Complexity of Infinite-Domain Constraint Satisfaction

Constraint Satisfaction Problems (CSPs) are natural computational problems that appear in many areas of theoretical computer science. Exploring which CSPs are solvable in polynomial time and which are NP-hard reveals a surprising link with central questions in universal algebra.

This monograph presents a self-contained introduction to the universal-algebraic approach to complexity classification, treating both finite and infinite-domain CSPs. It includes the required background from logic and combinatorics, particularly model theory and Ramsey theory, and explains the recently discovered link between Ramsey theory and topological dynamics and its implications for CSPs. The book will be of interest to graduate students and researchers in theoretical computer science and to mathematicians in logic, combinatorics, and dynamics who wish to learn about the applications of their work in complexity theory.

MANUEL BODIRSKY is Professor at the Institute for Algebra in the Faculty of Mathematics at TU Dresden.

LECTURE NOTES IN LOGIC

A Publication of The Association for Symbolic Logic

This series serves researchers, teachers, and students in the field of symbolic logic, broadly interpreted. The aim of the series is to bring publications to the logic community with the least possible delay and to provide rapid dissemination of the latest research. Scientific quality is the overriding criterion by which submissions are evaluated.

More information, including a list of the books in the series, can be found at http://aslonline.org/books/lecture-notes-in-logic/

LECTURE NOTES IN LOGIC 52

Complexity of Infinite-Domain Constraint Satisfaction

MANUEL BODIRSKY

Technische Universität Dresden

ASSOCIATION FOR SYMBOLIC LOGIC

CAMBRIDGE
UNIVERSITY PRESS

CAMBRIDGE
UNIVERSITY PRESS

University Printing House, Cambridge CB2 8BS, United Kingdom

One Liberty Plaza, 20th Floor, New York, NY 10006, USA

477 Williamstown Road, Port Melbourne, VIC 3207, Australia

314–321, 3rd Floor, Plot 3, Splendor Forum, Jasola District Centre, New Delhi – 110025, India

79 Anson Road, #06–04/06, Singapore 079906

Cambridge University Press is part of the University of Cambridge.

It furthers the University's mission by disseminating knowledge in the pursuit of education, learning, and research at the highest international levels of excellence.

www.cambridge.org
Information on this title: www.cambridge.org/9781107042841
DOI: 10.1017/9781107337534

Association for Symbolic Logic
Richard A. Shore, Publisher
Department of Mathematics, Cornell University, Ithaca, NY 14853
http://aslonline.org

First published 2021

A catalogue record for this publication is available from the British Library.

Library of Congress Cataloging-in-Publication Data
Names: Bodirsky, Manuel, 1976– author. | Association for Symbolic Logic, issuing body.
Title: Complexity of infinite-domain constraint satisfaction / Manuel Bodirsky.
Description: Cambridge, United Kingdom ; New York, NY : Cambridge University Press, 2021. | Series: Lecture notes in logic ; 52 | "A publication of the Association for Symbolic Logic." | Includes bibliographical references and index.
Identifiers: LCCN 2021010432 (print) | LCCN 2021010433 (ebook) | ISBN 9781107042841 (hardback) | ISBN 9781107337534 (epub)
Subjects: LCSH: Constraint programming (Computer science) | Constraints (Artificial intelligence)–Mathematics. | Linear programming. | Computational complexity.
Classification: LCC QA76.612 .B63 2021 (print) | LCC QA76.612 (ebook) | DDC 005.1/16–dc23
LC record available at https://lccn.loc.gov/2021010432
LC ebook record available at https://lccn.loc.gov/2021010433

ISBN 978-1-107-04284-1 Hardback

CONTENTS

ACKNOWLEDGEMENTS

I want to thank my current institution, TU Dresden, and my former institution, the CNRS, for the great freedom in research that allowed me to write this text. The research leading to the results presented here has received funding from the European Research Council (Grant Agreement 257039, CSP-Complexity, and and Grant Agreement 681988, CSP-Infinity) and from the German Research Foundation (DFG project number 622397). I also want to thank all of my constraint satisfaction co-authors for the good time we had with our joint work. Many thanks also to Andrés Aranda, François Bossière, Alex Kazda, Pablo Cubides Kovacsics, Johannes Greiner, Martin Hils, Simon Knäuer, Michael Kompatscher, Victor Lagerqvist, Erkko Lehtonen, Barnaby Martin, Antoine Mottet, Jakub Opršal, Thomas Quinn-Gregson, Christian Pech, Leopold Schlicht, Friedrich Martin Schneider, Florian Starke, Jakub Rydval, Trung Van Pham, Caterina Viola, Albert Vucaj, Michał Wrona, and the anonymous referees for their comments either on previous versions or my habilitation memoir which has been the basis for this book.

Special thanks to Christoph Dürr for the permission to include his wonderful pictures at the beginning of many chapters. More of his artwork can be found by searching the internet for *'La vie est Dürr'*. The pictures for Chapters 3, 7, and 12 are drawings that I kept from discussions with Jaroslav Nešetřil, Jan Kára, and Martin Kutz. The picture for Chapter 5 is by Lewin Bodirsky (2010), the pictures for Chapter 8 and Chapter 9 are by Otto Bodirsky (2011), and the pictures for Chapter 14 by Elsa Bodirsky (2019) and Michaela Metzger.

..POM, POM
OH, IL N'Y A PLUS D'ÉLECTRICITÉ AUJOURD'HUI?
PAS GRAVE, JE TRAVAILLERAIS SUR PAPIER
PAS DE LAMPE? JE M'APPROCHE DE LA FENÊTRE
PAS DE CHAUFFAGE? JE REMETS MA VESTE
CH'UIS PAS BIEN LÀ! IL NE MANQUE PLUS QU'UNE TASSE DE CAFÉ
QUOI? PAS DE CAFÉ?

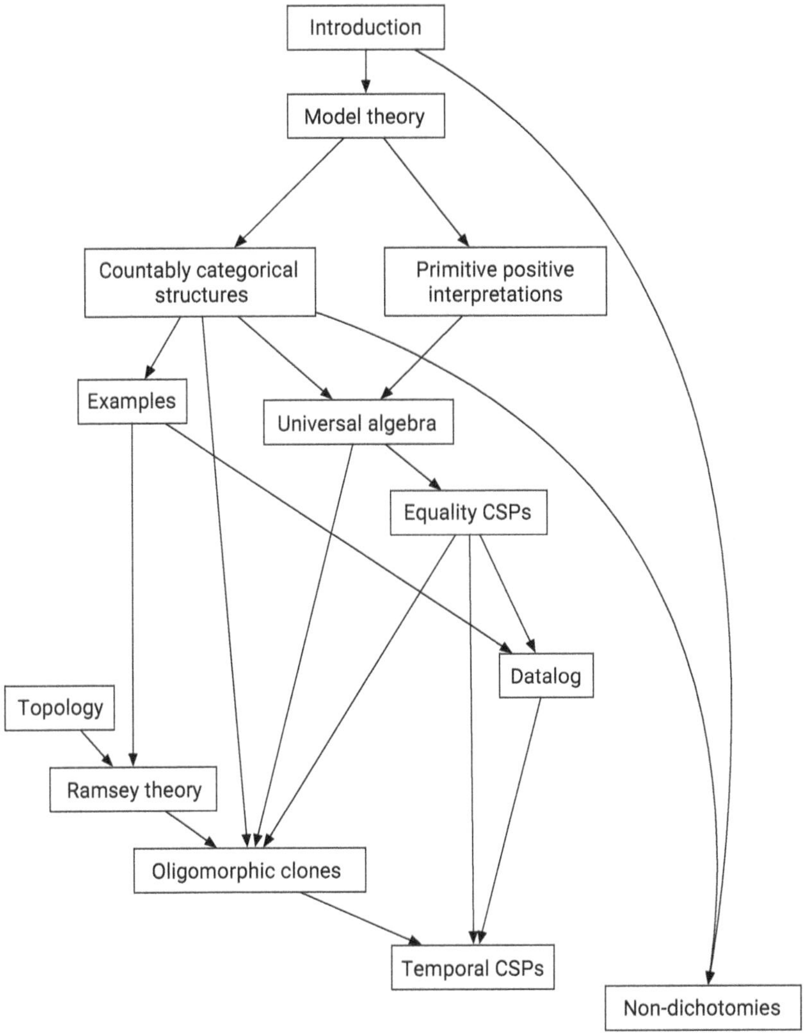
Introduction
Model theory
Countably categorical structures
Primitive positive interpretations
Examples
Universal algebra
Equality CSPs
Datalog
Topology
Ramsey theory
Oligomorphic clones
Temporal CSPs
Non-dichotomies

Chapter 1

INTRODUCTION TO CONSTRAINT SATISFACTION PROBLEMS

A *constraint satisfaction problem* (*CSP*) is a computational problem where we are given a finite set of variables and a finite set of *constraints* and where the task is to decide whether values can be assigned to the variables so that all the constraints are satisfied. Well-known examples of CSPs are the satisfiability problem for systems of linear equations over the two-element field, the satisfiability problem for systems of linear inequalities over the rational numbers, and the three-colouring problem for undirected graphs, just to name three. CSPs appear in almost every area of theoretical computer science, for instance in artificial intelligence, scheduling, computational linguistics, optimisation, computational biology, and verification. Many computational problems studied in those areas can be modelled by appropriately choosing a set of constraint types, the *constraint language* (or *template*), that are allowed in the input instance of a CSP. In the last decade, huge progress was made to find general criteria for constraint languages that imply that the corresponding CSP can be solved efficiently.

The complexity of CSPs became a topic that vitalises the field of universal algebra as it turned out that questions about the computational complexity of CSPs translate to important universal-algebraic questions about algebras that can be associated to CSPs. This approach is now known as the *algebraic approach* to constraint satisfaction complexity. The algebraic approach has raised questions that are of central importance in universal algebra. The so-called *dichotomy conjecture* of Feder and Vardi [170] stated that every CSP *with a finite domain* is either polynomial-time tractable (i.e., in P) or NP-complete. According to a well-known result by Ladner, it is known that there are NP-*intermediate* computational problems, i.e., problems in NP that are neither tractable nor NP-complete (unless P = NP). But the known NP-intermediate problems are extremely artificial. It would be interesting from a complexity theoretic perspective to discover more natural candidates for NP-intermediate problems. Two positive solutions to the dichotomy conjecture, both using the universal-algebraic approach, have been announced in 2017 by Bulatov and by Zhuk [111, 348].

The problems in the literature that can be described by specifying a constraint language over a finite domain, and that have been studied independently from the CSP framework, are quite limited, and mostly focussed on specialised graph theoretic problems or Boolean satisfiability problems. If we consider the class of all problems that can be formulated by specifying a constraint language over an *infinite* domain, the situation changes drastically. Many problems that have been studied independently in temporal reasoning, spatial reasoning, phylogenetic reconstruction, and computational linguistics can be directly formulated as CSPs over an infinite domain. Also feasibility problems in linear (and also non-linear, or tropical) programming (over the rationals, the integers, or other domains) can be cast as CSPs.

In this book we present a generalisation of the universal-algebraic approach to infinite domains. It turns out that this is possible when the constraint language, viewed as a relational structure $\mathfrak{B}$ with an infinite domain, is ω-*categorical* (or *countably categorical*), i.e., its first-order theory has at most one countable model up to isomorphism. An alternative characterisation of ω-categoricity of $\mathfrak{B}$ which is most useful in our context is in terms of the automorphism group of $\mathfrak{B}$: a structure $\mathfrak{B}$ is ω-categorical if and only if the automorphism group of $\mathfrak{B}$ is *oligomorphic*, i.e., for every $n \in \mathbb{N}$ the automorphism group of $\mathfrak{B}$ has only finitely many orbits in its componentwise action on n-tuples of elements of $\mathfrak{B}$. Many of the CSPs in the mentioned application areas can be formulated with ω-categorical constraint languages—in particular, problems coming from so-called *qualitative calculi* in artificial intelligence tend to have formulations with ω-categorical constraint languages. While ω-categoricity is quite a strong assumption from a model-theoretic point of view (and, for example, constraint languages for linear programming cannot be ω-categorical), the class of computational problems that can be formulated with ω-categorical constraint languages is still a very large generalisation of the class of finite-domain CSPs. This will be amply demonstrated by examples of ω-categorical constraint languages from many different areas in computer science in Chapter 5.

There are several general results for ω-categorical structures that are important when studying the computational complexity of the respective CSPs. We highlight some of these general results, providing pointers into the text where all the involved technical terms will be gently introduced.

- Every finite or countably infinite ω-categorical structure is homomorphically equivalent to a finite or countably infinite ω-categorical structure which is *model complete* and a *core* (Section 2.6). Model-complete cores have many good properties: for example, they have quantifier elimination once expanded by all primitively positively definable relations. Moreover, they can be expanded by finitely many singleton relations $\{a\}$ without changing the complexity of the CSP. Since homomorphically equivalent structures have the same CSP, we can therefore focus on structures having these properties.
- The so-called *polymorphism clone* of an ω-categorical structure $\mathfrak{B}$ fully captures the computational complexity of the corresponding CSP (Chapter 6). Every polymorphism clone gives rise to a *topological clone* with respect to the pointwise convergence topology. Topological clones are defined analogously to topological groups and are a promising new subject in mathematics. We will see that the complexity of the CSP is already captured by the polymorphism clone viewed as a topological clone (Section 9.4).
- Finally, if the expansion of an ω-categorical model-complete core by finitely many singleton relations does not interpret primitively positively all finite structures, then it must have a pseudo-Siggers polymorphism (see Section 10.2). For finite domains, the existence of such polymorphisms turned out to be the dividing line between the polynomial-time tractable and the NP-hard CSPs.

Despite all these general results for ω-categorical structures, the class of all ω-categorical structures is still too large to hope for a complete complexity classification. It is easy to see that there are ω-categorical structures $\mathfrak{B}$ whose CSP is undecidable, and, as we will see in Chapter 13, there are CSPs of various other complexities: for example there are ω-categorical structures with a coNP-complete CSP, or with a CSP that is contained in coNP but neither coNP-hard nor in P.

However, there is a natural subclass of the class of all ω-categorical structures that might exhibit a complexity dichotomy as well. Formally, we consider the CSPs for *reducts of finitely bounded homogeneous structures*; these notions will be introduced in the text. Such structures have a finite representation and their CSPs are always in NP. The class of CSPs of reducts of finitely bounded homogeneous structures contains all CSPs over finite domains, but also contains many additional CSPs from the mentioned application areas. Moreover, every CSP in the complexity class MMSNP (for Monotone Monadic Strict NP, treated in Section 5.6.2) can be formulated in this way. It follows from general principles that if the model-complete core template for such a CSP does not have a pseudo-Siggers polymorphism, then the CSP is NP-hard. We conjecture that otherwise, the CSP ought to be in P, and refer to this as the *infinite-domain tractability conjecture*. This conjecture has been confirmed in numerous special cases:

- For all finite-domain CSPs [348, 111];
- for all first-order reducts of
  - $(\mathbb{Q}; <)$ [69] (see Chapter 12),
  - the countable random graph [90],
  - the model companion of the class of all C-relations [64],
  - the universal homogeneous poset [241],
  - all homogeneous undirected graphs [80],
  - all unary structures [81];
- for all CSPs in the complexity class MMSNP [75].

Any outcome of the infinite-domain dichotomy conjecture for reducts of finitely bounded homogeneous structures is significant: a negative answer might provide relatively natural NP-intermediate problems, which would be of interest for complexity theorists. A positive answer probably comes with a criterion which describes the NP-hard CSPs, and it would probably provide algorithms for the polynomial-time tractable CSPs. But then we would have a fascinatingly rich catalogue of computational problems where the computational complexity is known. Such a catalogue would be a valuable tool for deciding the complexity of computational problems: since CSPs for finitely bounded homogeneous structures are abundant, one may derive algorithmic results by reducing the problem of interest to a known tractable CSP, and one may derive hardness results by reducing a known NP-hard CSP to the problem of interest.

An important feature of the universal-algebraic approach is that the tractability of a CSP can be linked to the existence of polymorphisms of the constraint language. This link can be exploited in several directions: first, if we already know that a constraint language of interest has a polymorphism satisfying good properties, then this polymorphism can guide the search for an efficient algorithm for the corresponding CSP. Another direction is that we already have an algorithm (or an algorithmic technique), and that we want to know for which CSPs the algorithm is a correct decision procedure: again, polymorphisms are the key tool for this task. Finally, we may use the absence of polymorphisms with good properties to prove that a CSP is

NP-hard. There are several instances where these three directions of the algebraic approach have been used very successfully for CSPs with finite domain constraint languages [113, 126, 212, 21, 111, 348] or ω-categorical constraint languages [69, 90, 64, 80].

Another tool that becomes useful specifically for polymorphisms over infinite domains is *Ramsey theory* (Chapter 11). The basic idea here is to apply Ramsey theory to show that polymorphisms must act *canonically* on large parts of their domain. If we are working over a homogeneous structure with finite relational signature, then there are only finitely many behaviours of canonical functions of a given arity, and so this technique allows to perform combinatorial analysis when proving classification results. With this approach we can also show that, under further assumptions on $\mathfrak{B}$, many questions about the expressive power of $\mathfrak{B}$ become decidable, such as the question whether a given quantifier-free first-order formula is in $\mathfrak{B}$ equivalent to a primitive positive formula.

In Chapter 12 we use polymorphisms to classify the computational complexity of a large family of constraint satisfaction problems, namely *temporal CSPs*, i.e., CSPs for first-order reducts of $(\mathbb{Q}; <)$ which includes many CSPs in qualitative temporal reasoning and scheduling. Our classification confirms the mentioned infinite-domain tractability conjecture for this family. The class of temporal CSPs is particularly important because it often provides counterexamples for naive generalisations of known facts for finite-domain constraint satisfaction; moreover, the polynomial-time algorithms are particularly interesting in this class. Similar classifications have also been obtained for the countable homogeneous universal poset [241], the model companion of the class of C-relations [64], and for all homogeneous graphs [80].

Finally, in Chapter 13, we present results that show that the CSPs for certain natural classes of infinite structures do *not* have a complexity dichotomy.

Chapter outline. Constraint satisfaction problems can appear in different forms, because there are several ways to formalise CSPs. The differences in formalising CSPs are related to the way that instances are coded and to the way that the problem itself is described. In the next sections we present four formalisations; each of them is attached to different lines of research. In later sections some arguments are more natural from one perspective than from the other, so it will be convenient to have them all discussed here. Table 1 shows the relationships among the four perspectives in tabular form.

1.1. The homomorphism perspective

A *relational signature* τ is a set of relation symbols R each of which has an associated finite *arity* $k \in \mathbb{N}$. A *relational structure* $\mathfrak{A}$ over the signature τ (also called τ-*structure*) consists of a set A (the *domain* or *base set*) together with

Perspective	*Instance*	*Problem description*	*Section*
Homomorphism	Structure	Structure	1.1
Sentence Evaluation	Sentence	Structure	1.2
Satisfiability	Sentence	Sentences	1.3
Existential Second-Order	Structure	Sentence	1.4

TABLE 1. The four perspectives on the definition of CSPs.

a relation $R^{\mathfrak{A}} \subseteq A^k$ for each relation symbol $R \in \tau$ of arity k. It causes no harm and will be convenient to allow structures whose domain is empty.

A *homomorphism* from a τ-structure $\mathfrak{A}$ with domain A to a τ-structure $\mathfrak{B}$ with domain B is a function $h: A \to B$ that *preserves* all relations; that is, if $R \in \tau$ has arity k and $(a_1, \dots, a_k) \in R^{\mathfrak{A}}$, then $(h(a_1), \dots, h(a_k)) \in R^{\mathfrak{B}}$. If a structure $\mathfrak{A}$ has a homomorphism to $\mathfrak{B}$ we also say that $\mathfrak{A}$ is *homomorphic to* $\mathfrak{B}$ or that $\mathfrak{A}$ *maps homomorphically* to $\mathfrak{B}$. An *isomorphism* is a bijective homomorphism h such that the inverse map $h^{-1}: B \to A$, which sends $h(x)$ to x, is a homomorphism, too.

In this text, a *constraint satisfaction problem* (*CSP*) is a computational problem that is specified by a single structure with a finite relational signature, called the *template* (or the *constraint language*; the name 'constraint language' is typically used in the context of the second perspective on CSPs that we present in Section 1.2). Such problems are sometimes also called *non-uniform CSPs* because $\mathfrak{B}$ is not part of the input, but fixed.

DEFINITION 1.1.1 (CSP($\mathfrak{B}$)). Let $\mathfrak{B}$ be a (possibly infinite) structure with a finite relational signature τ. Then CSP($\mathfrak{B}$) is the computational problem of deciding whether a given finite τ-structure $\mathfrak{A}$ maps homomorphically to $\mathfrak{B}$.

CSP($\mathfrak{B}$) can be considered to be a class—the class of all finite τ-structures that map homomorphically to $\mathfrak{B}$. A homomorphism from a given τ-structure $\mathfrak{A}$ to $\mathfrak{B}$ is called a *solution* of $\mathfrak{A}$ for CSP($\mathfrak{B}$). It is in general not clear how to represent solutions for CSP($\mathfrak{B}$) on a computer; however, for the definition of the problem CSP($\mathfrak{B}$) we do not need to represent solutions, since we only have to decide the *existence* of solutions. To represent an input structure $\mathfrak{A}$ of CSP($\mathfrak{B}$) we fix a representation of the relation symbols in the signature τ (the precise choice of the representation is irrelevant because of the assumption that τ is *finite*). Thus, CSP($\mathfrak{B}$) is a well-defined computational problem for *any* infinite structure $\mathfrak{B}$ with finite relational signature.

Example 1.1.2 (Digraph acyclicity). Next, consider the problem CSP($\mathbb{Z}$; $<$). Here, the relation $<$ denotes the strict linear order of the integers $\mathbb{Z}$. An instance $\mathfrak{A}$ of this problem is a finite $\{<\}$-structure, which can be viewed as a directed graph (also called *digraph*), potentially with loops. It is easy to see that $\mathfrak{A}$ maps homomorphically to $(\mathbb{Z}; <)$ if and only if there is no directed cycle in $\mathfrak{A}$ (loops are considered to be directed cycles, too). It is easy to see and well known that this can be tested in linear time, for example by performing a depth-first search on the digraph $\mathfrak{A}$.

Example 1.1.3 (Betweenness). The so-called *Betweenness Problem* [296] can be modelled as CSP($\mathbb{Z}$; Betw) where Betw is the ternary relation

$$\{(x, y, z) \in \mathbb{Z}^3 \mid (x < y < z) \vee (z < y < x)\}.$$

This problem is one of the NP-complete problems listed in the book of Garey and Johnson [176].

Example 1.1.4 (Cyclic ordering). The *Cyclic-ordering Problem* [174] can be modelled as CSP($\mathbb{Z}$; Cycl) where Cycl is the ternary relation

$$\{(x, y, z) \in \mathbb{Z}^3 \mid (x < y < z) \vee (y < z < x) \vee (z < x < y)\}.$$

This problem is again NP-complete and can be found in [176].

Example 1.1.5 ($\mathfrak{H}$-colouring problems). Let $\mathfrak{H}$ be an (undirected) graph. We view undirected graphs as τ-structures where τ contains a single binary relation symbol E, which denotes a symmetric relation; in our setting, it is natural and useful to allow *loops*, i.e., edges of the form (x, x). Then the *$\mathfrak{H}$-colouring problem* is the computational problem to decide for a given finite graph $\mathfrak{G}$ whether there exists a homomorphism from $\mathfrak{G}$ to $\mathfrak{H}$. For instance, if $\mathfrak{H}$ is the complete graph on three vertices K_3 without loops, then the $\mathfrak{H}$-colouring problem is the famous 3-*colorability problem* (see e.g. [176]). Similarly, for every fixed k, the k-colorability problem can be modelled as CSP(K_k), where K_k is the complete graph with k vertices, also called a *k-clique*.

The next lemma (Lemma 1.1.8) is a useful test to determine whether a computational problem can be formulated as CSP($\mathfrak{B}$) for some relational structure $\mathfrak{B}$.

DEFINITION 1.1.6. An (*induced*) *substructure* of a τ-structure $\mathfrak{A}$ is a τ-structure $\mathfrak{B}$ with $B \subseteq A$ and $R^{\mathfrak{B}} = R^{\mathfrak{A}} \cap B^n$ for each n-ary $R \in \tau$; we also say that $\mathfrak{B}$ is *induced on B by* $\mathfrak{A}$, and write $\mathfrak{A}[B]$ for $\mathfrak{B}$.

The *union* of two τ-structures $\mathfrak{A}, \mathfrak{B}$ is the τ-structure $\mathfrak{A} \cup \mathfrak{B}$ with domain $A \cup B$ and the relation $R^{\mathfrak{A} \cup \mathfrak{B}} := R^{\mathfrak{A}} \cup R^{\mathfrak{B}}$ for every $R \in \tau$. The *intersection* $\mathfrak{A} \cap \mathfrak{B}$ of $\mathfrak{A}$ and $\mathfrak{B}$ is defined analogously. A *disjoint union* of $\mathfrak{A}$ and $\mathfrak{B}$ is the union of isomorphic copies of $\mathfrak{A}$ and $\mathfrak{B}$ with disjoint domains. As disjoint unions are unique up to isomorphism, we usually speak of *the* disjoint union of $\mathfrak{A}$ and $\mathfrak{B}$, and denote it by $\mathfrak{A} \uplus \mathfrak{B}$. The disjoint union of a set of τ-structures C is defined analogously (and the disjoint union of an empty set of structures is the τ-structure with empty domain). A structure is called *connected* if it is not the disjoint union of two non-empty structures. A *maximal* connected substructure of $\mathfrak{B}$ (i.e., a connected substructure of $\mathfrak{B}$ such that every substructure of $\mathfrak{B}$ with a larger domain is not connected) is called a *connected component* of $\mathfrak{B}$.

DEFINITION 1.1.7. We say that a class C of relational structures is

- *closed under homomorphisms* if whenever $\mathfrak{A} \in C$ and $\mathfrak{A}$ maps homomorphically to $\mathfrak{B}$ we have $\mathfrak{B} \in C$;
- *closed under inverse homomorphisms* if whenever $\mathfrak{B} \in C$ and $\mathfrak{A}$ maps homomorphically to $\mathfrak{B}$ we have $\mathfrak{A} \in C$;
- *closed under (finite) disjoint unions* if whenever $\mathfrak{A}, \mathfrak{B} \in C$ the disjoint union of $\mathfrak{A}$ and $\mathfrak{B}$ is also in C.

Note that a class C of τ-structures is closed under inverse homomorphisms if and only if its complement in the class of all τ-structures is closed under homomorphisms. When a class is closed under inverse homomorphisms, or closed under homomorphisms, it is in particular closed under isomorphisms.

Let $\mathcal{F}$ be a class of τ-structures. We say that a structure $\mathfrak{A}$ is *$\mathcal{F}$-free* if no $\mathfrak{B} \in \mathcal{F}$ maps homomorphically to $\mathfrak{A}$ (so $\mathcal{F}$ denotes the homomorphically *forbidden* structures). The class of all finite $\mathcal{F}$-free structures we denote by $\mathrm{Forb}^{\mathrm{hom}}(\mathcal{F})$. The following is a simple, but fundamental lemma for CSPs.

LEMMA 1.1.8. *Let τ be a finite relational signature and let C a class of finite τ-structures. Then the following are equivalent.*

(1) $C = \mathrm{CSP}(\mathfrak{B})$ *for some τ-structure $\mathfrak{B}$.*
(2) $C = \mathrm{Forb}^{\mathrm{hom}}(\mathfrak{F})$ *for a class of finite connected τ-structures $\mathfrak{F}$.*
(3) *C is closed under disjoint unions and inverse homomorphisms.*
(4) $C = \mathrm{CSP}(\mathfrak{B})$ *for a countable τ-structure $\mathfrak{B}$.*

PROOF. It suffices to prove the implications (1) $\Rightarrow$ (2) $\Rightarrow$ (3) $\Rightarrow$ (4). For the implication from (1) to (2), let $\mathfrak{F}$ be the class of all finite connected τ-structures that do not map homomorphically to $\mathfrak{B}$. If a structure $\mathfrak{A}$ maps homomorphically to $\mathfrak{B}$ then no $\mathfrak{C} \in \mathfrak{F}$ can map homomorphically to $\mathfrak{A}$ because of the transitivity of the homomorphism relation. Conversely, if $\mathfrak{A}$ does not map homomorphically to $\mathfrak{B}$, then already one of the connected components of $\mathfrak{A}$ does not map homomorphically to $\mathfrak{B}$. This connected component is in $\mathfrak{F}$, which shows that a structure from $\mathfrak{F}$ maps homomorphically to $\mathfrak{A}$.

(2) $\Rightarrow$ (3). Suppose (2), and let $\mathfrak{A}_1$ and $\mathfrak{A}_2$ be two structures from $\mathrm{Forb}^{\mathrm{hom}}(\mathfrak{F})$. Since every $\mathfrak{C} \in \mathfrak{F}$ is connected, every homomorphism from $\mathfrak{C}$ to $\mathfrak{A}_1 \uplus \mathfrak{A}_2$ must already be a homomorphism to $\mathfrak{A}_1$ or to $\mathfrak{A}_2$, which is impossible. Hence, $\mathrm{Forb}^{\mathrm{hom}}(\mathfrak{F})$ is closed under disjoint unions. Closure under inverse homomorphisms follows straightforwardly from the transitivity of the homomorphism relation.

(3) $\Rightarrow$ (4). Suppose that C is a class of relational structures that is closed under disjoint unions and inverse homomorphisms. Let C' be a subclass of C where we select one structure from each isomorphism class of structures

Triangle-Freeness
Instance: An undirected graph $\mathfrak{G}$.
Question: Is $\mathfrak{G}$ triangle-free?

Acyclic-Bipartition
Instance: A digraph $\mathfrak{G}$.
Question: Is there a partition $V = V_1 \uplus V_2$ of the vertices V of $\mathfrak{G}$ such that both $\mathfrak{G}[V_1]$ and $\mathfrak{G}[V_2]$ are acyclic?

No-Mono-Tri
Instance: An undirected graph $\mathfrak{G}$.
Question: Is there a partition $V = V_1 \uplus V_2$ of the vertices V of $\mathfrak{G}$ such that both $\mathfrak{G}[V_1]$ and $\mathfrak{G}[V_2]$ are triangle-free?

TABLE 2. Three computational problems that are closed under disjoint unions and inverse homomorphisms.

in C. Let $\mathfrak{B}$ be the (countable) disjoint union over all structures in C' (if C' is empty then $\mathfrak{B}$ is by definition the empty structure[1]). Clearly, every structure in C maps homomorphically to $\mathfrak{B}$. Now, let $\mathfrak{A}$ be a finite structure with a homomorphism h to $\mathfrak{B}$. By the construction of $\mathfrak{B}$, the set $h(A)$ is contained in the disjoint union $\mathfrak{C}$ of a finite set of structures from C'. Since C is closed under disjoint unions, $\mathfrak{C}$ is in C. Clearly, $\mathfrak{A}$ maps homomorphically to $\mathfrak{C}$, and because C is closed under inverse homomorphisms, $\mathfrak{A}$ is in C as well. $\dashv$

Example 1.1.9. The computational problems in Table 2 are closed under disjoint unions and inverse homomorphisms. Hence, Lemma 1.1.8 shows that they can be formulated as CSP($\mathfrak{B}$) for some relational structure $\mathfrak{B}$. It is easy to see that none of those three problems can be formulated as CSP($\mathfrak{B}$) for a *finite* structure $\mathfrak{B}$.

We verify this for the problem of Triangle-freeness. For a fixed n, consider the graph that contains vertices $x_1, \dots, x_n$, and that contains for every pair i, j with $1 \leq i < j \leq n$ two additional vertices $u_{i,j}, v_{i,j}$ and the edges $(x_i, u_{i,j})$, $(u_{i,j}, v_{i,j})$, $(v_{i,j}, x_j)$. The resulting graph is clearly triangle-free. But note that every homomorphism f from this graph to a graph $\mathfrak{H}$ with strictly less than n vertices must identify at least two of the vertices $x_1, \dots, x_n$. So suppose that $f(x_i) = f(x_j)$. Then $(f(x_i), f(u_{i,j}))$, $(f(u_{i,j}), f(v_{i,j}))$, and $(f(v_{i,j}), f(x_j))$ are edges in $\mathfrak{H}$ because f is a homomorphism. Hence, $\mathfrak{H}$ either contains a triangle or a loop. In both cases, $\mathfrak{H}$ cannot be the template for Triangle-Freeness.

[1]Structures with an empty domain are often forbidden in model theory. Lemma 1.1.8 is one of the places that motivates our decision to allow them in this text.

We have thus ruled out all templates of size $n-1$, which concludes the proof since n was chosen arbitrarily.

The central conjecture for finite-domain constraint satisfaction problems was the dichotomy conjecture, due to Feder and Vardi [170], which states that for every structure $\mathfrak{B}$ with a finite relational signature and a finite domain, CSP($\mathfrak{B}$) is in P or NP-complete. A solution to the dichotomy conjecture has been announced by Bulatov [111] and, independently, by Zhuk [348]. Before, it has been verified for many classes of structures $\mathfrak{B}$, for instance

- for structures $\mathfrak{B}$ with a two-element domain (Schaefer's theorem [319]; see Section 6.2);
- structures over a 3-element domain [108];
- for finite undirected graphs $\mathfrak{B}$ (the theorem of Hell and Nešetřil [195]; see Section 6.8);
- finite digraphs without sources and sinks [25];
- finite structures $\mathfrak{B}$ that contain a unary relation symbol for each subset of the domain of $\mathfrak{B}$, due to [106] (see also [17] and [110]).

There are many equivalent descriptions of the border between polynomial-time tractable and NP-complete finite-domain CSPs; see Sections 3.7, 6.3.5, 6.6, 6.9.

We close this section with an important concept for finite structures $\mathfrak{B}$, the notion of a *core*; generalisations to infinite structures $\mathfrak{B}$ are presented in Section 2.6.2. The motivation for this concept is that for every finite-domain CSP has a template of minimal size which is unique up to isomorphism, and this template has many pleasant properties. To formalise this, we need the following definitions. Two structures $\mathfrak{A}$ and $\mathfrak{B}$ are called *homomorphically equivalent* if there exists a homomorphism from $\mathfrak{A}$ to $\mathfrak{B}$ and vice versa. An *embedding* of $\mathfrak{A}$ into $\mathfrak{B}$ is an injective map $f: A \to B$ such that $(a_1, \ldots, a_k)$ is in $R^{\mathfrak{A}}$ if and only if $(f(a_1), \ldots, f(a_k))$ is in $R^{\mathfrak{B}}$. An *endomorphism* of a structure $\mathfrak{B}$ is a homomorphism from $\mathfrak{B}$ to $\mathfrak{B}$.

DEFINITION 1.1.10. A structure $\mathfrak{B}$ is a *core* if all its endomorphisms are embeddings[2]. For structures $\mathfrak{A}, \mathfrak{B}$ of the same signature, the structure $\mathfrak{B}$ is called a *core of* $\mathfrak{A}$ if $\mathfrak{B}$ is a core and homomorphically equivalent to $\mathfrak{A}$.

In fact, we speak of *the core* of a finite structure $\mathfrak{A}$, due to the following fact, whose proof is easy and left to the reader.

PROPOSITION 1.1.11. *Let $\mathfrak{A}$ be a finite structure. Then $\mathfrak{A}$ has a core, and all cores of $\mathfrak{A}$ are isomorphic.*

Core structures $\mathfrak{B}$ have many pleasant properties when it comes to studying the computational complexity of CSP($\mathfrak{B}$) (see for instance Proposition 1.2.11

[2]For finite structures $\mathfrak{B}$, injective self-maps must be bijective, and in fact every injective endomorphism of a structure $\mathfrak{B}$ must be an automorphism. For infinite structures, however, this need not be true, and for reasons that become clear in Chapter 4 we choose the present formulation of the definition.

below). Clearly, when $\mathfrak{A}$ and $\mathfrak{B}$ are homomorphically equivalent, then $\mathrm{CSP}(\mathfrak{A}) = \mathrm{CSP}(\mathfrak{B})$. Therefore, and because of Proposition 1.1.11, we can assume without loss of generality that a finite structure $\mathfrak{B}$ is a core when studying $\mathrm{CSP}(\mathfrak{B})$. We finally remark that structures with a one-element core have a trivial CSP.

PROPOSITION 1.1.12. *Let $\mathfrak{B}$ be a relational structure with a finite relational signature and a one-element core. Then* $\mathrm{CSP}(\mathfrak{B})$ *is in P.*

PROOF. Let $\mathfrak{C}$ be the core of $\mathfrak{B}$, and let c be the unique element of $\mathfrak{C}$. The problem $\mathrm{CSP}(\mathfrak{B})$ can be solved as follows. Let $\mathfrak{A}$ be an input structure of $\mathrm{CSP}(\mathfrak{B})$. If there is $(t_1, \dots, t_n) \in R^{\mathfrak{A}}$ such that $(c, \dots, c) \notin R^{\mathfrak{C}}$, then reject. Otherwise accept. ⊣

1.2. The sentence evaluation perspective

Let τ be a relational signature. A first-order τ-formula $\phi(x_1, \dots, x_n)$ is called *primitive positive* if it is of the form

$$\exists x_{n+1}, \dots, x_m (\psi_1 \wedge \dots \wedge \psi_l)$$

where $\psi_1, \dots, \psi_l$ are *atomic τ-formulas*, i.e., formulas of the form $R(y_1, \dots, y_k)$ with $R \in \tau$ and $y_i \in \{x_1, \dots, x_m\}$, of the form $y = y'$ for $y, y' \in \{x_1, \dots, x_m\}$, or of the form $\bot$ or $\top$ for *false* and *true*.[3] As usual, formulas without free variables are called *sentences*. Primitive positive formulas also play an important role in database theory, where they are known under the name of *conjunctive queries*.

From a model-checking perspective, CSPs are defined as follows. We will see (in Propositions 1.2.4 and 1.2.5) that this definition is essentially the same definition as Definition 1.1.1, and that the differences are a matter of formalisation.[4]

DEFINITION 1.2.1. Let $\mathfrak{B}$ be a (possibly infinite) structure with a finite relational signature τ. Then $\mathrm{CSP}(\mathfrak{B})$ is the computational problem to decide whether a given primitive positive τ-sentence ϕ is true in $\mathfrak{B}$.

[3] Note that $\bot$ can be thought of as a symbol for the empty relation of arity 0, and $\top$ as a symbol for the non-empty relation of arity 0; this relation contains the element (), even if the domain is empty. Note that if the domain is non-empty then we do not need a symbol $\top$ for *true*, since we can use the primitive positive sentence $\exists x \colon x = x$ to express it.

[4] A small difference between the homomorphism perspective and the sentence evaluation problem results from the fact that we *do* allow equality in primitive positive formulas; as we will see in Lemma 1.2.6, adding equality to the constraint language does not affect the complexity of the CSP up to log-space reductions. There is research, though, that studies the complexity of CSPs up to an even finer level than log-space reducibility, and there equality might not automatically be allowed in the input to a constraint satisfaction problem (cf. [254]).

3SAT *Instance:* A propositional formula in conjunctive normal form (CNF) with at most three literals per clause. *Question:* Is there a Boolean assignment for the variables such that in each clause at least one literal is true?
Positive 1-in-3-3SAT *Instance:* A propositional 3SAT formula with only positive literals. *Question:* Is there a Boolean assignment for the variables such that in each clause exactly one literal is true?
Positive Not-All-Equal-3SAT *Instance:* A propositional 3SAT formula with only positive literals. *Question:* Is there a Boolean assignment for the variables such that in each clause neither all three literals are true nor all three are false?

TABLE 3. Three Boolean satisfiability problems from the list of NP-complete problems of [176] that can be formulated as CSP($\mathfrak{B}$) for appropriate $\mathfrak{B}$.

The given primitive positive τ-sentence ϕ is also called an *instance* of CSP($\mathfrak{B}$). The conjuncts of an instance ϕ are called the *constraints* of ϕ. A mapping from the variables of ϕ to the elements of B that is a satisfying assignment for the quantifier-free part of ϕ is also called a *solution* to ϕ.

Some authors omit the (existential) quantifier-prefix in instances ϕ of CSP($\mathfrak{B}$), and the question is then whether ϕ is *satisfiable* over $\mathfrak{B}$. Clearly, this is just re-phrasing the problem above, but it explains the terminology of *satisfiable* and *unsatisfiable* (rather than *true* and *false*) instances of CSP($\mathfrak{B}$).

Example 1.2.2 (Boolean satisfiability problems). There are many Boolean satisfiability problems that can be cast as CSPs. Well-known examples are 3SAT (see Table 3) and the restricted versions of 3SAT called 1-in-3-3SAT and Not-All-Equal-3SAT [176]. These three problems are NP-complete. An interesting feature of the last two problems is that they remain NP-complete even when all clauses in the input only contain positive literals. With this additional restriction, the problems are called Positive-1-in-3-3SAT and Positive-Not-All-Equal-3SAT, and their definition can be found in Table 3.

All of these problems can be formulated as CSP($\mathfrak{B}$), for an appropriate 2-element structure $\mathfrak{B}$. Positive-1-in-3-3SAT can be formulated as CSP($\mathfrak{B}$) for the template

$$\mathfrak{B} = (\{0,1\}; 1\text{IN}3), \text{ where } 1\text{IN}3 = \{(0,0,1),(0,1,0),(1,0,0)\},$$

and Positive-Not-All-Equal-3SAT as CSP($\mathfrak{B}$) for the template

$$\mathfrak{B} = (\{0,1\}, \text{NAE}), \text{ where } \text{NAE} = \{0,1\}^3 \setminus \{(0,0,0),(1,1,1)\}.$$

These problems can also be formulated as CSPs if we do *not* impose the restriction that all literals are positive; the corresponding problems are then called 1-*in*-3-3*SAT* and *Not-All-Equal*-3*SAT*, respectively. The idea is to use a different ternary relation for each of the eight ways that three distinct variables in a clause with three literals may be negated. In this way, we can also model the classical problem of 3SAT (again, see Figure 3) as a CSP. Clauses of the type $x \vee y \vee \neg z$ in the 3SAT problem will then be viewed as constraints $R^{++-}(x,y,z)$, where $R^{++-} = \{0,1\}^3 \setminus \{(0,0,1)\}$ (here, x, y, z are not necessarily distinct variables). Similarly, the well-known 2SAT problem can be viewed as CSP($\{0,1\}; R^{++}, R^{+-}, R^{-+}, R^{--}$) where

$$\begin{aligned} R^{++} &= \{(0,1),(1,0),(1,1)\}, \\ R^{+-} &= \{(0,0),(1,1),(1,0)\}, \\ R^{-+} &= \{(1,1),(0,0),(0,1)\}, \text{ and} \\ R^{--} &= \{(1,0),(0,1),(0,0)\}. \end{aligned}$$

Example 1.2.3 (Disequality constraints). Consider CSP($\mathbb{N}; =, \neq$). An instance of this problem can be viewed as an (existentially quantified) set of variables, some linked by equality, some by disequality[5] constraints. Such an instance is false in ($\mathbb{N}; =, \neq$) if and only if there is a path $x_1, \ldots, x_n$ from a variable x_1 to a variable x_n that uses only equality edges, i.e., '$x_i = x_{i+1}$' is a constraint in the instance for each $1 \leq i \leq n-1$, and additionally '$x_1 \neq x_n$' is a constraint in the instance. Clearly, it can be tested in linear time in the size of the input instance whether the instance contains such a path.

1.2.1. Canonical conjunctive queries. To every finite relational τ-structure $\mathfrak{A}$ we can associate a τ-formula, called the *canonical conjunctive query* of $\mathfrak{A}$, and denoted by Q($\mathfrak{A}$). The variables of this sentence are the elements of $\mathfrak{A}$; the formula is the conjunction of all atomic formulas of the form $R(a_1, \ldots, a_k)$ for $R \in \tau$ and $(a_1, \ldots, a_k) \in R^{\mathfrak{A}}$.

For example, the canonical conjunctive query Q(K_3) of the complete graph K_3 with the three vertices $\{u, v, w\}$ is the formula

$$E(u,v) \wedge E(v,u) \wedge E(v,w) \wedge E(w,v) \wedge E(u,w) \wedge E(w,u).$$

The following proposition follows straightforwardly from the definitions.

PROPOSITION 1.2.4. *Let $\mathfrak{B}$ be a structure with finite relational signature τ, and let $\mathfrak{A}$ be a finite τ-structure. Then there is a homomorphism from $\mathfrak{A}$ to $\mathfrak{B}$ if and only if* Q($\mathfrak{A}$) *is satisfiable in $\mathfrak{B}$.*

---

[5]We deliberately use the word *disequality* instead of *inequality*, since we reserve the word *inequality* for the relation $x \leq y$.

1.2.2. Canonical databases. To present a converse of Proposition 1.2.4, we define the *canonical database* $D(\phi)$ of a primitive positive τ-formula, which is a relational τ-structure defined as follows. We require that ϕ does not contain the symbols $\bot$ and $=$. Then the domain of $D(\phi)$ is the set of variables (both the free variables and the existentially quantified variables) that occur in ϕ. There is a tuple $(v_1, \ldots, v_k)$ in a relation R of $D(\phi)$ iff ϕ contains the conjunct $R(v_1, \ldots, v_k)$. The following is similarly straightforward as Proposition 1.2.4.

PROPOSITION 1.2.5. *Let $\mathfrak{B}$ be a structure with signature τ, and let ϕ be a primitive positive τ-sentence that does not contain the symbols $=$ and $\bot$. Then ϕ is true in $\mathfrak{B}$ if and only if $D(\phi)$ maps homomorphically to $\mathfrak{B}$.*

Note that the canonical database of the canonical query of a structure $\mathfrak{A}$ equals $\mathfrak{A}$. Conversely, the canonical query of the canonical database of a quantifier-free primitive positive formula ϕ is equivalent to ϕ. Also note that from an instance ϕ of $\mathrm{CSP}(\mathfrak{B})$ that contains the symbol $=$ we can easily compute an equivalent instance which does not, by replacing every occurrence of y by x if ϕ contains the conjunct $x = y$, and then removing this conjunct.

Due to Proposition 1.2.5 and Proposition 1.2.4, we may freely switch between the homomorphism and the logic perspective whenever this is convenient. In particular, instances of $\mathrm{CSP}(\mathfrak{B})$ can from now on be either finite structures $\mathfrak{A}$ or primitive positive sentences ϕ, whichever is more convenient.

1.2.3. Expansions. Let $\mathfrak{A}$ be a τ-structure, and let $\mathfrak{A}'$ be a τ'-structure with $\tau \subseteq \tau'$. If $\mathfrak{A}$ and $\mathfrak{A}'$ have the same domain and $R^{\mathfrak{A}} = R^{\mathfrak{A}'}$ for all $R \in \tau$, then $\mathfrak{A}$ is called the τ-*reduct* (or simply *reduct*) of $\mathfrak{A}'$, and $\mathfrak{A}'$ is called a τ'-*expansion* (or simply *expansion*) of $\mathfrak{A}$. If $\mathfrak{A}$ is a structure, and R is a relation over the domain of $\mathfrak{A}$, then we denote the expansion of $\mathfrak{A}$ by R by $(\mathfrak{A}, R)$.

When $\mathfrak{A}$ is a τ-structure, and $\phi(x_1, \ldots, x_k)$ is a formula with k free variables $x_1, \ldots, x_k$, then the *relation defined by* ϕ is the relation

$$\{(a_1, \ldots, a_k) \mid \mathfrak{A} \models \phi(a_1, \ldots, a_k)\}.$$

If the formula is primitive positive, then this relation is called *primitively positively definable*. Primitive positive definitions are an important concept in mathematics, either explicitly or implicitly; for instance, the famous theorem of Davis, Matiyasevich, Putnam, and Robinson [282] can be phrased as follows: a subset of $\mathbb{Z}$ is recursively enumerable if and only if it has a primitive positive definition in $(\mathbb{Z}; +, *, 1)$. Also note that if $\mathfrak{A}$ is a relational structure with domain $\mathbb{R}$ whose relations are convex and if R is primitively positively definable in $\mathfrak{A}$, then R must be convex, too.

The following lemma says that we can expand structures by primitively positively definable relations without changing the complexity of the corresponding CSP. Hence, primitive positive definitions are an important tool to prove NP-hardness: to show that $\mathrm{CSP}(\mathfrak{B})$ is NP-hard, it suffices to show that there is a primitive positive definition of a relation R such that $\mathrm{CSP}(\mathfrak{B}, R)$ is already known to be NP-hard. Stronger tools to prove NP-hardness of CSPs will be introduced in Section 6.3.5.

LEMMA 1.2.6 (Jeavons, Cohen, Gyssens [217]). *Let $\mathfrak{B}$ be a structure with finite relational signature, and let R be a relation that has a primitive positive definition in $\mathfrak{B}$. Then* CSP($\mathfrak{B}$) *and* CSP($\mathfrak{B}, R$) *are linear-time equivalent. They are also equivalent under log-space reductions.*

PROOF. It is clear that CSP($\mathfrak{B}$) reduces to the new problem. So suppose that ϕ is an instance of CSP($\mathfrak{B}, R$). Replace each conjunct $R(x_1, \ldots, x_l)$ of ϕ by its primitive positive definition $\psi(x_1, \ldots, x_l)$. Move all quantifiers to the front, so that the resulting formula is in *prenex normal form* and hence primitive positive. Finally, equalities can be eliminated one by one: for an equality $x = y$, remove y from the quantifier prefix, and replace all remaining occurrences of y by x. Let ψ be the formula obtained in this way.

It is straightforward to verify that ϕ is true in $(\mathfrak{B}, R)$ if and only if ψ is true in $\mathfrak{B}$, and it is also clear that ψ can be constructed in linear time in the representation size of ϕ. For the observation that the reduction is in log-space we need the recent result that undirected reachability can be decided in deterministic log-space [313]. ⊣

Example 1.2.7. Consider the 5-element clique $K_5 = (\{0, 1, 2, 3, 4\}; \neq)$. Its edge relation can be defined primitively positively over the undirected 5-element cycle C_5, i.e., the graph $(\{0, 1, 2, 3, 4\}; E)$ for $E = \{(x, y) \mid x - y \in \{1, -1\} \bmod 5\}$. The primitive positive definition is

$$\exists u_1, u_2 (E(x, u_1) \wedge E(u_1, u_2) \wedge E(u_2, y))$$

which states the existence of a path with three edges between x and y; note that from every vertex x of C_5 every other vertex is reachable via a path of length three, except for x itself. Hence, the relation defined by the formula is the edge relation of the 5-element clique.

A typical application of Lemma 1.2.6 is the following.

COROLLARY 1.2.8. *Let $\mathfrak{B}$ be a structure with domain $\{0, 1\}$ such that the relation* NAE *(see Example* 1.2.2*) has a primitive positive definition in $\mathfrak{B}$. Then* CSP($\mathfrak{B}$) *is NP-hard.*

PROOF. By Lemma 1.2.6, there is a polynomial-time (even log-space) reduction from CSP($\{0, 1\}$; NAE) to CSP($\mathfrak{B}$). The problem CSP($\{0, 1\}$; NAE) is positive Not-All-Equal-3SAT (see Example 1.2.2), which is NP-hard [176]. ⊣

An *automorphism* of a structure $\mathfrak{B}$ with domain B is an isomorphism between $\mathfrak{B}$ and itself. When applying an automorphism α to an element b from B we omit brackets, that is, we write αb instead of $\alpha(b)$. The set of all automorphisms α of $\mathfrak{B}$ is denoted by Aut($\mathfrak{B}$), and α^{-1} denotes the inverse map of α.

DEFINITION 1.2.9. Let $(b_1, \ldots, b_k)$ be a k-tuple of elements of $\mathfrak{B}$. A set of the form $S = \{(\alpha b_1, \ldots, \alpha b_k) \mid \alpha \in \text{Aut}(\mathfrak{B})\}$ is called an *orbit of k-tuples under* Aut($\mathfrak{B}$) (the *orbit of* $(b_1, \ldots, b_k)$).

LEMMA 1.2.10. *Let $\mathfrak{B}$ be a structure with a finite relational signature and domain B, and let $R = \{(b_1, \ldots, b_k)\}$ be a k-ary relation that only contains one tuple $(b_1, \ldots, b_k) \in B^k$. If the orbit of $(b_1, \ldots, b_k)$ under* Aut($\mathfrak{B}$) *is primitively positively definable, then there is a polynomial-time reduction from* CSP($\mathfrak{B}, R$) *to* CSP($\mathfrak{B}$).

PROOF. Let ϕ be an instance of CSP($\mathfrak{B}, R$) with variable set V. If ϕ contains two constraints $R(x_1, \ldots, x_k)$ and $R(y_1, \ldots, y_k)$, then replace each occurrence of y_1 in ϕ by x_1, then each occurrence of y_2 by x_2, and so on, and finally each occurrence of y_k by x_k. We repeat this step until all constraints that involve R are imposed on the same tuple of variables $(x_1, \ldots, x_k)$. Replace $R(x_1, \ldots, x_k)$ by the primitive positive definition θ of its orbit under Aut($\mathfrak{B}$). Finally, move all quantifiers to the front, so that the resulting formula ψ is in prenex normal form and thus an instance of CSP($\mathfrak{B}$). Clearly, ψ can be computed from ϕ in polynomial time. We claim that ϕ is true in $(\mathfrak{B}, R)$ if and only if ψ is true in $\mathfrak{B}$.

Suppose there is a map $s \colon V \to B$ showing that ϕ holds in $(\mathfrak{B}, R)$. Let s' be the restriction of s to the variables of V that also appear in ϕ. Since $(b_1, \ldots, b_n)$ satisfies θ, we can extend s' to the existentially quantified variables of θ to obtain a solution for ψ. In the opposite direction, suppose that s' is a solution to ψ over $\mathfrak{B}$. Let s be the restriction of s' to V. Because $(s(x_1), \ldots, s(x_k))$ satisfies θ it lies in the same orbit as $(b_1, \ldots, b_k)$. Thus, there exists an automorphism α of $\mathfrak{B}$ that maps $(s(x_1), \ldots, s(x_k))$ to $(b_1, \ldots, b_k)$. Then the extension of the map $x \mapsto \alpha s(x)$ that maps variables y_i of ϕ that have been replaced by x_i in ψ to the value b_i is a solution to ϕ over $(\mathfrak{B}, R)$. ⊣

Recall from Section 1.1 that every finite structure $\mathfrak{C}$ is homomorphically equivalent to a core structure $\mathfrak{B}$, which is unique up to isomorphism.

PROPOSITION 1.2.11. *Let $\mathfrak{B}$ be a finite core structure. Then orbits of k-tuples under* Aut($\mathfrak{B}$) *are primitively positively definable.*

Proposition 1.2.11 has a simple proof for finite structures $\mathfrak{B}$; however, the same fact is true for a large class of infinite structures, and is presented in Theorem 4.5.1, so we omit the proof of Proposition 1.2.11 at this point. Proposition 1.2.11 and Lemma 1.2.10 have the following well-known consequence.

COROLLARY 1.2.12. *Let $\mathfrak{B}$ be a finite core structure with elements $b_1, \ldots, b_n$ and finite signature. Then* CSP($\mathfrak{B}$) *and* CSP($\mathfrak{B}, \{b_1\}, \ldots, \{b_n\}$) *are polynomial-time equivalent.*

1.3. The satisfiability perspective

Yet another perspective on the constraint satisfaction problem translates not only the instances, but also the template of the CSP into logic. This leads to a natural perspective for various model-theoretic considerations in Chapter 2.

We use the opportunity to introduce some essential terminology from logic. We assume that the reader is already familiar with basic terminology of first-order logic; a highly recommended textbook is Hodges [207]. A (*first-order*) *theory* is a set of first-order sentences. If the first-order sentences are over the signature τ, we also say that T is a τ-*theory*. A *model* of a τ-theory T is a τ-structure $\mathfrak{B}$ such that $\mathfrak{B}$ satisfies all sentences in T. Theories that have a model are called *satisfiable*.

DEFINITION 1.3.1. Let τ be a finite relational signature, and let T be a τ-theory. Then CSP(T) is the computational problem to decide for a given primitive positive τ-sentence ϕ whether $T \cup \{\phi\}$ is satisfiable.

The satisfiability perspective on CSPs stresses the fact that the problem CSP$(\mathfrak{B})$ is fully determined by the (universal negative) first-order theory of $\mathfrak{B}$, that is, by the theory that consists of the (universal negative) sentences that are true in $\mathfrak{B}$. In fact, it is already determined by the primitive positive sentences that are false in $\mathfrak{B}$.

Example 1.3.2. Let T be the theory that consists of the following sentences.

$$\forall x, y, z \left((x < y \wedge y < z) \rightarrow x < z\right) \qquad \text{(transitivity)}$$
$$\forall x, y \neg(x < x) \qquad \text{(irreflexivity)}$$
$$\forall x, y, z \left((x < y) \vee (y < x) \vee (x = y)\right) \qquad \text{(totality)}$$

It is straightforward to verify that CSP(T) equals CSP$(\mathbb{Z}; <)$ (Example 1.1.2).

When T is a theory and ϕ a sentence, we say that T *entails* ϕ, in symbols $T \models \phi$, if every model of T satisfies ϕ. The following is clear from the definitions.

PROPOSITION 1.3.3. *Let τ be a finite relational signature, and let T be a τ-theory. Suppose that T entails exactly those negations of primitive positive sentences ϕ such that $\mathfrak{B} \not\models \phi$. Then* CSP$(T)$ *and* CSP$(\mathfrak{B})$ *are the same problem.*

We have already seen that two structures that are homomorphically equivalent have the same CSP; the following provides a necessary and sufficient condition that describes when two *theories* have the same CSP. Its proof is simple once the relevant notions from logic are introduced, and will be given in Section 2.1.9.

PROPOSITION 1.3.4. *Let T and T' be two first-order theories. Then the following are equivalent.*

- CSP(T) *equals* CSP(T').
- *Every model of T' has a homomorphism to some model of T, and every model of T has a homomorphism to some model of T'.*
- *T and T' entail the same negations of primitive positive sentences.*

We now present a couple of basic observations relating the definition of CSP(T) for a theory T with the definition of CSP($\mathfrak{B}$) for a relational structure $\mathfrak{B}$. We start with the observation that there are theories T such that CSP(T) cannot be formulated as CSP($\mathfrak{B}$) for a single structure $\mathfrak{B}$.

Example 1.3.5. Let τ be the signature $\{R, G\}$, where R and G are unary relation symbols, and let T be the τ-theory $\{\forall x, y \neg(R(x) \wedge G(y))\}$. There is no structure $\mathfrak{B}$ such that CSP($\mathfrak{B}$) equals CSP(T). To see this, observe that $T \cup \{\exists x \colon R(x)\}$ is satisfiable, and $T \cup \{\exists x \colon G(x)\}$ is satisfiable. But any structure $\mathfrak{B}$ that satisfies both $\exists x \colon R(x)$ and $\exists x \colon G(x)$ also satisfies $\exists x, y\, (R(x) \wedge R(y))$, which shows that CSP($\mathfrak{B}$) and CSP($T$) are different.

We next characterize those satisfiable theories T that have a model $\mathfrak{B}$ such that CSP($\mathfrak{B}$) and CSP(T) are the same problem.

PROPOSITION 1.3.6. *Let τ be a finite relational signature, and let T be a satisfiable first-order τ-theory. The following are equivalent.*

1. *There is a structure $\mathfrak{B}$ such that* CSP($\mathfrak{B}$) *and* CSP(T) *are the same problem.*
2. *There is a model $\mathfrak{B}$ of T such that* CSP($\mathfrak{B}$) *and* CSP(T) *are the same problem.*
3. *For all primitive positive τ-sentences ϕ_1 and ϕ_2, if $T \cup \{\phi_1\}$ is satisfiable and $T \cup \{\phi_2\}$ is satisfiable then $T \cup \{\phi_1, \phi_2\}$ is satisfiable as well.*
4. *T has the* Joint Homomorphism Property (JHP), *that is, when T has models $\mathfrak{A}$ and $\mathfrak{B}$, then it also has a model $\mathfrak{C}$ such that both $\mathfrak{A}$ and $\mathfrak{B}$ map homomorphically to $\mathfrak{C}$.*

We defer the proof of this fact to Section 2.1.9 when we have some more concepts from logic available.

1.4. The existential second-order perspective

By a famous result of Fagin, which will be reviewed below, the complexity class NP corresponds exactly to those problems that can be formulated in existential second-order logic (ESO). An important fragment of ESO that is particularly natural when it comes to the formulation of CSPs is the logic called *SNP* (for *strict NP*, see [298] and [170]), introduced by Kolaitis and Vardi under the name *strict* Σ^1_1 [236]. An existential second-order sentence is in SNP if its first-order part is universal. There are many links between constraint satisfaction and the complexity class SNP; many of those go back to [170] and [171], some others that we present here are new.

The logic SNP is often a convenient way to specify CSPs. However, not every problem in SNP is a CSP. In this section we present a natural syntactic condition that implies that an SNP sentence describes a problem of the form CSP($\mathfrak{B}$) for an infinite structure $\mathfrak{B}$ (Corollary 1.4.12). Conversely, if an SNP sentence describes a CSP, then there is an equivalent SNP sentence that satisfies our syntactic condition.

The special case in which all existentially quantified relations are unary, known as *monadic SNP*, deserves special attention, and will be discussed at the end of this section.

1.4.1. Fagin's theorem. We start by reviewing Fagin's theorem (see e.g. [162]). Fix a finite relational signature τ. Let C be a class of finite τ-structures that is closed under isomorphisms (that is, if $\mathfrak{B} \in C$, and $\mathfrak{A}$ is isomorphic to $\mathfrak{B}$, then $\mathfrak{A} \in C$). We say that C is in NP when there exists a nondeterministic polynomial-time algorithm that accepts exactly the structures from C (here we fix some standard way to code relational structures as finite strings so that they can be given as an input to a Turing machine, see again [162]).

A sentence of the form $\exists R_1, \ldots, R_m : \phi$ where ϕ is a first-order sentence with signature $\tau \cup \{R_1, \ldots, R_m\}$ is called an *existential second-order τ-sentence* (or simply *ESO-sentence*). When a structure $\mathfrak{A}$ satisfies Φ (and this is defined in the obvious way; see e.g. [162]), we write $\mathfrak{A} \models \Phi$.

THEOREM 1.4.1 (Fagin's Theorem; see e.g. [162]). *An isomorphism-closed class of finite τ-structures is in NP if and only if there exists an existential second-order sentence Φ that* describes C *in the sense that*

$$\mathfrak{A} \in C \text{ if and only if } \mathfrak{A} \models \Phi.$$

It would be great to have an algorithm that determined for a given ESO sentence Φ whether Φ describes a problem that can be solved in polynomial time. Alas, it turns out that such an algorithm cannot exist (unless, of course, P = NP).

THEOREM 1.4.2. *If P $\neq$ NP, then the problem of deciding whether a given ESO sentence describes a problem in P is undecidable.*

PROOF. Trakhtenbrot's theorem [340] asserts that the problem of deciding whether a given first-order sentence is valid over finite structures is undecidable. We reduce Trakhtenbrot's problem to the complement of our problem. Let φ be an arbitrary first-order sentence with signature $R_1, \ldots, R_r$. We first construct a universal second-order sentence Φ such that $\mathfrak{A} \models \Phi$ if and only if $\mathfrak{B} \models \forall R_1, \ldots, R_r : \phi$ for every induced substructure $\mathfrak{B}$ of $\mathfrak{A}$. The sentence Φ is

$$\forall U \forall R_1, \ldots, R_r : \psi$$

where ψ is obtained from φ by replacing each expression of the form $\exists x\colon \chi(x)$ by $\exists x\,(\neg U(x) \wedge \chi(x))$ and each expression of the form $\forall x\colon \chi(x)$ by $\forall x\,(U(x) \vee \chi(x))$.

Let Θ be an ESO sentence expressing SAT. By construction, $\Phi \Rightarrow \Theta$ is equivalent to an ESO sentence because $\neg\Phi$ is equivalent to an ESO sentence. We reduce φ to the computational question whether $\Phi \Rightarrow \Theta$ is in P or NP-complete (assuming P $\neq$ NP).

- If φ is true on all finite structures then $\Phi \Rightarrow \Theta$ expresses SAT and is therefore NP-complete.
- Otherwise, there exists a $c \in \mathbb{N}$ such that Φ is false for all structures with more than c elements. Hence, $\Phi \Rightarrow \Theta$ expresses the problem "the input structure encodes a satisfiable instance of SAT with strictly less than c elements or it has c or more elements". Since c is a constant, this problem is clearly in P.

We have therefore established a reduction from Trakhtenbrot's problem to the complement of our problem. It follows from Trakhtenbrot's theorem that our problem is also undecidable. ⊣

1.4.2. SNP. An *SNP* (τ-) *sentence* is an existential second-order sentence with a universal first-order part, i.e., a sentence of the form

$$\exists R_1, \dots, R_k \,\forall x_1, \dots, x_n\colon \phi$$

where ϕ is quantifier-free and over the signature $\tau \cup \{R_1, \dots, R_k\}$. The class of problems that can be described by SNP sentences is called SNP, too.

Example 1.4.3. The problem CSP($\mathbb{Z}$; <) can be described by the following SNP sentence.

$$\exists T\, \forall x, y, z\big((x < y \Rightarrow T(x, y)) \wedge \\ ((T(x, y) \wedge T(y, z)) \Rightarrow T(x, z)) \wedge \neg T(x, x)\big)$$

Example 1.4.4. The Betweenness Problem CSP($\mathbb{Z}$; Betw) from Example 1.1.3 can be described by the following SNP sentence.

$$\exists T\, \forall x, y, z\big(\neg T(x, x) \wedge ((T(x, y) \wedge T(y, z)) \Rightarrow T(x, z)) \wedge \\ (\mathrm{Betw}(x, y, z) \Rightarrow ((T(x, y) \wedge T(y, z)) \vee (T(z, y) \wedge T(y, x))))\big)$$

Example 1.4.5. The problem whether a given undirected graph $(V; E)$ can be partitioned into two triangle-free graphs (this problem has been called No-Mono-Tri in Example 1.1.9) can be described by the following SNP sentence.

$$\exists M\, \forall x, y, z\, \big(\neg(M(x) \wedge M(y) \wedge M(z) \wedge E(x, y) \wedge E(y, z) \wedge E(z, x)) \wedge \\ \neg(\neg M(x) \wedge \neg M(y) \wedge \neg M(z) \wedge E(x, y) \wedge E(y, z) \wedge E(z, x))\big)$$

The following fundamental lemma for SNP is due to Feder and Vardi [171], and an easy consequence of the compactness theorem (Theorem 2.1.6).

LEMMA 1.4.6 (Feder and Vardi [171]). *Let $\mathfrak{A}$ be an infinite structure, and Φ an SNP sentence. Then $\mathfrak{A} \models \Phi$ if and only if $\mathfrak{A}' \models \Phi$ for all finite substructures $\mathfrak{A}'$ of $\mathfrak{A}$.*

Since every finite substructure of $\mathfrak{B}$ maps homomorphically to $\mathfrak{B}$, and therefore satisfies Φ, we have the following consequence.

COROLLARY 1.4.7. *Let Φ be an SNP sentence that describes* CSP($\mathfrak{B}$) *for a structure $\mathfrak{B}$. Then $\mathfrak{B}$ itself satisfies Φ.*

1.4.3. SNP and CSPs. We say that two SNP sentences Φ and Ψ are *equivalent* if for all structures (equivalently: all finite structures) $\mathfrak{A}$ we have $\mathfrak{A} \models \Phi$ if and only if $\mathfrak{A} \models \Psi$. We assume in the following that the quantifier-free part ϕ of Φ is written in *conjunctive normal form* (*CNF*), i.e., it is a conjunction of disjunctions of *literals*, which are either atomic formulas or negated atomic formulas. The disjunctions of literals are also called *clauses* and often treated as finite *sets* of literals.

DEFINITION 1.4.8. Let Φ be an SNP τ-sentence with quantifier-free part ϕ in CNF. Then Φ is called *monotone* if each literal of ϕ with a symbol from $\tau \cup \{=\}$ is *negative*, that is, of the form $\neg R(\bar{x})$, for $R \in \tau \cup \{=\}$.

In particular, monotone SNP sentences do not contain literals of the form $x = y$.[6] We also assume that monotone SNP sentences do not contain literals of the form $x \neq y$. This is without loss of generality, since every monotone SNP sentence is equivalent to one which does not contain literals of the form $x \neq y$. To obtain the equivalent sentence, we remove literals of the form $x \neq y$ and replace all occurrences of y in the same clause by x. Note that the SNP sentences given in Example 1.4.3, 1.4.4, and 1.4.5 can be easily re-written into equivalent monotone SNP sentences.

The class of structures that satisfy a given monotone SNP sentence is clearly closed under inverse homomorphisms (Definition 1.1.7). The converse is a result by Feder and Vardi [171]; it shows that for SNP, the semantic restriction of closure under inverse homomorphisms and the syntactic restriction of monotonicity match.

THEOREM 1.4.9 (Feder and Vardi [171]). *Let Φ be an SNP sentence. Then the class of structures that satisfy Φ is closed under inverse homomorphisms if and only if Φ is equivalent to a monotone SNP sentence.*

PROOF. We only have to prove the forward direction. Introduce a new binary existentially quantified relation symbol E and conjuncts that imply that E is

[6] In the terminology of Feder and Vardi [170], we work here with *monotone SNP without inequality*; the reason why Feder and Vardi add the attribute *without inequalities* is that for them, the quantifier-free part of SNP sentences is written in negation normal form, so forbidding literals of the form $x = y$ amounts to forbidding inequalities in negation normal form.

an equivalence relation, and replace each subformula of the form $x = y$ by $E(x, y)$. For each existentially quantified relation symbol R of arity k in Φ let R' be a new relation symbol of arity k. Replace each occurrence of R in Φ by R', and then add the conjuncts $R(x_1, \ldots, x_n) \Rightarrow R'(x_1, \ldots, x_n)$ for new universally quantified variables $x_1, \ldots, x_n$. Moreover, for each $i \leq n$ we add the conjunct

$$\big(E(x_i, y) \wedge R'(x_1, \ldots, x_n)\big) \Rightarrow R'(x_1, \ldots, x_{i-1}, y, x_{i+1}, \ldots, x_n).$$

The resulting SNP sentence Ψ is clearly a monotone SNP sentence. If a structure $\mathfrak{A}$ satisfies Φ, then it also satisfies Ψ, because we may find an expansion of $\mathfrak{A}$ where E denotes the equality relation and R and R' denote the same relation, for every relation R in Φ, and which shows that $\mathfrak{A}$ satisfies Ψ. Conversely, suppose that $\mathfrak{A}$ satisfies Ψ. Then there exists an expansion $\mathfrak{A}'$ of $\mathfrak{A}$ that satisfies the quantifier-free part of Ψ. Let $\mathfrak{A}''$ be the relation obtained from $\mathfrak{A}'$ by

- factoring by $E^{\mathfrak{A}'}$; i.e., the elements of $\mathfrak{A}''$ are the equivalence classes of $E^{\mathfrak{A}'}$ and if $(a_1, \ldots, a_k) \in R^{\mathfrak{A}'}$ then $(a_1/E, \ldots, a_k/E) \in R^{\mathfrak{A}''}$.
- renaming each relation symbol R' to R and dropping the relation E.

Note that $\mathfrak{A}''$ satisfies Φ. Also note the map that sends an element $a \in A$ to its equivalence class $a/E \in A/E$ is a homomorphism from $\mathfrak{A}$ to $\mathfrak{A}''$. Since the class of all structures that satisfy Φ is closed under inverse homomorphisms, $\mathfrak{A}$ satisfies Φ as well. ⊣

We now introduce another syntactic restriction on SNP sentences which plays an important role in the context of CSPs.

DEFINITION 1.4.10 (Connected SNP). Let ψ be a clause of a first-order σ-formula ϕ in conjunctive normal form, and let $\mathfrak{C}$ be the σ-structure whose vertices are the variables of ψ, and where $(x_1, \ldots, x_n) \in R^{\mathfrak{C}}$ if and only if ψ contains a negative literal of the form $\neg R(x_1, \ldots, x_n)$. Then ψ is *connected* if $\mathfrak{C}$ is connected. An SNP sentence Φ is *connected* if every clause of the first-order part ϕ of Φ is connected.

Note that the SNP sentences in Example 1.4.3 and Example 1.4.4 are connected.

PROPOSITION 1.4.11. *Let Φ be an SNP sentence. Then the class of structures that satisfy Φ is closed under disjoint unions if and only if Φ is equivalent to a connected SNP sentence.*

PROOF. Let Φ be of the form $\exists R_1, \ldots, R_k\ \forall x_1, \ldots, x_l \colon \phi$ where ϕ is a quantifier-free first-order formula over the signature $\sigma = (\tau \cup \{R_1, \ldots, R_k\})$.

Suppose first that Φ is connected, and that $\mathfrak{A}_1$ and $\mathfrak{A}_2$ both satisfy Φ. In other words, there is a σ-expansion $\mathfrak{A}_1^*$ of $\mathfrak{A}_1$ and a σ-expansion $\mathfrak{A}_2^*$ of $\mathfrak{A}_2$ such that these expansions satisfy $\forall \bar{x} \colon \phi$. We claim that the disjoint union $\mathfrak{A}^*$ of $\mathfrak{A}_1^*$ and $\mathfrak{A}_2^*$ also satisfies $\forall \bar{x} \colon \phi$; otherwise, there would be a clause ψ in ϕ and elements $a_1, \ldots, a_q$ of $A_1 \cup A_2$ such that $\psi(a_1, \ldots, a_q)$ is false in $\mathfrak{A}^*$. Since $\mathfrak{A}_1^*$

and $\mathfrak{A}_2^*$ satisfy $\forall \bar{x} : \psi$, there must be i, j such that $a_i \in A_1$ and $a_j \in A_2$. But then ψ is disconnected, a contradiction.

For the opposite direction of the statement, assume that the class of structures that satisfy Φ is closed under disjoint unions. Consider the SNP sentence $\Psi = \exists R_1, \ldots, R_k, E \; \forall x_1, \ldots, x_l : \psi$ where ψ is the conjunction of the following clauses (we assume without loss of generality that $l \geq 3$).

- For each relation symbol $R \in \tau$, say of arity p, and each $i < j \leq p$, add the conjunct $\neg R(x_1, \ldots, x_p) \vee E(x_i, x_j)$ to ψ.
- Add the conjunct $\neg E(x_1, x_2) \vee \neg E(x_2, x_3) \vee E(x_1, x_3)$ to ψ.
- Add the conjunct $\neg E(x_1, x_2) \vee E(x_2, x_1)$ to ψ.
- For each clause ϕ' of ϕ with variables $y_1, \ldots, y_q \subseteq \{x_1, \ldots, x_l\}$, add to ψ the conjunct

$$\phi' \vee \bigvee_{i<j\leq q} \neg E(y_i, y_j).$$

We claim that the connected monotone SNP sentence Ψ is equivalent to Φ. Suppose first that $\mathfrak{A}$ is a finite structure that satisfies Φ. Then there is a σ-expansion $\mathfrak{A}'$ of $\mathfrak{A}$ that satisfies $\forall \bar{x} : \phi$. The expansion of $\mathfrak{A}'$ by the relation $E = A^2$ shows that $\mathfrak{A}$ also satisfies $\forall \bar{x} : \psi$.

Now suppose that $\mathfrak{A}$ is a finite structure with domain A that satisfies Ψ. Then there is a $(\sigma \cup \{E\})$-expansion $\mathfrak{A}'$ of $\mathfrak{A}$ that satisfies $\forall \bar{x} : \psi$. Write $\mathfrak{A}' = \mathfrak{A}'_1 \uplus \cdots \uplus \mathfrak{A}'_m$ for connected σ-structures $\mathfrak{A}'_1, \ldots, \mathfrak{A}'_m$. Note that the clauses of ψ force that the relation E denotes A_i^2 in the structure $\mathfrak{A}'_i$, for each $i \leq m$. Let $\mathfrak{A}_i$ be the σ-reduct of $\mathfrak{A}'_i$. Then $\mathfrak{A}_i$ satisfies $\forall \bar{x} : \phi$, because if there was a clause ϕ' from ϕ not satisfied in $\mathfrak{A}_i$ then the corresponding clause in ψ would not be satisfied in $\mathfrak{A}'_i$. Hence, $\mathfrak{A}_i \models \Phi$ for all $i \leq m$, and since Φ is closed under disjoint unions, we also have that $\mathfrak{A} \models \Phi$. ⊣

Theorem 1.4.9 combined with the previous result shows the following.

COROLLARY 1.4.12. *An SNP sentence Φ describes a problem of the form* CSP($\mathfrak{B}$) *for an infinite structure $\mathfrak{B}$ if and only if Φ is equivalent to a monotone and connected SNP sentence Ψ.*

PROOF. Suppose first that Φ is a monotone SNP sentence with connected clauses. To show that Φ describes a problem of the form CSP($\mathfrak{B}$) we can use Lemma 1.1.8. It thus suffices to show that the class of structures that satisfy Φ is closed under disjoint unions and inverse homomorphisms. But this has already been observed in Theorem 1.4.9 and Theorem 1.4.11.

For the implication in the opposite direction, suppose that Φ describes a problem of the form CSP($\mathfrak{B}$) for some infinite structure $\mathfrak{B}$. In particular, the class of structures that satisfy Φ is closed under inverse homomorphisms. By Theorem 1.4.9, Φ is equivalent to a monotone SNP sentence. Moreover, the class of structures that satisfy Φ is closed under disjoint unions, and hence Φ is also equivalent to a connected SNP sentence. By inspection of the proof of

Theorem 1.4.11, we see that if Φ is already monotone, then the connected SNP sentence in the proof of Theorem 1.4.11 will also be monotone. It follows that Φ is also equivalent to a connected monotone SNP sentence. ⊣

1.4.4. Monotone monadic SNP. If we further restrict monotone SNP by only allowing *unary* existentially quantified relations, the corresponding class of problems, called *montone monadic SNP* (or, short, *MMSNP*), gets very close to finite domain constraint satisfaction problems. Indeed, Feder and Vardi showed that the class MMSNP exhibits a complexity dichotomy if and only if the class of all finite domain CSPs exhibits a complexity dichotomy (that is, if the *dichotomy conjecture* mentioned in the introduction is true). In one direction, this is obvious since MMSNP obviously contains CSP($\mathfrak{B}$) for all finite structures $\mathfrak{B}$ (we may use a unary relation symbol for each element of $\mathfrak{B}$). In the other direction, Feder and Vardi showed that every problem in MMSNP is equivalent under randomised Turing reductions to a finite domain constraint satisfaction problem. The reduction has subsequently been derandomised by Kun [249].

In order to present the details of the link between MMSNP and CSPs, we start with the following fundamental observation.

PROPOSITION 1.4.13 (of [170]; see also [277]). *Let Φ be an MMSNP sentence. Then Φ is equivalent to a finite disjunction of connected MMSNP sentences.*

PROOF. Suppose that the first-order part ϕ of Φ contains a disconnected clause ψ. Since the positive literals of ψ must be unary, ψ is equivalent to $\psi_1 \vee \psi_2$ for non-empty clauses ψ_1 and ψ_2. Let ϕ_1 be the formula obtained from ϕ by replacing ψ by ψ_1, and let ϕ_2 be the formula obtained from ϕ by replacing ψ by ψ_2.

Let $P_1, \dots, P_k$ be the existential monadic predicates in Φ, and let τ be the input signature of Φ. It suffices to show that a $(\tau \cup \{P_1, \dots, P_k\})$-expansion $\mathfrak{A}'$ of $\mathfrak{A}$ satisfies ϕ if and only if $\mathfrak{A}'$ satisfies ϕ_1 or ϕ_2. If the universally quantified variables can be instantiated so that $\mathfrak{A}'$ falsifies a clause of ϕ, there is nothing to show since then $\mathfrak{A}'$ satisfies neither ϕ_1 nor ϕ_2. Conversely, suppose that $\mathfrak{A}'$ satisfies neither ϕ_1 nor ϕ_2. That is, the universally quantified variables can be instantiated so that $\mathfrak{A}'$ falsifies ψ_1 and they can be instantiated so that $\mathfrak{A}'$ falsifies ψ_2. Since the variables of ψ_1 and ψ_2 are distinct we can then also instantiate the variables so that $\mathfrak{A}'$ falsifies $\psi_1 \vee \psi_2$. This shows that $\mathfrak{A}'$ does not satisfy ϕ. Iterating this process for each disconnected clause of ϕ, we eventually arrive at a finite disjunction of connected MMSNP sentences. ⊣

We now show that we can reduce the complexity classification for MMSNP to the classification for *connected* MMSNP. We first prove a general result about finite unions of CSPs; the homomorphism perspective on the CSP is most natural to present this result.

PROPOSITION 1.4.14. *Let τ be a finite relational signature and let $\mathfrak{B}_1, \dots, \mathfrak{B}_k$ be τ-structures that have pairwise distinct CSPs. Then for each $i \leq k$ the problem* $\mathrm{CSP}(\mathfrak{B}_i)$ *has a polynomial-time reduction to* $\mathrm{CSP}(\mathfrak{B}_1) \cup \dots \cup \mathrm{CSP}(\mathfrak{B}_k)$. *Conversely, if each of* $\mathrm{CSP}(\mathfrak{B}_i)$ *can be solved in polynomial time, then* $\mathrm{CSP}(\mathfrak{B}_1) \cup \dots \cup \mathrm{CSP}(\mathfrak{B}_k)$ *can be solved in polynomial time as well.*

PROOF. Let $i \in \{1, \dots, k\}$. By assumption, for every $j \in \{1, \dots, k\} \setminus \{i\}$ there exists a finite τ-structure $\mathfrak{A}_j$ that maps homomorphically to $\mathfrak{B}_i$, but does not map to $\mathfrak{B}_j$. To reduce $\mathrm{CSP}(\mathfrak{B}_i)$ to $\mathrm{CSP}(\mathfrak{B}_1) \cup \dots \cup \mathrm{CSP}(\mathfrak{B}_k)$, execute for a given finite τ-structure $\mathfrak{A}$ an algorithm for $\mathrm{CSP}(\mathfrak{B}_1) \cup \dots \cup \mathrm{CSP}(\mathfrak{B}_k)$ on $\mathfrak{A}' := \mathfrak{A} \uplus \biguplus_{j \neq i} \mathfrak{A}_j$. We claim that $\mathfrak{A}' \in \mathrm{CSP}(\mathfrak{B}_1) \cup \dots \cup \mathrm{CSP}(\mathfrak{B}_k)$ if and only if $\mathfrak{A}$ maps homomorphically to $\mathfrak{B}_i$. First suppose that $\mathfrak{A}$ maps homomorphically to $\mathfrak{B}_i$. Since for every $j \neq i$ there is a homomorphism from $\mathfrak{A}_j$ to $\mathfrak{B}_i$, and since $\mathrm{CSP}(\mathfrak{B}_i)$ is closed under disjoint unions, we have that $\mathfrak{A}'$ has a homomorphism to $\mathfrak{B}_i$ as well. For the opposite direction, suppose that $\mathfrak{A}'$ is in $\mathrm{CSP}(\mathfrak{B}_1) \cup \dots \cup \mathrm{CSP}(\mathfrak{B}_k)$. Since $\mathfrak{A}_j$ does not map homomorphically to $\mathfrak{B}_j$ for all j distinct from i, there is no homomorphism from $\mathfrak{A}'$ to $\mathfrak{B}_j$ either. Hence, $\mathfrak{A}'$ and therefore also $\mathfrak{A}$ must be homomorphic to $\mathfrak{B}_i$.

If for every $i \leq k$ there is a polynomial-time algorithm A_i that solves $\mathrm{CSP}(\mathfrak{B}_i)$, then $\mathrm{CSP}(\mathfrak{B}_1) \cup \dots \cup \mathrm{CSP}(\mathfrak{B}_k)$ can be solved in polynomial time by running each of the algorithms $A_1, \dots, A_k$ on the input, and accepting if one of the algorithms accepts. ⊣

COROLLARY 1.4.15. *Every problem in MMSNP is in P or NP-complete if and only if every problem in* connected *MMSNP is in P or NP-complete.*

PROOF. The forward direction of the statement holds trivially. For the backwards direction, assume that every connected MMSNP sentence is either in P or NP-complete. Let Φ be an MMSNP sentence. By Proposition 1.4.13 there are connected MMSNP sentences $\Phi_1, \dots, \Phi_k$ so that Φ is logically equivalent to the disjunction $\Phi_1 \vee \dots \vee \Phi_k$. Each of the Φ_i describes a constraint satisfaction problem $\mathrm{CSP}(\mathfrak{B}_i)$, by Proposition 1.4.11; we may assume that the $\mathfrak{B}_i$ are pairwise homomorphically incomparable. The statement now follows from Proposition 1.4.14. ⊣

The following has been shown with *randomised* Turing-reductions by Feder and Vardi [170] (see also [275]); the reductions have been derandomised later by Kun [249].

THEOREM 1.4.16 (of [170] and [249]). *Every problem in monotone monadic SNP is polynomial-time Turing equivalent to* $\mathrm{CSP}(\mathfrak{B})$ *for a finite structure* $\mathfrak{B}$.

Similarly as in the previous section, we may ask for a syntactic characterisation of those monadic SNP sentences that describe a CSP. Note that this does not directly follow from the statement of Corollary 1.4.12, since the reductions used there introduce additional existentially quantified relations that are not monadic.

THEOREM 1.4.17 (Theorem 3 in [171]). *Let Φ be a monadic SNP sentence. Then the class of structures that satisfy Φ is closed under inverse homomorphisms if and only if Φ is equivalent to a monotone monadic SNP sentence.*

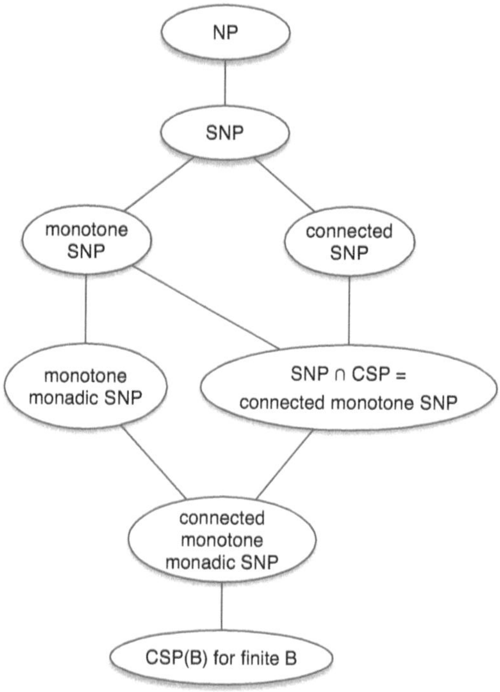

FIGURE 1. Fragments of NP related to CSPs.

Moreover, one can show the following monadic version of Proposition 1.4.11.

PROPOSITION 1.4.18. *Let Φ be a monadic SNP sentence. Then the class of structures that satisfy Φ is closed under disjoint unions if and only if Φ is equivalent to a connected monadic SNP sentence.*

PROOF. Let V be the set of variables of the first-order part ϕ of Φ, let $P_1, \dots, P_k$ be the existential monadic predicates in Φ, and let τ be the input signature so that ϕ has signature $\{P_1, \dots, P_k\} \cup \tau$. If Φ is connected, then it describes a problem that is closed under disjoint unions; this follows from Theorem 1.4.11.

For the opposite direction, suppose that Φ describes a problem that is closed under disjoint unions. We can assume without loss of generality that Φ is minimal in the sense that if we remove literals from some of the clauses the resulting SNP sentence is inequivalent. We shall show that then Φ must be connected. Let us suppose that this is not the case, and that there is a clause ψ in ϕ that is not connected. The clause ψ can be written as $\psi_1 \vee \psi_2$ where the set of variables $X \subset V$ of ψ_1 and the set of variables $Y \subset V$ of ψ_2 are non-empty and disjoint. Consider the formulas Φ_X and Φ_Y obtained from Φ by replacing ψ by ψ_1 and ψ by ψ_2, respectively. By minimality of Φ there is a τ-structure $\mathfrak{A}_1$ that satisfies Φ but not Φ_X, and similarly there exists a τ-structure $\mathfrak{A}_2$ that satisfies Φ but not Φ_Y. By assumption, the disjoint union $\mathfrak{A}$ of $\mathfrak{A}_1$ and $\mathfrak{A}_2$ satisfies Φ. So there exists a $\tau \cup \{P_1, \dots, P_k\}$-expansion $\mathfrak{A}'$ of $\mathfrak{A} = \mathfrak{A}_1 \uplus \mathfrak{A}_2$ that satisfies the first-order part of Φ. Consider the substructures $\mathfrak{A}'_1$ and $\mathfrak{A}'_2$ of $\mathfrak{A}'$ induced on A_1 and A_2, respectively. We have that $\mathfrak{A}'_1$ does not satisfy ψ_1 (otherwise $\mathfrak{A}_1$ would satisfy Φ_X). Consequently, there is an assignment $s_1 \colon V \to A_1$ of the universal variables that falsifies ψ_1. By similar reasoning we can infer that there is an assignment $s_2 \colon V \to A_2$ that falsifies ψ_2. Finally, fix any assignment $s \colon V \to A_1 \cup A_2$ that coincides with s_1 over X and with s_2 over Y (such an assignment exists because X and Y are disjoint). Clearly, s falsifies ψ and $\mathfrak{A}$ does not satisfy Φ, a contradiction. $\dashv$

Similarly as in Corollary 1.4.12 for SNP, the conditions of closure under inverse homomorphisms and closure under disjoint unions can be combined, arriving at the following.

COROLLARY 1.4.19. *A monadic SNP sentence Φ describes a problem of the form* CSP($\mathfrak{B}$) *for an infinite structure $\mathfrak{B}$ if and only if Φ is equivalent to a connected monotone monadic SNP sentence.*

The problems that can be described by connected monotone monadic SNP sentences are exactly the problems called *forbidden patterns problems* in the sense of Madelaine [273].

Example 1.4.20. CSP($\mathbb{Z}$; <) is not in MMSNP. Indeed, suppose that Φ is an MMSNP sentence which is true on all finite directed paths. We assume that the quantifier-free part of Φ is in conjunctive normal form. Let ρ be the existentially quantified unary relation symbols of Φ, let $k := |\rho|$, and let l be the number of variables in Φ. A directed path of length $(2^{kl}+1)l$, viewed as a $\{<\}$-structure, satisfies Φ, and therefore it has a $(\{<\} \cup \rho)$-expansion $\mathfrak{A}$ that satisfies the first-order part of Φ. Note that there are $L := 2^{kl}$ different $(\{<\} \cup \rho)$-expansions of a path of length l, and hence there must be $i, j \in \{0, \dots, L\}$ with $i < j$ such that the substructures of $\mathfrak{A}$ induced on $il+1, il+2, \dots, il+l$ and by $jl+1, jl+2, \dots, jl+l$ are isomorphic. We then claim that the directed cycle $(i+1)l+1, (i+1)l+2, \dots, jl+1, \dots, jl+l, (i+1)l+1$ satisfies Φ: this is witnessed by the $(\{<\} \cup \rho)$-expansion inherited from $\mathfrak{A}$ which satisfies the quantifier-free part of Φ. Hence, Φ does not express digraph acyclicity.

We summarise the landscape of classes of computational problems from this section in Figure 1.

1.5. CSPs and relation algebras

Many interesting infinite-domain CSPs, in particular in spatial and temporal reasoning, have been studied in the context of *relation algebras*; many examples will be given in Section 1.6 and Chapter 5. In Artificial Intelligence, relation algebras are used as a framework to formalise and study qualitative reasoning problems [251, 159, 203]. From the perspective of this book, the relation algebra approach does not bring substantially new tools, and Section 1.5 can be safely skipped. We give an introduction to this research direction in order to link the relation algebra terminology with the satisfiability perspective on the CSP (Section 1.5.3).

1.5.1. Proper relation algebras. Relation algebras are designed to handle binary relations in an algebraic way; we follow the presentation in [203].

DEFINITION 1.5.1. A *proper relation algebra* is a set B together with a set $\mathcal{R}$ of binary relations over B such that

1. $\mathrm{Id} := \{(x,x) \mid x \in D\} \in \mathcal{R}$;
2. if R_1 and R_2 are from $\mathcal{R}$, then $R_1 \vee R_2 := R_1 \cup R_2 \in \mathcal{R}$;
3. $1 := \bigcup_{B \in \mathcal{R}} B \in \mathcal{R}$;
4. $0 := \emptyset \in \mathcal{R}$;
5. if $B \in \mathcal{R}$, then $-B := 1 \setminus B \in \mathcal{R}$;
6. if $B \in \mathcal{R}$, then $B^{\smile} \in \mathcal{R}$, where $B^{\smile}$ is the *converse of* B defined by
$$B^{\smile} :- \{(x,y) \mid (y,x) \in B\};$$
7. If R_1 and R_2 are from $\mathcal{R}$, then $R_1 \circ R_2 \in \mathcal{R}$, where $R_1 \circ R_2$ is *the composition of* R_1 *and* R_2 defined by
$$R_1 \circ R_2 := \big\{(x,z) \mid \exists y\big((x,y) \in R_1 \wedge (y,z) \in R_2\big)\big\}.$$

We want to point out that in this standard definition of proper relation algebras it is *not* required that 1 denotes B^2. However, in most examples that we encounter, 1 indeed denotes B^2, and those proper relation algebras are called *square*. The minimal non-empty elements of $\mathcal{R}$ with respect to set-wise inclusion are called the *basic relations* of $\mathcal{R}$.

Example 1.5.2 (The Point Algebra). Let $B = \mathbb{Q}$ be the set of rational numbers, and consider
$$\mathcal{R} = \{\emptyset, =, <, >, \leq, \geq, \neq, \mathbb{Q}^2\}.$$
Those relations form a proper relation algebra (with atoms $<, >, =$, and where 1 denotes $\mathbb{Q}^2$) which is one of the most fundamental relation algebras and known under the name *point algebra*.

When $\mathcal{R}$ is finite, every relation in $\mathcal{R}$ can be written as a finite union of basic relations, and we abuse notation and sometimes write $R = \{B_1, \dots, B_k\}$ for

$R \in \mathcal{R}$ and $B_1, \ldots, B_k$ basic relations such that $R = B_1 \cup \cdots \cup B_k$. Note that composition of basic relations determines the composition of all relations in the relation algebra, since

$$R_1 \circ R_2 = \bigcup_{B_1 \in R_1, B_2 \in R_2} B_1 \circ B_2.$$

1.5.2. Abstract relation algebras. An abstract relation algebra is an algebra with signature $\{\vee, -, 0, 1, \circ, ^\smile, \mathrm{Id}\}$ that satisfies some of the laws that hold for these operators in every proper relation algebra.

DEFINITION 1.5.3 (Compare [203, 159, 251]). An (abstract) relation algebra $\mathbf{A}$ is an algebra with domain A and signature $\{\vee, -, 0, 1, \circ, ^\smile, \mathrm{Id}\}$ such that

- the structure $(A; \vee, \wedge, -, 0, 1)$ is a Boolean algebra where $\wedge$ is defined by $(x, y) \mapsto -(-x \vee -y)$ from $-$ and $\vee$;
- $\circ$ is an associative binary operation on A, called *composition*;
- $^\smile$ is a unary operation, called *converse*;
- $(a^\smile)^\smile = a$ for all $a \in A$;
- $a \circ (b \vee c) = a \circ b \vee a \circ c$;
- $(a \vee b)^\smile = a^\smile \vee b^\smile$;
- $(-a)^\smile = -(a^\smile)$;
- $(a \circ b)^\smile = b^\smile \circ a^\smile$;
- $(a \circ b) \wedge c^\smile = 0 \Leftrightarrow (b \circ c) \wedge a^\smile = 0$.

A *representation* $(\mathcal{R}, i)$ of $\mathbf{A}$ consists of a proper relation algebra $\mathcal{R}$ and a mapping i from the domain A of $\mathbf{A}$ to $\mathcal{R}$ such that i is an isomorphism with respect to the functions (and constants) $\{\vee, -, 0, 1, \circ, ^\smile, \mathrm{Id}\}$. In this case, we also say that $\mathbf{A}$ is the *abstract relation algebra of* $\mathcal{R}$. The representation is called *countable* if the base set of $\mathcal{R}$ is countable. There are finite relation algebras that do not have a representation [268]. A relation algebra that has a representation is called *representable*.

We define $x \leq y$ by $x \wedge y = x$. Note that if $(\mathcal{R}, i)$ is a representation of $\mathbf{A}$, then $i(a)$ is a basic relation of $\mathcal{R}$ if and only if $a \neq 0$, and for every $b \leq a$ we have $b = a$ or $b = 0$; we call a an *atom* of $\mathbf{A}$. Using the axioms of relation algebras, it can be shown that the composition operator is uniquely determined by the composition operator on the atoms. Similarly, the converse of an element $a \in A$ is the disjunction of the converses of all the atoms below a.

Example 1.5.4. The (abstract) point algebra is a relation algebra with 8 elements and 3 atoms, $=$, $<$, and $>$, and can be described as follows. The composition operator of the basic relations of the point algebra is shown in the table of Table 4. By the observation we just made, this table determines the full composition table. The converse of $<$ is $>$, and Id denotes $=$ which is its own converse. This fully determines the relation algebra. The point algebra with domain $\mathbb{Q}$ presented in Example 1.5.2 gives a representation of this relation algebra.

$\circ$	$=$	$<$	$>$
$=$	$=$	$<$	$>$
$<$	$<$	$<$	1
$>$	$>$	1	$>$

TABLE 4. The composition table for the basic relations in the point algebra.

1.5.3. Network satisfaction problems. The central computational problems that have been studied for relation algebras are *network satisfaction problems* [251, 159, 203], defined as follows. Let $\mathbf{A}$ be a relation algebra. An $(A$-) *network* $N = (V; f)$ consists of a finite set of nodes V and a function $f: V \times V \to A$. Two types of network satisfaction problems have been studied for A-networks. The first is the *network satisfaction problem for a (fixed) representation of* $\mathbf{A}$, defined as follows.

DEFINITION 1.5.5. Let $(\mathcal{R}, i)$ be a representation of a finite relation algebra $\mathbf{A}$ where $\mathcal{R}$ has the domain B. Then the *network satisfaction problem for* $(\mathcal{R}, i)$ is the computational problem to decide whether a given A-network $N = (V; f)$ is *satisfiable in* $(\mathcal{R}, i)$, that is, whether there exists a mapping $s: V \to D$ such that for all $u, v \in V$

$$(s(u), s(v)) \in i(f(u, v)).$$

The second problem is the *(general) network satisfaction problem for* $\mathbf{A}$.

DEFINITION 1.5.6. Let $\mathbf{A}$ be a finite relation algebra. Then the *network satisfaction problem for* $\mathbf{A}$ is the computational problem to decide whether for a given A-network N there *exists* a representation $(\mathcal{R}, i)$ of $\mathbf{A}$ such that N is satisfiable in $(\mathcal{R}, i)$.

Note that the network satisfaction problem for relation algebras that are not representable is trivial. Every network satisfaction problem for a fixed representation, but also the general network satisfaction problem, is closely related to a corresponding constraint satisfaction problem. To present this link between network satisfaction problems and CSPs we need the following notation. Let τ_A be a signature consisting of a binary relation symbol R_a for each element $a \in A$. If $(\mathcal{R}, i)$ is a representation of $\mathbf{A}$ where $\mathcal{R}$ is over the domain B, then this gives rise to a τ_A-structure $\mathfrak{B}_{\mathcal{R},i}$ in a natural way:

- the domain of the structure is B, and
- the relation symbol R_a is interpreted by $i(a)$.

We can associate to each A-network $N = (V; f)$ a primitive positive τ_A-sentence ϕ_N in the following straightforward way:

- the variables of ϕ_N are V, and

- ϕ_N contains the conjunct $R_a(u, v)$ if and only if $f(u, v) = a$.

Conversely, we associate to each primitive positive τ_A-sentence ϕ with variables V finitely many A-networks as follows. Let $D_1, \dots, D_{k(\phi)}$ be the connected components of the canonical database $D(\phi)$ of ϕ. For $j \in \{1, \dots, k(\phi)\}$, let $N_{\phi,j}$ be the network whose nodes V_j are the elements of D_j. Let $u, v \in V_j$, and list by $a_1, \dots, a_m$ all those $a \in A$ such that ϕ contains the conjunct $R_a(u, v)$. Then define $f(u, v) = a$ for $a = (a_1 \wedge a_2 \wedge \cdots \wedge a_m)$; if $m = 0$, then $f(u, v) = 1$.

The following link between the network satisfaction problem for a fixed representation $(\mathcal{R}, i)$ of $\mathbf{A}$, and the constraint satisfaction problem for $\mathfrak{B}_{\mathcal{R},i}$ is straightforward from the definitions.

PROPOSITION 1.5.7. *Let* $\mathbf{A}$ *be a relation algebra with a representation* $(\mathcal{R}, i)$.

1. *An* A*-network* N *is satisfiable in* $(\mathcal{R}, i)$ *if and only if* $\mathfrak{B}_{\mathcal{R},i} \models \phi_N$.
2. *Conversely,* $\mathfrak{B}_{\mathcal{R},i}$ *satisfies a primitive positive* τ_A*-sentence* ϕ *if and only if each of the networks* $N_{\phi,1}, \dots, N_{\phi,k(\phi)}$ *defined above is satisfiable in* $(\mathcal{R}, i)$.

PROOF. A map $s\colon V \to B$ shows that the A-network $N = (V; f)$ is satisfiable in $(\mathcal{R}, i)$ if and only if s shows that the sentence ϕ_N is satisfiable in $\mathfrak{B}_{\mathcal{R},i}$. For the second statement, let ϕ be a primitive positive τ_A-sentence with variables V and let $s\colon V \to B$ be a map. Then s shows that ϕ is true in $\mathfrak{B}_{\mathcal{R},i}$ if and only if for each $j \in \{1, \dots, k\}$ the restriction of s to the nodes of $N_{\phi,j}$ shows that $N_{\phi,j}$ is satisfiable in $(\mathcal{R}, i)$. ⊣

Note that the canonical database of ϕ_N has only one component, and hence $N_{\phi_N,1} = N$.

COROLLARY 1.5.8. *Let* $\mathbf{A}$ *be a finite relation algebra with a representation* $(\mathcal{R}, i)$. *Then there is a polynomial-time many-one reduction from the network satisfaction problem for* $(\mathcal{R}, i)$ *to* CSP$(\mathfrak{B}_{\mathcal{R},i})$. *Conversely, there is a polynomial-time Turing reduction from* CSP$(\mathfrak{B}_{\mathcal{R},i})$ *to the network satisfaction problem for* $(\mathcal{R}, i)$. *If* $(\mathcal{R}, i)$ *is a square representation, then the Turing reduction can be replaced by a polynomial-time many-one reduction.*

PROOF. All statements except the last follow from Proposition 1.5.7. For the last statement, it suffices to observe that $N_{\phi,1}, \dots, N_{\phi,k(\phi)}$ are satisfiable in a square representation $(\mathcal{R}, i)$ if and only if the network N_ϕ obtained from the disjoint union of $N_{\phi,1}, \dots, N_{\phi,k(\phi)}$ by setting $f(u, v) = 1$ for $u \in V_{j_1}$, $v \in V_{j_2}$, and $j_1 \neq j_2$, is satisfiable in $(\mathcal{R}, i)$. ⊣

Proposition 1.5.7 and Corollary 1.5.8 show that network satisfaction problems for fixed square representations essentially *are* constraint satisfaction problems, and that the differences are only a matter of formalisation. To also relate the *general* network satisfaction problem for a finite relation algebra $\mathbf{A}$ to a constraint satisfaction problem, we need the following definitions.

DEFINITION 1.5.9. Let $\mathbf{A}$ be a relation algebra. An A-network $N = (V; f)$ is called

- *closed* (here we follow the terminology of Hirsch [202]) if

$$f(a,c) \le f(a,b) \circ f(b,c) \text{ for all } a,b,c \in V; \tag{1}$$

- *atomic*[7] if the image of f only contains atoms of $\mathbf{A}$.

A representation $(\mathcal{R}, i)$ of $\mathbf{A}$ is called

- *universal* (again we follow the terminology of Hirsch [202], which is different from the one of Christiani and Hirsch [143]) if every satisfiable $\mathbf{A}$-network is satisfiable in $(\mathcal{R}, i)$;
- *fully universal* if every closed atomic $\mathbf{A}$-network is satisfiable in $(\mathcal{R}, i)$.

Clearly, if $(\mathcal{R}, i)$ is a universal representation of $\mathbf{A}$, then the network satisfaction problem for $(\mathcal{R}, i)$ equals the general network satisfaction problem for $\mathbf{A}$.

PROPOSITION 1.5.10. *If a representation $(\mathcal{R}, i)$ of a relation algebra $\mathbf{A}$ is fully universal, then it is also universal.*

PROOF. Let $N = (V; f)$ be an A-network. Suppose that N is satisfiable in some representation $(\mathcal{S}, j)$ of $\mathbf{A}$, where $\mathcal{S}$ is over the base set B, and let $s \colon V \to B$ be a mapping witnessing this. Let $N' = (V, f')$ be the network given by $f'(u,v) = a$, where a is an atom of $\mathbf{A}$ such that $(s(u), s(v)) \in j(a)$. Then N' is atomic, and satisfiable in $(\mathcal{S}, j)$ and therefore closed. Hence, N' is satisfiable in $(\mathcal{R}, i)$. Since $f'(u,v) \le f(u,v)$ for all $u, v \in V$ it follows that N is satisfiable in $(\mathcal{R}, i)$, too. This proves the statement. ⊣

The point algebra is an example of a relation algebra with a fully universal square representation. Note that if $\mathbf{A}$ has a fully universal representation, then the network satisfaction problem for $\mathbf{A}$ is decidable in NP: for a given network (V, f), simply select for each $x \in V^2$ an atom $a \in A$ with $a \le f(x)$, replace $f(x)$ by a, and then exhaustively check condition (1). We mention that a finite relation algebra has a fully universal representation if and only if the so-called path consistency procedure (see Section 8.2.1) decides satisfiability of atomic $\mathbf{A}$-networks (see, e.g., [62]). However, not every finite relation algebra has a fully universal representation, as the following example shows.

Example 1.5.11. There is a relation algebra with 4 atoms, called $\mathbf{B}_9$ in [262], which is universal but not fully universal. A representation of $\mathbf{B}_9$ with domain $\{0, 1, \dots, 6\}$ is given by the basic relations $\{R_0, R_1, R_2, R_3\}$ where $R_i = \{(x,y) : |x - y| = i \mod 7\}$, for $i \in \{0,1,2,3\}$. In fact, every representation of $\mathbf{B}_9$ is isomorphic to this representation. Let N be the network (V, f) with $V = \{a,b,c,d\}$, $f(a,b) = f(c,d) = R_3$, $f(a,d) = f(b,c) = R_2$, $f(a,c) = f(b,d) = R_1$, $f(i,i) = R_0$ for all $i \in V$, and $f(i,j) = f(j,i)$ for all $i, j \in V$. Then N is atomic and closed but not satisfiable.

[7]For Hirsch [202], an atomic network satisfies additionally the property of being closed.

$$T_{\mathbf{A}} := \{\forall x, y(\neg 0(x,y) \wedge (\mathrm{Id}(x,y) \Leftrightarrow x = y))\} \tag{2}$$

$$\cup\{\forall x, y(1(x,y) \Leftrightarrow \bigvee_{a\in A} R_a(x,y))\} \tag{3}$$

$$\cup \bigcup_{a\in A} \{\forall x, y(R_{a^\smile}(x,y) \Leftrightarrow R_a(y,x) \wedge (R_{-a}(x,y) \Leftrightarrow \neg R_a(x,y))\} \tag{4}$$

$$\cup \bigcup_{a,b\in A} \{\forall x, y(R_{a\vee b}(x,y) \Leftrightarrow (R_a(x,y) \vee R_b(x,y)))\} \tag{5}$$

$$\cup \bigcup_{a,b\in A} \{\forall x, z\big(R_{a\circ b}(x,z) \Leftrightarrow \exists y(R_a(x,y) \wedge R_b(y,z))\big)\} \tag{6}$$

FIGURE 2. The definition of the τ_A-theory $T_{\mathbf{A}}$.

We will now show that every representable relation algebra has a universal representation. Figure 2 contains the definition of a first-order τ_A-theory $T_{\mathbf{A}}$ (as in [203], Section 2.3). The models of $T_{\mathbf{A}}$ correspond to the representations of **A**, as described in the following.

PROPOSITION 1.5.12. *Let* **A** *be a finite relation algebra and let* $\mathfrak{B}$ *be a model of* $T_{\mathbf{A}}$. *Let* $\mathcal{R}$ *be the relations of* $\mathfrak{B}$ *and let* $i\colon A \to \mathcal{R}$ *be given by* $i(a) := R_a^{\mathfrak{B}}$. *Then* $(\mathcal{R}; i)$ *is a representation of* **A**. *Conversely, for every representation* $(\mathcal{R}, i)$ *of* **A** *the* τ_A*-structure* $\mathfrak{B}_{\mathcal{R},i}$ *is a model of* $T_{\mathbf{A}}$.

PROOF. The proof is straightforward by matching the sentences in $T_{\mathbf{A}}$ with the items of Definition 1.5.1. ⊣

It is easy to see that $T_{\mathbf{A}}$ has the Joint Homomorphism Property (JHP, introduced in Proposition 2.1.16); in fact, the disjoint union of two models of $T_{\mathbf{A}}$ is again a model of $T_{\mathbf{A}}$.

PROPOSITION 1.5.13. *Every finite representable relation algebra has a countable universal representation.*

PROOF. Let **A** be a finite representable relation algebra. Then by Proposition 1.5.12 the theory $T_{\mathbf{A}}$ is satisfiable. Since $T_{\mathbf{A}}$ also has the JHP, we can apply Proposition 2.1.16 to obtain a countable model $\mathfrak{B}$ of $T_{\mathbf{A}}$ such that CSP($\mathfrak{B}$) and CSP($T_{\mathbf{A}}$) are the same problem. Then Proposition 1.5.12 shows that **A** has a representation $(\mathcal{R}, i)$. We claim that $(\mathcal{R}, i)$ is universal: if N is an A-network that is satisfiable in some representation $(\mathcal{R}', i')$, then ϕ_N holds in $\mathfrak{B}_{\mathcal{R}',i'} \models T_{\mathbf{A}}$ (Proposition 1.5.7), and hence $\mathfrak{B} \models \phi_N$. This in turn implies that N is satisfiable in $(\mathcal{R}, i)$ (see Proposition 1.5.7 and the comment after this proposition). ⊣

COROLLARY 1.5.14. *Let* **A** *be a finite representable relation algebra. Then there exists a countable A-structure* $\mathfrak{B}$ *such that the network satisfaction problem for* **A** *is polynomial-time Turing equivalent to* CSP($\mathfrak{B}$).

PROOF. By Proposition 1.5.13 the relation algebra **A** has a countable universal representation $(\mathcal{R}, i)$, and hence the network satisfaction problem for **A** equals the network satisfaction problem for $(\mathcal{R}, i)$. By Corollary 1.5.8, this problem is polynomial-time Turing equivalent to CSP($\mathfrak{B}$) for the countable structure $\mathfrak{B} := \mathfrak{B}_{\mathcal{R},i}$. ⊣

When we later discuss examples of constraint satisfaction problems that are introduced as the network satisfaction problem for some representation $(\mathcal{R}, i)$ of a finite relation algebra **A** (for example in Section 1.6) then we identify this representation with the relational A-structure $\mathfrak{B}_{\mathcal{R},i}$.

1.5.4. Discussion. We close this section by discussing the weaknesses of the relation algebra approach to constraint satisfaction. First of all, the class of problems that can be formulated as a network satisfiability problem for some finite relation algebra **A** is severely restricted. The relations that we allow in the input networks are closed under unions; this introduces a sort of restricted disjunction that quickly leads to NP-hardness, and indeed only a few exceptional situations have a polynomial-time tractable network satisfiability problem [203]. The typical work-around here is to introduce another parameter, which is a subset A' of the domain of **A**, and to study the network satisfaction problem for networks $N = (V; f)$ where the image of f is contained in A'. Such subsets A' are often called a *fragment* of **A**. Note that such an additional parameter is not necessary for CSPs as studied here: with the techniques of this section, we can also formulate the network satisfaction problems for fragments of **A** as CSPs. Also note that the network satisfaction problem is restricted to *binary* relations, whereas many important CSPs can only be formulated in a natural way with relations of higher arity (see e.g. Section 1.6.2 or Section 1.6.8).

Every network satisfaction problem can be formulated as CSP($\mathfrak{B}$) for an appropriate infinite structure $\mathfrak{B}$ (as we have seen in Proposition 1.5.13), but as the above remarks show, only a very small fraction of CSPs can be formulated as a network satisfaction problem for a finite relation algebra **A**. Despite this, there are hardly any additional techniques available for studying network satisfaction problems. The tools we have for network satisfaction also apply more generally to constraint satisfaction problems.

The study of composition of relations in the context of the network satisfiability problem is usually justified by the fact that a network with constraints that involve the relation $R \circ S$ can be simulated by networks using constraints that involve the relations R and S and adding extra nodes. To study the computational complexity of the network satisfaction problem for a fragment B of a relation algebra **A**, one therefore typically computes the closure of B

under the operations of the relation algebra. But note that every binary relation in the closure of B is also primitively positively definable in any representation of $\mathbf{A}$ and that the converse of this statement is false.[8] Since the computational complexity is preserved also for expansions by primitively positively definable relations (see Lemma 1.2.6), primitive positive definitions therefore appear to be the more appropriate tool for studying network satisfaction problems. Apart from being more powerful, primitive positive definability has another advantage in comparison to closure in relation algebras: while the latter is intricate and not well understood, we can offer a powerful Galois theory to study primitive positive definability of relations (see Chapter 6).

1.6. Examples

This section presents computational problems that have been studied in various areas of theoretical computer science, and that can be formulated as constraint satisfaction problems in the sense of Section 1.1, 1.2, 1.3, or 1.4. Each problem is described from the perspective in which the computational first appeared in the literature. Our list is by far not exhaustive; computational problems that can be exactly formulated as CSP($\mathfrak{B}$) for an infinite structure $\mathfrak{B}$ are abundant in almost every area of theoretical computer science.

1.6.1. Allen's interval algebra. Allen's Interval Algebra [10] is a formalism that is famous in artificial intelligence, and which has been introduced to reason about intervals and about the relationships between intervals.

Formally, Allen's interval algebra is a proper relation algebra (see Section 1.5); we can also view it as a structure with a binary relational signature. The domain is the set $\mathbb{I}$ of all closed intervals $[a, b]$ of rational numbers, where $a, b \in \mathbb{Q}$ and $a < b$. When $x = [a, b]$ is an interval, then $-x$ denotes the interval $[-b, -a]$. For $R \subseteq \mathbb{I}^2$ we write R^- for the relation $\{(-x, -y) \mid (x, y) \in R\}$. Recall that in proper relation algebras $R^\smile$ denotes the relation $\{(y, x) \mid (x, y) \in R\}$.

The basic relations of Allen's interval algebra are the 13 relations P, M, O, S, D, E (defined in Table 5), P^-, M^-, O^-, S^-, and the converse of S, D, and S^-, denoted by $S^\smile$, $D^\smile$, and $(S^-)^\smile$, respectively. Note that these 13 relations are pairwise disjoint, that their union equals $\mathbb{I}^2$, and that they are closed under composition. Recall our convention that when $\mathcal{R}$ is a subset of the basic relations, we write $x\mathcal{R}y$ if $(x, y) \in \bigcup_{R \in \mathcal{R}} R$. For example, $x\{P, P^-\}y$ signifies that the intervals x and y are disjoint. The 2^{13} relations that arise in this way will be called the *relations of Allen's interval algebra*.

[8] An example that shows this for a fragment of Allen's Interval Algebra (cf. Section 1.6.1) is described by Krokhin, Jeavons, and Jonsson [248], Section 4.2.2).

Relation symbol	*Definition*	*Explanation*
P	$\{([a,b],[c,d]) \mid b < c\}$	$[a,b]$ precedes $[c,d]$
M	$\{([a,b],[c,d]) \mid b = c\}$	$[a,b]$ meets $[c,d]$
O	$\{([a,b],[c,d]) \mid a < c < b < d\}$	$[a,b]$ overlaps with $[c,d]$
S	$\{([a,b],[c,d]) \mid a = c \text{ and } b < d\}$	$[a,b]$ starts $[c,d]$
D	$\{([a,b],[c,d]) \mid c < a < b < d\}$	$[a,b]$ is during $[c,d]$
E	$\{([a,b],[c,d]) \mid a = c, b = d\}$	$[a,b]$ equals $[c,d]$

TABLE 5. The definitions for the basic relations of Allen's interval algebra.

An important computational problem for Allen's interval algebra is the network satisfaction problem for Allen's interval algebra. We will see in Exampe 5.5.5 that Allen's interval algebra is a fully universal representation of its abstract relation algebra **A**, so the network satisfaction problem for Allen's Interval Algebra equals the general network satisfaction for **A** (cf. Proposition 1.5.10). As explained in Corollary 1.5.8 this problem can be viewed as CSP($\mathfrak{A}$) where $\mathfrak{A}$ is a structure with domain $\mathbb{I}$ and a signature containing 2^{13} binary relation symbols. More on this structure can be found in Chapter 2, Example 2.4.2. The name *Allen's interval algebra* may refer to either $\mathfrak{A}$ or **A**, depending on the context.

The problem CSP($\mathfrak{A}$) is NP-complete [10] (this follows also from Theorem 3.4.3). The complexity of the CSP for (binary) reducts of Allen's interval algebra has been completely classified in [248].

1.6.2. Phylogenetic reconstruction problems. In modern biology it is believed that in the course of their evolution, the various species have developed in a tree-like fashion. That is, any species derives predominantly from a single prior species, and all species share a common ancestor. The goal of phylogenetic reconstruction is to determine the evolutionary tree from given partial information about the tree. This motivates the computational problem of *rooted triple satisfiability* (also called *rooted triple consistency*), defined below. In 1981, Aho, Sagiv, Szymanski, and Ullman [9] presented a quadratic time algorithm for this problem, motivated independently from computational biology by questions in database theory.

A *tree* (in the sense of graph theory) is a connected acyclic graph. Let $\mathfrak{T}$ be a tree with vertex set T and with a distinguished vertex r, the *root* of $\mathfrak{T}$. Vertices that are distinct from r and have *degree* one, i.e., that have exactly one neighbour in $\mathfrak{T}$, are called *leaves*. For $u, v \in T$, we say that u *lies below* v if the path from r to u passes through v. We say that u *lies strictly below* v if u lies below v and $u \neq v$. The *youngest common ancestor* (*yca*) of two vertices $u, v \in T$ is the node w such that both u and v lie below w and w has maximal distance from the root r.

Rooted-Triple Satisfiability
Instance: A finite set of variables V, and a set of triples $xy|z$ for $x, y, z \in V$.
Question: Is there a rooted tree $\mathfrak{T}$ with leaves L and a mapping $s\colon V \to L$ such that for every triple $xy|z$ the yca of $s(x)$ and $s(y)$ lies strictly below the yca of $s(x)$ and $s(z)$ in $\mathfrak{T}$?

Another famous problem that has been studied in this context is the quartet satisfiability problem, which is NP-complete [332].

Quartet Satisfiability
Instance: A finite set of variables V, and a set of quartets $xy{:}uv$ with $x, y, u, v \in V$.
Question: Is there a tree $\mathfrak{T}$ with leaves L and a mapping $s\colon V \to L$ such that for every quartet $xy{:}uv \in R^{yca}$ the shortest path from x to y is disjoint from the shortest path from u to v?

It is straightforward to check that the class of positive instances (viewed as relational structures) of each of these two computational problems is closed under disjoint unions and inverse homomorphisms. By Lemma 1.1.8, both the rooted triple satisfaction problem and the quartet satisfaction problem can be formulated as CSP($\mathfrak{B}$) for an infinite structure $\mathfrak{B}$. We come back to these CSPs in Section 5.1.

1.6.3. Branching-time constraints. An important model in temporal reasoning is *branching time*, where for every point in time the past is linearly ordered, but the future is only partially ordered.

This motivates the so-called *left-linear point algebra* [203, 159], which is a relation algebra with four basic relations, denoted by $=$, $<$, $>$, and $|$. Here, $x|y$ signifies that x and y are incomparable in time, and '$x < y$' signifies that x is *earlier in time than* y, and *to the left of* y when we draw points in the plane; this motivates the name *left linear point algebra.* The composition operator on these four basic relations is given in Table 6. The converse of $<$ is $>$, Id denotes $=$, and $|$ is its own converse. The relation algebra is uniquely given by this data.

As explained in Section 1.5, the network satisfaction problem for a representable relation algebra can be viewed as CSP($\mathfrak{B}$) for an appropriate infinite structure. In any such structure, $<$ must denote a dense partial order which is *semilinear*[9], i.e., for every x the set $\{y \mid y < x\}$ is linearly ordered by $<$. The network satisfaction problem of the left-linear point algebra is polynomial-time

[9]In some publications, a semilinear order is also required to have for all elements x, y an element z such that $z < x$ and $z < y$; we do not require this.

$\circ$	$=$	$<$	$>$	$\mid$
$=$	$=$	$<$	$>$	$\mid$
$<$	$<$	$<$	$\{<,=,>\}$	$\{<,\mid\}$
$>$	$>$	1	$>$	$\mid$
$\mid$	$\mid$	$\mid$	$\{>,\mid\}$	1

TABLE 6. The composition table for the basic relations in the left-linear point algebra from Section 1.6.3.

equivalent to the following problem, called the *branching-time satisfiability problem*.

Branching-Time Satisfiability
Instance: A finite relational structure $\mathfrak{A} = (A; \leq, \|, \neq)$ where $\leq$, $\|$, and $\neq$ are binary relations.
Question: Is there a rooted tree $\mathfrak{T}$ and a mapping $s\colon A \to T$ such that in $\mathfrak{T}$ the following is satisfied:

- If $(x, y) \in \leq^{\mathfrak{A}}$, then $s(x)$ lies above $s(y)$;
- if $(x, y) \in \|^{\mathfrak{A}}$, then neither $s(x)$ lies strictly above $s(y)$, nor $s(y)$ strictly above $s(x)$;
- if $(x, y) \in \neq^{\mathfrak{A}}$, then $s(x) \neq s(y)$.

The polynomial-time equivalence of branching-time satisfiability and the network satisfaction problem of the left-linear point algebra can be explained by the fact that in any representation of the left-linear point algebra, the relation $x\{<,>,=\}y$ has the primitive positive definition

$$\exists z\ (x\{<,=\}z \wedge y\{<,=\}z)$$

and the relation $x\{<,|,=\}y$ has the primitive positive definition

$$\exists z\ (x\{<,=\}z \wedge z\{|,=\}y);$$

we can then use Theorem 1.2.6 (for details, see [73]).

The branching-time satisfiability problem can be formulated as CSP($\mathfrak{C}$) for the structure with domain $C := \{0,1\}^*$ and relations $\leq$, $\|$, and $\neq$, where $\leq$ denotes the relation

$$\{(u,v) \in C^2 \mid u \text{ is a prefix of } v\}.$$

The relation $\neq$ is the disequality relation, and $u \parallel v$ holds if u and v are equal or incomparable with respect to $\leq$. Let $<$ denote the intersection of $\leq$ and $\neq$. Other templates for the branching-time satisfiability problem that have better model-theoretic properties will be discussed in Section 5.2. Note that the structure $\mathfrak{C}$ *cannot* be used to obtain a representation of the left-linear point algebra, since $(<) \circ (<)$ does not equal $<$. A fully universal

$\circ$	$=$	$<$	$>$	$\prec$	$\succ$
$=$	$=$	$<$	$>$	$\prec$	$\succ$
$<$	$<$	$<$	$\{<,>\}$	$\{<,\prec\}$	$\{<,\succ\}$
$>$	$>$	1	$>$	$\prec$	$\succ$
$\prec$	$\prec$	$\prec$	$\{>,\prec\}$	$\prec$	1
$\succ$	$\succ$	$\succ$	$\{>,\succ\}$	1	$\succ$

TABLE 7. The composition table for the basic relations of Cornell's tree algebra **C** from Section 1.6.4.

square representation of the left-linear point algebra will be presented in Example 5.5.6.

The first polynomial-time algorithm for the branching-time satisfiability problem (and therefore also for the network satisfaction problem of the left-linear point algebra) is due to Hirsch [203], and has a worst-case running time in $O(n^5)$. This has been improved by Broxvall and Jonsson [104], who presented an algorithm running in $O(n^{3.376})$ (this algorithm uses an $O(n^{2.376})$ algorithm for fast integer matrix multiplication). A simpler algorithm which does not use fast matrix multiplication and runs in $O(nm)$ has been found in [72].

1.6.4. Cornell's tree description constraints. Motivated by problems in computational linguistics, Cornell [138] introduced a computational problem[10] which is equivalent to the general network satisfaction problem for the relation algebra **C** with atoms $=, <, >, \prec$, and $\succ$ which is given by the composition table in Table 7. This is a strictly more expressive problem than the branching-time satisfiability problem from the previous section, and was introduced independently from [203] and [104]. The idea is that $<$ denotes a dense semilinear order (see Section 1.6.3), and $\prec \cup <$ denotes a linear order. The way these relations arise in natural language grammar formalisms like dependency grammars is that $<$ represents the syntactic structure of a natural language sentence whereas $\prec \cup <$ stands for the word order.

Similarly as in Section 1.6.3, all 2^5 relations of **C** can be obtained by repeated compositions and intersections of the four relations $\{<,=\}$, $\{\prec,=\}$, $\{\prec,\succ,=\}$, and $\{<,>,\prec,\succ\}$; for details, see [73]. The algorithm presented for the general network satisfaction problem for **C** in [138] (which is the so-called path consistency algorithm; see Section 8.2.1) is not complete (in fact, as a consequence of Lemma 8.6.6 the problem cannot even be solved by Datalog). A polynomial-time algorithm has been given in [73].

[10] I feel personally committed to Cornell's problem since it was the first CSP with an ω-categorical template I encountered.

1.6.5. Set constraints. Many fundamental problems in artificial intelligence, knowledge representation, and verification involve reasoning problems about relations between sets that can be modelled as constraint satisfaction problems. A fundamental problem of this type is the following. We denote the set of all subsets of $\mathbb{N}$ by $\mathcal{P}(\mathbb{N})$.

Basic Set Constraint Satisfiability
Instance: A finite set of variables V, and a set ϕ of constraints of the form $x \subseteq y$, $x \parallel y$, or $x \neq y$, for $x, y \in V$.
Question: Is there a mapping $s \colon V \to \mathcal{P}(\mathbb{N})$ such that

(a) if $x \subseteq y$ is in ϕ, then $s(x)$ is contained in $s(y)$;
(b) if $x \parallel y$ is in ϕ, then $s(x)$ and $s(y)$ are disjoint sets;
(c) if $x \neq y$ is in ϕ, then $s(x)$ and $s(y)$ are distinct sets.

This problem has the shorter description CSP($\mathcal{P}(\mathbb{N}); \subseteq, \parallel, \neq$) where $\subseteq, \parallel, \neq$ are binary relations over $\mathcal{P}(\mathbb{N})$, standing for the binary relations containment, disjointness, and inequality between sets. Drakengren and Jonsson [153] showed that basic set constraint satisfiability can be decided in polynomial time. They also showed that a generalisation of this problem can be solved in polynomial time in which each constraint has the form

$$x_1 \neq y_1 \vee \cdots \vee x_k \neq y_k \vee x_0 R y_0$$

where R is either $\subseteq$, $\parallel$, or $\neq$, and where $x_0, \ldots, x_k, y_0, \ldots, y_k$ are not necessarily distinct variables. Set constraint languages will be revisited in Section 5.3.

1.6.6. Spatial reasoning. Qualitative spatial reasoning is concerned with representation formalisms that are considered close to conceptual schemata used by humans for reasoning about their physical environment—in particular, about processes or events, and about the spatial environment in which they are situated. An approach in qualitative spatial reasoning is to develop relational schemata that abstract from concrete metrical data of entities (for example coordinate positions or distances) by subsuming similar metric or topological configurations of entities under one qualitative representation.

There are many formalisms for qualitative spatial reasoning. In particular, several relation algebras (see Section 1.5) have been studied in this context. A basic example is the RCC5 relation algebra (with 5 atoms; the RCC5 relation algebra is also known under the name *containment algebra* [37, 159]), and the RCC8 relation algebra (with 8 atoms). In both formalisms, the variables denote 'non-empty regions'. In RCC5, the five atoms are denoted by DR, PO, PP, PPI, EQ, and they stand for *disjointness*, *proper overlap*, *proper containment* (*proper-part-of*), its converse (also called *inverse* in this context), and equality, respectively. In RCC8, we further distinguish the way the 'boundaries' of two regions relate to each other. We do not further discuss RCC8; for details, see [159, 99].

∘	DR	PO	PP	PPI	EQ
DR	1	PP ∪ DR ∪ PO	PP∪DR∪PO	DR	DR
PO	PPI ∪ DR ∪ PO	1	PP ∪ PO	PPI ∪ PO	PO
PP	DR	PP ∪ DR ∪ PO	PP	1	PP
PPI	PPI ∪ DR ∪ PO	PPI ∪ PO	PP ∪ PPI ∪ PO ∪ EQ	PPI	PPI
EQ	DR	PO	PP	PPI	EQ

TABLE 8. The composition table for the basic relations of RCC5.

There are many equivalent ways to formally define RCC5. Every relation algebra is uniquely given by the composition table for its atomic relations. The table for RCC5 is given in Table 8. More elegant descriptions require concepts from model theory and will be presented in Section 5.4.

The network satisfaction problem for RCC5 is NP-complete [314]; the computational complexity of the CSPs for the (binary) reducts of $\mathfrak{B}$ has been classified in [220]. A polynomial-time tractable case of particular interest is the *network satisfaction problem for the basic relations of RCC5* [314], i.e., the network satisfaction problem for RCC5 if the input is restricted to networks $N = (V; f)$ where the image of f only contains atoms of RCC5.

In any representation of RCC5, the atomic relations satisfy the following set of axioms T. We use $\mathrm{P}(x, y)$ as a shortcut for $\mathrm{PP}(x, y) \vee \mathrm{EQ}(x, y)$ and $\mathrm{PI}(x, y)$ as a shortcut for $\mathrm{PPI}(x, y) \vee \mathrm{EQ}(x, y)$.

$$T := \{\forall x, y, z \; (\mathrm{DR}(x, y) \wedge \mathrm{P}(z, y) \rightarrow \mathrm{DR}(x, z)) \tag{7}$$

$$\forall x, y, z \; (\mathrm{PO}(x, y) \wedge \mathrm{P}(y, z) \rightarrow (\mathrm{PO}(x, z) \vee \mathrm{PP}(x, z))) \tag{8}$$

$$\forall x, y, z \; (\mathrm{PP}(x, y) \wedge \mathrm{PP}(y, z) \rightarrow \mathrm{PP}(x, z)) \tag{9}$$

$$\forall x, y, z \; (\mathrm{PI}(x, y) \wedge \mathrm{P}(y, z) \rightarrow \neg\mathrm{DR}(x, y))\} \tag{10}$$

It follows from the results in Section 5.4 that the network satisfaction problem for the basic relations of RCC5 is the same problem as CSP(T), where T is the first-order theory defined as above. It can be checked that T satisfies item (2) in the statement of Proposition 1.3.6, and hence there exists an infinite structure $\mathfrak{B}$ such that CSP($\mathfrak{B}$) equals the satisfiability problem for RCC5. We will give more explicit descriptions of such an infinite structure $\mathfrak{B}$ in Section 5.4 (and it turns out that there are close links with the basic set constraints problem from Section 1.6.5).

1.6.7. Horn-SAT. The Horn-SAT problem is an important computational problem that can be solved in linear time in the size of the input [152]. It is complete for polynomial time under log-space reductions [118]. A propositional formula in conjunctive normal form is called *Horn* if each clause is a *Horn clause*, i.e., has at most one positive literal.

Horn-SAT
Instance: A propositional Horn formula.
Question: Is there a Boolean assignment for the variables such that in each clause at least one literal is true?

We cannot model this problem as CSP($\mathfrak{B}$) for a finite signature structure $\mathfrak{B}$; however, note that a clause $\neg x_1 \vee \cdots \vee \neg x_k \vee x_0$ is equivalent to

$$\exists y_1, \ldots, y_{k-1} \big((\neg x_1 \vee \neg x_2 \vee y_1) \wedge (\neg y_1 \vee \neg x_3 \vee y_2) \wedge \cdots \wedge (\neg y_{k-2} \vee \neg x_k \vee x_0) \big).$$

Hence, by introducing new variables, there is a straightforward reduction of Horn-SAT to the restriction of Horn-SAT where every clause has at most three literals. This restricted problem, which we call Horn-3SAT, can be formulated as CSP($\mathfrak{B}$) for

$$\mathfrak{B} = \big(\{0,1\}; \{(x,y,z) \mid (x \wedge y) \Rightarrow z\}, \{(x,y,z) \mid \neg x \vee \neg y \vee \neg z\}, \{(x,y) \mid x \Rightarrow y\}, \{(x,y) \mid \neg x \vee \neg y\}, \{0\}, \{1\}\big).$$

1.6.8. Precedence constraints in scheduling. The following problem has been studied in scheduling [285]: given is a finite set of variables V, and a finite set of constraints of the form

$$\bigvee_{i \in \{1,\ldots,k\}} x_i < x_0 \tag{11}$$

for $x_0, x_1, \ldots, x_k \in V$. The question is whether there exists an assignment $V \to \mathbb{Q}$ (equivalently, we can replace $\mathbb{Q}$ by $\mathbb{Z}$, or any other infinite linearly ordered set) such that all these constraints are satisfied.

As in the case of Horn-SAT, we cannot directly model this problem as CSP($\mathfrak{B}$) for a finite signature structure $\mathfrak{B}$. However, note that Formula (11) is equivalent to

$$\exists y_1, \ldots, y_{k-1} \big((x_1 < x_0 \vee y_1 < x_0) \wedge (x_2 < y_1 \vee y_2 < y_1) \wedge \cdots \wedge (x_{k-1} < y_{k-1} \vee x_k < y_{k-1}) \big).$$

This shows that and/or precedence constraints can be translated into conjunctions of constraints of the form $x_1 < x_0 \vee x_2 < x_0$ by introducing new existentially quantified variables. Hence, the problem whether a given set of and/or precedence constraints is satisfiable reduces naturally to CSP($\mathbb{Q}; R^{\min}_<$) where

$$R^{\min}_< := \{(a,b,c) \mid b < a \vee c < a\}.$$

Note that $R_<^{\min}$ holds on exactly those triples (a, b, c) where a is larger than the minimum of b and c. The problem CSP($\mathbb{Q}$; $R_<^{\min}$) can be solved in polynomial time; this is essentially due to [285]. For more expressive constraint languages over $\mathbb{Q}$ that contain the relation $R_<^{\min}$ and whose CSP can still be solved in polynomial time, see Section 12.4.1, Section 12.5.2, and Section 12.5.3.

1.6.9. Ord-Horn constraints. In this section we work with first-order formulas over the signature $\{<\}$. We write $x \leq y$ as a shortcut for $(x < y) \vee (x = y)$ (recall our convention that equality is part of first-order logic). A formula over the signature $\{<\}$ and with variables V is called *Ord-Horn* if it is a conjunction of disjunctions of the form

$$x_1 \neq y_1 \vee \cdots \vee x_k \neq y_k \vee R(x_0, y_0)$$

where $x_0, x_1, \ldots, x_k, y_0, y_1, \ldots, y_k \in V$, and R is either $\leq$, $<$, $\neq$, or $=$.

Ord-Horn Satisfiability
Instance: A finite set of variables V, and a finite set of Ord-Horn formulas with variables from V.
Question: Is there an assignment $V \to \mathbb{Q}$ that satisfies all the given formulas over $(\mathbb{Q}; <)$?

Nebel and Bürckert [288] showed that Ord-Horn Satisfiability can be solved in polynomial time. A relation $R \subseteq \mathbb{Q}^k$ is called *Ord-Horn* if it is definable by an Ord-Horn formula over $(\mathbb{Q}; <)$. As in the case of Horn-SAT and of and/or precedence constraints, there are structures $\mathfrak{B}$ with finitely many Ord-Horn relations such that all Ord-Horn relations have a primitive positive definition in $\mathfrak{B}$. Similarly as in the previous two sections, it is easy to see that the following structure has this property:

$$(\mathbb{Q}; \leq, \neq, \{(x, y, u, v) \mid (x = y) \Rightarrow (u = v)\})$$

In Section 12.4.1 we present a constraint language that contains and/or precedence constraints *and* Ord-Horn constraints and that can still be solved in polynomial time.

1.6.10. Ord-Horn interval constraints. For some (binary) reducts $\mathfrak{B}$ of Allen's interval algebra CSP($\mathfrak{B}$) can be solved in polynomial time. The most important of these reducts has been introduced by Nebel and Bürkert [288] under the name *Ord-Horn interval constraints*. It consists of all the relations R of Allen's interval algebra such that the relation $\{(x, y, u, v) \mid ([x, y], [u, v]) \in R\}$ is Ord-Horn (see Section 1.6.9). Now it is not hard to see that satisfiability for Ord-Horn interval constraints has a polynomial-time reduction to Ord-Horn satisfiability. This type of reduction will be studied in Section 6.3.5.

1.6.11. Linear program feasibility. Linear Programming is a computational problem of outstanding theoretical and practical importance (see e.g. [326]). It is known to be computationally equivalent to the problem to decide whether a given set of linear (non-strict) inequalities is *feasible*, i.e., defines a non-empty set.

Linear Program Feasibility
Instance: A finite set of variables V; a finite set of linear inequalities of the form $a_1x_1 + \cdots + a_kx_k \leq a_0$ where $x_1, \ldots, x_k \in V$ and $a_0, \ldots, a_k$ are rational numbers where numerator and denominator are represented in binary.
Question: Does there exist an $x \in \mathbb{R}^{|V|}$ that satisfies all inequalities?

Khachyian showed in [232] that Linear Program Feasibility can be solved in polynomial time. It is clearly not possible to formulate this problem as CSP($\mathfrak{B}$) for a structure $\mathfrak{B}$ with a finite relational signature. However, it is polynomial-time equivalent to CSP($\mathbb{R}; \{(x, y, z) \mid x + y = z\}, \{1\}, \leq$). For this, we need the following lemma.

LEMMA 1.6.1 (Lemma 2.11 in [66]). *Let $n_0, \ldots, n_l \in \mathbb{Q}$ be arbitrary rational numbers. Then the relation $\{(x_1, \ldots, x_l) \mid n_1x_1 + \cdots + n_lx_l = n_0\}$ is primitively positively definable in $(\mathbb{R}; \{(x, y, z) \mid x + y = z\}, \{1\})$. Furthermore, the primitive positive formula that defines the relation can be computed in polynomial time.*

The idea of the proof is the use of iterated doubling to define large numbers by small primitive positive formulas. By extending the previous result to inequalities, one can prove the following.

PROPOSITION 1.6.2 (from [66]). *The linear program feasibility problem for linear programs is polynomial-time equivalent to* CSP($\mathbb{R}; \{(x, y, z) \mid x + y = z\}, \{1\}, \leq$).

Clearly, we could have chosen $\mathbb{Q}$ instead of $\mathbb{R}$ to formulate the linear program feasibility problem. However, if we replace the domain by $\mathbb{Z}$ we obtain the NP-complete integer linear program feasibility problem [326] (cf. Section 1.6.14).

1.6.12. The Max-atom problem. In our list of problems from the literature that can be formulated as CSP($\mathfrak{B}$), we also want to include one problem in NP for which it is not known whether CSP($\mathfrak{B}$) is in P. The problem in question is a variant of the following problem, which has been introduced in [39] and, independently, in [285], and which is not itself of the form CSP($\mathfrak{B}$) for a structure $\mathfrak{B}$ with a finite relational signature.

The Max-atom Problem
Instance: A finite set of variables V; a finite set of constraints of the form

$$x_0 \leq \max(a_1 + x_1, \ldots, a_k + x_k)$$

where $x_0, x_1, \ldots, x_k \in V$ are variables and $a_0, \ldots, a_k \in \mathbb{Z}$ are coefficients that are represented in binary.
Question: Does there exist an $x \in \mathbb{Q}^{|V|}$ that satisfies all inequalities?

The Max-atom problem is known to be contained in NP $\cap$ coNP; this was proved by Möhring, Skutella, and Stork [285], with another proof given by Bezem, Nieuwenhuis and Rodríguez-Carbonell [39]. They also show that the problem is in P when the coefficients in the input are represented in unary. It is not known whether the problem is in P. The problem is polynomial-time equivalent to the famous problem of deciding the winner in mean-payoff games [285]. Mean-payoff games generalise so-called *Parity games* (see [183]), which in turn are equivalent to the satisfiability problem for the propositional μ-calculus [165]. A recent breakthrough result is that the winner in a Parity game can be decided in quasi-polynomial time [120]; it appears to be difficult to generalise this complexity result to prove that the Max-atom problem can be decided in quasi-polynomial time. The Max-atom problem is also polynomial-time equivalent to an intensively studied problem in an area called *max*/+ *algebra* [39].

As in the case of linear program feasibility, the Max-atom problem cannot be formulated as CSP($\mathfrak{B}$) for a structure $\mathfrak{B}$ with a finite relational signature. The problem we consider in its place is

$$\mathrm{CSP}(\mathbb{Q}; \{(x,y) \mid y = x+1\}, \{(x,y) \mid y = 2x\}, R^{\min}_{\leq})$$

where $R^{\min}_{\leq} = \{(x,y,z) \mid y \leq x \vee z \leq x\}$ is a variant of the relation $R^{\min}_{<}$ from Section 1.6.8. The Max-atom problem can be reduced to the latter problem: namely, we replace expressions of the form $x_i + a_i$ by a new variable y_i, and add a primitive positive formula $\phi(x,y)$ that defines $y_i = x_i + a_i$ and can be computed in polynomial time in the input size of the Max-atom problem. We do not know how to prove hardness for the CSP above, and rather think that the problem might well be in P.

1.6.13. Unification. Unification (and unification modulo equational theories) is a field in its own right in computational logic, and the complexity of the unification problem has been studied in numerous variants [15]. Many unification problems can be viewed as CSP($\mathfrak{B}$), for an appropriate infinite structure $\mathfrak{B}$, as we will see in the following. We start with the most fundamental unification problem.

Let $\tau := \{f_1, \dots, f_k\}$ be a finite set of function symbols, and let x be a variable symbol. Then $\mathcal{F}(x)$ denotes the set of all terms that can be constructed from τ and the variable x. The *unnested unification problem over* τ is the following problem[11].

Unnested Unification Problem over τ

Instance: a finite set of variables V, and a finite set of 'un-nested' term equations, i.e., expressions of the form $y_0 \approx f(y_1, \dots, y_k)$ for $y_0, y_1, \dots, y_k \in V$ and $f \in \tau$.

[11] This problem is known to be equivalent to the standard unification problem where the input is a single equation $t_1 \approx t_2$ for 'nested' terms $t_1, t_2 \in \mathcal{F}(x)$.

Question: is there an assignment $s \colon V \to \mathcal{F}(x)$ such that for every expression $y_0 \approx f(y_1, \ldots, y_k)$ in the input we have $s(y_0) = f(s(y_1), \ldots, s(y_k))$?

For fixed τ as above, let $\mathfrak{T} = (\mathcal{F}(x); F_1, \ldots, F_k)$ be the structure where F_i is the relation $\{(t_0, t_1, \ldots, t_r) \in (\mathcal{F}(x))^{r+1} \mid t_0 = f_i(t_1, \ldots, t_r)\}$ (here, r is the arity of f_i). It is clear that the unnested unification problem over τ can be described as CSP($\mathfrak{T}$). In a similar way, equational unification problems (see [15]) can be viewed as CSPs.

1.6.14. CSPs over the integers. The structure ($\mathbb{Z}$; Succ) of the integers with the successor relation Succ $= \{(x, y) \in \mathbb{Z}^2 \mid x = y + 1\}$ constitutes one of the simplest structures with a finite signature that is not ω-categorical. In Section 5.8 we will show that CSP($\mathbb{Z}$; Succ) can be expressed in connected monotone SNP. It is natural to ask which expansions of ($\mathbb{Z}$; Succ) have a CSP that can still be solved in polynomial time.

Difference Logic
Instance: A finite set of variables V; a finite set of constraints, each of the form

$$x - y \leq c$$

where $x, y \in V$ are variables and $c \in \mathbb{Z}$ is a constant represented in binary.
Question: Does there exist an $x \in \mathbb{Q}^{|V|}$ that satisfies all inequalities?

This problem can be solved in polynomial time by the algorithm of Bellman-Ford [137]. More expressive fragments of integer linear programming quickly become NP-hard: for example, satisfiability of systems of linear inequalities with at most two variables per inequality over the integers is NP-complete [253]. Even more expressive is the following famous problem.

Integer Program Feasibility
Instance: A finite set of variables V; a finite set of linear inequalities of the form

$$a_1 x_1 + \cdots + a_k x_k \leq 0$$

where $x_1, \ldots, x_k \in V$ are variables and $a_1, \ldots, a_k \in \mathbb{Z}$ are integer constants represented in binary.
Question: Does there exist an $x \in \mathbb{Z}^{|V|}$ that satisfies all inequalities?

Note that all of the problems mentioned above cannot be modelled as CSP($\mathfrak{B}$) for some fixed structure $\mathfrak{B}$ with a finite relational signature (see the discussion on infinite signatures in Section 1.7.1). However, similarly as for Linear Program Feasibility, there exists a structure with finite relational signature whose CSP is polynomial-time equivalent to Integer Program Feasibility, namely

$$(\mathbb{Z}; \{(x, y, z) \mid x + y = z\}, \{1\}, \leq).$$

1.6.15. Expansions of linear program feasibility. Linear Programming is a computational problem of outstanding theoretical and practical importance. As we have seen in Section 1.6.11, it is computationally equivalent to CSP($\mathbb{R}; \{(x, y, z) \mid x + y = z\}, \{1\}, \leq$). It is natural to ask which expansions of this structure still have a polynomial-time solvable CSP. It has been shown in [66] that every *first-order expansion* of this structure is contained in a class called Horn-DLR [219] and polynomial-time tractable, or otherwise NP-hard.

An important class of relations over the real numbers that generalises the class of relations defined by linear inequalities is the class of all *semi-algebraic* relations, i.e., relations with a first-order definition in ($\mathbb{R}; +, *$). By the fundamental theorem of Tarski and Seidenberg it is known that a relation $S \subseteq \mathbb{R}^n$ is semi-algebraic if and only if it has a quantifier-free first-order definition in ($\mathbb{R}; +, *, 0, 1, \leq$). Geometrically, we can view semi-algebraic sets as unions of intersections of the solution sets of strict and non-strict polynomial inequalities. The classification of the computational complexity of CSPs for real-valued semi-algebraic constraint languages is an ambitious research project, and has important links to semidefinite programming: every semidefinite representable set is semi-algebraic and *convex*. There are many fundamental questions in this area that are wide open, for instance the complexity of semidefinite programming feasibility (see e.g. Section 6.4.4 in [345], or [312]). The conjecture that every convex semi-algebraic set is *semidefinite representable* [198], i.e., primitively positively definable over the structure that has as its relations all the solution spaces of semidefinite programs, has been refuted recently [321]. These important questions from real algebraic geometry are outside the scope of this text.

1.7. Topics not covered

When choosing the material for this book, certain restrictive choices had to be made. We comment on some related lines of research or facets of the area that we had to omit. There are for example the following fields of research.

1. Finding optimal solutions to *valued constraint satisfaction problems* [239, 338, 240, 245], even over infinite domains [78].
2. Studying the complexity of CSPs for *random instances* [136, 2]; often it is interesting to determine the threshold of the density where such instances are with high probability satisfiable or with high probability unsatisfiable. Also this topic has also been studied for constraint satisfaction problems over infinite domains [189, 180, 181].
3. *Counting the number of solutions to a CSP* [112, 160, 161, 109], exactly or approximately.

4. *Enumerating the solutions to a CSP*, for instance with the goal to find a polynomial-delay algorithm [114],
5. Studying the complexity of CSPs not only relative to the set of allowed constraints, but also with restrictions on the possible input instances. This is modelled by fixing a class of structures $\mathcal{K}$ and allowing only structures from $\mathcal{K}$ in the input, sometimes referred to as *left-hand side restriction* (since we restrict the left-hand side in the homomorphism problem $\mathfrak{A} \xrightarrow{?} \mathfrak{B}$); see [188].
6. The recent topic of *promise CSPs*, where the task is to distinguish an instance which is satisfiable in $\mathfrak{A}$ from an instance which is not even satisfiable in $\mathfrak{B}$, for two relational structures $\mathfrak{A}, \mathfrak{B}$ with the same signature [117].
7. *Exact time complexity* of the (NP-hard) CSPs. In this context, primitive positive definability is too powerful, but weaker forms of definability exist that are still useful for complexity classification. The universal-algebraic approach can be adapted to this setting using *partial clones* instead of clones; we refer to the recent survey of Couceiro, Haddad, and Lagerkvist on this topic [139].

In each of these fields, techniques that will be presented in this book have been applied fruitfully to obtain systematic understanding of the complexity of the respective computational problems. In the following, we comment in more detail on three further topics: *infinite signatures*, *complexity classes within NP*, and *quantified CSPs*.

1.7.1. Infinite signatures. Several natural computational problems could be formulated as CSP($\mathfrak{B}$) if we allow the structure $\mathfrak{B}$ to have a countably infinite signature. For example, we may view the feasibility problem for linear programs (Section 1.6.11) as CSP($\mathfrak{B}$) where $\mathfrak{B}$ contains all relations of the form $\{(x_1, \dots, x_k) \mid a_1x_1 + \cdots + a_kx_k \leq a_0\}$, for all rational numbers $a_0, a_1, \dots, a_k$. Indeed, some general results for constraint satisfaction that we present here carry over to infinite signatures. However, if we wanted to extend the present definition of CSP($\mathfrak{B}$) to structures $\mathfrak{B}$ with an infinite signature, we are faced with the problem of specifying how the constraints in input instances of CSP($\mathfrak{B}$) are *represented*. When $\mathfrak{B}$ has a finite signature, this causes no difficulties, since we can fix any representation for the finite number of relation symbols; as $\mathfrak{B}$ is fixed, the precise choice of the representation is irrelevant.

When $\mathfrak{B}$ has an infinite signature, a good choice of encoding for the constraints in the input very much depends on the structure $\mathfrak{B}$. In the example of linear programming feasibility, for instance, we may represent the constraint $a_1x_1 + \cdots + a_kx_k \leq a_0$ by specifying the coefficients $a_0, a_1, \dots, a_k$ in binary. Note that the issue of finite versus infinite constraint languages is not specific to infinite domains, but becomes relevant already for finite domains. Typically, for infinite constraint languages over a finite domain each constraint in the input is represented by listing all tuples of the corresponding relation in the constraint language. But this is not the only, and sometimes not even the most natural way to represent the constraints. For instance for the Horn-SAT problem (see Section 1.6.7), the most natural way to present the constraints is by writing them as conjunctions of Horn clauses. In the general setting, several representations have been proposed, some of which are more concise than listing all tuples [128], and some of which are less concise [281]. It turns out that typically when a CSP with an infinite constraint language is computationally hard, then there is a finite set of relations in this language such that the CSP for this sub-language is already NP-hard. For infinite constraint languages over a finite domain, and assuming that each constraint is represented by explicitly listing all satisfying assignments for the variables, it has even been conjectured [115] that this might be true in general; that is, when CSP($\mathfrak{B}$) is NP-hard under this representation, then $\mathfrak{B}$ has a finite signature reduct with an NP-hard CSP.

We have decided to keep the focus on CSPs for *finite* constraint languages. The main reason is that we can then work with the same definition of the computational problem CSP($\mathfrak{B}$) for *all* infinite structures $\mathfrak{B}$; for finite languages, there is no need for discussions of how the input to the constraint satisfaction problem is represented. Moreover, for all of the algorithms presented in this book it will be immediately clear for which representation of the input they can be generalised to infinite languages; we will illustrate this with the algorithms given in Chapter 12. Working with finite signatures does not prevent us from stating relevant mathematical facts in full generality when they also hold for structures with an infinite signature; only when it comes to statements about CSP($\mathfrak{B}$), do we insist that $\mathfrak{B}$ has finite relational signature.

1.7.2. Complexity classes below P. Besides the mentioned progress on the complexity dichotomy for finite domain CSPs, there has been considerable research activity to localise the exact complexity of CSPs inside the complexity class P, or with respect to definability in certain logics. By *definability* of CSP($\mathfrak{B}$) we mean that there exists a sentence Φ in some logic (typically extensions of first-order logic and restrictions of second-order logics; some appear in Chapter 8) such that $\mathfrak{A} \models \Phi$ if and only if $\mathfrak{A}$ maps homomorphically to $\mathfrak{B}$ (in this case it is most natural to work with the definitions of the CSP presented in Section 1.1 and in Section 1.4).

One motivation for studying the computational complexity within P is the question whether it is possible to solve problems faster in parallel models of computation. Another motivation is the goal to better understand the scope of existing algorithmic techniques to solve CSPs (such as Datalog, or restrictions of Datalog). In this line of research, the computational complexity of CSP($\mathfrak{B}$) has been completely classified if $\mathfrak{B}$ is a two-element structure [11]. Each problem in this class is complete for one of the complexity classes NP, P, $\oplus L$, NL, L, and AC^0 under AC^0 isomorphisms. For general finite domains, several universal-algebraic conditions are known that imply hardness for various complexity classes [26, 14, 255]. Concerning definability of CSPs, there are precise characterisations of those CSPs that are definable by a first-order sentence [254, 13]. Moreover, if CSP($\mathfrak{B}$) is not first-order definable, then it is L-complete under AC^0-reductions [255] (see also [145, 164]). For infinite-domain constraint satisfaction, apart from a characterisation of first-order definable CSPs [317, 61], there are no general results about pinpointing the complexity of CSPs within the complexity class P yet. We would like to mention that already for some concrete and model-theoretically well-behaved structures $\mathfrak{B}$ the precise complexity of CSP($\mathfrak{B}$) within P is open.

Example 1.7.1. Consider the problem

$$\text{CSP}(\mathbb{Q}; \neq, \{(x, y, z) \mid (x = y \Rightarrow y \leq z) \wedge x \leq y).$$

This problem is hard for the complexity class *non-deterministic logspace* (*NL*) since there is an easy reduction from directed reachability to this problem, and directed reachability is an NL-complete problem. However, the precise complexity of this problem is not known; it might be that the problem is contained in NL, but it might also be P-hard.

1.7.3. Quantified CSPs. Let $\mathfrak{B}$ be a structure with a finite relational signature. Then the *quantified constraint satisfaction problem for* $\mathfrak{B}$, denoted QCSP($\mathfrak{B}$), is the computational problem to decide for a given first-order sentence ϕ in prenex normal form and without disjunction and negation symbols whether ϕ is true in $\mathfrak{B}$. The difference of QCSP($\mathfrak{B}$) from CSP($\mathfrak{B}$) as we have presented it in Section 1.1 is that *universal quantification* is permitted in the input sentences ϕ.

The additional expressiveness often comes at the price of higher computational complexity; whereas for finite structures $\mathfrak{B}$, the CSP for $\mathfrak{B}$ is always in NP, there are finite structures $\mathfrak{B}$ where QCSP($\mathfrak{B}$) is PSPACE-complete. But quite surprisingly, several constraint languages with a polynomial-time tractable CSP also have a polynomial-time tractable QCSP. This is for instance the case for 2SAT [12] (see Example 1.2.2), or for Horn-3SAT [225] (see Section 1.6.7). Similarly, it can be shown that the temporal constraint languages presented in Section 12.5.2 and Section 12.5.4 are not only tractable for the CSP, but also for the QCSP. These are attractive results, since they assert that we can solve an even more expressive computational problem than the CSP for the same constraint language without loosing polynomial-time tractability. From a methodological point of view, we remark that the universal-algebraic approach can also be applied to study the complexity of the QCSP [102]; as in the case of the CSP, the computational complexity of QCSP($\mathfrak{B}$) is captured by the (*surjective*) *polymorphisms* of $\mathfrak{B}$ (see Chapter 6). Classifications of the QCSP often rely on the corresponding classification for the CSP. In particular, any hardness result for the CSP immediately translates into a hardness result for the

QCSP. Moreover, in the cases where the CSP($\mathfrak{B}$) is tractable, the algorithmic insight is often the starting point for further investigations of QCSP($\mathfrak{B}$).

However, complexity classifications for QCSPs are typically harder to obtain than the corresponding complexity classifications for CSPs. One of the reasons is that several relevant universal-algebraic facts require the assumption that the algebra be *idempotent*. The complexity of the QCSP, however, is not preserved by homomorphic equivalence, thus when we study QCSP($\mathfrak{B}$) we cannot pass to the core of $\mathfrak{B}$. Hence, in general we cannot make the assumption that the polymorphism clone of $\mathfrak{B}$ is idempotent. For CSP($\mathfrak{B}$), a powerful way of proving NP-hardness is to give a primitive positive interpretation of a Boolean template with a hard CSP (see Section 6.3.5). For the QCSP, there are other sources of hardness. There are for example 3-element templates $\mathfrak{B}$ such that QCSP($\mathfrak{B}$) is PSPACE-complete [102] and $\mathfrak{B}$ has a semilattice polymorphism so that no NP-hard Boolean CSP can be interpreted in $\mathfrak{B}$ (not even up to homomorphic equivalence; see Section 6.7). Finally, we would like to mention that PSPACE-hardness proofs for the QCSP are often more difficult than NP-hardness proofs for the CSP [102, 52]. In view of the above it is not surprising that a full classification of the QCSP complexity for three-element structures is still open. Similarly, there is no classification of the QCSP for the class of temporal constraint languages presented in Chapter 12. There are concrete temporal constraint languages for which the QCSP is of unknown computational complexity, for instance the QCSP for

$$(\mathbb{Q}; \{(x, y, z) \mid x = y \Rightarrow y \geq z\}).$$

For this problem, we do not know hardness for any complexity class above P, and do not know containment in any complexity class below PSPACE.

Chapter 2

MODEL THEORY

Hodges [207] writes that *"model theory is about the classification of mathematical structures, maps and sets by means of logical formulas"*. This book is about the computational complexity of constraint satisfaction problems for infinite structures $\mathfrak{B}$—and since the constraint satisfaction problem for $\mathfrak{B}$ (and in particular its complexity) is fully determined by the first-order theory of $\mathfrak{B}$, it is not surprising that model theory has a great deal to say about constraint satisfaction problems.

We focus in this chapter on those classical themes from model theory that become relevant for constraint satisfaction; these are in particular saturation, various preservation theorems, and the direct limit construction. Compared to 'classical model theory' we have to make one important twist here: since negation is not permitted in constraint satisfaction, the results we present here will be concerned with *positive* logic. Consequently, the prominent notion of embedding in model theory will be replaced by homomorphism, existential definability by existential positive definability, limits of chains by direct limits, etc. Positive logic is topic of independent interest in model theory [35]. It is important to note that the positive results in model theory imply the corresponding classical results: we only have to add a symbol for the negation of each atomic formula to the signature, and apply the result for positive logic, to obtain the corresponding classical result.

When we present applications of the findings of this chapter to the study of CSPs, it is natural to take the *Satisfiability Perspective* on the CSP, that is, we consider problems of the form CSP(T) for a first-order theory T (see Section 1.3). The main theme of this chapter is methods for constructing models of T with good properties such that CSP($\mathfrak{B}$) = CSP(T). Important in this context is the notion of *(first-order) types* and *primitive positive types* (Section 2.2). Models with a rich automorphism group and good model-theoretic properties can be constructed from classes of finite structures via *Fraïssé-amalgamation* (Section 2.3), or via *first-order interpretations* from known structures (Section 2.4). Section 2.5 presents algebraic characterisations of good model-theoretic properties (so-called *preservation theorems*), Section 2.6 studies theories that have good model-theoretic properties (namely *model-complete theories* and *core theories*), and Section 2.6 explains how in some situations a theory may be replaced by a theory that has good model-theoretic properties (*model companions* and *core compansions*).

2.1. Preliminaries in logic

This section collects some basic terminology and facts from logic. The notation mostly follows Hodges' textbook [207, 206], so many readers may safely skip this section. The section can be consulted later, if needed, for particular concepts that we introduce here. At some rare occasions, we use the concepts *ordinal*, *cardinal*, *well-order*, *transfinite induction*, and *axiom of choice* (cf. [207, 206]); however, these concepts are only needed for strongest possible formulations of the results or in some side remarks. The readers who feel uneasy about set theoretic foundations can resort to the countable setting, which is sufficient for most applications. The smallest infinite ordinal is denoted by ω and the smallest infinite cardinal by $\aleph_0$.

2.1.1. Basic conventions. In this text, $\mathbb{N} = \{0, 1, 2, \dots\}$ denotes the set of natural numbers including 0. If A is a set, $a = (a_1, \dots, a_n) \in A^n$, and

$i \in \{1, \ldots, n\}$, then we also use the notation $a[i]$ to denote the i-th entry a_i of the tuple a.

2.1.2. Structures. In Section 1.1 we have already defined *relational structures*; we now give the general definition of structures that may also contain functions, since we need those later. One occasion where we need functions rather than relations is in Chapter 6 where we consider *algebras*, i.e., structures with a purely functional signature, which are an important tool in the study of the complexity of CSPs. Another occasion where we need function symbols is to conveniently define several important templates of CSPs, e.g. in Section 5.3. Most definitions go in parallel for functional and relational signatures, so we give them together in this section.

A *signature* τ is a set of relation and function symbols, each equipped with an arity. A τ-*structure* $\mathfrak{A}$ is a set A (the *domain* of $\mathfrak{A}$) together with a relation $R^{\mathfrak{A}} \subseteq A^k$ for each k-ary relation symbol in τ and a function $f^{\mathfrak{A}} \colon A^k \to A$ for each k-ary function symbol in τ; here we allow the case $k = 0$ to model constant symbols. Unless stated otherwise, $A, B, C, \ldots$ denote the domains of the structures $\mathfrak{A}, \mathfrak{B}, \mathfrak{C}, \ldots$, respectively. We sometimes write $(A; R_1^{\mathfrak{A}}, R_2^{\mathfrak{A}}, \ldots, f_1^{\mathfrak{A}}, f_2^{\mathfrak{A}}, \ldots)$ for the relational structure $\mathfrak{A}$ with relations $R_1^{\mathfrak{A}}, R_2^{\mathfrak{A}}, \ldots$ and functions $f_1^{\mathfrak{A}}, f_2^{\mathfrak{A}}, \ldots$. When there is no danger of confusion, we use the same symbol for a function and its function symbol, and for a relation and its relation symbol. The cardinality of a structure is defined to be the cardinality of its domain. By *countable* we mean at most countable; if we want to exclude finite sets, we use the formulation *countably infinite*.

The most important special cases of structures that appear in this text are *relational structures*, that is, structures with a purely relational signature, and *algebras*, that is, structures with a purely functional signature. Algebras with domain $A, B, C, \ldots$ are typically denoted by $\mathbf{A}, \mathbf{B}, \mathbf{C}, \ldots$.

When working with function symbols, it is sometimes convenient to work with *multi-sorted* structures, where we have distinguished unary predicates, called *sorts*, that define a partition of the domain, and where function symbols may only be defined on some of the sorts (that is, the function symbols may not be defined on some of the elements). We are sloppy with the formal details since they can always be worked out easily. Two-sorted structures will be denoted by $(\mathfrak{A}, \mathfrak{B})$—here one sort induces the structure $\mathfrak{A}$, and the other sort induces the structure $\mathfrak{B}$.

2.1.3. Expansions and reducts. Let σ, τ be signatures with $\sigma \subseteq \tau$. If $\mathfrak{A}$ is a σ-structure and $\mathfrak{B}$ is a τ-structure, both with the same domain, such that $R^{\mathfrak{A}} = R^{\mathfrak{B}}$ for all relations $R \in \sigma$ and $f^{\mathfrak{A}} = f^{\mathfrak{B}}$ for all functions and constants $f \in \sigma$, then $\mathfrak{A}$ is called a *reduct* of $\mathfrak{B}$, and $\mathfrak{B}$ is called an *expansion* of $\mathfrak{A}$. We also write $\mathfrak{B}^{\sigma}$ for the reduct of $\mathfrak{B}$ with signature σ. An expansion $\mathfrak{B}$ of $\mathfrak{A}$ is called *first-order* if all new relations in $\mathfrak{B}$ are first-order definable over $\mathfrak{A}$. A reduct of a first-order expansion of $\mathfrak{A}$ is called a *first-order reduct* of $\mathfrak{A}$ (in some articles, first-order reducts are simply called reducts; we do not follow this convention, since we also need the classical notion of reduct in this text).

We also write $(\mathfrak{A}, R)$ (and, similarly, $(\mathfrak{A}, f)$) for the expansion of $\mathfrak{A}$ by a new relation R (a new function or constant f, respectively). If $\mathfrak{A}$ is a τ-structure and $(a_i)_{i\in I}$ a sequence of elements of A indexed by I, then $(\mathfrak{A}; (a_i)_{i\in I})$ is the natural $(\tau \cup \{c_i \mid i \in I\})$-expansion of $\mathfrak{A}$ with $|I|$ new constants, where c_i is interpreted by a_i for every $i \in I$. If $S \subseteq A$ we also write $\mathfrak{A}_S$ for any expansion of $\mathfrak{A}$ by constant symbols such that for every $a \in S$ there is a constant symbol c in the expansion that denotes a.

2.1.4. Extensions and substructures. A τ-structure $\mathfrak{A}$ is a *substructure* of a τ-structure $\mathfrak{B}$ iff

- $A \subseteq B$,
- for every $R \in \tau$ of arity n and $\bar{a} \in A^n$ we have that $\bar{a} \in R^{\mathfrak{A}}$ iff $\bar{a} \in R^{\mathfrak{B}}$, and
- for every $f \in \tau$ of arity n and $\bar{a} \in A^n$ we have that $f^{\mathfrak{A}}(\bar{a}) = f^{\mathfrak{B}}(\bar{a})$.

In this case, we also say that $\mathfrak{B}$ is an *extension* of $\mathfrak{A}$, and write $\mathfrak{A} \leq \mathfrak{B}$. Substructures $\mathfrak{A}$ of $\mathfrak{B}$ and extensions $\mathfrak{B}$ of $\mathfrak{A}$ are called *proper* if the domains of $\mathfrak{A}$ and $\mathfrak{B}$ are distinct.

Note that for every subset S of the domain of $\mathfrak{B}$ there is a unique smallest substructure of $\mathfrak{B}$ whose domain contains S, which is called the *substructure of $\mathfrak{B}$ generated by S*, and which is denoted by $\mathfrak{B}[S]$. Note that every element of $\mathfrak{B}[S]$ can be written as $t^{\mathfrak{B}}(b_1, \ldots, b_k)$ for some $k \geq 1$, a k-ary τ-term t, and $b_1, \ldots, b_k \in S$. We say that $\mathfrak{B}$ is *finitely generated* if $\mathfrak{B} = \mathfrak{B}[S]$ for a finite set S of elements. Recall that we have defined algebras as structures with a purely functional signature. In particular, we may use the notation $\mathbf{B}[S]$ and $\mathbf{A} \leq \mathbf{B}$ for algebras. A *subalgebra* $\mathbf{A}$ of an algebra $\mathbf{B}$ (generated by S) is simply a substructure of $\mathbf{B}$ (generated by S).

Example 2.1.1. A *group* is an algebra $\mathbf{G}$ with a binary function symbol $\cdot$ for composition, a unary function symbol $^{-1}$ for the inverse, and a constant 1 for the identity element of $\mathbf{G}$, satisfying the sentences

$$\forall x, y, z\colon\ x \cdot (y \cdot z) = (x \cdot y) \cdot z, \qquad \text{(associativity)}$$
$$\forall x\colon\ x \cdot x^{-1} = 1,$$
$$\text{and} \quad \forall x\colon\ 1 \cdot x = x \cdot 1 = x.$$

In this signature, the subgroups of $\mathbf{G}$ are precisely the subalgebras of $\mathbf{G}$ as defined above. The subgroup of $\mathbf{G}$ consisting of the identity element only is called *trivial*, and subgroups of $\mathbf{G}$ that are distinct from $\mathbf{G}$ are called *proper*. We typically omit the function symbol $\cdot$ and write fg for the product of elements f, g of $\mathbf{G}$. Such groups will also be called *abstract groups* to distinguish them from *permutation groups*; a permutation group (over a set X) is a set of permutations of X closed under composition and inverse, and containing the identity.

Example 2.1.2. A *meet-semilattice* $\mathfrak{S}$ is a $\{\leq\}$-structure with domain S such that $\leq^{\mathfrak{S}}$ denotes a partial order where any two $u, v \in S$ have a (unique) *greatest lower bound* $u \wedge v$, i.e., an element w such that $w \leq u$, $w \leq v$, and for all w' with $w' \leq u$ and $w \leq v$ we have $w' \leq w$. Dually, a *join-semilattice* is a partial order with least upper bounds, denoted by $u \vee v$. A *semilattice* is a meet-semilattice or a join-semilattice where the distinction between meet and join is either not essential or clear from the context.

Semilattices can also be characterised as $\{\wedge\}$-algebras where $\wedge$ is a binary operation that must satisfy the following axioms

$$\forall x, y, z\colon\ x \wedge (y \wedge z) = (x \wedge y) \wedge z \qquad (\textit{associativity})$$
$$\forall x, y\colon\ x \wedge y = y \wedge x \qquad (\textit{commutativity})$$
$$\forall x\colon x \wedge x = x \qquad (\textit{idempotency, or idempotence}).$$

Clearly, the operation $\wedge^{\mathfrak{S}}$, defined as above in a semilattice $\mathfrak{S}$ viewed as a partially ordered set (*poset*), satisfies these axioms. Conversely, if $(S; \wedge)$ is a semilattice, then the formula $x \wedge y = x$ defines a partial order on S which is a meet-semilattice (and $x \wedge y = y$ defines a partial order on S which is a join-semilattice).

Note that the two ways of formalising semilattices differ when it comes to the notion of a substructure; a *subsemilattice* is referring to the substructure of a semilattice when formalised as an algebraic structure.

Example 2.1.3. A *lattice* $\mathfrak{L}$ is a $\{\leq\}$-structure with domain L such that $\leq^{\mathfrak{L}}$ denotes a partial order such that any two $u, v \in L$ have a largest lower bound $u \wedge v$ and a least upper bound, denoted by $u \vee v$. Lattices can also be characterised as $\{\wedge, \vee\}$-algebras where $\wedge$ and $\vee$ are semilattice operations (Example 2.1.2) that additionally satisfy

$$\forall x, y\colon\ x \wedge (x \vee y) = x \text{ and } x \vee (x \wedge y) = x \qquad (\textit{absorption}).$$

If $\mathfrak{L}$ is a lattice and the operations $\wedge$ and $\vee$ are defined as above for semilattices, then these two operations also satisfy the absorption axiom. Conversely, if $(L; \wedge, \vee)$ satisfies the axioms described above, then the formula $x \wedge y = x$ (equivalently, the formula $x \vee y = y$) defines a partial order on L which is a lattice. Of course, there is potential danger of confusion of the symbols for lattice operations $\wedge$ and $\vee$ with the propositional connectives $\wedge$ for conjunction and $\vee$ for disjunction (which can be seen as lattice operations on the set $\{0, 1\}$) which luckily should not cause trouble here. A lattice $\mathfrak{L} = (L; \wedge, \vee)$ is called *distributive* if it satisfies

$$\forall x, y\colon\ x \wedge (y \vee z) = (x \wedge y) \vee (x \wedge z) \qquad \text{(distributivity).}$$

Example 2.1.4. A *Boolean algebra* $\mathfrak{B}$ is a $\{\wedge, \vee, \neg, 0, 1\}$-structure with domain B such that $(B; \wedge, \vee)$ is a distributive lattice and $(B; \wedge, \vee, \neg, 0, 1)$ satisfies

$$\forall x\colon\ x \vee \neg x = 1$$
$$\forall x\colon\ x \wedge \neg x = 0.$$

The following concept is only defined for *relational* structures.

DEFINITION 2.1.5. The *Gaifman graph* of a relational structure $\mathfrak{B}$ is the undirected graph whose vertex set equals the domain B of $\mathfrak{B}$, and which has an edge between distinct elements $x, y \in B$ if and only if there is a tuple in one of the relations of $\mathfrak{B}$ that has both x and y as entries.

A relational structure $\mathfrak{B}$ is readily seen to be connected (in the sense of Section 1.1) if and only if its Gaifman graph is connected (in the usual graph-theoretic sense).

2.1.5. Products. Let $\mathfrak{A}$ and $\mathfrak{B}$ be two structures with domains A and B, and the same signature τ. Then the (*direct*, or *categorical*) *product* $\mathfrak{C} = \mathfrak{A} \times \mathfrak{B}$ is the τ-structure with domain $A \times B$, which has for each k-ary $R \in \tau$ the relation that contains a tuple $((a_1, b_1), \dots, (a_k, b_k))$ if and only if $R(a_1, \dots, a_k)$ holds in $\mathfrak{A}$ and $R(b_1, \dots, b_k)$ holds in $\mathfrak{B}$. For each k-ary $f \in \tau$ the structure $\mathfrak{C}$ has the operation that maps $((a_1, b_1), \dots, (a_k, b_k))$ to $(f(a_1, \dots, a_k), f(b_1, \dots, b_k))$. The direct product $\mathfrak{A} \times \mathfrak{A}$ is also denoted by $\mathfrak{A}^2$, and the k-fold product $\mathfrak{A} \times \cdots \times \mathfrak{A}$, defined analogously, by $\mathfrak{A}^k$.

We generalise the definition of products in the obvious way to infinite products. For a sequence of τ-structures $(\mathfrak{A}_i)_{i \in I}$, the direct product $\mathfrak{B} = \prod_{i \in I} \mathfrak{A}_i$ is the τ-structure on the domain $\prod_{i \in I} A_i$ such that for $R \in \tau$ of arity k

$$((a_i^1)_{i \in I}, \dots, (a_i^k)_{i \in I}) \in R^{\mathfrak{B}} \text{ iff } (a_i^1, \dots, a_i^k) \in R^{\mathfrak{A}_i} \text{ for each } i \in I\ ,$$

and for $f \in \tau$ of arity k, we have

$$f^{\mathfrak{B}}((a_i^1)_{i \in I}, \dots, (a_i^k)_{i \in I}) = (f^{\mathfrak{A}_i}(a_i^1, \dots, a_i^k))_{i \in I}.$$

If $\mathbf{A}_i = \mathbf{A}$ for all $i \in I$, we also write $\mathbf{A}^I$ instead of $\prod_{i \in I} \mathfrak{A}_i$, and call it the I-*th power of* $\mathfrak{A}$.

2.1.6. Functions and preservation. Throughout the text, we use the following conventions. If $f\colon A \to B$ is a function, and S is a subset of A, then $f(S)$ denotes the set $\{f(s) \mid s \in S\} \subseteq B$. If $t = (t_1, \dots, t_k)$ is a k-tuple of elements of A, then $f(t)$ denotes the tuple $(f(t_1), \dots, f(t_k))$. Moreover, we use the same convention for functions of higher arity $f\colon B^m \to B$: if $t^1, \dots, t^m$ are k-tuples of elements of B, then $f(t^1, \dots, t^m)$ denotes the k-tuple $(f(t_1^1, \dots, t_1^m), \dots, f(t_k^1, \dots, t_k^m))$ (that is, the k-tuple is computed componentwise).

If $h\colon A \to B$ is a map, then the *kernel* of h is the equivalence relation E on A where $(a, a') \in E$ if $h(a) = h(a')$. For $a \in A$, we denote by a/E the equivalence class of a in E, and by A/E the set of all equivalence classes of elements of A.

In the following, let $\mathfrak{A}$ be a τ-structure with domain A and $\mathfrak{B}$ a τ-structure with domain B. A *homomorphism* h from $\mathfrak{A}$ to $\mathfrak{B}$ is a function from A to B that *preserves* each function and each relation for the symbols in τ; that is,

- if $(a_1, \dots, a_k)$ is in $R^{\mathfrak{A}}$, then $(h(a_1), \dots, h(a_k))$ must be in $R^{\mathfrak{B}}$;
- $f^{\mathfrak{B}}(h(a_1), \dots, h(a_k)) = h(f^{\mathfrak{A}}(a_1, \dots, a_k))$.

When $\mathbf{A}, \mathbf{B}$ are algebras with the same signature and domain A, B, respectively, and f is a homomorphism from $\mathbf{A}$ to $\mathbf{B}$, then $f(A)$ induces a subalgebra of $\mathbf{B}$, and this subalgebra is called a *homomorphic image* of $\mathbf{A}$.

When a mapping h preserves a relation R, we also say that R is *invariant* under h. A homomorphism from $\mathfrak{A}$ to $\mathfrak{B}$ is called a *strong homomorphism* if it also preserves the complements of the relations from $\mathfrak{A}$. An injective strong homomorphism e from $\mathfrak{A}$ to $\mathfrak{B}$ is called an *embedding*, and denoted as $e \colon \mathfrak{A} \hookrightarrow \mathfrak{B}$. Surjective embeddings are called *isomorphisms*. A homomorphism from a substructure of $\mathfrak{A}$ to $\mathfrak{B}$ is called a *partial homomorphism* from $\mathfrak{A}$ to $\mathfrak{B}$. An embedding from a substructure of $\mathfrak{A}$ into $\mathfrak{B}$ is called a *partial isomorphism* between $\mathfrak{A}$ and $\mathfrak{B}$.

Homomorphisms and isomorphisms from $\mathfrak{B}$ to itself are called *endomorphisms* and *automorphisms*, respectively. When $f \colon A \to B$ and $g \colon B \to C$, then $g \circ f$ denotes the composed function $x \mapsto g(f(x))$. Clearly, the composition of two homomorphisms (embeddings, automorphisms) is again a homomorphism (embedding, automorphism). Let $\mathrm{Aut}(\mathfrak{A})$ and $\mathrm{End}(\mathfrak{A})$ be the sets of automorphisms and endomorphisms, respectively, of $\mathfrak{A}$. The set $\mathrm{Aut}(\mathfrak{A})$ can be viewed as a group, and $\mathrm{End}(\mathfrak{A})$ as a monoid with respect to composition; more on that can be found in Section 4.2 and Section 4.4.

2.1.7. Formulas and theories. We assume familiarity with basic concepts of classical first-order logic; see for example [163]. In particular, we will use the concepts of *atomic τ-formula* (where τ is a fixed signature), *free and bound variable*, and *τ-term*.

We always allow the first-order formula $x = y$ (for equality), $\bot$ (for 'false'), and $\top$ (for 'true') independently of the signature, unless stated otherwise. A formula without free variables will be called a *sentence*. A *(first-order) τ-theory* is a set of (first-order) τ-sentences. A τ-structure $\mathfrak{B}$ is a *model* of a τ-sentence ϕ (or a theory T) if ϕ (all sentences in T, respectively) holds true in $\mathfrak{B}$; in this case we write $\mathfrak{B} \models \phi$ ($\mathfrak{B} \models T$). If C is a class of τ-structures and ϕ is a τ-sentence, we say that ϕ holds in C if every structure in C is a model of ϕ.

The set of all first-order τ-sentences that are true in a given τ-structure $\mathfrak{B}$ is called the *first-order theory of* $\mathfrak{B}$, and denoted by $\mathrm{Th}(\mathfrak{B})$. Similarly, the *first-order theory of a class of τ-structures* is the set of all τ-sentences that holds in all structures in the class. When the reference to a particular signature τ

is clear or not important, we omit the specification of τ in the terminology introduced above. If a sentence or a theory has a model, we call it *satisfiable*.

THEOREM 2.1.6 (Compactness; see Theorem 5.1.1 in [207]). *Let T be a first-order theory. If every finite subset of T is satisfiable then T is satisfiable.*

If T is a theory and ϕ a sentence, we say that T *entails* ϕ, in symbols $T \models \phi$, if every model of T satisfies ϕ. Analogously we define $T_1 \models T_2$ for two theories. Two theories T_1, T_2 are said to be (*logically*) *equivalent* if $T_1 \models T_2$ and $T_2 \models T_1$; we apply analogous definitions for logical equivalence of single sentences and for other logics than first-order logic.

LEMMA 2.1.7 (Lemma 2.3.2 in [207]). *Let T be a first-order τ-theory, and ϕ a first-order τ-formula with free variables $x_1, \dots, x_n$. Let $c_1, \dots, c_n$ be distinct constants that are not in τ. Then $T \models \phi(c_1, \dots, c_k)$ if and only if $T \models \forall x_1, \dots, x_n \colon \phi$.*

2.1.8. Syntactic restrictions. There is a series of syntactic restrictions of first-order logic that are important for the applications of model theory in this text. A first-order τ-formula ϕ is said to be

- *primitive positive* (pp) if it is of the form $\exists x_1, \dots, x_n \ (\psi_1 \wedge \cdots \wedge \psi_m)$, where $\psi_1, \dots, \psi_m$ are atomic (primitive positive formulas are of central importance in the entire text);
- *quantifier-free* (qf) if it does not contain any quantifiers; that is, it is built from the logical connectives $\wedge, \vee, \neg$, the binary relation $=$, the (free) variables, and the symbols from τ only;
- *quantifier-free primitive positive* (qfpp) if it contains neither quantifiers, nor negation, nor disjunction; that is, it is built from the logical connectives $\wedge$, the binary relation $=$, the (free) variables, and the symbols from τ only;
- *positive* if ϕ does not contain any negation symbol $\neg$;
- *in prenex normal form* if it is of the form $Q_1 x_1 \dots Q_n x_n \colon \psi$ where $Q_i \in \{\forall, \exists\}$ and ψ is quantifier-free;
- *Horn* if it is written in conjunctive normal form and every clause has at most one positive literal (Horn formulas appear e.g. in Section 7.3);
- *existential* if it is of the form $\exists x_1, \dots, x_n \colon \psi$ where ψ is quantifier-free (existential formulas appear e.g. in Section 2.6.1);
- *universal* if is of the form $\forall x_1, \dots, x_n \colon \psi$ where ψ is quantifier-free;
- *existential positive* ($\exists^+$) if it is existential and positive;
- *universal negative* ($\forall^-$) if it is of the form $\forall x_1, \dots, x_n \colon \neg\psi$ where ψ is positive quantifier-free (such formulas have also been called *h-universal* in [35]);
- *universal conjunctive* if it is universal and if the quantifier-free part of ϕ does not contain any negation or disjunction symbols (universal conjunctive formulas appear e.g. in Section 6.5);

- *forall-exists* ($\forall\exists$) if it is of the form $\forall y_1, \dots, y_m \colon \psi$ where ψ is existential (forall-exists formulas appear e.g. in Section 2.6.1; they are sometimes also called *inductive* [207]);
- *h-inductive* ($\forall\exists^+$) if they are conjunctions of formulas of the form

$$\forall \bar{x} \big(\exists \bar{y} \colon \phi(\bar{x}, \bar{y}) \Rightarrow \exists \bar{z} \colon \psi(\bar{x}, \bar{z})\big)$$

where ϕ and ψ are positive quantifier-free formulas. Such formulas appear throughout Section 2.6; they have been studied first by Ben Yaacov and Poizat [35].

Clearly, every existential positive formula is equivalent to a finite disjunction of primitive positive formulas. An important property of primitive positive sentences ϕ is that $\mathfrak{A} \times \mathfrak{B} \models \phi$ iff $\mathfrak{A} \models \phi$ and $\mathfrak{B} \models \phi$.

There are several descriptions of h-inductive formulas up to equivalence; the name comes from a semantic characterisation of such formulas that will be presented later (Theorem 2.5.5).

PROPOSITION 2.1.8. *Let ϕ be a formula. Then the following are equivalent.*

- *ϕ is h-inductive.*
- *ϕ is equivalent to a conjunction of universally quantified disjunctions of primitive positive formulas and negated atomic formulas.*
- *ϕ is equivalent to a formula of the form $\forall \bar{y} \colon \phi(\bar{y})$ where $\phi(\bar{y})$ is a positive boolean combination of quantifier-free formulas and existential positive formulas (such formulas were called* positively restricted forall-exists formulas *in* [61]).

A first-order theory T is said to be *existential* if all sentences in T are existential. Analogously, we define universal, existential positive, and universal negative theories. If T is a theory, then $T_\exists$ denotes the set of all existential sentences implied by T. The theories $T_{\exists^+}$, $T_\forall$, $T_{\forall^-}$, $T_{\forall\exists^+}$, T_{qf}, T_{pp}, T_{qfpp} are defined analogously.

2.1.9. Formulas and preservation. Let $\mathfrak{B}$ be a τ-structure. If ϕ is a first-order τ-formula and $x_1, \dots, x_n$ is an ordered list such that all free variables in ϕ come from $x_1, \dots, x_n$, then $\phi(x_1, \dots, x_n)$ *defines* over $\mathfrak{B}$ the relation

$$\{(b_1, \dots, b_n) \in B^n \mid \mathfrak{B} \models \phi(b_1, \dots, b_n)\}.$$

If two structures $\mathfrak{A}$ and $\mathfrak{B}$ have the same domain and all relations and functions from $\mathfrak{A}$ have a first-order definition in $\mathfrak{B}$, then we say that $\mathfrak{A}$ is (*first-order*) *definable* in $\mathfrak{B}$. Two structures $\mathfrak{A}, \mathfrak{B}$ over the same domain are (*first-order*) *interdefinable* if $\mathfrak{A}$ is definable in $\mathfrak{B}$ and vice versa.

DEFINITION 2.1.9. Let $\mathfrak{A}$ and $\mathfrak{B}$ be τ-structures and let $\phi(x_1, \dots, x_n)$ be a τ-formula. Then a function $h \colon A \to B$ *preserves* ϕ if $\mathfrak{B} \models \phi(h(a_1), \dots, h(a_n))$ whenever $\mathfrak{A} \models \phi(a_1, \dots, a_n)$.

Note that homomorphisms preserve all existential positive formulas. Also note that partial isomorphisms preserve quantifier-free formulas, embeddings preserve existential formulas, and isomorphisms preserve all first-order formulas. The following is straightforward from the definitions, and sometimes called the *diagram lemma* (see Lemma 1.4.2 in [207]).

LEMMA 2.1.10. *Let $\mathfrak{A}$ and $\mathfrak{B}$ be τ-structures. Then the following are equivalent.*

1. *There is an expansion $\mathfrak{B}'$ of $\mathfrak{B}$ such that $\mathfrak{B}' \models \mathrm{Th}(\mathfrak{A}_A)_{\mathrm{qfpp}}$.*
2. *There is a homomorphism from $\mathfrak{A}$ to $\mathfrak{B}$.*

Embeddings that preserve all first-order formulas are called *elementary*. If $\mathfrak{B}$ is an extension of $\mathfrak{A}$ such that the identity map from $\mathfrak{A}$ to $\mathfrak{B}$ is an elementary embedding, we say that $\mathfrak{B}$ is an *elementary extension* of $\mathfrak{A}$, and that $\mathfrak{A}$ is an *elementary substructure* of $\mathfrak{B}$.

THEOREM 2.1.11 (Löwenheim-Skolem; see Corollary 3.1.4 in [207]). *Let $\mathfrak{A}$ be a τ-structure, X a set of elements of $\mathfrak{A}$, and λ an infinite cardinal such that $|\tau| + |X| \leq \lambda \leq |A|$. Then $\mathfrak{A}$ has an elementary substructure $\mathfrak{B}$ of cardinality λ with $X \subseteq B$.*

We will need the following lemma.

LEMMA 2.1.12. *Let $\mathfrak{A}$ and $\mathfrak{B}$ be τ-structures and suppose that every primitive positive sentence true in $\mathfrak{A}$ is also true in $\mathfrak{B}$. Then there exists an elementary extension $\mathfrak{C}$ of $\mathfrak{B}$ and a homomorphism $g \colon \mathfrak{A} \to \mathfrak{C}$.*

PROOF. It suffices to show that the theory $T := \mathrm{Th}(\mathfrak{A}_A)_{\mathrm{qfpp}} \cup \mathrm{Th}(\mathfrak{B}_B)$ has a model $\mathfrak{C}'$, since Lemma 2.1.10 then asserts the existence of a homomorphism from $\mathfrak{A}$ to the τ-reduct $\mathfrak{C}$ of $\mathfrak{C}'$, which will be an elementary extension of $\mathfrak{B}$.

If T has no model, then by the compactness theorem there is a τ-formula ϕ such that $\phi(\bar{c}) \in \mathrm{Th}(\mathfrak{A}_A)_{\mathrm{qfpp}}$ and $\{\phi(\bar{c})\} \cup \mathrm{Th}(\mathfrak{B}_B)$ has no model, and in particular $\mathfrak{B} \models \neg\exists \bar{y} \colon \phi(\bar{y})$. Since $\exists \bar{y} \colon \phi(\bar{y})$ is primitive positive, the assumptions imply that $\mathfrak{A} \models \neg\exists \bar{y} \colon \phi(\bar{y})$. This contradicts that $\mathfrak{A}_A \models \phi(\bar{c})$. ⊣

The following example shows that moving to an elementary extension of $\mathfrak{B}$ is necessary, because in general there need not be a homomorphism from $\mathfrak{A}$ to $\mathfrak{B}$.

Example 2.1.13. Let $\mathfrak{A}$ be a countably infinite clique, and $\mathfrak{B}$ a countable disjoint union of finite cliques of arbitrary large size. Then every primitive positive sentence over the signature of graphs which is true in $\mathfrak{A}$ is also true in $\mathfrak{B}$, but there is no homomorphism from $\mathfrak{A}$ to $\mathfrak{B}$.

PROPOSITION 2.1.14. *Let S and T be τ-theories. The following are equivalent.*

1. *Every model of S has a homomorphism to a model of T.*
2. *Every universal negative consequence of T is also a consequence of S, i.e., $T_{\forall-} \subseteq S_{\forall-}$.*
3. *For every primitive positive sentence ϕ, if $S \cup \{\phi\}$ is satisfiable then $T \cup \{\phi\}$ is satisfiable.*

4. $T \cup \{\phi \mid S \cup \{\phi\}$ *is satisfiable*$\}$ *is satisfiable.*

PROOF. To prove the implication from (1) to (2) let ϕ be a universal negative sentence implied by T, and let $\mathfrak{C}$ be a model of S. By (1), there is a homomorphism from $\mathfrak{C}$ to a model $\mathfrak{B}$ of T, which satisfies in particular ϕ. Since $\neg\phi$ is equivalent to an existential positive sentence, it is preserved by homomorphisms, so we must have $\mathfrak{C} \models \phi$. Thus, $\phi \in S_{\forall-}$.

(2) $\Rightarrow$ (3): if $S \cup \{\phi\}$ is satisfiable, then $\neg\phi$ is universal negative, so can not be a consequence of T by (2). Hence, $T \cup \{\phi\}$ has a model.

(3) $\Rightarrow$ (4) is by compactness.

(4) $\Rightarrow$ (1). Let $\mathfrak{A}$ be a model of S, and let S' be the existential positive theory of $\mathfrak{A}$. By assumption, $T \cup S'$ has a model $\mathfrak{B}$. Then $\mathfrak{B}$ satisfies the assumptions from Lemma 2.1.12, so there exists an elementary extension $\mathfrak{C}$ of $\mathfrak{B}$ and a homomorphism $h \colon \mathfrak{A} \to \mathfrak{C}$. $\dashv$

We can now prove a generalisation of the condition given in Proposition 1.3.4 from Section 1.3 that characterises when two theories have the same CSP.

COROLLARY 2.1.15. *Let S and T be τ-theories. The following are equivalent.*

1. *Every model of S has a homomorphism to a model of T, and every model of T has a homomorphism to a model of S.*
2. *S and T imply the same universal negative sentences, i.e., $S_{\forall-} = T_{\forall-}$.*

This indeed proves Proposition 1.3.4, because theories that imply the same universal negative sentences obviously have the same CSP. It is now also easy to prove Proposition 1.3.6 from Section 1.3, characterising those theories T for which there exists a structure $\mathfrak{B}$ such that $\mathrm{CSP}(T) = \mathrm{CSP}(\mathfrak{B})$. We first show the following.

PROPOSITION 2.1.16. *For any satisfiable theory T, the following are equivalent.*

1. *There exists a structure $\mathfrak{B}$ such that for every existential positive sentence ϕ the theory $T \cup \{\phi\}$ is satisfiable if and only if $\mathfrak{B} \models \phi$.*
2. *T has a model $\mathfrak{B}$ that satisfies every existential positive sentence ϕ for which $T \cup \{\phi\}$ is satisfiable.*
3. *For all existential positive sentences ϕ_1 and ϕ_2, if $T \cup \{\phi_1\}$ is satisfiable and $T \cup \{\phi_2\}$ is satisfiable, then $T \cup \{\phi_1, \phi_2\}$ is satisfiable as well.*
4. *T has the Joint Homomorphism Property (JHP; cf. Proposition 1.3.6).*

PROOF. We prove (1) $\Leftrightarrow$ (2), (2) $\Leftrightarrow$ (3), (3) $\Leftrightarrow$ (4). The implications (2) $\Rightarrow$ (1) and (2) $\Rightarrow$ (3) are obvious.

We first prove (1) $\Rightarrow$ (2). If (1) holds, then the implication (3) $\Rightarrow$ (1) in Proposition 2.1.14 shows that there is a homomorphism from $\mathfrak{B}$ to a model $\mathfrak{C}$ of T. This model $\mathfrak{C}$ has the desired property: if ϕ is existential positive such that $T \cup \{\phi\}$ is satisfiable, then $\mathfrak{B}$ satisfies ϕ, and since homomorphisms preserve existential positive formulas, $\mathfrak{C}$ satisfies ϕ as well.

(3) $\Rightarrow$ (2). Let P be the set of all existential positive sentences ϕ such that $T \cup \{\phi\}$ is satisfiable. By the assumption that T is satisfiable, and by (3),

all finite subsets of $T \cup P$ are satisfiable, so by the compactness theorem of first-order logic (Theorem 2.1.6) we have that $T \cup P$ has a model $\mathfrak{B}$.

(3) $\Rightarrow$ (4). Let τ be the signature of T. Let $\mathfrak{A}_1$ and $\mathfrak{A}_2$ be models of T. We have to show that there exists a model $\mathfrak{B}$ of T that admits homomorphisms from $\mathfrak{A}_1$ and $\mathfrak{A}_2$. Let $\mathfrak{A}'_1$ and $\mathfrak{A}'_2$ be expansions of $\mathfrak{A}_1$ and $\mathfrak{A}_2$, respectively, where every element is denoted by a distinct constant symbol. By Lemma 2.1.10 the theories $S_1 := T \cup \mathrm{Th}(\mathfrak{A}'_1)_{\mathrm{qfpp}}$ and $S_2 := T \cup \mathrm{Th}(\mathfrak{A}'_2)_{\mathrm{qfpp}}$ are satisfiable. Let $S'_1 \subseteq S_1$ and $S'_2 \subseteq S_2$ be finite. We obtain a single qfpp formula ϕ_i by forming a conjunction over all elements of S'_i, for $i = 1$ and $i = 2$. By (3), the theory $T \cup \{\phi_1, \phi_2\}$ is satisfiable as well, showing that $T \cup S'_1 \cup S'_2$ is satisfiable. Again, the compactness theorem of first-order logic implies that $T' := T \cup S_1 \cup S_2$. Let $\mathfrak{B}'$ be a model of T', and let $\mathfrak{B}$ be the τ-reduct of $\mathfrak{B}'$. Then another application of Lemma 2.1.10 asserts the existence of a homomorphism from $\mathfrak{A}_1$ to $\mathfrak{B}$ and from $\mathfrak{A}_2$ to $\mathfrak{B}$, which proves (4).

(4) $\Rightarrow$ (3). Let ϕ_1 and ϕ_2 be existential positive sentences such that $T \cup \{\phi_1\}$ has a model $\mathfrak{A}_1$ and $T \cup \{\phi_2\}$ has a model $\mathfrak{A}_2$. By (4), there exists a model $\mathfrak{B}$ of T such that $\mathfrak{A}_1$ and $\mathfrak{A}_2$ map homomorphically to $\mathfrak{B}$. Then $\mathfrak{B}$ clearly satisfies $T \cup \{\phi_1, \phi_2\}$ since homomorphisms preserve existential positive sentences. $\dashv$

Note that in the statement and the proof above, the phrase *existential positive* can be used interchangeably with the phrase *primitive positive*. With the additional assumption that T has a finite relational signature, item (1) in Proposition 2.1.16 becomes the statement that there exists a structure $\mathfrak{B}$ such that CSP($\mathfrak{B}$) and CSP(T) are the same problem, so we indeed proved in particular Proposition 1.3.6.

2.1.10. Chains and direct limits. Chains and direct limits of sequences of τ-structures are an important method of constructing models of first-order theories, and will be used for instance in Section 2.5 and Section 2.7.

Let $(\mathfrak{A}_i)_{i<\kappa}$ be a sequence of τ-structures for a relational signature τ. Then $(\mathfrak{A}_i)_{i<\kappa}$ is called a *chain* if $\mathfrak{A}_i$ is a substructure of $\mathfrak{A}_j$ for all $i < j < \kappa$. A chain is called an *elementary chain* if for all $i, j < \kappa$ the extension $\mathfrak{A}_j$ of $\mathfrak{A}_i$ is elementary.

DEFINITION 2.1.17. The *union* of the chain $(\mathfrak{A}_i)_{i<\gamma}$ is a τ-structure $\mathfrak{B}$ defined as follows. The domain of $\mathfrak{B}$ is $\bigcup_{i\leq\gamma} A_i$, and for each relation symbol $R \in \tau$ we put $\bar{a} \in R^{\mathfrak{B}}$ if $\bar{a} \in R^{\mathfrak{A}_i}$ for some (or all) $\mathfrak{A}_i$ containing $\bar{a}$.

THEOREM 2.1.18 (Tarski-Vaught; Theorem 2.5.2 in [207]). *Let $(\mathfrak{A}_i)_{i<\kappa}$ be an elementary chain. Then $\bigcup_{i<\kappa} \mathfrak{A}_i$ is an elementary extension of $\mathfrak{A}_i$ for each $i < \kappa$.*

We say that a formula ϕ is *preserved by chains* (*of models of T*) if for all chains $(\mathfrak{A}_i)_{0\leq i<\kappa}$ of τ-structures (where all the $\mathfrak{A}_i$ and $\mathfrak{A} := \bigcup_{i<\kappa} \mathfrak{A}_i$ are models of T), and every finite n-tuple $\bar{a}$ of elements of A_0 we have $\mathfrak{A} \models \phi(\bar{a})$ whenever $\mathfrak{A}_i \models \phi(\bar{a})$ for every (equivalently, for some) $i < \kappa$.

PROPOSITION 2.1.19 (Theorem 2.4.4 in [207]). *Every* $\forall\exists$*-formula is preserved by unions of chains.*

Proposition 2.1.19 is a direct consequence of the more general Lemma 2.1.21 below. Direct limits can be seen as a positive variant of the notion of a union of chains; we essentially replace the identity embeddings of $\mathfrak{A}_i$ into $\mathfrak{A}_j$ in the chain by homomorphisms. Let τ be a relational signature, and let $\mathfrak{A}_0, \mathfrak{A}_1, \dots$ be a sequence of τ-structures such that there are homomorphisms $f_{ij} \colon \mathfrak{A}_i \to \mathfrak{A}_j$. Those homomorphisms are called *coherent* if $f_{jk} \circ f_{ij} = f_{ik}$ for every $i \leq j \leq k$.

DEFINITION 2.1.20. Let $(\mathfrak{A}_i)_{i<\kappa}$ be a sequence of τ-structures with coherent homomorphisms $f_{ij} \colon \mathfrak{A}_i \to \mathfrak{A}_j$. Then the *direct limit* $\lim_{i<\kappa} \mathfrak{A}_i$ is the τ-structure $\mathfrak{A}$ defined as follows. Let $X := \bigcup_{i<\kappa} A_i$ where we assume for simplicity that A_i and A_j are disjoint; if they are not, replace them by isomorphic copies. The domain A of $\mathfrak{A}$ comprises the equivalence classes of the equivalence relation $\sim$ defined on X by setting $x_i \sim x_j$ for $x_i \in A_i, x_j \in A_j$ iff there is a k such that $f_{ik}(x_i) = f_{jk}(x_j)$. Let $g_i \colon A_i \to A$ be the function that maps $a \in A_i$ to the equivalence class of a in A. For $R \in \tau$ and a tuple $\bar{a}$ over A, define $\mathfrak{A} \models R(\bar{a})$ iff there is a $k < \kappa$ and a tuple $\bar{b}$ over A_k such that $\mathfrak{A}_k \models R(\bar{b})$ and $\bar{a} = g_k(\bar{b})$.

Note that the definition of $\lim_{i<\kappa} \mathfrak{A}_i$ also depends on the coherent family f_{ij}, but this is left implicit and will be clear from the context. Also note that g_i defines a homomorphism from $\mathfrak{A}_i$ to $\mathfrak{A}$; this function is called the *limit homomorphism* from $\mathfrak{A}_i$ to the direct limit $\mathfrak{A}$. Note that $g_i = g_j \circ f_{ij}$ for all $i < j < \kappa$.

We have seen that unions of chains preserve $\forall\exists$-formulas; the analogous statement for direct limits is as follows. We say that a first-order formula $\phi(x_1, \dots, x_n)$ is *preserved by direct limits (of models of* T*)* if for every direct limit $\mathfrak{A} := \lim_{i<\kappa} \mathfrak{A}_i$ as defined above where all the $\mathfrak{A}_i$ and $\mathfrak{A}$ are models of T, we have $\mathfrak{A} \models \phi(\bar{a})$ for $\bar{a} \in A^n$ whenever there exists $i < \kappa$ and $\bar{a}_i \in A_i^n$ whose j-th entry is a representative of the j-th entry of $\bar{a}$ such that $\mathfrak{A}_i \models \phi(\bar{a}_i)$. The following is Theorem 2.4.6 in [206]; it is given there without proof but we present the proof here for completeness.

LEMMA 2.1.21. *Every* $\forall\exists^+$*-formula is preserved by direct limits of models of* T.

PROOF. Let $(\mathfrak{A}_i)_{i<\kappa}$ be a sequence of models of T with coherent homomorphisms $h_{ij} \colon \mathfrak{A}_i \to \mathfrak{A}_j$, for $i, j < \kappa$, such that $\mathfrak{A} := \lim_{i<\kappa} \mathfrak{A}_i$ is a model of T. Let g_i be the limit homomorphism from $\mathfrak{A}_i$ to $\mathfrak{A}$. Let ϕ be a $\forall\exists^+$-formula, $\bar{a} \in A^n$, and $i < \kappa$ be such that there exists $\bar{a}_i \in A_i^n$ with $g_i(\bar{a}_i) = \bar{a}$ and $\mathfrak{A}_i \models \phi(\bar{a}_i)$. We have to show that $\mathfrak{A} \models \phi(\bar{a})$. Since ϕ is $\forall\exists^+$, we can assume that $\phi(\bar{x})$ is of the form $\forall \bar{y} \colon \phi'(\bar{x}, \bar{y})$ where ϕ' is a disjunction of negated atomic formulas and existential positive formulas. Suppose that $\phi'(\bar{a}, \bar{b})$ is false in $\mathfrak{A}$ for some tuple $\bar{b}$ of elements of $\mathfrak{A}$. Every disjunct ψ of $\phi'(\bar{a}, \bar{b})$ is false in $\mathfrak{A}$. Then there exists an $p < \kappa$ such that all entries of $\bar{b}$ have representatives in $\mathfrak{A}_p$, and the negated atomic disjuncts of ϕ' are already false in $\mathfrak{A}_p$, by the definition of direct limits. Let $q := \max(p, i)$. Let $\bar{b}_q$ be a tuple of elements of $\mathfrak{A}_q$ where the j-th entry is a representative of the j-th entry in $\bar{b}$. Define $\bar{a}_q := h_{ip}(\bar{a}_i)$ and $b_q := h_{pq}(\bar{b}_q)$. Since $\mathfrak{A}_q \models \phi(a_q)$, there must exist a disjunct ψ of ϕ' such that $\psi(\bar{a}_q, \bar{b}_q)$ holds in $\mathfrak{A}_q$. The limit homomorphism g_q maps $(\bar{a}_q, \bar{b}_q)$ to $(\bar{a}, \bar{b})$ and is a homomorphism from $\mathfrak{A}_q$ to $\mathfrak{A}$, and therefore preserves existential positive formulas, contradicting the assumption that $\phi'(\bar{a}, \bar{b})$ is false in $\mathfrak{A}$. It follows that $\mathfrak{A} \models \forall \bar{y} \colon \phi'(\bar{a}, \bar{y})$ and hence $\mathfrak{A} \models \phi(\bar{a})$. ⊣

2.2. Types

Loosely speaking, types are used to talk about elements in a structure that are not necessarily there; a *type* of a structure $\mathfrak{B}$ is a set of formulas that is satisfied by a real or by a 'virtual' element of B, that is, an element of some structure that has the same theory as $\mathfrak{B}$. In this section we present fundamental results about the existence of models that realise many types, and constructing models that avoid certain types; these results will be needed in elegant proofs of logical preservation theorems that are relevant for constraint satisfaction.

We now introduce types formally. A set Φ of formulas with free variables $x_1, \dots, x_n$ is called *satisfiable* over a structure $\mathfrak{B}$ if there are elements $b_1, \dots, b_n$ of $\mathfrak{B}$ such that for all sentences $\phi \in \Phi$ we have $\mathfrak{B} \models \phi(b_1, \dots, b_n)$. We say that Φ is *satisfiable* if there exists a structure $\mathfrak{B}$ such that Φ is satisfiable over $\mathfrak{B}$. For $n \geq 0$, an *n-type* of a theory T is a set p of formulas with free variables $x_1, \dots, x_n$ such that $p \cup T$ is satisfiable. An *n-type over a structure* $\mathfrak{B}$ is an n-type of the first-order theory of $\mathfrak{B}$; note that we do not admit parameters in these formulas (i.e., we allow constants to appear in the formulas of the type only if they belong to the signature). An n-type p of T is *maximal* or *complete* if $T \cup p \cup \{\phi(x_1, \dots, x_n)\}$ is unsatisfiable for any formula $\phi \notin p$. The set of all maximal n-types of $\mathrm{Th}(\mathfrak{A})$ is denoted by $S_n^{\mathfrak{A}}$.

A *quantifier-free* (*existential positive*) *type* is a type whose formulas are quantifier-free (existential positive). Maximality for existential positive types and for quantifier-free types is defined analogously with the obvious modifications. If p is an n-type, and $I \subseteq \{1, \dots, n\}$ with $|I| = k$, then the *subtype of p induced on I* is the k-type obtained from p by existentially quantifying in all formulas in p the variables x_i for $i \in \{1, \dots, n\} \setminus I$, and then renaming the variables in the resulting set of formulas to $x_1, \dots, x_k$ in a way that preserves the order on the indices of the variables.

An n-type p of $\mathfrak{A}$ is *realised* in $\mathfrak{A}$ if there exist $a_1, \dots, a_n \in A$ such that $\mathfrak{A} \models \phi(a_1, \dots, a_n)$ for each $\phi \in p$; otherwise, p is *omitted* in $\mathfrak{A}$. The set of all first-order formulas with free variables $x_1, \dots, x_n$ satisfied by an n-tuple

$\bar{a} = (a_1, \ldots, a_n)$ in $\mathfrak{A}$ is a maximal type of $\mathfrak{A}$, called the *type of* $\bar{a}$, and denoted by $\mathrm{tp}^{\mathfrak{A}}(\bar{a})$.

2.2.1. Realising types and saturated structures. A structure $\mathfrak{B}$ is *saturated* if, informally, 'as many types as possible' come from real elements of $\mathfrak{B}$. Formally, for an infinite cardinal κ, a structure $\mathfrak{A}$ is κ-*saturated* if, for all $\beta < \kappa$ and expansions $\mathfrak{A}'$ of $\mathfrak{A}$ by at most β constants, every 1-type of $\mathfrak{A}'$ is realised in $\mathfrak{A}'$. We say that an infinite structure $\mathfrak{A}$ is *saturated* if it is $|A|$-saturated.

THEOREM 2.2.1 (Corollary 8.2.2 in [207]; Theorem 4.3.12 in [278]). *Let τ be a signature and $\lambda \geq |\tau|$. Then every τ-structure $\mathfrak{B}$ has a λ^+-saturated elementary extension of cardinality $\leq |B|^\lambda$.*

The proof of this theorem is very similar to the proof of Lemma 2.7.3, and we refer to the cited textbooks for a proof. The proof of the following result uses an important technique, called a *back-and-forth argument*. For a back-and-forth argument in a more concrete setting, see Proposition 4.1.1.

LEMMA 2.2.2. *Let $\mathfrak{A}$ and $\mathfrak{B}$ be two saturated structures with the same theory and the same cardinality. Then $\mathfrak{A}$ and $\mathfrak{B}$ are isomorphic.*

PROOF. Let $(a_\alpha)_{\alpha<\kappa}$ be an enumeration of A and $(b_\alpha)_{\alpha<\kappa}$ an enumeration of B. We inductively construct a sequence $(c_\alpha)_{\alpha<\kappa}$ of elements of A and a sequence$(d_\alpha)_{\alpha<\kappa}$ of elements of B such that for all $\beta < \kappa$

$$\mathrm{Th}(\mathfrak{A}; (a_\alpha)_{\alpha<\beta}, (d_\alpha)_{\alpha<\beta}) = \mathrm{Th}(\mathfrak{B}; (c_\alpha)_{\alpha<\beta}, (b_\alpha)_{\alpha<\beta}).$$

The base case $\beta = 0$ holds by the assumptions of the theorem. Suppose that $(c_\alpha)_{\alpha<\beta}$ and $(d_\alpha)_{\alpha<\beta}$ have already been constructed. If β is a limit ordinal, there is nothing to be done. Otherwise, $\beta = \gamma + 1$ and $(c_\alpha)_{\alpha<\gamma}$ and $(d_\alpha)_{\alpha<\gamma}$ have already been constructed. We use saturation of $\mathfrak{B}$ to find c_β such that

$$\mathrm{Th}(\mathfrak{A}; (a_\alpha)_{\alpha\leq\beta}, (d_\alpha)_{\alpha<\beta}) = \mathrm{Th}(\mathfrak{B}; (c_\alpha)_{\alpha\leq\beta}, (b_\alpha)_{\alpha<\beta}).$$

Then we use saturation of $\mathfrak{A}$ to find d_β such that

$$\mathrm{Th}(\mathfrak{A}; (a_\alpha)_{\alpha\leq\beta}, (d_\alpha)_{\alpha\leq\beta}) = \mathrm{Th}(\mathfrak{B}; (c_\alpha)_{\alpha\leq\beta}, (b_\alpha)_{\alpha\leq\beta}).$$

At the end of the day, the map $f : A \to B$ defined by $f(a_\alpha) := c_\alpha$ for all $\alpha < \kappa$ is an embedding $\mathfrak{A} \hookrightarrow \mathfrak{B}$, and the map $b_\alpha \mapsto d_\alpha$ is an embedding $\mathfrak{B} \hookrightarrow \mathfrak{A}$ which is the inverse of f. ⊣

2.2.2. Omitting types and atomic structures. A formula $\phi(x_1, \ldots, x_n)$ *isolates* an n-type p over T if

- $T \cup \{\exists \bar{x} : \phi(\bar{x})\}$ is satisfiable, and
- for every formula $\psi \in p$ we have that $T \models \forall \bar{x}(\phi(x) \Rightarrow \psi(\bar{x}))$.

A type p of a theory T is called *principal* if it is isolated by some formula. If T is complete, then every principal type p is realised in every model of T, because $\mathfrak{B} \models \exists \bar{x} : \phi(\bar{x})$, and if $\mathfrak{B} \models \phi(\bar{a})$, then $\bar{a}$ realises p. The omitting types theorem can be viewed as a converse of this observation; for a proof, see [207] or [278].

THEOREM 2.2.3 (Countable omitting types theorem). *Let τ be a countable signature, let T be a satisfiable τ-theory, and let $p_1, p_2, \dots$ be non-principal types of T. Then T has a countable model that omits all of the p_i's.*

We say that a structure $\mathfrak{B}$ is *atomic* if for every $a \in B^n$, the type of a in $\mathfrak{B}$ is principal.

THEOREM 2.2.4 (Theorem 6.2.2 in [207]). *Let T be a complete satisfiable theory with countably many n-types for every $n \in \mathbb{N}$. Then T has a countable atomic model.*

PROOF. There are only countably many non-principal complete types in T, so by the countable omitting types theorem (Theorem 2.2.3) there is a countable model $\mathfrak{B}$ of T that omits all of them. ⊣

LEMMA 2.2.5. *Let $\mathfrak{A}$ and $\mathfrak{B}$ be atomic countable structures with the same theory. Then $\mathfrak{A}$ and $\mathfrak{B}$ are isomorphic. In particular, if $\mathfrak{B}$ is atomic, and $a, b \in B^n$ have the same type, then a and b lie in the same orbit under* $\mathrm{Aut}(\mathfrak{B})$.

PROOF. As in the proof of Lemma 2.2.2, this can be shown by a back-and-forth argument. Where we previously used saturation of $\mathfrak{B}$ to show the existence of an element c_β that has the same type over $(\mathfrak{B}; (c_\alpha)_{\alpha<\beta}, (b_\alpha)_{\alpha<\beta})$, we now use that this type is principal (since the type of $(c_1, \dots, c_\beta)$ in $\mathfrak{B}$ is principal by assumption), and hence must be realised. ⊣

2.2.3. Existential positive types. Recall that an *existential positive n-type* (*ep-n-type*) of a theory T is a set p of existential positive formulas with free variables $x_1, \dots, x_n$ such that $p \cup T$ is satisfiable, and that an ep-n-type p of T is *maximal* if $T \cup p \cup \{\phi(x_1, \dots, x_n)\}$ is unsatisfiable for any existential positive formula $\phi \notin p$. Note that $p \cup \mathrm{Th}(\mathfrak{A})$ is satisfiable if and only if $p \cup \mathrm{Th}(\mathfrak{A})_{\forall-}$ is satisfiable; thus we could equivalently have defined existential positive n-types with respect to the latter theory.

A structure $\mathfrak{A}$ is *existentially positively κ-saturated* if for all $\beta < \kappa$ and expansions $\mathfrak{A}'$ of $\mathfrak{A}$ by at most β constants every existential positive 1-type of $\mathfrak{A}'$ is realised in $\mathfrak{A}'$. Note that a structure that is κ-saturated is a fortiori existentially positively κ-saturated.

LEMMA 2.2.6. *Let $\mathfrak{A}$ and $\mathfrak{B}$ be τ-structures, where $\mathfrak{B}$ is existentially positively $|A|$-saturated. Suppose that every primitive positive sentence that is true in $\mathfrak{A}$ is also true in $\mathfrak{B}$. Then there is a homomorphism from $\mathfrak{A}$ to $\mathfrak{B}$.*

PROOF. Let $(a_i)_{0 \le i < |A|}$ well-order A. We claim that for every $\mu \le |A|$ there is a sequence $(b_i)_{i<\mu}$ of elements of B such that every primitive positive $(\tau \cup \{c_i \mid i < \mu\})$-sentence true on $(\mathfrak{A}, (a_i)_{i<\mu})$ is true on $(\mathfrak{B}, (b_i)_{i<\mu})$. The proof is by transfinite induction on μ.

- The base case, $\mu = 0$, follows from the hypothesis of the lemma.
- The inductive step for limit ordinals μ follows from the observation that a sentence can only mention a finite collection of constants, whose indices must all be less than some $\nu < \mu$.

- For the inductive step for successor ordinals $\mu = \nu + 1 < |A|$, set

$$\Sigma := \{\phi(x) \mid \phi \text{ is a ep-}(\tau \cup \{c_i \mid i < \nu\})\text{-formula such that } (\mathfrak{A}, (a_i)_{i<\nu}) \models \phi(a_\nu)\}.$$

  By the inductive assumption $(\mathfrak{B}, (b_i)_{i<\nu}) \models \exists x \colon \phi(x)$ for every $\phi \in \Sigma$. By compactness, since Σ is closed under conjunction, we have that Σ is an ep-1-type of $(\mathfrak{B}, (b_i)_{i<\nu})$. Then Σ is realised by some element $b_\gamma \in B$ because $\mathfrak{B}$ is ep-$|A|$-saturated. By construction we maintain that all primitive positive $(\tau \cup \{c_i \mid i < \mu\})$-sentences true on $(\mathfrak{A}, (a_i)_{i<\mu})$ are true on $(\mathfrak{B}, (b_i)_{i<\mu})$.

The function that maps a_i to b_i for all $i < |A|$ is a homomorphism from $\mathfrak{A}$ to $\mathfrak{B}$. ⊣

2.3. Fraïssé-amalgamation

A versatile tool for the construction of beautiful countably infinite structures from classes of finite structures is *amalgamation* à la Fraïssé. The resulting infinite structures can be used as templates to model various computational problems as CSPs. Such a structure has a large automorphism group which will be important for the universal-algebraic approach in Chapter 6. We present Fraïssé-amalgamation for the special case of *relational structures*; this is all that is needed in the examples that we present in this text. For a stronger version of Fraïssé-amalgamation for classes of structures that may involve function symbols, see [207]. In this section, τ always denotes a countable relational signature.

2.3.1. The age of a structure. Let C be a class of τ-structures. We say that C has the *joint embedding property* (*JEP*) if for all structures $\mathfrak{B}_1, \mathfrak{B}_2 \in C$ there exists a structure $\mathfrak{C} \in C$ that embeds both $\mathfrak{B}_1$ and $\mathfrak{B}_2$. The *age* of a τ-structure $\mathfrak{A}$ is the class of all finite τ-structures that embed into $\mathfrak{A}$; it is denoted by $\mathrm{Age}(\mathfrak{A})$. The proof of the following lemma is similar to that of Lemma 1.1.8.

PROPOSITION 2.3.1. *A class C of finite τ-structures is the age of a countably infinite τ-structure $\mathfrak{B}$ if and only if C contains countably many isomorphism types, is closed under isomorphisms, taking substructures, and if it has the JEP.*

2.3.2. The amalgamation property. Let $\mathfrak{B}_1, \mathfrak{B}_2$ be τ-structures such that $\mathfrak{A}$ is an induced substructure of both $\mathfrak{B}_1$ and $\mathfrak{B}_2$ and all common elements of $\mathfrak{B}_1$ and $\mathfrak{B}_2$ are elements of $\mathfrak{A}$; note that in this case $\mathfrak{A} = \mathfrak{B}_1 \cap \mathfrak{B}_2$. We call $(\mathfrak{B}_1, \mathfrak{B}_2)$ an *amalgamation diagram* (*over C*). The structure $\mathfrak{B}_1 \cup \mathfrak{B}_2$ is called the *free amalgam* of $\mathfrak{B}_1, \mathfrak{B}_2$ over $\mathfrak{A}$. More generally, a τ-structure $\mathfrak{C}$ is an *amalgam of* $(\mathfrak{B}_1, \mathfrak{B}_2)$ if for $i = 1, 2$ there are embeddings $f_i \colon \mathfrak{B}_i \hookrightarrow \mathfrak{C}$ such that $f_1(a) = f_2(a)$ for every $a \in \mathfrak{A}$. A *strong amalgam* of $(\mathfrak{B}_1, \mathfrak{B}_2)$ is an amalgam of $(\mathfrak{B}_1, \mathfrak{B}_2)$ where $f_1(B_1) \cap f_2(B_2) = f_1(A)$ $(= f_2(A))$.

DEFINITION 2.3.2. Let C be an isomorphism-closed class of relational τ-structures. Then C has the (*free, strong*) *amalgamation property* (*AP*) if all amalgamation diagrams over C have a (free, strong) amalgam in C. A class of finite τ-structures that contains at most countably many non-isomorphic structures, has the amalgamation property, and is closed under taking induced substructures and isomorphisms is called an *amalgamation class*.

Note that since we only consider relational structures here (and since we allow structures to have an empty domain), the amalgamation property of C implies the joint embedding property.

Example 2.3.3. Let C be the class of all finite structures over the signature $\{<\}$ where $<$ denotes a linear order. Then C is clearly closed under isomorphisms and induced substructures, and has countably many isomorphism types (one for each domain size). To show that it also has the amalgamation property, let $\mathfrak{B}_1, \mathfrak{B}_2 \in C$, and let $\mathfrak{A}$ be an induced substructure of both $\mathfrak{B}_1$ and $\mathfrak{B}_2$. Let $\mathfrak{C}$ be the free amalgam of $\mathfrak{B}_1$ and $\mathfrak{B}_2$ over $\mathfrak{A}$. Then $\mathfrak{C}$ is an acyclic finite graph; therefore, any linear extension of $\mathfrak{C}$ is an amalgam (even a strong amalgam, but not a free amalgam) in C of $\mathfrak{B}_1$ and $\mathfrak{B}_2$ over $\mathfrak{A}$. It follows that C is an amalgamation class.

Example 2.3.4. Let $n \geq 3$. Let C be the class of all finite graphs that do not embed K_n, viewed as structures over the signature $\{E\}$. Such graphs are also called *K_n-free*. Again it is clear that C is closed under isomorphisms and induced substructures, has countably many isomorphism types, and the free amalgamation property.

Example 2.3.5. Let $k \in \mathbb{N} \setminus \{0\}$. Let C be the class of all finite structures over the signature $\{E\}$ where E denotes an equivalence relation with at most k classes. Again it is clear that C is closed under isomorphisms and induced substructures, has countably many isomorphism types and the strong amalgamation property, but not the free amalgamation property.

Example 2.3.6. Let $k \in \mathbb{N} \setminus \{0\}$. Let C be the class of all finite structures over the signature $\{E\}$ where E denotes an equivalence relation where each equivalence class has size at most k. Again it is clear that C is closed under isomorphisms and induced substructures, and has countably many isomorphism types. Moreover, C has the amalgamation property, but this time not the strong amalgamation property.

A τ-structure $\mathfrak{A}$ is *homogeneous* (sometimes also called *ultra-homogeneous* [207]) if every isomorphism between finite substructures of $\mathfrak{A}$ can be extended to an automorphism of $\mathfrak{A}$.

PROPOSITION 2.3.7. *If* $\mathfrak{C}$ *is a homogeneous structure, then* Age($\mathfrak{C}$) *is an amalgamation class.*

PROOF. Let $(\mathfrak{B}_1, \mathfrak{B}_2)$ be an amalgamation diagram over Age($\mathfrak{C}$). Let g_1 and g_2 be embeddings of $\mathfrak{B}_1$ and $\mathfrak{B}_2$ into $\mathfrak{C}$. For $i \in \{1,2\}$, let $\mathfrak{A}_i$ be the substructure of $\mathfrak{C}$ induced on $g_i(B_1 \cap B_2)$. Then $g_2 \circ g_1^{-1}|_{A_1}$ is an isomorphism between the finite substructures $\mathfrak{A}_1$ and $\mathfrak{A}_2$ of $\mathfrak{C}$, and hence can be extended to an automorphism α of $\mathfrak{C}$. The substructure of $\mathfrak{C}$ induced on $\alpha(B_1) \cup B_2$ is the required amalgam of $(\mathfrak{B}_1, \mathfrak{B}_2)$ witnessed by the embeddings $f_1 := \alpha g_1$ and $f_2 := g_2$. ⊣

Fraïssé's theorem can be seen as a converse to this observation.

THEOREM 2.3.8 (Fraïssé [172, 173]). *Let* τ *be a countable relational signature and let* C *be an amalgamation class of* τ*-structures. Then there is a homogeneous and at most countable* τ*-structure* $\mathfrak{C}$ *whose age equals* C. *The structure* $\mathfrak{C}$ *is unique up to isomorphism, and called the* Fraïssé-limit *of* C.

For the proof of Theorem 2.3.8, we refer to [207]; the proof of Lemma 2.7.3 below is similar.

Example 2.3.9. Let C be the class of all finite graphs. It is even easier than in the previous examples to verify that C is an amalgamation class, since here the free amalgam itself shows the amalgamation property. The Fraïssé-limit of C is also known as the *countable random graph* or the *Rado graph*.

Example 2.3.10. Let C be the class of all finite K_n-free graphs from Example 2.3.4, for some $n \geq 3$. Then the Fraïssé-limit of C is called the *K_n-free Henson graph* [199].

Example 2.3.11. Let C be the class of all finite partially ordered sets. Amalgamation can be shown by computing the transitive closure: if $\mathfrak{C}$ is the free amalgam of $\mathfrak{B}_1$ and $\mathfrak{B}_2$ over $\mathfrak{A}$, then the transitive closure of $\mathfrak{C}$ gives an amalgam in C. The Fraïssé-limit of C is called the *homogeneous universal partial order*.

All countable homogeneous graphs have been classified in [250], all countable homogeneous posets in [322], and all countable homogeneous directed graphs in [131].

2.3.3. Forbidden substructures. In the following, it will be convenient to specify classes of finite τ-structures by specifying forbidden substructures. If $\mathcal{F}$ is a set of finite τ-structures, we write $\mathrm{Forb}^{\mathrm{emb}}(\mathcal{F})$ for the set of finite τ-structures $\mathfrak{A}$ such that no structure in $\mathcal{F}$ embeds into $\mathfrak{A}$.

Example 2.3.12. Henson [200] used Fraïssé limits to construct 2^{ω} many pairwise non-isomorphic homogeneous directed graphs. A *tournament* is a directed graph without loops such that for all pairs x, y of distinct vertices exactly one of the pairs $(x, y), (y, x)$ is an arc in the graph. Let $\mathfrak{L}$ be the directed graph with just one vertex 0 and the edge $(0,0)$). Note that for all classes $\mathcal{N}$ of finite tournaments, $\text{Forb}^{\text{emb}}(\mathcal{N} \cup \{\mathfrak{L}\})$ is an amalgamation class, because if $\mathfrak{A}_1$ and $\mathfrak{A}_2$ are directed graphs in $\text{Forb}^{\text{emb}}(\mathcal{N} \cup \{\mathfrak{L}\})$ such that $\mathfrak{A} = \mathfrak{A}_1 \cap \mathfrak{A}_2$ is an induced substructure of both $\mathfrak{A}_1$ and $\mathfrak{A}_2$, then the free amalgam $\mathfrak{A}_1 \cup \mathfrak{A}_2$ is also in $\text{Forb}^{\text{emb}}(\mathcal{N} \cup \{\mathfrak{L}\})$.

In his proof, Henson specified an infinite set $\mathcal{T}$ of tournaments $\mathfrak{T}_1, \mathfrak{T}_2, \ldots$ with the property that $\mathfrak{T}_i$ does not embed into $\mathfrak{T}_j$ if $i \neq j$; the set $\mathcal{T}$ will be described in Section 13.3. Note that this property implies that for two distinct subsets $\mathcal{N}_1$ and $\mathcal{N}_2$ of $\mathcal{T}$ the two free amalgamation classes $\text{Forb}^{\text{emb}}(\mathcal{N}_1 \cup \{\mathfrak{L}\})$ and $\text{Forb}^{\text{emb}}(\mathcal{N}_2 \cup \{\mathfrak{L}\})$ are distinct as well. Since there are 2^{ω} many subsets of the infinite set $\mathcal{T}$, there are also that many distinct homogeneous directed graphs; they are often referred to as *Henson digraphs*. Note that non-isomorphic Henson digraphs have distinct CSPs. Since there are only countably many algorithms, this proves that there exist homogeneous digraphs with an undecidable CSP [85].

The structures from Example 2.3.12 can be used to prove various negative results about homogeneous structures with finite signature, for instance in Section 11.6 and in Section 13.3. A better behaved class of homogeneous structures are those whose age is *finitely bounded* (we use the same terminology as in [270]).

DEFINITION 2.3.13. A class $\mathcal{C}$ of finite τ-structures (or a structure with age $\mathcal{C}$) is called *finitely bounded* if τ is finite and there exists a finite set of finite τ-structures $\mathcal{F}$ such that $\mathcal{C} = \text{Forb}^{\text{emb}}(\mathcal{F})$. We sometimes refer to the elements of $\mathcal{F}$ as the *bounds* of $\mathcal{C}$. A structure $\mathfrak{B}$ is called *finitely bounded* if $\text{Age}(\mathfrak{B})$ is finitely bounded.

LEMMA 2.3.14. *Let τ be a relational signature. A class $\mathcal{C}$ of τ-structures is finitely bounded if and only if $\mathcal{C}$ has a* universal axiomatisation, *i.e., there exists a universal τ-sentence ϕ such that $\mathcal{C}$ is precisely the class of finite models of ϕ.*

PROOF. If $\mathcal{F}$ is such that $\mathcal{C} = \text{Forb}^{\text{emb}}(\mathcal{F})$ then the conjunction of all sentences of the form $\forall \bar{x}: \neg Q_{\mathfrak{F}}$ for $\mathfrak{F} \in \mathcal{F}$ is a universal axiomatisation of $\mathcal{C}$ (here $Q_{\mathfrak{F}}$ is the canonical query of $\mathfrak{F}$ from Section 1.2.1), Conversely, if ϕ is universal axiomatisation with n variables, we choose a representative for each isomorphism class of τ-structures with at most n elements that does not belong to $\mathcal{C}$. If $\mathcal{F}$ is the set of chosen representatives, then $\text{Forb}^{\text{emb}}(\mathcal{F}) = \mathcal{C}$. ⊣

PROPOSITION 2.3.15. *If $\mathfrak{B}$ is a reduct of a finitely bounded structure, then* $\text{CSP}(\mathfrak{B})$ *is in NP.*

PROOF. It is easy to see that $\text{CSP}(\mathfrak{B})$ is in monotone SNP (Section 1.4). ⊣

PROPOSITION 2.3.16. *A structure $\mathfrak{B}$ is a finite-signature first-order reduct of a finitely bounded homogeneous structure if and only if $\mathfrak{B}$ is a reduct of a finitely bounded homogeneous structure.*

PROOF. Clearly, if $\mathfrak{B}$ is a reduct of a finitely bounded homogenous structure $\mathfrak{C}$, then $\mathfrak{B}$ is also a first-order reduct of $\mathfrak{C}$ and has a finite signature. Conversely, suppose that $\mathfrak{B}$ is a finite-signature first-order reduct of a finitely bounded homogeneous structure $\mathfrak{C}$, i.e., there exists a finite set of finite τ-structures $\mathcal{F}$ such that $\text{Forb}^{\text{emb}}(\mathcal{F}) = \text{Age}(\mathfrak{C})$. We claim that the expansion $\mathfrak{C}'$ of $\mathfrak{C}$ by all relations from $\mathfrak{C}$ is still homogeneous and finitely bounded. This is clear for homogeneity. For finite boundedness, let m be the maximal arity of the relations of $\mathfrak{B}$. Up to isomorphism, there are finitely many structures of size at most m in the signature of $\mathfrak{C}'$. Let $\mathcal{F}'$ be obtained from $\mathcal{F}$ by adding these structures. Then $\text{Forb}^{\text{emb}}(\mathcal{F}') = \text{Age}(\mathfrak{C}')$. Hence, $\mathfrak{B}$ is a reduct of the finitely bounded homogeneous structure $\mathfrak{C}'$. ⊣

The following conjecture was proposed by Michael Pinsker and the author at the Fields Institute in Toronto in 2011 (at the time, conditional on the finite-domain dichotomy conjecture).

CONJECTURE 2.1 (Infinite-domain dichotomy conjecture). *For every reduct $\mathfrak{B}$ of a finitely bounded homogeneous structure,* CSP($\mathfrak{B}$) *is in P or NP-complete.*

2.3.4. One-point amalgamation. In some situations, in place of the full amalgamation property it is more convenient to work with the 1*-point amalgamation property* (1*-AP*), which only requires amalgams if the given structures $\mathfrak{B}_1, \mathfrak{B}_2 \in C$ over a common substructure $\mathfrak{A}$ are such that $|B_1| = |B_2| = |A| + 1$. By Proposition 2.3.17 below, this ostensibly weaker property is equivalent to full amalgamation. Similarly, the strong amalgamation property may be replaced by a *strong* 1*-point amalgamation property* which requires the existence of an amalgam $\mathfrak{C}$ that additionally satisfies $|C| = |A| + 2$.

When using the 1-AP another formulation is more practical, namely item (4) in the proposition below. Similarly, when using the AP item (1) is a more practical formulation (so that for many authors, item (1) is the official definition of the amalgamation property).

PROPOSITION 2.3.17. *Let C be a class of finite τ-structures that is closed under isomorphisms and substructures. Then the following are equivalent.*

1. *for all $\mathfrak{A}, \mathfrak{B}_1, \mathfrak{B}_2 \in C$ and embeddings $e_i \colon \mathfrak{A} \hookrightarrow \mathfrak{B}_i$, for $i \in \{1, 2\}$, there exists a structure $\mathfrak{C} \in C$ and embeddings $f_i \colon \mathfrak{B}_i \hookrightarrow \mathfrak{C}$ such that $f_1 \circ e_1 = f_2 \circ e_2$.*
2. *C has the amalgamation property.*
3. *C has the* 1*-point amalgamation property.*
4. *for all $\mathfrak{A}, \mathfrak{B}_1, \mathfrak{B}_2 \in C$ and embeddings $e_i \colon \mathfrak{A} \hookrightarrow \mathfrak{B}_i$ for $i \in \{1, 2\}$ such that $|B_1| = |A| + 1 = |B_2|$, there exist $\mathfrak{C} \in C$ and embeddings $f_i \colon \mathfrak{B}_i \hookrightarrow \mathfrak{C}$ such that $f_1 \circ e_1 = f_2 \circ e_2$.*

PROOF. The implication from (1) to (2) is easy (choosing e_1 and e_2 to be id_A), and the converse implication follows easily from the assumption that C is closed under isomorphisms. Similarly, (3) and (4) are equivalent.

The implication from (2) to (3) is trivial. The missing implication from (4) to (2) is more interesting. Let $\mathfrak{A}, \mathfrak{B}_1, \mathfrak{B}_2 \in C$ be such that $\mathfrak{B}_1 \cap \mathfrak{B}_2 = \mathfrak{A}$. We prove that $\mathfrak{B}_1, \mathfrak{B}_2$ has an amalgam $\mathfrak{C}$ over $\mathfrak{A}$ in C by induction on $|B_1 \setminus A| + |B_2 \setminus A|$. The statement is trivial for $|B_1| = |A|$ since in this case B_2 is an amalgam. Likewise, the statement is trivial if $|B_2| = |A|$. Let $b_1 \in B_1 \setminus A$ and $b_2 \in B_2 \setminus A$. By assumption, $\mathfrak{B}'_1 := \mathfrak{B}_1[A \cup \{b_1\}]$ and $\mathfrak{B}'_2 := \mathfrak{B}_2[A \cup \{b_2\}]$ have an amalgam $\mathfrak{C}'$ over $\mathfrak{A}$. By the inductive assumption, for $i \in \{1,2\}$ the structures $\mathfrak{B}_i$ and $\mathfrak{C}'$ have an amalgam $\mathfrak{C}_i$ over $\mathfrak{B}'_i$ via the embeddings $f_i \colon \mathfrak{B}_i \to \mathfrak{C}_i$ and $f'_i \colon \mathfrak{C}' \to \mathfrak{C}_i$. Finally, applying the inductive assumption again we obtain that $\mathfrak{C}_1$ and $\mathfrak{C}_2$ have an amalgam $\mathfrak{C}$ over $\mathfrak{C}'$ via the embeddings $g_i \colon \mathfrak{C}_i \to \mathfrak{C}$. Then $\mathfrak{C}$ is also an amalgam of $\mathfrak{B}_1$ and $\mathfrak{B}_2$ via the embeddings $g_i \circ f_i \colon \mathfrak{B}_i \to \mathfrak{C}$; indeed,

$$g_1 \circ f_1|_A = g_1 \circ f'_1|_A = g_2 \circ f'_2|_A = g_2 \circ f_2|_A. \qquad \dashv$$

The following variant of Proposition 2.3.17 for strong amalgamation can be shown analogously.

PROPOSITION 2.3.18. *Let C be a class of finite τ-structures that is closed under isomorphisms and substructures. Then the following are equivalent.*

1. *for all $\mathfrak{A}, \mathfrak{B}_1, \mathfrak{B}_2 \in C$ and embeddings $e_i \colon \mathfrak{A} \hookrightarrow \mathfrak{B}_i$, for $i \in \{1,2\}$, there exists a structure $\mathfrak{C} \in C$ and embeddings $f_i \colon \mathfrak{B}_i \hookrightarrow \mathfrak{C}$ such that $f_1 \circ e_1|_A = f_2 \circ e_2|_A$ and such that $|f_1(B_1) \cap f_2(B_2)| = |A|$.*
2. *C has the strong amalgamation property.*
3. *C has the strong 1-point amalgamation property.*
4. *for all $\mathfrak{A}, \mathfrak{B}_1, \mathfrak{B}_2 \in C$ and embeddings $e_i \colon \mathfrak{A} \hookrightarrow \mathfrak{B}_i$, for $i \in \{1,2\}$ such that $|B_1| = |A| + 1 = |B_2|$, there exist $\mathfrak{C} \in C$ and embeddings $f_i \colon \mathfrak{B}_i \hookrightarrow \mathfrak{C}$ such that $f_1 \circ e_1|_A = f_2 \circ e_2|_A$ and such that $|f_1(B_1) \cap f_2(B_2)| = |A|$.*

2.3.5. Deciding amalgamation. Let $\mathcal{F}$ be a finite set of finite τ-structures. We are interested in the computational problem of deciding whether $\mathrm{Forb}^{\mathrm{emb}}(\mathcal{F})$ is an amalgamation class. It is open whether this problem is decidable (Question 53).

Recall that for relational signatures, the Amalgamation Property implies the Joint Embedding Property. It is known that whether $\mathrm{Forb}^{\mathrm{emb}}(\mathcal{F})$ has the JEP is undecidable [103]; however, this does not answer our question. In this section we use 1-point amalgamation to prove that if the signature is binary then it is algorithmically decidable whether $\mathrm{Forb}^{\mathrm{emb}}(\mathcal{F})$ has the amalgamation property; this follows from the following proposition.[12]

PROPOSITION 2.3.19. *Let τ be a finite relational signature where all relation symbols have arity at most 2. Let $\mathcal{F}$ be a finite set of finite τ-structures and let $C := \mathrm{Forb}^{\mathrm{emb}}(\mathcal{F})$. Let $m \geq 3$ be larger than the maximal number of elements of the structures in $\mathcal{F}$ and let ℓ be the number of 2-element structures in C. Then C*

[12]The author has learned about this fact from Gregory Cherlin but is not aware of a proper reference in the literature.

has the amalgamation property if and only if all 1-point amalgamation diagrams of size at most $(m-2)\ell$ have an amalgam.

PROOF. By Proposition 2.3.17, C has the amalgamation property if and only if it has 1-point amalgamation. Let $(\mathfrak{B}_1, \mathfrak{B}_2)$ be a 1-point amalgamation diagram without an amalgam. In particular, no τ-structure $\mathfrak{C}$ with domain $B_1 \cup B_2$ such that $\mathfrak{B}_1$ and $\mathfrak{B}_2$ are substructures of $\mathfrak{C}$ can be an amalgam for $(\mathfrak{B}_1, \mathfrak{B}_2)$. Hence, there must exist $A = \{a_1, \ldots, a_{m-2}\} \in B_0 := B_1 \cap B_2$ such that the substructure induced on $\{a_1, \ldots, a_{m-2}, p, q\}$ embeds a structure from $\mathcal{F}$. Note that since the maximal arity is 2, the number of such τ-structures $\mathfrak{C}$ is bounded by ℓ since they only differ by the substructure induced on $\{p, q\}$. So let $A_1, \ldots, A_\ell \subseteq B_0$ be a list of sets witnessing that all of these structures $\mathfrak{C}$ embed a structure from $\mathcal{F}$. Let $\mathfrak{B}_1'$ be the substructure of $\mathfrak{B}_1$ induced on $\{p\} \cup A_1 \cup \cdots \cup A_\ell$ and let $\mathfrak{B}_2'$ be the substructure of $\mathfrak{B}_2$ induced on $\{q\} \cup A_1 \cup \cdots \cup A_\ell$. Suppose for contradiction that $(\mathfrak{B}_1', \mathfrak{B}_2')$ has an amalgam $\mathfrak{D}$; we may assume that this amalgam is of size at most $(m-2) \cdot \ell$. Depending on the two-element structure induced on $\{p, q\}$ in $\mathfrak{D}$, there exists an $i \leq \ell$ such that the structure induced on $\{p, q\} \cup A_i$ in $\mathfrak{D}$ embeds a structure from $\mathcal{F}$, a contradiction. ⊣

2.3.6. Generic superpositions. For strong amalgamation classes there is a powerful construction to obtain new strong amalgamation classes from known ones. If τ_1 and τ_2 are disjoint relational signatures, and for $i \in \{1, 2\}$ let $\mathfrak{A}_i$ be a τ_i-structure such that $\mathfrak{A}_1$ and $\mathfrak{A}_2$ have the same domain. Then the $(\tau_1 \cup \tau_2)$-structure whose τ_i-reduct equals $\mathfrak{A}_i$ is called the *superposition* of $\mathfrak{A}_1$ and $\mathfrak{A}_2$.

DEFINITION 2.3.20. Let C_1 and C_2 be classes of finite structures with disjoint relational signatures τ_1 and τ_2, respectively. Then the *superposition of C_1 and C_2*, denoted by $C_1 * C_2$, is the class of all superpositions of structures from C_1 with structures from C_2.

The following lemma has a straightforward proof by combining amalgamation in C_1 with amalgamation in C_2.

LEMMA 2.3.21. *Let C_1 and C_2 be strong amalgamation classes with disjoint relational signatures. Then $C_1 * C_2$ is also a strong amalgamation class.*

DEFINITION 2.3.22 (Generic superposition). Let $\mathfrak{B}_1$ and $\mathfrak{B}_2$ be homogeneous structures with disjoint relational signatures whose ages $\mathcal{C}_1$ and $\mathcal{C}_2$ have strong amalgamation. Then $\mathfrak{B}_1 * \mathfrak{B}_2$ denotes the (up to isomorphism unique) Fraïssé-limit of $C_1 * C_2$. We refer to $\mathfrak{B}_1 * \mathfrak{B}_2$ as the *generic superposition of $\mathfrak{B}_1$ and $\mathfrak{B}_2$*.

It can be shown by a back and forth argument (an example of such an argument will be given in Proposition 4.1.1) that for $i \in \{1, 2\}$ the τ_i-reduct of $\mathfrak{B}_1 * \mathfrak{B}_2$ is isomorphic to $\mathfrak{B}_i$. In the following, we identify the domain of $\mathfrak{B}_i$

with the domain of $\mathfrak{B}_1 * \mathfrak{B}_2$ along this isomorphism, so that $\mathfrak{B}_1 * \mathfrak{B}_2$ is indeed a superposition of $\mathfrak{B}_1$ and $\mathfrak{B}_2$ as defined in the beginning of this section.

Example 2.3.23. For $i \in \{1, 2\}$, let $\tau_i = \{<_i\}$, let $\mathcal{C}_i$ be the class of all finite τ_i-structures where $<_i$ denotes a linear order, and let $\mathfrak{B}_i$ be the Fraïssé-limit of $\mathcal{C}_i$. is known as the *random permutation* (see e.g. [124]).

More facts about generic superpositions in a slightly more general setting can be found in Section 4.7.1.

2.4. First-order interpretations

First-order interpretations are a powerful tool to derive new structures from known structures. To keep the presentation simple, we define first-order interpretations for relational structures only; but the formalism can be extended to general signatures without any problems.

DEFINITION 2.4.1. Let $\mathfrak{A}$ and $\mathfrak{B}$ be structures with the relational signatures τ and σ and let $d \in \mathbb{N}$. A (*first-order*) *interpretation of dimension* d of $\mathfrak{B}$ in $\mathfrak{A}$ is a partial surjection $I \colon A^d \to B$ (also called the *coordinate map*) such that for every relation R defined by an atomic σ-formula ϕ, say of arity k, the dk-ary relation

$$I^{-1}(R) := \{(a_1^1, \dots, a_d^1, \dots, a_1^k, \dots, a_d^k) \mid (I(a_1^1, \dots, a_d^1), \dots, I(a_1^k, \dots, a_d^k)) \in R\}$$

has a first-order definition ϕ_I in $\mathfrak{A}$.

Since equality and $\top$ are always allowed as atomic formulas, there must in particular exist

- a τ-formula $\top_I$, called the *domain formula*, such that $\top_I(x_1, \dots, x_d)$ holds if and only if $(x_1, \dots, x_d)$ is in the domain $\mathrm{dom}(I)$ of I;
- a τ-formula $=_I$ such that $=_I (x_{1,1}, \dots, x_{1,d}, x_{2,1}, \dots, x_{2,d})$ holds if and only if $((x_{1,1}, \dots, x_{1,d}), (x_{2,1}, \dots, x_{2,d}))$ lies in the kernel of I.

In order to specify a σ-structure $\mathfrak{B}$ with a first-order interpretation in a given τ-structure $\mathfrak{A}$ up to isomorphism, it suffices to specify the interpreting formulas for the atomic σ-formulas of $\mathfrak{A}$; in particular, if the signature of $\mathfrak{A}$ is relational and finite, then an interpretation has a finite presentation.

We say that $\mathfrak{B}$ is *interpretable in* $\mathfrak{A}$ *with finitely many parameters* if there are $c_1, \dots, c_n \in A$ such that $\mathfrak{B}$ is interpretable in the expansion of $\mathfrak{A}$ by the constants c_i for all $1 \le i \le n$.

Example 2.4.2. In Section 1.6 we have described *Allen's Interval Algebra* for temporal reasoning in Artificial Intelligence [10], and the corresponding CSP. Formally, it is easiest to describe the template $\mathfrak{A}$ for this CSP by a first-order interpretation I in $(\mathbb{Q};<)$. The dimension of the interpretation is two, and the domain formula $\top_I(x,y)$ is $x<y$. Hence, the elements of $\mathfrak{A}$ can indeed be viewed as non-empty closed bounded intervals $[x,y]$ over $\mathbb{Q}$. The template $\mathfrak{A}$ contains for each inequivalent $\{<\}$-formula ϕ with four variables a binary relation R such that (a_1,a_2,a_3,a_4) satisfies ϕ if and only if $((a_1,a_2),(a_3,a_4)) \in R$. In particular, $\mathfrak{A}$ has relations for equality of intervals, containment of intervals, and so forth.

Example 2.4.3. Let $\mathfrak{G}=(V;E)$ be an undirected graph (viewed as a symmetric digraph). Then the *line graph* $L_{\mathfrak{G}}$ of $\mathfrak{G}$ is the (undirected) graph with vertex set

$$V(L_{\mathfrak{G}}) := \{\{u,v\} \mid (u,v) \in E\}$$

and the edge set

$$E(L_{\mathfrak{G}}) := \{(\{u,v\},\{v,w\}) \mid \{u,v\},\{v,w\} \in E\}.$$

The line graph has the 2-dimensional first-order interpretation $I\colon E \to V(L_{\mathfrak{G}})$ in $\mathfrak{G}$ given by $I(x,y) := \{x,y\}$:

- $I^{-1}(L_{\mathfrak{G}})$ has the first-order definition $E(x_1,x_2)$.
- $I^{-1}(\{(u,u) \mid u \in V(L_{\mathfrak{G}})\})$ has the first-order definition

$$\begin{aligned}&(x_{1,1}=x_{2,1} \wedge x_{1,2}=x_{2,2})\\ \vee\ &(x_{1,1}=x_{2,2} \wedge x_{1,2}=x_{2,1}).\end{aligned}$$

- $I^{-1}(E(L_{\mathfrak{G}}))$ has the first-order definition

$$\begin{aligned}&x_{1,1}=x_{2,1} \vee x_{1,1}=x_{2,2}\\ \vee\ &x_{1,2}=x_{2,1} \vee x_{2,2}=x_{2,1}.\end{aligned}$$

LEMMA 2.4.4. *Let $\mathfrak{B}$ be a structure with at least two elements. Then every finite structure has a first-order interpretation in $\mathfrak{B}$.*

PROOF. Let $\mathfrak{A}$ be a τ-structure with domain $\{1,\dots,n\}$. The statement is trivial if $n=1$; so let us assume that $n>1$ in the following. Our first-order interpretation I of $\mathfrak{A}$ in $\mathfrak{B}$ is n-dimensional. For $k \in \{1,\dots,n-1\}$, define

$$\rho_k(x_1,\dots,x_n) := \left(x_k \neq x_{k+1} \wedge \bigwedge_{i=1}^{k} x_1 = x_i\right)$$

$$\rho_n := (x_1 = \cdots = x_n).$$

The domain formula of our interpretation is *true*. Equality is interpreted by the formula

$$=_I (x_1,\dots,x_n,y_1,\dots,y_n) := \bigvee_{k<n} \big(\rho_k(x_1,\dots,x_n) \wedge \rho_k(y_1,\dots,y_n)\big).$$

Note that the equivalence relation defined by $=_I$ on A^n has exactly n equivalence classes. If $R \in \tau$ is m-ary, then the formula $R(x_1,\dots,x_m)_I$ is a disjunction of conjunctions with the nm variables $x_{1,1}, \dots, x_{m,n}$. For each tuple $(t_1, \dots, t_m)$ from $R^{\mathfrak{A}}$ the disjunction contains the conjunct

$$\bigwedge_{i \leq m} \rho_{t_i}(x_{i,1}, \dots, x_{i,n}).$$

⊣

2.4.1. Composing interpretations. First-order interpretations can be composed. In order to conveniently treat these compositions, we first describe how an interpretation of a σ-structure $\mathfrak{B}$ gives rise to interpreting formulas for arbitrary σ-formulas $\psi(x_1, \dots, x_n)$. Replace each atomic σ-formula $\phi(y_1, \dots, y_n)$ in ψ by $\phi_I(y_{1,1}, \dots, y_{1,d}, \dots, y_{n,1}, \dots, y_{n,d})$; we write $\psi_I(x_{1,1}, \dots, x_{1,d}, \dots, x_{n,1}, \dots, x_{n,d})$ for the resulting τ-formula, and call it the *interpreting formula* for ψ. Note that if ψ defines the relation R in $\mathfrak{B}$, then ψ_I defines $I^{-1}(R)$ in $\mathfrak{A}$. For all d-tuples $a_1, \dots, a_n \in I^{-1}(B)$

$$\mathfrak{B} \models \psi(I(a_1), \dots, I(a_n)) \Leftrightarrow \mathfrak{A} \models \psi_I(a_1, \dots, a_n).$$

DEFINITION 2.4.5. Let $\mathfrak{C}$, $\mathfrak{B}$, $\mathfrak{A}$ be structures with the relational signatures ρ, σ, and τ. Suppose that

- $\mathfrak{C}$ has a d-dimensional interpretation I in $\mathfrak{B}$, and
- $\mathfrak{B}$ has an e-dimensional interpretation J in $\mathfrak{A}$.

Then $\mathfrak{C}$ has a natural de-dimensional first-order interpretation $I \circ J$ in $\mathfrak{A}$: the domain of $I \circ J$ is the set of all de-tuples in A that satisfy the τ-formula $(\top_I)_J$, and we define

$$I \circ J(a_{1,1}, \dots, a_{1,e}, \dots, a_{d,1}, \dots, e_{d,e}) := I(J(a_{1,1}, \dots, a_{1,e}), \dots, J(a_{d,1}, \dots, e_{d,e})).$$

Let ϕ be a τ-formula which defines a relation R over $\mathfrak{A}$. Then the formula $(\phi_I)_J$ defines in $\mathfrak{A}$ the preimage of R under $I \circ J$.

2.4.2. Bi-interpretations. Let I_1 and I_2 be two interpretations of $\mathfrak{B}$ in $\mathfrak{A}$ of dimension d_1 and d_2, respectively. Then I_1 and I_2 are called *homotopic*[13] if the relation $\{(\bar{x}, \bar{y}) \mid I_1(\bar{x}) = I_2(\bar{y})\}$ of arity $d_1 + d_2$ is first-order definable in $\mathfrak{A}$. Note that id_C is an interpretation of $\mathfrak{C}$ in $\mathfrak{C}$, called the *identity interpretation* of $\mathfrak{C}$ (in $\mathfrak{C}$).

DEFINITION 2.4.6. Two structures $\mathfrak{A}$ and $\mathfrak{B}$ with an interpretation I of $\mathfrak{B}$ in $\mathfrak{A}$ and an interpretation J of $\mathfrak{A}$ in $\mathfrak{B}$ are called *mutually interpretable*. If both $I \circ J$ and $J \circ I$ are homotopic to the identity interpretation (of $\mathfrak{A}$ and of $\mathfrak{B}$, respectively), then we say that $\mathfrak{A}$ and $\mathfrak{B}$ are *bi-interpretable* (*via I and J*).

Example 2.4.7. It is easy to see that Allen's interval algebra from Section 1.6.1 is bi-interpretable with $(\mathbb{Q}; <)$ (a detailed proof can be found in Example 3.3.3).

[13] We follow the terminology from [8].

2.4.3. Full interpretations. An interpretation I of $\mathfrak{B}$ in $\mathfrak{A}$ is called *full* if for every $R \subseteq B^k$ we have that R is first-order definable in $\mathfrak{B}$ if and only if the relation $I^{-1}(R)$ is first-order definable in $\mathfrak{A}$. Note that every structure with an interpretation in $\mathfrak{A}$ is a first-order reduct of a structure with a full interpretation in $\mathfrak{A}$. Full interpretations play an important role in Section 6.3.5 since they are the counterpart for certain standard algebraic constructions that will be studied later. The interpretation in Example 2.4.2 is full: this follows easily from previous observations and the following lemma.

LEMMA 2.4.8. *Suppose that $\mathfrak{A}$ and $\mathfrak{B}$ are bi-interpretable via I and J. Then I (and, symmetrically, J) is a full interpretation.*

PROOF. Let τ be the signature of $\mathfrak{A}$ and ρ the signature of $\mathfrak{B}$. Let d be the dimension of I and e the dimension of J. Let $R \subseteq B^k$ be a relation such that the dk-ary relation $I^{-1}(R)$ has a first-order definition ϕ in $\mathfrak{A}$. Let ψ be the τ-formula that defines in $\mathfrak{B}$ the relation

$$\{(x_{1,1},\dots,x_{1,e},\dots,x_{d,1},\dots,x_{d,e},y) \mid I(J(x_{1,1},\dots,x_{1,e}),\dots, J(x_{d,1},\dots,x_{d,e})) = y\}$$

witnessing homotopy of $I \circ J$ with id_B. Then the ρ-formula $\chi(y_1,\dots,y_k)$ given by

$$\exists y_{1,1,1},\dots,y_{k,d,e}\Big(\phi_J(y_{1,1,1},\dots,y_{k,d,e}) \wedge \bigwedge_{i\le k} \psi(y_{i,1,1},\dots,y_{i,d,e},y_i)\Big)$$

is a first-order definition of R in $\mathfrak{B}$: if $(b_1,\dots,b_k) \in R$, choose $b_{1,1,1},\dots,b_{k,d,e}$ such that $J(I(b_{i,1,1},\dots,b_{i,1,e}),\dots,I(b_{i,d,1},\dots,b_{i,d,e})) = b_i$ which exist by the surjectivity of I and J. Then

$$\begin{aligned}(b_1,\dots,b_k) \in R &\Leftrightarrow \mathfrak{A} \models \phi(J(b_{1,1,1},\dots,b_{1,1,e}),\dots,J(b_{k,d,1},\dots,b_{k,d,e}))\\ &\Leftrightarrow \mathfrak{B} \models \phi_J(b_{1,1,1},\dots,b_{k,d,e})\\ &\Leftrightarrow \mathfrak{B} \models \chi(b_1,\dots,b_k).\end{aligned}$$

⊣

2.5. Preservation theorems

Preservation theorems in model theory establish links between definability in (a syntactically restricted fragment of) a given logic with certain 'semantic' closure properties. For the syntactic restrictions on first-order formulas that we have introduced in Section 2.1.7 we have already made remarks about various types of mappings that automatically preserve the respective formulas. Interestingly, these mappings can be used to obtain an exact characterisation of definability in the corresponding fragment of first-order logic.

When studying CSP(T) for a given theory T, preservation theorems become relevant in two contexts. The first is that they can be used to give exact characterisations for a first-order formula to be equivalent to an existential, existential positive, or quantifier-free formula over T. The second context in which we encounter model-theoretic preservation theorems is in connection with syntactic restrictions for theories T in the study of CSP(T), for example when proving Proposition 2.6.13.

DEFINITION 2.5.1. When T is a first-order theory and $\phi(\bar{x})$ and $\psi(\bar{x})$ are formulas, we say that ϕ and ψ are *equivalent modulo T* if $T \models \forall \bar{x}(\phi(\bar{x}) \Leftrightarrow \psi(\bar{x}))$.

THEOREM 2.5.2 (Homomorphism Preservation Theorem). *Let T be a first-order theory. A first-order formula ϕ is equivalent to an existential positive formula modulo T if and only if ϕ is preserved by all homomorphisms between models of T.*

PROOF. It is clear that homomorphisms preserve existential positive formulas. For the converse, let ϕ be first-order, with free variables $x_1, \dots, x_n$, and preserved by homomorphisms between models of T. Let τ be the signature of T and ϕ, and let $\bar{c} = (c_1, \dots, c_n)$ be a sequence of constant symbols that do not appear in τ. If $T \cup \{\phi(\bar{c})\}$ is unsatisfiable, the statement is clearly true, so assume otherwise. Let Ψ be the set of all existential positive $(\tau \cup \{c_1, \dots, c_n\})$-sentences ψ such that $T \cup \{\phi(\bar{c})\} \models \psi$. Let $\mathfrak{A}$ be a model of $T \cup \Psi$. Let U be the set of all primitive positive sentences θ such that $\mathfrak{A} \models \neg\theta$.

We claim that $T \cup \{\neg\theta \mid \theta \in U\} \cup \{\phi(\bar{c})\}$ is satisfiable. For otherwise, by compactness, there would be a finite subset U' of U such that $T \cup \{\neg\theta \mid \theta \in U'\} \cup \{\phi(\bar{c})\}$ is unsatisfiable. But then $\psi := \bigvee_{\theta \in U'} \theta$ is an existential positive sentence such that $T \cup \{\phi(\bar{c})\} \models \psi$, and hence $\psi \in \Psi$. This is in contradiction to the assumption that $\mathfrak{A} \models \neg\theta$ for all $\theta \in U$. We conclude that there exists a model $\mathfrak{B}$ of $T \cup \{\neg\theta \mid \theta \in U\} \cup \{\phi(\bar{c})\}$.

By Theorem 2.2.1, $\mathfrak{A}$ has an elementary extension $\mathfrak{A}'$ which is $|B|$-saturated. Every primitive positive $(\tau \cup \{c_1, \dots, c_n\})$-sentence θ that is true in $\mathfrak{B}$ is also true in $\mathfrak{A}'$: for if otherwise θ were false in $\mathfrak{A}'$, then it would be also false in $\mathfrak{A}$, and hence $\theta \in U$ in contradiction to the assumption that $\mathfrak{B} \models \{\neg\theta \mid \theta \in U\}$. Hence, by Lemma 2.2.6, there exists a homomorphism from $\mathfrak{B}$ to $\mathfrak{A}'$. Since $\mathfrak{B} \models \phi(\bar{c})$, and ϕ is preserved by homomorphisms between models of T, we have $\mathfrak{A}' \models \phi(\bar{c})$.

We conclude that $T \cup \Psi \cup \{\neg\phi(\bar{c})\}$ is unsatisfiable, and again by compactness there exists a finite subset Ψ' of Ψ such that $T \cup \Psi' \cup \{\neg\phi(\bar{c})\}$ is unsatisfiable. Then $\bigwedge \Psi'$ is an existential positive sentence; let ψ be the formula obtained from this sentence by replacing for all $i \leq n$ all occurrences of c_i by x_i. Then $T \models \forall \bar{x}(\psi(\bar{x}) \Leftrightarrow \phi(\bar{x}))$ (see Lemma 2.1.7), which is what we wanted to show. $\dashv$

Note that here the assumption that $\bot$ is always part of first-order logic is important: the first-order formula $\exists x \colon x \neq x$ is preserved by all homomorphisms between models of T, but without $\bot$ it may not be equivalent to an existential positive formula modulo T (for instance when T is the empty theory).

The classical theorem of Łoś-Tarski for preservation under embeddings of models of a theory is a direct consequence of the homomorphism preservation theorem.

COROLLARY 2.5.3 (Łoś-Tarski; see e.g. Corollary in 5.4.5 of [207]). *Let T be a first-order theory. A first-order formula ϕ is equivalent to an existential formula modulo T if and only if ϕ is preserved by all embeddings between models of T.*

PROOF. For each atomic formula ψ add a new relation symbol N_ψ to the signature of T, and add the sentence $\forall \bar{x}(N_\psi(\bar{x}) \Leftrightarrow \neg\psi(\bar{x}))$; let T' be the resulting theory. Then every existential formula ϕ is equivalent to an existential positive formula in T', which can be obtained from ϕ by replacing negative literals $\neg\psi(\bar{x})$ in ϕ by $N_\psi(\bar{x})$. Similarly, homomorphisms between models of T' must be embeddings. Hence, the statement follows from Theorem 2.5.2. ⊣

Our next preservation theorem, Theorem 2.5.5, is a positive variant of the well-known Chang-Łoś-Suszko preservation theorem, which characterises $\forall\exists$-definability. Much as we have just seen in the case of the homomorphism preservation theorem, our positive variant easily implies the classical one. The proof we give for the positive version is similar to the proof of the Chang-Łoś-Suszko theorem given in [207]. We need the following lemma.

LEMMA 2.5.4. *Let T be a first-order theory and let $\mathfrak{A}$ be a model of $T_{\forall\exists^+}$. Then $\mathfrak{A}$ can be extended to a model $\mathfrak{B}$ of T such that for every $\bar{a} \in A^n$ and every existential positive formula ϕ if $\mathfrak{B} \models \phi(\bar{a})$ then $\mathfrak{A} \models \phi(\bar{a})$.*

PROOF. Let $\mathfrak{A}'$ be an expansion of $\mathfrak{A}$ by constants such that each element of $\mathfrak{A}'$ is denoted by a constant symbol. It suffices to prove that $T \cup \mathrm{Th}(\mathfrak{A}')_{\mathrm{qf}} \cup \mathrm{Th}(\mathfrak{A}')_{\forall-}$ has a model $\mathfrak{B}$. Suppose for contradiction that it is inconsistent; then by compactness, there exists a finite subset U of $\mathrm{Th}(\mathfrak{A}')_{\mathrm{qf}} \cup \mathrm{Th}(\mathfrak{A}')_{\forall-}$ such that $T \cup U$ is inconsistent. Let ϕ be the conjunction over U where all new constant symbols are existentially quantified. Then $T \cup \{\phi\}$ is inconsistent as well. But $\neg\phi$ is equivalent to a $\forall\exists^+$ formula, and a consequence of T. Hence, $\mathfrak{A} \models \neg\phi$, a contradiction. ⊣

Theorem 2.5.5 below is essentially Theorem 23 in [35]; the formulation below is taken from [61].

THEOREM 2.5.5 (Positive Chang-Łoś-Suszko). *Let T be a first-order τ-theory, and Φ a set of τ-formulas. Then the following are equivalent.*

1. *Φ is equivalent modulo T to a set of $\forall\exists^+$-formulas Ψ.*
2. *Φ is preserved in direct limits of sequences of models of T.*
3. *Φ is preserved in direct limits of countable sequences of models of T.*

PROOF. The implication from (1) to (2) is Lemma 2.1.21. The implication from (2) to (3) is trivial. For the implication from (3) to (1), assume that Φ

is preserved by direct limits of sequences $(\mathfrak{A}_i)_{i<\omega}$ as in the statement of the proposition. We can assume that Φ is a set of sentences (by adding constants, Lemma 2.1.7). Let Ψ be the set of all $\forall\exists^+$-sentences that are consequences of $T \cup \Phi$. To show that $T \cup \Psi$ implies Φ, it suffices to show that every model of $T \cup \Psi$ is elementary equivalent to the direct limit of a sequence $(\mathfrak{B}_i)_{i<\omega}$ of models of $T \cup \Phi$. To construct this sequence, we define an elementary chain of models $(\mathfrak{A}_i)_{i<\omega}$ of $T \cup \Psi$ such that there are

- $\mathfrak{A}_i$ has an extension to a model $\mathfrak{B}_i$ of $T \cup \Phi$ such that for every tuple $\bar{a}_i$ of elements from $\mathfrak{A}_i$ and every existential positive formula θ, if $\mathfrak{B}_i \models \theta\bar{a}_i)$, then $\mathfrak{A}_i \models \theta(\bar{a}_i)$, and
- homomorphisms $g_i : \mathfrak{B}_i \to \mathfrak{A}_{i+1}$ such that g_i is the identity on $\mathfrak{A}_i$.

Let $\mathfrak{A}_0$ be a model of $T \cup \Psi$. To construct the rest of the sequence, suppose that $\mathfrak{A}_i$ has been chosen. Since $\mathfrak{A}_0$ is an elementary substructure of $\mathfrak{A}_i$, in particular all the $\forall\exists^+$-consequences of $T \cup \Phi$ hold in $\mathfrak{A}_i$. By Lemma 2.5.4, the structure $\mathfrak{A}_i$ can be extended to a model $\mathfrak{B}_i$ of $T \cup \Phi$ such that every ep-sentence that holds in $(\mathfrak{B}_i, \bar{a}_i)$ also holds in $(\mathfrak{A}_i, \bar{a}_i)$. By Lemma 2.1.12 there are an elementary extension $\mathfrak{A}_{i+1}$ of $\mathfrak{A}_i$ and a homomorphism $g_i : \mathfrak{B}_i \to \mathfrak{A}_{i+1}$. Then $\mathfrak{C} := \bigcup_{i<\omega} \mathfrak{A}_i$ equals $\lim_{i<\omega} \mathfrak{B}_i$, and by the Tarski-Vaught elementary chain theorem (Theorem 2.1.18) $\mathfrak{A}_0$ is an elementary substructure of $\mathfrak{C}$. So $\mathfrak{C}$ is a model of T, and the direct limit of models $\mathfrak{B}_i$ of $T \cup \Phi$, and hence by assumption $\mathfrak{C} \models \Phi$. This shows that $T \cup \Psi$ implies Φ. ⊣

By compactness one can show that if Φ is finite, then the set of formulas Ψ from item (1) in Theorem 2.5.5 above can be chosen to be a single formula.

COROLLARY 2.5.6. (Chang-Łoś-Suszko Theorem; Theorem 5.4.9 in [207] and remarks after the proof). *Let T be a first-order τ-theory.*

- *A set of first-order τ-formulas Φ is equivalent to a set of $\forall\exists$-formulas Ψ modulo T if and only if Φ is preserved in unions of chains of models of T.*
- *A first-order τ-formula ϕ is equivalent to a $\forall\exists$-formula ψ modulo T if and only if ϕ is preserved in unions of chains (A_i) of models of T.*

PROOF. The statement can be derived by adding relation symbols for the negations of all atomic formulas, and applying Theorem 2.5.5 to the corresponding theory, as in the proof of Corollary 2.5.3. ⊣

2.6. Model-completeness and cores

This section is concerned with theories T where various fragments of first-order logic have equal expressive power. In particular, we consider the situation that modulo T

(a) every first-order formula is equivalent to an existential formula (Section 2.6.1),

(b) every existential formula is equivalent to an existential positive formula (Section 2.6.2),

(a) & (b) every first-order formula is equivalent to an existential positive formula (Section 2.6.3.

In fact, the most important case in later sections will be the third, so we treat it separately.

These collapse results will be useful when studying the complexity of CSPs. For example, they clarify when the so-called *constraint entailment problem* is inter-reducible with a corresponding CSP (see Section 2.6.2). They are also important for the universal-algebraic approach for CSPs of ω-categorical structures that we present in Chapter 6. While the collapse properties listed above appear to be quite strong assumptions on T, we will see in Section 2.7 that often (for instance if T is ω-categorical) there exists a theory T' such that $\mathrm{CSP}(T) = \mathrm{CSP}(T')$ and T' satisfies the collapse properties listed above.

2.6.1. Model-complete theories. We start by recalling the classical concept of model-completeness of theories, since it inspired the new results of the next section about core theories.

DEFINITION 2.6.1. A theory T is *model complete* if every embedding between models of T is elementary, i.e., preserves all first-order formulas.

An equivalent characterisation of model-completeness is as follows.

THEOREM 2.6.2 (Theorem 7.3.1 in [207]). *Let T be a theory. Then the following are equivalent.*

1. *T is model complete;*
2. *every first-order formula is equivalent modulo T to an existential formula;*
3. *for every embedding e of a model $\mathfrak{A}$ of T into a model $\mathfrak{B}$ of T, every tuple $\bar{a}$ of elements of A, and every existential formula ϕ, if $\mathfrak{B} \models \phi(e(\bar{a}))$ then $\mathfrak{A} \models \phi(\bar{a})$.*
4. *every existential formula is equivalent modulo T to a universal formula;*
5. *every first-order formula is equivalent modulo T to a universal formula.*

PROOF. (1) $\Rightarrow$ (2). Suppose that T is model complete, and let ϕ be a first-order formula. Since T is model complete, ϕ is preserved by all embeddings between models of T. It follows from Theorem 2.5.3 that ϕ is equivalent to an existential formula.

(2) $\Rightarrow$ (3). Let e be an embedding of a model $\mathfrak{A}$ of T into a model $\mathfrak{B}$ of T. Let $\bar{a}$ be a tuple of $\mathfrak{A}$, and ϕ an existential formula such that $\mathfrak{B} \models \phi(e(\bar{a}))$. By (2), $\neg\phi$ is equivalent to an existential formula. Therefore, e preserves $\neg\phi$. Since $\mathfrak{B} \models \phi(e(\bar{a}))$ we therefore must have $\mathfrak{A} \models \phi(\bar{a})$.

(3) $\Rightarrow$ (4). Let ϕ be an existential formula. We have to show that $\neg\phi$ is equivalent to an existential formula. But (3) implies that $\neg\phi$ is preserved by embeddings between models of T, so the statement follows from Theorem 2.5.3

(4) $\Rightarrow$ (5). Let ϕ be a first-order formula, written in prenex normal form $Q_1 x_1 \cdots Q_n x_n : \psi$ for ψ quantifier-free. Let $i \leq n$ be smallest so that $Q_i = \cdots = Q_n$. If $i = 1$ then either ϕ is already universal, or equivalent to a universal formula by (4), and we are done. Otherwise, if $Q_i = \cdots = Q_n = \exists$ then by (4) the formula $\exists x_i \ldots \exists x_n : \psi$ is equivalent modulo T to a universal formula ψ'. We proceed with the formula $Q_1 x_1 \cdots Q_{i-1} x_{i-1} : \psi'$ which has fewer quantifier alternations than ϕ. Finally, suppose that $Q_i = \cdots = Q_n = \forall$. By (4) the formula $\exists x_i \ldots \exists x_n : \neg\psi$ is equivalent modulo T to a universal formula ψ'. Then the formula $Q_1 x_1 \cdots Q_{i-1} x_{i-1} : \neg\psi'$ is clearly equivalent to ϕ, but has fewer quantifier alternations. The claim follows by induction on the number of quantifier alternations of ϕ.

(5) $\Rightarrow$ (1). Let ϕ be a first-order formula. Then $\neg\phi$ is equivalent to a universal formula, therefore ϕ is equivalent to an existential formula, and hence preserved by all embeddings between models of T. $\dashv$

For many theories, model completeness can be shown by proving an even stronger property, namely *quantifier elimination*.

DEFINITION 2.6.3. A τ-theory admits *quantifier elimination* if for every first-order τ-formula there is a quantifier-free τ-formula which is equivalent modulo T.

In this context, our assumption that we allow $\bot$ as a first-order formula, becomes relevant.[14] Theorem 2.6.2 shows that theories with quantifier elimination are model complete.

Example 2.6.4. Let T be the first-order theory of $(\mathbb{Z}; \text{Succ})$ where Succ is the binary relation $\{(x, y) \mid y = x + 1\}$). Then T does not have quantifier elimination, but is model complete.

We say that a *structure* $\mathfrak{A}$ *is model complete* if and only if the first-order theory $\text{Th}(\mathfrak{A})$ of $\mathfrak{A}$ is model complete.

Example 2.6.5. The structure $(\mathbb{Q}_0^+; <)$, where $\mathbb{Q}_0^+$ denotes the non-negative rational numbers, is not model complete, because the self-embedding $x \mapsto x+1$ of $(\mathbb{Q}_0^+; <)$ does not preserve the formula $\phi(x) = \forall y\, (x < y \vee x = y)$ (which is satisfied only by 0).

In the following we use the easy fact that the first-order theory of a finite structure determines the structure up to isomorphism.

Example 2.6.6. All finite structures $\mathfrak{A}$ are model complete: self-embeddings of $\mathfrak{A}$ are automorphisms, and hence they are elementary. Every relation that is first-order definable in a finite structure also has an existential definition.

[14] Hodges [206] does not make this assumption, and therefore has to distinguish between *quantifier elimination* and what he calls *quantifier elimination for non-sentences*.

PROPOSITION 2.6.7. *Every model-complete theory T is equivalent to a $\forall\exists$-theory.*

PROOF. This is an immediate consequence of the Chang-Łoś-Suszko theorem (Corollary 2.5.6; where the theory denoted by T in Corollary 2.5.6 is empty and Φ from Corollary 2.5.6 equals the theory T from the statement here) because for any sequence $(\mathfrak{B}_i)_{i<\kappa}$ of models of T with embeddings $g_{ij}\colon \mathfrak{B}_i \hookrightarrow \mathfrak{B}_j$, the g_{ij} are elementary. By the Tarski-Vaught theorem (Theorem 2.1.18), we have that $(\lim_{i<\kappa} \mathfrak{B}_i) \models T$. ⊣

2.6.2. Core theories. We have already encountered the concept of a *core* of a finite structure in Section 1.1. To recall, a finite structure $\mathfrak{B}$ is called a *core* if all endomorphisms of $\mathfrak{B}$ are embeddings. Cores play an important role in the classification program for finite-domain CSPs. There are many equivalent definitions of the notion of a core of a finite structure: for example, a finite structure is a core if and only if all endomorphisms are surjective, or injective, or bijective, or automorphisms. For infinite structures these definitions are in general not equivalent, even when if they are ω-categorical; see [32, 31, 44]. To motivate our general definition of cores, let us review some important properties of finite cores:

- Existence: every finite structure $\mathfrak{A}$ has a core $\mathfrak{B}$ (Proposition 1.1.11).
- Uniqueness: all core structures $\mathfrak{B}$ of $\mathfrak{A}$ are isomorphic (Proposition 1.1.11).
- Definability: orbits of k-tuples under $\mathrm{Aut}(\mathfrak{B})$ for finite cores $\mathfrak{B}$ are primitively positively definable in $\mathfrak{B}$ (Proposition 1.2.11).
- Definability property 2: every first-order formula is equivalent over $\mathfrak{B}$ to an existential positive formula (a direct consequence of the previous item).

Note that the last property only depends on the first-order theory of $\mathfrak{B}$, and not on $\mathfrak{B}$ itself. We make the following definition.

DEFINITION 2.6.8. A theory T is called a *core theory* if every homomorphism between models of T is an embedding.

Note that a finite structure $\mathfrak{B}$ is a core if and only if the first-order theory of $\mathfrak{B}$ is a core theory (again we use that the first-order theory of a finite structure $\mathfrak{B}$ determines the structure up to isomorphism). In fact, the same is true for all ω-categorical structures $\mathfrak{B}$; we will revisit the ω-categorical case in Chapter 4.

Example 2.6.9. The first-order theory T of $(\mathbb{Q}; <)$ is easily seen to be a core theory: homomorphisms between models of T must be injective and also preserve the complement of $<$. In contrast, the first-order theory of $(\mathbb{Q}; \leq)$ is not a core theory, since $(\mathbb{Q}; \leq)$ has a constant endomorphism.

PROPOSITION 2.6.10. *Let T be a first-order τ-theory. The following are equivalent.*

1. *T is a core theory.*
2. *Every existential formula is equivalent modulo T to an existential positive formula.*
3. *For every atomic τ-formula ψ, the formula $\neg\psi$ is equivalent to an existential positive formula modulo T.*

PROOF. (1) $\Rightarrow$ (2). Let T be a core theory, and let ϕ be an existential formula. Then ϕ is preserved by all embeddings between models of T. Since all homomorphisms between models of T are embeddings, ϕ is also preserved by all homomorphisms between models of T. Hence, Theorem 2.5.2 implies that ϕ is equivalent modulo T to an existential positive formula.

(2) $\Rightarrow$ (3) is clear since negations of atomic formulas are existential formulas.

(3) $\Rightarrow$ (1) is immediate because homomorphisms between models of T must preserve existential positive formulas, so by (3) they must preserve the negations of atomic formulas, and hence are embeddings. ⊣

We would like to point out a computational corollary. Let T be a theory with finite relational signature τ. The *constraint entailment problem for T* is the following computational problem. The input consists of a primitive positive τ-formula ϕ, and a single atomic τ-formula ψ, both ϕ and ψ with free variables $x_1, \ldots, x_n$. The question is whether ϕ implies (*entails*) ψ over T, i.e., whether

$$T \models \forall x_1, \ldots, x_n \, (\phi \Rightarrow \psi).$$

COROLLARY 2.6.11. *Let τ be a finite relational signature, and let T be a core τ-theory. Then the constraint entailment problem for T is equivalent to* CSP(T) *under polynomial-time Turing reductions.*

PROOF. The reduction from CSP(T) to the constraint entailment problem for T is trivial, because in order to decide satisfiability of $T \cup \{\phi\}$, we can test whether ϕ entails *false* over T.

For the converse reduction, let ϕ, ψ be an input to the constraint entailment problem for T. Since T is a core theory, $\neg\psi$ is by Proposition 2.6.10 equivalent to an existential positive τ-formula, and hence equivalent to a disjunction $\psi_1 \vee \cdots \vee \psi_m$ of primitive positive formulas. We may assume that the size of this disjunction is bounded by a constant, for all possible inputs, because T is fixed and the signature τ is finite. Then ϕ entails ψ if and only if for all $1 \leq i \leq m$, we have that $\exists x_1, \ldots, x_k \, (\phi \wedge \psi_i)$ is false in T. In this way we have reduced the entailment problem to solving a constant number of constraint satisfaction problems over T. ⊣

2.6.3. Model-complete core theories. The results from the previous two sections can be combined to obtain alternative characterisations of model-complete core theories. We mostly work with such theories in later sections for reasons that will become apparent in Section 2.7, so we explicitly combine these results here for easy reference.

THEOREM 2.6.12. *Let T be a first-order theory. Then the following are equivalent*:

1. *T is a model-complete core theory*;
2. *every first-order formula is equivalent modulo T to an existential positive formula*;
3. *for every homomorphism h from a model $\mathfrak{A}$ of T to a model $\mathfrak{B}$ of T, every tuple $\bar{a}$ of elements of A, and every existential positive formula ϕ, if $\mathfrak{B} \models \phi(h(\bar{a}))$ then $\mathfrak{A} \models \phi(\bar{a})$.*
4. *every existential positive formula is equivalent modulo T to a universal negative formula*;
5. *every first-order formula is equivalent modulo T to a universal negative formula.*

PROOF. Analogous to the proof of Theorem 2.6.2, replacing Theorem 2.5.2 by Corollary 2.5.3, existential by existential positive definability, universal by universal negative definability, and embeddings by homomorphisms. ⊣

PROPOSITION 2.6.13. *Let T be a model-complete core theory. Then T is equivalent to a $\forall\exists^+$-theory.*

PROOF. Analogous to the proof of Proposition 2.6.7, replacing the Chang-Łoś-Suszko theorem by its positive variant, Theorem 2.5.5. ⊣

2.7. Companions

When we are interested in CSP(T), it is often useful to first pass to a 'nicer' theory T' such that CSP(T) = CSP(T'). Ideally, we would like that T' is a model-complete core theory—but such a theory T' may not exist. However, if such a model-complete core theory T' does exist, then it turns out to be unique up to equivalence of theories, and will be called the *core companion* of T.

If the core companion exists, it can be constructed using the concept of *existential positive closure*, which we develop in Section 2.7.1. Another closely related concept is the *positive Kaiser hull* of a theory T, defined in Section 2.7.2. The positive Kaiser hull may not be model complete or not a core theory, but if T has a core companion, then the core companion equals the positive Kaiser hull of T. In particular, if the core companion exists, then the core companion is unique—all this up to logical equivalence of theories. Using these concepts, we prove in Section 2.7.3 the central statements about core companions. Our results about core companions imply the corresponding classical results about existential closure, the Kaiser hull, and model companions; this will be the topic of Section 2.7.4.

2.7.1. Existential positive closure. The direct limit construction from Section 2.1.10 can be used to build models with a certain desirable property, *existential positive closure*, which we will introduce in this section. Much of the material presented here is from [61] and analogous to, but more powerful than, the classical facts about *existential closure*.

DEFINITION 2.7.1. Let T be a theory. A structure $\mathfrak{A}$ is called *existentially positively closed for* T (short: T*-epc*) if there is a homomorphism from $\mathfrak{A}$ to a model of T, and if for any homomorphism h from $\mathfrak{A}$ into a model $\mathfrak{B}$ of T, any tuple $\bar{a}$ from A, and any existential positive formula ϕ with $\mathfrak{B} \models \phi(h(\bar{a}))$ we have $\mathfrak{A} \models \phi(\bar{a})$.

Note that we can equivalently replace 'existential positive' by 'primitive positive' in the previous definition.

LEMMA 2.7.2. *A structure is T-epc if and only if it is $T_{\forall-}$-epc.*

PROOF. Suppose that $\mathfrak{A}$ is T-epc. Then $\mathfrak{A}$ maps homomorphically to a model $\mathfrak{B}$ of $T_{\forall-}$. Let h be any such homomorphism, let $\bar{a}$ be a tuple in A, and ϕ an existential positive formula such that $\mathfrak{B} \models \phi(h(\bar{a}))$. By Proposition 2.1.14, $\mathfrak{B}$ has a homomorphism g to a model of T. Then $g(h(\bar{a}))$ holds since g preserves existential positive formulas. Apply the assumption that $\mathfrak{A}$ is T-epc to the homomorphism $g \circ h$ to conclude that $\phi(\bar{a})$, which proves that $\mathfrak{A}$ is $T_{\forall-}$-epc.

Now, let $\mathfrak{A}$ be $T_{\forall-}$-epc. Proposition 2.1.14 shows that $\mathfrak{A}$ has a homomorphism to a model of T. Any such homomorphism is in particular a homomorphism to a model of $T_{\forall-}$, and hence the assumption that $\mathfrak{A}$ is $T_{\forall-}$-epc implies that $\mathfrak{A}$ is T-epc. ⊣

To show the existence of T-epc structures we apply the direct limit construction from Section 2.1.10 (see also [36]).

LEMMA 2.7.3. *Let T be a τ-theory and $\kappa \geq \max(|\tau|, \aleph_0)$. Then any model $\mathfrak{A}$ of T of cardinality at most κ admits a homomorphism to a T-epc structure $\mathfrak{B}$ of cardinality at most κ.*

PROOF. Set $\mathfrak{B}_0 := \mathfrak{A}$. Suppose that we have already constructed $\mathfrak{B}_i$ of cardinality at most κ, for $i < \omega$. Let $\{(\phi_\alpha, \bar{a}_\alpha) \mid \alpha < \kappa\}$ be an enumeration of all pairs $(\phi, \bar{a})$ where ϕ is existential positive with free variables $x_1, \ldots, x_n$, and $\bar{a}$ is an n-tuple from B_i. We construct a sequence $(\mathfrak{B}_i^\alpha)_{0 \leq \alpha < \kappa}$ of models of $T_{\forall-}$ of cardinality at most κ and a coherent sequence $(f_i^{\mu,\alpha})_{0 \leq \mu < \alpha < \kappa}$ where $f_i^{\mu,\alpha}$ is a homomorphism from $\mathfrak{B}_i^\mu$ to $\mathfrak{B}_i^\alpha$, as follows.

Set $\mathfrak{B}_i^0 = \mathfrak{B}_{i-1}$. Now let $\alpha = \beta + 1 < \kappa$ be a successor ordinal. Let $\bar{b}_\beta$ be the image of $\bar{a}_\beta$ in $\mathfrak{B}_i^\beta$ under $f_i^{0,\beta}$. If there is a model $\mathfrak{C}$ of $T_{\forall-}$ and a homomorphism $h \colon \mathfrak{B}_i^\beta \to \mathfrak{C}$ such that $\mathfrak{C} \models \phi_\beta(h(\bar{b}_\beta))$, then by the theorem of Löwenheim-Skolem (Theorem 2.1.11) there is also a model $\mathfrak{C}'$ of $T_{\forall-}$ of cardinality at most κ and a homomorphism $h' \colon \mathfrak{B}_i^\beta \to \mathfrak{C}'$ such that $\mathfrak{C}' \models \phi_\beta(h'(\bar{b}_\beta))$. Set $\mathfrak{B}_i^\alpha := \mathfrak{C}'$ and $f_i^{\mu,\alpha} := h' \circ f_i^{\mu,\beta}$ for all $\mu < \alpha$. Otherwise, if there is no such model $\mathfrak{C}$, we set $\mathfrak{B}_i^\alpha := \mathfrak{B}_i^\beta$ and $f_i^{\beta,\alpha} :=$ id (the identity mapping) and $f_i^{\mu,\alpha} := f_i^{\mu,\beta}$. Finally, for limit ordinals $\alpha < \kappa$, set

$\mathfrak{B}_i^\alpha := \lim_{\mu<\alpha} \mathfrak{B}_i^\mu$ and let $f_i^{\mu,\alpha}$ be the corresponding limit homomorphism from $\mathfrak{B}_i^\mu$ to $\mathfrak{B}_i^\alpha$.

Let $\mathfrak{B}_i$ be $\lim_{\alpha<\kappa} \mathfrak{B}_i^\alpha$ and let $g_i : \mathfrak{B}_{i-1} \to \mathfrak{B}_i$ be the limit homomorphism mapping each element of $\mathfrak{B}_{i-1} = \mathfrak{B}_i^0$ to its equivalence class in $\mathfrak{B}_i$. In the natural way, the g_i give rise to a coherent sequence of homomorphisms, and by Lemma 2.1.21, $\mathfrak{B} := \lim_{i<\omega} \mathfrak{B}_i$ is a model of $T_{\forall-}$; let $h_i : \mathfrak{B}_i \to \mathfrak{B}$ for $i < \omega$ be the corresponding limit homomorphisms.

To verify that $\mathfrak{B}$ is T-epc it suffices to show that it is $T_{\forall-}$-epc, by Lemma 2.7.2. Let g be a homomorphism from $\mathfrak{B}$ to a model $\mathfrak{C}$ of $T_{\forall-}$, and suppose that there is a tuple $\bar{b}$ over B and an existential positive formula ϕ such that $\mathfrak{C} \models \phi(g(\bar{b}))$. There is an $i < \omega$ and an $\bar{a} \in B_i$ such that $h_i(\bar{a}) = \bar{b}$. Then $g \circ h_i$ is a homomorphism from $\mathfrak{B}_i$ to $\mathfrak{C}$, and by construction we have that $\mathfrak{B}_{i+1} \models \phi(g_{i+1}(\bar{a}))$. Note that $h_{i+1} \circ g_{i+1} = h_i$. Thus, since h_{i+1} preserves existential positive formulas, we also have that $\mathfrak{B} \models \phi(\bar{b})$, which is what we had to show. ⊣

Existential positive closure can be characterised using maximal ep-types.

PROPOSITION 2.7.4. *Let T be a theory and let $\mathfrak{A}$ be a model of T. Then $\mathfrak{A}$ is T-epc if and only if for every $n \in \mathbb{N}$ every ep-n-type of a tuple in $\mathfrak{A}$ is a maximal ep-type of T.*

PROOF. (Forwards.) Let $(a_1, \ldots, a_n)$ be a tuple of elements of $\mathfrak{A}$ and let p be the ep-n-type of $(a_1, \ldots, a_n)$. Let $c_1, \ldots, c_n$ be new constant symbols that denote $a_1, \ldots, a_n$ in $\mathfrak{A}$. Let $\phi(x_1, \ldots, x_n)$ be an existential positive formula such that $T \cup p(c_1, \ldots, c_n) \cup \{\phi(c_1, \ldots, c_n)\}$ has a model $\mathfrak{C}$. Then $\mathfrak{C}$ has an $|A|$-saturated elementary extension $\mathfrak{C}_{\text{sat}}$ by Theorem 2.2.1. Clearly, $\mathfrak{C}_{\text{sat}}$ is in particular ep-$|A|$-saturated, and all existential positive sentences true on $(\mathfrak{A}, c_1, \ldots, c_n)$ are true on $\mathfrak{C}_{\text{sat}}$. By Lemma 2.2.6, there is a homomorphism h from $(\mathfrak{A}, c_1, \ldots, c_n)$ to $\mathfrak{C}_{\text{sat}}$. Now, since $\mathfrak{C}_{\text{sat}} \models \phi(c_1, \ldots, c_n)$ and $\mathfrak{A}$ is T-epc, we find that $(\mathfrak{A}, c_1, \ldots, c_n) \models \phi(c_1, \ldots, c_n)$. Hence, p is a maximal ep-type of T.

(Backwards.) Let $\mathfrak{B}$ be a model of T and $h : \mathfrak{A} \to \mathfrak{B}$ a homomorphism, $\bar{a} \in A^n$, and $\phi(x_1, \ldots, x_n)$ an existential positive formula such that $\mathfrak{B} \models \phi(h(\bar{a}))$. Let p be the ep-type of $\bar{a}$ in $\mathfrak{A}$. Since $\mathfrak{B}$ is a model of T and h preserves all existential positive formulas, it follows that $T \cup p \cup \{\phi\}$ is satisfiable. By the maximality of p, we have that $\phi \in p$, and therefore $\mathfrak{A} \models \phi(\bar{a})$. ⊣

The classical results about existential closure can be seen as a special case of the results about existential positive closure. Let T be a first-order theory. A structure $\mathfrak{A}$ is called *existentially closed for T* (short, *T-ec*) if $\mathfrak{A}$ embeds into a model of T and $\mathfrak{A} \models \phi(\bar{a})$ for any embedding e from $\mathfrak{A}$ into another model $\mathfrak{B}$ of T, any tuple $\bar{a}$ from A, and any existential formula ϕ with $\mathfrak{B} \models \phi(e(\bar{a}))$. The following is Lemma 3.2.11 in [337], a variant of Corollary 7.2.2 in [207].

COROLLARY 2.7.5. *Let T be a τ-theory and $\kappa \geq \max(|\tau|, \aleph_0)$. Then any model $\mathfrak{A}$ of T of cardinality at most κ embeds into a T-ec structure of cardinality at most κ.*

PROOF. A direct consequence of Lemma 2.7.3, by appropriately choosing the signature (as in the proof of Corollary 2.5.3). ⊣

2.7.2. The positive Kaiser hull. In this section we introduce a positive variant of the classical notion of the *Kaiser Hull* of a theory. The results about the Kaiser Hull are direct consequences of their positive analogues that we present here. Our presentation follows the presentation of the classical case given in [337].

LEMMA 2.7.6. *For every theory T there exists a unique largest $\forall\exists^+$-theory T' such that $T'_{\forall-} = T_{\forall-}$.*

PROOF. Suppose for contradiction that the set of all $\forall\exists^+$-theories S such that $S_{\forall-} = T_{\forall-}$ is not closed under unions. This is equivalent to the existence of $\forall\exists^+$-theories S, S' such that

- $S_{\forall-} = S'_{\forall-} = T_{\forall-}$,
- T has a model $\mathfrak{A}$, and
- $S \cup S'$ is unsatisfiable.

By Proposition 2.1.14 there exists a homomorphism from $\mathfrak{A}$ to a model $\mathfrak{A}_0$ of S, and a homomorphism from $\mathfrak{A}_0$ to a model $\mathfrak{A}_1$ of S'. Repeating this step we construct a sequence of structures $(\mathfrak{A}_i)_{i \in \mathbb{N}}$, with coherent homomorphisms $f_{ij} : \mathfrak{A}_i \to \mathfrak{A}_j$, such that $\mathfrak{A}_i$ is a model of S for even i and a model of S' for odd i. Then by Lemma 2.1.21 the direct limit $\mathfrak{B} := \lim_{i<\omega} \mathfrak{A}_i$ is a model of $S \cup S'$, a contradiction. ⊣

The theory T' from Lemma 2.7.6 will be called the *positive Kaiser hull of T*, denoted in the following by T^{KH^+}.

LEMMA 2.7.7. *The positive Kaiser hull of T equals the set of $\forall\exists^+$-sentences that hold in all T-epc structures.*

PROOF. Let T^* be the set of all $\forall\exists^+$-sentences satisfied by all T-epc structures. To show that $T^* \subseteq T^{\mathrm{KH}^+}$ it suffices to show that $(T^*)_{\forall-} = T_{\forall-}$. By Lemma 2.7.3, every model of $T_{\forall-}$ has a homomorphism to a T^*-epc structure. Therefore, every universal-negative sentence that holds in all T^*-epc structures must be in $T_{\forall-}$, i.e., $(T^*)_{\forall-} \subseteq T_{\forall-}$. Conversely, every T-epc structure homomorphically maps to a model of $T_{\forall-}$, and therefore satisfies $T_{\forall-}$, so $T_{\forall-} \subseteq (T^*)_{\forall-}$.

We now show that $T^{\mathrm{KH}^+} \subseteq T^*$. For this, we have to show that every T-epc structure $\mathfrak{A}$ satisfies all $\phi \in T^{\mathrm{KH}^+}$. Since T^{KH^+} is a $\forall\exists^+$-theory, ϕ is of the form $\forall \bar{y} : \psi(\bar{y})$ where ψ is a disjunction of existential positive and negative atomic τ-formulas. Let $\bar{a}$ be a tuple of elements of $\mathfrak{A}$. We have to show that $\mathfrak{A} \models \psi(\bar{a})$. Since $(T^*)_{\forall-} = T_{\forall-} = T^{\mathrm{KH}^+}_{\forall-}$, by Proposition 2.1.14 there is a homomorphism h from $\mathfrak{A}$ to a model $\mathfrak{B}$ of $T_{\forall-}^{\mathrm{KH}^+}$. Since $\mathfrak{B} \models \forall \bar{y} \colon \psi(\bar{y})$, at least one disjunct $\theta(h(\bar{a}))$ of ψ is true in $\mathfrak{B}$. If θ is a negative atomic formula, then $\theta(\bar{a})$ is also true in $\mathfrak{A}$ since h is a homomorphism. Now suppose that θ is an existential positive formula. Since $T_{\forall-}^{\mathrm{KH}^+} = T_{\forall-}$, by Proposition 2.1.14 there is a homomorphism g from $\mathfrak{B}$ to a model $\mathfrak{C}$ of $T_{\forall-}$. Since g preserves θ we have that $\mathfrak{C} \models \theta(g(h(\bar{a})))$. Now $\mathfrak{A} \models \theta(\bar{a})$ since $\mathfrak{A}$ is T-epc. In both cases we can conclude that $\mathfrak{A} \models \psi(\bar{a})$. Hence, $\mathfrak{A} \models T^{\mathrm{KH}^+}$. ⊣

2.7.3. Core companions. In this section we study conditions that imply that we can pass from a theory T to a model-complete core theory T' satisfying the same universal negative sentences (and therefore, when T has a finite relational signature, T and T' have the same CSP).

DEFINITION 2.7.8 (from [61]). Let T be a first-order τ-theory. Then a τ-theory S is called a *core companion* of T if

- S is a model-complete core theory;
- S and T have the same universal negative consequences, i.e., $S_{\forall-} = T_{\forall-}$.

Recall from Corollary 2.1.15 that the second item in Definition 2.7.8 is equivalent to requiring that every model of S maps homomorphically to a model of T, and conversely every model of T maps homomorphically to a model of S.

Example 2.7.9. Let T be the first-order theory of $(\mathbb{Z}; <)$. Then the core companion T is the first-order theory of $(\mathbb{Q}; <)$.

Example 2.7.10. Let T be the first-order theory of all undirected and loopless graphs. Then the core companion of T is the first-order theory of $(\mathbb{N}; \neq)$.

The core companion, if it exists, is unique; we choose a formulation of this fact that follows the presentation of the classical results on model companions given by Tent and Ziegler (Theorem 3.2.14 in [337]; some parts were generalised to the positive setting in [35], Corollary 24).

THEOREM 2.7.11. *Let T be a theory. Then the following are equivalent.*

1. *T has a core companion;*
2. *all models of the positive Kaiser hull of T are T-epc;*
3. *the class of T-epc structures has a first-order axiomatisation.*

In particular, if T has a core companion T^*, then T^* is the theory of all T-epc structures, and T^* is equivalent to T^{KH^+}.

PROOF. (1) ⇒ (2). Let U be the core companion of T. By Proposition 2.6.13, U is equivalent to a $\forall\exists^+$-theory. Since $U_{\forall-} = T_{\forall-}$ we therefore have $U \subseteq T^{\mathrm{KH}^+}$. So it suffices to show that every model $\mathfrak{A}$ of U is T-epc. The structure $\mathfrak{A}$ has a homomorphism h to a model of T since $\mathfrak{A} \models U_{\forall-} (= T_{\forall-})$ by Proposition 2.1.14. Let h be an arbitrary homomorphism from $\mathfrak{A}$ to a model $\mathfrak{B}$ of T, let $\bar{a}$ be a tuple from $\mathfrak{A}$, and ϕ an existential positive formula with

$\mathfrak{B} \models \phi(h(\bar{a}))$. Then $\mathfrak{B}$ has a homomorphism g to a model $\mathfrak{C}$ of U. Since U is a model-complete core theory, the homomorphism $g \circ h$ is elementary. Since g preserves existential positive formulas, $\mathfrak{C} \models \phi(g(h(\bar{a})))$. Since $g \circ h$ is elementary, $\mathfrak{A} \models \phi(\bar{a})$.

(2) $\Rightarrow$ (3). Lemma 2.7.7 implies that T^{KH^+} is satisfied by all T-epc structures. Together with (2) this implies that the first-order theory T^{KH^+} axiomatises the class of all T-epc structures.

(3) $\Rightarrow$ (1). Suppose that the class of T-epc structures equals the class of all models of a first-order theory U. We claim that U is the core companion of T. Every model of U is in particular a model of $T_{\forall-}$, and every model of $T_{\forall-}$ maps homomorphically to a model of U by Lemma 2.7.3. So we only have to verify that U is a model-complete core theory. It suffices to verify item (3) in Theorem 2.6.12. Let h be a homomorphism between models $\mathfrak{A}$ and $\mathfrak{B}$ of U, let $\bar{a}$ be a tuple from A, and ϕ an existential positive formula such that $\mathfrak{B} \models \phi(h(a))$. Since $\mathfrak{B}$ satisfies $T_{\forall-}$ and $\mathfrak{A}$ is $T_{\forall-}$-epc by Lemma 2.7.2, we have that $\mathfrak{B} \models \phi(\bar{a})$, as desired.

The final statement of the theorem is a clear consequence of the proof above. ⊣

Example 2.7.12. Let T be the first-order theory of $(\mathbb{Z}; \mathrm{Succ})$, as in Example 2.6.4. Then T is not a core theory since there are models of T consisting of several disjoint copies of $(\mathbb{Z}; \mathrm{Succ})$, and homomorphisms from such models to $(\mathbb{Z}; \mathrm{Succ})$ are non-injective. We claim that T has no core companion. To see this, observe that all structures that are T-epc are isomorphic to $(\mathbb{Z}; \mathrm{Succ})$; hence, by Theorem 2.7.11 if there were a core companion, it would have to be T itself, contradicting our observation above that T is not a core theory.

2.7.4. Model companions. In this section we state classical results about model companions and derive them from the more general results about core companions from the previous section.

DEFINITION 2.7.13. A theory S is a *model companion* of a theory T if

- S is model complete;
- S and T have the same universal consequences, i.e., $S_\forall = T_\forall$.

Note that the second item in this definition is equivalent to requiring that every model of S embeds into a model of T, and vice versa (this follows from Corollary 2.1.15 by choosing an appropriate signature, as explained in the proof of Corollary 2.5.3).

Example 2.7.14. In Example 2.6.5 we have seen that $(\mathbb{Q}_0^+; <)$ is not model-complete. However, it has a model companion: the first-order theory of $(\mathbb{Q}; <)$.

Example 2.7.15. We give an example of a theory without model companion. Let B be a binary relation symbol, and let T be the first-order theory of

$(\mathbb{Q}; \leq)$ together with sentences that assert that B is the graph of a bijection that preserves $\leq$. Then T does not have a model companion; this follows from a much more general result of Kikyo [233] about theories without model companions. But T has a core companion: the theory of the structure $(\{0\}; \leq, \{(0,0)\})$.

It follows from Lemma 2.7.7 that there exists a $\forall\exists$-theory T^{KH} which is *largest* with respect to containment and such that $T_\forall = T^{\mathrm{KH}}_\forall$; the theory T^{KH} is called the *Kaiser Hull* of T. And similarly as in the positive case, if T has a model companion, then the model companion equals T^{KH}.

THEOREM 2.7.16 (Theorem 3.2.14 in [337]). *Let T be a theory. Then the following are equivalent.*

1. *T has a model companion;*
2. *all models of the Kaiser hull of T are T-ec;*
3. *the class of T-ec structures has a first-order axiomatisation.*

In particular, if T has a model companion T^*, then it is unique up to equivalence of theories, and $T^* = T^{\mathrm{KH}}$ is the theory of all T-ec structures.

PROOF. An immediate consequence of Theorem 2.7.11 by appropriately changing the signature if necessary. ⊣

Chapter 3

PRIMITIVE POSITIVE INTERPRETATIONS

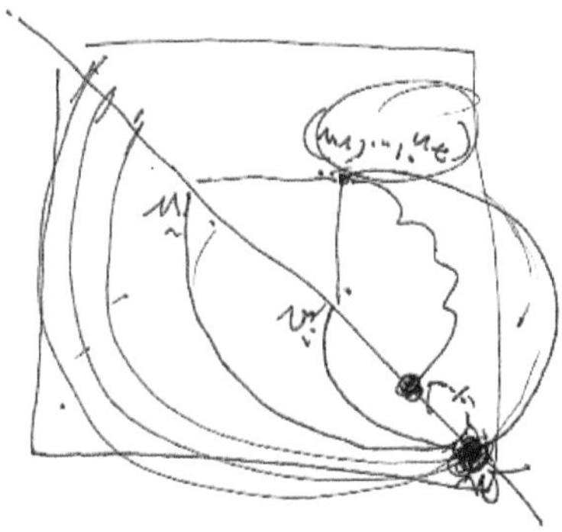

Jaroslav Nešetřil, 2004

Primitive positive definability from Section 1.2 is a strong tool to prove that certain CSPs are hard, but in some cases this tool is not strong enough. In this chapter we discuss the more flexible concept of *primitive positive interpretations*, a restriction of first-order interpretations that were introduced in Section 2.4. Later, in Chapter 6, we will see an important tight connection between primitive positive interpretability in finite or countably infinite ω-categorical structures and pseudo-varieties from universal algebra.

3.1. Introducing primitive positive interpretations

Primitive positive interpretations are interpretations for which all the defining formulas are primitive positive. As we will see, such interpretations can be used to study the computational complexity of constraint satisfaction problems.

DEFINITION 3.1.1. Let I be an interpretation of $\mathfrak{A}$ in $\mathfrak{B}$. If all the interpreting formulas of I can be chosen to be primitive positive then we say that I is a *primitive positive interpretation*. A d-dimensional primitive positive interpretation I is called *full* if a relation $R \subseteq A^k$ is primitively positively definable in $\mathfrak{A}$ whenever

$$I^{-1}(R) = \{(b_1^1, \dots, b_1^d, \dots, b_k^1, \dots, b_k^d) \in B^{dk} \mid (b_1^1, \dots, b_1^d), \dots, (b_k^1, \dots, b_k^d) \in \mathrm{dom}(I), (I(b_1^1, \dots, b_1^d), \dots, I(b_k^1, \dots, b_k^d)) \in R\}$$

is primitively positively definable in $\mathfrak{B}$.

Example 3.1.2. Let G be a directed graph and let F be an equivalence relation on $V(G)$. Then G/F is the directed graph whose vertices are the equivalence classes of F, and where (S, T) is an arc if there are $s \in S$ and $t \in T$ such that $(s, t) \in E(G)$. If F has a primitive positive definition in G, then G/F has a primitive positive interpretation in G.

This interpretation is not full in general: consider for example the digraph

$$G = (\{-1, 0, 1\}; \{(-1, 0), (0, 1), (-1, 1)\})$$

and the equivalence relation F with equivalence classes $\{0\}$ and $\{-1, 1\}$. Then G/F is isomorphic to K_2 and has an automorphism that exchanges the vertex $\{0\}$ with the vertex $\{-1, 1\}$, so the unary relation that just contains the class $\{0\}$ is not primitively positively definable in G/F. On the other hand, the relation $\{0\}$ has the primitive positive definition $\phi(x) = \exists u, v\big(E(u, x) \wedge E(x, v)\big)$ in G.

For an example where this interpretation is full, consider the digraph

$$G = (\{-1, 0, 1\}; \{(0, -1), (-1, 0), (0, 1), (1, 0)\})$$

and the equivalence relation F with equivalence classes $\{0\}$ and $\{-1, 1\}$. Again, G/F is isomorphic to K_2. Let ϕ be a primitive positive formula over the signature of graphs, defining a relation R over G. Then the same primitive positive formula ϕ defines $I^{-1}(R)$ over G/F.

Example 3.1.3. The field of rational numbers $(\mathbb{Q}; 0, 1, +, *)$ has a primitive positive 2-dimensional interpretation I in $(\mathbb{Z}; 0, 1, +, *)$. First observe that the non-negative integers are primitively positively definable in $(\mathbb{Z}; 0, 1, +, *)$, namely by the following formula $\phi(x)$ which states that x is the sum of four squares:

$$\exists x_1, x_2, x_3, x_4 (x = x_1^2 + x_2^2 + x_3^2 + x_4^2).$$

Clearly, every integer that satisfies $\phi(x)$ is non-negative; the converse is the famous four-square theorem of Lagrange [192]. The interpretation is now given as follows.

- The domain formula $\top_I(x, y)$ is $y \geq 1$ (using $\phi(x)$, it is straightforward to express this with a primitive positive formula);
- the formula $=_I (x_1, y_1, x_2, y_2)$ is $x_1 y_2 = x_2 y_1$;
- the formula $0_I(x, y)$ is $x = 0$, the formula $1_I(x, y)$ is $x = y$;

- the formula $+_I(x_1, y_1, x_2, y_2, x_3, y_3)$ is $y_3 * (x_1 * y_2 + x_2 * y_1) = x_3 * y_1 * y_2$;
- the formula $*_I(x_1, y_1, x_2, y_2, x_3, y_3)$ is $x_1 * x_2 * y_3 = x_3 * y_1 * y_2$.

THEOREM 3.1.4. *Let $\mathfrak{A}$ and $\mathfrak{B}$ be structures with finite relational signatures. If there is a primitive positive interpretation of $\mathfrak{A}$ in $\mathfrak{B}$, then there is a polynomial-time reduction from* CSP($\mathfrak{A}$) *to* CSP($\mathfrak{B}$).

PROOF. Let d be the dimension of the primitive positive interpretation I of the τ-structure $\mathfrak{A}$ in the σ-structure $\mathfrak{B}$. Let ϕ be an instance of CSP($\mathfrak{A}$) with variable set $U = \{x_1, \dots, x_n\}$. We construct an instance ψ of CSP($\mathfrak{B}$) as follows. For distinct variables $V := \{y_1^1, \dots, y_n^d\}$, we set ψ_1 to be the formula

$$\bigwedge_{1 \le i \le n} \top_I(y_i^1, \dots, y_i^d).$$

Let ψ_2 be the conjunction of the formulas $\theta_I(y_{i_1}^1, \dots, y_{i_1}^d, \dots, y_{i_k}^1, \dots, y_{i_k}^d)$ over all conjuncts $\theta = R(x_{i_1}, \dots, x_{i_k})$ of ϕ. By moving existential quantifiers to the front, the sentence

$$\exists y_1^1, \dots, y_n^d \, (\psi_1 \wedge \psi_2)$$

can be re-written to a primitive positive σ-formula ψ, and clearly ψ can be constructed in polynomial time in the size of $\mathfrak{A}$.

We claim that ϕ is true in $\mathfrak{A}$ if and only if ψ is true in $\mathfrak{B}$. Let V be the variables of ψ. Suppose that $f: V \to B$ satisfies all conjuncts of ψ in $\mathfrak{B}$. Hence, by construction of ψ, if ϕ has a conjunct $\theta = R(x_{i_1}, \dots, x_{i_k})$, then

$$\mathfrak{B} \models \theta_I((f(y_{i_1}^1), \dots, f(y_{i_1}^d)), \dots, (f(y_{i_k}^1), \dots, f(y_{i_k}^d))).$$

By the definition of interpretations, this implies that

$$\mathfrak{A} \models R(I(f(y_{i_1}^1), \dots, f(y_{i_1}^d)), \dots, I(f(y_{i_k}^1), \dots, f(y_{i_k}^d))).$$

Hence, the mapping $g: U \to A$ that sends x_i to $I(f(y_i^1), \dots, f(y_i^d))$ satisfies all conjuncts of ϕ in $\mathfrak{A}$.

Now, suppose that $f: U \to A$ satisfies all conjuncts of ϕ over $\mathfrak{A}$. Since h is a surjective mapping from dom(I) to A, there are $b_i^1, \dots, b_i^d \in B$ such that $I(b_i^1, \dots, b_i^d) = f(x_i)$ for all $i \in \{1, \dots, n\}$. We claim that the mapping $g: V \to B$ that maps y_i^j to b_i^j satisfies ψ in $\mathfrak{B}$. By construction, any constraint in ψ either comes from ψ_1 or from ψ_2. If it comes from ψ_1 then it must be of the form $\top_I(y_i^1, \dots, y_i^d)$, and is satisfied since $\top_I$ defines the pre-image of h. If the constraint comes from ψ_2, then it must be a conjunct of a formula $\theta_I(y_{i_1}^1, \dots, y_{i_1}^d, \dots, y_{i_k}^1, \dots, y_{i_k}^d)$ that was introduced for $\theta = R(x_{i_1}, \dots, x_{i_k})$ in ϕ. It therefore suffices to show that

$$\mathfrak{B} \models \theta_I(g(y_{i_1}^1), \dots, g(y_{i_1}^d), \dots, g(y_{i_k}^1), \dots, g(y_{i_k}^d)).$$

By assumption, $R(f(x_{i_1}), \dots, f(x_{i_k}))$ holds in $\mathfrak{A}$. By the choice of $b_1^1, \dots, b_n^d$, this shows that $R(I(b_{i_1}^1, \dots, b_{i_1}^d), \dots, I(b_{i_k}^1, \dots, b_{i_k}^d))$ holds in $\mathfrak{B}$. By the definition of interpretations, this is the case if and only if $\theta_I(b^1_{i_1}, \ldots, b^d_{i_1}, \ldots, b^1_{i_k}, \ldots, b^d_{i_k})$ holds in $\mathfrak{B}$, which is what we had to show. ⊣

Remark 3.1.5. As for first-order interpretations, primitive positive interpretations can be composed. Recall from Section 2.4 that if a τ_2-structure $\mathfrak{C}_2$ has a first-order interpretation I_1 in a τ_1-structure $\mathfrak{C}_1$, and the τ_3-structure $\mathfrak{C}_3$ has a first-order interpretation I_2 in $\mathfrak{C}_2$, then $\mathfrak{C}_3$ has a natural $(d_1 d_2)$-dimensional first-order interpretation $I_2 \circ I_1$ in $\mathfrak{C}_1$. Note that if I_1 and I_2 are primitive positive, then $I_2 \circ I_1$ is primitive positive, too. This comes from the observation that if ϕ is a primitive positive τ_2-formula, then the τ_1-formula ϕ_{I_1} is primitive positive, too. The analogous statement holds for existential positive definability as well, but not for existential definability: there are existential formulas ϕ and existential interpretations I_1 such that ϕ_{I_1} is no longer existential.

In many hardness proofs we use Theorem 3.1.4 in the following way.

COROLLARY 3.1.6. *Let $\mathfrak{B}$ be a relational structure such that K_3, $(\{0,1\}; 1\text{IN}3)$, or $(\{0,1\}; \text{NAE})$ has a primitive positive interpretation in $\mathfrak{B}$. Then $\mathfrak{B}$ has a finite-signature reduct with an NP-hard* CSP.

PROOF. The primitive positive formulas involved in the primitive positive interpretation can mention only finitely many relations from $\mathfrak{B}$. Let $\mathfrak{B}'$ be the reduct of $\mathfrak{B}$ that contains exactly those relations. Then the NP-hardness of CSP($\mathfrak{B}'$) follows via Theorem 3.1.4 from the NP-hardness of CSP(K_3) (see, e.g., [176]) and of CSP($\{0,1\}$; 1IN3) and CSP($\{0,1\}$; NAE) (see Section 1.2 and Example 1.2.2). ⊣

There are situations where Theorem 3.1.4 can be combined with Lemma 1.2.10 to prove hardness of CSPs, as described in the following.

PROPOSITION 3.1.7. *Let $\mathfrak{A}$ be a structure with finite relational signature, and let $\mathfrak{B}$ be a structure with elements $c_1, \ldots, c_k$ such that*

- *the orbit of $(c_1, \ldots, c_k)$ under* Aut($\mathfrak{B}$) *is primitively positively definable, and*
- *$\mathfrak{A}$ has a primitive positive interpretation in $(\mathfrak{B}, c_1, \ldots, c_k)$.*

Then there is a finite-signature reduct $\mathfrak{B}'$ of $\mathfrak{B}$ and a polynomial-time reduction from CSP($\mathfrak{A}$) *to* CSP($\mathfrak{B}'$).

PROOF. Let $\mathfrak{B}'$ be the finite-signature reduct of $\mathfrak{B}$ that contains exactly the relations mentioned in the primitive positive interpretation of $\mathfrak{A}$ in $\mathfrak{B}$ and the relations mentioned in a primitive positive definition ϕ of the orbit of $(c_1, \ldots, c_k)$ under Aut($\mathfrak{B}$). Since $\mathfrak{C} := (\mathfrak{B}', \{c_1\}, \ldots, \{c_k\})$ interprets $\mathfrak{A}$ primitively positively, there is a polynomial-time reduction from CSP($\mathfrak{A}$) to CSP($\mathfrak{C}$) by Theorem 3.1.4. Since ϕ is a primitive positive definition of the orbit of $(c_1, \ldots, c_k)$ over the signature of $\mathfrak{B}'$, we can apply Lemma 1.2.10 and obtain a polynomial-time reduction from CSP($\mathfrak{C}$) to CSP($\mathfrak{B}'$). Composing reductions, we conclude that there is a polynomial-time reduction from CSP($\mathfrak{A}$) to CSP($\mathfrak{B}'$). ⊣

We give a concrete example of this technique of proving hardness.

DEFINITION 3.1.8. Let T_3 be the ternary relation

$$T_3 := \{(x, y, z) \in \mathbb{Q}^3 \mid (x = y < z) \vee (x = z < y)\}.$$

PROPOSITION 3.1.9. *The structure* $(\{0, 1\}; \text{1IN3})$ *has a primitive positive interpretation in* $(\mathbb{Q}; T_3, 0, 1)$. *The problem* CSP$(\mathbb{Q}; T_3)$ *is NP-hard.*

PROOF. We give a 2-dimensional primitive positive interpretation I of the structure $(\{0, 1\}; \text{1IN3})$ in $(\mathbb{Q}; T_3, 0, 1)$. The domain formula $\top_I(x_1, x_2)$ is $T_3(0, x_1, x_2)$; the formula $\text{1IN3}_I(x_1, x_2, y_1, y_2, z_1, z_2)$ is

$$\exists u, v \left(T_3(0, u, z_1) \wedge T_3(u, v, y_1) \wedge T_3(v, 1, x_1)\right);$$

the formula $=_I(x_1, x_2, y_1, y_2)$ is $T_3(0, x_1, y_2)$. Let (b_1, b_2) be a pair of elements of $\mathfrak{B}$ that satisfies $\top_I$. Then exactly one of b_1, b_2 must have value 0, and the other element is strictly greater than 0. We define $I(b_1, b_2)$ to be 1 if $b_1 = 0$, and to be 0 otherwise.

To see that this is the intended interpretation, let $(x_1, x_2), (y_1, y_2), (z_1, z_2)$ be pairs that satisfy $\top_I$ in $\mathfrak{B}$ and suppose that

$$t := (I(x_1, x_2), I(y_1, y_2), I(z_1, z_2)) = (1, 0, 0) \in \text{1IN3}\,.$$

We have to verify that $(x_1, x_2, y_1, y_2, z_1, z_2)$ satisfies 1IN3_I in $\mathfrak{B}$. Since $I(x_1, x_2) = 1$, we have $x_1 = 0$, and similarly we get that $y_1, z_1 > 0$. We can then set u and v to 0 and satisfy $T_3(0, u, z_1)$, $T_3(u, v, y_1)$, and $T_3(v, 1, x_1)$. In the case that $t = (0, 1, 0)$ we set u to 0 and v to $\min(1, x_1)$ and again satisfy the three atomic formulas. Finally, if $t = (0, 0, 1) \in \text{1IN3}$ then $x_1, y_1 > 0$ and $z_1 = 0$ and we can set v to $\min(1, x_1)$ and u to $\min(v, y_1)$.

Conversely, suppose that the tuple $(x_1, x_2, y_1, y_2, z_1, z_2)$ satisfies 1IN3_I in $\mathfrak{B}$. Since $T_3(0, u, z_1)$, exactly one out of u and z_1 equals 0. If $u = 0$, then exactly one out of v and y_1 equals 0 because of $T_3(u, v, y_1)$. If $v = 0$ then $x_1 = 0$ because of $T_3(v, 1, x_1)$, and we get that

$$(I(x_1, x_2), I(y_1, y_2), I(z_1, z_2)) = (1, 0, 0) \in \text{1IN3}\,.$$

If $v > 0$ then $x_1 > 0$ because of $T_3(v, 1, x_1)$ and $(I(x_1, x_2), I(y_1, y_2), I(z_1, z_2)) = (0, 1, 0) \in \text{1IN3}$. If $u > 0$, then $y_1 > 0$ and $v > 0$ and consequently $x_1 > 0$, so

$$(I(x_1, x_2), I(y_1, y_2), I(z_1, z_2)) = (0, 0, 1) \in \text{1IN3}\,.$$

The orbit of the pair $(0, 1)$ under $\text{Aut}(\mathbb{Q}; T_3)$ is primitively positively definable in $(\mathbb{Q}; T_3)$ by the formula $T_3(x, x, y)$. Hence, the NP-hardness of CSP$(\mathbb{Q}; T_3)$ follows from the NP-hardness of CSP$(\{0, 1\}; \text{1IN3})$ via Proposition 3.1.7. ⊣

We give another concrete application. We have defined in Example 1.1.3 the relation Betw on $\mathbb{Z}$; we use the analogous definition for Betw over $\mathbb{Q}$, that is,

$$\text{Betw} := \{(x, y, z) \in \mathbb{Q}^3 \mid x < y < z \vee z < y < x\}.$$

PROPOSITION 3.1.10. *The structure* $(\{0, 1\}; \text{NAE})$ *has a primitive positive interpretation in* $(\mathbb{Q}; \text{Betw}, 0)$, *and* $\text{CSP}(\mathbb{Q}; \text{Betw})$ *is NP-hard.*

PROOF. Recall that the relation NAE is $\{0, 1\}^3 \setminus \{(0, 0, 0), (1, 1, 1)\}$. The dimension of our interpretation I is one, and the domain formula is $\exists z\colon \text{Betw}(x, 0, z)$, which is equivalent to $x \neq 0$. The formula $=_I (x_1, y_1)$ is

$$\exists z \big(\text{Betw}(x_1, 0, z) \wedge \text{Betw}(y_1, 0, z)\big).$$

Note that $=_I$ is over $(\mathbb{Q}; \text{Betw}, 0)$ equivalent to $(x_1 > 0 \Leftrightarrow y_1 > 0)$. Finally, the formula $\text{NAE}(x_1, y_1, z_1)_I$ is

$$\exists u \big(\text{Betw}(x_1, u, y_1) \wedge \text{Betw}(u, 0, z_1)\big).$$

The coordinate map sends positive points to 1, and all other points from $\mathbb{Q}$ to 0.

Since the orbit of 0 under $\text{Aut}(\mathbb{Q}; \text{Betw})$ is the entire set $\mathbb{Q}$ it is in particular primitively positively definable, and we can show the NP-hardness of $\text{CSP}(\mathbb{Q}; \text{Betw})$ using Proposition 3.1.7 and the fact that $\text{CSP}(\{0, 1\}; \text{NAE})$ is NP-hard. ⊣

More applications of Proposition 3.1.7 can be found in Section 12.2.

3.2. Interpreting all finite structures primitively positively

Some structures $\mathfrak{B}$ have the remarkable property that they can interpret all finite structures primitively positively. We have already seen that we can always find *first-order* interpretations of all finite structures in $\mathfrak{B}$ if $\mathfrak{B}$ has at least two elements (Lemma 2.4.4). This is relevant here because there are many structures where every first-order formula is equivalent to a primitive positive formula. An example for such a structure is K_n for $n \geq 3$, and the Boolean structures $(\{0, 1\}; \text{1IN3})$ and $(\{0, 1\}; \text{NAE})$; the proof has to wait until Section 6.1.8.

COROLLARY 3.2.1. K_3 *interprets all finite structures primitively positively.*

PROOF. In Proposition 6.1.43 we will see that for $n \geq 3$, every first-order formula is equivalent over K_n to a primitive positive formula. Therefore the statement follows from Lemma 2.4.4. ⊣

The class of structures that admit primitive positive interpretations of *all* finite structures can be characterised in many different ways. We write $\text{I}(\mathfrak{B})$ for the class of all structures with a primitive positive interpretation in $\mathfrak{B}$.

Theorem 3.2.2. *Let $\mathfrak{B}$ be any structure. Then the following are equivalent.*

1. $(\{0,1\}; 1\text{IN}3) \in \text{I}(\mathfrak{B})$.
2. $(\{0,1\}; \text{NAE}) \in \text{I}(\mathfrak{B})$.
3. $K_n \in \text{I}(\mathfrak{B})$, *for some* $n \geq 3$.
4. $\text{I}(\mathfrak{B})$ *contains a structure with at least two elements for which all first-order formulas are equivalent to primitive positive formulas.*
5. $\text{I}(\mathfrak{B})$ *contains all finite structures.*

If these equivalent conditions apply, then $\mathfrak{B}$ *has a finite-signature reduct whose* CSP *is NP-hard.*

Proof. Clearly, (5) implies (1), (2), and (3). We have given a primitive positive definition of NAE in $(\{0,1\}; 1\text{IN}3)$ in the proof of Theorem 6.2.7, which proves that (1) $\Rightarrow$ (2) by composing primitive positive interpretations (see Section 3.1). The implication (2) $\Rightarrow$ (4) is by Proposition 6.2.8. The implication (3) $\Rightarrow$ (4) is Proposition 6.1.43. (4) $\Rightarrow$ (5) is immediate from Lemma 2.4.4 and composing primitive positive interpretations (see Section 3.1).

For the final statement of the theorem, we use the fact that $K_3 \in \text{I}(\mathfrak{B})$; we consider the reduct $\mathfrak{B}'$ of $\mathfrak{B}$ that consists of all relations of $\mathfrak{B}$ that appear in the interpretation of K_3. Then the NP-hardness of CSP($\mathfrak{B}'$) follows from the NP-hardness of CSP(K_3) via Corollary 3.1.6. ⊣

The equivalent conditions in Theorem 3.2.2 capture many of the structures $\mathfrak{B}$ such that CSP($\mathfrak{B}$) is NP-hard, but not all. A concrete example is the structure $\mathfrak{B} = (\mathbb{Q}; T_3)$ from Proposition 3.1.9; in Example 9.6.3 we will see that the conditions from Theorem 3.2.2 fail for this structure. A finite core with an NP-hard CSP that cannot interpret all finite structures primitively positively will be discussed in Example 6.7.1.

3.3. Bi-interpretations

The notion of *primitive positive homotopy* (*pp-homotopy*) between two interpretations I_1, I_2 of $\mathfrak{C}$ in $\mathfrak{D}$ is defined analogously to homotopy in Section 2.4: we require that the relation $\{(\bar{x}, \bar{y}) \mid I_1(\bar{x}) = I_2(\bar{y})\}$ is primitively positively definable in $\mathfrak{D}$. Note that the identity interpretation is primitive positive.

Definition 3.3.1. Two structures $\mathfrak{C}$ and $\mathfrak{D}$ with a primitive positive interpretation I of $\mathfrak{C}$ in $\mathfrak{D}$ and a primitive positive interpretation J of $\mathfrak{D}$ in $\mathfrak{C}$ are called *mutually primitively positively interpretable*. If both $I \circ J$ and $J \circ I$ are pp-homotopic to the identity interpretation (of $\mathfrak{D}$ and of $\mathfrak{C}$, respectively), then we say that $\mathfrak{C}$ and $\mathfrak{D}$ are *primitively positively bi-interpretable*.

Example 3.3.2. The directed graph $\mathfrak{C} := (\mathbb{N}^2; M)$ where

$$M := \{((u_1, u_2), (v_1, v_2)) \mid u_2 = v_1\}$$

and the structure $\mathfrak{D} := (\mathbb{N}; =)$ are primitive positive bi-interpretable. The primitive positive interpretation I of $\mathfrak{C}$ in $\mathfrak{D}$ is 2-dimensional, the domain formula is true, and $I(u_1, u_2) = (u_1, u_2)$. The primitive positive interpretation J of $\mathfrak{D}$ in $\mathfrak{C}$ is 1-dimensional, the domain formula is true, and $J(x, y) = x$. Both interpretations are clearly primitive positive.

Then $J(I(x, y)) = z$ is definable by the formula $x = z$, and hence $I \circ J$ is pp-homotopic to the identity interpretation of $\mathfrak{D}$. Moreover, $I(J(u), J(v)) = w$ is primitively positively definable by

$$M(w, v) \wedge \exists p\, \big(M(p, u) \wedge M(p, w)\big),$$

so $J \circ I$ is also pp-homotopic to the identity interpretation of $\mathfrak{C}$.

Example 3.3.3. Let I be the 2-dimensional interpretation of $(\mathbb{I}; m)$ in $(\mathbb{Q}; <)$ with domain formula $x < y$, mapping $(x, y) \in \mathbb{Q}^2$ with $x < y$ to the interval $[x, y] \in \mathbb{I}$. The formula $(y_1 = y_2)_I$ is true, and the formula $(m(y_1, y_2))_I$ has variables $x_1^1, x_2^1, x_1^2, x_2^2$ and is given by $x_2^1 = x_1^2$.

Let J be the 1-dimensional interpretation with the domain formula true and $J([x, y]) := x$. The formula $(x < y)_I$ is the primitive positive formula

$$\exists u, v\, \big(m(u, x_1) \wedge m(u, v) \wedge m(v, x_2)\big).$$

We show that $J \circ I$ and $J \circ I$ are pp-homotopic to the identity interpretation. The relation $\big\{(x_1, x_2, y) \mid J(I(x_1, x_2)) = y\big\}$ has the primitive positive definition $x_1 = y$. To see that the relation $R := \big\{(u, v, w) \mid I(J(u), J(v)) = w\big\}$ has a primitive positive definition in $(\mathbb{I}; m)$, first note that the relation

$$\big\{(u, v) \mid u = [u_1, u_2], v = [v_1, v_2], u_1 = v_1\big\}$$

has the primitive positive definition $\phi_1(u, v) = \exists w\, \big(m(w, u_1) \wedge m(w, u_2)\big)$ in $(\mathbb{I}; m)$. Similarly, $\{(u, v) \mid u = [u_1, u_2], v = [v_1, v_2], u_2 = v_2\}$ has a primitive positive definition $\phi_2(u, v)$. Then the formula $\phi_1(u, w) \wedge \phi_2(v, w)$ is equivalent to a primitive positive formula over $(\mathbb{I}; m)$, and defines R.

Example 3.3.4. The structures $\mathfrak{C} := \big(\mathbb{N}^2; \{(x, y), (u, v) \mid x = u\}\big)$ and $\mathfrak{D} := (\mathbb{N}; =)$ are mutually primitive positive interpretable, but *not* primitive positive bi-interpretable. There is a primitive positive interpretation I_1 of $\mathfrak{D}$ in $\mathfrak{C}$, and a primitive positive interpretation of $\mathfrak{C}$ in $\mathfrak{D}$ such that $I_2 \circ I_1$ is pp-homotopic to the identity interpretation. However, the two structures are not even *first-order* bi-interpretable, as we will see in Example 9.5.27 in Section 9.5.3.

DEFINITION 3.3.5. A structure $\mathfrak{B}$ has *essentially infinite signature* if every relational structure $\mathfrak{C}$ that is primitively positively interdefinable with $\mathfrak{B}$ has an infinite signature.

Examples of finite and countably infinite ω-categorical structures with essentially infinite signature will be presented in Section 8.5.5. We show that the property of having essentially infinite signature is preserved by bi-interpretability.

PROPOSITION 3.3.6. *Let $\mathfrak{B}$ and $\mathfrak{C}$ be structures that are primitive positive bi-interpretable. Then $\mathfrak{B}$ has essentially infinite signature if and only if $\mathfrak{C}$ has essentially infinite signature.*

PROOF. Let τ be the signature of $\mathfrak{B}$. Suppose that the interpretation I_1 of $\mathfrak{C}$ in $\mathfrak{B}$ is d_1-dimensional, and that the interpretation I_2 of $\mathfrak{B}$ in $\mathfrak{C}$ is d_2-dimensional. Let $\theta(x, y_{1,1}, \dots, y_{d_1,d_2})$ be the τ-formula that shows that $I_2 \circ I_1$ is pp-homotopic to the identity interpretation of $\mathfrak{B}$. That is, θ defines in $\mathfrak{B}$ the $(d_1 d_2 + 1)$-ary relation that contains a tuple $(a, b_{1,1}, \dots, b_{d_1,d_2})$ iff

$$a = h_2(h_1(b_{1,1}, \dots, b_{1,d_2}), \dots, h_1(b_{d_1,1}, \dots, b_{d_1,d_2})).$$

We have to show that if $\mathfrak{C}$ has a finite signature, then $\mathfrak{B}$ is primitively positively interdefinable with a structure $\mathfrak{B}'$ with a finite signature. Let $\sigma \subseteq \tau$ be the set of all relation symbols that appear in θ and in all the formulas of the interpretation of $\mathfrak{C}$ in $\mathfrak{B}$. Since the signature of $\mathfrak{C}$ is finite, the cardinality of σ is finite as well. We will show that there is a primitive positive definition of $\mathfrak{B}$ in the σ-reduct $\mathfrak{B}'$ of $\mathfrak{B}$.

Let ϕ be an atomic τ-formula with k free variables $x_1, \dots, x_k$. Then the primitive positive σ-formula

$$\exists y^1_{1,1}, \dots, y^k_{d_1,d_2} \Big(\bigwedge_{i \le k} \theta(x_i, y^i_{1,1}, \dots, y^i_{d_1,d_2}) \\ \wedge \; \phi_{I_1 I_2}(y^1_{1,1}, \dots, y^k_{1,d_2}, y^1_{2,d_2}, \dots, y^k_{2,d_2}, \dots, y^k_{d_1,d_2}) \Big)$$

is equivalent to $\phi(x_1, \dots, x_k)$ over $\mathfrak{B}'$. Indeed, by the surjectivity of h_2, for every element a_i of $\mathfrak{B}$ there are elements $c^i_1, \dots, c^i_{d_2}$ of $\mathfrak{C}$ such that $h_2(c^i_1, \dots, c^i_{d_2}) = a_i$, and by the surjectivity of h_1, for every element c^i_j of $\mathfrak{C}$ there are elements $b^i_{1,j}, \dots, b^i_{d_1,j}$ of $\mathfrak{B}$ such that $h_1(b^i_{1,j}, \dots, b^i_{d_1,j}) = c^i_j$. Then

$$\begin{aligned} \mathfrak{B} \models R(a_1, \dots, a_k) &\Leftrightarrow \mathfrak{C} \models \phi_{I_2}(c^1_1, \dots, c^1_{d_2}, \dots, c^k_1, \dots, c^k_{d_2}) \\ &\Leftrightarrow \mathfrak{B}' \models \phi_{I_1 I_2}(b^1_{1,1}, \dots, b^k_{1,d_2}, b^1_{2,d_2}, \dots, b^k_{2,d_2}, \dots, b^k_{d_1,d_2}). \end{aligned}$$ ⊣

3.4. Classification transfer

Let $\mathfrak{C}$ be a structure with finite relational signature. By the *classification project for* $\mathfrak{C}$ we mean a complexity classification for CSP($\mathfrak{B}$) for all first-order expansions $\mathfrak{B}$ of $\mathfrak{C}$ with finite relational signature. For instance, the classification project for $(\mathbb{N}; =)$ is treated in Chapter 7, and the classification project for $(\mathbb{Q}; \le)$ is treated in Chapter 12.

Sometimes, it is possible to derive the classification project for $\mathfrak{C}$ from the classification project for $\mathfrak{D}$, for another structure $\mathfrak{D}$. For instance, we will show below how to derive the classification project for the directed graph

$$\mathfrak{C} := (\mathbb{N}^2; \{(x, y), (u, v) \mid y = u\})$$

from the classification project for $\mathfrak{D} := (\mathbb{N}; =)$; a more advanced application of such a classification transfer can be found in Theorem 3.4.3 below. Here is the central lemma for complexity classification transfer.

LEMMA 3.4.1. *Suppose $\mathfrak{D}$ has a primitive positive interpretation I in $\mathfrak{C}$, and $\mathfrak{C}$ has a primitive positive interpretation J in $\mathfrak{D}$ such that $J \circ I$ is pp-homotopic to the identity interpretation of $\mathfrak{C}$. Then for every first-order expansion $\mathfrak{C}'$ of $\mathfrak{C}$ there is a first-order expansion $\mathfrak{D}'$ of $\mathfrak{D}$ such that $\mathfrak{C}'$ and $\mathfrak{D}'$ are mutually pp-interpretable.*

PROOF. Let $\mathfrak{C}'$ be a first-order expansion of $\mathfrak{C}$. Let d be the dimension of J and c the dimension of I. Then we set $\mathfrak{D}'$ to be the expansion of $\mathfrak{D}$ that contains for every k-ary R in the signature of $\mathfrak{C}'$ the (dk)-ary relation S defined as follows. If ϕ is the first-order definition of R in $\mathfrak{C}$, then S is the relation defined by ϕ_J in $\mathfrak{D}$ (see Section 3.1).

We claim that J is also a primitive positive interpretation of $\mathfrak{C}'$ in $\mathfrak{D}'$. First note that $J^{-1}(C') = J^{-1}(C)$ is primitively positively definable in $\mathfrak{D}'$ since $J^{-1}(C)$ is primitively positively definable in $\mathfrak{D}$ and $\mathfrak{D}'$ is an expansion of $\mathfrak{D}$. An atomic formula ψ with free variables $x_1, \dots, x_k$ in the signature of $\mathfrak{C}'$ can be interpreted in $\mathfrak{D}'$ as follows. We replace the relation symbol in ψ by its definition in $\mathfrak{C}$, and obtain a formula ϕ in the signature of $\mathfrak{C}$. Let S be the symbol in the signature of $\mathfrak{D}'$ for the relation defined by $\phi_J(x_1^1, \dots, x_1^d, \dots, x_k^1, \dots, x_k^d)$ over $\mathfrak{D}'$. Then indeed $S(x_1^1, \dots, x_1^d, \dots, x_k^1, \dots, x_k^d)$ is a defining formula for ψ, because

$$\mathfrak{C}' \models \psi(J(a_1^1, \dots, a_1^d), \dots, J(a_k^1, \dots, a_k^d)) \Leftrightarrow \\ \mathfrak{D}' \models S(a_1^1, \dots, a_1^d, \dots, a_k^1, \dots, a_k^d)$$

for all $a_1, \dots, a_k \in J^{-1}(C)$.

Conversely, we claim that I is a primitive positive interpretation of $\mathfrak{D}'$ in $\mathfrak{C}'$. Again, $I^{-1}(D') = I^{-1}(D)$ is primitively positively definable in $\mathfrak{C}'$ since $\mathfrak{C}'$ is an expansion of $\mathfrak{C}$. Let ϕ be an atomic formula over the (relational) signature of $\mathfrak{D}'$. If the relation symbol in ϕ is already in the signature of $\mathfrak{D}$, then there is a primitive positive interpreting formula in $\mathfrak{C}$ and therefore also in $\mathfrak{C}'$. Otherwise, by definition of $\mathfrak{D}'$, the relation symbol in ϕ has arity dk for some $k \in \mathbb{N}$ and has been introduced for a k-ary relation R from $\mathfrak{C}'$. We have to find a defining formula with kcd variables. Let $\theta(x_0, x_{1,1}, \dots, x_{1,c}, \dots, x_{d,1}, \dots, x_{c,d})$ be the primitive positive formula of arity $cd + 1$ that defines $J\big(I(x_{1,1}, \dots, x_{c,1}), \dots, I(x_{1,d}, \dots, x_{c,d})\big) = x_0$ in $\mathfrak{C}$. Then the defining formula ϕ_I for the atomic formula $\phi(x_1^1, \dots, x_d^k)$ has free variables $x_{1,1}^1, \dots, x_{c,d}^k$ and equals

$$\exists x^1, \dots, x^k \left(R(x^1, \dots, x^k) \wedge \bigwedge_{i=1}^{k} \theta(x^i, x_{1,1}^i, \dots, x_{c,d}^i)\right).$$

We have to verify that $(a_{1,1}^1, \dots, a_{c,d}^k) \in C^{cdk}$ satisfies ϕ_I in $\mathfrak{C}'$ if and only if

$$(I(a_{1,1}^1, \dots, a_{c,1}^1), \dots, I(a_{1,d}^k, \dots, a_{c,d}^k))$$

satisfies ϕ in $\mathfrak{D}'$. Suppose that $\mathfrak{C} \models \phi_I(a_{1,1}^1, \dots, a_{c,d}^k) \in C^{cdk}$ and let for all $i \leq k$

$$a^i := J\big(I(a_{1,1}^i, \dots, a_{c,1}^i), \dots, I(a_{1,d}^i, \dots, a_{c,d}^i)\big).$$

Then $R(a^1, \dots, a^k)$. Since J is an interpretation this shows that

$$\mathfrak{D}' \models \phi(I(a_{1,1}^1, \dots, a_{c,1}^1), \dots, I(a_{1,d}^k, \dots, a_{c,d}^k)).$$

All implications in this final argument can be reversed, which concludes the proof. ⊣

In particular, when $\mathfrak{C}$, $\mathfrak{D}$, $\mathfrak{C}'$ and $\mathfrak{D}'$ are as in Lemma 3.4.1, and $\mathfrak{C}'$ and $\mathfrak{D}'$ have a finite relational signature, then CSP($\mathfrak{C}'$) and CSP($\mathfrak{D}'$) have the same computational complexity, by Theorem 3.1.4. Hence, Lemma 3.4.1 shows that the classification project for $\mathfrak{C}$ can be reduced to the classification project for $\mathfrak{D}$. With a slightly stronger assumption we can get the following consequence.

COROLLARY 3.4.2. *Let $\mathfrak{C}$ and $\mathfrak{D}$ be primitive positive bi-interpretable structures. Then every first-order expansion of $\mathfrak{C}$ is primitive positive bi-interpretable with a first-order expansion of $\mathfrak{D}$.*

Let us conclude with a concrete application of Corollary 3.4.2.

THEOREM 3.4.3. *Let $\mathfrak{B}$ be a reduct of Allen's interval algebra (Example 2.4.2) that contains the relation $m = \{((u_1, u_2), (v_1, v_2)) \mid u_2 = v_1\}$. Then* CSP($\mathfrak{B}$) *is either in P or NP-complete.*

PROOF. In Example 3.3.3 we have shown that the structure $(\mathbb{I}; m)$ is primitive positive bi-interpretable with $(\mathbb{Q}; <)$. The result follows from the main result of Chapter 12 and Corollary 3.4.2. ⊣

3.5. Binary signatures and the dual encoding

In this section we prove that every structure $\mathfrak{C}$ with a relational signature of maximal arity $m \in \mathbb{N}$ is primitively positively bi-interpretable with a *binary structure* $\mathfrak{B}$, i.e., a relational structure where every relation symbol has arity at most two. Moreover, if $\mathfrak{C}$ has a finite signature, then $\mathfrak{B}$ can be chosen to have a finite signature as well. It follows from Theorem 3.1.4 that every CSP is polynomial-time equivalent to a binary CSP. This transformation is known under the name *dual encoding*[148, 133]. We want to stress that the transformation works for relational structures with domains of arbitrary cardinality.

DEFINITION 3.5.1. Let $\mathfrak{C}$ be a structure and $d \in \mathbb{N}$. Then a *d-th full power of* $\mathfrak{C}$ is a structure $\mathfrak{D}$ with domain C^d such that the identity map on C^d is a full d-dimensional primitive positive interpretation of $\mathfrak{D}$ in $\mathfrak{C}$.

In particular, for all $i, j \in \{1, \dots, d\}$ the relation

$$E_{i,j} := \{((x_1, \dots, x_d), (y_1, \dots, y_d)) \mid x_1, \dots, x_d, y_1, \dots, y_d \in C \text{ and } x_i = y_j\}$$

is primitively positively definable in $\mathfrak{D}$.

PROPOSITION 3.5.2. *Let $\mathfrak{C}$ be a structure and $\mathfrak{D}$ a d-th full power of $\mathfrak{C}$ for $d \geq 1$. Then $\mathfrak{C}$ and $\mathfrak{D}$ are primitively positively bi-interpretable.*

PROOF. Let I be the identity map on C^d which is a full interpretation of $\mathfrak{D}$ in $\mathfrak{C}$. Our interpretation J of $\mathfrak{C}$ in $\mathfrak{D}$ is one-dimensional and the coordinate map is the first projection. The domain formula is *true* and the pre-image of the equality relation in $\mathfrak{C}$ under the coordinate map has the primitive positive definition $E_{1,1}(x, y)$. To define the pre-image of a k-ary relation R of $\mathfrak{C}$ under the coordinate map it suffices to observe that the k-ary relation

$$S := \{((a_{1,1}, \dots, a_{d,1}), \dots, (a_{1,k}, \dots, a_{d,k})) \mid (a_{1,1}, \dots, a_{1,k}) \in R\}$$

is primitively positively definable in $\mathfrak{D}$ and $J(S) = R$.

To show that $\mathfrak{C}$ and $\mathfrak{D}$ are primitively positive bi-interpretable we prove that $I \circ J$ and $J \circ I$ are pp-homotopic to the identity interpretation. Then the relation

$$\{(u_0, u_1, \dots, u_k) \mid u_0 = I(J(u_1), \dots, J(u_k)), u_1, \dots, u_k \in C^{k+1}\}$$

has the primitive positive definition $\bigwedge_{i \in \{1,\dots,k\}} E_{i,1}(u_0, u_i)$ and the relation

$$\{(v_0, v_1, \dots, v_k) \mid v_0 = J(I(v_1, \dots, v_k)), v_1, \dots, v_k \in D^{k+1}\}$$

has the primitive positive definition $v_0 = v_1$. ⊣

Note that for every relation R of arity $k \leq d$ of $\mathfrak{C}$, in a d-th full power $\mathfrak{D}$ of $\mathfrak{C}$ the unary relation

$$R' := \{(a_1, \dots, a_d) \mid (a_1, \dots, a_k) \in R\}$$

must be primitively positively definable. We now define a particular full power.

DEFINITION 3.5.3. Let $\mathfrak{C}$ be a relational structure with maximal arity m and let $d \geq m$. Then the structure $\mathfrak{B} := \mathfrak{C}^{[d]}$ with domain C^d is defined as follows:

- for every relation $R \subseteq C^k$ of $\mathfrak{C}$ the structure $\mathfrak{B}$ has the unary relation $R' \subseteq B = C^d$ defined above, and
- for all $i, j \in \{1, \dots, d\}$ the structure $\mathfrak{B}$ has the binary relation symbol $E_{i,j}$.

It is clear that the signature of $\mathfrak{B}$ is finite if the signature of $\mathfrak{C}$ is finite. Also note that the signature of $\mathfrak{C}^{[d]}$ is always binary.

LEMMA 3.5.4. *Let $\mathfrak{C}$ be a relational structure with maximal arity m and let $d \geq m$. Then the binary structure $\mathfrak{C}^{[d]}$ is a full power of $\mathfrak{C}$. If $\mathfrak{C}$ is finitely bounded, then $\mathfrak{C}^{[d]}$ is finitely bounded. Moreover, if $\mathrm{Age}(\mathfrak{C}) = \mathrm{Forb}^{\mathrm{emb}}(\mathcal{F})$ for a finite set of finite τ-structures $\mathcal{F}$, then we can compute from $\mathcal{F}$ a finite set of finite structures $\mathcal{F}'$ (in polynomial time in the representation size of $\mathcal{F}$) such that $\mathrm{Age}(\mathfrak{C}^{[d]}) = \mathrm{Forb}^{\mathrm{emb}}(\mathcal{F}')$.*

PROOF. The identity map is a d-dimensional primitive positive interpretation I of $\mathfrak{B} := \mathfrak{C}^{[d]}$ in $\mathfrak{C}$. Our interpretation J of $\mathfrak{C}$ in $\mathfrak{B}$ is one-dimensional and the coordinate map is the first projection. The domain formula is *true* and the pre-image of the equality relation in $\mathfrak{C}$ under the coordinate map has the primitive positive definition $E_{1,1}(x, y)$. The pre-image of the relation R of $\mathfrak{C}$ under the coordinate map is defined by the primitive positive formula

$$\exists y \Big(\bigwedge_{i \in \{1,\dots,k\}} E_{1,i}(x_i, y) \wedge R'(y) \Big).$$

The proof that $I \circ J$ and $J \circ I$ are pp-homotopic to the identity interpretation is as in the proof of Proposition 3.5.2.

Now suppose that $\mathfrak{C}$ is finitely bounded with signature τ, i.e., $\mathrm{Age}(\mathfrak{C}) = \mathrm{Forb}^{\mathrm{emb}}(\mathcal{F})$ for some finite set of finite τ-structures. For $\mathfrak{F} \in \mathcal{F}$, note that $\mathfrak{F}^{[d]}$ does not embed into $\mathfrak{C}^{[d]}$. Otherwise, suppose that e is such an embedding. Then the map $x \mapsto e(x, \dots, x)_1$ is an embedding of $\mathfrak{F}$ into $\mathfrak{C}$: if $(x_1, \dots, x_k) \in R^{\mathfrak{F}}$ for $R \in \tau$, let $R' \in \rho$ be the unary relation symbol introduced in $\mathfrak{F}^d$ for R. Pick any $x_{k+1}, \dots, x_d \in F$ and note that $(x_1, \dots, x_d) \in (R')^{\mathfrak{F}^{[d]}}$. Hence, $e(x_1, \dots, x_d) \in (R')^{\mathfrak{B}}$. Also note that for every $i \leq d$

$$\mathfrak{F}^{[d]} \models R_{1,i}\big((x_i, \dots, x_i), (x_1, \dots, x_d)\big)$$

and hence $\mathfrak{B} \models R_{1,i}\big(e(x_i, \dots, x_i), e(x_1, \dots, x_d)\big)$. By the definition of $\mathfrak{B}$, this implies that $(e(x_1, \dots, x_1)_1, \dots, e(x_d, \dots, x_d)_1) \in (R')^{\mathfrak{B}}$, and we conclude that

$$\big(e(x_1, \dots, x_1)_1, \dots, e(x_k, \dots, x_k)_1\big) \in R^{\mathfrak{C}}.$$

Let $\mathcal{F}'$ be the finite set of structures of the form $\mathfrak{F}^{[d]}$ for $\mathfrak{F} \in \mathcal{F}$ together with finitely many structures in the signature of $\mathfrak{B}$ (of size at most three) that ensure that for every structure $\mathfrak{A} \in \mathrm{Forb}^{\mathrm{emb}}(\mathcal{F}')$

1. $E_{i,i}(x, x)$ holds for all $i \leq d$ and $x \in A$;
2. $E_{i,j}(x, y)$ implies that $E_{j,i}(y, x)$ for all $i, j \leq d$ and $x, y \in A$;
3. $E_{i,j}(x, y)$ and $E_{j,k}(y, z)$ imply that $E_{i,k}(x, z)$ for all $i, j, k \leq d$ and $x, y, z \in A$.
4. $\bigwedge_{i \in \{1,\dots,d\}} E_{i,i}(x, y)$ implies that $x = y$.

In particular, every relation $E_{i,i}$ is an equivalence relation on A; if a is an element, then we write $[a]_i$ for the equivalence class of a with respect to the equivalence relation $E_{i,i}$. Note that every element $a \in A$ is uniquely determined by $[a]_1, \ldots, [a]_d$. Clearly, no structure in $\mathcal{F}'$ embeds into $\mathfrak{B}$. We claim that every structure $\mathfrak{A} \in \text{Forb}^{\text{emb}}(\mathcal{F}')$ embeds into $\mathfrak{B}$. Let $\mathfrak{A}'$ be the following τ-structure. The elements of A' are the equivalence classes of all the equivalence relations $R_{i,i}$ for $i \leq d$, where $[a]_i$ is identified with $[b]_j$ if $R_{i,j}(a,b)$ holds in $\mathfrak{A}$. We define $([a_1]_1, \ldots, [a_k]_k) \in R^{\mathfrak{A}'}$ if $(a_1, \ldots, a_k) \in R^{\mathfrak{A}}$. Note that $(a_1, \ldots, a_d) \mapsto ([a_1]_1, \ldots, [a_d]_d)$ is an embedding from $\mathfrak{A}$ into $(\mathfrak{A}')^{[d]}$.

The structure $\mathfrak{A}'$ embeds into $\mathfrak{C}$: suppose otherwise that there is an embedding f from $\mathfrak{F} \in \mathcal{F}$ into $\mathfrak{A}$. Then $(x_1, \ldots, x_d) \mapsto (f(x_1), \ldots, f(x_d))$ is an embedding from $\mathfrak{F}^{[d]}$ into $\mathfrak{A}$, a contradiction. Hence, $\mathfrak{A}' \hookrightarrow \mathfrak{C}$ and it follows that $\mathfrak{A} \hookrightarrow (\mathfrak{A}')^{[d]} \hookrightarrow \mathfrak{C}^{[d]}$ which concludes the proof. ⊣

COROLLARY 3.5.5. *For every structure $\mathfrak{C}$ with maximal arity m there exists a structure $\mathfrak{B}$ with maximal arity 2 such that $\mathfrak{B}$ and $\mathfrak{C}$ are primitively positively bi-interpretable. If the signature of $\mathfrak{C}$ is finite, then the signature of $\mathfrak{B}$ can be chosen to be finite as well.*

PROOF. An immediate consequence of Lemma 3.5.4 and Proposition 3.5.2. ⊣

In the case that the structure $\mathfrak{C}$ is ω-categorical then a universal-algebraic interpretation of this result can be found in Section 6.3.5. Note that if $\mathfrak{C}$ has a finite relational signature, then Corollary 3.5.5 implies that CSP($\mathfrak{C}$) is polynomial-time equivalent to CSP($\mathfrak{B}$) for some structure $\mathfrak{B}$ with a finite *binary* signature. A quite different reduction to a binary signature has been given by Feder and Vardi [170]. Their reduction even produces a constraint language with a *single* binary relation. However, this approach has only been described for finite domains.

3.6. Primitive positive constructions

In the previous sections we have seen several conditions on $\mathfrak{A}$ and $\mathfrak{B}$ that imply that CSP($\mathfrak{A}$) reduces to CSP($\mathfrak{B}$); in this section we compare them. We will do this by introducing operators on classes of structures. For finite or countably infinite ω-categorical structures, these operators have algebraic counterparts that will be introduced in Section 6.4. Let C be a class of structures. We write

1. H(C) for the class of structures that are homomorphically equivalent to structures in C;
2. C(C) for the class of all structures obtained by expanding a structure $\mathfrak{B} \in C$ with a singleton relation $\{b\}$, for some $b \in B^n$ and $n \in \mathbb{N}$, if the orbit of b under Aut($\mathfrak{B}$) is primitively positively definable in $\mathfrak{B}$;

3. $\mathrm{P}^{\mathrm{fin}}_{\mathrm{full}}(\mathcal{C})$ for the class of all structures that are a full finite power of a structure in $\mathcal{C}$ (Definition 3.5.1);
4. $\mathrm{Red}(\mathcal{C})$ for the class of all *primitive positive reducts* $\mathfrak{A}$ of structures $\mathfrak{B}$ in $\mathcal{C}$, i.e., $\mathfrak{A}$ has the same domain as $\mathfrak{B}$ and all relations of $\mathfrak{A}$ are primitively positively definable in $\mathfrak{B}$;
5. $\mathrm{I}(\mathcal{C})$ for the class of structures with a primitive positive interpretation in a structure from $\mathcal{C}$.

We also write $\mathrm{H}(\mathfrak{B})$ instead of $\mathrm{H}(\{\mathfrak{B}\})$, and similarly for the operators C, $\mathrm{P}^{\mathrm{fin}}_{\mathrm{full}}$, Red, and I. Brackets can be omitted when composing those operators. Clearly, $\mathrm{Red}(\mathrm{P}^{\mathrm{fin}}_{\mathrm{full}}(\mathcal{C})) = \mathrm{I}(\mathcal{C})$ and $\mathrm{I}(\mathrm{I}(\mathcal{C})) = \mathrm{I}(\mathcal{C})$ (see Remark 3.1.5).

LEMMA 3.6.1. *For any class of structures we have* $\mathrm{C}(\mathrm{C}(\mathcal{C})) = \mathrm{C}(\mathcal{C})$.

PROOF. Let $\mathfrak{B} \in \mathcal{C}$ and $b \in B^n$ be such that the orbit of b under $\mathrm{Aut}(\mathfrak{B})$ has the primitive positive definition $\phi(\bar{x})$ in $\mathfrak{B}$. Let $c \in B^m$ be such that the orbit of c under $\mathrm{Aut}(\mathfrak{B}, b)$ has the primitive positive definition $\psi(\bar{y})$ in $(\mathfrak{B}, b)$. Write $\psi(\bar{y})$ as $\psi'(b, \bar{y})$ where ψ' is now a formula over the signature of $\mathfrak{B}$. We claim that $\chi(\bar{x}, \bar{y}) := \psi(\bar{x}, \bar{y}) \wedge \phi(\bar{x})$ is a primitive positive definition of the orbit of (b, c) under $\mathrm{Aut}(\mathfrak{B})$. Clearly, $\mathfrak{B} \models \chi(b, c)$. Conversely, suppose that $\mathfrak{B} \models (b', c')$. Then there exists a $\beta \in \mathrm{Aut}(\mathfrak{B})$ such that $\beta(b') = b$. Hence, $\mathfrak{B} \models \chi(b, \beta(c'))$ and in particular $\mathfrak{B} \models \psi'(b, \beta(c'))$. Since $\psi(\bar{y})$ defines the orbit of c under $\mathrm{Aut}(\mathfrak{B}, b)$ there exists a $\gamma \in \mathrm{Aut}(\mathfrak{B}, b)$ such that $\gamma(b, \beta(c')) = (b, c)$. Hence, $\gamma \circ \beta(b', c') = \gamma(b, \beta(c')) = (b, c)$. It follows that $(\mathfrak{B}, b, c) \in \mathrm{C}(\mathfrak{B})$. $\dashv$

Barto, Opršal, and Pinsker [27] showed that if there is an arbitrary chain of applications of the operators H, C, and I to derive $\mathfrak{A}$ from $\mathfrak{B}$, then there is also a three-step chain to derive $\mathfrak{A}$ from $\mathfrak{B}$, namely by finding $\mathfrak{A}$ in $\mathrm{H}\,\mathrm{Red}\,\mathrm{P}^{\mathrm{fin}}_{\mathrm{full}}(\mathfrak{B})$. If $\mathfrak{A} \in \mathrm{H}\,\mathrm{Red}\,\mathrm{P}^{\mathrm{fin}}_{\mathrm{full}}(\mathfrak{B})$ then $\mathfrak{A}$ is also called *pp-constructible in* $\mathfrak{B}$ (see [27]).

THEOREM 3.6.2 (Barto, Opršal, and Pinsker [27]). *Let $\mathcal{C}$ be a class of finite structures, and let $\mathcal{D}$ be the smallest class that contains $\mathcal{C}$ and is closed under* H, C, *and* I. *Then*

$$\mathcal{D} = \mathrm{H}\,\mathrm{Red}\,\mathrm{P}^{\mathrm{fin}}_{\mathrm{full}}(\mathcal{C}) = \mathrm{HI}(\mathcal{C}).$$

This insight is conceptually important for the CSP since it leads to a better understanding of the power of the available tools. We split the proof into several propositions.

PROPOSITION 3.6.3. *Let $\mathfrak{B}$ be a τ-structure and $c \in B$ be such that the orbit of c under* $\mathrm{Aut}(\mathfrak{B})$ *is primitively positively definable in* $\mathfrak{B}$. *Then*

$$\mathfrak{C} := (\mathfrak{B}, \{c\}) \in \mathrm{HI}(\mathfrak{B}).$$

PROOF. Let O be the orbit of c under $\mathrm{Aut}(\mathfrak{B})$ and let ϕ be a primitive positive definition of O in $\mathfrak{B}$. We give a 2-dimensional primitive positive interpretation I of a structure $\mathfrak{A}$ with the same signature as $\mathfrak{C}$ and with domain $B \times O$. The domain formula $\top_I(x_1, x_2)$ is $\phi(x_2)$ and the coordinate map is the identity on

$B \times O$. If $R \in \tau$ has arity k then

$$R^{\mathfrak{A}} := \{((a_1, b_1), \dots, (a_k, b_k)) \in A^k \mid (a_1, \dots, a_k) \in R^{\mathfrak{B}} \text{ and } b_1 = \dots = b_k \in O\}.$$

If S is the symbol for the relation $\{c\}$ of $\mathfrak{C}$, then we define $S^{\mathfrak{A}} := \{(a, a) \mid a \in O\}$.

We claim that $\mathfrak{A}$ and $\mathfrak{C}$ are homomorphically equivalent. The homomorphism from $\mathfrak{C}$ to $\mathfrak{A}$ is given by $a \mapsto (a, c)$:

- if $(a_1, \dots, a_k) \in R^{\mathfrak{C}} = R^{\mathfrak{B}}$ then $((a_1, c), \dots, (a_k, c)) \in R^{\mathfrak{A}}$;
- the relation $S^{\mathfrak{C}} = \{c\}$ is preserved since $(c, c) \in S^{\mathfrak{A}}$.

To define a homomorphism h from $\mathfrak{A}$ to $\mathfrak{C}$ we pick for each $a \in O$ an automorphism $\alpha_a \in \text{Aut}(\mathfrak{B})$ such that $\alpha_a(a) = c$. Note that $b \in O$ since $\mathfrak{B} \models \top_I(a, b)$, and we define $h(a, b) := \alpha_b(a)$. To check that this is indeed a homomorphism, let $R \in \tau$ be k-ary, and let $t = ((a_1, b_1), \dots, (a_k, b_k)) \in R^{\mathfrak{A}}$. Then $b_1 = \dots = b_k =: b \in O$ and we have that $h(t) = (\alpha_b(a_1), \dots, \alpha_b(a_k))$ is in $R^{\mathfrak{C}}$ since $(a_1, \dots, a_k) \in R^{\mathfrak{B}} = R^{\mathfrak{C}}$ and α_b preserves $R^{\mathfrak{B}} = R^{\mathfrak{C}}$. If $S^{\mathfrak{A}} = \{(a, a) \mid a \in O\}$, then S is preserved as well, because $h((a, a)) = \alpha_a(a) = c \in \{c\} = S^{\mathfrak{C}}$. ⊣

PROPOSITION 3.6.4. *For every structure* $\mathfrak{B}$ *we have* $\text{I}(\mathfrak{B}) \subseteq \text{H}\,\text{Red}\,\text{P}^{\text{fin}}_{\text{full}}(\mathfrak{B})$.

PROOF. Let $\mathfrak{C}$ be a τ-structure with a d-dimensional primitive positive interpretation I in $\mathfrak{B}$. Let $\mathfrak{D}$ be a d-th full power of $\mathfrak{B}$. Let $\mathfrak{D}'$ be the τ-structure where $R \in \tau$ of arity k denotes the relation

$$\{((a_{1,1}, \dots, a_{1,d}), \dots, (a_{k,1}, \dots, a_{k,d})) \mid (I(a_{1,1}, \dots, a_{1,d}), \dots, I(a_{k,1}, \dots, a_{k,d})) \in R^{\mathfrak{C}}\}.$$

We have $\mathfrak{D}' \in \text{Red}(\mathfrak{D})$ by the definition of full powers. We prove that $\mathfrak{D}'$ is homomorphically equivalent to $\mathfrak{C}$. Let f be any map from D to C extending I. Let g be any mapping from C to D such that $f \circ g$ is the identity on C. Then f and g witness that $\mathfrak{D}'$ and $\mathfrak{C}$ are homomorphically equivalent. ⊣

PROPOSITION 3.6.5. *For any class of structures* $\mathcal{C}$ *we have*

$$\text{H}(\text{H}(\mathcal{C})) \subseteq \text{H}(\mathcal{C}) \tag{1}$$
$$\text{Red}(\text{Red}(\mathcal{C})) \subseteq \text{Red}(\mathcal{C}) \tag{2}$$
$$\text{P}^{\text{fin}}_{\text{full}}(\text{Red}(\mathcal{C})) \subseteq \text{Red}(\text{P}^{\text{fin}}_{\text{full}}(\mathcal{C})) \tag{3}$$
$$\text{H}(\text{Red}(\text{H}(\text{Red}(\mathcal{C})))) \subseteq \text{H}(\text{Red}(\mathcal{C})) \tag{4}$$
$$\text{P}^{\text{fin}}_{\text{full}}(\text{H}(\mathcal{C})) \subseteq \text{H}(\text{Red}(\text{P}^{\text{fin}}_{\text{full}}(\mathcal{C}))) \tag{5}$$
$$\text{P}^{\text{fin}}_{\text{full}}(\text{P}^{\text{fin}}_{\text{full}}(\mathcal{C})) \subseteq \text{P}^{\text{fin}}_{\text{full}}(\mathcal{C}) \tag{6}$$

where (6) *holds if we consider structures up to isomorphism.*

PROOF. The proof of (1) follows from the transitivity of homomorphic equivalence. Also the other statements are simple consequences of the definitions; we only verify that $\mathrm{P}^{\mathrm{fin}}_{\mathrm{full}}(\mathrm{H}(\mathcal{C})) \subseteq \mathrm{H}(\mathrm{Red}(\mathrm{P}^{\mathrm{fin}}_{\mathrm{full}}(\mathcal{C})))$. Let $\mathfrak{B} \in \mathcal{C}$ and $\mathfrak{C}$ be a structure such that there is a homomorphism f from $\mathfrak{B}$ to $\mathfrak{C}$ and a homomorphism g from $\mathfrak{C}$ to $\mathfrak{B}$. Let $\mathfrak{D}$ be a d-th full power of $\mathfrak{C}$. If R is a k-ary relation of $\mathfrak{D}$, then there is a primitive positive formula ϕ_R with dk free variables that defines R over $\mathfrak{C}$. Let $\mathfrak{D}'$ be the reduct of a d-th full power of $\mathfrak{B}$ where we keep for each relation R of $\mathfrak{D}$ the relation defined by ϕ_R over $\mathfrak{B}$ (rather than over $\mathfrak{C}$). Then the map $(a_1, \dots, a_d) \mapsto (f(a_1), \dots, g(a_d))$ is a homomorphism from $\mathfrak{D}'$ to $\mathfrak{D}$ and the map $(a_1, \dots, a_d) \mapsto (g(a_1), \dots, g(a_d))$ is a homomorphism from $\mathfrak{D}$ to $\mathfrak{D}'$. ⊣

We can finally prove Theorem 3.6.2.

PROOF. We have to show that $\mathrm{H}\,\mathrm{Red}\,\mathrm{P}^{\mathrm{fin}}_{\mathrm{full}}(\mathcal{C})$ is closed under H, C, and I. For H this follows from (1) above. For closure under interpretations, we have

$$\begin{aligned}
\mathrm{I}(\mathrm{H}\,\mathrm{Red}\,\mathrm{P}^{\mathrm{fin}}_{\mathrm{full}}(\mathcal{C})) &\subseteq \mathrm{H}\,\mathrm{Red}\,\mathrm{P}^{\mathrm{fin}}_{\mathrm{full}}(\mathrm{H}\,\mathrm{Red}\,\mathrm{P}^{\mathrm{fin}}_{\mathrm{full}}(\mathcal{C}))) && \text{(Proposition 3.6.4)}\\
&\subseteq \mathrm{H}\,\mathrm{Red}\,\mathrm{H}\,\mathrm{Red}\,\mathrm{P}^{\mathrm{fin}}_{\mathrm{full}}\,\mathrm{Red}\,\mathrm{P}^{\mathrm{fin}}_{\mathrm{full}}(\mathcal{C}))) && \text{(Proposition 3.6.5, (5))}\\
&\subseteq \mathrm{H}\,\mathrm{Red}\,\mathrm{H}\,\mathrm{Red}\,\mathrm{P}^{\mathrm{fin}}_{\mathrm{full}}\,\mathrm{P}^{\mathrm{fin}}_{\mathrm{full}}(\mathcal{C}))) && \text{(Prop. 3.6.5, (3,2))}\\
&\subseteq \mathrm{H}\,\mathrm{Red}\,\mathrm{P}^{\mathrm{fin}}_{\mathrm{full}}(\mathcal{C}) && \text{(Prop. 3.6.5, (4,6))}.
\end{aligned}$$

For closure under C, Proposition 3.6.3 implies that

$$\mathrm{C}(\mathrm{H}\,\mathrm{Red}\,\mathrm{P}^{\mathrm{fin}}_{\mathrm{full}}(\mathcal{C})) \subseteq \mathrm{HI}(\mathrm{H}\,\mathrm{Red}\,\mathrm{P}^{\mathrm{fin}}_{\mathrm{full}}(\mathcal{C}))$$

which equals $\mathrm{H}\,\mathrm{Red}\,\mathrm{P}^{\mathrm{fin}}_{\mathrm{full}}(\mathcal{C})$ by the above. ⊣

There are finite core structures $\mathfrak{A}, \mathfrak{B}$ such that $\mathfrak{A} \in \mathrm{HI}(\mathfrak{B}) \setminus \mathrm{H}(\mathfrak{B})$. The following example is taken from [27].

Example 3.6.6. Let $\mathfrak{B}$ be the structure with the domain $(\mathbb{Z}_2)^2$ and the signature $\{R_a \mid a \in (\mathbb{Z}_2)^2\}$ such that

$$R_a^{\mathfrak{B}} := \{(x, y, z) \in ((\mathbb{Z}_2)^2)^3 \mid x + y + z = a\}.$$

Note that $\mathfrak{B}$ is a core. Let $\mathfrak{B}'$ be the reduct of $\mathfrak{B}$ with the signature $\tau := \{R_{(0,0)}, R_{(1,0)}\}$. Let $\mathfrak{A}$ be the τ-structure with domain $\mathbb{Z}_2$ such that for $b \in \{0, 1\}$

$$R_{(b,0)}^{\mathfrak{A}} := \{(x, y, z) \in (\mathbb{Z}_2)^3 \mid x + y + z = b\}.$$

Now observe that $(x_1, x_2) \mapsto x_1$ is a homomorphism from $\mathfrak{B}'$ to $\mathfrak{A}$, and $x \mapsto (x, 0)$ is a homomorphism from $\mathfrak{A}$ to $\mathfrak{B}'$. Therefore $\mathfrak{A} \in \mathrm{H}(\mathfrak{B}')$. Moreover, $\mathfrak{B}' \in \mathrm{Red}(\mathfrak{B})$ and consequently $\mathfrak{A} \in \mathrm{H}\,\mathrm{Red}(\mathfrak{B})$. We finally show that $\mathfrak{A} \notin \mathrm{I}(\mathfrak{B})$. Suppose for contradiction that there is a pp-interpretation of $\mathfrak{A}$ in $\mathfrak{B}$ with coordinate map $I\colon C \to A$ such that $C \subseteq B^n$ is primitively positively definable in $\mathfrak{B}$. The kernel K of I has a primitive positive definition ϕ in $\mathfrak{B}$. The two equivalence classes of K are pp-definable relations over $\mathfrak{B}$ as well: the formula $\exists x(\phi(x, y) \wedge R_a(x))$ defines the equivalence class of a. But the relations with a primitive positive definition in $\mathfrak{B}$ are precisely the affine linear subspaces of the vector space $(\mathbb{Z}_2)^2$, so their cardinality must be a power of 4. And two powers of 4 cannot add up to a power of 4.

3.7. The tractability conjecture

The previous section suggests that the NP-hardness condition for CSPs from Theorem 3.2.2 can be generalised.

COROLLARY 3.7.1. *Let $\mathfrak{B}$ be a relational structure such that $K_3 \in \mathrm{HI}(\mathfrak{B})$. Then $\mathfrak{B}$ has a finite-signature reduct whose CSP is NP-hard.*

PROOF. Follows from the NP-hardness of CSP(K_3), the fact that homomorphic equivalence preserves the CSP, and Theorem 3.1.4. ⊣

For finite templates, the negation of the condition in the previous result ensures polynomial-time tractability of the CSP. This result has been obtained by Bulatov [111] and, independently, Zhuk [348], providing a solution to the Feder-Vardi dichotomy conjecture.

THEOREM 3.7.2 (Finite-domain tractability [111, 348]). *Let $\mathfrak{B}$ be a finite structure with a finite signature. If $K_3 \notin \mathrm{HI}(\mathfrak{B})$ then* CSP($\mathfrak{B}$) *is in P.*

Also all the known reducts $\mathfrak{B}$ of finitely bounded homogeneous structures such that CSP($\mathfrak{B}$) is NP-hard satisfy $K_3 \in \mathrm{HI}(\mathfrak{B})$. The infinite-domain tractability conjecture (Conjecture 3.1 below) generalises Theorem 3.7.2 and states that otherwise CSP($\mathfrak{B}$) is in P.

CONJECTURE 3.1 (Infinite-domain tractability conjecture). *Let $\mathfrak{B}$ be a reduct of a finitely bounded homogeneous structure. If $K_3 \notin \mathrm{HI}(\mathfrak{B})$ then* CSP($\mathfrak{B}$) *is in P.*

The infinite-domain tractability conjecture was originally stated in a different form [95] (see Conjecture 4.1) which is closer to the condition that has been used in the tractability conjecture as formulated in [116] and proved by Bulatov and Zhuk. For finite structures $\mathfrak{B}$, and more generally for reducts $\mathfrak{B}$ of finitely bounded homogeneous structures (but not for general ω-categorical structures), the conditions of Conjecture 3.1 and Conjecture 4.1 are equivalent. The equivalence of the two conjectures will be shown in Section 6.6 for finite structures and in Section 10.3 for reducts of homogeneous structures with finite relational signature.

The conjecture has been verified for many finitely bounded homogeneous structures, e.g., for $(\mathbb{Q}; <)$ (see Chapter 12), for the countable homogeneous universal poset [241], for the structure $(\mathbb{L}; |)$ for phylogenetic reconstruction from Section 5.1, for all unary structures [81], and for all homogeneous graphs [80].

Chapter 4

COUNTABLY CATEGORICAL STRUCTURES

Many important infinite-domain constraint satisfaction problems can be formulated with templates that are *ω-categorical* (the adjectives 'ω-categorical' and '$\aleph_0$-categorical' are used interchangeably). The concept of ω-categoricity is central in model theory, and for reasons that will become clear in Section 4.2, also in permutation group theory. From a model-theoretic perspective, ω-categoricity is a very strong assumption—but still many problems that have been studied in the literature, in particular constraint satisfaction problems for *qualitative* reasoning formalisms in artificial intelligence[15], can be formulated as CSPs with ω-categorical templates. We will also see that every CSP that can be expressed by a first-order sentence describes a constraint satisfaction

[15]The question which reasoning formalisms in Artificial Intelligence should be called *qualitative* has been the topic of scientific discussion [264]. My own response to this question is: it is qualitative if and only if it can be formulated with an ω-categorical template.

problem of an ω-categorical structure (Section 5.6.1). This even holds for the larger class of CSPs that can be expressed in monadic SNP (Section 5.6.2). The corresponding computational problems have also been called called *forbidden patterns problems* and studied in [275, 273, 276]); we will see that each of these problems is a constraint satisfaction problem for an ω-categorical structure.

In this chapter we present fundamental results about ω-categorical structures: for example how to algebraically characterize syntactically restricted forms of definability of relations over ω-categorical structures (Sections 4.2, 4.4, and 4.5), and how to construct ω-categorical structures (Sections 4.3, 4.7.2, and 4.6). We also give an exact model-theoretic characterisation of those constraint satisfaction problems that can be formulated with ω-categorical templates (Section 4.6).

There is already an excellent literature on ω-categoricty: most notably, the book by Cameron [123], the recent survey by Macpherson [270], and the collection [226]. Moreover, classical textbooks on model theory, such as [337, 207, 278, 231], always treat ω-categority, and use ω-categorical structures as a rich source of examples. The present chapter is different from those in that it focusses on techniques and facts that will be relevant for complexity classification for the corresponding constraint satisfaction problems. It contains many results that are not contained in any of the sources mentioned above (and which have been published in [61, 85, 196, 69, 68, 90, 89, 67]).

4.1. Introducing countable categoricity

"Every statement about all ω-categorical structures is either trivial, or false."
Martin Ziegler, 2005

A satisfiable first-order theory T is called *ω-categorical* (or *$\aleph_0$-categorical*) if all countable models of T are isomorphic. A structure is called *ω-categorical* if its first-order theory is ω-categorical. Note that the theory of a finite structure does not have countable models, and hence is ω-categorical.

Cantor [125] showed that the linear order of the rational numbers $(\mathbb{Q}; <)$ is ω-categorical; we use this as a running example in this section. We will see many more examples of ω-categorical structures in this section and in Chapter 5. One of the standard approaches to verify that a structure is ω-categorical is via a so-called *back-and-forth argument* [207, 306]. Such an argument already appeared in the proof of Lemma 2.2.2, but we would like to illustrate this technique again in a concrete situation, for proving the ω-categoricity of the structure $(\mathbb{Q}; <)$.

PROPOSITION 4.1.1. *The structure $(\mathbb{Q}; <)$ is ω-categorical.*

PROOF. Let $\mathfrak{A}$ be a countable model of the first-order theory T of $(\mathbb{Q}; <)$. It is easy to verify that T contains (and, as this argument will show, is uniquely given by)

- $\exists x: x = x$ (no empty model)
- $\forall x, y, z\ \big((x < y \land y < z) \Rightarrow x < z\big)$ (transitivity)
- $\forall x: \neg(x < x)$ (irreflexivity)
- $\forall x, y\ (x < y \lor y < x \lor x = y)$ (totality)
- $\forall x\, \exists y: x < y$ (no largest element)
- $\forall x\, \exists y: y < x$ (no smallest element)
- $\forall x, z\, \exists y\ (x < y \land y < z)$ (density).

An isomorphism between $\mathfrak{A}$ and $(\mathbb{Q}; <)$ can be defined inductively as follows. Suppose that we have already defined f on a finite subset S of $\mathbb{Q}$ and that f is an embedding of the substructure of $(\mathbb{Q}; <)$ induced on S into $\mathfrak{A}$. Since $<^{\mathfrak{A}}$ is dense and unbounded, we can extend f to any other element of $\mathbb{Q}$ such that the extension is still an embedding from a substructure of $\mathbb{Q}$ into $\mathfrak{A}$ (*going forth*). Symmetrically, for every element v of $\mathfrak{A}$ we can find an element $u \in \mathbb{Q}$ such that the extension of f that maps u to v is also an embedding (*going back*). We now alternate between going forth and going back; when going forth, we extend the domain of f by the *next* element of $\mathbb{Q}$, according to some fixed enumeration of the elements in $\mathbb{Q}$. When going back, we extend f such that the image of A contains the *next* element of $\mathfrak{A}$, according to some fixed enumeration of the elements of $\mathfrak{A}$. If we continue in this way, we have defined the value of f on all elements of $\mathbb{Q}$. Moreover, f will be surjective, and an embedding, and hence an isomorphism between $\mathfrak{A}$ and $(\mathbb{Q}; <)$. ⊣

Remark 4.1.2. The proof in fact shows that $(\mathbb{Q}; <)$ is homogeneous, since taking $\mathfrak{A} = (\mathbb{Q}; <)$ we extend an arbitrary embedding from a finite substructure of $(\mathbb{Q}; <)$ into $(\mathbb{Q}; <)$ to an automorphism of $(\mathbb{Q}; <)$.

A second important running example of this section is the *countable random graph* $(\mathbb{V}; E)$, introduced in Section 2.3, Example 2.3.9. This (simple and undirected) graph has the following *extension property*: for all finite disjoint subsets U, U' of $\mathbb{V}$ there exists a vertex $v \in \mathbb{V} \setminus (U \cup U')$ such that v is adjacent to all vertices in U and to no vertex in U'.

PROPOSITION 4.1.3. *The random graph* $(\mathbb{V}; E)$ *is* ω*-categorical.*

PROOF. Note that the extension property of $(\mathbb{V}; E)$ given above is a first-order property; a simple back-and-forth argument shows that every countably infinite graph with this property is isomorphic to $(\mathbb{V}; E)$. ⊣

Another important ω-categorical structure that turns out to be useful in many different contexts in this text (see Section 5.3, Example 5.7.2, Example 6.7.5, Example 7.3.5) is introduced in the following example.

Example 4.1.4. An *atom* in a Boolean algebra $(B; \wedge, \vee, -, 0, 1)$ is an element $x \neq 0$ such that for all $y \in B$ with $(x \wedge y) = y$ and $x \neq y$ we have $y = 0$. If a Boolean algebra does not contain atoms, it is called *atomless*. Note that the axioms of Boolean algebras and the property of not having atoms can be written as first-order sentences. Let T be the first-order theory of all atomless Boolean algebras. Clearly, atomless Boolean algebras exist, so T is satisfiable, and T also has a countable model by Theorem 2.1.11. It is well known and easy to show by a back-and-forth argument that all countable atomless Boolean algebras are isomorphic (Corollary 5.16 in [242]; also see Example 4 on page 100 in [206]). Hence, T is ω-categorical, and we refer to its countable model as *the* atomless Boolean algebra. We also mention that the structure $\mathfrak{A}$ has quantifier elimination (see Exercise 17 on Page 391 in [206]).

4.1.1. The theorem of Engeler, Svenonius, and Ryll-Nardzewski. There are many equivalent characterisations of ω-categoricity, the most important one being in terms of the automorphism group of $\mathfrak{B}$. In the following, let $\mathcal{G}$ be a set of permutations of a set X. We say that $\mathcal{G}$ is a *permutation group* if $\mathcal{G}$ contains the identity id_X, and for $\alpha, \beta \in \mathcal{G}$ the functions $x \mapsto \alpha\beta x$ and $x \mapsto \alpha^{-1}x$ are also in $\mathcal{G}$. For $n \geq 1$ the *orbit* of $t = (t_1, \ldots, t_n) \in X^n$ under $\mathcal{G}$ is the set $\{(\alpha t_1, \ldots, \alpha t_n) \mid \alpha \in \mathcal{G}\}$, and is sometimes denoted by $\mathcal{G}t$. An *orbit* of $\mathcal{G}$ refers to an orbit for a tuple of length $n = 1$. An *orbital* of $\mathcal{G}$ is an orbit of pairs, that is, a set of the form $\{(\alpha a, \alpha b) \mid \alpha \in \mathcal{G}\}$ for $a, b \in B$.

DEFINITION 4.1.5. A permutation group $\mathcal{G}$ on a countably infinite set X is *oligomorphic* if $\mathcal{G}$ has only finitely many orbits of n-tuples for each $n \geq 1$.

The following theorem can be found in Hodges' book (Theorem 6.3.1 in [207]).

THEOREM 4.1.6 (Engeler, Ryll-Nardzewski, Svenonius). *For a countable structure $\mathfrak{B}$ with countable signature, the following are equivalent:*

1. *$\mathfrak{B}$ is ω-categorical;*
2. *all types of $\mathfrak{B}$ are principal;*
3. *all models of* $\mathrm{Th}(\mathfrak{B})$ *are atomic;*
4. *for all $n \geq 1$, every set of n-tuples that is preserved by all automorphisms of $\mathfrak{B}$ is first-order definable in $\mathfrak{B}$;*
5. *the automorphism group* $\mathrm{Aut}(\mathfrak{B})$ *of $\mathfrak{B}$ is oligomorphic;*
6. *for each $n \geq 1$, there are finitely many inequivalent formulas with free variables $x_1, \ldots, x_n$ over $\mathfrak{B}$;*
7. *$\mathfrak{B}$ has finitely many n-types, for all $n \geq 1$.*

PROOF. We first show the implications $(1) \Rightarrow (2) \Rightarrow (3) \Rightarrow (1)$, and then show $(3) \Rightarrow (4) \Rightarrow (5) \Rightarrow (6) \Rightarrow (7) \Rightarrow (2)$.

$(1) \Rightarrow (2)$. If $\mathfrak{B}$ has a non-principal type p, then $\mathrm{Th}(\mathfrak{B})$ has a countable model where p is realised, and by the omitting types theorem (Theorem 2.2.3;

here we use the assumption that the signature is countable) a countable model where p is omitted, so $\mathfrak{B}$ is not ω-categorical.

(2) $\Rightarrow$ (3). This is immediate from the definitions, because if all types of T are principal, then all types of n-tuples in models of T are principal.

(3) $\Rightarrow$ (1) follows from the fact that all countable atomic structures with the same theory are isomorphic (Lemma 2.2.5).

(3) $\Rightarrow$ (4). If all types of T are principal, then $\mathfrak{B}$ is atomic. If two n-tuples of an atomic structure have the same type, then there is an automorphism of $\mathfrak{B}$ that maps one to the other, by Lemma 2.2.5. So types define orbits of n-tuples under $\text{Aut}(\mathfrak{B})$, and since n-types of $\mathfrak{B}$ are principal, it follows that the orbits of n-tuples are first-order definable in $\mathfrak{B}$.

(4) $\Rightarrow$ (5). Suppose that $\text{Aut}(\mathfrak{B})$ has infinitely many orbits of n-tuples, for some n. Then the union of any subset of the set of all orbits of n-tuples is preserved by all automorphisms of $\mathfrak{B}$; but there are only countably many first-order formulas over a countable signature, so not all the invariant sets of n-tuples can be first-order definable in $\mathfrak{B}$.

(5) $\Rightarrow$ (6) is immediate since automorphisms preserve first-order formulas.

(6) $\Rightarrow$ (7). Suppose that for each $n \in \mathbb{N}$ there are formulas $\psi_1, \dots, \psi_m$ such that every first-order formula $\phi(x_1, \dots, x_n)$ is equivalent to one of the ψ_i's. Then every n-type of $\mathfrak{B}$ is completely determined by which ψ_i it contains.

(7) $\Rightarrow$ (2). Let $p_1, \dots, p_m$ be the finitely many distinct n-types of $\mathfrak{B}$. For every pair (i, j) with distinct entries $i, j \in \{1, \dots, m\}$ there is a formula $\phi_{i,j}$ such that $\phi_{i,j} \in p_i \setminus p_j$. Then $\bigwedge_{j \in \{1,\dots,m\} \setminus \{i\}} \phi_{i,j}$ isolates p_i. ⊣

The second condition in Theorem 4.1.6 provides another way to verify that a structure is ω-categorical. We again illustrate this with the structure $(\mathbb{Q}; <)$; suppose that all we know about this structure is that it is the (homogeneous) Fraïssé-limit of the class $\mathcal{C}$ of all finite linear orders from Example 2.3.3. The homogeneity of $(\mathbb{Q}; <)$ implies that the orbit of an n-tuple $(t_1, \dots, t_n)$ from $\mathbb{Q}^n$ under $\text{Aut}(\mathbb{Q}; <)$ is determined by the weak linear order induced on $(t_1, \dots, t_n)$ in $(\mathbb{Q}; <)$. We write *weak linear order*, and not *linear order*, because some of the elements $t_1, \dots, t_n$ may be equal (that is, a *weak linear order* is a total quasiorder). The number of weak linear orders on n elements is bounded by n^n, and hence the automorphism group of $(\mathbb{Q}; <)$ has a finite number of orbits of n-tuples, for all $n \geq 1$. See Lemma 4.3.1 for a generalisation of this argument.

4.1.2. Compactness and ω-categoricity. Lemma 4.1.7 below states a useful property that ω-categorical structures have in common with finite structures, and is an easy consequence of Königs tree lemma.

LEMMA 4.1.7. *Let $\mathfrak{B}$ be a finite or countably infinite ω-categorical structure with relational signature τ, and let $\mathfrak{A}$ be a countable τ-structure. If there is no homomorphism (embedding) from $\mathfrak{A}$ to $\mathfrak{B}$, then there is a finite substructure of $\mathfrak{A}$ that does not map homomorphically (embed) to $\mathfrak{B}$.*

PROOF. We present the proof for homomorphisms; the proof for embeddings is analogous. Suppose every finite substructure of $\mathfrak{A}$ maps homomorphically to $\mathfrak{B}$. We show the contraposition of the lemma, and prove the existence of a homomorphism from $\mathfrak{A}$ to $\mathfrak{B}$. Let $a_1, a_2, \ldots$ be an enumeration of $\mathfrak{A}$. We construct a rooted tree with finite out-degree, where each node lies on some level $n \geq 0$. The nodes on level n are equivalence classes of homomorphisms from the substructure of $\mathfrak{A}$ induced on $a_1, \ldots, a_n$ to $\mathfrak{B}$. Hence, there is only one vertex on level 0, which will be the root of the tree. Two such homomorphisms f and g are equivalent if there is an automorphism α of $\mathfrak{B}$ such that $\alpha f = g$. Two equivalence classes of homomorphisms on level n and $n+1$ are adjacent if there are representatives of the classes such that one is a restriction of the other. Theorem 4.1.6 asserts that $\text{Aut}(\mathfrak{A})$ has only finitely many orbits of k-tuples, for all $k \geq 0$ (clearly, this also holds if $\mathfrak{B}$ is finite). Hence, the constructed tree has finite out-degree. By assumption, there is a homomorphism from the substructure of $\mathfrak{A}$ induced on $a_1, a_2, \ldots, a_n$ to $\mathfrak{B}$ for all $n \geq 0$, and hence the tree has vertices on all levels. König's lemma asserts the existence of an infinite path in the tree, which can be used inductively to define a homomorphism h from $\mathfrak{A}$ to $\mathfrak{B}$ as follows.

The restriction of h to $\{a_1, \ldots, a_n\}$ will be an element from the n-th node of the infinite path. Initially, this is trivially true if h is restricted to the empty set. Suppose h is already defined on $a_1, \ldots, a_n$, for $n \geq 0$. By construction of the infinite path, we find representatives h_n and h_{n+1} of the n-th and the $(n+1)$-st element on the path such that h_n is a restriction of h_{n+1}. The inductive assumption gives us an automorphism α of $\mathfrak{A}$ such that $\alpha h_n(x) = h(x)$ for all $x \in \{a_1, \ldots, a_n\}$. We set $h(a_{n+1})$ to be $\alpha h_{n+1}(a_{n+1})$. The restriction of h to $a_1, \ldots, a_{n+1}$ will therefore be a member of the $(n+1)$-st element of the infinite path. The operation h defined in this way is indeed a homomorphism from $\mathfrak{A}$ to $\mathfrak{B}$. ⊣

The assumption that $\mathfrak{A}$ is countable is necessary in Lemma 4.1.7; consider for example $\mathfrak{A} := (\mathbb{R}; <)$, which does not admit a homomorphism to $\mathfrak{B} := (\mathbb{Q}; <)$ for cardinality reasons, even though any finite substructure of $\mathfrak{A}$ does.

COROLLARY 4.1.8. *For any structure $\mathfrak{C}$, there is a finite structure $\mathfrak{B}$ with the same* CSP *as $\mathfrak{C}$ if and only if $\mathfrak{C}$ has a finite core.*

PROOF. If there exists a finite structure $\mathfrak{B}$ with the same CSP as $\mathfrak{C}$, then every finite substructure of $\mathfrak{C}$ maps homomorphically to the core $\mathfrak{B}'$ of $\mathfrak{B}$, and by Lemma 4.1.7 there exists a homomorphism from $\mathfrak{C}$ to the finite core structure $\mathfrak{B}'$ (which is unique up to isomorphism); since $\mathfrak{B}'$ also maps to $\mathfrak{C}$, it is a core of $\mathfrak{C}$. The converse is trivial. ⊣

COROLLARY 4.1.9. *Two countable ω-categorical relational τ-structures $\mathfrak{A}$ and $\mathfrak{B}$ have the same* CSP *if and only if there is a homomorphism from $\mathfrak{A}$ to $\mathfrak{B}$ and a homomorphism from $\mathfrak{B}$ to $\mathfrak{A}$.*

For general countable relational structures Corollary 4.1.9 is false. Consider for example the structure $(\mathbb{Z}; \{(x, y) \mid y = x + 1\})$—the 'infinite line', and the structure $(\mathbb{N}; \{(x, y) \mid y = x + 1\})$—the 'infinite ray'. Clearly, these two structures give rise to the same CSP, but there is no homomorphism from the line to the ray.

Several times we need variants of Lemma 4.1.7 that can be proved in the same way. For instance, we can replace *homomorphism* in the statement and the proof by *strong homomorphism*, or *injective homomorphism*. Sometimes, we are also looking for functions f from $\mathfrak{A}^2$ to $\mathfrak{B}$ that satisfy universal identities such as $\forall x, y\colon f(x, y) = f(y, x)$. What is common for all those statements is that the respective property of the function can be expressed by *universal first-order sentences*. We make this more precise and derive the following generalisation of Lemma 4.1.7 based on the compactness theorem.

LEMMA 4.1.10. *Let $\mathfrak{B}$ be a countable ω-categorical structure with countable relational signature τ, let $\mathfrak{A}$ be a countably infinite τ-structure, and let σ be a countable set of function symbols. Then for any universal $(\tau \cup \sigma)$-theory T the following are equivalent.*

1. *The two-sorted structure $(\mathfrak{A}, \mathfrak{B})$ has a $(\tau \cup \sigma)$-expansion that satisfies T such that every $f \in \sigma$ denotes a function from $\mathfrak{A}$ to $\mathfrak{B}$.*
2. *For every finite induced substructure $\mathfrak{C}$ of $\mathfrak{A}$ the two-sorted structure $(\mathfrak{C}, \mathfrak{B})$ has a $(\tau \cup \sigma)$-expansion that satisfies T such that every $f \in \sigma$ denotes a function from $\mathfrak{C}$ to $\mathfrak{B}$.*

PROOF. Any substructure of a model of a universal theory is again a model of the theory, so (1) implies (2). For the converse, let P_1, P_2 be unary relation symbols not contained in τ. Let $\mathfrak{A}'$ be an expansion of $\mathfrak{A}$ by countably many constants such that every element of $\mathfrak{A}$ is named in $\mathfrak{A}'$ by a constant symbol; let τ' be the (countable) signature of $\mathfrak{A}'$. Let $D := \mathrm{Th}(\mathfrak{A}')_{\mathrm{qf}}$, and let S be a set of first-order sentences that expresses that

- P_1 and P_2 are disjoint and denote two distinct sorts such that all function symbols from σ denote functions from P_1 to P_2, and
- the τ-reduct of the structure induced by the elements in P_2 has the same first-order theory as $\mathfrak{B}$.

We first prove that $D \cup S \cup T$ is satisfiable. By compactness, it suffices to prove the satisfiability of $D' \cup S \cup T$ for all finite subsets D' of D. Let $c_1, \dots, c_n$ be the constant symbols mentioned in D'. Let $\mathfrak{C}'$ be the substructure of $\mathfrak{A}'$ induced on $\{c_1, \dots, c_n\}$. Clearly, $\mathfrak{C}' \models D'$. Let $\mathfrak{C}$ be the τ-reduct of $\mathfrak{C}'$. By assumption, the two-sorted structure $(\mathfrak{C}, \mathfrak{B})$ with sorts P_1 and P_2 can be expanded to a two-sorted $(\tau \cup \sigma)$-structure $\mathfrak{D}$ that satisfies T; this structure also satisfies S. If we additionally denote the constants $c_1, \dots, c_n$ as in $\mathfrak{A}'$, then the expansion satisfies also D', and so we have found a model of $D' \cup S \cup T$.

By compactness, there exists an (infinite) model of $D \cup S \cup T$, and by Theorem 2.1.11 and since $\tau' \cup \sigma$ is countable there is also a countably infinite model $\mathfrak{M}$ of $D \cup S \cup T$. Consider the substructure $\mathfrak{M}'$ of $\mathfrak{M}$ generated by the constants from τ' and the elements in $P_2^{\mathfrak{M}}$. Then $\mathfrak{M}'$ still satisfies D and S, and since universal sentences are preserved by taking substructures, $\mathfrak{M}'$ also satisfies T. Then $\mathfrak{M}' \models D$ implies that in $\mathfrak{M}'$, the constants from τ' induce a copy of $\mathfrak{A}$. Since $\mathfrak{M} \models S$ the elements of $P_2^{\mathfrak{M}}$ induce a structure that is isomorphic to $\mathfrak{B}$, because $\mathfrak{B}$ is ω-categorical. So the functions of $\mathfrak{M}'$ denoted by the function symbols from σ provide the required $(\tau \cup \sigma)$-expansion of $(\mathfrak{A}, \mathfrak{B})$. ⊣

Lemma 4.1.10 is indeed a generalisation of Lemma 4.1.7: to make sure that f is a homomorphism, T contains for every relation symbol $R \in \tau$ the sentence $\forall \bar{x}\, (R(\bar{x}) \Rightarrow R(f(\bar{x}))$.

4.2. Oligomorphic permutation groups

We have seen in Section 4.1.1 that a countably infinite structure is ω-categorical if and only if its automorphism group is *oligomorphic*, i.e., has for each $n \geq 1$ only finitely many orbits of n-tuples. This section describes the connection between logic and permutation groups in more detail.

4.2.1. Topology. Automorphism groups of relational structures $\mathfrak{B}$ are naturally equipped with a topology, given by the following definition (a more general topological treatment can be found in Chapter 9). We write $\mathrm{Sym}(B)$ for the set of all permutations of the set B; this set forms a permutation group and is called the *symmetric group over X*.

DEFINITION 4.2.1. A subset $\mathcal{P}$ of $\mathrm{Sym}(B)$ is called *closed in* $\mathrm{Sym}(B)$ if $\mathcal{P}$ contains all $\alpha \in \mathrm{Sym}(B)$ with the property that for every finite $A \subseteq B$ there exists $\beta \in \mathcal{P}$ such that $\alpha x = \beta x$ for all $x \in A$.

PROPOSITION 4.2.2. *For $\mathcal{P} \subseteq \mathrm{Sym}(B)$, the following are equivalent.*

1. *$\mathcal{P}$ is the automorphism group of a relational structure with domain B;*
2. *$\mathcal{P}$ is a closed subgroup of* $\mathrm{Sym}(B)$;
3. *$\mathcal{P}$ is the automorphism group of a homogeneous relational structure with domain B.*

In the proof of this proposition, the following concept is useful.

DEFINITION 4.2.3. If $\mathcal{P} \subseteq \mathrm{Sym}(B)$ be a set of permutations, then $\mathrm{sInv}(\mathcal{P})$ denotes the *strong invariants* of $\mathcal{P}$, i.e., the set of all relations R over B such that for every $\alpha \in \mathcal{P}$, both α and α^{-1} preserve R. Occasionally, we may view $\mathrm{sInv}(\mathcal{P})$ as a relational structure with domain B whose relations are precisely the orbits of n-tuples, for all $n \geq 1$.

PROOF OF PROPOSITION 4.2.2. For the implication from (1) to (2), let $\mathcal{P}$ be the set of automorphisms of a relational structure $\mathfrak{B}$ with domain B. It is clear that $\mathcal{P}$ is closed under composition and taking inverses. To show that $\mathcal{P}$ is closed in $\mathrm{Sym}(B)$, let $\alpha \in \mathrm{Sym}(B)$ be such that for every finite $A \subseteq B$ there exists $\beta \in \mathcal{P}$ such that $\alpha x = \beta x$ for all $x \in A$. Then α must preserve all relations from $\mathfrak{B}$, because if α does not preserve a relation from $\mathfrak{B}$, then this can be seen from the restriction of α to a finite subset of B. Hence, $\alpha \in \mathcal{P}$.

For the implication from (2) to (3), first note that $\mathrm{sInv}(\mathcal{P})$ is homogeneous: suppose that i is an isomorphism between finite substructures of $\mathrm{sInv}(\mathcal{P})$ with domain $\{a_1, \dots, a_n\}$. Then (2) implies that $\{(\alpha a_1, \dots, \alpha a_n) \mid \alpha \in \mathcal{P}\} \in \mathrm{sInv}(\mathcal{P})$. Hence, this relation is preserved by i and $(i(a_1), \dots, i(a_n)) = (\alpha a_1, \dots, \alpha a_n)$ for some $\alpha \in \mathcal{P}$. This shows that there is an automorphism of $\mathrm{sInv}(\mathcal{P})$ that extends i. In fact, $\mathcal{P}$ is closed in $\mathrm{Sym}(B)$, so this also shows that every automorphism of $\mathrm{sInv}(\mathcal{P})$ is from $\mathcal{P}$. Since clearly $\mathcal{P} \subseteq \mathrm{Aut}(\mathrm{sInv}(\mathcal{P}))$, we therefore have that $\mathcal{P} = \mathrm{Aut}(\mathrm{sInv}(\mathcal{P}))$ is the automorphism of a homogeneous relational structure.

The implication from (3) to (1) is trivial. ⊣

4.2.2. The sInv-Aut Galois connection. The set of all relations that are first-order definable in a structure $\mathfrak{B}$ is denoted by $\langle \mathfrak{B} \rangle_{\mathrm{fo}}$. We will see in this section that for every ω-categorical structure $\mathfrak{C}$ the set

$$\{\langle \mathfrak{B} \rangle_{\mathrm{fo}} \mid \mathfrak{B} \text{ first-order definable in } \mathfrak{C}\},$$

partially ordered by inclusion, is closely connected to the set of all closed subgroups of $\mathrm{Sym}(B)$ that contain $\mathrm{Aut}(\mathfrak{C})$, again partially ordered by inclusion.

Recall that the automorphism group of a relational structure $\mathfrak{B}$, i.e., the set of all automorphisms of $\mathfrak{B}$, is denoted by $\mathrm{Aut}(\mathfrak{B})$. In the following it will be convenient to define the operator Aut also on sets $\mathcal{R}$ of relations over the same domain B, in which case $\mathrm{Aut}(\mathcal{R})$ denotes the set of all permutations α of B such that α and its inverse α^{-1} preserve all relations from $\mathcal{R}$.

DEFINITION 4.2.4. An (*anti-tone*) *Galois connection* is a pair of functions $F \colon U \to V$ and $G \colon V \to U$ between two posets U and V, such that $v \leq F(u)$ if and only if $u \leq G(v)$ for all $u \in U, v \in V$.

If F, G form a Galois connection then $u \leq G(F(u))$ for every $u \in U$: it follows immediately from Definition 4.2.4 for $v := F(u)$ that $u \leq G(v) = G(F(u))$, because $F(u) \leq F(u)$. Similarly, we get that $v \leq F(G(v))$ for every $v \in V$. Moreover, $F(u) = F(G(F(u)))$ and $G(v) = G(F(G(v)))$ for all $u \in U$ and $v \in V$. See Figure 1.

PROPOSITION 4.2.5. *Let B be a set. The operators* sInv *and* Aut *form a Galois connection between sets of relations over B and sets of permutations of B, both partially ordered by inclusion.*

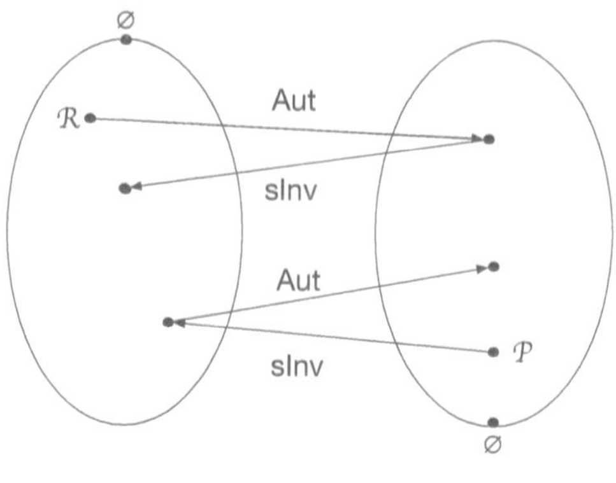

FIGURE 1. Illustration for the Galois connection Aut-sInv between sets of relations $\mathcal{R}$ over the base set B, ordered by $\supseteq$, and and sets of permutations $\mathcal{P}$ of B, ordered by $\subseteq$.

PROOF. Let $\mathcal{R}$ be a set of relations over B and let $\mathcal{P}$ be a set of permutations of B. First suppose that $\mathcal{P} \subseteq \text{Aut}(\mathcal{R})$, and let $R \in \mathcal{R}$ and $g \in \mathcal{P}$. Then $g \in \text{Aut}(\mathcal{R})$ and hence g and g^{-1} preserve $\mathcal{R}$. Thus, $\mathcal{R} \subseteq \text{sInv}(\mathcal{P})$.

Conversely, suppose that $\mathcal{R} \subseteq \text{sInv}(\mathcal{P})$, and again let $g \in \mathcal{P}$ and $R \in \mathcal{R}$. Then $R \in \text{sInv}(\mathcal{P})$, and hence g preserves R. Since $g^{-1} \in \mathcal{P}$, and g^{-1} also preserves R, we have that $g \in \text{Aut}(\mathcal{R})$. Thus, $\mathcal{P} \subseteq \text{Aut}(\mathcal{R})$. ⊣

For sets of permutations $\mathcal{P} \subseteq \text{Sym}(B)$ and sets $\mathcal{R}$ of relations over the domain B, we now present descriptions of the closure operators

$$\mathcal{P} \mapsto \text{Aut}(\text{sInv}(\mathcal{P}))$$
$$\mathcal{R} \mapsto \text{sInv}(\text{Aut}(\mathcal{R})).$$

We start with the former, for which we need the following definitions.

DEFINITION 4.2.6. For $\mathcal{P} \subseteq \text{Sym}(B)$, we define

- $\langle \mathcal{P} \rangle$, the *permutation group generated by* $\mathcal{P}$, to be the smallest permutation group that contains $\mathcal{P}$.
- $\overline{\mathcal{P}}$, the *closure of* $\mathcal{P}$ *in* $\text{Sym}(B)$, to be the smallest closed subset of $\text{Sym}(B)$ that contains $\mathcal{P}$.

PROPOSITION 4.2.7. *Let* $\mathcal{P} \subseteq \text{Sym}(B)$ *and let* $\mathcal{P}'$ *be the smallest permutation group that contains* $\mathcal{P}$ *and is closed in* $\text{Sym}(B)$. *Then* $\mathcal{P}' = \text{Aut}(\text{sInv}(\mathcal{P})) = \overline{\langle \mathcal{P} \rangle}$.

PROOF. Since $\mathcal{P} \subseteq \mathcal{P}'$ and $\mathcal{P}'$ is a permutation group, we must have $\langle \mathcal{P} \rangle \subseteq \mathcal{P}'$, and therefore also $\overline{\langle \mathcal{P} \rangle} \subseteq \mathcal{P}'$ since $\mathcal{P}'$ is closed in $\text{Sym}(B)$. To show the converse inclusion $\mathcal{P}' \subseteq \overline{\langle \mathcal{P} \rangle}$, it suffices to verify that $\overline{\langle \mathcal{P} \rangle}$ is a closed subgroup of $\text{Sym}(B)$. Since $\overline{\langle \mathcal{P} \rangle}$ is clearly closed in $\text{Sym}(B)$ we only have to show that $\overline{\langle \mathcal{P} \rangle}$ is closed under compositions and contains inverses, which is both straightforward and left to the reader.

We now show that $\overline{\langle \mathcal{P} \rangle} \subseteq \mathrm{Aut}(\mathrm{sInv}(\mathcal{P}))$. Let $\alpha \in \overline{\langle \mathcal{P} \rangle}$ be arbitrary, and let R be from $\mathrm{sInv}(\mathcal{P})$. We have to show that α and α^{-1} preserve R. Let $t \in R$; since $\alpha \in \overline{\langle \mathcal{P} \rangle}$, we have that $\alpha t = \beta_1 \cdots \beta_k t$ for some permutations $\beta_1, \dots, \beta_k$ from $\mathcal{P} \cup \mathcal{P}^{-1}$. Since $\beta_1, \dots, \beta_k$ preserve R, we have that $\alpha t \in R$. The argument for α^{-1} is analogous.

Finally, we show $\mathrm{Aut}(\mathrm{sInv}(\mathcal{P})) \subseteq \overline{\langle \mathcal{P} \rangle}$. Let α be from $\mathrm{Aut}(\mathrm{sInv}(\mathcal{P}))$. It suffices to show that for every finite subset $\{a_1, \dots, a_n\}$ of B there is a $\beta \in \langle \mathcal{P} \rangle$ such that $\alpha a_i = \beta a_i$ for all $i \leq n$. Consider the relation $\{(\beta t_1, \dots, \beta t_n) \mid \beta \in \langle \mathcal{P} \rangle\}$. It is preserved by all permutations in $\mathcal{P}$. Therefore, α preserves this relation, and so there exists $\beta \in \langle \mathcal{P} \rangle$ as required. ⊣

We now turn to characterisations of the hull operator $\mathcal{R} \mapsto \mathrm{sInv}(\mathrm{Aut}(\mathcal{R}))$. First observe the following, which is straightforward.

PROPOSITION 4.2.8. *Let $\mathfrak{B}$ be any structure. Then* $\mathrm{sInv}(\mathrm{Aut}(\mathfrak{B}))$ *contains* $\langle \mathfrak{B} \rangle_{\mathrm{fo}}$, *the set of all relations that are first-order definable in $\mathfrak{B}$.*

An exact characterisation of $\mathrm{sInv}(\mathrm{Aut}(\mathfrak{B}))$ can be given if $\mathfrak{B}$ is ω-categorical. The analogous statement of Theorem 4.2.9 below has been observed for finite structures by Krasner [246]. The fact that it extends to ω-categorical structures is a direct consequence of the theorem of Ryll-Nardzewski (Theorem 4.1.6).

THEOREM 4.2.9. *Let $\mathfrak{B}$ be a countable ω-categorical structure with countable signature and let $R \subseteq B^k$ be a relation. Then R is first-order definable in $\mathfrak{B}$ if and only if R is preserved by the automorphisms of $\mathfrak{B}$, in symbols*: $\mathrm{sInv}(\mathrm{Aut}(\mathfrak{B})) = \langle \mathfrak{B} \rangle_{\mathrm{fo}}$.

As we have seen in the proof of Proposition 4.2.2, it follows in particular that the expansion of every ω-categorical structure by all *first-order* definable relations is homogeneous. Observe that the equivalence of (1) ad (3) in Theorem 4.1.6 implies that the conclusion in Theorem 4.2.9 holds *if and only if* $\mathfrak{B}$ is ω-categorical.

We have the following consequence of Proposition 4.2.7 and Theorem 4.2.9. An *anti-isomorphism* between two posets U and V is a bijection f from the elements of U to the elements of V such that $u \leq v$ in U if and only if $f(u) \geq f(v)$ in V.

COROLLARY 4.2.10. *Let $\mathfrak{B}$ be a countable ω-categorical structure. Then we have the following.*

- *The sets of the form $\langle \mathfrak{A} \rangle_{\mathrm{fo}}$, where $\mathfrak{A}$ is a first-order reduct of $\mathfrak{B}$, ordered by inclusion, form a lattice.*
- *The closed subgroups of* $\mathrm{Sym}(B)$ *containing* $\mathrm{Aut}(\mathfrak{B})$, *ordered by inclusion, form a lattice.*
- *The operator* sInv *is an anti-isomorphism between those two lattices, and* Aut *is its inverse.*

Recall that two structures $\mathfrak{B}$, $\mathfrak{C}$ on the same domain are said to be *first-order interdefinable* iff all relations of $\mathfrak{B}$ have a first-order definition in $\mathfrak{C}$ and vice-versa. Then it follows from the above that two ω-categorical structures are first-order interdefinable if and only if they have the same automorphisms.

4.2.3. Primitivity and transitivity. We further illustrate the connection between ω-categorical structures and oligomorphic permutation groups by translating several concepts from permutation group theory into model-theoretic terminology, and vice versa.

A *congruence* of a permutation group on a set B is an equivalence relation on B that is preserved by all permutations in the permutation group. Subsets of B that are equivalence classes of some congruence are also called *blocks*.

LEMMA 4.2.11. *Let $\mathcal{G}$ be a permutation group on a set B. Then $S \subseteq B$ is a block if and only if $g(S) = S$ or $g(S) \cap S = \emptyset$ for all $g \in \mathcal{G}$.*

PROOF. Suppose that S is a block of the congruence C, and suppose that $g(S) \cap S$ contains an element t. That is, there is an element $s \in S$ such that $g(s) = t$. We will show that $g(S) = S$. Let $r \in S$ be arbitrary. Then $(r, s) \in C$, and hence $(g(r), g(s)) = (g(r), t) \in C$. Since $t \in S$, it follows that $g(r) \in S$.

Now suppose that $S \subseteq B$ is such that $g(S) = S$ or $g(S) \cap S = \emptyset$ for all $g \in \mathcal{G}$. Define

$$C := \{(x, y) \mid x = y \text{ or } \exists g \in \mathcal{G} : g(x), g(y) \in S\}.$$

This relation is clearly reflexive, symmetric, and preserved by $\mathcal{G}$. In order to prove that C is a congruence it remains to verify transitivity. Let $(x, y), (y, z) \in C$. If $x = y$ or $y = z$ then $(x, z) \in C$. Otherwise, there are $g_1, g_2 \in \mathcal{G}$ such that $g_1(x), g_1(y), g_2(y), g_2(z) \in S$. Then $S \cap g_2(g_1^{-1}(S))$ contains $g_2(y)$, and hence $(g_2 \circ g_1^{-1})(S) = S$ contains both $g_2(x)$ and $g_2(z)$, and hence $(x, z) \in E$. ⊣

A congruence is *trivial* if each block contains only one element (and *non-trivial* otherwise), and it is called *proper* if it is distinct from the equivalence relation that has only one block.

PROPOSITION 4.2.12. *Let $\mathfrak{B}$ be an ω-categorical structure and $\mathcal{G}$ its (oligomorphic) automorphism group. Then the congruences of $\mathcal{G}$ are exactly the first-order definable equivalence relations in $\mathfrak{B}$.*

PROOF. An immediate consequence of Theorem 4.2.9. ⊣

A permutation group $\mathcal{G}$ on a set B is called

- *primitive* if every proper congruence of $\mathcal{G}$ is trivial;
- *k-transitive* if for any two k-tuples s, t of distinct elements from B there is an $\alpha \in \mathcal{G}$ such that $\alpha s = t$, where the action of α on tuples is componentwise, i.e., $\alpha(s_1, \dots, s_k) = (\alpha s_1, \dots, \alpha s_k)$;
- *transitive* if it is 1-transitive;
- *k-set transitive* if for any two sets $S, T \subseteq B$ of cardinality k there is an $\alpha \in \mathcal{G}$ such that $\alpha S = \{\alpha s \mid s \in S\} = T$.

Clearly, 2-transitive structures are always primitive.

It is easy to see that every 2-set transitive permutation group $\mathcal{G}$ on a set B of size at least three is also transitive. We prove the contraposition: assume that c_1 and c_2 are elements of B in distinct orbits. Let $c_3 \in B$ be distinct from c_1 and c_2. There is no permutation in $\mathcal{G}$ that maps $\{c_1, c_3\}$ to $\{c_2, c_3\}$, and hence $\mathcal{G}$ is not 2-set transitive. More generally, it can be shown that the number of orbits of n-element subsets of B under $\mathcal{G}$ is a non-decreasing sequence [123]; here, by an *orbit of a set* $S \subseteq B$ under $\mathcal{G}$ we mean $\{\alpha(S) \mid \alpha \in \mathcal{G}\}$.

4.2.4. Group actions. It will be convenient to take a more general perspective on permutation groups (and this will again be used in Chapter 9). We now consider *abstract groups* (Example 2.1.1). The link to permutation groups is given by the concept of an *action* of such a group on a set, which is described below.

DEFINITION 4.2.13. A *(left) group action* of an (abstract) group $\boldsymbol{G}$ on a set X is a binary function $\cdot : G \times X \to X$ which satisfies that $(gh) \cdot x = g \cdot (h \cdot x)$ for all $g, h \in G$ and $x \in S$, and $e \cdot x = x$ for every $x \in X$. The action is *faithful* if for any two distinct $g, h \in G$ there exists an $x \in X$ such that $g \cdot x \neq h \cdot x$.

Equivalently, a group action of $\mathbf{G}$ on a set X is a homomorphism from $\mathbf{G}$ into $\mathrm{Sym}(X)$, and a faithful group action is an isomorphism between $\mathbf{G}$ and a subgroup of $\mathrm{Sym}(X)$.

Example 4.2.14 (The componentwise action on X^n). If $\mathbf{G}$ is a permutation group on a set X, then *the componentwise action of* $\mathbf{G}$ *on* X^n is given by

$$\xi(g)(x_1, \ldots, x_n) := (g(x_1), \ldots, g(x_n)).$$

This action is faithful.

Example 4.2.15 (The setwise action on $\binom{X}{k}$). If $\mathbf{G}$ is a permutation group on a set X, then *the setwise action of* $\mathbf{G}$ *on* $\binom{X}{n}$ is given by

$$\xi(g)(\{x_1, \ldots, x_n\}) := \{g(x_1), \ldots, g(x_n)\}.$$

Again, this action is faithful.

The *orbit* of x under an action of $\boldsymbol{G}$ on X is the set $\{g \cdot x \mid g \in G\}$. Clearly, to every action of $\mathbf{G}$ on B we can associate a permutation group $\mathcal{G}$ as considered before, namely the image of the action in $\mathrm{Sym}(B)$. We see that an orbit of n-tuples under $\mathrm{Aut}(\mathfrak{B})$, as defined in Definition 1.2.9, is the same as an orbit under the componentwise action of $\mathrm{Aut}(\mathfrak{B})$ on B (Example 4.2.14), and an orbit of a k-element subset of B as defined in the previous section is an orbit under the setwise action of $\mathrm{Aut}(\mathfrak{B})$ on $\binom{B}{k}$. Conversely, to every permutation group $\mathcal{G}$ on a set X we can associate an abstract group $\boldsymbol{G}$ whose domain is $\mathcal{G}$, where composition and inverse are defined in the obvious way, and which acts on X faithfully by $g \cdot x = gx$.

An action is called *oligomorphic* if the associated permutation group is oligomorphic. In this way we can also use other terminology introduced for permutation groups (such a transitivity, congruences, primitivity, etc.) for group actions.

4.2.5. Products. In this section we review the classical theory of the analysis of a permutation group in terms of transitive ones. The same idea can be used to construct new oligomorphic permutation groups from known ones.

The *product* of a sequence of groups $(\mathbf{G}_i)_{i \in I}$ is the product of this sequence as defined in general in Section 2.1; note that the product is again a group. Products appear in several ways when studying permutation groups; the first in connection with the relation between a permutation group and its 'transitive constituents', described in the following.

4.2.5.1. *The intransitive action of a group product.* If $\mathbf{G}$ acts on a set X and $O \subseteq X$ is an orbit under this action, then $\mathbf{G}$ naturally acts transitively (but not necessarily faithfully) on O by restriction.

PROPOSITION 4.2.16 (see [123]). *Let* $\mathbf{G}$ *be a group acting faithfully on a set* X *and let* $\mathcal{O}$ *be the set of all orbits under the action. Then for each* $O \in \mathcal{O}$ *there exists a group* $\mathbf{G}_O$ *acting transitively on* O *such that* $\mathbf{G}$ *has a surjective homomorphism to* $\mathbf{G}_O$ *and* $\mathbf{G}$ *is a subgroup of* $\prod_{O \in \mathcal{O}} \mathbf{G}_O$.

We can use the same idea to construct new oligomorphic permutation groups from known ones.

DEFINITION 4.2.17. Let $\mathbf{G}_1$ and $\mathbf{G}_2$ be groups acting on disjoint sets X and Y, respectively. Then the action of $\mathbf{G}_1 \times \mathbf{G}_2$ on $X \cup Y$ defined by $(g_1, g_2) \cdot z = g_1 z$ if $z \in X$, and $g_2 z$ if $y \in Y$, is called the (*natural*) *intransitive action* of $\mathbf{G}_1 \times \mathbf{G}_2$ on $X \cup Y$.

Note that if $\mathbf{G}_1$ and $\mathbf{G}_2$ act oligomorphically on X and Y, respectively, then the natural intransitive action of $\mathbf{G}_1 \times \mathbf{G}_2$ is also oligomorphic: If $f_1(n)$ is the number of orbits under the setwise action of $\mathbf{G}_1$ on $\binom{X}{n}$ and $f_2(n)$ is the number of orbits under the setwise action of $\mathbf{G}_2$ on $\binom{Y}{n}$ (Example 4.2.15), then the number $f(n)$ of orbits under the setwise of $\mathbf{G}_1 \times \mathbf{G}_2$ on $X \cup Y$ is $\sum_{0 \leq i \leq n} f_1(i) f_2(n-i)$, and hence finite for all n. Since $n^n f(n)$ is an upper bound on the number of orbits under the componentwise action of $\mathbf{G}_1 \times \mathbf{G}_2$ on $X \cup Y$, it follows that the action of $\mathbf{G}_1 \times \mathbf{G}_2$ on $X \cup Y$ is oligomorphic.

If $\mathbf{G}_1$ and $\mathbf{G}_2$ are the automorphism groups of ω-categorical relational structures $\mathfrak{A}$ and $\mathfrak{B}$ with disjoint domains A and B, respectively, then the natural intransitive action on $A \cup B$ can also be described as the automorphism group of a relational structure $\mathfrak{C}$. If τ is the signature of $\mathfrak{A}$ and σ the signature of $\mathfrak{B}$, then we can take for $\mathfrak{C}$ the structure

- whose signature is $\tau \cup \sigma \cup \{P\}$, where P is a new unary relation symbol,

- whose domain is $A \cup B$, and
- whose relations are $R^{\mathfrak{C}} := R^{\mathfrak{A}}$ for $R \in \tau$, $R^{\mathfrak{C}} := R^{\mathfrak{B}}$ for $R \in \sigma$, and $P^{\mathfrak{C}} := A$.

Since reducts of ω-categorical structures are again ω-categorical, this shows in particular that the disjoint union of two ω-categorical structures is again ω-categorical. Note that if $\mathfrak{A}$ and $\mathfrak{B}$ are finitely bounded, then so is $\mathfrak{C}$.

4.2.5.2. *The product action.* When $\mathbf{G}_1$ is a group acting on a set X, and $\mathbf{G}_2$ a group acting on a set Y, there is another important natural action of $\mathbf{G}_1 \times \mathbf{G}_2$ besides the intransitive natural action of $\mathbf{G}_1 \times \mathbf{G}_2$, namely the *product action*. In this action, $\mathbf{G}_1 \times \mathbf{G}_2$ acts on $X \times Y$ by $(g_1, g_2) \cdot (x, y) = (g_1 x, g_2 y)$. If $\mathcal{G}$ is the permutation group on X induced by the action of $\mathbf{G}_1$ on X and $\mathcal{H}$ is the permutation group on Y induced by the action of $\mathbf{G}_2$ on Y then we also write $\mathcal{G} \times \mathcal{H}$ for the permutation group induced by the product action of $\mathbf{G}_1 \times \mathbf{G}_2$ on $X \times Y$.

If the actions of $\mathbf{G}_1$ and $\mathbf{G}_2$ are transitive, then the product action is clearly transitive, too. We claim that if the actions of $\mathbf{G}_1$ and $\mathbf{G}_2$ are oligomorphic, then the product action is also oligomorphic. Let $F_1(n)$ and $F_2(n)$ be the number of orbits under the componentwise action of G_1 on X^n and Y^n, respectively. Then the number of orbits under the componentwise action of $\boldsymbol{G}$ on $(X \times Y)^n$ is $F_1(n)F_2(n)$, and in particular finite, which proves the claim.

If $\mathbf{G}_1$ and $\mathbf{G}_2$ are the automorphism groups of ω-categorical structures $\mathfrak{A}$ and $\mathfrak{B}$, then the image of the product action of $\boldsymbol{G}$ in $\mathrm{Sym}(A \times B)$ is the automorphism group of the following structure $\mathfrak{A} \boxtimes \mathfrak{B}$, which we call the *algebraic product* of $\mathfrak{A}$ and $\mathfrak{B}$.

DEFINITION 4.2.18. Let $\mathfrak{A}$ be a relational σ-structure and $\mathfrak{B}$ be a relational τ-structure. We assume that σ and τ are disjoint, otherwise we rename the relations so that the assumption is satisfied. We also assume that each of σ and τ has a symbol for equality. Then the structure $\mathfrak{A} \boxtimes \mathfrak{B}$ contains the relations

$$\{((a_1, b_1), \dots, (a_k, b_k)) \mid (a_1, \dots, a_k) \in R^{\mathfrak{A}}, b_1, \dots, b_k \in B\} \qquad \text{for each } k\text{-ary } R \in \sigma,$$

$$\{((a_1, b_1), \dots, (a_k, b_k)) \mid (b_1, \dots, b_k) \in R^{\mathfrak{B}}, a_1, \dots, a_k \in A\} \qquad \text{for each } k\text{-ary } R \in \tau.$$

By our assumption that σ and τ have a symbol for equality, $\mathfrak{A} \boxtimes \mathfrak{B}$ carries in particular the following two relations.

$$E_1 := \{((a_1, b_1), (a_2, b_2)) \mid a_1, a_2 \in A, b_1, b_2, \in B, a_1 = a_2\};$$
$$E_2 := \{((a_1, b_1), (a_2, b_2)) \mid a_1, a_2 \in A, b_1, b_2, \in B, b_1 = b_2\}.$$

PROPOSITION 4.2.19. *Let $\mathfrak{A}$ and $\mathfrak{B}$ be structures. Then the automorphism group of $\mathfrak{C} := \mathfrak{A} \boxtimes \mathfrak{B}$ equals* $\text{Aut}(\mathfrak{A}) \times \text{Aut}(\mathfrak{B})$ *in its product action on $A \times B$. Moreover,*

- *if $\mathfrak{A}$ and $\mathfrak{B}$ are homogeneous, then so is $\mathfrak{A} \boxtimes \mathfrak{B}$;*
- *if $\mathfrak{A}$ and $\mathfrak{B}$ are finitely bounded, then so is $\mathfrak{A} \boxtimes \mathfrak{B}$.*

PROOF. Let h be the product action of $\mathbf{G} := \text{Aut}(\mathfrak{A}) \times \text{Aut}(\mathfrak{B})$ on $A \times B$, viewed as a homomorphism from G to $\text{Sym}(A \times B)$. Let (g_1, g_2) be an element of G. Then $h((g_1, g_2))$ is the permutation $(x, y) \mapsto (g_1 x, g_2 y)$ of $A \times B$, and this map preserves $\mathfrak{C}$: if $((a_1, b_1), \ldots, (a_k, b_k)) \in R^{\mathfrak{C}}$, for $R \in \sigma$, then $(a_1, \ldots, a_k) \in R^{\mathfrak{A}}$, and so $(g_1 a_1, \ldots, g_1 a_k) \in R^{\mathfrak{A}}$. Therefore, $((g_1 a_1, g_2 b_1), \ldots, (g_1 a_k, g_2 b_k)) \in R^{\mathfrak{C}}$. The proof for the relation symbols $R \in \tau$ is analogous.

We now show that conversely, every automorphism g of $\mathfrak{C}$ is in the image of h. Note that E_1 and E_2 are congruences of the automorphism group of $\mathfrak{C}$. Fix elements $a_0 \in A, b_0 \in B$. Let g_1 be the permutation of A that maps $a \in A$ to the point a' such that $g((a, b_0)) = (a', b_0)$. Similarly, let g_2 be the permutation of B that maps $b \in B$ to the point b' such that $g((a_0, b)) = (a_0, b')$. Since g preserves E_1 and E_2, the definition of g_1 and g_2 does not depend on the choice of a_0 and b_0. Moreover, g_1 is from $\text{Aut}(\mathfrak{A})$, since g preserves the relations for the symbols from σ. Similarly, g_2 is from $\text{Aut}(\mathfrak{B})$. Then $h((g_1, g_2))$ equals g. Hence, g is a permutation of $A \times B$ that lies in the image of h. We leave the proof of the statements about homogeneity and finite boundedness of $\mathfrak{A} \boxtimes \mathfrak{B}$ to the reader. ⊣

Note that Proposition 4.2.19 becomes false in general when we omit the relations E_1 and E_2 in $\mathfrak{A} \boxtimes \mathfrak{B}$. Consider for example a structure $\mathfrak{B}$ with empty signature and at least two elements. Then the automorphism group of $\mathfrak{B} \boxtimes \mathfrak{B}$ is not primitive, but without the relations E_1 and E_2, the structure is primitive. Note that $(\mathfrak{A} \boxtimes \mathfrak{B}) \boxtimes \mathfrak{C}$ and $\mathfrak{A} \boxtimes (\mathfrak{B} \boxtimes \mathfrak{C})$ have the same automorphism group (on the domain $A \times B \times C$). We explicitly define the d-fold algebraic product as follows.

DEFINITION 4.2.20 (Algebraic product of d structures). Let $\mathfrak{B}_1, \ldots, \mathfrak{B}_d$ be structures with disjoint relational signatures $\tau_1, \ldots, \tau_d$. We denote by $\mathfrak{B}_1 \boxtimes \cdots \boxtimes \mathfrak{B}_d$ the structure with domain $B := B_1 \times \cdots \times B_d$ that contains for every $i \leq d$, and every m-ary $R \in (\tau_i \cup \{=\})$ an m-ary relation defined by

$$\{((x_1^1, \ldots, x_1^d), \ldots, (x_m^1, \ldots, x_m^d)) \in B^m \mid (x_1^i, \ldots, x_m^i) \in R^{\mathfrak{B}_i}\}.$$

If $\mathfrak{B} := \mathfrak{B}_1 = \cdots = \mathfrak{B}_k$, then we first rename $R \in \tau_i$ into R_i so that the factors have pairwise disjoint signatures. Then $\mathfrak{B}_1 \boxtimes \cdots \boxtimes \mathfrak{B}_d$ is well-defined, and called the d-th algebraic power of $\mathfrak{B}$, written $\mathfrak{B}^{(d)}$.

The structure $\mathfrak{B}^{(d)}$ should not be confused with the full power structure $\mathfrak{B}^{[d]}$ from Definition 3.5.3.

Remark 4.2.21. If $\mathfrak{A}$ and $\mathfrak{B}$ have the same signature τ, then the automorphism group of the τ-structure $\mathfrak{A} \times \mathfrak{B}$ *contains* the automorphism group of $\mathfrak{A} \boxtimes \mathfrak{B}$, and hence $\mathfrak{A} \times \mathfrak{B}$ is ω-categorical, by Theorem 4.1.6. As a consequence, the class of all ω-categorical structures of some fixed signature, considered up to homomorphic equivalence, forms a lattice with respect to the homomorphism order (where disjoint union gives the join, and product the meet of two ω-categorical structures).

4.3. Countable categoricity from Fraïssé-amalgamation

Fraïssé's theorem can be used to construct ω-categorical structures, because homogeneous structures with finite relational signature are ω-categorical. More generally, we have the following.

LEMMA 4.3.1. *Let $\mathfrak{C}$ be a countably infinite homogeneous structure such that for each k only a finite number of distinct k-ary relations can be defined by atomic formulas. Then $\mathfrak{C}$ is ω-categorical.*

PROOF. By the homogeneity of $\mathfrak{C}$, the atomic formulas that hold on the elements of t in $\mathfrak{C}$ determine the orbit of t under Aut($\mathfrak{C}$). Since there are only finitely many such atomic formulas, it follows that there are finitely many orbits of k-tuples under Aut($\mathfrak{C}$). The claim follows by Theorem 4.1.6. ⊣

4.3.1. Homogeneity and quantifier elimination. There is an exact characterisation of those ω-categorical structures that are homogeneous. A structure has *quantifier elimination* if its first-order theory has quantifier elimination (Definition 2.6.3). We will prove shortly that an ω-categorical structure $\mathfrak{B}$ admits quantifier elimination if and only if it is homogeneous. This is an immediate corollary of more general results about definability by *positive* Boolean combinations in ω-categorical structures. Following Truss and Lockett [267], a structure $\mathfrak{B}$ is called

- *HA-homogeneous* if every finite partial endomorphism of $\mathfrak{B}$ extends to an automorphism.
- *HI-homogeneous* if every finite partial endomorphism of $\mathfrak{B}$ extends to a self-embedding of $\mathfrak{B}$.

LEMMA 4.3.2. *Let $\mathfrak{B}$ be a countable ω-categorical τ-structure. Then the following are equivalent.*

1. *$\mathfrak{B}$ is HI-homogeneous.*
2. *$\mathfrak{B}$ is HA-homogeneous.*
3. *Every orbit of k-tuples under* Aut($\mathfrak{B}$) *can be defined by a conjunction of atomic τ-formulas.*
4. *Every first-order formula is equivalent over $\mathfrak{B}$ to a positive Boolean combination of atomic τ-formulas.*

PROOF. The implication (2) ⇒ (1) is trivial.

(1) ⇒ (2). If h is a homomorphism from a finite substructure $\mathfrak{A}$ of $\mathfrak{B}$ to $\mathfrak{B}$, then by assumption h can be extended to a self-embedding of $\mathfrak{B}$. We have to show that h can even be extended to an automorphism of $\mathfrak{B}$. We construct the automorphism by a back-and-forth argument. Suppose we have already constructed an isomorphism f between two finite substructures $\mathfrak{A}$ and $\mathfrak{A}'$ of $\mathfrak{B}$, and let b be any element of B. (Going forth) Then there is a self-embedding of $\mathfrak{B}$ that extends f, and the restriction of this self-embedding to $A \cup \{b\}$ is an embedding of the substructure of $\mathfrak{B}$ induced on $A \cup \{b\}$ that extends f. (Going back) Apply the same reasoning to f^{-1} to define an extension of f whose image includes b. Since $\mathfrak{B}$ is countable we eventually obtain an automorphism of $\mathfrak{B}$ that extends h.

(2) ⇒ (3). Let $a = (a_1, \ldots, a_k) \in B^k$; we claim that the conjunction ψ of all atomic τ-formulas that hold on a defines in $\mathfrak{B}$ the orbit of a under Aut($\mathfrak{B}$). To see this, suppose that $b \in B^k$ is such that $\mathfrak{B} \models \psi(b)$. Then the mapping that sends a_i to b_i is a partial endomorphism, and by HI-homogeneity can be extended to an automorphism of $\mathfrak{B}$. So b is in the same orbit as a, which proves the claim.

(3) ⇒ (4). Every first-order definable relation R is a finite union of orbits of n-tuples under Aut($\mathfrak{B}$). The assumption implies that each orbit can be defined by a conjunction of atomic τ-formulas. The disjunction over all those conjunctions is a positive Boolean combination of atomic τ-formulas defining R.

(4) ⇒ (2). Let $\bar{a} = (a_1, \ldots, a_k)$ and $\bar{b} = (b_1, \ldots, b_k)$ be k-tuples of elements of $\mathfrak{B}$ such that the mapping f that sends a_i to b_i, for $1 \leq i \leq n$, is a homomorphism between the substructures of $\mathfrak{B}$ induced on $\{a_1, \ldots, a_k\}$ and on $\{b_1, \ldots, b_k\}$. By Theorem 4.1.6, the orbit of $(a_1, \ldots, a_k)$ under Aut($\mathfrak{B}$) has a first-order definition ϕ, and by assumption ϕ is equivalent to a positive Boolean combination of atomic τ-formulas. Such formulas are preserved by homomorphisms, and hence $(b_1, \ldots, b_k)$ lies in the same orbit as $(a_1, \ldots, a_k)$. It follows that f can be extended to an automorphism of $\mathfrak{B}$. ⊣

The equivalence of HA-homgeneity and HI-homogeneity holds for countable structures in general [267]. By expanding an ω-categorical τ-structure $\mathfrak{B}$ by all negations of atomic τ-formulas, Lemma 4.3.2 implies the following.

COROLLARY 4.3.3 (Statement 2.22 in [123]). *An ω-categorical structure $\mathfrak{B}$ admits quantifier elimination if and only if it is homogeneous.*

4.3.2. Strong amalgamation and algebraic closure. When C is a strong amalgamation class, then the Fraïssé-limit of C has a remarkable property. Let $\mathfrak{B}$ be a structure, and A a finite set of elements of $\mathfrak{B}$. Then $\operatorname{acl}_{\mathfrak{B}}(A)$ denotes the (model-theoretic) *algebraic closure of A in $\mathfrak{B}$*, i.e., the elements of $\mathfrak{B}$ that lie in finite sets that are first-order definable over $\mathfrak{B}$ with parameters from A. In ω-categorical structures, this is precisely the set of elements of $\mathfrak{B}$ that lie in finite orbits under $\mathrm{Aut}(\mathfrak{B})_{(A)}$. We say that $\mathfrak{B}$ *has no algebraicity* if $\mathrm{acl}_{\mathfrak{B}}(A) = A$ for all finite sets of parameters A. Otherwise, if $\mathrm{acl}_{\mathfrak{B}}(A) \setminus A$ is non-empty, we say that $\mathfrak{B}$ *has algebraicity*. Note that $\mathfrak{B}$ has algebraicity if and only if there are finitely many elements $a_1, \ldots, a_n$ in $\mathfrak{B}$ such that a single element distinct from $a_1, \ldots, a_n$ is definable in $(\mathfrak{B}, a_1, \ldots, a_n)$.

Example 4.3.4. Let E be an equivalence relation on a countably infinite set B whose classes have size $k \in \mathbb{N}$. Then the structure $(B; E)$ is homogeneous (the Fraïssé-limit of Example 2.3.6), $\mathrm{Aut}(B; E)$ has no finite orbits, but algebraicity.

THEOREM 4.3.5 (See (2.15) in [123]). *A homogeneous ω-categorical structure $\mathfrak{B}$ has no algebraicity if and only if the age of $\mathfrak{B}$ has strong amalgamation.*

PROOF. Suppose first that $\mathrm{Age}(\mathfrak{B})$ has strong amalgamation, and suppose for contradiction that there are elements $a_1, \ldots, a_n$ of $\mathfrak{B}$ such that the set $\{a_0\}$ is definable in $(\mathfrak{B}, a_1, \ldots, a_n)$. Let $\mathfrak{A}$ be the substructure of $\mathfrak{B}$ induced on $A := \{a_1, \ldots, a_n\}$ and let $\mathfrak{C}$ be the substructure of $\mathfrak{B}$ induced on $A \cup \{a_0\}$. Then $\mathrm{Age}(\mathfrak{B})$ contains a strong amalgam $\mathfrak{D}$ of $\mathfrak{C}$ and $\mathfrak{C}$ over $\mathfrak{A}$, so there are embeddings e_1 and e_2 of $\mathfrak{C}$ into $\mathfrak{D}$ such that $e_1(C) \cap e_2(C) = A$. By the homogeneity of $\mathfrak{B}$ we can assume that $\mathfrak{D}$ is a substructure of $\mathfrak{B}$ that contains $\mathfrak{A}$ as a substructure, and that e_1 and e_2 fix the elements of A. Again by the homogeneity of $\mathfrak{B}$ the tuples $(a_1, \ldots, a_n, e(a_0))$ and $(a_1, \ldots, a_n, e(a_0))$ lie in the same orbit of $(n+1)$-tuples under $\mathrm{Aut}(\mathfrak{B})$, and it follows that the elements $e_1(a_0)$ and $e_2(a_0)$ lie in the same orbit under $\mathrm{Aut}(\mathfrak{B}, a_1, \ldots, a_n)$, contradicting the assumption that $\{a_0\}$ is definable in $(\mathfrak{B}, a_1, \ldots, a_n)$.

Now suppose that $\mathfrak{B}$ has no algebraicity. To prove that $\mathrm{Age}(\mathfrak{B})$ has strong amalgamation, it suffices to verify the strong 1-point amalgamation property (Proposition 2.3.18). Let $\mathfrak{A}, \mathfrak{B}_1, \mathfrak{B}_2 \in \mathrm{Age}(\mathfrak{B})$ be such that $\mathfrak{B}_1 \cap \mathfrak{B}_2 = \mathfrak{A}$ and $|B_1| = |B_2| = |A| + 1$. By the homogeneity of $\mathfrak{B}$, there exists an amalgam $\mathfrak{C} \in \mathrm{Age}(\mathfrak{B})$ of $\mathfrak{B}_1$ and $\mathfrak{B}_2$ over $\mathfrak{A}$ with embeddings $e_1 \colon \mathfrak{B}_1 \hookrightarrow \mathfrak{C}$ and $e_2 \colon \mathfrak{B}_2 \hookrightarrow \mathfrak{C}$. Without loss of generality we assume that $\mathfrak{C}$ is a substructure of $\mathfrak{B}$. Let $b \in B_1 \setminus A$ and let $b_1, \ldots, b_k$ be the elements of B_2 such that $b_1 = B_2 \setminus A$. Since $\mathfrak{B}$ has no algebraicity and is ω-categorical, the orbit of $e_1(b)$ under $\mathrm{Aut}(\mathfrak{B}, e_2(b_1), \ldots, e_2(b_k))$ contains more than one element, so pick c in this orbit distinct from $e_1(b)$. Then the substructure of $\mathfrak{B}$ induced on $C \cup \{c\}$ is a strong amalgam of $\mathfrak{B}_1$ and $\mathfrak{B}_2$ over $\mathfrak{A}$ via e_1 and the embedding obtained from e_2 by mapping b_1 to c. ⊣

Note that if $\mathfrak{B}$ has no algebraicity, and $\mathfrak{C}$ is a first-order reduct of $\mathfrak{B}$, then $\mathfrak{C}$ has no algebraicity, too. We give an application of algebraicity in the context of constraint satisfaction.

LEMMA 4.3.6. *Let $\mathfrak{B}$ be an ω-categorical relational structure without algebraicity all of whose relations R are* injective, *i.e., every tuple in R has pairwise* *distinct entries. Then every finite structure that maps homomorphically to $\mathfrak{B}$ also has an injective homomorphism to $\mathfrak{B}$.*

PROOF. Let τ be the signature of $\mathfrak{B}$. Let f be a homomorphism from a finite τ-structure $\mathfrak{F}$ to $\mathfrak{B}$ such that $f(F)$ is maximal. If $|f(F)| = |F|$ then f is injective and we are done. Otherwise, there are $u, v \in F$ such that $f(u) = f(v)$. Let $\mathfrak{A}$ be the substructure of $\mathfrak{B}$ induced on $f(F) \setminus \{f(u)\}$. By Theorem 4.3.5, Age($\mathfrak{B}$) has strong amalgamation, and hence there exist embeddings $e_1, e_2 \colon \mathfrak{B}[f(F)] \hookrightarrow \mathfrak{B}$ such that $e_1 \circ f|_A = e_2 \circ f|_A$ and $e_1(f(u)) \neq e_2(f(v))$. Then the mapping $f' \colon F \to B$ defined by

$$f'(w) := \begin{cases} e_2(f(w)) & \text{if } w \neq u \\ e_1(f(w)) & \text{if } w \neq v \end{cases}$$

is well defined. To see that f' is a homomorphism from $\mathfrak{F}$ to $\mathfrak{B}$, let $(x_1, \dots, x_n) \in R^{\mathfrak{F}}$. Since f is a homomorphism we have $(f(x_1), \dots, f(x_n)) \in R^{\mathfrak{B}}$, and at most one of the $f(x_i)$ equals $f(u) = f(v)$ since $R^{\mathfrak{B}}$ is injective. Hence, $f'|_{\{x_1,\dots,x_n\}} = e_1 \circ f|_{\{x_1,\dots,x_n\}}$ or $f'|_{\{x_1,\dots,x_n\}} = e_2 \circ f|_{\{x_1,\dots,x_n\}}$, proving that $(f'(x_1), \dots, f'(x_n)) \in R^{\mathfrak{B}}$. Since $f'(u) \neq f'(v)$, we have $|f'(F)| > |f(F)|$, a contradiction. ⊣

4.3.3. Homogenisation. It is sometimes convenient to define an ω-categorical τ-structure $\mathfrak{B}$ by specifying an amalgamation class $\mathcal{C}$ with a larger signature than τ such that $\mathfrak{B}$ is a reduct of the Fraïssé-limit of $\mathcal{C}$. If the Fraïssé-limit of $\mathcal{C}$ satisfies the condition of Lemma 4.3.1, it will be ω-categorical, and therefore also all its reducts are ω-categorical (Lemma 4.7.3).

This method will be used in Section 5.1 to construct ω-categorical templates for computational problems that have been studied in phylogenetic analysis. It has also been used in [210] to give another proof of a theorem due to Cherlin, Shelah, and Shi (Theorem 4.3.7). The result appears in [129] for the special case where τ has a single binary relation denoting the edge relationship of undirected graphs. The statement for general relational signatures τ also follows from a result of [141]. The theorem of Cherlin, Shelah, and Shi will be useful to formulate problems in *monotone monadic SNP* as CSPs with ω-categorical templates (Section 5.6.2).

Let $\mathcal{F}$ be a finite set of finite structures with a finite relational signature τ. Recall that a τ-structure $\mathfrak{B}$ is called *$\mathcal{F}$-free* if there is no homomorphism from any structure in $\mathcal{F}$ to $\mathfrak{B}$. A structure $\mathfrak{A}$ in a class of structures $\mathcal{C}$ is called *universal* for $\mathcal{C}$ if the class of structures that embed into $\mathfrak{A}$ is precisely $\mathcal{C}$. Recall that a structure is *connected* if it cannot be given as the disjoint union of non-empty structures.

THEOREM 4.3.7 (Cherlin, Shelah, and Shi [129]). *Let $\mathcal{F}$ be a finite set of finite connected τ-structures; then there exists a countable model-complete τ-structure $\mathfrak{B}$ which is universal for the class of at most countable $\mathcal{F}$-free structures. Moreover, $\mathfrak{B}$ is ω-categorical and without algebraicity.*

Observe that

- the structure $\mathfrak{B}$ from Theorem 4.3.7 is unique up to isomorphism: this follows from the uniqueness of the model companion (Theorem 2.7.16), and ω-categoricity;
- in countable ω-categorical structures, being universal for the class of all countable $\mathcal{F}$-free structures is equivalent to the requirement that $\text{Age}(\mathfrak{B}) = \text{Forb}^{\text{hom}}(\mathcal{F})$, by Lemma 4.1.7.

If we are interested in an ω-categorical structure $\mathfrak{B}$ with $\text{Age}(\mathfrak{B}) = \text{Forb}^{\text{hom}}(\mathcal{F})$, then there is another candidate for $\mathfrak{B}$, which is easier to construct than the structure in Theorem 4.3.7. The structure $\mathfrak{B}$ that we construct in Theorem 4.3.8 below is in general not isomorphic to the structure from Theorem 4.3.7 (see Example 4.3.9). However, we will see in Section 4.7.4 that Theorem 4.3.8 below implies Theorem 4.3.7 using general principles. A primitive positive formula is called *connected* if its canonical database $D(\phi)$ (see Section 1.2.2) is connected (as a structure, in the sense of Section 1.1). If $\mathfrak{B}$ is a structure, we write $\mathfrak{B}_{\text{pp}(m)}$ for the expansion of $\mathfrak{B}$ by all relations that can be defined with a connected primitive positive formula with at most m variables, at least one free variable, and without equality. The following theorem is a rephrasing of results of Hubička and Nešetřil [210] and [209].

THEOREM 4.3.8. *Let $\mathcal{F}$ be a finite set of finite connected τ-structures and let m be the maximal domain size of the structures in $\mathcal{F}$. Then there exists a countable ω-categorical τ-structure $\mathfrak{B}$ without algebraicity such that*

- $\text{Age}(\mathfrak{B}) = \text{Forb}^{\text{hom}}(\mathcal{F})$;
- $\mathfrak{B}_{\text{pp}(m)}$ *is homogeneous;*
- $\text{Age}(\mathfrak{B}_{\text{pp}(m)})$ *is the class $\mathcal{P}$ of all substructures of structures of the form $\mathfrak{A}_{\text{pp}(m)}$ for $\mathfrak{A} \in \text{Forb}^{\text{hom}}(\mathcal{F})$.*

PROOF. The class $\mathcal{P}$ is clearly closed under isomorphisms and substructures, and contains only countably many non-isomorphic structures. We verify that it has the strong amalgamation property. Let $\mathfrak{A}, \mathfrak{B}_1, \mathfrak{B}_2 \in \mathcal{P}$ with $\mathfrak{A} = \mathfrak{B}_1 \cap \mathfrak{B}_2$. Let ϕ_i be the canonical query of $\mathfrak{B}_i$, for $i \in \{1, 2\}$. If $R_\psi(\bar{x})$ is a conjunct of ϕ_i where R_ψ is the relation that has been added for a primitive positive formula ψ, then replace the conjunct $R_\psi(\bar{x})$ in ϕ_i by the formula $\psi(\bar{x})$ (using fresh variable names for the existentially quantified variables in ψ). Let $\mathfrak{C}$ be the canonical database of $\phi_1 \wedge \phi_2$ (a τ-structure). We will show that $\mathfrak{C}_{\text{pp}(m)}$ is the required (strong) amalgam, where $f_1 \colon \mathfrak{B}_1 \to \mathfrak{C}$ and $f_2 \colon \mathfrak{B}_2 \to \mathfrak{C}$ are the identity map.

Claim 1. $\mathfrak{C}$ is $\mathcal{F}$-free. Suppose for contradiction that there was a homomorphism $h \colon \mathfrak{F} \to \mathfrak{C}$ for some $\mathfrak{F} \in \mathcal{F}$. Let χ be the canonical query of the substructure of $\mathfrak{C}$ induced on $h(F)$; note that χ has at most m variables. Let χ_i be the subset of the conjuncts of χ that is contained in ϕ_i, for $i \in \{1, 2\}$. Let

$x_1, \dots, x_n$ be the variables of χ_1 contained in A, and let $y_1, \dots, y_k$ be the remaining variables. Note that $k \geq 1$, since otherwise χ_2 would hold in the canonical query of $\mathfrak{B}_2$, a contradiction to the assumption that $\mathfrak{B}_2$ is $\mathcal{F}$-free. Similarly, there is at least one variable of χ_2 which is not contained in A. Therefore, $n \geq 1$ by the assumption that $\mathfrak{F}$ is connected. The primitive positive formula obtained by existentially quantifying $y_1, \dots, y_k$ is connected, has at most m variables and at least one free variable, so the structure $\mathfrak{A} \in \mathcal{P}$ contains a relation symbol R for the relation defined by this formula, and $\mathfrak{A} \models R(x_1, \dots, x_n)$. By the definition of $\mathcal{P}$, there exists a structure $\mathfrak{A}' \in \text{Forb}^{\text{hom}}(\mathcal{F})$ such that $\mathfrak{A}$ is a substructure of $\mathfrak{A}'_{\text{pp}(m)}$. The structure $\mathfrak{A}'$ must have witnesses for the variables that we quantified in χ_1 (here we use that A is non-empty), and the variables that we quantified in χ_2, so $\mathfrak{F}$ maps homomorphically to $\mathfrak{A}'$, a contradiction.

Claim 2. The structures $\mathfrak{B}_1$ and $\mathfrak{B}_2$ are substructures of $\mathfrak{C}_{\text{pp}(m)}$. We show this for $\mathfrak{B}_1$; the proof for $\mathfrak{B}_2$ is analogous. Let $\bar{a} \in A^k$ and $\bar{b} \in B_1^\ell$ and let R_ψ be a $(k+\ell)$-ary relation from the signature of $\mathfrak{B}_1$. First note that if $\mathfrak{B}_1 \models R_\psi(\bar{a}, \bar{b})$ then the conjunct $\psi(\bar{a}, \bar{b})$ is part of ϕ_1, and hence $\mathfrak{C} \models \psi(\bar{a}, \bar{b})$. Conversely, suppose that $\mathfrak{C} \models \psi(\bar{a}, \bar{b})$ for some primitive positive formula ψ with at most m variables. The definition of $\mathfrak{C}$ implies that $\psi(\bar{a}, \bar{b})$ can be written as $\exists \bar{x}, \bar{y}, \bar{z} \big(\psi_1(\bar{a}, \bar{b}, \bar{x}, \bar{y}) \wedge \psi_2(\bar{a}, \bar{y}, \bar{z})\big)$ where ψ_1 and ψ_2 are quantifier-free primitive positive formulas such that

- $\bar{x}$ lists the variables from ϕ_1 that are not variables of ϕ_2;
- $\bar{y}$ lists the variables that appear both in ϕ_1 and in ϕ_2;
- $\bar{z}$ lists the variables that appear in ϕ_2 but not in ϕ_1.

Note that the variables in $\bar{y}$ are in fact elements of A (recall that ϕ_1 and ϕ_2 were constructed from the canonical database of $\mathfrak{B}_1$ and $\mathfrak{B}_2$, respectively, so they contain the elements of A as variables). Let $\theta(\bar{a}, \bar{y})$ be the formula $\exists \bar{z} \colon \psi_2(\bar{a}, \bar{y}, \bar{z})$ which is primitive positive with at most m variables. Let $\mathfrak{B}_1'$ be a structure in $\text{Forb}^{\text{hom}}(\mathcal{F})$ such that $\mathfrak{B}_1$ is a substructure of $(\mathfrak{B}_1')_{\text{pp}(m)}$ (such a structure exists since $\mathfrak{B}_1 \in \mathcal{P}$). The above implies that there are elements $\bar{v}$ of A such that $\mathfrak{B}_1' \models \exists \bar{x} \colon \psi_1(\bar{a}, \bar{b}, \bar{x}, \bar{v})$ and $\mathfrak{A} \models R_\theta(\bar{a}, \bar{v})$. In particular, $\mathfrak{B}_1 \models R_\theta(\bar{a}, \bar{v})$. It follows that

$$\mathfrak{B}_1' \models \exists \bar{x}, \bar{z} \big(\psi_1(\bar{a}, \bar{b}, \bar{x}, \bar{v}) \wedge \psi_2(\bar{a}, \bar{b}, \bar{v}, \bar{z})\big)$$

and hence $\mathfrak{B}_1' \models \exists \bar{x}, \bar{y}, \bar{z} \big(\psi_1(\bar{a}, \bar{b}, \bar{x}, \bar{y}) \wedge \psi_2(\bar{a}, \bar{b}, \bar{y}, \bar{z})\big)$. Thus, $\mathfrak{B}_1 \models R_\psi(\bar{a}, \bar{b})$ as required.

Claims 1 and 2 show that $\mathfrak{C}_{\text{pp}(m)} \in \mathcal{P}$ is an amalgam of $\mathfrak{B}_1$ and $\mathfrak{B}_2$ over $\mathfrak{A}$. So $\mathcal{P}$ is a Fraïssé-class. Let $\mathfrak{B}'$ be its Fraïssé-limit, and let $\mathfrak{B}$ be the τ-reduct of $\mathfrak{B}'$. Then $\mathfrak{B}$ has the desired properties:

- it is ω-categorical, because it is a reduct of a homogeneous structure with a finite relational signature;

- it has no algebraicity, because it is a reduct of a structure without algebraicity: $\mathfrak{B}'$ is the Fraïssé-limit of a strong amalgamation class, and hence has no algebraicity by Theorem 4.3.5;
- clearly, $\mathrm{Age}(\mathfrak{B}) = \mathrm{Forb}^{\mathrm{hom}}(\mathcal{F})$;
- as we prove below, the structure $\mathfrak{B}_{\mathrm{pp}(m)}$ equals $\mathfrak{B}'$, which shows the last two items of the statement.

Let $\psi(x_1, \dots, x_k)$ be a primitive positive τ-formula with at most m variables and without equality conjuncts. We have to prove that $\mathfrak{B}' \models R_\psi(a_1, \dots, a_k)$ if and only if $\mathfrak{B}_{\mathrm{pp}(m)} \models R_\psi(a_1, \dots, a_k)$ for all $a \in B^k$. By definition, the latter is the case if and only if $\mathfrak{B} \models \psi(a_1, \dots, a_k)$. The substructure $\mathfrak{A}'$ of $\mathfrak{B}'$ induced on $a_1, \dots, a_k$ is contained in $\mathcal{P}$, and hence there exists a superstructure of $\mathfrak{A}'$ of the form $\mathfrak{A}_{\mathrm{pp}(m)} \in \mathcal{P}$ for a structure $\mathfrak{A} \in \mathrm{Forb}^{\mathrm{hom}}(\mathcal{F})$. Note that

$$\begin{aligned} \mathfrak{B}' \models R_\psi(a_1, \dots, a_k) &\Leftrightarrow \mathfrak{A}' \models R_\psi(a_1, \dots, a_k) \quad && (\mathfrak{A}' \in \mathrm{Age}(\mathfrak{B}')) \\ &\Leftrightarrow \mathfrak{A} \models \psi(a_1, \dots, a_k) && (\text{by definition of } \mathfrak{A}_{\mathrm{pp}(m)}). \end{aligned}$$

The structure $\mathfrak{A}_{\mathrm{pp}(m)}$ embeds into $\mathfrak{B}'$ and by the homogeneity of $\mathfrak{B}'$ we obtain that $\mathfrak{B}' \models R_\psi(a_1, \dots, a_k)$ if and only if $\mathfrak{B} \models \psi(a_1, \dots, a_k)$, which is what we had to show. ⊣

Example 4.3.9. Let $\mathcal{F}$ be $\{C_5\}$. Let $\mathfrak{B}$ be the model-complete structure with $\mathrm{Age}(\mathfrak{B}) = \mathrm{Forb}^{\mathrm{hom}}(\mathcal{F})$. We claim that $\mathfrak{B}$ is not isomorphic to the structure $\mathfrak{B}'$ that we constructed in Theorem 4.3.8. To see this, observe that the formula

$$\neg \exists y (E(x, y) \wedge E(y, z))$$

is over $\mathfrak{B}$ equivalent to an existential formula (Theorem 2.6.2); in our case, this is the existential formula

$$\varphi = E(x, z) \vee \exists y_1, y_2 (E(x, y_1) \wedge E(y_1, y_2) \wedge E(y_2, z)).$$

But note that the class $\mathcal{P}$ from the proof of Theorem 4.3.8 contains $\mathfrak{A}_{\mathrm{pp}(m)}$ for the structure $\mathfrak{A} \in \mathrm{Forb}^{\mathrm{hom}}(\mathcal{F})$ that contains two vertices x, z and where all relations are empty. In the τ-reduct $\mathfrak{B}'$ of the Fraïssé-limit of $\mathcal{P}$, any path from x to z must have at least six vertices. So the formula $\neg E(x, y)$ is in $\mathfrak{B}'$ not equivalent to φ, so $\mathfrak{B}$ and $\mathfrak{B}'$ cannot be isomorphic.

4.3.4. Expansions by constants. It is easy to see that every expansion of a homogeneous structure $\mathfrak{B}$ by constant symbols is again homogeneous. Also note that if $c \in B$ then $\mathrm{Aut}(\mathfrak{B}, c) = \mathrm{Aut}(\mathfrak{B}, \{c\})$. However, the relational structure $(\mathfrak{B}, \{c\})$ is in general not homogeneous, as the following example shows.

Example 4.3.10. Let $(V; E)$ be the random graph (Example 2.3.9) and let $c \in \mathbb{V}$. Let $p \in V \setminus \{c\}$ be such that $E(p, c)$ and $q \in V \setminus \{c\}$ be such that $\neg E(c, q)$. Then the mapping that sends p to q is an isomorphism between (one-element) substructures of $(V; E, \{c\})$ which cannot be extended to an automorphism of $(V; E, \{c\})$. (The difference to $(V; E, c)$ is that all substructures of $(V; E, c)$ must contain c.)

It will be convenient later to work with homogeneous structures with finite relational signature, and we therefore define the following structure.

DEFINITION 4.3.11. Let $\mathfrak{B}$ be a relational structure with signature τ and let $b_1, \dots, b_n \in B$. Then $\mathfrak{B}_{b_1,\dots,b_n}$ denotes the expansion of $\mathfrak{B}$ which contains all relations that are defined by atomic formulas over the structure $(\mathfrak{B}, b_1, \dots, b_n)$.

Note that if the signature of $\mathfrak{B}$ is finite then the signature of $\mathfrak{B}_{b_1,\dots,b_n}$ is also finite, and the maximal arity is unaltered.

PROPOSITION 4.3.12. *Let $\mathfrak{B}$ be a homogeneous relational structure and $b_1, \dots, b_n \in B$. Then $\mathfrak{B}_{b_1,\dots,b_n}$ is homogeneous and* $\mathrm{Aut}(\mathfrak{B}, b_1, \dots, b_n) = \mathrm{Aut}(\mathfrak{B}_{b_1,\dots,b_n})$.

PROOF. The claim about the automorphism groups follows from the observation that $\mathfrak{B}_{b_1,\dots,b_n}$ has in particular the unary relations $\{b_1\}, \dots, \{b_n\}$. To show the homogeneity of $\mathfrak{B}_{b_1,\dots,b_n}$, let a be an isomorphism between two finite substructures A_1, A_2 of $\mathfrak{B}_{b_1,\dots,b_n}$. Since $\mathfrak{B}_{b_1,\dots,b_n}$ contains for all $i \leq n$ the relation $\{b_i\}$ which is preserved by a, it follows that if A_1 or A_2 contains b_i, then both A_1 and A_2 must contain b_i, and $a(b_i) = b_i$. If b_i is contained in neither A_1 nor A_2, then a can be extended to a partial isomorphism a' of $\mathfrak{B}_{b_1,\dots,b_n}$ with domain $A_1 \cup \{b_i\}$ by setting $a(b_i) = b_i$: this follows directly from the definition of $\mathfrak{B}_{b_1,\dots,b_n}$. By the homogeneity of $\mathfrak{B}$, the map a' can be extended to an automorphism of $\mathfrak{B}$. This automorphism fixes each of $b_1, \dots, b_n$, and hence is an automorphism of $\mathfrak{B}_{b_1,\dots,b_n}$. ⊣

4.4. Oligomorphic endomorphism monoids

We apply the model-theoretic preservation theorems from Section 2.5 to characterise existential and existential positive definability of relations in ω-categorical structures. For *finite* structures, a characterisation of existential positive definability has already been noted by Krasner [246]. Note that existential definability over finite structures equals first-order definability (for finite structures, self-embeddings are necessarily automorphisms).

THEOREM 4.4.1 (from [43] and [89, 67]). *Let $\mathfrak{B}$ be an ω-categorical structure with base set B, and $R \subseteq B^k$ be a relation.*

1. *R has an existential positive definition in $\mathfrak{B}$ if and only if R is preserved by all endomorphisms of $\mathfrak{B}$.*
2. *R has an existential definition in $\mathfrak{B}$ if and only if R is preserved by all self-embeddings of $\mathfrak{B}$.*

PROOF. We have already remarked in Section 2.1 that existential positive formulas are preserved by homomorphisms, and existential formulas are preserved by embeddings.

For the other direction, note that the endomorphisms and self-embeddings of $\mathfrak{B}$ contain the automorphisms of $\mathfrak{B}$, and hence Theorem 4.1.6 shows that R has a first-order definition ϕ in $\mathfrak{B}$. Suppose for contradiction that R is preserved by all endomorphisms of $\mathfrak{B}$ but has no existential positive definition in $\mathfrak{B}$. We use the homomorphism preservation theorem (Theorem 2.5.2). Since by assumption ϕ is not equivalent to an existential positive formula in $\mathfrak{B}$, there are models $\mathfrak{B}_1$ and $\mathfrak{B}_2$ of the first-order theory of $\mathfrak{B}$ and a homomorphism h from $\mathfrak{B}_1$ to $\mathfrak{B}_2$ that does not preserve ϕ. By the theorem of Löwenheim-Skolem (Theorem 2.1.11; by Theorem 4.1.6 we may assume that the signature is countable) the first-order theory of the two-sorted structure $(\mathfrak{B}_1, \mathfrak{B}_2; h)$ has a countable model $(\mathfrak{B}'_1, \mathfrak{B}'_2; h')$. Since both $\mathfrak{B}'_1$ and $\mathfrak{B}'_2$ must be countably infinite, and because $\mathfrak{B}$ is ω-categorical, we have that $\mathfrak{B}'_1$ and $\mathfrak{B}'_2$ are isomorphic to $\mathfrak{B}$, and h' can be seen as an endomorphism of $\mathfrak{B}$ that does not preserve ϕ; a contradiction.

The argument for existential definitions is similar, but instead of the homomorphism preservation theorem we use the theorem of Łos-Tarski (Corollary 2.5.3). ⊣

We now present a Galois connection between sets of relations on a base set B and sets of functions from B to B for which the Galois-closed sets correspond to existential positive definability and to closed transformation monoids, respectively. This is similar to the Galois connection presented in Section 4.2 for first-order definability and closed permutation groups.

For a structure $\mathfrak{B}$, we denote the set of relations with an existential positive definition in $\mathfrak{B}$ by $\langle\mathfrak{B}\rangle_{\text{ep}}$. A set $\mathcal{F} \subseteq B^B$ is *closed (in B^B)* if it contains every $f \in B^B$ such that for every finite subset A of B there exists a $g \in \mathcal{F}$ such that $f(a) = g(a)$ for all $a \in A$. The set of endomorphisms of a relational structure $\mathfrak{B}$ is denoted by $\text{End}(\mathfrak{B})$. The set of functions from B^B that preserve a set of relations $\mathcal{R}$ over the domain B is also denoted by $\text{End}(\mathcal{R})$. The following can be shown in a similarly straightforward way as Proposition 4.2.2.

PROPOSITION 4.4.2. *For every $\mathcal{F} \subseteq B^B$, the following are equivalent.*

1. *$\mathcal{F}$ is the transformation monoid of a relational structure;*
2. *$\mathcal{F}$ is a transformation monoid that is closed in B^B.*

Proceeding as in Section 4.1.1, we make the following definition. Note the differences from Definition 4.2.6; however, it will always be clear from the context which of the two definitions applies.

DEFINITION 4.4.3. Let $\mathcal{F}$ be a subset of B^B.

- $\langle\mathcal{F}\rangle$ denotes the smallest transformation monoid that contains $\mathcal{F}$.
- The *closure* $\overline{\mathcal{F}}$ of $\mathcal{F}$ is the smallest closed subset of B^B that contains $\mathcal{F}$.

The set of all relations that are preserved by all functions in $\mathcal{F} \subseteq B^B$ is denoted by $\text{Inv}(\mathcal{F})$. The proof of the following statement is similar to the proof of Proposition 4.2.7.

PROPOSITION 4.4.4. *Let* $\mathfrak{F} \subseteq B^B$. *Then* $\text{End}(\text{Inv}(\mathfrak{F})) = \overline{\langle \mathfrak{F} \rangle}$ *equals the smallest transformation monoid that contains* $\mathfrak{F}$ *and is closed in* B^B.

Theorem 4.4.1 implies the following analog to Corollary 4.2.10.

COROLLARY 4.4.5. *Let* $\mathfrak{C}$ *be an* ω*-categorical structure. Then the lattice of closed submonoids of* C^C *that contain* $\text{Aut}(\mathfrak{C})$ *is anti-isomorphic to the lattice of sets of the form* $\langle \mathfrak{B} \rangle_{\text{ep}}$, *ordered by inclusion, where* $\mathfrak{B}$ *is first-order definable in* $\mathfrak{C}$.

To illustrate we present a simple and typical application.

LEMMA 4.4.6. *Let* $\mathfrak{B}$ *be such that* $\text{Aut}(\mathfrak{B})$ *is 2-set transitive. If* $\mathfrak{B}$ *has a non-injective endomorphism* f, *then* $\mathfrak{B}$ *also has a constant endomorphism.*

PROOF. Let f be an endomorphism of $\mathfrak{B}$ such that $f(b) = f(b')$ for two distinct values $b, b' \in B$. Let $b_1, b_2, \dots$ be an enumeration of B. We construct a sequence of endomorphisms $e_1, e_2, \dots,$ where e_i is an endomorphism that maps all of the values $b_1, \dots, b_i$ to b_1. This suffices, since then the mapping defined by $e(x) := b_1$ for all x is an endomorphism of $\mathfrak{B}$ as well since $\text{End}(\mathfrak{B})$ is closed.

For e_1, we take the identity map, which clearly is an endomorphism with the desired properties. To define e_i for $i \geq 2$, let α be an automorphism of $\mathfrak{B}$ that maps $\{b_1, e_{i-1}(b_i)\}$ to $\{b, b'\}$; such an automorphism exists because $\text{Aut}(\mathfrak{B})$ is 2-set transitive. Then the endomorphism $f(\alpha e_{i-1}(x))$ is constant on $b_1, \dots, b_i$; recall that $b_1 = e_{i-1}(b_1) = \dots = e_{i-1}(b_{i-1})$. Since $\mathfrak{B}$ is 2-transitive, it is in particular transitive, and there is an automorphism β that maps $f(b)$ to b_1. Then $e_i : x \mapsto \beta f(\alpha e_{i-1}(x))$ is an endomorphism of $\mathfrak{B}$ with the desired properties. ⊣

There are structures $\mathfrak{B}$ such that $\text{Aut}(\mathfrak{B}) = \text{End}(\mathfrak{B})$, for example the structure $\mathfrak{B} = (\mathbb{Z}; \{(x, y) \mid x = y + 1\})$. This cannot happen if $\mathfrak{B}$ is countably infinite ω-categorical! The most natural proof of this fact seems to be a model-theoretic one.

PROPOSITION 4.4.7. *Let* $\mathfrak{B}$ *be an* ω*-categorical structure. Then there exists an elementary self-embedding* $e \in \text{End}(\mathfrak{B}) \setminus \text{Aut}(\mathfrak{B})$.

PROOF. Let τ be the signature of $\mathfrak{B}$ and let $b_1, b_2, \dots$ be constant symbols and $\mathfrak{C}$ an expansion of $\mathfrak{B}$ with signature $\tau \cup \{b_1, b_2, \dots\}$ so that $b_1^{\mathfrak{C}}, b_2^{\mathfrak{C}}, \dots$ enumerates all elements of B. Let c be a new constant symbol. Note that every finite subset of the theory $T := \text{Th}(\mathfrak{B}') \cup \{c \neq b_i \mid i \in \mathbb{N}\}$ has an infinite model, and hence by compactness of first-order logic (Theorem 2.1.6) T has a infinite model $\mathfrak{A}$; by Theorem 2.1.11, we may assume that $\mathfrak{A}$ is countably infinite. Note that the τ-reduct of $\mathfrak{A}$ satisfies the same first-order sentences as $\mathfrak{B}$, and by the ω-categoricity of $\mathfrak{B}$ there exists an isomorphism i between this reduct and $\mathfrak{B}$. Then the map $e : B \to B$ that sends $b_n^{\mathfrak{B}}$ to $i(b_n^{\mathfrak{A}})$ is an elementary self-embedding such that $e(B)$ does not contain $i(c^{\mathfrak{A}})$. ⊣

4.5. Countably categorical model-complete cores

We have seen that first-order definability in a countable ω-categorical structure $\mathfrak{B}$ is captured by the automorphism group of $\mathfrak{B}$ (Section 4.2) and that existential-positive definability in $\mathfrak{B}$ is captured by the endomorphism monoid of $\mathfrak{B}$ (Section 4.4). Particularly interesting is the situation where existential positive and first-order definability is the same, which happens precisely if $\mathfrak{B}$ is model complete and a core. In the more general terminology of Section 2.6, this is the case if and only if $\mathfrak{B}$ has a model-complete core theory. We will see later that *every* ω-categorical structure $\mathfrak{B}$ is homomorphically equivalent to an ω-categorical structure that has this property (Theorem 4.7.4); so when it comes to the classification of the computational complexity of CSPs for ω-categorical structures $\mathfrak{B}$, we usually work with model-complete cores. In this section we prove various equivalent conditions that characterise ω-categorical model-complete cores and illustrate them with examples.

THEOREM 4.5.1. *Let $\mathfrak{B}$ be a countable ω-categorical structure. Then the following are equivalent.*

1. *$\mathfrak{B}$ is a model-complete core.*
2. *$\mathfrak{B}$ has a homogeneous expansion such that all relations of the expansion and their complements have an existential positive definition in $\mathfrak{B}$.*
3. *Every first-order formula is equivalent to an existential positive one over $\mathfrak{B}$.*
4. *The orbits of n-tuples under* $\mathrm{Aut}(\mathfrak{B})$ *are primitively positively definable in $\mathfrak{B}$.*
5. $\mathrm{Aut}(\mathfrak{B})$ *is dense in* $\mathrm{End}(\mathfrak{B})$.
6. *Every $e \in$* $\mathrm{End}(\mathfrak{B})$ *is* locally invertible, *i.e., for every finite tuple t of elements of B there exists $i \in$* $\mathrm{End}(\mathfrak{B})$ *such that $i(e(t)) = t$.*
7. *Every existential positive formula is equivalent to a universal-negative formula over $\mathfrak{B}$.*
8. *$\mathfrak{B}$ has a model-complete core theory.*
9. *The theory of $\mathfrak{B}$ is equivalent to a $\forall\exists^+$-theory.*

PROOF. We show these statements in cyclic order. For the implication from (1) to (2), consider the expansion $\mathfrak{B}'$ of $\mathfrak{B}$ by all relations with an existential positive definition in $\mathfrak{B}$. Since every endomorphism of $\mathfrak{B}$ preserves all first-order formulas, it also preserves all relations in $\mathfrak{B}'$ and their complements. It follows that the complements also have an existential positive definition by Theorem 4.4.1, and hence the expansion is of the desired type. Since the endomorphism of $\mathfrak{B}$ preserve by assumption all first-order formulas, and since by the ω-categoricity of $\mathfrak{B}$ and Theorem 4.1.6 the orbits of n-tuples under $\mathrm{Aut}(\mathfrak{B})$ are first-order definable in $\mathfrak{B}$, it follows from Theorem 4.4.1 that the orbits of n-tuples under $\mathrm{Aut}(\mathfrak{B})$ (and its expansion) have an existential positive definition in $\mathfrak{B}$. Then, the expansion $\mathfrak{B}'$ by all existentially positively definable relations in homogeneous.

For the implication from (2) to (3), let ϕ be a first-order formula. Then ϕ has in the homogeneous expansion of $\mathfrak{B}$ from Item (2) a quantifier-free definition ψ (Corollary 4.3.3); assume without loss of generality that ψ is written in conjunctive normal form. If we replace all literals by their existential positive definition in $\mathfrak{B}$ we arrive at an equivalent formula which is existential positive in the signature of $\mathfrak{B}$.

For the implication from (3) to (4), let O be an orbit of n-tuples under $\mathrm{Aut}(\mathfrak{B})$. By Theorem 4.1.6, O is first-order definable, so by assumption, O even has an existential positive definition. Note that every existential positive formula can be written as a disjunction of primitive positive formulas, so let ϕ be such a definition of O. We can also assume without loss of generality that none of the disjuncts in ϕ implies another (otherwise, we simply omit it). Since O is a minimal non-empty first-order definable relation, ϕ can only contain a single disjunct, and therefore is primitive positive.

(4) $\Rightarrow$ (5). Let $e \in \mathrm{End}(\mathfrak{B})$. To show that $e \in \overline{\mathrm{Aut}(\mathfrak{B})}$, let t be a finite tuple of elements of $\mathfrak{B}$. We have to show that there is an automorphism α of $\mathfrak{B}$ such that $e(t) = \alpha t$. The orbit of t is by assumption primitively positively definable, and hence preserved by e. So $e(t)$ is in the same orbit as t, and we are done.

(5) $\Rightarrow$ (6). Let $\bar{x}$ be a finite tuple of elements of B, and $e \in \mathrm{End}(\mathfrak{B})$. By (5) there exists an $\alpha \in \mathrm{Aut}(\mathfrak{B})$ such that $e(\bar{x}) = \alpha\bar{x}$. Choose $g = \alpha^{-1}$.

(6) $\Rightarrow$ (7). Let ϕ be an existential positive formula. In order to show that ϕ is equivalent to an universal negative formula, we use Theorem 4.4.1 and show that $\neg\phi$ is preserved by all endomorphisms f of $\mathfrak{B}$, and hence equivalent to an existential positive formula. Let $\bar{a}$ be a finite tuple of elements of B such that $\mathfrak{B} \models \phi(f(\bar{a}))$. By assumption, there exists a $g \in \mathrm{End}(\mathfrak{B})$ such that $g(f(\bar{a})) = \bar{a}$, and hence $\mathfrak{B} \models \phi(\bar{a})$. This shows that f preserves $\neg\phi$.

The implication from (7) to (8) follows from the implication from (4) to (1) in Theorem 2.6.12.

The implication from (8) to (9) follows from Theorem 2.7.11: The theory of $\mathfrak{B}$ is its own core companion and hence is equivalent to its positive Kaiser hull, which is by definition a $\forall\exists^+$-theory.

For (9) $\Rightarrow$ (1), note that by definition the theory T of $\mathfrak{B}$ equals its positive Kaiser hull; see Theorem 2.7.11. Lemma 2.7.7 states that T equals the set of all $\forall\exists^+$-sentences that hold in all T-epc structures, and then Theorem 2.7.11 implies that T is its own core companion and therefore a model-complete core theory. $\dashv$

Note that in (2), we have to insist on the complements of all relations being existentially positively definable; in other words, it is not true that an ω-categorical structure $\mathfrak{B}$ is a model-complete core if and only if the expansion by all relations with an existential positive definition is homogeneous, as illustrated by the following example.

Example 4.5.2. The existential positive expansion of the structure $(\mathbb{Q}_0^+; <)$ by the unary predicate P such that $P(x) \Leftrightarrow \exists y\,(y < x)$ is homogeneous, but $(\mathbb{Q}_0^+; <)$ is not even model complete (Example 2.6.5).

4.5.1. Adding constants. The fact that in countable ω-categorical model-complete cores the orbits of n-tuples are primitively positively definable is one of the key facts for finite cores $\mathfrak{B}$ mentioned earlier. This can be exploited in the study of the CSP in many ways. For instance, it can be combined with Proposition 3.1.7 to obtain the following.

COROLLARY 4.5.3. *Let $\mathfrak{B}$ be an ω-categorical model-complete core, and suppose that $\mathfrak{A}$ is a structure with finite relational signature and a primitive positive interpretation in $\mathfrak{B}$ with parameters from $\mathfrak{B}$. Then there exists a finite signature reduct $\mathfrak{B}'$ of $\mathfrak{B}$ such that* CSP$(\mathfrak{A})$ *has a polynomial-time reduction to* CSP$(\mathfrak{B}')$.

This in turn immediately yields the following hardness condition.

COROLLARY 4.5.4. *Let $\mathfrak{B}$ be an ω-categorical structure, and let $\mathfrak{C}$ be its model-complete core. If K_3 has a primitive positive interpretation in $\mathfrak{C}$ with parameters from $\mathfrak{C}$ then $\mathfrak{B}$ has a finite-signature reduct with an NP-hard* CSP.

We can now state the infinite-domain tractability conjecture (Conjecture 3.1) in its historically first formulation. We will see in Section 10.3 that the two conjectures are equivalent.

CONJECTURE 4.1. *Let $\mathfrak{B}$ be a reduct of a finitely bounded homogeneous structure, and let $\mathfrak{C}$ be its model-complete core. If K_3 does not have a primitive positive interpretation in $\mathfrak{C}$ with parameters from $\mathfrak{C}$ then* CSP$(\mathfrak{B})$ *is in P.*

This conjecture involves two operators on structures that we have already introduced in Section 3.6: the operator I for primitive positive interpretations and the operator C for adding constants for tuples whose orbit is primitively positively definable. We revisit these operators for ω-categorical model-complete cores.

LEMMA 4.5.5. *Let $\mathfrak{B}$ be an ω-categorical model-complete core. Then any structure in* C$(\mathfrak{B})$ *is an ω-categorical model-complete core as well.*

PROOF. Let $c \in B$, let e be an endomorphism of $(\mathfrak{B}, c)$, and let $\phi(\bar{x})$ be a first-order formula over the signature of $(\mathfrak{B}; c)$. Let $\psi(\bar{x}, y)$ be the first-order formula over the signature of $\mathfrak{B}$ such that $\psi(\bar{x}, c)$ equals $\phi(\bar{x})$. As $\mathfrak{B}$ is an ω-categorical model-complete core, e preserves ψ. Since e also preserves c we find that e preserves $\phi(\bar{x})$ and it follows that $(\mathfrak{B}, c)$ is a model-complete core. The ω-categoricity of $(\mathfrak{B}, c)$ is an easy consequence of Theorem 4.1.6; see Lemma 4.7.3. ⊣

Also note the following relation between expansions by constants and primitive positive interpretations in ω-categorical model-complete core.

LEMMA 4.5.6. *Let $\mathfrak{B}$ be an ω-categorical model-complete core. Suppose that $\mathfrak{A}$ has a primitive positive interpretation in $\mathfrak{B}$. Then any expansion of $\mathfrak{A}$ by constants is in* $\mathrm{IC}(\mathfrak{B})$.

PROOF. Suppose that $\mathfrak{A}$ has a d-dimensional primitive positive interpretation I in $\mathfrak{B}$ and let $a \in A^n$. Arbitrarily choose $b \in I^{-1}(a)$; note that $b = (b_{1,1}, \dots, b_{n,d}) \in B^{dn}$. Let $\chi(\bar{x}, \bar{y})$ be the primitive positive definition of $I^{-1}(=^{\mathfrak{A}})$. We claim that I is also an interpretation of $(\mathfrak{A}, a)$ in $(\mathfrak{B}, b)$: the pre-image $I^{-1}(a)$ has the primitive positive definition $\bigwedge_{i \leq n} \chi(x_{i,1}, \dots, x_{i,d}, b_{i,1}, \dots, b_{i,d})$. Moreover, the orbit of b under $\mathrm{Aut}(\mathfrak{B})$ has a primitive positive definition in $\mathfrak{B}$ because $\mathfrak{B}$ is a model-complete core. Hence, $(\mathfrak{A}, a) \in \mathrm{C}(\mathfrak{B})$. ⊣

Combining this observation with Lemma 3.6.1, we obtain the following.

COROLLARY 4.5.7. *Let $\mathfrak{B}$ be an ω-categorical model-complete core, and let $\mathcal{C}$ be the smallest class that contains $\mathfrak{B}$ and is closed under* I *and expansions by finitely many constants. Then $\mathcal{C} = \mathrm{IC}(\mathfrak{B})$.*

4.5.2. Model completeness. The description of ω-categorical model-complete cores from Theorem 4.5.1 implies the following characterisation of model-completeness for countable ω-categorical structures.

THEOREM 4.5.8. *Let $\mathfrak{B}$ be a countable ω-categorical structure. Then the following are equivalent.*

1. *The structure $\mathfrak{B}$ is model complete.*
2. *$\mathfrak{B}$ has a homogeneous expansion by relations $R_1, R_2, \dots$ such that both the R_i and their complements have existential definitions in $\mathfrak{B}$.*
3. *Every self-embedding of $\mathfrak{B}$ is in* $\overline{\mathrm{Aut}(\mathfrak{B})}$.
4. $\mathrm{Th}(\mathfrak{B})$ *is equivalent to a $\forall\exists$-theory.*

PROOF. Follows from Theorem 4.5.1. ⊣

We give an overview of the various restrictions of first-order logic that were considered in this text in Figure 2. The following example shows that being a core *and* homogeneous (and therefore model-complete) does not imply that every first-order formula is equivalent to a formula which is both quantifier-free and existential positive.

Example 4.5.9. The structure $(\{0, 1\}; Z, \neq)$, for $Z = \{0\}$, is clearly homogeneous and a core. The relation $\{1\}$ has the quantifier-free definition $\neg Z(x)$, and it also has the existential positive definition $\exists y (Z(y) \wedge x \neq y)$, but it is clearly not definable by a formula which is both quantifier-free and existential positive.

4.6. Existential positive Ryll-Nardzewski

In this section we present an existential positive variant of the Ryll-Nardzewski theorem (Theorem 4.6.1 below). This also answers the question which CSPs can be formulated with an ω-categorical template, and it implies an existential

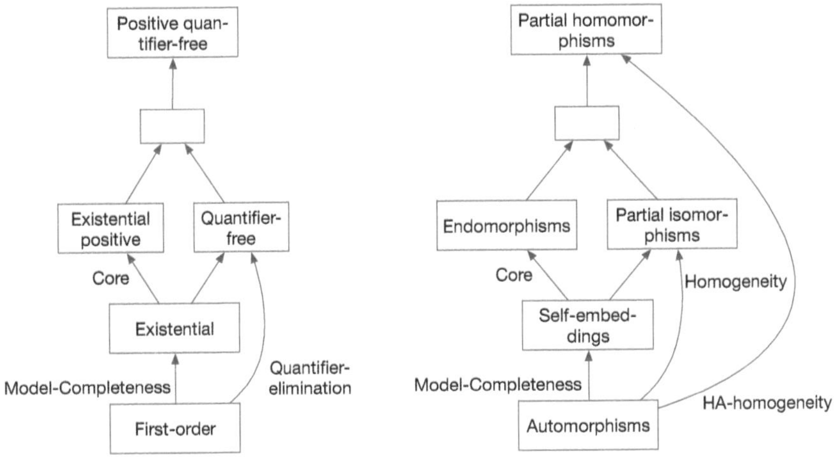

FIGURE 2. Various forms of syntactically restricted definability (left side), ordered by relative strength, and the corresponding class of operations (right side). The labels on the arrows indicate the condition on the structure when the corresponding two forms of definability coincide (left side), and correspondingly when one set of operations generates the other (right side).

variant of Ryll-Nardzewski due to Simmons [330]. Another consequence of Theorem 4.6.1 is that every ω-categorical theory has a model companion due to Saracino [318] and a core companion (a result from [44]).

For a satisfiable theory T, let $\sim_n^T$ be the equivalence relation defined on existential positive formulas with n free variables $x_1, \dots, x_n$ (we could have equivalently used primitive positive formulas here) as follows. For two such formulas ϕ_1 and ϕ_2, let $\phi_1 \sim_n^T \phi_2$ if for all existential positive formulas ψ with free variables $x_1, \dots, x_n$ we have that $\{\phi_1, \psi\} \cup T$ is satisfiable if and only if $\{\phi_2, \psi\} \cup T$ is satisfiable. The *index* of an equivalence relation is the number of its equivalence classes.

THEOREM 4.6.1 (Theorem 4.27 in [61]). *Let T be a theory with countable relational signature and the joint homomorphism property* (*JHP*; *cf. Proposition* 2.1.16). *Then the following are equivalent.*

(*i*) *T has an ω-categorical core companion.*
(*ii*) *$\sim_n^T$ has for each n finite index.*
(*iii*) *T has finitely many maximal existential positive n-types for each n.*
(*iv*) *There is a* (*finite or countably infinite*) *ω-categorical model-complete core $\mathfrak{B}$ that satisfies an existential positive sentence ϕ if and only if $T \cup \{\phi\}$ is satisfiable.*

PROOF. We show (i) $\Rightarrow$ (ii) $\Rightarrow$ (iii) $\Rightarrow$ (iv) $\Rightarrow$ (i).

(i) $\Rightarrow$ (ii). Let U be the core companion of T. Since U and T entail the same universal negative sentences, we can deduce that for every existential positive formula ψ the theory $U \cup \{\psi\}$ is satisfiable if and only if $T \cup \{\psi\}$ is satisfiable; from which it follows that the indices of $\sim_n^U$ and $\sim_n^T$ coincide.

For a proof by contraposition, assume that $\sim_n^U$ has infinite index for some n. Let ψ_1 and ψ_2 be two existential positive formulas from different equivalence classes of $\sim_n^U$. Hence, there is an existential positive formula ψ_3 such that exactly one of $\{\psi_1, \psi_3\} \cup U$ and $\{\psi_2, \psi_3\} \cup U$ is satisfiable. This shows that ψ_1 and ψ_2 are inequivalent modulo U. Therefore there are infinitely many first-order formulas with free variables $x_1, \ldots, x_n$ that are inequivalent modulo U, and U cannot be ω-categorical by Theorem 4.1.6.

(ii) $\Rightarrow$ (iii). We show that every maximal ep-n-type p is determined completely by the $\sim_n^T$ equivalence classes of the existential positive formulas contained in p. Since there are finitely many such classes, the result follows. Let p and q be maximal ep-n-types such that for every $\phi_1 \in p$ there exists $\phi_1' \in q$ such that $\phi_1 \sim_n^T \phi_1'$ and for every $\phi_2 \in q$, there exists $\phi_2' \in p$ such that $\phi_2 \sim_n^T \phi_2'$. We aim to prove that $p = q$. If not then there exists, without loss of generality, $\psi \in p$ such that $\psi \notin q$. Since q is maximal, $T \cup q \cup \{\psi\}$ is not satisfiable. By compactness, $T \cup \{\theta, \psi\}$ is not satisfiable for some finite conjunction θ of formulas from q. Now, $\theta \in q$ by maximality and there exists by assumption $\theta' \in p$ such that $\theta \sim_n^T \theta'$. By the definition of $\sim_n^T$ we deduce that $T \cup \{\theta', \psi\}$ is satisfiable if and only if $T \cup \{\theta, \psi\}$ is satisfiable. Since the latter is not satisfiable, we deduce that neither is the former, which yields the contradiction that $T \cup p \cup \{\psi\}$ is not satisfiable.

(iii) $\Rightarrow$ (iv). Let S be the union of the set of all existential positive sentences ϕ such that $T \cup \{\phi\}$ is satisfiable and the set of all universal negative consequences of T. By Proposition 2.1.16, S has a model $\mathfrak{C}$, and by Theorem 2.1.11 we can assume that $\mathfrak{C}$ is either finite or countable. Lemma 2.7.3 gives a homomorphism from $\mathfrak{C}$ to a finite or countable epc τ-model $\mathfrak{B}$ of S. Note that also $\mathfrak{B}$ satisfies exactly those existential positive sentences that are satisfiable together with T. Let σ be a signature that contains a relation symbol for each maximal ep-n-types of T, and let $\tau' := \tau \cup \sigma$. Then $\mathfrak{B}$ has a canonical τ'-expansion $\mathfrak{B}'$ where a new relation symbol denotes the set of all tuples that attain the respective maximal ep-type.

Claim. $\mathfrak{B}'$ is homogeneous. Let $b, a_1, \ldots, a_m \in B'$ and let f be an embedding of the substructure of $\mathfrak{B}'$ induced on $\{a_1, \ldots, a_m\}$ into $\mathfrak{B}'$. By Proposition 2.7.4, the ep-type p of $(f(a_1), \ldots, f(a_m), b)$ in $\mathfrak{B}$ is maximal. For each of the finitely many other ep-$(m+1)$-types q in $\mathfrak{B}$ pick a primitive positive formula in p which is not in q; let $\phi(x_1, \ldots, x_m, x)$ be the conjunction of all these formulas. Then the ep-type of $(f(a_1), \ldots, f(a_m))$ contains $\exists x \colon \phi(x_1, \ldots, x_m, x)$. Since the ep-type of $(a_1, \ldots, a_m)$ is maximal (again by

Proposition 2.7.4) it must also contain this primitive positive formula. We deduce the existence of $a \in B$ such that $\mathfrak{B} \models \phi(a_1, \dots, a_m, a)$ and consequently $\mathfrak{B} \models q(a_1, \dots, a_m, a)$. It follows that the extension of f given by $f(a) := b$ is still an isomorphism between substructures of $\mathfrak{B}'$. This was the step of *going back* in a back-and-forth argument that shows that f can be extended to an automorphism of $\mathfrak{B}'$, and the claim follows.

The homogeneity of $\mathfrak{B}'$ implies that $\mathfrak{B}'$ and $\mathfrak{B}$ are ω-categorical by Lemma 4.3.1 (since variable identifications are existential positive, there is only a finite number of inequivalent atomic formulas with n free variables). To prove that $\mathfrak{B}$ is a model-complete core, note that the argument above has shown that all relations of the homogeneous expansion $\mathfrak{B}'$ of $\mathfrak{B}$ and their complements have an existential positive definition in $\mathfrak{B}$. The statement then follows from Theorem 4.5.1 (2).

For the implication (iv) $\Rightarrow$ (i), observe that a (finite or countably infinite) ω-categorical structure $\mathfrak{B}$ is a model-complete core if and only if it has a model-complete core theory; this is an easy consequence of the Löwenheim-Skolem theorem (Theorem 2.1.11). So it suffices to show that the first-order theory of $\mathfrak{B}$ and T have the same universal negative consequences, by Corollary 2.1.15. A universal negative sentence ϕ is implied by T if and only if $T \cup \{\neg\phi\}$ is unsatisfiable, which is the case if and only if $\mathfrak{B}$ does not satisfy $\neg\phi$ (and hence satisfies ϕ). ⊣

Theorem 4.6.1 implies a necessary and sufficient condition for whether a CSP can be formulated with an ω-categorical template.

COROLLARY 4.6.2. *Let $\mathfrak{A}$ be a structure with a finite relational signature. Then the following are equivalent.*

1. *There is an ω-categorical template $\mathfrak{B}$ such that* $\mathrm{CSP}(\mathfrak{B}) = \mathrm{CSP}(\mathfrak{A})$.
2. $\sim_n^{\mathrm{Th}(\mathfrak{A})}$ *has for each n finite index.*
3. *There exists a structure $\mathfrak{B}$ with* $\mathrm{CSP}(\mathfrak{B}) = \mathrm{CSP}(\mathfrak{A})$ *which has for all $n \geq 1$ finitely many primitively positively definable relations of arity n.*

PROOF. The implications from (1) to (3) and from (3) to (2) are easy. The implication from (2) to (1) follows from $(ii) \Rightarrow (iv)$ in Theorem 4.6.1. ⊣

A related existential positive variant of the theorem of Ryll-Nardzewski has been found by Pech and Pech [300]; their version is less general than the variant presented here since it does not cover the following example.

Example 4.6.3. The structure $(\mathbb{Z}; <)$ has infinitely many 2-types, but only three maximal existential positive n-types for each n. And indeed, the ω-categorical model-complete core $(\mathbb{Q}; <)$ satisfies the same existential positive sentences as $(\mathbb{Z}; <)$.

Simmons' theorem. The following is due to Simmons [330], and another immediate consequence of Theorem 4.6.1 Recall the *joint embedding property*, which has been defined for classes of structures in Section 2.3; a *theory* T has the joint embedding property (JEP) if for any two models $\mathfrak{B}_1, \mathfrak{B}_2$ of T there exists a model $\mathfrak{C}$ of T that embeds both $\mathfrak{B}_1$ and $\mathfrak{B}_2$.

THEOREM 4.6.4 (Simmons [330]). *Let T be a theory with the JEP. Then the following are equivalent.*

- *T has an ω-categorical model companion.*
- *For every n, the theory T has finitely many maximal existential n-types.*

In particular, every ω-categorical theory has an ω-categorical model companion.

PROOF. By appropriately choosing the signature and applying Theorem 4.6.1 to the corresponding theory, as in Corollary 2.5.3. The second part of the statement clearly follows from the first. ⊣

The consequence stated for ω-categorical theories T at the end of Theorem 4.6.4 has first been shown by Saracino [318].

COROLLARY 4.6.5. *An ω-categorical structure is model-complete if and only if its first-order theory is equivalent to a $\forall\exists$-theory.*

PROOF. Follows from Corollary 4.7.6. ⊣

4.7. Constructing countably categorical structures

In this section we present powerful techniques to construct ω-categorical structures from known ones:

- generic superpositions (Section 4.7.1),
- first-order interpretations (Section 4.7.2),
- forming model companions, and, more generally,
- forming model-complete cores (Section 4.7.3).

In Section 4.7.4 we present an application of constructing ω-categorical structures by taking model-complete cores.

4.7.1. Generic superpositions. If two ω-categorical structures $\mathfrak{A}$ and $\mathfrak{B}$ have disjoint signatures and no algebraicity, then their expansions $\mathfrak{A}'$ and $\mathfrak{B}'$ by all first-order definable relations are homogeneous and still have no algebraicity. Hence, $\mathfrak{A}'$ and $\mathfrak{B}'$ have a generic superposition $(\mathfrak{A}') * (\mathfrak{B}')$ as defined in Section 2.3.6, and the generic superposition is still ω-categorical. Suppose that $\mathfrak{A}$ has the signature τ and $\mathfrak{B}$ has the signature ρ and that τ and ρ are disjoint. Then the $(\rho \cup \tau)$-reduct $\mathfrak{C}$ of $(\mathfrak{A}') * (\mathfrak{B}')$ is called the *generic superposition of* $\mathfrak{A}$ *and* $\mathfrak{B}$, and denoted by $\mathfrak{A} * \mathfrak{B}$, too (this is a slight generalisation of the definitions given in Section 2.3.6). The following can be shown by an easy back-and-forth argument.

LEMMA 4.7.1. *Let $\mathfrak{C}$ be the generic superposition of $\mathfrak{A}$ and $\mathfrak{B}$. Then the reduct of $\mathfrak{C}$ in the signature of $\mathfrak{A}$ is isomorphic to $\mathfrak{A}$, and the reduct of $\mathfrak{C}$ in the signature of $\mathfrak{B}$ is isomorphic to $\mathfrak{B}$.*

We mention that the generic superposition $\mathfrak{C}$ is uniquely given by the following properties (see [58]).

- For all $a, b \in C^k$, $k \in \mathbb{N}$ the orbit of a under $\mathrm{Aut}(\mathfrak{C}^\tau)$ and the orbit of b under $\mathrm{Aut}(\mathfrak{C}^\rho)$ have a non-empty intersection.
- For all $a \in C^k$, $k \in \mathbb{N}$ the orbit of a under $\mathrm{Aut}(\mathfrak{C})$ is the intersection of the orbit of a under $\mathrm{Aut}(\mathfrak{C}^\tau)$ and the orbit of a under $\mathrm{Aut}(\mathfrak{C}^\rho)$.

Generic superpositions inherit some of the properties from their constituents.

LEMMA 4.7.2. *Let $\mathfrak{A}$ and $\mathfrak{B}$ be ω-categorical model-complete cores. Then $\mathfrak{A} * \mathfrak{B}$ is an ω-categorical model-complete core as well.*

PROOF. The generic superposition is a reduct of an ω-categorical structure and hence ω-categorical. To show that $\mathfrak{A} * \mathfrak{B}$ is a model-complete core we verify item (2) in Theorem 4.5.1. Let $\mathfrak{A}'$ and $\mathfrak{B}'$ be the expansions of $\mathfrak{A}$ and $\mathfrak{B}'$ by all first-order definable relations so that $\mathfrak{A} * \mathfrak{B}$ is a reduct of the homogeneous structure $\mathfrak{A}' * \mathfrak{B}'$. Since $\mathfrak{A}$ is a model-complete core, all relations of $\mathfrak{A}'$ have an existential positive definition in $\mathfrak{A}$, and likewise all relations of $\mathfrak{B}'$ have an existential positive definition in $\mathfrak{B}$. Therefore, $\mathfrak{A}' * \mathfrak{B}'$ is a homogeneous expansion of $\mathfrak{A} * \mathfrak{B}$ such that all relations of $\mathfrak{A}' * \mathfrak{B}'$ and their complements have an existential positive definition $\mathfrak{A} * \mathfrak{B}$, which is item (2) in Theorem 4.5.1. ⊣

4.7.2. Countable categoricity from interpretations. Many ω-categorical structures can be derived from other ω-categorical structures via first-order interpretations.

LEMMA 4.7.3. *Let $\mathfrak{A}$ be an ω-categorical structure. Then every structure $\mathfrak{B}$ that is first-order interpretable in $\mathfrak{A}$ with finitely many parameters is ω-categorical.*

PROOF. By the theorem of Ryll-Nardzewski (Theorem 4.1.6) it suffices to show that the number $o(n)$ of orbits of n-tuples under $\mathrm{Aut}(\mathfrak{B})$ is finite, for every n. If $\mathfrak{B}$ is the expansion of $\mathfrak{A}$ by a constant c, then $o_\mathfrak{B}(n) \leq o_\mathfrak{A}(n+1)$ since the map that sends the orbit of t to the orbit of (c, t) is an injection. If $\mathfrak{B}$ has a d-dimensional interpretation in $\mathfrak{A}$ then $o_\mathfrak{B}(n) \leq o_\mathfrak{A}(dn)$ and hence is finite, too. ⊣

Note that in particular all first-order reducts of an ω-categorical structure and all expansions of an ω-categorical structure by finitely many constants are again ω-categorical.

4.7.3. Forming model-complete cores. Recall that a structure $\mathfrak{C}$ is called a *core* if and only if all endomorphisms of $\mathfrak{B}$ are embeddings. If $\mathfrak{C}$ is ω-categorical, this is equivalent to saying that the first-order theory of $\mathfrak{C}$ is a core theory (an easy consequence of the Löwenheim-Skolem theorem, Theorem 2.1.11). We present a simple proof of the following result from [44].

THEOREM 4.7.4 (Theorem 16 in [44]). *Every ω-categorical structure $\mathfrak{B}$ is homomorphically equivalent to an ω-categorical model-complete core $\mathfrak{C}$. All model-complete cores of $\mathfrak{B}$ are isomorphic to $\mathfrak{C}$.*

PROOF. Let T be the first-order theory of $\mathfrak{B}$; clearly, T has the JHP. Since T is ω-categorical, $\sim_n^T$ has finite index for each n (Theorem 4.1.6), and Theorem 4.6.1 implies that T has a core companion S which is either ω-categorical or the theory of a finite structure. Let $\mathfrak{C}$ be the unique countable model of S. By Theorem 2.7.11, the theory S is unique up to equivalence of first-order theories, so it follows that $\mathfrak{C}$ is unique up to isomorphism. Finally, $\mathfrak{B}$ and $\mathfrak{C}$ have the same universal negative theory and are ω-categorical, and therefore Lemma 4.1.7 implies that they are homomorphically equivalent, which proves the statement. ⊣

Since the model-complete core $\mathfrak{C}$ of $\mathfrak{B}$ from the previous theorem is unique up to isomorphism, we call it *the* model-complete core of $\mathfrak{B}$.

Remark 4.7.5. If h is a homomorphism from an ω-categorical structure $\mathfrak{B}$ to its model-complete core $\mathfrak{C}$, and i is a homomorphism from $\mathfrak{C}$ to $\mathfrak{B}$, then i must in fact be an embedding: $h \circ i$ is an endomorphism of $\mathfrak{C}$, and hence must be an embedding because $\mathfrak{C}$ is a model-complete core. This implies that i must be an embedding, too. So the model-complete core of $\mathfrak{B}$ embeds into $\mathfrak{B}$.

The existence of model-complete cores for ω-categorical structures also yields yet another characterisation of ω-categorical model-complete cores.

COROLLARY 4.7.6. *An ω-categorical structure is a model-complete core if and only if its first-order theory T is equivalent to a $\forall\exists^+$-theory.*

PROOF. The forward implication is Proposition 2.6.13. For the backwards implication, assume that T is equivalent to a $\forall\exists^+$-theory. Then T is equivalent to T^{KH^+}. Theorem 4.7.4 implies that T has a core companion, and by Theorem 2.7.11, T is its own model companion, and hence model complete. ⊣

The following gives an indication that the model-complete core of an ω-categorical structure $\mathfrak{B}$ is typically 'simpler' than $\mathfrak{B}$.

PROPOSITION 4.7.7. *Let $\mathfrak{B}$ be a countable ω-categorical structure, and let $\mathfrak{C}$ be its model-complete core. Then the following statements hold.*

1. *If $\mathfrak{B}$ is homogeneous, then $\mathfrak{C}$ is homogeneous as well.*
2. *If i is a homomorphism from $\mathfrak{C}$ to $\mathfrak{B}$ and $t_1, t_2 \in C^n$ lie in the same orbit under* $\mathrm{Aut}(\mathfrak{C})$, *then there exist endomorphisms $e_1, e_2 \in \mathrm{End}(\mathfrak{B})$ such that $e_1(i(t_1)) = i(t_2)$ and $e_2(i(t_2)) = i(t_1)$.*
3. *If i is a homomorphism from $\mathfrak{C}$ to $\mathfrak{B}$ and $t_1, t_2 \in C^n$ are such that $i(t_1)$ and $i(t_2)$ lie in the same orbit under* $\mathrm{Aut}(\mathfrak{B})$, *then t_1 and t_2 lie in the same orbit under* $\mathrm{Aut}(\mathfrak{C})$.
4. *For every n, the number of orbits of n-tuples under* $\mathrm{Aut}(\mathfrak{C})$ *is at most the number of orbits of n-tuples under* $\mathrm{Aut}(\mathfrak{B})$.
5. *if for every n, the number of orbits of n-tuples under* $\mathrm{Aut}(\mathfrak{C})$ *equals the number of orbits of n-tuples under* $\mathrm{Aut}(\mathfrak{B})$, *then $\mathfrak{B}$ and $\mathfrak{C}$ are isomorphic.*

PROOF. Let h be a homomorphism from $\mathfrak{B}$ to $\mathfrak{C}$, and i be a homomorphism from $\mathfrak{C}$ to $\mathfrak{B}$. To prove (2), suppose that t_1 and t_2 lie in the same orbit. Then $t_3 := h(i(t_1))$ also lies in the same orbit as t_2 because $\mathfrak{C}$ is a model-complete core; let $\gamma \in \mathrm{Aut}(\mathfrak{C})$ be such that it maps t_3 to t_2. Then $i\gamma h$ is an endomorphism of $\mathfrak{B}$ that maps $i(t_1)$ to $i(t_2)$. Symmetrically, there exists an endomorphism of $\mathfrak{B}$ that maps $i(t_2)$ to $i(t_1)$.

We now prove (3). Since $\mathfrak{C}$ is an ω-categorical model-complete core, there are primitive positive definitions ϕ_1 and ϕ_2 of the orbits of t_1 and t_2. Suppose that there exists $\beta \in \mathrm{Aut}(\mathfrak{B})$ that maps $i(t_1)$ to $i(t_2)$. Since h, β, and i preserve primitive positive formulas, the tuple $t_3 := h(\beta i(t_1))$ satisfies ϕ_1. As $\beta i(t_1) = i(t_2)$, the tuple t_3 can also be written as $h(i(t_2))$, and hence also satisfies ϕ_2. Thus, ϕ_1 and ϕ_2 define the same orbit, and t_1 and t_2 are in the same orbit.

(1) and (4) are an immediate consequence of (3).

For (5), let $n \in \mathbb{N}$, $t \in B^n$, and $e \in \mathrm{End}(\mathfrak{B})$. Select from each orbit O of n-tuples under $\mathrm{Aut}(\mathfrak{C})$ a tuple s_O; then (3) shows that the map I from the orbits of n-tuples under $\mathrm{Aut}(\mathfrak{C})$ to the orbits of n-tuples under $\mathrm{Aut}(\mathfrak{B})$ that maps O to the orbit of $i(s_O)$ is an injection; by assumption, it must be a bijection. Arbitrarily choose s from the preimage of the orbit of t under the map I, and s' from the preimage of the orbit of $e(t)$. Then there are $\alpha, \beta \in \mathrm{Aut}(\mathfrak{B})$ such that $\alpha i(s) = t$ and $\beta i(s') = e(t)$. Since $\mathfrak{C}$ is a model-complete core, the endomorphism $f \circ e \circ \alpha \circ i$ of $\mathfrak{C}$ is in $\overline{\mathrm{Aut}(\mathfrak{C})}$, so s and $f \circ e \circ \alpha \circ i(s)$ lie in the same orbit under $\mathrm{Aut}(\mathfrak{C})$. Note that $f \circ e \circ \alpha \circ i(s) = f \circ e(t) = f \circ \beta \circ i(s')$, and that $f \circ \beta \circ i$ is an endomorphism of $\mathfrak{C}$ and hence in $\overline{\mathrm{Aut}(\mathfrak{C})}$. So s' lies in the same orbit as s, and $e(t)$ lies in the same orbit as t. This shows that $\mathfrak{B}$ is a model-complete core, and hence $\mathfrak{B}$ and $\mathfrak{C}$ are isomorphic by Theorem 4.7.4. ⊣

Note that item (2) of Proposition 4.7.7 cannot be strengthened by requiring that $i(t_2)$ and $i(t_1)$ lie in the same orbit, as the following example shows.

Example 4.7.8. Let $\mathfrak{B}$ be the substructure of $(\mathbb{Q}; <)$ induced on $\mathbb{Q} \setminus (-1, 1)$. Clearly, $(\mathbb{Q}; <)$ is the model-complete core of $\mathfrak{B}$. Let i be a homomorphism from $(\mathbb{Q}; <)$ to $\mathfrak{B}$ that maps the negative rationals to the rationals smaller than -1, and that maps to non-negative rationals to rationals larger than 1. Then -1 and 1 lie in the same orbit under $\mathrm{Aut}(\mathbb{Q}; <)$, but $i(-1)$ and $i(1)$ lie in different orbits under $\mathrm{Aut}(\mathfrak{B})$. However, there are endomorphisms of $\mathfrak{B}$ mapping $i(-1)$ to $i(1)$ and vice versa.

4.7.4. Proving Cherlin, Shelah, and Shi. Using model companions we can now also derive the theorem of Cherlin, Shelah, and Shi (Theorem 4.3.7) from Theorem 4.3.8. We also obtain an even more powerful existential positive variant of Theorem 4.3.7; both are combined in the following corollary.

COROLLARY 4.7.9. *Let $\mathcal{F}$ be a finite set of finite connected τ-structures. Then there are*

1. *a countable ω-categorical model-complete structure $\mathfrak{B}$ such that* $\text{Age}(\mathfrak{B}) = \text{Forb}^{\text{hom}}(\mathcal{F})$ (*this is the structure from Theorem* 4.3.7);
2. *a countable ω-categorical model-complete core structure $\mathfrak{B}$ such that a countable structure maps* homomorphically *to $\mathfrak{B}$ if and only if it is $\mathcal{F}$-free.*

PROOF. By Theorem 4.3.8 there exists an ω-categorical structure $\mathfrak{C}$ such that $\text{Age}(\mathfrak{C}) = \text{Forb}^{\text{hom}}(\mathcal{F})$. By Theorem 4.6.4, the model companion $\mathfrak{B}$ of $\mathfrak{C}$ exists, satisfies $\text{Age}(\mathfrak{B}) = \text{Forb}^{\text{hom}}(\mathcal{F})$, and is ω-categorical, showing (1). Likewise, we can use Theorem 4.7.4 and obtain that the model-complete core of $\mathfrak{C}$ has the required properties in (2). $\dashv$

Chapter 5

EXAMPLES

The structure ($\mathbb{Q}; <$) was an important running example in the previous section. First-order reducts of ($\mathbb{Q}; <$) provide further examples of ω-categorical structures, and they will be studied in great detail in Chapter 12. In this chapter, we present other ω-categorical structures $\mathfrak{A}$ that will not be treated at the same level of detail. For example, we treat homogeneous C-relations (Section 5.1), dense semilinear orders (Section 5.2), and the atomless Boolean algebra (Section 5.3). In each case, we give a brief discussion on what is known about CSPs for first-order reducts of these ω-categorical structures. Thereby, we revisit many problems from Section 1.6. We also discuss constructions of ω-categorical structures that serve as templates for network satisfaction problems for certain finite relation algebras (Section 5.5) or for problems in fragments of existential second-order logic (Section 5.6).

The ω-categorical structures presented in this chapter are chosen to illustrate the diversity of the class of all ω-categorical structures, and many computational problems and classes of computational problems from the literature can be formulated as CSPs for these structures.

5.1. Phylogeny constraints and homogeneous C-relations

The rooted-triple satisfiability problem from Section 1.6.2 can be formulated as CSP($\mathfrak{B}$) for an ω-categorical template $\mathfrak{B}$ (an observation from [44]). There are various different ways to define such a structure $\mathfrak{B}$; the most convenient for us is via Fraïssé-amalgamation (Section 5.1.1). The resulting homogeneous structure can be found in [122] and appears explicitly in the model theory literature about so-called C-relations; this is discussed in Section 5.1.2.

5.1.1. Leaf structures. A rooted tree (see Section 1.6.2) is called *binary* if the root has degree zero (in which case the tree consists of just one vertex) or two, and all other vertices have degree one (i.e., they are *leaves*) or three. Let $\mathcal{T}$ be the class of all finite rooted binary trees $\mathfrak{T}$. Clearly, $\mathcal{T}$ is not closed under substructures. The closure of $\mathcal{T}$ under substructures does not satisfy the amalgamation property (Figure 1 shows two trees with the common set of leaves $\{a, b, u, v\}$ that cannot be amalgamated; in this figure, we did not indicate the roots but draw directed arcs instead). Generalising our notation from Section 1.6.2, we define the *youngest common ancestor* $yca(S)$ of a subset S of the nodes of $\mathfrak{T}$ to be the node w that lies in $\mathfrak{T}$ above each vertex in S and has maximal distance from the root of $\mathfrak{T}$.

DEFINITION 5.1.1. The *leaf structure* $\mathfrak{C}$ of a tree $\mathfrak{T} \in \mathcal{T}$ with leaves L is the relational structure $(L; |)$ where $|$ is a ternary relation symbol, and $ab|c$ holds in $\mathfrak{C}$ iff $yca(\{a, b\})$ lies strictly below $yca(\{a, b, c\})$ in $\mathfrak{T}$.

We also call $\mathfrak{T}$ the *underlying tree* of $\mathfrak{C}$. Let C be the class of all leaf structures for trees from $\mathcal{T}$.

PROPOSITION 5.1.2. *The class C is an amalgamation class.*

PROOF. Closure under isomorphisms holds by definition, and closure under substructures is easy to see. For the amalgamation property, let $\mathfrak{B}_1, \mathfrak{B}_2 \in C$ be such that $\mathfrak{A} = \mathfrak{B}_1 \cap \mathfrak{B}_2$ is a substructure of both $\mathfrak{B}_1$ and $\mathfrak{B}_2$. To show that there is an amalgam of $\mathfrak{B}_1$ and $\mathfrak{B}_2$ over $\mathfrak{A}$ in C, we inductively assume that the statement has been shown for all triples $(\mathfrak{A}, \mathfrak{B}'_1, \mathfrak{B}'_2)$ where $B'_1 \cup B'_2$ is a proper subset of $B_1 \cup B_2$.

Let $\mathfrak{T}_1$ be the rooted binary tree underlying $\mathfrak{B}_1$, and $\mathfrak{T}_2$ the rooted binary tree underlying $\mathfrak{B}_2$. Let $\mathfrak{B}_1^1 \in C$ be the substructure of $\mathfrak{B}_1$ induced on the vertices below one of the children of the root of $\mathfrak{T}_1$, and $\mathfrak{B}_1^2 \in C$ be the substructure of $\mathfrak{B}_1$ induced on the vertices below the other child of the root of $\mathfrak{T}_1$. The structures $\mathfrak{B}_2^1$ and $\mathfrak{B}_2^2$ are defined analogously for $\mathfrak{B}_2$ instead of $\mathfrak{B}_1$.

First consider the case that there is a vertex u that lies in both $\mathfrak{B}_1^1$ and $\mathfrak{B}_2^1$, and a vertex v that lies in both $\mathfrak{B}_2^1$ and $\mathfrak{B}_1^2$. In this case no vertex w from $\mathfrak{B}_2^2$ can lie inside $\mathfrak{B}_1$: for otherwise, w is either in $\mathfrak{B}_1^1$, in which case we have $uw|v$ in $\mathfrak{B}_1$, or in $\mathfrak{B}_1^2$, in which case we have $vw|u$ in $\mathfrak{B}_1$. But since u, v, w are in A, this is in contradiction to the fact that $uv|w$ holds in $\mathfrak{B}_2$. Let $\mathfrak{C}' \in C$ be the

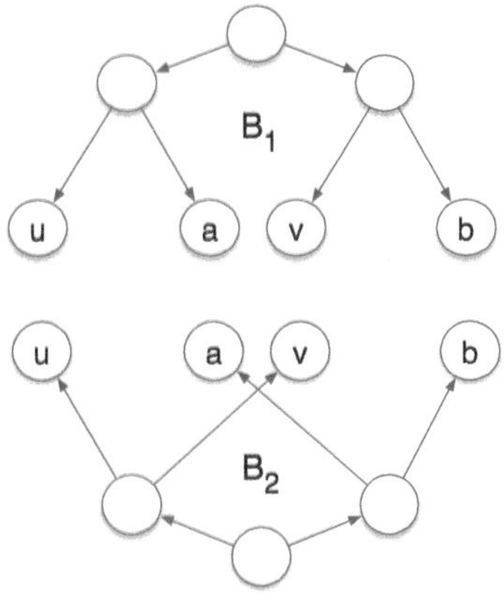

FIGURE 1. Illustration for the proof of Proposition 5.1.2 and for Section 5.2.3.

amalgam of $\mathfrak{B}_1$ and $\mathfrak{B}_2^1$ over $\mathfrak{A}$, which exists by inductive assumption, and let $\mathfrak{T}' \in \mathcal{T}$ be its underlying tree. Now let $\mathfrak{T}$ be the tree with root r and $\mathfrak{T}'$ as a subtree, and the underlying tree of $\mathfrak{B}_2^2$ as the other subtree. It is straightforward to verify that the leaf structure of $\mathfrak{T}$ is in $\mathcal{C}$, and that it is an amalgam of $\mathfrak{B}_1$ and $\mathfrak{B}_2$ over $\mathfrak{A}$ (via the identity embeddings).

Up to symmetry, the only remaining essentially different case we have to consider is that $B_1^1 \cup B_2^1$ and $B_1^2 \cup B_2^2$ are disjoint. In this case it is similarly straightforward to first amalgamate $\mathfrak{B}_1^1$ with $\mathfrak{B}_2^1$ and $\mathfrak{B}_1^2$ with $\mathfrak{B}_2^2$ to obtain an amalgam of $\mathfrak{B}_1$ and $\mathfrak{B}_2$; the details are left to the reader. $\dashv$

Let $(\mathbb{L};|)$ denote the Fraïssé-limit of $\mathcal{C}$. The structure $(\mathbb{L};|)$ is homogeneous in a finite relational signature, so it is a fortiori model complete. Clearly, $(\mathbb{L};|)$ is a core: every endomorphism of $(\mathbb{L};|)$ preserves the binary relation defined by $xx|y$, which is equivalent to $\neq$, and hence must be injective. Moreover, the formula $\neg(xy|z)$ is equivalent to $yz|x \vee zx|y \vee x = y = z$, and hence is preserved by all endomorphisms as well. Since Age$(\mathbb{L};|)$ is the class of all leaf structures for structures from $\mathcal{T}$, it is clear that CSP$(\mathbb{L};|)$ is the rooted triple satisfiability problem (Section 1.6.2).

5.1.2. C-relations. The relation defined by $xy|z$ on $\mathbb{L}$ from the previous section is a so-called *C-relation*, following the terminology of Adeleke and Neumann [3], which became an important concept in model theory [193, 271]. For a discussion of C-relations in the context of the constraint satisfaction problem, see [64].

A ternary relation C is said to be a *C-relation* on a set L if for all $a, b, c, d \in L$ the following conditions hold:

C1 $C(a; b, c) \rightarrow C(a; c, b)$;

C2 $C(a;b,c) \to \neg C(b;a,c)$;
C3 $C(a;b,c) \to C(a;d,c) \vee C(d;b,c)$;
C4 $a \neq b \to C(a;b,b)$.

A C-relation is called *proper* if it satisfies

C5 $\forall b, c \exists a : C(a;b,c)$;
C6 $\forall a, b \big(a \neq b \Rightarrow \exists c (b \neq c \wedge C(a;b,c)\big)$.

A C-relation is called *dense* if it satisfies

C7 $\forall a, b, c \big(C(a;b,c) \to \exists e \,(C(e;b,c) \wedge C(a;b,e))\big)$

and it is called *uniform with branching number* 2, or short *binary branching*, if it satisfies

$$\forall x, y, z \,\big((x \neq y \vee x \neq z \vee y \neq z) \Rightarrow \\ (C(x;y,z) \vee C(y;x,z) \vee C(z;x,y))\big).$$

The structure $(L;C)$ is also called a *C-set*. A structure $\mathfrak{B}$ is said to be *relatively* k*-transitive* if for every partial isomorphism f between substructures of $\mathfrak{B}$ of size k there exists an automorphism of $\mathfrak{B}$ that extends f.

THEOREM 5.1.3 (Theorem 14.7 in [4] and comments thereafter). *Let $(L;C)$ be a relatively 3-transitive C-set. Then $(L;C)$ is ω-categorical. There is a relatively 3-transitive countable C-set which is dense and binary branching, and unique up to isomorphism.*

Note that the relation $|$ on $\mathbb{L}$ satisfies all the properties in Theorem 5.1.3; also this is discussed in [4]. In fact, the properties C1-C4 together with the property to be binary branching provide a universal axiomatisation of Age$(\mathbb{L};|)$ (Part III in [4]), showing that $(\mathbb{L};|)$ is finitely bounded (Lemma 2.3.14). It also follows that the theory of $(\mathbb{L};|)$ is the model companion of the theory of binary branching C-relations.

5.1.3. The quartet satisfiability problem. A similar amalgamation approach can be used to construct a homogeneous template for the quartet satisfiability problem from Section 1.6.2. Alternatively, an ω-categorical template for the quartet satisfiability problem can be given via a first-order reduct $(\mathbb{L};Q)$ of $(\mathbb{L};|)$.

DEFINITION 5.1.4. If $u_1, \ldots, u_k$ and $v_1, \ldots, v_l$ are leaves in a rooted tree $\mathfrak{T}$, then we write $u_1 \ldots u_k | v_1 \ldots, v_l$ if $u := yca(\{u_1, \ldots, u_k\})$ and $v := yca(\{v_1, \ldots, v_l\})$ are *disjoint* in $\mathfrak{T}$, i.e., neither u lies above v nor v lies above u in $\mathfrak{T}$.

The first-order definition of $Q(x,y,u,v)$ is

$$(xy|uv) \;\vee\; (uv|x \wedge vx|y) \;\vee\; (xy|u \wedge yu|v).$$

Indeed, if $u, v, x, y \in \mathbb{L}$, and $\mathfrak{T}$ is the tree underlying the substructure of $(\mathbb{L};|)$ induced on $\{u,v,x,y\}$, then the given formula describes the situation that the shortest path from x to y in $\mathfrak{T}$ does not intersect the shortest path from u to v in $\mathfrak{T}$. Note that whether this is true is in fact independent from the position of the root of $\mathfrak{T}$. We leave the verification to the reader that CSP($\mathbb{L}$; Q) indeed describes the quartet satisfiability problem. Lemma 4.7.3 implies the ω-categoricity of $(\mathbb{L}; Q)$. Similarly as for the C-relation given above, an axiomatic treatment of $(\mathbb{L}; Q)$ has been given in [4]; there, the relation Q was called a *D-relation*, and this is standard terminology in model theory. As we have mentioned above, the structure $(\mathbb{L}; Q)$ can also be defined as a Fraïssé-limit of D-relations on finite sets (see Cameron [123]).

5.2. Branching-time constraints and semilinear orders

The branching-time satisfiability problem from Section 1.6.3 can be formulated as CSP($\mathfrak{B}$) for an ω-categorical structure $\mathfrak{B}$. This has been observed in [85] and can be shown in various ways.

5.2.1. An explicit construction. Let $\mathbb{S}$ be the set of all non-empty finite sequences of rational numbers. For

$$a = (q_1, q_1, \dots, q_n), b = (q'_1, q'_1, \dots, q'_m), n \leq m$$

we write $a < b$ if one of the following conditions holds:

- a is a proper initial subsequence of b, i.e., $n < m$ and $q_i = q'_i$ for $1 \leq i \leq n$;
- $q_i = q'_i$ for $1 \leq i < n$, and $q_n < q'_n$.

We use $a \leq b$ to denote $(a < b) \vee (a = b)$, and $\|$ denotes the binary relation that contains all pairs of elements that are equal or incomparable with respect to $\leq$. A proof that $(\mathbb{S}; \leq, \|, \neq)$ is ω-categorical with a transitive automorphism group can be found in [4] (Section 5). The reduct $(\mathbb{S}; \leq)$ of this structure is a *semilinear order*, i.e., for all $x \in \mathbb{S}$, the set $\{y \mid y \leq x\}$ is linearly ordered by $<$. Such structures have been studied systematically in the context of infinite permutation groups; see [155, 123]. We warn the reader that in some publications the order is reversed, so that $\{y \mid y \geq x\}$ is linearly ordered by $<$. Since the structure $(\mathbb{S}; \leq, \|, \neq)$ has the same age as the structure $\mathfrak{C}$ constructed in Section 1.6.3, we obtain the following.

PROPOSITION 5.2.1. *The branching-time satisfiability problem is*

$$\mathrm{CSP}(\mathbb{S}; \leq, \|, \neq).$$

Note that the structure $(\mathbb{S}; \leq, \|, \neq)$ is *not* model-complete. To see this, first observe that $(\mathbb{S}; \leq)$ is a meet-semilattice (Example 2.1.2). Indeed, for any two $a, b \in \mathbb{S}$, there exists least upper bound $a \wedge b$ of a and b: if a and b are comparable, then $a \wedge b \in \{a, b\}$. Otherwise, let c be the longest common prefix of a and b. Note that $a = (c, a')$ for $a' \in \mathbb{S}$, and let $a'' \in \mathbb{Q}$ be the first entry of a. Similarly, $b = (c, b')$ for $b' \in \mathbb{S}$, and let $b'' \in \mathbb{Q}$ be the first entry of a. Then $(c, \min(a', b'))$ is the least upper bound of a and b.

Now observe that for any two $x, y \in \mathbb{S}$ that are incomparable with respect to $\leq$ there exists an embedding e of $(\mathbb{S}; \leq)$ which fixes all elements of $\mathbb{S}$ except for $C := \{u \mid u \geq (x \wedge y)\}$, and which maps $u = (x \wedge y, u') \in C$ to $(x \wedge y, u', 0)$. Then e preserves $\leq$, but it does not preserve $\wedge$ because $e(x) \wedge e(y) = x \wedge y \neq (x \wedge y, 0) = e(x \wedge y)$ This shows that $(\mathbb{S}; \leq)$ is not model-complete.

However, by Theorem 4.6.4 the structure $(\mathbb{S}; \leq, \|, \neq)$ has an (up to isomorphism unique) ω-categorical model companion $\mathfrak{B}$, which is also a template for the branching-time satisfiability problem. In the following section this template will be constructed more directly.

5.2.2. Construction via existential closure. In this section we essentially follow the presentation in [49]. Let T be the first-order theory of semilinear orders. By Corollary 2.7.5 there exists a countable semilinear order $(\mathbb{T}; \leq)$ which is existentially closed for T. Clearly, $(\mathbb{T}; \leq)$ is

- *downwards directed*: for all $x, y \in \mathbb{T}$ there exists $z \in \mathbb{T}$ such that $z \leq x$ and $z \leq y$;
- *dense*: for all $x, y \in \mathbb{T}$ such that $x < y$ there exists $z \in \mathbb{T}$ such that $x < z < y$;
- *unbounded*: for every $x \in \mathbb{T}$ there are $y, z \in \mathbb{T}$ such that $y < x < z$;
- *binary branching*: (a) for all $x, y \in \mathbb{T}$ such that $x < y$ there exists $u \in \mathbb{T}$ such that $u < y$ and $u|x$, and (b) for any three incomparable elements of $\mathbb{T}$ there is an element in $\mathbb{T}$ that is larger than two out of the three, and incomparable to the third;
- *nice* (adopting the terminology from [158]): for every $x, y \in \mathbb{T}$ such that $x|y$ there exists $z \in \mathbb{T}$ such that $z > x$ and $z|y$;
- *without joins*: for all $x, y, z \in \mathbb{T}$ with $x, y \leq z$ and x, y incomparable, there exists a $u \in \mathbb{T}$ such that $x, y \leq u$ and $u < z$.

It can be shown by a back-and-forth argument that all countable, downwards directed, dense, unbounded, nice, and binary branching semilinear orders without joins are isomorphic to $(\mathbb{T}; \leq)$. For a proof, see [49] (Proposition 3.4); this also follows from results of Droste [156] and Droste, Holland, and Macpherson [157]. Since all of the properties of $(\mathbb{T}; \leq)$ listed above can be expressed by first-order sentences, it follows that $(\mathbb{T}; \leq)$ is ω-categorical. Moreover, the theory of $(\mathbb{T}; \leq)$ is the model companion of the theory of semilinear orders (Theorem 2.7.16). Again writing $a \leq b$ for $(a < b) \vee (a = b)$ and $\|$ for the binary relation that contains all pairs of elements that are equal or incomparable with respect to $\leq$, we obtain that $\mathrm{CSP}(\mathbb{T}; \leq, \|, \neq)$ equals the branching-time satisfiability problem.

5.2.3. Construction via Fraïssé amalgamation. The age of $(\mathbb{T}; \leq)$ is not an amalgamation class: an illustration of an amalgamation diagram that fails can be found in Figure 1. Let $|$ be the ternary relation with the following primitive positive definition over $(\mathbb{T}; \leq)$.

$$xy|z :\Leftrightarrow \exists u\, \big((u \leq x) \wedge (u \leq y) \wedge \neg(u \leq z) \wedge \neg(z \leq u)\big).$$

PROPOSITION 5.2.2 (Proposition 3.4 in [49]). $(\mathbb{T}; \leq, |)$ *is homogeneous.*

We are not aware of a reference for the following fact.

PROPOSITION 5.2.3. $(\mathbb{T}; \leq, |)$ *is finitely bounded.*

PROOF. The definition of semilinear orders can be expressed by a universal sentence ϕ (with three variables). Writing $x|y$ as a shortcut for $\neg(x \leq y) \wedge \neg(y \leq x)$, we claim that the additional axioms

$$\forall x, y, z, u\colon\ (u \leq x \wedge u \leq y \wedge u|z) \Rightarrow xy|z \tag{1}$$

$$\forall x, y, z\colon\ (xy|z \Rightarrow yx|z) \tag{2}$$

$$\forall x, y, z\colon\ \big(xy|z \Rightarrow x|z \wedge y|z\big) \tag{3}$$

$$\forall x, y, z\colon\ \neg(xy|z \wedge yz|x) \tag{4}$$

$$\forall x, y, z\colon\ (x|y \wedge y|z \wedge x|y) \Rightarrow (xy|z \vee yz|x \vee zx|y) \tag{5}$$

$$\forall x, y, z, u\colon\ (xy|z \wedge yz|u) \Rightarrow xz|u \tag{6}$$

provide a universal axiomatisation of $\mathrm{Age}(\mathbb{T}; \leq, |)$. The claim proves the statement by Lemma 2.3.14. Clearly, $(\mathbb{T}; \leq, |)$ satisfies all the above axioms.

Let $\mathfrak{A}$ be a finite $\{\leq, |\}$-structure which satisfies the above axioms We prove by induction on $|A|$ that $\mathfrak{A}$ embeds into $(\mathbb{T}; \leq, |)$. We already know that the $\{\leq\}$-reduct of $\mathfrak{A}$ must be a semilinear order. If $|A| \leq 2$ then the statement is immediate, so suppose that $|A| \geq 3$. We first consider the case that there exists $r \in A$ with $r \leq a$ for all $a \in A$. Then by the inductive assumption there exists an embedding $e\colon \mathfrak{A}[A \setminus \{r\}] \hookrightarrow (\mathbb{T}; \leq, |)$. Since $(\mathbb{T}; \leq)$ is unbounded and downwards directed there exists $t \in \mathbb{T}$ such that $t \leq e(a)$ for all $a \in A$. If $rx|y$, $xr|y$, or $xy|r$ held in $\mathfrak{A}$ for some $x, y \in A$, then x, y, r show that $\mathfrak{A}$ does not satisfy (3), a contradiction. Since neither $te(x)|e(y)$, $e(x)t|e(y)$, nor $e(x)e(y)|t$ holds in $(\mathbb{T}; \leq, |)$, it follows that the extension of e that maps r to t is an embedding of $\mathfrak{A}$ into $(\mathbb{T}; \leq, |)$.

Otherwise, there are $r_1, r_2 \in A$ with $r_1|r_2$ such that there is no $a \in A$ with $a < r_1$ or $a < r_2$. Note that (1) implies that $ab|c$ for all $a, b \in B_i := \{u \mid r_i \leq u\}$ and $c \in A \setminus B_i$. Also note that (4) implies that $\neg ab|c$ for $a \in B_1$, $b \in B_2$, and $c \in B_1 \cup B_2$. We claim that there exists a partition of A into two non-empty sets B_1, B_2 such that $ab|c$ for all $a, b \in B_1$ and $c \in B_2$ and $ab|c$ for all $a, b \in B_2$ and $c \in B_1$. Otherwise, choose B_1 and B_2 such that $ab|c$ for all $a, b \in B_1$ and $c \in B_2$ and $ab|c$ for all $a, b \in B_2$ and $c \in B_1$ and such that $|B_1| + |B_2|$ is maximal. Suppose for contradiction that there exists $d \in A \setminus (B_1 \cup B_2)$. If $d \leq b$ for $b \in B_1$, then (1) implies that $db|a$ for every $a \in B_2$. Axioms (6) and (2) imply that $dc|a$ for all $c \in B_1$, contradicting the maximality of $|B_1| + |B_2|$. With similar reasoning we can infer that we must have that $b_1|d$ and $b_2|d$ for all $b_1 \in B_1$ and $b_2 \in B_2$. Therefore, (5) implies that $b_1b_2|d$, $b_2d|b_1$, or $d|b_1b_2$. In each of the cases we may use (6) to derive a contradiction to the maximality of $|B_1 + B_2|$.

By the inductive assumption there exist embeddings $e_i : \mathfrak{A}[B_i] \hookrightarrow (\mathbb{T}; \leq, |)$. Since $(\mathbb{T}; \leq)$ is downwards directed and unbounded there exist $u_1, u_2 \in \mathbb{T}$ such that $u_i < e_i(a)$ for all $a \in B_i$. Using the homogeneity of $(\mathbb{T}; \leq, |)$ we may assume that $u_1|u_2$ and consequently that $e_1(a_1)|e_2(a_2)$ for all $a_1 \in B_1$ and $a_2 \in B_2$. It is now straightforward to verify that the common extension e of e_1 and e_2 to all of A is an embedding of $\mathfrak{A}$ into $(\mathbb{T}; \leq, |)$. $\dashv$

Hence, we could have introduced $(\mathbb{T}; \leq)$ as the reduct of the Fraïssé-limit of the class of all finite $\{\leq, |\}$-structures that satisfy the universal sentences from the proof of Proposition 5.2.3. We refrain from doing so since the proof of the amalgamation property of this class is tedious.

Note that $\mathrm{Aut}(\mathbb{T}; \leq)$ has four orbitals, with the primitive positive definitions $x \leq y \wedge x \neq y$, $y \leq x \wedge x \neq y$, $x \| y \wedge x \neq y$, and $x = y$ in $(\mathbb{T}; \leq, \|, \neq)$. Since all relations of $(\mathbb{T}; \leq, \|, \neq)$ are binary, this implies that every endomorphism of $(\mathbb{T}; \leq, \|, \neq)$ must be an embedding, and hence $(\mathbb{T}; \leq, \|, \neq)$ is a core.

5.3. Set constraints and the atomless Boolean algebra

In Section 1.6.5 we defined the problem of *Basic Set Constraint Satisfiability*. This problem, and many other set constraint satisfaction problems, can also be formulated as a CSP for an ω-categorical structure. We first define what we mean by a *set constraint satisfaction problem* and later explain how to construct ω-categorical templates.

Let $\mathfrak{S}$ be the structure with the domain $\mathcal{P}(\mathbb{N})$, the set of all subsets of natural numbers, and with the signature $\{\sqcap, \sqcup, c, \mathbf{0}, \mathbf{1}\}$, where

- $\sqcap$ is a binary function symbol that denotes intersection, i.e., $\sqcap^{\mathfrak{S}} = \cap$;
- $\sqcup$ is a binary function symbol for union, i.e., $\sqcup^{\mathfrak{S}} = \cup$;
- c is a unary function symbol for complementation, i.e., $c^{\mathfrak{S}}$ is the function that maps $S \subseteq \mathbb{N}$ to $\mathbb{N} \setminus S$;
- $\mathbf{0}$ and $\mathbf{1}$ are constants (treated as 0-ary function symbols) denoting the empty set $\emptyset$ and the full set $\mathbb{N}$, respectively.

A *set constraint language* is a relational structure with the domain $\mathcal{P}(\mathbb{N})$ whose relations have a quantifier-free first-order definition in $\mathfrak{S}$. For example, the relation $\{(x, y, z) \in \mathcal{P}(\mathbb{N})^3 \mid x \cap y \subseteq z\}$ has the quantifier-free first-order definition $z \sqcap (x \sqcap y) = x \sqcap y$ over $\mathfrak{S}$.

The first-order theory of the structure $\mathfrak{S}$ is certainly not ω-categorical; it is easy to verify that there are infinitely many pairwise inequivalent first-order formulas with one free variable. However, all set constraint satisfaction problems can be formulated with an ω-categorical template.

PROPOSITION 5.3.1. *Let $\mathfrak{C}$ be a set constraint language. Then there exists an ω-categorical structure $\mathfrak{B}$ such that $\mathfrak{B}$ and $\mathfrak{C}$ have the same existential theory. In particular, if $\mathfrak{C}$ has finite signature, then $\mathfrak{B}$ and $\mathfrak{C}$ have the same* CSP.

PROOF. Let $\phi_1, \phi_2, \dots$ be quantifier-free first-order formulas that define the relations $R_1^{\mathfrak{C}}, R_2^{\mathfrak{C}}, \dots$ of $\mathfrak{C}$ over $\mathfrak{S} = (\mathcal{P}(\mathbb{N}); \cup, \cap, c, \mathbf{0}, \mathbf{1})$. Let $\mathfrak{A}$ be the countable atomless Boolean algebra (Example 4.1.4) and let $R_1^{\mathfrak{A}}, R_2^{\mathfrak{A}}, \dots$ be the relations defined by $\phi_1, \phi_2, \dots$ over the atomless Boolean algebra $\mathfrak{A}$, where $\wedge, \vee, \neg, 0, 1$ play the role of $\cap, \cup, c, \mathbf{0}, \mathbf{1}$. The structure $\mathfrak{B} = (\mathbb{A}; R_1^{\mathfrak{A}}, R_2^{\mathfrak{A}}, \dots)$ is ω-categorical (see the comment after Lemma 4.7.3). To verify that $\mathfrak{B}$ and $\mathfrak{C}$ have the same existential theory, let ψ be an existential sentence over the signature $\{R_1, R_2, \dots\}$. Replace each atomic formula of the form $R_i(x_1, \dots, x_k)$ in ψ by the formula $\phi_i(x_1, \dots, x_k)$. The resulting formula ψ' is a quantifier-free first-order formula in the signature of Boolean algebras, $\{\cup, \cap, c, \mathbf{0}, \mathbf{1}\}$. Clearly, $\mathfrak{C} \models \psi$ if and only if $\mathfrak{S} \models \psi'$ and $\mathfrak{B} \models \psi$ if and only if $\mathfrak{A} \models \psi'$. It therefore suffices to show that $\mathfrak{S} \models \psi'$ if and only if $\mathfrak{A} \models \psi'$. To see this, first note that the structure $(\mathcal{P}(\mathbb{N}); \cup, \cap, c, \mathbf{0}, \mathbf{1})$ is a Boolean algebra, with

- $\cap$ and $\cup$ playing the role of $\wedge$ and $\vee$, respectively;
- c playing the role of $-$;
- $\mathbf{0}$ and $\mathbf{1}$ playing the role of 0 and 1.

Now the statement follows from the well-known fact that a quantifier-free formula is satisfiable in some infinite Boolean algebra if and only if it is satisfiable in all infinite Boolean algebras (see, e.g., Corollary 5.7 in [280]). $\dashv$

All finite set constraint languages have a CSP in NP [60]. A large class of polynomial-time tractable set constraint languages has been described in [60]; the class given there is *maximal tractable* in the sense that every strictly larger class of set constraint languages contains a finite subset with an NP-hard CSP.

5.4. Spatial reasoning

In this section we present a homogeneous structure $\mathfrak{R}$ whose CSP equals the network satisfaction problem for RCC5 from Section 1.6.6 (for an explanation of the identification of network satisfaction problems with CSPs see Section 1.5.3). We present two different ways of constructing $\mathfrak{R}$.

5.4.1. Construction via Fraïssé-amalgamation. Our first construction follows the presentation in [51]. Let $\mathfrak{S}$ be the structure with domain $S := \mathcal{P}(\mathbb{N}) \setminus \{\emptyset\}$, i.e., the set of all non-empty subsets of the natural numbers $\mathbb{N}$. The signature of $\mathfrak{S}$ consists of the binary relation symbols P, DR, PO, and for $x, y \subseteq \mathbb{N}$ we have

$$(x, y) \in \mathrm{P}^{\mathfrak{S}} \text{ iff } x \subseteq y, \qquad \text{“}x \text{ is contained in } y\text{”}$$

$$(x, y) \in \mathrm{DR}^{\mathfrak{S}} \text{ iff } x \cap y = \emptyset, \qquad \text{“}x \text{ and } y \text{ are disjoint”}$$

$(x, y) \in \mathrm{PO}^{\mathfrak{S}}$ iff $x \nsubseteq y \wedge y \nsubseteq x \wedge x \cap y \neq \emptyset$, "$x$ and y properly overlap".

Note that every pair (x, y) of distinct elements of S is contained in precisely one of the relations $\mathrm{P}^{\mathfrak{S}}$, $\{(x, y) \mid (y, x) \in \mathrm{P}^{\mathfrak{S}}\}$, $\mathrm{DR}^{\mathfrak{S}}$, and $\mathrm{PO}^{\mathfrak{S}}$. Also note that the structure $\mathfrak{S}$ is neither ω-categorical nor homogeneous. But there exists a homogeneous structure with the same age as $\mathfrak{S}$; this follows from Fraïssé's theorem (Theorem 2.3.8) and the following proposition.

PROPOSITION 5.4.1 (Theorem 30 in [51]). Age($\mathfrak{S}$) *is a strong amalgamation class.*

PROOF. It suffices to verify the strong amalgamation property. Let $\mathfrak{A}, \mathfrak{B}_0, \mathfrak{B}_1 \in \mathrm{Age}(\mathfrak{S})$ be such that $A = B_0 \cap B_1$. We have to show that there exist embeddings $f_i : \mathfrak{B}_i \hookrightarrow \mathfrak{S}$, for $i \in \{0, 1\}$, such that $f_0(a) = f_1(a)$ for all $a \in A$ and $f_0(B_0) \cap f_1(B_1) = f_0(A) = f_1(A)$. For $i \in \{0, 1\}$, let $e_i : \mathfrak{B}_i \hookrightarrow \mathfrak{S}$ be an embedding. We may choose e_0 and e_1 such that $\bigcup_{b \in B_0} e_0(b) \cap \bigcup_{b \in B_1} e_1(b) = \emptyset$. For $b \in B_i$ and $X \subseteq B_i$, define

$$N_i(b) := \{b' \in B_i \mid (b', b) \in \mathrm{P}^{\mathfrak{B}_i}\}$$
$$N_i(X) := \{b' \in B_i \mid \exists b \in X : (b', b) \in \mathrm{P}^{\mathfrak{B}_i}\}.$$

We define $f_i : B_i \to \mathfrak{S}$ by

$$f_i(b) := e_i(b) \cup \bigcup_{b' \in N_{1-i}(N_i(b))} e_{1-i}(b').$$

Note that if $b \in A$ then $N_1(N_0(b)) = N_1(b)$ and $N_0(N_1(b)) = N_0(b)$, and $f_0(b) = f_1(b) = e_i(b) \cup e_{1-i}(b)$. We claim that f_i is an embedding. Let $v, v' \in B_i$.

1. If $(v', v) \in \mathrm{P}^{\mathfrak{B}_i}$ then $e_i(v') \subseteq e_i(v)$. Since $\mathrm{P}^{\mathfrak{B}_i}$ is transitive, $N_{1-i}(N_i(v')) \subseteq N_{1-i}(N_i(v))$, and hence $f_i(v') \subseteq f_i(v)$. This shows that f_i preserves P.
2. To prove that f_i is injective suppose that $f_i(v) = f_i(v')$. Then by the definition of f_i, every element $x \in e_i(v') \subseteq f_i(v') = f_i(v)$ must be contained in $e_i(v)$ or in $e_{1-i}(u)$ for some $u \in N_{1-i}(N_i(v))$. Since $e_i(v') \cap e_{1-i}(u) = \emptyset$, we must have that $x \in e_i(v)$. We conclude that $e_i(v') \subseteq e_i(v)$. Symmetrically, one can show that $e_i(v) \subseteq e_i(v')$. Since e_i is an embedding this implies that $v = v'$ and proves the injectivity of f_i.
3. Now suppose that $(v, v') \in \mathrm{DR}^{\mathfrak{B}_i}$. Suppose for contradiction that there exists $x \in f_i(v) \cap f_i(v')$. By the definition of f_i this means that $x \in e_i(v)$ or there exists $u \in N_{1-i}(N_i(v))$ such that $x \in e_{1-i}(u)$.
   - Suppose that $x \in e_i(v)$. By the definition of f_i either $x \in e_i(v')$ or there exists $u' \in N_{1-i}(N_i(v'))$ such that $x \in e_{1-i}(u')$. The latter is impossible because $e_i(v) \cap e_{1-i}(u') = \emptyset$. Thus, $x \in e_i(v) \cap e_i(v')$, in contradiction to the assumption that e_i is an embedding and $(v, v') \in \mathrm{DR}^{\mathfrak{B}_i}$.
   - Suppose that there exists $u \in N_{1-i}(N_i(v))$ such that $x \in e_{1-i}(u)$. By the definition of f_i either $x \in e_i(v')$ or there exists $u' \in N_{1-i}(N_i(v'))$

such that $x \in e_{1-i}(u')$. The former is impossible because $e_{1-i}(u) \cap e_i(v') = \emptyset$. We thus have $u \in N_{1-i}(N_i(v))$ and $u' \in N_{1-i}(N_i(v'))$ such that $x \in e_{1-i}(u) \cap e_{1-i}(u')$. Let $w \in N_i(v)$ be such that $u \in N_{1-i}(w)$ and let $w' \in N_i(v')$ such that $u' \in N_{1-i}(w')$. Since e_{1-i} is an embedding and $x \in e_{1-i}(u) \cap e_{1-i}(u')$ we have that $(w, w') \notin \mathrm{DR}^{\mathfrak{B}_{1-i}}$, and since w, w must be in A we have that $(w, w') \notin \mathrm{DR}^{\mathfrak{B}_i}$. This in turn implies that $(v, v') \notin \mathrm{DR}^{\mathfrak{B}_i}$, in contradiction to the assumptions.

4. Finally, suppose that $(v, v') \in \mathrm{PO}^{\mathfrak{B}_i}$. Then there exist $x \in e_i(v) \cap e_i(v')$, $y \in e_i(v) \setminus e_i(v')$, and $y' \in e_i(v') \setminus e_i(v)$. Note that $x \in f_i(v) \cap f_i(v')$, $y \in f_i(v)$, and $y' \in f_i(v')$. Also note that $y \notin f_i(v')$ since $y \notin e_i(v')$ and $e_i(v)$ and $e_{1-i}(u)$ are disjoint for every $u \in N_{i-1}(N_i(v'))$. Similarly, $y' \notin f_i(v)$. This shows that f_i preserves PO.

Moreover, $f_0(u_0) \neq f_1(u_1)$ for $u_i \in B_i \setminus A$; this can be shown similarly as in (2). Hence, $f_0(B_0) \cap f_1(B_1) = f_0(A) = f_1(A)$. This concludes the proof of the strong amalgamation property of $\mathrm{Age}(\mathfrak{S})$. ⊣

Proposition 5.4.1 and Theorem 2.3.8 imply that there exists a countable homogeneous structure with the same age as $\mathfrak{S}$. Let $\mathfrak{R}$ be the first-order expansion of this structure where

- $\mathrm{PP}(x, y)$ is defined by $\mathrm{P}(x, y) \wedge x \neq y$,
- $\mathrm{PPI}(x, y)$ is defined by $\mathrm{P}(y, x) \wedge x \neq y$,
- $\mathrm{EQ}(x, y)$ is defined by $x = y$, and
- all other elements of the relation algebra RCC5 are defined as the respective unions of PP, PPI, DR, PO, EQ.

It is now straightforward to verify that the relations of this structure satisfy the composition table given in Figure 8. It follows that $\mathfrak{R}$ is a square representation of RCC5.

PROPOSITION 5.4.2. *$\mathfrak{R}$ is a fully universal representation of RCC5.*

PROOF. Let $\mathfrak{A}$ be a finite $\{\mathrm{PP}, \mathrm{PPI}, \mathrm{DR}, \mathrm{PO}, \mathrm{EQ}\}$-structure that satisfies the axioms (7),(8), (9), and (10) from Section 1.6.6. It suffices to verify that the structure $(A; \mathrm{PP}^{\mathfrak{A}} \cup \mathrm{EQ}^{\mathfrak{A}}, \mathrm{DR}^{\mathfrak{A}}, \mathrm{PO}^{\mathfrak{A}})$ embeds into $\mathfrak{S}$. Suppose without loss of generality that $A = \{1, \dots, n\}$. Then $e \colon A \to \mathcal{P}(\mathbb{N})$ given by $e(a) := \{b \mid (a, b) \in \mathrm{PP}^{\mathfrak{A}} \cup \mathrm{EQ}^{\mathfrak{A}}\}$ is such an embedding. ⊣

5.4.2. Construction via the atomless Boolean algebra. The structure $\mathfrak{R}$ can also be obtained from the atomless Boolean algebra (Example 4.1.4), following the presentation in [62]. Let PP, PPI, DR, PO, and EQ be the binary relations with the following first-order definitions in $\mathfrak{A}$ (and their intuitive meaning in quotes).

$\mathrm{PP}(x, y)$ iff $(x \cap y = x) \wedge x \neq y \wedge x, y \notin \{0\}$ 'y properly contains x'

$\mathrm{PPI}(x, y)$ iff $(x \cap y = y) \wedge x \neq y \wedge x, y \notin \{0\}$ 'x properly contains y'

$\mathrm{DR}(x, y)$ iff $(x \cap y = 0) \wedge x \neq y \wedge x, y \notin \{0\}$ 'x and y are disjoint'

$$\begin{aligned} \mathrm{PO}(x,y) \text{ iff } & \neg\mathrm{DR}(x,y) \wedge \neg\mathrm{PP}(x,y) \wedge \\ & \neg\mathrm{PPI}(x,y) \wedge x \neq y \wedge x,y \notin \{0\} && \text{`$x$ and $y$ properly overlap'} \\ \mathrm{EQ}(x,y) \text{ iff } & x = y \wedge x,y \notin \{0\} && \text{`x equals y'} \end{aligned}$$

The structure $(A; \mathrm{PP}, \mathrm{PPI}, \mathrm{DR}, \mathrm{PO}, \mathrm{EQ})$ is a first-order reduct of $\mathfrak{A}$ and hence ω-categorical, and by Theorem 4.6.4 it has an ω-categorical model companion. We claim that the expansion of this structure by all binary first-order definable relations is isomorphic to $\mathfrak{R}$ as defined in Section 5.4.1. To prove this, first observe that $\mathfrak{A}$ satisfies the axioms (7),(8), (9), and (10) from Section 1.6.6. Hence, every finite substructure of $(A; \mathrm{PP} \cup \mathrm{EQ}, \mathrm{DR}, \mathrm{PO})$ embeds into $\mathfrak{S}$ as we have seen in the proof of Proposition 5.4.2. Conversely, every finite substructure of $\mathfrak{S}$ embeds into $(A; \mathrm{PP} \cup \mathrm{EQ}, \mathrm{DR}, \mathrm{PO})$. From this it follows that $\mathfrak{R}$ and the structure constructed above have the same age. Moreover, both structures are model-complete and ω-categorical, so they must be isomorphic.

5.5. Finite relation algebras from countably categorical structures

Every ω-categorical structure $\mathfrak{B}$ gives rise to a finite relation algebra, which we call the *orbital relation algebra for* $\mathfrak{B}$, and which is described by the following proposition.

PROPOSITION 5.5.1. *Let* $\mathfrak{B}$ *be a structure such that* $\mathrm{Aut}(\mathfrak{B})$ *has finitely many orbitals (defined in Section* 4.1.1*). Then the unions of these orbitals form a proper square relation algebra* $\mathbf{A}$.

PROOF. Clearly, the orbitals of $\mathrm{Aut}(\mathfrak{B})$ partition B^2. Since composition is first-order definable, it follows that unions of orbitals of $\mathrm{Aut}(\mathfrak{B})$ are preserved under composition. Also the other properties of proper relation algebras in Definition 1.5.1 are straightforward to verify. ⊣

Note that not every ω-categorical structure provides a universal representation of its orbital relation algebra: for example the K_4-free Henson graph (Example 2.3.10) has the same orbital relation algebra as the random graph (Example 2.3.9), but only the latter provides a (fully) universal representation.

PROPOSITION 5.5.2. *Let* $\mathbf{A}$ *be a finite relation algebra with a fully universal square representation* $\mathfrak{B}$. *Then* $\mathfrak{B}$ *is finitely bounded and the network satisfaction problem for* $\mathbf{A}$ *equals the constraint satisfaction problem* $\mathfrak{B}$.

PROOF. Besides some bounds of size at most two that make sure that the atomic relations partition B^2, it suffices to include appropriate three-element structures into $\mathcal{F}$ that can be read off from the composition table of $\mathbf{A}$. The statement about the network satisfaction problem holds because fully universal representations are in particular universal (Proposition 1.5.10). ⊣

The connection between (network satisfaction problems of) finite relation algebras $\mathfrak{A}$ and ω-categorical structures is clearest for a certain class of finite relation algebras, namely those that have a normal representation [202].

DEFINITION 5.5.3. A representation for $\mathfrak{A}$ is called *normal* if it is square, fully universal, and (as a relational structure, see Section 1.5.3) homogeneous.

Normal representations of finite relation algebras are ω-categorical (Lemma 4.3.1) and clearly they are model-complete cores.

COROLLARY 5.5.4. *Let* **A** *be a finite relation algebra with a normal representation* $\mathfrak{B}$. *Then the orbital relation algebra of* $\mathfrak{B}$ *is isomorphic to* **A**.

PROOF. The statement follows directly from the definition of the orbital relation algebra from Proposition 5.5.1. ⊣

Examples of finite relation algebras with a normal representation are the point algebra (Example 1.5.2), RCC5 (Sections 1.6.6 and 5.4), and Allen's Interval Algebra which we revisit in the following example.

Example 5.5.5. Allen's Interval Algebra $\mathfrak{A}$ from Section 1.6.1 is a normal representation of its orbital relation algebra. It suffices to show that $\mathfrak{A}$ is homogeneous. Let $\mathfrak{B}$ be the expansion of $(\mathbb{Q};<)^{[2]}$, the full second power of $(\mathbb{Q};<)$ (Definition 3.5.3), by all binary first-order definable relations. We claim that $\mathfrak{B}$ is homogeneous: if $g\colon \mathfrak{B}_1 \to \mathfrak{B}_2$ is an isomorphism between finite substructures of $\mathfrak{A}$, let $C_i := \{a \in \mathbb{Q} \mid (a,b) \in B_i \text{ or } (b,a) \in B_i\}$. The preservation of the relations $E_{1,1}$, $E_{1,2}$, and $E_{2,2}$ implies that g induces an function $f\colon C_1 \to C_2$; this map must preserve $<$, and $(a,b) \mapsto (f(a), f(b))$ is an automorphism of $\mathfrak{B}$ that extends g. Note that $\mathfrak{A}$ is the substructure of $\mathfrak{B}$ with domain $\{(a,b) \mid a < b\}$; it is easy to see that substructures of homogeneous structures whose domain is first-order definable are homogeneous as well.

We finally present an example of a finite relation algebra with an ω-categorical fully universal representation, but without normal representation.

Example 5.5.6. An example of a finite relation algebra with no normal representation is the left-linear point algebra (Sections 1.6.3 and 5.2). The results in Section 5.2.3 imply that the left-linear point algebra is the orbital relation algebra of the ω-categorical structure $(\mathbb{T}; \leq)$ and that it is a fully universal and square.

5.6. Fragments of existential second-order logic

Already in Section 1.4 we have pointed out the important role of existential second-order logic in complexity theory, and the role of fragment of ESO, in particular of SNP, in constraint satisfaction. If CSP($\mathfrak{B}$) is expressible in one of these fragments, then this often means that $\mathfrak{B}$ can be chosen to satisfy strong model-theoretic properties (here we will use concepts that we have introduced in Chapter 2 and Chapter 4).

We start from first-order logic (in Section 5.6.1), then re-visit the logic MMSNP from Section 1.4.4 (in Section 5.6.2), and then consider a further extension MMSNP_2 (in Section 5.6.3). Throughout this section, if Φ is an ESO sentence, we write $[\![\Phi]\!]$ for the class of all finite models of Φ.

5.6.1. Finitely many connected obstructions. Every CSP can be described by homomorphically forbidding a set of finite connected structures (Lemma 1.1.8). In this section we study CSPs that can be described by forbidding a *finite* set of finite connected structures. It turns out that every such CSP can be formulated with an ω-categorical template.

The same class of CSPs arises naturally in a different context. Recall that if $\mathfrak{B}$ is a structure with a finite relational signature τ, then $\text{CSP}(\mathfrak{B})$ can be viewed as a class of finite τ-structures that homomorphically maps to $\mathfrak{B}$ (as described in Section 1.1 and Section 1.4). Consider the situation in which this class can be *described* by a first-order sentence Φ in the sense that $[\![\Phi]\!] = \text{CSP}(\mathfrak{B})$. If there is such a first-order sentence Φ then we say that $\text{CSP}(\mathfrak{B})$ *is in FO*. It is clear that all CSPs that can be described by homomorphically forbidding a finite set of finite connected structures are in FO; we can in fact construct from the homomorphic obstructions a universal-negative sentence describing the CSP. The converse implication follows from the following result, which was a famous open problem in finite model theory until its solution by Rossman [317]. The theorem will not be needed further in this text and we do not present its proof.

THEOREM 5.6.1 (Homomorphism Preservation in the Finite). *Let τ be a finite relational signature, and let Φ be a first-order τ-sentence. Then Φ is equivalent to an existential positive sentence on all finite τ-structures if and only if the class of all finite τ-models of Φ is closed under homomorphisms* (*Definition* 1.1.7).

In the remainder of this section, τ is a finite relational signature and $\mathfrak{B}$ is a τ-structure. Recall that $\text{CSP}(\mathfrak{B})$, viewed as a class of finite τ-structures, is closed under inverse homomorphisms and disjoint unions. In particular, the class of all finite τ-structures that do *not* map homomorphically to $\mathfrak{B}$ is closed under homomorphisms, and by Theorem 5.6.1 it can be described by an existential positive τ-sentence Ψ. This leads us to the following.

THEOREM 5.6.2. *Let τ be a finite relational signature and let C be a class of finite τ-structures. Then the following are equivalent.*

1. $C = \text{CSP}(\mathfrak{B})$ *for some τ-structure $\mathfrak{B}$, and* $\text{CSP}(\mathfrak{B})$ *is in FO*;
2. $C = \text{Forb}^{\text{hom}}(\mathcal{F})$ *for a finite set of finite connected τ-structures $\mathcal{F}$*;
3. $C = \text{CSP}(\mathfrak{B})$ *for an ω-categorical τ-structure $\mathfrak{B}$, and there exists a universal-negative sentence Ψ such that $\mathfrak{A} \in C$ if and only if $\mathfrak{A} \models \Psi$.*

PROOF. (1) $\Rightarrow$ (2). The remarks above explain how to deduce from Theorem 5.6.1 that there is an existential positive τ-sentence Ψ such that $\mathfrak{A}$ maps homomorphically to $\mathfrak{B}$ if and only if $\mathfrak{A}$ satisfies $\neg\Psi$. The sentence $\neg\Psi$ can be re-written as a universal negative sentence in conjunctive normal form; let Φ be such a universal negative sentence of minimal size. We claim that the canonical database $\mathfrak{C}$ for each conjunct in Φ is connected. To see this, suppose that $\mathfrak{C}$ has several connected components. If one of them does not map homomorphically to $\mathfrak{B}$, then Φ was not of minimal size, since the corresponding conjunct could have been replaced by the (smaller) conjunctive query for the component. If all components map homomorphically to $\mathfrak{B}$, then so does $\mathfrak{C}$, a contradiction to the fact that $\mathfrak{C}$ is the canonical database of a conjunct of Φ. Therefore, we obtain finitely many connected τ-structures $\mathfrak{C}$, and they form the desired set $\mathcal{F}$.

(2) $\Rightarrow$ (3). We can apply Theorem 4.3.8 to the finite set $\mathcal{F}$ of finite connected τ-structures, and obtain an ω-categorical τ-structure $\mathfrak{B}$ such that $\mathrm{CSP}(\mathfrak{B}) = \mathrm{Forb}^{\mathrm{hom}}(\mathcal{F})$. When ψ is the canonical query for a structure $\mathfrak{C} \in \mathcal{F}$, then $\neg\psi$ is equivalent to a universal-negative sentence. The conjunction over the all those universal-negative formulas for all structures in $\mathcal{F}$ provides a universal-negative sentence Ψ with the required properties.

(3) $\Rightarrow$ (1). Trivial. ⊣

5.6.2. CSPs in Monadic SNP. In this section we show that every CSP in monadic SNP can be formulated using an ω-categorical template.

THEOREM 5.6.3 (Theorem 7 in [56]). *Let $\mathfrak{C}$ be a structure with a finite relational signature. If $\mathrm{CSP}(\mathfrak{C})$ can be described by a monadic SNP sentence Φ, then there is an ω-categorical $\mathfrak{B}$ such that $\mathrm{CSP}(\mathfrak{B}) = \mathrm{CSP}(\mathfrak{C})$.*

PROOF. By Corollary 1.4.19, we can assume without loss of generality that Φ is a *connected* and *monotone* monadic SNP sentence. Let $P_1, \dots, P_k$ be the existentially quantified monadic predicates in Φ. Let τ' be the signature containing the input relations from τ, the monadic relation symbols $P_1, \dots, P_k$, and new monadic relation symbols $P'_1, \dots, P'_k$.

We replace positive literals of the form $P_i(x)$ in Φ by $\neg P'_i(x)$. We shall denote the τ'-formula obtained from Φ after this transformation by Φ'. Observe that each clause ψ' of Φ' can be written as $\neg\psi$ where ψ is quantifier-free primitive positive. We define $\mathcal{F}$ to be the set of τ'-structures containing for each clause $\psi = \neg\psi'$ in Φ' the canonical database of ψ'. We shall use the fact that a τ'-structure $\mathfrak{A}$ satisfies a clause ψ' if and only if the canonical database of ψ is not homomorphic to $\mathfrak{A}$. Since Φ is connected, all structures in $\mathcal{F}$ are connected.

Then Theorem 4.3.8 asserts the existence of an $\mathcal{F}$-free ω-categorical τ'-structure $\mathfrak{B}'$ that is universal for all $\mathcal{F}$-free structures. We use $\mathfrak{B}'$ to define the template $\mathfrak{B}$ with the properties required in the statement of the theorem we are about to prove. The structure $\mathfrak{B}$ is the τ-reduct of the restriction of $\mathfrak{B}'$ to the points with the property that for all existential monadic predicates

P_i, $1 \leq i \leq k$, either P_i or P'_i holds (but not both P_i and P'_i). It follows from Theorem 4.7.3 that reducts of ω-categorical structures and restrictions to first-order definable subsets of ω-categorical structures are again ω-categorical. Hence, the resulting τ-structure $\mathfrak{B}$ is ω-categorical.

We claim that $[\![\Phi]\!] = \mathrm{CSP}(\mathfrak{B})$. First, let $\mathfrak{A}$ be a finite structure that has a homomorphism h to $\mathfrak{B}$. Let $\mathfrak{A}'$ be the τ'-expansion of $\mathfrak{A}$ such that for all $i \leq k$ and $a \in A$ the relation $P_i(a)$ holds in $\mathfrak{A}'$ if and only if $P_i(h(a))$ holds in $\mathfrak{B}'$, and $P'_i(a)$ holds in $\mathfrak{A}'$ if and only if $P'_i(h(a))$ holds in $\mathfrak{B}'$. Clearly, h defines a homomorphism from $\mathfrak{A}'$ to $\mathfrak{B}'$. In consequence, none of the structures from $\mathcal{F}$ maps to $\mathfrak{A}'$. Hence, the τ-reduct $\mathfrak{A}$ of $\mathfrak{A}'$ satisfies Φ.

Conversely, let $\mathfrak{A}$ be a finite τ-structure satisfying Φ. Consequently, there exists a τ'-expansion $\mathfrak{A}'$ of $\mathfrak{A}$ that satisfies the first-order part of Φ', and where for every $a \in A$ exactly one of $P_i(a)$ or $P'_i(a)$ holds. Clearly, no structure in $\mathcal{F}$ is homomorphic to $\mathfrak{A}'$, and by the universality of $\mathfrak{B}'$ the τ'-structure $\mathfrak{A}'$ is a substructure of $\mathfrak{B}'$. Since for every $a \in A$ exactly one of $P_i(a)$ and $P'_i(a)$ holds, $\mathfrak{A}'$ is also a substructure of the restriction of $\mathfrak{B}'$ to B. Consequently, $\mathfrak{A}$ is homomorphic to the τ-reduct of this restriction. This completes the proof. ⊣

5.6.3. Guarded monotone SNP. In this section we consider an expressive generalisation of MMSNP introduced by Bienvenu, ten Cate, Lutz, and Wolter [40] in the context of ontology-based data access, called *guarded monotone SNP* (also called guarded disjunctive Datalog [40]). It has the same expressive power as the logic MMSNP_2 introduced by Madelaine [274], which is the extension of the class MMSNP from Section 1.4.4 where we are also allowed to quantify over subsets of (extensional) relations, rather than just over subsets of the domain [75]. See Figure 1.

DEFINITION 5.6.4. Let Φ be a monotone SNP τ-sentence with existentially quantified predicates ρ. Then Φ is called *guarded* if each conjunct of Φ can be written in the form

$$\alpha_1 \wedge \cdots \wedge \alpha_n \Rightarrow \beta_1 \vee \cdots \vee \beta_m,$$

where

- $\alpha_1, \ldots, \alpha_n$ are atomic $(\tau \cup \rho)$-formulas, called *body atoms*,
- $\beta_1, \ldots, \beta_m$ are atomic ρ-formulas, called *head atoms*,
- for every head atom β_i there is a body atom α_j such that α_j contains all variables from β_i (such clauses are called *guarded*).

We do allow the case that $m = 0$, i.e., the case where the head consists of the empty disjunction, which is equivalent to $\perp$ (false).

The following proposition can be shown similarly as Proposition 1.4.11.

PROPOSITION 5.6.5. *Every guarded monotone SNP sentence Φ is equivalent to a finite disjunction $\Phi_1 \vee \cdots \vee \Phi_k$ of connected guarded monotone SNP sentences.*

Similarly as in Corollary 1.4.15 is can be shown that if connected guarded monotone SNP has a complexity dichotomy into P and NP-complete, then so has guarded monotone SNP.

THEOREM 5.6.6 (Theorem 2 in [71]). *For every sentence Φ in connected guarded monotone SNP there exists a reduct $\mathfrak{C}$ of a finitely bounded homogeneous structures such that $[\![\Phi]\!] = \text{CSP}(\mathfrak{C})$.*

PROOF. Let Φ be a τ-sentence in connected guarded monotone SNP with existentially quantified relation symbols $\{E_1, \dots, E_k\}$. Let σ be the signature which contains for every relation symbol $R \in \{E_1, \dots, E_k\}$ two new relation symbols R^+ and R^- of the same arity and for every relation symbol $R \in \tau$ a new relation symbol R'. Let ϕ be the first-order part of Φ, written in conjunctive normal form, and let n be the number of variables in the largest clause of ϕ. Let ϕ' be the sentence obtained from ϕ by replacing each occurrence of $R \in \{E_1, \dots, E_k\}$ by R^+ and each occurrence of $\neg R$ by R^-, and finally each occurrence of $R \in \tau$ by R'. Let $\mathcal{F}$ be the (finite) class of all finite σ-structures with at most n elements that do not satisfy ϕ'. By Theorem 4.3.8, there exists a finitely bounded homogeneous $(\sigma \cup \rho)$-structure $\mathfrak{B}$ such that the age of the σ-reduct $\mathfrak{C}$ of $\mathfrak{B}$ equals $\text{Forb}^{\text{emb}}(\mathcal{N})$. We say that $S \subseteq B$ is *correctly labelled* if for every $R \in \{E_1, \dots, E_k\}$ of arity m and $s_1, \dots, s_m \in S$ we have $R^-(s_1, \dots, s_m)$ if and only if $\neg R(s_1, \dots, s_m)$. Let $\mathfrak{B}'$ the $(\tau \cup \sigma \cup \rho)$-expansion of $\mathfrak{B}$ where $R \in \tau$ of arity m denotes

$$\{(t_1, \dots, t_m) \in (R')^{\mathfrak{B}} \mid \{t_1, \dots, t_m\} \text{ is correctly labelled}\}.$$

Since $\mathfrak{B}$ is finitely bounded homogeneous, $\mathfrak{B}'$ is finitely bounded homogeneous, too. Let $\mathfrak{C}$ be the τ-reduct of $\mathfrak{B}'$. We claim that $[\![\Phi]\!] = \text{CSP}(\mathfrak{C})$. First suppose that $\mathfrak{A}$ is a finite τ-structure that satisfies Φ. Then it has an $\{E_1, \dots, E_k\}$-expansion $\mathfrak{A}'$ that satisfies ϕ. Let $\mathfrak{A}''$ be the σ-structure with the same domain as $\mathfrak{A}'$ where

- R' denotes $R^{\mathfrak{A}'}$ for each $R \in \tau$;
- R^+ denotes $R^{\mathfrak{A}'}$ for each $R \in \{E_1, \dots, E_k\}$;
- R^- denotes $\neg R^{\mathfrak{A}'}$ for each $R \in \{E_1, \dots, E_k\}$.

Then $\mathfrak{A}''$ satisfies ϕ', and hence embeds into $\mathfrak{B}$. This embedding is a homomorphism from $\mathfrak{A}$ to $\mathfrak{C}$ since the image of the embedding is correctly labelled by the construction of $\mathfrak{A}''$.

Conversely, suppose that $\mathfrak{A}$ has a homomorphism h to $\mathfrak{C}$. Let $\mathfrak{A}'$ be the $(\tau \cup \{E_1, \dots, E_k\})$-expansion of $\mathfrak{A}$ by defining for every n-ary $R \in \{E_1, \dots, E_k\}$ that $(a_1, \dots, a_n) \in R^{\mathfrak{A}}$ if and only if $(h(a_1), \dots, h(a_n)) \in R^{\mathfrak{B}'}$. Then each clause of ϕ is satisfied, because each clause of ϕ is guarded: let $x_1, \dots, x_m$ be the variables of some clause of ϕ. If $a_1, \dots, a_m \in A$ satisfy the body of this clause, and $\psi(a_{i_1}, \dots, a_{i_l})$ is a head atom of such a clause, then the set $\{h(a_{i_1}), \dots, h(a_{i_l})\}$ is correctly labelled. This implies that some of the head atoms of the clause must be true in $\mathfrak{A}'$ because $\mathfrak{B}'$ satisfies ϕ'. ⊣

Not all CSPs of finitely bounded homogeneous structures can be expressed in GMSNP: similarly as in Example 1.4.20 it can be shown that CSP($\mathbb{Q}$; $<$) is not in GMSNP [71].

5.7. Examples with doubly exponential orbit growth

This section presents an example of an ω-categorical structure $\mathfrak{B}$ such that CSP($\mathfrak{B}$) is in NP, but there is no first-order reduct $\mathfrak{C}$ of a homogeneous structure with a finite relational signature such that CSP($\mathfrak{B}$) = CSP($\mathfrak{C}$). For this purpose, the number of maximal ep-n-types of $\mathfrak{B}$ is useful (see Section 4.6). Note that when two structures have the same CSP, then for all $n \geq 1$ they have the same number of maximal ep-n-types. The number of orbits of n-tuples under Aut($\mathfrak{B}$) is un upper bound for the number of maximal ep-n-types of $\mathfrak{B}$; equality holds if $\mathfrak{B}$ is an ω-categorical model-complete core.

PROPOSITION 5.7.1. *Let $\mathfrak{B}$ be a reduct of a homogeneous relational τ-structure $\mathfrak{C}$ with maximal arity m. Then* Aut($\mathfrak{B}$) *has at most $2^{|\tau|n^m}$ orbits of n-tuples.*

PROOF. An orbit of n-tuples under Aut($\mathfrak{C}$) is uniquely described by the atomic formulas that hold on a (equivalently, all) tuples from this orbit. Since each atomic formula has at most m variables, there are at most $|\tau|n^m$ possible atomic formulas that can hold on such an n-tuple, which yields the bound. ⊣

To construct the announced example we use a set constraint language.

Example 5.7.2. Let $\mathfrak{B}$ be the structure that contains all relations of arity at most three with a quantifier-free first-order definition in the atomless Boolean algebra $\mathfrak{A}$ (Example 4.1.4). Since $\mathfrak{A}$ is ω-categorical, the signature of $\mathfrak{B}$ is finite. We have already mentioned in Section 5.3 that the CSP for all finite set constraint languages, and in particular for CSP($\mathfrak{B}$), is in NP [60].

PROPOSITION 5.7.3. *Let $\mathfrak{B}$ be the structure from Example* 5.7.2. *Then there is no first-order reduct $\mathfrak{B}'$ of a homogeneous structure with finite signature such that* CSP($\mathfrak{B}'$) = CSP($\mathfrak{B}$).

PROOF. We first show that $\mathfrak{B}$ is a model-complete core. Trivially, $\mathfrak{B}$ is a core, since with each relation also the complement of the relation is a relation of $\mathfrak{B}$. To see that $\mathfrak{B}$ is model complete, let ϕ be a first-order formula that defines a first-order relation R over $\mathfrak{B}$; we have to show that R also has an existential definition over $\mathfrak{B}$. By quantifier elimination of $\mathfrak{A}$ (recall that $\mathfrak{A}$ has function symbols $\cup, \cap, c, \mathbf{0}, \mathbf{1}$), there is a quantifier-free first-order formula ψ that defines R over $\mathfrak{A}$. By un-nesting terms in ψ with the help of new existentially quantified variables, and replacing occurrences of atomic formulas by the corresponding formulas in the signature of $\mathfrak{B}$ (for instance replacing formulas of the form $x \cap y = z$ by $S(x, y, z)$ where S is the relation of $\mathfrak{B}$ defined by $x \cap y = z$), we find the required existential definition of R in $\mathfrak{B}$. By Theorem 4.5.1, the orbits of n-tuples under $\mathrm{Aut}(\mathfrak{B})$ are primitively positively definable, and so the number of maximal ep-n-types equals the number of orbits of n-tuples under $\mathrm{Aut}(\mathfrak{B})$. Because of Proposition 5.7.1, it therefore suffices to show that for every $m \geq 1$, the number of orbits of n-tuples under $\mathrm{Aut}(\mathfrak{B})$ is not in $O(2^{n^m})$.

We show that this number is at least $2^{2^{n-1}}$. Let $X := \{x_1, \ldots, x_{n-1}\}$ be such that for each $S \subseteq \{1, \ldots, n-1\}$ we have

$$x_S := \bigcap_{i \in S} x_i \cap \bigcap_{i \in \{1,\ldots,n\} \setminus S} c(x_i) \neq \emptyset.$$

Note that if $S, T \subseteq \{1, \ldots, n-1\}$ are distinct, then $x_S \cap x_T = \emptyset$. Hence, there are $2^{2^{n-1}}$ many different elements b that can be obtained from the elements of the form x_S by applying $\cup$. Since the relations $\{(x, y) \mid x = c(y)\}$, $\{(x, y, z) \mid x \cap y = z\}$, and $\{(x, y, z) \mid x \cup y = z\}$ are in $\mathfrak{B}$, for each of these elements b the tuple $(x_1, \ldots, x_{n-1}, b)$ will lie in a different orbit, which shows the claim. $\dashv$

5.8. CSPs in SNP without a countably categorical template

In this section we present a simple example of a connected monotone SNP sentence that cannot be formulated with an ω-categorical template. Recall from Section 1.6.14 that Succ denotes the relation $\{(x, y) \in \mathbb{Z}^2 \mid x = y + 1\}$.

PROPOSITION 5.8.1. $\mathrm{CSP}(\mathbb{Z}; \mathrm{Succ})$ *can be expressed in connected monotone SNP.*

PROOF. Let Φ be the following connected monotone SNP sentence over the signature $\{\mathrm{Succ}\}$.

$$\begin{aligned} \exists E, S, T\, \forall x, y, z\, \big(&\text{`}E \text{ is an equivalence relation'} \\ &\wedge \text{`}T \text{ is transitive, irreflexive, and extends } S, \\ &\qquad\qquad \text{and } S \text{ extends Succ'} \\ &\wedge \big((S(x, y) \wedge S(x, z)) \Rightarrow E(y, z)\big) \\ &\wedge \big((S(x, y) \wedge E(x, x')) \Rightarrow S(x', y)\big) \end{aligned}$$

First note that if a finite $\{\mathrm{Succ}\}$-structure has a homomorphism h to $(\mathbb{Z}; \mathrm{Succ}\}$ then there exists an $\{\mathrm{Succ}, E, S, T\}$-expansion that satisfies the quantifier-free part ϕ of Φ, given as follows:

- $E(x, y)$ holds if $h(x) = h(y)$,
- $S(x, y)$ holds and if $\mathrm{Succ}(h(x), h(y))$, and
- $T(x, y)$ holds if $h(x) < h(y)$.

Conversely, suppose that a $\{\mathrm{Succ}, S, E, T\}$-structure satisfies ϕ. Then the $\{\mathrm{Succ}\}$-reduct must be an acyclic directed graph (i.e, the digraph contains no directed cycles); we have to show that this digraph homomorphically maps to

$(\mathbb{Z}; \text{Succ})$. It is easy to see that such a homomorphism exists if and only if the *net length* of all (not necessarily directed) cycles in the digraph is 0, i.e., if we traverse a cycle in one direction, then the number of traversed forward edges equals the number of traversed backwards edges. Suppose for contradiction that the digraph contains a cycle of net length different from 0. We now consider cycles where we are additionally allowed to traverse E-edges and S-edges. The E-edges are ignored when computing the net length of a cycle, while the S edges are treated as Succ-edges. Let C be the smallest such cycle of positive net length in the $\{\text{Succ}, S, E\}$-structure. Not all vertices on the cycle can have no incoming S-edge because of the acyclicity of the digraph. If for some vertex on the cycle there are no incoming S-edges, then the preconditions of the implications in the last two conjuncts of ϕ can be applied to find a shorter cycle with the same net length, a contradiction. ⊣

PROPOSITION 5.8.2. $\text{CSP}(\mathbb{Z}; \text{Succ})$ *cannot be formulated with an* ω*-categorical template.*

PROOF. The number of maximal ep-n-types is the same in any structure $\mathfrak{B}$ where $\text{CSP}(\mathfrak{B}) = \text{CSP}(\mathbb{Z}; \text{Succ})$, so by Corollary 4.6.2 it suffices to check that $(\mathbb{Z}; \text{Succ})$ has an infinite number of maximal pp-2-types. But this is clear since for each n the formula $\phi_n(x_0, x_n)$ defined by $\exists x_1, \dots, x_{n-1} \bigwedge_{i=1}^{n} \text{Succ}(x_{i-1}, x_i)$ is in a different pp-2-type. ⊣

Chapter 6

UNIVERSAL ALGEBRA

One of the central concerns of universal algebra, as in model-theory, is the *classification* of mathematical structures. Often, model-theory is considered to be an extension of universal algebra, as formulated by Chang and Keisler in

$$\text{model-theory} = \text{univeral algebra} + \text{logic}.$$

We have a different perspective. Universal algebra leads to classification results with finer distinctions: while model theory often considers two relational structures to be equivalent if they are first-order (or perhaps existentially) interdefinable, universal algebra provides methods that allow to distinguish relational structures up to primitive positive definability. To do so, we study higher-dimensional generalisations of endomorphism monoids, called *polymorphism clones*; from the perspective of this text, we therefore have

$$\text{model-theory} = \text{one-dimensional universal algebra}.$$

The strongest universal-algebraic classification results are available on *finite* domains [205]. In recent years, strong links between deep and central questions in universal algebra and the Feder-Vardi conjecture have led to renewed activity. In fact, several important new and purely algebraic results, for example from [329, 22, 212, 279], were originally motivated by questions about CSPs.

There has been less work on algebras over infinite domains. However, a considerable body of universal-algebraic techniques applies when the algebra under consideration contains as operations all the permutations from an oligomorphic permutation group; we will call such algebras *oligomorphic*. The assumption that the algebra be oligomorphic seems to provide the appropriate amount of 'finiteness' that we need in order to apply universal-algebraic methods.

The step from algebras with only unary functions to algebras that contain higher-ary functions is the point where universal algebra becomes interesting. At the same time, the step from studying automorphisms and embeddings to studying polymorphisms is the step that is new to model-theorists, so we found it natural to divide the background material into a chapter on model theory and a chapter on universal algebra (with a slight overlap).

In Section 6.1 we give a brief introduction to clones, to the study of structures via their polymorphism clones, and to the study of algebras via invariant relations. In particular, we will see that for a finite or ω-categorical structure $\mathfrak{B}$ the complexity of CSP($\mathfrak{B}$) only depends on the polymorphism clone of $\mathfrak{B}$; we refer to this step as the *first abstraction step*. The Boolean case is particularly well understood and provides many examples in the field, so we have dedicated an entire section to clones over a two-element set (Section 6.2). In Section 6.3 we introduce algebras (in the sense of universal algebra), pseudo-varieties, and varieties. For finite and ω-categorical structures $\mathfrak{B}$, the pseudo-variety generated by the polymorphism algebra of $\mathfrak{B}$ is closely linked to the concept of primitive positive interpretability from Chapter 3. Again, the computational complexity of CSP($\mathfrak{B}$) only depends on this pseudo-variety; we refer to the passage to the associated pseudo-variety as the *second abstraction step*. To even study primitive positive interpretability *modulo homomorphic equivalence* we need a recent (but rather natural) concept from universal algebra, the concept of a *reflection* of an algebra (Section 6.4).

For finite structures $\mathfrak{B}$, there is even a third abstraction step: the complexity of CSP($\mathfrak{B}$) depends in fact only on the set of *identities* that are satisfied by the polymorphism algebra of $\mathfrak{B}$, which corresponds, by the fundamental theorem of Birkhoff, to the variety generated by the polymorphism algebra of $\mathfrak{B}$, and to the polymorphism clone of $\mathfrak{B}$ as an abstract clone (Section 6.5). This third abstraction step is harder to take in the context of ω-categorical structures $\mathfrak{B}$; it requires some topological considerations, and is deferred until Chapter 9. In Section 6.6 we present some important results for *idempotent* algebras.

The third abstraction step has a variant for pseudo-varieties that are additionally closed under reflections; this variant shows that certain identities are particularly important, namely identities of height one, where nesting of functions in terms is forbidden (Section 6.7). We then show that every finite algebra that satisfies a non-trivial height-one identity also satisfies a single non-trivial identity involving one function symbol of arity six, a so-called *Siggers term* (Section 6.8.2); also this result will be generalised for oligomorphic algebras later (in Chapter 10). Finally, in Section 6.9 we present some universal-algebraic results for finite algebras that have not yet been generalised to oligomorphic algebras.

6.1. Operation clones

Let B be a set and $n \geq 1$. We denote by $\mathcal{O}_B^{(n)} := B^{B^n}$ the set of *operations* of arity n on B, i.e., the set of functions from B^n to B. The set B will be called the *domain* or *base set*. The set of all operations on B of finite arity will be denoted by $\mathcal{O}_B := \bigcup_{n \in \mathbb{N}} \mathcal{O}_B^{(n)}$. An *operation clone* (over B) is a subset $\mathcal{C}$ of $\mathcal{O}_B$ satisfying the following two properties:

- $\mathcal{C}$ contains all projections, that is, for all $1 \leq k \leq n$ it contains the operation $\pi_k^n \in \mathcal{O}_B^{(n)}$ defined by $\pi_k^n(x_1, \ldots, x_n) = x_k$, and
- $\mathcal{C}$ is *closed under composition*, that is, for all $f \in \mathcal{C} \cap \mathcal{O}_B^{(n)}$ and $g_1, \ldots, g_n \in \mathcal{C} \cap \mathcal{O}_B^{(m)}$ it contains the operation $f(g_1, \ldots, g_n) \in \mathcal{O}_B^{(m)}$ defined by

$$(x_1, \ldots, x_m) \mapsto f(g_1(x_1, \ldots, x_m), \ldots, g_n(x_1, \ldots, x_m)).$$

A *clone* is an abstraction of an operation clone that will be introduced in Section 6.5. In the literature, operation clones are often called clones, or *concrete clones*; we prefer to use the terms 'operation clone' and 'clone' in analogy to 'permutation group' and 'group'.

A *subclone* of $\mathcal{C}$ is a clone that is contained in $\mathcal{C}$. If $\mathcal{S} \subseteq \mathcal{O}_B$ is a set of operations we write $\langle \mathcal{S} \rangle$ for the smallest operation clone $\mathcal{C}$ which contains $\mathcal{S}$, and call $\mathcal{C}$ the clone *generated* by $\mathcal{S}$. The most important source of operation clones in this text are *polymorphism clones* of structures, which we introduce next.

DEFINITION 6.1.1. A *polymorphism* of a structure $\mathfrak{B}$ is a homomorphism from a finite power of $\mathfrak{B}$ to $\mathfrak{B}$; the set of all polymorphisms of $\mathfrak{B}$ is denoted by $\mathrm{Pol}(\mathfrak{B})$.

It is easy to verify that $\mathrm{Pol}(\mathfrak{B})$ is an operation clone, called the *polymorphism clone* of $\mathfrak{B}$. Let f be from $\mathcal{O}_B^{(n)}$, and let $R \subseteq B^m$ be a relation. Then we say that f *preserves* R (and that R is *invariant under* f) iff $f(r_1, \ldots, r_n) \in R$ whenever $r_1, \ldots, r_n \in R$, where $f(r_1, \ldots, r_n)$ is calculated componentwise.

DEFINITION 6.1.2. An operation clone $\mathcal{C} \subseteq \mathcal{O}_B$ is called *finitely related* if there exists a structure $\mathfrak{B}$ with finite relational signature such that $\mathcal{C} = \mathrm{Pol}(\mathfrak{B})$.

It will be convenient to define the operator Pol not only for relational structures $\mathfrak{B}$, but also for sets of relations (since the polymorphisms do not depend on the choice of the signature, but only on the set of relations of $\mathfrak{B}$). We use the following notational conventions. For $n \geq 1$ and a set B, we write $\mathcal{R}_B^{(n)}$ for the set of relations $R \subseteq B^n$ of arity n over B (also called the *n-ary relations over B*). The set of all relations over B will be denoted by $\mathcal{R}_B := \bigcup_{n \geq 1} \mathcal{R}_B^{(n)}$. For $\mathcal{R} \subseteq \mathcal{R}_B$, we write $\mathrm{Pol}(\mathcal{R})$ for the set of operations of $\mathcal{O}$ that preserve all relations from $\mathcal{R}$. Conversely, given a set of operations $\mathcal{S} \subseteq \mathcal{O}$, we write $\mathrm{Inv}(\mathcal{S})$ for the set of all relations which are invariant under all $f \in \mathcal{S}$.

Primitive positive definability has been introduced in Section 2.1.7. The following is straightforward.

PROPOSITION 6.1.3. *Let $\mathfrak{B}$ be any structure. Then* $\mathrm{Inv}(\mathrm{Pol}(\mathfrak{B}))$ *contains* $\langle \mathfrak{B} \rangle_{\mathrm{pp}}$, *the set of all relations that are primitively positively definable in* $\mathfrak{B}$.

PROOF. Suppose that R is k-ary, has a primitive positive definition ψ, and let f be an l-ary polymorphism of $\mathfrak{B}$. To show that f preserves R, let $t_1, \ldots, t_l$ be tuples from R. Then there must be witnesses for the existentially quantified variables $x_{l+1}, \ldots, x_n$ of ψ that show that $\psi(t_i)$ holds in $\mathfrak{B}$, for all $1 \leq i \leq n$. Write s_i for the extension of t_i such that s_i satisfies the quantifier-free part $\psi'(x_1, \ldots, x_l, x_{l+1}, \ldots, x_n)$ of ψ (we assume that ψ is written in prenex normal form). Then the tuple $f(s_1, \ldots, s_l)$ satisfies ψ' as well. This shows that $f(t_1, \ldots, t_l)$ satisfies ψ in $\mathfrak{B}$, which is what we had to show. $\dashv$

6.1.1. Pol-Inv. In this section we fix a countably infinite base set B and write $\mathcal{O}$ instead of $\mathcal{O}_B$ and $\mathcal{O}^{(n)}$ instead of $\mathcal{O}_B^{(n)}$. Operation clones are naturally equipped with a topology, the *topology of pointwise convergence*, which is given by the following definition (for a proper topological treatment, see Chapter 9.)

DEFINITION 6.1.4. Let $\mathcal{S}$ be a subset of $\mathcal{O}$. Suppose that for all $n \geq 1$ and all $g \in \mathcal{O}^{(n)}$, if for all finite $A \subseteq B^n$ there exists an n-ary $f \in \mathcal{S}$ which agrees with g on A, then $g \in \mathcal{S}$. Then we say that $\mathcal{S}$ is *locally closed*. The *local closure* of $\mathcal{S}$, denoted by $\overline{\mathcal{S}}$, is the smallest locally closed subset of $\mathcal{O}$ that contains $\mathcal{S}$. If $f \in \overline{\langle S \rangle}$ then we also say that f is *locally generated* by S.

The following proposition is a description of the hull operator $\mathcal{S} \mapsto \mathrm{Pol}(\mathrm{Inv}(\mathcal{S}))$ (cf. [334], in particular Corollary 1.9) in the Galois connection defined by the operators Pol and Inv.

PROPOSITION 6.1.5. *Let $\mathcal{C} \subseteq \mathcal{O}$ be the smallest locally closed clone that contains $\mathcal{S} \subseteq \mathcal{O}$. Then $\mathcal{C} = \overline{\langle \mathcal{S} \rangle} = \mathrm{Pol}(\mathrm{Inv}(\mathcal{S}))$.*

PROOF. Clearly, $\overline{\langle \mathcal{S} \rangle} \subseteq \mathcal{C}$ since $\mathcal{C}$ contains $\mathcal{S}$, is a clone, and is locally closed. To show that conversely $\mathcal{C} \subseteq \overline{\langle \mathcal{S} \rangle}$ it suffices to show that $\overline{\langle \mathcal{S} \rangle}$ is a locally closed clone that contains $\mathcal{S}$, because $\mathcal{C}$ is the smallest such clone. Clearly, $\overline{\langle \mathcal{S} \rangle}$ contains $\mathcal{S}$ and is locally closed; it is straightforward to verify that $\overline{\langle \mathcal{S} \rangle}$ is a clone.

To show that $\overline{\langle \mathcal{S} \rangle} \subseteq \mathrm{Pol}(\mathrm{Inv}(\mathcal{S}))$, let $f \in \overline{\langle \mathcal{S} \rangle}$ be k-ary. Let R be from $\mathrm{Inv}(\mathcal{S})$. We have to show that f preserves R. Let $t_1, \dots, t_k$ be from R. By assumption $f(t_1, \dots, t_k) = g(t_1, \dots, t_k)$ for some operation g generated from operations in $\mathcal{S}$ and projections. Note that $\mathrm{Inv}(\mathcal{S}) = \mathrm{Inv}(\langle \mathcal{S} \rangle)$ and hence all those operations preserve R. So we conclude that $f(t_1, \dots, t_k) \in R$.

We finally show that $\mathrm{Pol}(\mathrm{Inv}(\mathcal{S})) \subseteq \overline{\langle \mathcal{S} \rangle}$. Let $f \in \mathrm{Pol}(\mathrm{Inv}(\mathcal{S}))$ be n-ary. It suffices to show that for every finite subset A of B there is an operation $g \in \langle \mathcal{S} \rangle$ such that $f(\bar{a}) = g(\bar{a})$ for every $\bar{a} \in A^n$. List all elements of A^n by $a_1, \dots, a_m$ and consider the relation $R := \{(g(a_1), \dots, g(a_m)) \mid g \in \langle \mathcal{S} \rangle^{(n)}\}$. Note that R is preserved by all operations in $\mathcal{S}$. By assumption, f preserves R. Also note that since $\pi_i^n \in \langle \mathcal{S} \rangle$ for all $i \leq n$ we have that $(a_1[i], \dots, a_m[i]) \in R$. Therefore, $(f(a_1), \dots, f(a_n)) \in R$, and hence there exists a $g \in \langle \mathcal{S} \rangle$ such that $(f(a_1), \dots, f(a_n)) = (g(a_1), \dots, g(a_n))$, as required. $\dashv$

Theorem 6.1.5 has the following folklore consequence in universal algebra [334].

COROLLARY 6.1.6. *For $\mathcal{S} \subseteq \mathcal{O}$, the following are equivalent.*

1. *$\mathcal{S}$ is the polymorphism clone of a relational structure;*
2. *$\mathcal{S}$ is a locally closed clone.*

Arbitrary intersections of subclones of $\mathcal{O}$ are operation clones, and arbitrary intersections of locally closed subclones of $\mathcal{O}$ are locally closed. In fact, the set of all locally closed subclones of $\mathcal{O}$, partially ordered by inclusion, forms a complete lattice. The join of a family $(\mathcal{C}_i)_{i \in I}$ in this lattice equals $\overline{\langle \bigcup_{i \in I} \mathcal{C}_i \rangle}$, and its meet is $\bigcap_{i \in I} \mathcal{C}_i$.

For every permutation group $\mathcal{G}$ there is a unique largest operation clone $\mathcal{C}$ on the same domain such that $\mathcal{C}^{(1)} = \langle \mathcal{G} \rangle$, namely the polymorphism clone of the structure introduced in the following definition.

DEFINITION 6.1.7. If $\mathcal{G}$ is a permutation group on the set B, we write $\mathrm{Orb}(\mathcal{G})$ for a relational structure with domain B whose relations are precisely the orbits of n-tuples under $\mathcal{G}$, for all $n \geq 1$.

The structure $\mathrm{Orb}(\mathcal{P})$ is not unique because there is no restriction on the exact choice of the symbols in the signature; however, this choice never matters in what follows. Note that $\mathrm{End}(\mathrm{Orb}(\mathcal{G})) = \overline{\mathcal{G}}$. Also note that if $\mathcal{G}$ is oligomorphic, then $\mathrm{Orb}(\mathcal{G})$ and all its first-order expansions are ω-categorical model-complete cores.

Remark 6.1.8. The n-ary operations in $\mathrm{Pol}(\mathrm{Orb}(\mathcal{G}))$ are precisely the operations $f \in \mathcal{O}_B^{(n)}$ such that for all $g_1, \dots, g_n \in \mathcal{G}$ the operation

$$x \mapsto f(g_1(x), \dots, g_n(x))$$

is contained in $\overline{\mathcal{G}}$. If f preserves all orbits of n-tuples under $\mathcal{G}$ then the operation above preserves them and hence is in $\overline{\mathcal{G}}$; if f does not preserve an orbit O of n-tuples, then we can find a $t \in O$ and $g_1, \dots, g_n \in \mathcal{G}$ such that $f(g_1(t), \dots, g_n(t)) \notin O$, and hence the operation above is not in $\overline{\mathcal{G}}$.

6.1.2. Inv-Pol. In this section we present a characterisation of $\mathrm{Inv}(\mathrm{Pol}(\mathfrak{B}))$ for arbitrary countable structures $\mathfrak{B}$. The characterisation specialises for ω-categorical structures $\mathfrak{B}$ to a characterisation of primitive positive definability, using the theorem of Ryll-Nardzewski. A somewhat similar statement is due to Romov (Theorem 3.5 in [315]); other characterisations of the operator Inv-Pol have been proved by Szabó [333], Geiger [178], and Pöschel [308], but these characterisations involve infinitary relations or infinitary superposition, which are not needed in the theorem below. The proof below is due to Marcello Mamino (personal communication); we are grateful for the permission to present it here.

The class of *infinitary primitive positive formulas* is inductively defined as follows.

- atomic formulas belong to the class;
- the class contains all finite and infinite conjunctions of formulas $\phi(x_1, \dots, x_n)$ in the class that have the same free variables $x_1, \dots, x_n$;
- if $\phi(x_1, \dots, x_n)$ is in the class, then $\exists x_i \colon \phi(x_1, \dots, x_n)$ is in the class.

The following example shows that there are relations that can be defined by infinitary primitive positive formulas that are not definable as infinite conjunctions of primitive positive formulas.

Example 6.1.9. Consider $\mathfrak{B} := (\mathbb{Z} \cup \{\infty\}; <)$ where $n < \infty$ holds if and only if $n \neq \infty$. Clearly, for each $i \in \mathbb{N}$ the relation $\{(n, m) \in B^2 \mid n + i < m\}$ has a primitive positive definition

$$\phi_i(x, y) := \exists x_1, \dots, x_i \left(x < x_1 \wedge x_i < y \wedge \bigwedge_{j=1}^{i-1} x_j < x_{j+1} \right)$$

in $\mathfrak{B}$. Then the unary relation $\{\infty\}$ can be defined by the infinitary primitive positive formula $\exists x \bigwedge_{i=1}^{\infty} \phi_i(x, y)$. But note that if a primitive positive formula $\phi(y)$ is satisfiable, it is also satisfiable by an element from $\mathbb{Z}$, and since $\mathbb{Z}$ is an orbit under $\mathrm{Aut}(\mathfrak{B})$, it is satisfiable by all elements from $\mathbb{Z}$. Infinite intersections of relations that contain $\mathbb{Z}$ again contain $\mathbb{Z}$, and in particular it follows that the relation $\{\infty\}$ is not definable by an infinite intersection of primitively positively definable relations.

A *chain* of relations is a sequence of relations $(R_i)_{i \in \mathbb{N}}$ of the same arity with the property that $R_i \subseteq R_{i+1}$ for all $i \in \mathbb{N}$.

THEOREM 6.1.10. *Let $\mathfrak{B}$ be a countable structure. Then $R \subseteq B^n$ is preserved by all polymorphisms of $\mathfrak{B}$ if and only if R is the union of a chain of relations that have infinitary primitive positive definitions in $\mathfrak{B}$.*

PROOF. We have already seen in Proposition 6.1.3 that every relation with a primitive positive definition in $\mathfrak{B}$ is preserved by all polymorphisms of $\mathfrak{B}$. The proof that $\mathrm{Inv}(\mathrm{Pol}(\mathfrak{B}))$ is closed under infinite conjunctions and unions of chains is similarly straightforward.

Now suppose that $R \in \mathrm{Inv}(\mathrm{Pol}(\mathfrak{B}))$ is a relation of arity n. Let $a_1, a_2, \dots$ be an enumeration of R. For $i \in \mathbb{N}$, let R_i be the intersection of all relations that contain the tuples $a_1, \dots, a_i$ and that have an infinitary primitive positive definition in $\mathfrak{B}$. We have $R_i \subseteq R_{i+1}$ for all $i \in \mathbb{N}$ since each relation in the intersection that defines R_{i+1} also appears as a relation in the intersection that defines R_i. So it suffices to prove that $R = \bigcup_{i \in \mathbb{N}} R_i$. The inclusion $R \subseteq \bigcup_{i \in \mathbb{N}} R_i$ is clear. To prove the reverse inclusion, we have to prove that $R_i \subseteq R$ for all $i \in \mathbb{N}$. Let $t = (t_1, \dots, t_n) \in R_i$. We claim that there exists an $f \in \mathrm{Pol}(\mathfrak{B})$ of arity i such that $f(a_1, \dots, a_i) = t$. Let $b_1, b_2, \dots$ be an enumeration of B^i starting with the i-tuples $b_1 := (a_1[1], \dots, a_i[1]), \dots, b_n := (a_1[n], \dots, a_i[n])$. Note that if $b_r = b_s$, for some $r, s \in \{1, \dots, i\}$, then the primitive positive formula $x_r = x_s$ defines a relation that contains $a_1, \dots, a_i$ and hence contains R_i, and therefore $t_r = t_s$. Define $f(b_1) := t_1, \dots, f(b_n) := t_n$ and note that the partially defined function f preserves all infinitary primitive positive formulas since $t \in R_i$. We extend f to the next element in the enumeration of B^i while preserving the property that f preserves all infinitary primitive positive formulas. Suppose that we already have defined f for $b_1, \dots, b_m$, for $m \geq n$, and that we want to define f for b_{m+1}. Let $\phi(x_1, \dots, x_m, x_{m+1})$ be the conjunction of all infinitary primitive positive formulas satisfied by $b_1, \dots, b_{m+1}$. Then f preserves $\phi'(x_1, \dots, x_m) := \exists x_{m+1} \colon \phi(x_1, \dots, x_m, x_{m+1})$, and hence $\mathfrak{B} \models \phi'(f(b_1), \dots, f(b_m))$. Therefore, there exists an element t_{m+1} such that $\mathfrak{B} \models \phi(f(b_1), \dots, f(b_m), t_{m+1})$. Define $f(b_{m+1}) := t_{m+1}$. And indeed, by construction this extension still has the property that it preserves every infinitary primitive positive formula. In particular, the function f defined on all of B^i is a polymorphism of $\mathfrak{B}$. This shows that $t \in R$. Hence, $R \in \mathrm{Inv}(\mathrm{Pol}(\mathfrak{B}))$, as required. ⊣

Note that Example 6.1.9 shows that in general, unions of chains of infinite intersections of primitive positive definable relations are not strong enough to express all relations in $\text{Inv}(\text{Pol}(\mathfrak{B}))$: in the example, the unary relation $\{\infty\}$ cannot be expressed in this way.

6.1.3. Oligomorphic clones. Let $\mathcal{C} \subseteq \mathcal{O}$ be an operation clone. A unary operation $e \in \mathcal{C}$ is called *invertible in* $\mathcal{C}$ if there exists a unary $i \in \mathcal{C}$ such that $i(e(x)) = e(i(x)) = x$ for all $x \in B$. When $\mathcal{C}$ is the polymorphism clone of a structure $\mathfrak{B}$, then the invertible operations of $\mathcal{C}$ are precisely the automorphisms of $\mathfrak{B}$.

DEFINITION 6.1.11. An operation clone $\mathcal{C} \subseteq \mathcal{O}$ is *oligomorphic* if the set of invertible operations in $\mathcal{C}$ forms an oligomorphic permutation group.

It is immediate from Theorem 4.1.6 and Corollary 6.1.6 that a locally closed subclone of $\mathcal{O}$ is oligomorphic if and only if it is the polymorphism clone of an ω-categorical structure.

Geiger [178] and independently Bodnarčuk, Kalužnin, Kotov, and Romov [101] have shown that a relation R has a primitive positive definition in a *finite* structure $\mathfrak{B}$ if and only if R is preserved by all polymorphisms of $\mathfrak{B}$. This characterisation of primitive positive definability holds in fact for all ω-categorical structures $\mathfrak{B}$.

THEOREM 6.1.12 (Theorem 5.1 in [84]). *Let $\mathfrak{B}$ be a countable ω-categorical structure. A relation R has a primitive positive definition in $\mathfrak{B}$ if and only if R is preserved by all polymorphisms of $\mathfrak{B}$; in symbols,* $\text{Inv}(\text{Pol}(\mathfrak{B})) = \langle\mathfrak{B}\rangle_{\text{pp}}$.

PROOF. One direction has been shown in Proposition 6.1.3. For the other direction, recall that there are finitely many primitively positively definable relations for each fixed arity. Hence, every infinite conjunction of primitive positive formulas is equivalent to a primitive positive formula. For the same reason, every union of a chain of primitively positively definable relations is primitively positively definable. Therefore, Theorem 6.1.10 implies that every relation in $\text{Inv}(\text{Pol}(\mathfrak{B}))$ is primitively positively definable in $\mathfrak{B}$. ⊣

Analogously to Corollary 4.2.10 for permutation groups and to Corollary 4.4.5 for transformation monoids, we obtain a Galois connection between structures with a first-order definition in an ω-categorical structure $\mathfrak{B}$, considered up to primitive positive interdefinability, and subsets of $\mathcal{O}_B$ containing $\text{Aut}(\mathfrak{B})$.

THEOREM 6.1.13. *Let $\mathfrak{B}$ be a countable ω-categorical structure. Then:*

1. *for first-order reducts $\mathfrak{C}$ of $\mathfrak{B}$, the sets of the form $\langle\mathfrak{C}\rangle_{\text{pp}}$, ordered by inclusion, form a lattice;*
2. *the closed subclones of $\mathcal{O}_B$ containing* $\text{Aut}(\mathfrak{B})$, *ordered by inclusion, form a lattice;*
3. *the operator* Inv *is an anti-isomorphism between those two lattices, and* Pol *is its inverse.*

Theorem 6.1.13 tells us that classifying the first-order reducts of an ω-categorical structure $\mathfrak{B}$ up to primitive positive interdefinability amounts to understanding the lattice of closed subclones of $\mathcal{O}_B$ that contain $\mathrm{Aut}(\mathfrak{B})$. To further illustrate this connection and to facilitate later use we state some consequences.

COROLLARY 6.1.14. *Let $\mathfrak{B}$ be ω-categorical and let $\mathcal{F} \subseteq \mathcal{O}_B$. Then:*

- $\overline{\langle \mathcal{F} \rangle} = \mathrm{Pol}(\mathfrak{B})$ *if and only if* $\langle \mathfrak{B} \rangle_{\mathrm{pp}} = \mathrm{Inv}(\mathcal{F})$.
- *The smallest relation in $\langle \mathfrak{B} \rangle_{\mathrm{pp}}$ that contains $R \subseteq B^k$ equals*

$$\{f(a_1, \dots, a_k) \mid k \in \mathbb{N}, f \in \mathrm{Pol}^{(k)}(\mathfrak{B}), a_1, \dots, a_k \in R\}.$$

If $\mathfrak{B}$ is not ω-categorical, then $\mathrm{Inv}(\mathrm{Pol}(\mathfrak{B}))$ may or may not be equal to $\langle \mathfrak{B} \rangle_{\mathrm{pp}}$, as the following examples illustrate.

Example 6.1.15. Let $\mathfrak{B}$ be the structure $(\mathbb{Z}; \{(x, y) \mid x = y + 1\})$. We use Theorem 6.1.10 to prove that $\mathrm{Inv}(\mathrm{Pol}(\mathfrak{B})) = \langle \mathfrak{B} \rangle_{\mathrm{pp}}$. By Theorem 6.1.10 it suffices to show that $\langle \mathfrak{B} \rangle_{\mathrm{pp}}$ is closed under taking infinite conjunctions and unions of chains. It can be shown that every relation in $\langle \mathfrak{B} \rangle_{\mathrm{pp}}$ can be defined as a conjunction of binary relations in $\langle \mathfrak{B} \rangle_{\mathrm{pp}}$, which are precisely the empty relation, the full relation $\mathbb{Z}^2$, and the relations of the form $\{(x, y) \mid x = y + c\}$ for $c \in \mathbb{Z}$. Clearly, infinite intersections and unions of chains of such relations are again of this form, which shows the statement.

6.1.4. Essentially unary operations. Let $k \in \mathbb{N}$ and $i \in \{1, \dots, k\}$. We say that $f \in \mathcal{O}_B^{(k)}$ *depends on the i-th argument* if there is no $(k-1)$-ary operation f' such that $f(x_1, \dots, x_k) = f'(x_1, \dots, x_{i-1}, x_{i+1}, \dots, x_k)$ for all $x_1, \dots, x_k \in B$. If f does not depend on the i-th argument, then we also say that the i-th argument of f is *fictitious*. We can equivalently characterise k-ary operations that depend on the i-th argument by requiring that there are $x_1, \dots, x_k \in B$ and $x_i' \in B$ such that

$$f(x_1, \dots, x_k) \neq f(x_1, \dots, x_{i-1}, x_i', x_{i+1}, \dots, x_k).$$

We say that an operation f is *essentially unary* iff there is an $i \in \{1, \dots, k\}$ and a unary operation f_0 such that $f(x_1, \dots, x_k) = f_0(x_i)$. Operations that are not essentially unary are called *essential*.[16]

DEFINITION 6.1.16. For any set B, the relations P_B^3 and P_B^4 over B are defined as follows.

$$P_B^3 := \{(a, b, c) \in B^3 \mid a = b \text{ or } b = c\};$$
$$P_B^4 := \{(a, b, c, d) \in B^4 \mid a = b \text{ or } c = d\}.$$

[16]This is standard in clone theory, and it makes sense also when studying the complexity of CSPs, since the essential operations are those that are essential for complexity classification.

LEMMA 6.1.17. *Let $f \in \mathscr{O}$ be an operation on a set B. Then the following are equivalent.*

1. *f is essentially unary.*
2. *f preserves P_B^3.*
3. *f preserves P_B^4.*
4. *f depends on at most one argument.*

PROOF. Let k be the arity of f. The implication from (1) to (2) is obvious, since unary operations clearly preserve P_B^3.

To show the implication from (2) to (3), we show the contrapositive, and assume that f does not preseve P_B^4. By permuting arguments of f, we can assume that there are 4-tuples $a^1, \ldots, a^k \in P_B^4$ with $f(a^1, \ldots, a^k) \notin P_B^4$ and $l \leq k$ such that in $a^1, \ldots, a^l$ the first two coordinates are equal, and in $a^{l+1}, \ldots, a^k$ the last two coordinates are equal. Let c be the tuple $(a_1^1, \ldots, a_1^l, a_4^{l+1}, \ldots, a_4^k)$. Since $f(a^1, \ldots, a^k) \notin P_B^4$ we have $f(a_1^1, \ldots, a_1^k) \neq f(a_2^1, \ldots, a_2^k)$, and therefore $f(c) \neq f(a_1^1, \ldots, a_1^k)$ or $f(c) \neq f(a_2^1, \ldots, a_2^k)$. Let $d = (a_1^1, \ldots, a_1^k)$ in the first case, and $d = (a_2^1, \ldots, a_2^k)$ in the second case. Likewise, we have $f(c) \neq f(a_3^1, \ldots, a_3^k)$ or $f(c) \neq f(a_4^1, \ldots, a_4^k)$, and let $e = (a_3^1, \ldots, a_3^k)$ in the first, and $e = (a_4^1, \ldots, a_4^k)$ in the second case. Then for each $i \leq k$, the tuple (d_i, c_i, e_i) is from P_B^3, but $(f(d), f(c), f(e)) \notin P_B^3$.

The proof of the implication from (3) to (4) is again by contraposition. Suppose f depends on the i-th and j-th argument, $1 \leq i \neq j \leq k$. Hence there exist tuples $a_1, b_1, a_2, b_2 \in B^k$ such that a_1, b_1 and a_2, b_2 only differ at the entries i and j, respectively, and such that $f(a_1) \neq f(b_1)$ and $f(a_2) \neq f(b_2)$. Then $(a_1(l), b_1(l), a_2(l), b_2(l)) \in P_B^4$ for all $l \leq k$, but $(f(a_1), f(b_1), f(a_2), f(b_2)) \notin P_B^4$, which shows that f does not preserve P_B^4.

For the implication from (4) to (1), suppose that f depends only on the first argument. Let $i \leq k$ be minimal such that there is an operation g with $f(x_1, \ldots, x_k) = g(x_1, \ldots, x_i)$. If $i = 1$ then f is essentially unary and we are done. Otherwise, observe that since f does not depend on the i-th argument, neither does g, and so there is an $(i-1)$-ary operation g' such that for all $x_1, \ldots, x_n \in B$ we have $f(x_1, \ldots, x_n) = g(x_1, \ldots, x_i) = g'(x_1, \ldots, x_{i-1})$, contradicting the choice of i. ⊣

Example 6.1.18. We claim that all polymorphisms of the structure

$$\mathfrak{B} = (\mathbb{Z}; 0, \{(x, y) \mid x = y + 1\}, \{(u, v, x, y) \mid u = v \vee x = y\})$$

are projections. Lemma 6.1.17 implies that all polymorphisms are essentially unary; all unary maps that preserve $\{(x, y) \mid x = y + 1\}$ must be of the form $z \mapsto z + 1$, and since they also have to preserve 0 all endomorphisms must be the identity, which proves the claim. Hence, $\mathrm{Inv}(\mathrm{Pol}(\mathfrak{B}))$ is uncountable (every subset of $\mathbb{Z}$ is preserved by all polymorphisms), but $\langle \mathfrak{B} \rangle_{\mathrm{pp}}$ is countable, so $\mathrm{Inv}(\mathrm{Pol}(\mathfrak{B})) \neq \langle \mathfrak{B} \rangle_{\mathrm{pp}}$.

For ω-categorical structures we can combine Lemma 6.1.17 with Theorem 4.4.1 to obtain an algebraic characterisation of the situation where disjunction can be eliminated from existential positive formulas.

PROPOSITION 6.1.19. *Let $\mathfrak{B}$ be a countable ω-categorical structure. Then the following are equivalent.*

1. *All relations with an existential positive definition in $\mathfrak{B}$ also have a primitive positive definition in $\mathfrak{B}$.*
2. *The relation P_B^3 is primitively positively definable in $\mathfrak{B}$.*
3. *All polymorphisms of $\mathfrak{B}$ are essentially unary.*

PROOF. (1) $\Rightarrow$ (2). The formula $(x = y) \vee (y = z)$ is existential positive, and thus has a primitive positive definition in $\mathfrak{B}$.

(2) $\Rightarrow$ (3). All polymorphisms of $\mathfrak{B}$ preserve all primitive positive formulas, so the preserve P_B^3 and the statement follows from Lemma 6.1.17.

(3) $\Rightarrow$ (1). Unary operations preserve all existential positive formulas. Hence, if ϕ is an existential positive formula, by assumption all polymorphisms of $\mathfrak{B}$ preserve ϕ, and thus ϕ is equivalent to a primitive positive formula by Theorem 4.4.1. ⊣

6.1.5. Elementary clones. Recall the definition of *elementary embedding* from Section 2.1: an embedding is elementary if it preserves all first-order formulas. Analogously, we say that a polymorphism f of a structure $\mathfrak{B}$ is *elementary* if it preserves all first-order formulas. By Lemma 6.1.17, such a polymorphism must be essentially unary, and in fact is contained in the smallest closed subclone of $\mathscr{O}$ that contains the automorphisms of $\mathfrak{B}$. If every polymorphism of $\mathfrak{B}$ is elementary, we say that $\text{Pol}(\mathfrak{B})$ is *elementary*.

If $\text{Pol}(\mathfrak{B})$ is elementary, then $\mathfrak{B}$ has the remarkable property that every first-order formula is equivalent to a primitive positive formula over $\mathfrak{B}$. The following corollary is a straightforward combination of results from previous sections. We state it here for future use.

COROLLARY 6.1.20. *Let $\mathfrak{B}$ be a countable ω-categorical structure. Then the following are equivalent.*

1. *Every relation with a first-order definition also has a primitive positive definition in $\mathfrak{B}$.*
2. *$\mathfrak{B}$ is a model-complete core, and P_B^3 is primitively positively definable in $\mathfrak{B}$.*
3. $\text{Pol}(\mathfrak{B})$ *is generated by the unary operations that are invertible in* $\text{Pol}(\mathfrak{B})$.
4. $\text{Pol}(\mathfrak{B})$ *is elementary.*

PROOF. (1) $\Rightarrow$ (2). We assume that every first-order definable relation has a primitive positive definition, and hence is preserved by all polymorphisms of $\mathfrak{B}$. In particular, the endomorphisms of $\mathfrak{B}$ preserve all first-order definable relations, and hence $\mathfrak{B}$ is a model-complete core. Moreover, the relation P_B^3 is clearly first-order definable, and therefore also primitively positively definable in $\mathfrak{B}$.

(2) $\Rightarrow$ (3). Assume (2). Then Proposition 6.1.19 implies that all polymorphisms of $\mathfrak{B}$ are essentially unary. Thus, for every n-ary polymorphism f of $\mathfrak{B}$ there is an endomorphism g of $\mathfrak{B}$ and an $j \leq n$ such that $f(x_1, \ldots, x_n) = g(x_j)$. Since $\mathfrak{B}$ is a model-complete core, Theorem 4.5.1 implies that $g \in \overline{\text{Aut}(\mathfrak{B})}$, which proves (3).

(3) $\Rightarrow$ (4). Invertible operations of $\mathcal{C}$ are automorphisms of $\mathfrak{B}$ and therefore preserve all first-order definable relations in $\mathfrak{B}$. Hence, the implication follows from Proposition 6.1.5.

(4) $\Rightarrow$ (1). By Theorem 6.1.12. $\dashv$

COROLLARY 6.1.21. *Every countable ω-categorical structure $\mathfrak{B}$ with more than one element and an elementary polymorphism clone interprets all finite structures primitively positively.*

PROOF. Lemma 2.4.4 states that every finite structure has a first-order definition in $\mathfrak{B}$. The statement follows because every first-order formula is equivalent over $\mathfrak{B}$ to a primitive positive formula. $\dashv$

LEMMA 6.1.22. *Let $\mathfrak{B}$ be a countable ω-categorical structure all of whose polymorphisms are essentially unary. Then the model-complete core of $\mathfrak{B}$ has an elementary polymorphism clone.*

PROOF. Let h be a homomorphism from $\mathfrak{B}$ to the model-complete core $\mathfrak{C}$ of $\mathfrak{B}$, and let i be a homomorphism from $\mathfrak{C}$ to $\mathfrak{B}$. By Corollary 6.1.20 it suffices to verify that the relation P_C^3 is primitively positively definable in $\mathfrak{C}$. Since all polymorphisms of $\mathfrak{B}$ are essentially unary, Proposition 6.1.19 implies that the relation P_B^3 has a primitive positive definition $\phi(x, y, z)$ in $\mathfrak{B}$. We claim that ϕ is also a primitive positive definition of P_C^3 in $\mathfrak{C}$. Suppose that $\mathfrak{C} \models \phi(a)$ for $a \in C^3$. Then $\mathfrak{B} \models \phi(i(a))$ since the homomorphism i preserves primitive positive formulas. This in turn means that $i(a) \in P_B^3$, and so $h(i(a)) \in P_C^3$. Since $h \circ i$ is an embedding, we get that $a \in P_C^3$. Now suppose that conversely $a \in P_C^3$. Then $i(a) \in P_B^3$, and hence $\mathfrak{B} \models \phi(i(a))$. This in turn implies that $\mathfrak{C} \models \phi(h(i(a)))$ since homomorphisms preserve primitive positive formulas, so $h(i(a)) \in P_C^3$. Since $h \circ i$ is an embedding because $\mathfrak{C}$ is a model-complete core we get that $a \in P_C^3$. $\dashv$

COROLLARY 6.1.23. *Let $\mathfrak{B}$ be a countable ω-categorical structure with no constant endomorphism such that all polymorphisms are essentially unary. Then* HI($\mathfrak{B}$) *contains all finite structures, and $\mathfrak{B}$ has a finite-signature reduct whose CSP is NP-hard.*

PROOF. Let $\mathfrak{C}$ be the model-complete core of $\mathfrak{B}$. Since $\mathfrak{B}$ does not have a constant endomorphism, the structure $\mathfrak{C}$ has at least two elements. Moreover, $\mathfrak{C}$ is elementary by Lemma 6.1.22. By Corollary 6.1.21 we have that I($\mathfrak{C}$) contains all finite structures. Since I($\mathfrak{C}$) $\subseteq$ I(H($\mathfrak{B}$)) $\subseteq$ HI($\mathfrak{B}$) by Theorem 3.6.2, the statement follows from Corollary 3.7.1. $\dashv$

6.1.6. Arity reduction. For many combinatorial arguments with polymorphism clones it is crucial to have bounds on the arity of polymorphisms that have certain properties. A basic, yet very useful observation to obtain such bounds is the following (which holds for arbitrary structures $\mathfrak{B}$).

LEMMA 6.1.24. *Let $\mathfrak{B}$ be a relational structure and let R be a k-ary relation contained in m orbits of k-tuples under* $\mathrm{Aut}(\mathfrak{B})$. *If $\mathfrak{B}$ has a polymorphism f that does not preserve R, then $\mathfrak{B}$ also has an m-ary polymorphism that does not preserve R.*

PROOF. Let f' be an polymorphism of $\mathfrak{B}$ of smallest arity l that does not preserve R. Then there are k-tuples $t_1, \ldots, t_l \in R$ such that $f'(t_1, \ldots, t_l) \notin R$. For $l > m$ there are two tuples t_i and t_j that lie in the same orbit of k-tuples, and therefore $\mathfrak{B}$ has an automorphism α such that $\alpha t_j = t_i$. By permuting the arguments of f', we can assume that $i = 1$ and $j = 2$. Then the $(l-1)$-ary operation g defined as

$$g(x_2, \ldots, x_l) := f'(\alpha x_2, x_2, \ldots, x_l)$$

is also a polymorphism of $\mathfrak{B}$, and also does not preserve R, a contradiction. Hence, $l \leq m$. An operation of arity exactly m that does not preserve R can then be obtained from f' by adding fictitious variables. ⊣

We present some applications of Lemma 6.1.24; other applications can be found in Section 6.2 and in Chapter 12. Recall that $r(\mathcal{G})$ denotes the *rank* of $\mathcal{G}$, i.e., the number of orbitals of $\mathcal{G}$ (see Section 4.2).

COROLLARY 6.1.25. *Let $\mathfrak{B}$ be a structure with an automorphism group $\mathcal{G}$. If $\mathfrak{B}$ has an essential polymorphism, then it must also have an essential polymorphism of arity at most $2r(\mathcal{G}) - 1$.*

PROOF. The structure $\mathfrak{B}$ has an essential polymorphism if and only if it has a polymorphism that does not preserve the relation P^3_B, where B is the domain of $\mathfrak{B}$, by Proposition 6.1.19. The relation P^3_B consists of at most $2r(\mathcal{G}) - 1$ orbits of triples: there are at most $r(\mathcal{G})$ orbits of triples (t_1, t_2, t_3) where $t_1 = t_2 \neq t_3$, and at most that many where $t_1 \neq t_2 = t_3$. Only the orbit of the tuple where $t_1 = t_2 = t_3$ is counted twice. The statement follows from Lemma 6.1.24. ⊣

We give another simple example of how Lemma 6.1.24 may be used.

COROLLARY 6.1.26. *Let $\mathfrak{B}$ be first-order definable in $(\mathbb{Q}; <)$, and suppose there is a polymorphism of $\mathfrak{B}$ that does not preserve $<$. Then there is also an endomorphism of $\mathfrak{B}$ that does not preserve $<$.*

PROOF. Observe that $<$ consists of a single orbit of pairs under $\mathrm{Aut}(\mathbb{Q}; <)$, and therefore also in $\mathrm{Aut}(\mathfrak{B})$. ⊣

COROLLARY 6.1.27. *Suppose that $\mathfrak{B}$ has a 2-transitive automorphism group with a polymorphism that does not preserve $\neq$. Then $\mathfrak{B}$ has a constant endomorphism.*

PROOF. The relation $\neq$ consists of a single orbit of pairs under $\mathrm{Aut}(\mathfrak{B})$. Hence, there is an endomorphism of $\mathfrak{B}$ that does not preserve $\neq$, by Lemma 6.1.24. The rest follows by Lemma 4.4.6. ⊣

6.1.7. Kára's method. If $\mathfrak{B}$ is a structure with a 2-transitive automorphism group, then Corollary 6.1.25 implies that if $\mathfrak{B}$ has an essential polymorphism, then $\mathfrak{B}$ also has a binary essential polymorphism. In this section we present another method for showing that an oligomorphic clone with essential operations must contain a binary essential operation. The idea is taken from [68], where it was used to prove a result that only applies to structures preserved by all permutations of the domain. The result has been generalised slightly in [89]. To state it in full generality, we introduce the following, apparently new, concept.

DEFINITION 6.1.28. A permutation group $\mathcal{G}$ on a set B has the *orbital extension property* (*OEP*) if there is an orbital O such that for all $b_1, b_2 \in B$ there is an element $c \in B$ where $(c, b_1) \in O$ and $(c, b_2) \in O$.

Examples of oligomorphic permutation groups with the orbital extension property are the automorphism group of the Random graph, $(\mathbb{Q}; <)$, the countable universal homogeneous poset, the universal homogeneous C-relation, and many more. An example of a structure whose automorphism group does not have the OEP is $K_{\omega,\omega}$, the complete bipartite graph where both parts are countably infinite. An example of an oligomorphic permutation group which is not primitive but has the OEP is the automorphism group of an equivalence relation on an infinite set with infinitely many infinite classes.

LEMMA 6.1.29. *Let $\mathcal{C}$ be a clone with an essential operation that contains a permutation group $\mathcal{G}$ with the orbital extension property. Then $\mathcal{C}$ must also contain a binary essential operation.*

PROOF. Let f be an essential operation in $\mathcal{C}$, and let k be the arity of f. Assume without loss of generality that f depends on all its arguments and is at least ternary. In particular, there are $a_1, \ldots, a_k$ and a_1' such that $f(a_1, \ldots, a_k) \neq f(a_1', a_2, \ldots, a_k)$. Let O be the orbital which exists due to the orbital extension property of $\mathcal{G}$. We distinguish two cases.

Case 1. There are elements $b_1, \ldots, b_k$ such that $(b_i, a_i) \in O$ for $2 \leq i \leq k$ and $f(b_1, a_2, \ldots, a_k) \neq f(b_1, \ldots, b_k)$. Then there are $\alpha_3, \ldots, \alpha_k \in \mathcal{G}$ such that $\alpha_i(a_2) = a_i$ and $\alpha_i(b_2) = b_i$. We define

$$g(x, y) := f(x, y, \alpha_3(y), \ldots, \alpha_k(y)),$$

which depends on both arguments:

$$\begin{aligned} g(a_1, a_2) = f(a_1, a_2, \alpha_3(a_2), \ldots, \alpha_k(a_2)) &= f(a_1, a_2, a_3, \ldots, a_k) \\ &\neq f(a_1', a_2, a_3, \ldots, a_k) = g(a_1', a_2) \end{aligned}$$

shows that g depends on the first argument, and

$$\begin{aligned} g(b_1, a_2) &= f(b_1, a_2, a_3, \ldots, a_k) \\ &\neq f(b_1, b_2, b_3, \ldots, b_k) = g(b_1, b_2) \end{aligned}$$

shows that g depends on the second argument.

Case 2. For all $b_1, \ldots, b_k$, if $(a_i, b_i) \in O$ for $2 \leq i \leq k$, then $f(b_1, a_2, \ldots, a_k) = f(b_1, b_2, \ldots, b_k)$. Since f depends on its second coordinate, there are $c_1, \ldots, c_k$ and c_2' such that $f(c_1, c_2, c_3, \ldots, c_k) \neq f(c_1, c_2', c_3, \ldots, c_k)$. Then $f(c_1, a_2, \ldots, a_k)$ can be equal to either $f(c_1, c_2, c_3, \ldots, c_k)$, or to $f(c_1, c_2', c_3, \ldots, c_k)$, but not to both. We assume without loss of generality that $f(c_1, a_2, \ldots, a_k) \neq f(c_1, c_2, c_3, \ldots, c_k)$. Since $\mathcal{G}$ has the orbital extension property, we can choose $d_2, \ldots, d_k$ such that for each $2 \leq i \leq k$, the pairs (d_i, a_i) and the pairs (d_i, c_i) all lie in O. Hence, there are $\alpha_3, \ldots, \alpha_k \in \mathcal{G}$ such that $\alpha_i(c_2) = c_i$ and $\alpha_i(d_2) = d_i$. We claim that the operation g defined by

$$g(x, y) := f(x, y, \alpha_3(y), \ldots, \alpha_k(y))$$

depends on both arguments. Indeed,

$$\begin{aligned} g(a_1, d_2) &= f(a_1, d_2, \ldots, d_k) = f(a_1, \ldots, a_k) \\ \text{and } g(a_1', d_2) &= f(a_1', d_2, \ldots, d_k) = f(a_1', a_2, \ldots, a_k). \end{aligned}$$

and by the choice of the values $a_1, \ldots, a_k$ and a_1' these two values are distinct. Thus, g depends on the first argument. Finally, g also depends on the second argument, because

$$\begin{aligned} g(c_1, d_2) &= f(c_1, d_2, \ldots, d_k) = f(c_1, a_2, \ldots, a_k) \\ \text{and } g(c_1, c_2) &= f(c_1, c_2, \ldots, c_k). \end{aligned}$$

adn $f(c_1, a_2, \ldots, a_k)$ and $f(c_1, c_2, \ldots, c_k)$ are distinct. ⊣

Note that every permutation group with the orbital extension property is transitive, and that every 2-transitive infinite permutation group has the orbital extension property. We even have the following.

LEMMA 6.1.30 (Lemma 3.7 in [58]). *Every 2-set-transitive permutation group on a set with at least 4 elements has the orbital extension property.*

PROOF. Let $\mathcal{G}$ be a 2-set-transitive permutation group on a set B with $|B| \geq 4$. If $\mathcal{G}$ is even 2-transitive then the statement is obvious. Otherwise, the relation defined by $x \neq y$ is the union of exactly two orbitals O and P. Thus, the relation O defines a tournament on B. It is easy to see that $|B| \geq 4$ implies that there are $u, v, w \in B$ with $(u, v), (v, w), (u, w) \in O$. We claim that O witnesses the OEP. Let $b_1, b_2 \in B$. If $b_1 = b_2$ the statement is trivial; so suppose without loss of generality that $(b_1, b_2) \in O$. There exists $\alpha \in \mathcal{G}$ that maps $\{b_1, b_2\}$ to $\{v, w\}$ because $\mathcal{G}$ is 2-set transitive. We then have $(\alpha^{-1}(u), b_1) \in O$ and $(\alpha^{-1}(u), b_2) \in O$. ⊣

COROLLARY 6.1.31. *Let $\mathfrak{B}$ be an infinite 2-set-transitive structure with an essential polymorphism. Then $\mathfrak{B}$ also has a binary essential polymorphism.*

6.1.8. Minimal clones. When classifying closed clones that contain a given oligomorphic permutation group, e.g. in Chapter 12, it often turns out that a *bottom-up* approach works: we first classify all the minimal (with respect to set inclusion) closed subclones of $\mathcal{O}$ that strictly contain $\mathrm{Aut}(\mathbb{Q};<)$, and then the remaining classification argument is organised according to those *minimal* clones. Of course, for this method to work, one must at least have the existence of such minimal clones (cf. Theorem 6.1.37). To present the results in this section in their strongest formulation, we define a relative notion of minimality of a closed clone above some other closed clone.

DEFINITION 6.1.32. Let $\mathcal{C}$ and $\mathcal{D}$ be closed subclones of $\mathcal{O}$. We say that $\mathcal{D}$ is *minimal above* $\mathcal{C}$ if $\mathcal{C} \subsetneq \mathcal{D}$ and for all closed subclones $\mathcal{E}$ of $\mathcal{O}$

$$\mathcal{C} \subsetneq \mathcal{E} \subseteq \mathcal{D} \text{ implies that } \mathcal{E} = \mathcal{D}.$$

DEFINITION 6.1.33. An operation $f \in \mathcal{O} \setminus \mathcal{C}$ is *minimal above* $\mathcal{C}$ if f is of minimal arity such that for every $g \in \mathcal{O} \setminus \mathcal{C}$ locally generated by $\{f\} \cup \mathcal{C}$ we have that $\{g\} \cup \mathcal{C}$ locally generates f.

The following is straightforward from the definitions.

PROPOSITION 6.1.34. *Let $\mathcal{C}$ be a closed subclone of $\mathcal{O}$, and let f be minimal above $\mathcal{C}$. Then $\{f\} \cup \mathcal{C}$ locally generates an operation clone that is minimal above $\mathcal{C}$. Conversely, every closed subclone $\mathcal{D}$ of $\mathcal{O}$ that is minimal above $\mathcal{C}$ contains an operation that is minimal above $\mathcal{C}$ such that $\{f\} \cup \mathcal{C}$ locally generates $\mathcal{D}$.*

For oligomorphic operation clones we obtain the following.

PROPOSITION 6.1.35. *Let $\mathfrak{B}$ be an ω-categorical structure, and let $\mathfrak{C}$ be primitively positively definable in $\mathfrak{B}$. Then* $\mathrm{Pol}(\mathfrak{C})$ *is minimal above* $\mathrm{Pol}(\mathfrak{B})$ *if and only if for every $R \in \langle \mathfrak{B} \rangle_{\mathrm{pp}} \setminus \langle \mathfrak{C} \rangle_{\mathrm{pp}}$ the structure $\mathfrak{B}$ has a primitive positive definition in $(\mathfrak{C}, R)$.*

PROOF. The equivalence follows from Proposition 6.1.13. ⊣

It is well known that every operation clone over a finite domain contains an operation clone $\mathcal{D}$ which is minimal above the trivial clone that just contains the projections [144]. The following example shows that same is not true in general over infinite domains.

Example 6.1.36. Let $\mathcal{D}$ be the closed operation clone over the domain $\mathbb{N}$ which is generated by the operation $x \mapsto x + 1$. Then $\mathcal{D}$ does not contain an operation clone that is minimal above the closed clone $\mathcal{C}$ that just consists of the set of all projections over $\mathbb{N}$. To see this, note that every operation in $\mathcal{D}$ is essentially unary, and that every unary operation $f \in \mathcal{D}$ is of the form $x \mapsto x + c$ for some $c \in \mathbb{N}$. If $c > 0$ then $\{f\}$ generates the operation g given by $x \mapsto f(f(x)) = x + 2c$, but $\{g\} \cup \mathcal{C}$ only generates operations of the form $x \mapsto x + kc$ for some $k \in \mathbb{N}$, and hence does not locally generate f.

The situation is better if $\mathscr{C}$ is oligomorphic.

THEOREM 6.1.37 (Theorem 4.6 in [50]). *Let $\mathfrak{B}$ be a countable ω-categorical structure with a finite relational signature. Then any locally closed operation clone $\mathscr{C}$ that properly contains* $\mathrm{Pol}(\mathfrak{B})$ *contains a closed operation clone $\mathscr{D}$ that is minimal above* $\mathrm{Pol}(\mathfrak{B})$.

PROOF. Let $\mathscr{B}$ be the polymorphism clone of $\mathfrak{B}$. Consider the partially ordered set of all locally closed operation clones that contain $\mathscr{B}$ and that are contained in $\mathscr{C}$, ordered by inclusion. From this poset we remove $\mathscr{B}$, which is the unique minimal element; the resulting poset will be denoted by P. We claim that in P, all descending chains $\mathscr{S}_1 \supseteq \mathscr{S}_2 \supseteq \cdots$ are bounded, i.e., for all such chains there exists an $\mathscr{S} \in P$ such that $\mathscr{S}_i \supseteq \mathscr{S}$ for all $i \geq 1$. To see this, observe that $\bigcup_{i\geq 1} \mathrm{Inv}(\mathscr{S}_i)$ is *closed under primitive positive definability* in the sense that it can be written as $\langle \mathfrak{S} \rangle_{\mathrm{pp}}$ for some relational structure $\mathfrak{S}$, because only a finite number of relations can be mentioned in a formula, and because $\mathrm{Inv}(\mathfrak{S}_i)$ is closed under primitive positive definability, for each $i \geq 1$. Moreover, there must be a relation $R \in \mathfrak{B}$ that is not contained in $\bigcup_{i\geq 1} \mathrm{Inv}(\mathscr{S}_i)$; otherwise, since $\mathfrak{B}$ has finitely many relations, there is a $j < \omega$ such that $\mathrm{Inv}(\mathscr{S}_j)$ contains all relations from $\mathfrak{B}$, and hence equals $\mathscr{B}$, which is impossible since $\mathscr{B}$ is not an element of P. Hence, by Corollary 4.2.10, the structure $\mathfrak{S}$ has a polymorphism f that is not from $\mathscr{B}$. Then $\mathrm{Pol}(\mathfrak{S}) = \bigcap_{i\geq 1} \mathscr{S}_i$ is a lower bound in P of the descending chain $(\mathscr{S}_i)_{i\geq 0}$. We can thus apply Zorn's lemma and conclude that P contains a minimal element $\mathscr{S}$. ⊣

For essentially unary oligomorpic clones $\mathscr{C}$, we can bound the arity of minimal functions above $\mathscr{C}$.

PROPOSITION 6.1.38 (Proposition 24 in [96]). *Let $\mathfrak{B}$ be an arbitrary structure such that* $\mathrm{Aut}(\mathfrak{B})$ *has r orbitals and s orbits. Let $\mathscr{B}$ be the (locally closed) clone generated by the endomorphisms of $\mathfrak{B}$. Then every minimal closed clone above $\mathscr{B}$ is locally generated by* $\mathrm{End}(\mathfrak{B}) \cup \{f\}$ *for some function f of arity at most* $2r - s$.

PROOF. Let $\mathscr{C}$ be a minimal closed clone above $\mathscr{B}$. If all the functions in $\mathscr{C}$ are essentially unary, then $\mathscr{C}$ is generated by a unary operation together with $\mathrm{End}(\mathfrak{B})$ and we are done. Otherwise, let f be an essential operation in $\mathscr{C}$. By Lemma 6.1.17 the operation f does not preserve P^3_B over the domain B of $\mathfrak{B}$; recall that P^3_B is defined by the formula $(x = y) \vee (y = z)$. The subset of P^3_B that contains all tuples of the form (a, a, b), for $a, b \in B$, clearly consists of r orbits under $\mathrm{Aut}(\mathfrak{B})$. Similarly, the subset of P^3_B that contains all tuples of the form (a, b, b), for $a, b \in B$, consists of the same number of orbits. The intersection of these two relations consists of exactly s orbits (namely, the triples with three equal entries), and therefore P^3_B is the union of $2r - s$ different orbits. The assertion now follows from Lemma 6.1.24. ⊣

In Section 11.4 we will see that under further Ramsey-theoretic assumptions on the structure $\mathfrak{B}$, there are only *finitely many* minimal closed clones above $\text{End}(\mathfrak{B})$. In the remainder of this section, we present Rosenberg's five types theorem for clones on finite domains [316]; one of the five Rosenberg cases can be ruled out for oligomorphic clones. An operation is called *idempotent* if $f(x,\dots,x) = x$ for all domain elements x. We define several important properties of operations.

DEFINITION 6.1.39. A k-ary operation f is

- *symmetric* (or *commutative*) if f is binary and satisfies
$$\forall x,y\colon f(x,y) = f(y,x);$$
- a *quasi near-unanimity operation* iff $k \geq 3$ and f satisfies
$$\begin{aligned}\forall x,y\colon f(x,\dots,x,y) &= f(x,\dots,y,x)\\ &= \cdots = f(y,x,\dots,x) = f(x,\dots,x);\end{aligned}$$
- a *quasi majority operation* if f is a ternary quasi near-unanimity operation;
- a *quasi minority operation* if f is ternary and satisfies
$$\forall x,y\colon f(x,y,y) = f(y,x,y) = f(y,y,x) = f(x,x,x);$$
- a *quasi Maltsev operation* if f is ternary and satisfies
$$\forall x,y\colon f(x,y,y) = f(y,y,x) = f(x,x,x);$$
- a *quasi semiprojection* if there exists an $i \in \{1,\dots,k\}$ and a unary operation g such that for every sequence of variables $x_1,\dots,x_k$ with $|\{x_1,\dots,x_k\}| < k$ (i.e., at least one variable is repeated) we have the operations f and g satisfy
$$\forall x_1,\dots,x_k\colon f(x_1,\dots,x_k) = g(x_i).$$

An idempotent quasi near-unanimity is called a *near-unanimity*; we make an analogous convention for all other properties involving the qualifier 'quasi-'. That is, a *majority operation* is an idempotent quasi majority operation, etc.

DEFINITION 6.1.40. If f is an operation, then $\hat{f}$ denotes the unary operation given by $x \mapsto f(x,\dots,x)$.

Note that an operation f is a quasi semiprojection if and only if there is an $i \in \{1,\dots,k\}$ such that f satisfies the equations
$$\begin{aligned}f(x_1,x_1,x_3,\dots,x_k) &= f(x_1,x_2,x_1,x_4,\dots,x_k)\\ &= \dots = f(x_1,\dots,x_{k-1},x_{k-1}) = \hat{f}(x_i).\end{aligned}$$

The following is a slight generalisation of a lemma of Świerczkowski (which is Satz 4.4.6 in [309]), and a special case of a result of Couceiro and Lehtonen [140].

For the purpose of proving the next lemma, we call an n-ary operation f a *weak semiprojection* if for all distinct $i,j \in \{1,\dots,n\}$ there exists an index $s(i,j)$

and a unary non-constant operation $g_{i,j}$ such that $f(x_1, \ldots, x_n) = g_{i,j}(x_{s(i,j)})$ whenever $x_i = x_j$. In the proof of the following lemma the following notation for weak semiprojections will be practical. Let $S \subseteq \{1, \ldots, n\}$ be of cardinality at least two, and let $(x_1, \ldots, x_n)$ be a tuple such that $x_i = x_j$ for all $i, j \in S$. Then $f(x_1, \ldots, x_n) = g_{i,j}(x_i)$ for some $i \in \{1, \ldots, n\}$. If $i \in S$ define $E(S) := S$. Otherwise, define $E(S) := \{i\}$. Note that if $S \subseteq T \subseteq \{1, \ldots, n\}$, then $E(S) \subseteq E(T)$. Also note that if there exists an $i \in \{1, \ldots, n\}$ such that $i \in E(S)$ for every $S \subseteq \{1, \ldots, n\}$ with at least two elements, then f is a quasi semiprojection.

LEMMA 6.1.41. *Let f be a weak semiprojection of arity at least $n \geq 4$. Then f is a quasi semiprojection.*

PROOF. We first show that $E(\{1,2\}) \cap E(\{3,4\}) \neq \emptyset$. If $E(\{1,2,3,4\}) = \{\ell\}$ for some $\ell \notin \{1,2,3,4\})$, then $E(\{1,2\}) = \{\ell\} = E(\{3,4\})$ and we are done. So we assume that $E(\{1,2,3,4\}) = \{1,2,3,4\})$. If $E(\{1,2\}) = \{1,2\}$ then $E(\{1,2,3\}) = \{1,2,3\}$ and hence $E(\{2,3\}) = \{2,3\}$ and $E(\{2,3,4\}) = \{2,3,4\}$ which implies that $E(\{3,4\}) = \{3,4\}$. But then $f(x, x, y, y, x_5, \ldots, x_n) = g_{1,2}(x)$ because of $E(\{1,2\}) \subseteq \{1,2\}$ and $f(x, x, y, y, x_5, \ldots, x_n) = g_{3,4}(y)$ because of $E(\{3,4\}) \subseteq \{3,4\}$, which is a contradiction since $g_{1,2}$ and $g_{3,4}$ are non-constant.

Let $i \in E(\{1,2\}) \cap E(\{3,4\})$. Note that if $i \notin \{1,2\}$, then $E(\{1,2\}) = \{i\}$. Similarly, if $i \notin \{3,4\}$ then $E(\{3,4\}) = \{i\}$. We therefore have a set $S \subseteq \{1, \ldots, n\}$ with $E(S) = \{i\}$. Let $T \subseteq \{1, \ldots, n\}$ be of cardinality at least two. We will show that $i \in E(T)$. Observe that if $T \subseteq \{1, \ldots, n\} \setminus \{i\}$, then $E(T) = E(\{1, \ldots, n\} \setminus \{i\}) = E(S) = \{i\}$. Now suppose that $T = \{i, j\}$ for some $j \in \{1, \ldots, n\} \setminus \{i\}$. Then $\{1, \ldots, n\} \setminus T$ has at least two elements (since $n \geq 4$). We can therefore apply the argument from the first paragraph, up to renaming argument, to conclude that $E(\{i,j\}) \cap E(\{1, \ldots, n\} \setminus \{i,j\}) \neq \emptyset$, and hence $E(\{i,j\}) = \{i\}$. This implies that $E(T) = \{i\}$ for all $T \subseteq \{1, \ldots, n\}$ of cardinality at least 2 containing i. ⊣

The following result was shown for finite *idempotent* clones $\mathcal{C}$ in [316]; a clone is called *idempotent* if all its operations are idempotent. The infinite non-idempotent version below appeared in slightly different form in [50]. For oligomorphic clones, the statement will be strengthened in Theorem 6.1.45.

THEOREM 6.1.42. *Let $\mathcal{C}$ be an essentially unary clone and let f be a minimal operation above $\mathcal{C}$. Then f is of one of the following types:*

1. *A unary operation.*
2. *A binary essential operation such that $\hat{f} \in \mathcal{C}$.*
3. *A quasi Maltsev operation.*
4. *A quasi majority operation.*
5. *A k-ary quasi semiprojection for some $k \geq 3$.*

PROOF. There is nothing to show when f is unary or binary. If f is ternary, we have to show that f satisfies the equations of quasi majorities, quasi Maltsev operations, or quasi semiprojections. By minimality of f, the operation $f_1(x, y) := f(y, x, x)$ is in $\mathcal{C}$, and hence $f_1(x, y) = \hat{f}(x)$ or $f_1(x, y) = \hat{f}(y)$. Similarly, the other operations $f_2(x, y) := f(x, y, x)$, and $f_3(x, y) := f(x, x, y)$ obtained by identifications of two variables are essentially unary, and each of f_1, f_2, f_3 is either equal to $\hat{f}(x)$ or to $\hat{f}(y)$. We therefore distinguish eight cases, depicted in the following table; in each case, f must be a quasi majority, a quasi semiprojection, or a quasi Maltsev operation, as indicated in the final column of the table.

f_1	f_2	f_3	type
$\hat{f}(x)$	$\hat{f}(x)$	$\hat{f}(x)$	quasi majority
$\hat{f}(x)$	$\hat{f}(x)$	$\hat{f}(y)$	quasi semiprojection
$\hat{f}(x)$	$\hat{f}(y)$	$\hat{f}(x)$	quasi semiprojection
$\hat{f}(x)$	$\hat{f}(y)$	$\hat{f}(y)$	quasi Maltsev
$\hat{f}(y)$	$\hat{f}(x)$	$\hat{f}(x)$	quasi semiprojection
$\hat{f}(y)$	$\hat{f}(x)$	$\hat{f}(y)$	quasi Maltsev
$\hat{f}(y)$	$\hat{f}(y)$	$\hat{f}(x)$	quasi Maltsev
$\hat{f}(y)$	$\hat{f}(y)$	$\hat{f}(y)$	quasi Maltsev

If f is k-ary for $k \geq 4$ then Lemma 6.1.41 implies that f is a quasi semiprojection, and we are done. ⊣

We finally present a proof, based on Theorem 6.1.42, of a statement that we have already used in the proof of Corollary 3.2.1 in Section 3.2. The first part of the statement also follows from Lemma 6.8.4.

PROPOSITION 6.1.43. *For each $n \geq 3$ all idempotent polymorphisms of K_n are projections. Every relation that is first-order definable in K_n is primitively positively definable in K_n; these are precisely the relations that are preserved by all permutations of the domain.*

PROOF. By Theorem 6.1.37, it suffices to show that the clone of idempotent polymorphisms of K_n does not contain a minimal operation. Hence, by Theorem 6.1.42, we have to verify that $\mathrm{Pol}(K_n)$ does not contain any essentially binary idempotent operation, Maltsev operation, majority operation, or k-ary semiprojection for $k \geq 3$.

1. Let f be a binary idempotent polymorphism of K_n.

   Observation 1. $f(u, v) \in \{u, v\}$: otherwise, $i := f(u, v)$ is adjacent to both u and v, but $f(i, i) = i$ is not adjacent to i, in contradiction to f being a polymorphism.

Observation 2. If $f(u,v) = u$, then $f(v,u) = v$: this is clear if $u = v$, and if $u \neq v$ then note that $f(u,v) \neq f(v,u)$ because f preserves $\neq$, so $f(v,u) \neq u$. Moreover, $f(v,u) \in \{u,v\}$ by Observation 1, so $f(v,u) = v$.

To prove that f is a projection, it therefore suffices to show that there cannot be distinct u, v and distinct u', v' such that $f(u,v) = u$ and $f(u',v') = v'$. Suppose for contradiction that there are such u, v, u', v'.
Case 1. $u = u'$. Since $f(u,v') = f(u',v') = v'$, we have $f(v',u) = u$ by Observation 2. This is in contradiction to $f(u,v) = u$ since v' is adjacent to $u = u'$, and u is adjacent to v.
Case 2. $u \neq u'$.
Case 2.1. $f(u',u) = u$. This is impossible because $f(u,v) = u$, $E(u,u')$, and $E(u,v)$.
Case 2.2. $f(u',u) = u'$. This is impossible because $f(v',u') = u'$, $E(u',v')$, and $E(u',u)$.

2. Since $(1,0), (1,2), (0,2) \in E(K_n)$, but $(0,0) \notin E(K_n)$, the graph K_n has no Maltsev polymorphism.
3. If f is a majority, note that $f(0,1,2) = f(x_0, x_1, x_2)$ where x_i is some element distinct from i if $f(0,1,2) = i$, and $x_i := f(0,1,2)$ otherwise. But $(i, x_i) \in E(K_n)$, so f is not a polymorphism of K_n.
4. Finally, let f be a k-ary semiprojection for $k \geq 3$ which is not a projection. Suppose without loss of generality that $f(x_1, \ldots, x_k) = x_1$ whenever $|\{x_1, \ldots, x_k\}| < k$ (otherwise, permute the arguments of f). Since f is not a projection, there exist pairwise distinct $a_1, \ldots, a_k \in V(K_n)$ such that $c := f(a_1, \ldots, a_k) \neq a_1$. Let $b_1, \ldots, b_k$ be such that b_i is any element of $V(K_n) \setminus \{c\}$ if $c = a_i$, and $b_i := c$ otherwise. Note that $b_1 = a_1$ since $c \neq a_1$, and that $f(b_1, \ldots, b_k) = b_1 = a_1$ because f is a semiprojection. But $(a_i, b_i) \in E(K_n)$ for all $i \leq k$, so f is not a polymorphism of K_n.

The second part of the statement follows from Theorem 6.1.12. ⊣

When the clone $\mathcal{C}$ is generated by an oligomorphic permutation group $\mathcal{G}$, the characterisation of minimal operations f above $\mathcal{C}$ can be strengthened. First of all, we can exclude the case that f is a quasi Maltsev operation. This follows from the following more general fact, which we prove using a consequence of Ramsey's theorem (which is presented in Section 11.1).

PROPOSITION 6.1.44. *Let $\mathcal{D}$ be an oligomorphic clone on a countably infinite set B and suppose that $\mathcal{D}$ contains a quasi Maltsev operation. Then $\mathcal{D}$ must contain a unary non-injective operation.*

PROOF. Since $\mathcal{D}$ is oligomorphic, the group of invertible unary operations in $\mathcal{D}$ has a finite number r of orbitals. Let $v_1, v_2, \ldots$ be an enumeration of B. For $i < j$, we assign to a pair (v_i, v_j) the orbital of (v_i, v_j) as a colour. Ramsey's theorem (Theorem 11.1.1) implies that there is a monochromatic triangle with respect to this colouring, i.e., there exist three distinct elements $a, b, c \in B$

such that (a, b), (a, c), and (b, c) lie in the same orbital O. Let $\alpha, \beta \in \mathcal{D}^{(1)}$ be such that $\alpha(a, b) = (a, c)$ and $\beta(a, b) = (b, c)$. Now, suppose f is a quasi Maltsev operation. We claim that $g \in \mathcal{D}^{(1)}$ given by $g(x) := f(x, \alpha(x), \beta(x))$ is non-injective:

$$\begin{aligned} g(a) = f(a, \alpha(a), \beta(a)) &= f(a, a, b) = \hat{f}(b) \\ &= f(b, c, c) = g(b, \alpha(b), \beta(b)). \end{aligned} \quad \dashv$$

Hence, in particular, ω-categorical model-complete cores cannot have quasi Maltsev polymorphisms.

THEOREM 6.1.45. *Let $\mathcal{G}$ be an oligomorphic permutation group on a countably infinite set B with r orbitals and s orbits, and let f be minimal above $\langle \mathcal{G} \rangle$. Then f is of one of the following types:*

1. *A unary operation.*
2. *A binary operation.*
3. *A ternary quasi majority operation.*
4. *A k-ary quasi semiprojection, for $3 \leq k \leq 2r - s$.*

PROOF. The statement follows from Theorem 6.1.42 as follows. By Proposition 6.1.44 the operation f cannot be a quasi Maltsev operation, since in this case $\langle \{f\} \cup \mathcal{G} \rangle$ contains a unary operation g that is non-injective and therefore not in $\langle \mathcal{G} \rangle$, in contradiction to minimality of f.

Next, observe that if the arity of a quasi semiprojection is larger than $2r - s$, then it generates an essential operation of arity at most $2r - s$ by Corollary 6.1.25; hence, minimal quasi semiprojections have arity at most $2r - s$. $\dashv$

6.2. The Boolean domain

Schaefer's theorem states that for every 2-element template the CSP is either in P or NP-hard (Theorem 6.2.7). Via the Inv-Pol Galois connection (Section 6.1.2), most of the classification arguments in Schaefer's article follow from earlier work of Post [310], who classified all clones over the Boolean domain $\{0, 1\}$. We present a short proof of Schaefer's theorem here, using the results and ideas from Section 6.1.6 and Section 6.1.8.

Note that over the Boolean domain, there is precisely one minority operation, and precisely one majority operation.

THEOREM 6.2.1 (Post [310]). *Let $\mathcal{C}$ be an essentially unary operation clone with domain $\{0, 1\}$. Then every minimal operation above $\mathcal{C}$ is among one of the following:*

- *a unary operation;*
- *the binary operation $(x, y) \mapsto \min(x, y)$;*
- *the binary operation $(x, y) \mapsto \max(x, y)$;*

- *the minority operation;*
- *the majority operation.*

PROOF. Note that $\mathcal{C}$ either consists of the projections only or equals the operation clone generated by $\neg: x \mapsto \neg x$. Let f be a minimal operation above $\mathcal{C}$ of arity at least two. Note that $\hat{f} \in \mathcal{C}$. Hence, either $\hat{f}$ is the identity, in which case f is idempotent, or $\hat{f}$ equals $\neg$, in which case $\neg f$ is idempotent and minimal above $\mathcal{C}$ as well. So we can assume without loss of generality that f is idempotent.

There are only four binary idempotent operations on $\{0, 1\}$, two of which are projections and therefore cannot be minimal. The other two operations are min and max. Next, note that a semiprojection of arity at least three on a Boolean domain must be a projection. Thus, Theorem 6.1.42 implies that f is the majority or a Maltsev operation. In the former case, we are done. In the latter case, if $f(x, y, x) = y$ then f is the minority operation, and we are also done. Otherwise, the minimality of f implies that $f(x, y, x) = x$, and we will define the majority operation in terms of f. Indeed, the operation g defined by

$$g(x, y, z) := f(x, f(x, y, z), z)$$

satisfies

$$\begin{aligned} g(x, x, z) &= f(x, f(x, x, z), z) = f(x, z, z) = x \\ g(x, y, y) &= f(x, f(x, y, y), y) = f(x, x, y) = y \\ g(x, y, x) &= f(x, f(x, y, x), x) = f(x, x, x) = x. \end{aligned}$$ ⊣

Using Theorem 6.2.1, the following result about the structure $(\{0, 1\}; \text{1IN3})$ introduced in Example 1.2.2 is straightforward to check.

COROLLARY 6.2.2. *All polymorphisms of* $(\{0, 1\}; \text{1IN3})$ *are projections.*

PROOF. The Boolean relation 1IN3 is preserved neither by min, max, minority, nor majority. The only unary polymorphism is the identity. Hence, the statement follows from Theorem 6.2.1. ⊣

In the following propositions we present alternative descriptions of the relations which are preserved by min, max, minority, and majority.

PROPOSITION 6.2.3. *A Boolean relation is preserved by the minority operation if and only if R has a definition by a conjunction of linear equations modulo* 2.

PROOF. This statement follows from basic facts in linear algebra. Let R be n-ary. We view R as a subset of the Boolean vector space $\{0, 1\}^n$. By definition, R is called *affine* if it is the solution space of a system of linear equations, and it is a well-known fact from linear algebra that affine spaces are precisely those that are closed under *affine combinations*, i.e., linear combinations of the form $\alpha_1 x_1 + \cdots + \alpha_k x_k$ such that $\alpha_1 + \cdots + \alpha_k = 1$. In particular, if R is affine it is preserved by $(x_1, x_2, x_3) \mapsto x_1 + x_2 + x_3$, which is the minority operation.

Conversely, if R is preserved by the minority operation, then $x_1 + \cdots + x_k$, for odd k, can be written as

$$\text{minority}(x_1, x_2, \text{minority}(x_3, x_4, \ldots \text{minority}(x_{n-2}, x_{k-1}, x_k)))$$

and hence is contained in R. ⊣

The following definition is very useful for proving that certain Boolean relations R can be defined in syntactically restricted propositional logic.

DEFINITION 6.2.4. A propositional formula ϕ in conjunctive normal form is called *reduced* if whenever we remove a literal from a clause in ϕ, the resulting formula is not equivalent to ϕ.

Clearly, every Boolean relation has a reduced definition: simply remove literals from any definition in CNF until the formula becomes reduced.

PROPOSITION 6.2.5. *A Boolean relation R has a definition by a Horn formula (Section* 1.6.7) *if and only if R is preserved by* min.

PROOF. It is easy to see that min preserves every relation defined by a clause that contains at most one positive literal, and hence every relation with a Horn definition. Conversely, let R be a Boolean relation preserved by min. Let ϕ be a reduced propositional formula in CNF that defines ϕ. Now suppose for contradiction that ϕ contains a clause C with two positive literals x and y. Since ϕ is reduced, there is an assignment s_1 that satisfies ϕ such that $s_1(x) = 1$, and such that all other literals of C evaluatc to 0. Similarly, there is a satisfying assignment s_2 for ϕ such that $s_2(y) = 1$ and all other literal s of C evaluate to 0. Then $s_0 \colon x \mapsto \min(s_1(x), s_2(y))$ does not satisfy C, and does not satisfy ϕ, in contradiction to the assumption that min preserves R. ⊣

A binary relation is called *bijunctive* if it can be defined by a propositional formula that is a conjunction of clauses of size two.

PROPOSITION 6.2.6. *A Boolean relation R is preserved by the majority operation if and only if it is bijunctive.*

PROOF. It is easy to see that the majority operation preserves every Boolean relation of arity two, and hence every bijunctive Boolean relation. For the converse, it suffices to show that if R is preserved by the majority operation, and ϕ is a reduced definition of R, then all clauses C of ϕ have at most two literals. Suppose for contradiction that C has three literals l_1, l_2, l_3. Since ϕ is reduced, there must be satisfying assignments s_1, s_2, s_3 to ϕ such that under s_i all literals of C evaluate to 0 except for l_i. Then the mapping $s_0 \colon x \mapsto \text{majority}(s_1(x), s_2(x), s_3(x))$ does not satisfy C and therefore does not satisfy ϕ, in contradiction to the assumption that majority preserves R. ⊣

Recall that NAE was defined as the Boolean relation $\{0,1\}^3 \setminus \{(0,0,0), (1,1,1)\}$. We combine the findings above in the following theorem.

THEOREM 6.2.7 (Schaefer [319]). *Let $\mathfrak{B}$ be a structure with finite signature over a two-element universe. Then either* $(\{0, 1\}; \text{NAE})$ *has a primitive positive definition in* $\mathfrak{B}$, *and* CSP($\mathfrak{B}$) *is NP-complete, or*

1. $\mathfrak{B}$ *is preserved by a constant operation.*
2. $\mathfrak{B}$ *is preserved by* min. *In this case, every relation of* $\mathfrak{B}$ *has a definition by a propositional Horn formula.*
3. $\mathfrak{B}$ *is preserved by* max. *In this case, every relation of* $\mathfrak{B}$ *has a definition by a* dual-Horn *formula, that is, by a propositional formula in CNF where every clause contains at most one negative literal.*
4. $\mathfrak{B}$ *is preserved by the majority operation. In this case, every relation of* $\mathfrak{B}$ *is bijunctive.*
5. $\mathfrak{B}$ *is preserved by the minority operation. In this case, every relation of* $\mathfrak{B}$ *can be defined by a conjunction of linear equations modulo* 2.

In case $(1 - 5)$ *the problem* CSP($\mathfrak{B}$) *can be solved in polynomial time.*

PROOF. If Pol($\mathfrak{B}$) contains a constant operation, then we are in case one, and the statement follows from Proposition 1.1.12; so suppose in the following that this is not the case.

If NAE is primitively positively definable in $\mathfrak{B}$, then CSP($\mathfrak{B}$) is NP-hard by Corollary 1.2.8. Otherwise, by Theorem 6.1.12 there is an operation $f \in \text{Pol}(\mathfrak{B})$ that does not preserve NAE. If $\hat{f}$ equals the identity then f is idempotent. Otherwise, $\hat{f}$ equals $\neg$. But then $\neg f \in \text{Pol}(\mathfrak{B})$ is idempotent and also does not preserve NAE. So let us assume in the following that f is idempotent. Then f generates a minimal operation $g \in \text{Pol}(\mathfrak{B})$ of arity at least two.

By Theorem 6.2.1, the operation g equals min, max, the Boolean minority, or the Boolean majority function.

- $g = \min$ or $g = \max$. If $\mathfrak{B}$ is preserved by min, then by Proposition 6.2.5 all relations of $\mathfrak{B}$ can be defined by propositional Horn formulas. It is well-known that the satisfiability of propositional Horn formulas can be solved in linear time [152].[17] A general tractability condition generalising the algorithmic content of this results will be presented in Section 8.4. The case that $g = \max$ is dual to this case.
- $g =$ majority. By Proposition 6.2.6, all relations of $\mathfrak{B}$ are bijunctive. Hence, in this case the instances of CSP($\mathfrak{B}$) can be viewed as instances of the 2SAT problem, and can be solved in linear time [12]. A general tractability condition generalising the algorithmic content of this results will be presented in Section 8.5.2.

[17] A well-known algorithmic technique to decide the satisfiability of a given propositional Horn formula is *positive unit-resolution* [324].

- $g =$ minority. By Proposition 6.2.3 every relation of $\mathfrak{B}$ has a definition by a conjunction of linear equalities modulo 2. Then CSP($\mathfrak{B}$) can be solved in polynomial time by Gaussian elimination.

This concludes the proof of the statement. ⊣

NP-hard Boolean constraint languages can be characterised in many equivalent ways via Corollary 6.1.20, as we will see in the following.

PROPOSITION 6.2.8. *Let $\mathfrak{B}$ be a structure over a two-element universe. Then the following are equivalent.*

1. *The relation* NAE *has a primitive positive definition in $\mathfrak{B}$.*
2. *$\mathfrak{B}$ is preserved neither by* min, max, *minority, majority, nor the constant operations.*
3. *Either the polymorphism clone of $\mathfrak{B}$ contains only projections, or it is generated by the unary operation $x \mapsto \neg x$.*
4. *Every first-order formula is equivalent over $\mathfrak{B}$ to a primitive positive formula.*

PROOF. The implication from (1) to (2) follows from the fact that NAE is not preserved by min, max, minority, majority, or constant operations, which is straightforward to verify. The implication (2) $\Rightarrow$ (3) follows from Theorem 6.2.1. The implication from (3) to (4) follows from Corollary 6.1.20. For the implication from (4) to (1), note that NAE has the first-order definition $\text{NAE}(x, y, z) \Leftrightarrow (x \neq y \vee y \neq z)$. So (4) implies that NAE also has a primitive positive definition in $\mathfrak{B}$. ⊣

6.3. Algebras and pseudo-varieties

In the previous sections of this chapter we have seen a useful characterisation of primitive positive definability in an ω-categorical structure $\mathfrak{B}$ in terms of the polymorphism clone of $\mathfrak{B}$. In this section we present a similarly useful universal-algebraic characterisation of the notion of (*full*) *primitive positive interpretability in* $\mathfrak{B}$, which we introduced in Section 3.1, based on *pseudo-varieties*.

6.3.1. Algebras. Algebras have been defined in Section 2.1: they are simply structures with a purely functional signature. When $\mathbf{A}$ is an algebra with signature τ and domain A, we denote by $\text{Clo}(\mathbf{A})$ the set of all term functions of $\mathbf{A}$, that is, functions with domain A of the form $(x_1, \ldots, x_n) \mapsto t(x_1, \ldots, x_n)$ where t is any term over the signature τ whose set of variables is contained in $\{x_1, \ldots, x_n\}$. Clearly, $\text{Clo}(\mathbf{A})$ is an operation clone since it is closed under compositions, and contains the projections, and in fact it is the smallest operation clone that contains $\{f^{\mathbf{A}} \mid f \in \tau\}$. An algebra $\mathbf{A}$ is called *oligomorphic* if $\text{Clo}(\mathbf{A})$ is oligomorphic.

Conversely, it is clear that for any operation clone $\mathfrak{C}$ one can find an algebra $\mathbf{A}$ such that $\text{Clo}(\mathbf{A}) = \mathfrak{C}$. In the context of complexity classification of CSPs, algebras arise as follows.

DEFINITION 6.3.1. Let $\mathfrak{B}$ be a structure with domain B. An algebra $\mathbf{B}$ with domain B such that $\text{Clo}(\mathbf{B}) = \text{Pol}(\mathfrak{B})$ is called a *polymorphism algebra of* $\mathfrak{B}$.

A relational structure has many different polymorphism algebras, since Definition 6.3.1 does not prescribe how to assign function symbols to the polymorphisms of $\mathfrak{B}$. We mention that there is a canonical way of choosing a signature, namely by using the functions in the operation clone themselves as the symbols of the signature (we can use any set as a signature for an algebra). This is why it also makes sense to speak about *the* polymorphism algebra, which is the algebra $\mathbf{B}$ with signature $\text{Pol}(\mathfrak{B})$ such that $f^{\mathbf{B}} := f$. In later sections (at the latest from Chapter 9 onwards) we often work directly with operation clones rather than algebras, which allows elegant presentation of many results. However, we believe it to be necessary for this text to present the treatment via algebras as well, for several reasons. One reason is that we would like to relate the results in this text with results in their traditional presentation in universal algebra. Another reason is that when we discuss properties of a τ-algebra $\mathbf{A}$ it will be convenient to use universal-conjunctive first-order τ-sentences in order to express properties of the operation clone $\text{Clo}(\mathbf{A})$. Later we will see that there is another, essentially equally expressive way for talking about these properties, namely primitive positive sentences over the signature of abstract clones (see Section 6.5). But these sentences are syntactically cumbersome, and we often prefer to work with τ-terms instead.

6.3.2. Subalgebras of powers. Let $f : A^n \to A$ be an operation on A and let $R \subseteq A^m$ be a relation on A, for some $n, m \in \mathbb{N}$. The equivalence of the following statements is immediate from the definitions.

1. $f \in \text{Pol}(A; R)$.
2. f preserves R.
3. f is a homomorphism from $(A; R)^n \to (A; R)$.
4. R is a subalgebra of $(A; f)^m$.

6.3.3. Congruences and quotients. A *congruence* K of an algebra $\mathbf{A}$ is an equivalence relation on A that is preserved by all operations in $\mathbf{A}$ (so K can be viewed as a subalgebra of $\mathbf{A}^2$). The results in Section 6.1.2 show that for ω-categorical structures $\mathfrak{A}$ with polymorphism algebra $\mathbf{A}$, the congruences of $\mathbf{A}$ are exactly the primitively positively definable equivalence relations over $\mathfrak{A}$.

PROPOSITION 6.3.2 (see [119]). *Let* $\mathbf{A}$ *be an algebra. Then* E *is a congruence of* $\mathbf{A}$ *if and only if* E *is the kernel of a homomorphism from* $\mathbf{A}$ *to some other algebra* $\mathbf{B}$.

Note that the notion of a congruence relates to normal subgroups in group theory (the normal subgroups are precisely the equivalence classes of the neutral element with respect to some congruence of the group; see Proposition 9.2.16) but also generalises the notion of a congruence of a permutation group (we view the permutation group as an algebra which has a unary function for each permutation of $\mathcal{G}$; see Section 4.2.3).

DEFINITION 6.3.3. If K is a congruence of a τ-algebra $\mathbf{A}$, then the *quotient algebra* $\mathbf{A}/K$ denotes τ-algebra with domain A/K where

$$f^{\mathbf{A}/K}(a_1/K, \dots, a_k/K) = f^{\mathbf{A}}(a_1, \dots, a_k)/K$$

where $a_1, \dots, a_k \in A$ and $f \in \tau$ is k-ary. This is well defined since K is preserved by all operations of $\mathbf{A}$. If K is the kernel of a homomorphism h then we also write $\mathbf{A}/h$ instead of $\mathbf{A}/K$.

The following is well known.

LEMMA 6.3.4. *Let* $\mathbf{A}$ *and* $\mathbf{B}$ *be algebras with the same signature, and let* $h\colon \mathbf{A} \to \mathbf{B}$ *be a homomorphism. Then the image of any subalgebra* $\mathbf{A}'$ *of* $\mathbf{A}$ *under* h *is a subalgebra of* $\mathbf{B}$, *and the preimage of any subalgebra* $\mathbf{B}'$ *of* $\mathbf{B}$ *under* h *is a subalgebra of* $\mathbf{A}$.

PROOF. Let $f \in \tau$ be k-ary. Then for all $a_1, \dots, a_k \in A'$,

$$f^{\mathbf{B}}(h(a_1), \dots, h(a_k)) = h(f^{\mathbf{A}}(a_1, \dots, a_k)) \in h(A') ,$$

so $h(A')$ is a subalgebra of $\mathbf{B}$. Now suppose that $h(a_1), \dots, h(a_k)$ are in B'; then $f^{\mathbf{B}}(h(a_1), \dots, h(a_k)) \in B'$ and hence $h(f^{\mathbf{A}}(a_1, \dots, a_k)) \in B'$. So, $f^{\mathbf{A}}(a_1, \dots, a_k) \in h^{-1}(B')$ which shows that $h^{-1}(B')$ induces a subalgebra of $\mathbf{A}$. ⊣

6.3.4. Homomorphic images, subalgebras, products. In this section we recall some basic universal-algebraic operators on classes of algebras that will be used in the following subsections. When $\mathcal{K}$ is a class of algebras of the same signature, then

- $\mathrm{H}(\mathcal{K})$ denotes the class of all homomorphic images of algebras from $\mathcal{K}$.
- $\mathrm{S}(\mathcal{K})$ denotes the class of all subalgebras of algebras from $\mathcal{K}$.
- $\mathrm{P}(\mathcal{K})$ denotes the class of all products of algebras from $\mathcal{K}$.
- $\mathrm{P}^{\mathrm{fin}}(\mathcal{K})$ denotes the class of all finite products of algebras from $\mathcal{K}$.
- $\mathrm{Exp}(\mathcal{K})$ denotes the class of all expansions of algebras from $\mathcal{K}$.

(Homomorphic images, subalgebras, products, and expansions have been defined in Section 2.1.) Note that closure under homomorphic images implies in particular closure under isomorphisms. For the operators H, S, P, $\mathrm{P}^{\mathrm{fin}}$, and Exp we often omit the brackets when applying them to singletons $\mathcal{K} = \{\mathbf{A}\}$, i.e., we write $\mathrm{H}(\mathbf{A})$ instead of $\mathrm{H}(\{\mathbf{A}\})$. A class $\mathcal{V}$ of algebras with the same signature τ is called

- a *pseudo-variety* if $\mathcal{V}$ contains all homomorphic images, subalgebras, and finite direct products of algebras in $\mathcal{V}$, i.e., $\mathrm{H}(\mathcal{V}) = \mathrm{S}(\mathcal{V}) = \mathrm{P}^{\mathrm{fin}}(\mathcal{V})$;

- a *variety* if $\mathcal{V}$ also contains all (finite and infinite) products of algebras in $\mathcal{V}$.

So the only difference between pseudo-varieties and varieties is that pseudo-varieties need not be closed under direct products of infinite cardinality. The smallest pseudo-variety (variety) that contains a class $\mathcal{C}$ of τ-algebras is called the pseudo-variety (variety) *generated* by $\mathcal{C}$.

LEMMA 6.3.5 (HSP lemma). *Let $\mathcal{C}$ be a class of τ-algebras.*

- *The pseudo-variety generated by $\mathcal{C}$ equals* $\mathrm{HSP}^{\mathrm{fin}}(\mathcal{C})$.
- *The variety generated by $\mathcal{C}$ equals* $\mathrm{HSP}(\mathcal{C})$.

PROOF. Clearly, $\mathrm{HSP}^{\mathrm{fin}}(\mathcal{C})$ is contained in the pseudo-variety generated by $\mathcal{C}$, and $\mathrm{HSP}(\mathcal{C})$ is contained in the variety generated by $\mathcal{C}$. We have to verify that $\mathrm{HSP}^{\mathrm{fin}}(\mathcal{C})$ is closed under H, S, and $\mathrm{P}^{\mathrm{fin}}$. It is clear that $\mathrm{H}(\mathrm{HSP}^{\mathrm{fin}}(\mathcal{C})) = \mathrm{HSP}^{\mathrm{fin}}(\mathcal{C})$. Lemma 6.3.4 implies that $\mathrm{S}(\mathrm{HSP}^{\mathrm{fin}}(\mathcal{C})) \subseteq \mathrm{HS}(\mathrm{SP}^{\mathrm{fin}}(\mathcal{C})) = \mathrm{HSP}^{\mathrm{fin}}(\mathcal{C})$. Finally,

$$\mathrm{P}^{\mathrm{fin}}(\mathrm{HSP}^{\mathrm{fin}}(\mathcal{C})) \subseteq \mathrm{H}\,\mathrm{P}^{\mathrm{fin}}\,\mathrm{S}\,\mathrm{P}^{\mathrm{fin}}(\mathcal{C}) \subseteq \mathrm{HSP}^{\mathrm{fin}}\,\mathrm{P}^{\mathrm{fin}}(\mathcal{C}) = \mathrm{HSP}^{\mathrm{fin}}(\mathcal{C}).$$

The proof that $\mathrm{HSP}(\mathcal{C})$ is closed under H, S, and P is analogous. ⊣

6.3.5. Interpretations and pseudo-varieties. We present the aforementioned connection between primitive positive interpretations and pseudo-varieties.

PROPOSITION 6.3.6. *Let $\mathfrak{C}$ be a structure and let $\mathbf{C}$ be a polymorphism algebra of $\mathfrak{C}$. If $\mathfrak{B} \in \mathrm{I}(\mathfrak{C})$ then* $\mathrm{Exp}\,\mathrm{HSP}^{\mathrm{fin}}(\mathbf{C})$ *contains a polymorphism algebra of $\mathfrak{B}$.*

PROOF. Suppose that $\mathfrak{B}$ has a d-dimensional primitive positive interpretation I in $\mathfrak{C}$. Since $I^{-1}(B)$ is primitively positively definable in $\mathfrak{C}$, it is preserved by all operations in $\mathbf{C}$, and therefore induces a subalgebra $\mathbf{D}$ of $\mathbf{C}^d$. Let K be the kernel of I. Since $I^{-1}(=_B)$ is primitively positively definable in $\mathfrak{C}$, all operations of $\mathbf{C}$ preserve the equivalence relation $K = I^{-1}(=_B)$, so K is a congruence of $\mathbf{D}$. Thus, I is a surjective homomorphism from $\mathbf{D}$ to $\mathbf{B} := \mathbf{D}/K$. It is straightforward to verify that $\mathrm{Clo}(\mathbf{B}) \subseteq \mathrm{Pol}(\mathfrak{B})$. ⊣

Recall that any structure with an interpretation in a structure $\mathfrak{C}$ is a reduct of a structure with a full interpretation in $\mathfrak{C}$. If $\mathcal{C}$ is a class of structures, we write $\mathrm{I}_{\mathrm{full}}(\mathcal{C})$ for the class of structures with a full primitive positive interpretation in a structure from $\mathcal{C}$.

THEOREM 6.3.7. *Let $\mathfrak{C}$ be a countable ω-categorical structure and let $\mathbf{C}$ be a polymorphism algebra of $\mathfrak{C}$. Then*

1. $\mathfrak{B} \in \mathrm{I}_{\mathrm{full}}(\mathfrak{C})$ *if and only if there exists* $\mathbf{B} \in \mathrm{HSP}^{\mathrm{fin}}(\mathbf{C})$ *such that*
$$\overline{\mathrm{Clo}(\mathbf{B})} = \mathrm{Pol}(\mathfrak{B}).$$
2. $\mathfrak{B} \in \mathrm{Red}(\mathfrak{C})$ *if and only if there exists* $\mathbf{B} \in \mathrm{Exp}(\mathbf{C})$ *such that*
$$\mathrm{Clo}(\mathbf{B}) = \mathrm{Pol}(\mathfrak{B}).$$

3. $\mathfrak{B} \in \mathrm{I}(\mathfrak{C})$ *if and only if there exists* $\mathbf{B} \in \operatorname{Exp} \mathrm{HSP}^{\text{fin}}(\mathbf{C})$ *such that*

$$\overline{\mathrm{Clo}(\mathbf{B})} = \mathrm{Pol}(\mathfrak{B}).$$

PROOF. (2) follows from Theorem 6.1.12. Moreover, (3) follows from (1) and (2).

To show (1), suppose that $\mathfrak{B}$ has a d-dimensional full primitive positive interpretation I in $\mathfrak{C}$. Since $I^{-1}(B)$ is primitively positively definable in $\mathfrak{C}$, it is preserved by all operations in $\mathbf{C}$, and therefore induces a subalgebra $\mathbf{D}$ of $\mathbf{C}^d$. Let K be the kernel of I. Since $I^{-1}(=_B)$ is primitively positively definable in $\mathfrak{C}$, all operations of $\mathbf{C}$ preserve $K = I^{-1}(=_B)$, so K is a congruence of $\mathbf{D}$. Thus, I is a surjective homomorphism from $\mathbf{D}$ to $\mathbf{B} := \mathbf{D}/K$. We verify that $\overline{\mathrm{Clo}(\mathbf{B})} = \mathrm{Pol}(\mathfrak{B})$. By Corollary 6.1.14 it suffices to show that a relation $R \subseteq B^k$ is primitively positively definable in $\mathfrak{B}$ if and only if it is preserved by all operations of $\mathbf{B}$. For every $f \in \tau$, the relation R is preserved by $f^{\mathbf{B}}$ if and only if $f^{\mathbf{C}}$ preserves $I^{-1}(R)$, which is the case if and only if $I^{-1}(R)$ is primitively positively definable in $\mathfrak{C}$. This in turn is the case if and only if R is primitively positively definable in $\mathfrak{B}$ by the assumption that the primitive positive interpretation I is full.

Now suppose that there is an algebra $\mathbf{B} \in \mathrm{HSP}^{\text{fin}}(\mathbf{C})$ such that $\overline{\mathrm{Clo}(\mathbf{B})} = \mathrm{Pol}(\mathfrak{B})$. So there exists a finite number $d \geq 1$, a subalgebra $\boldsymbol{D}$ of $\mathbf{C}^d$, and a surjective homomorphism h from $\mathbf{D}$ to $\mathbf{B}$. We claim that h is a d-dimensional primitive positive interpretation of $\mathfrak{B}$ in $\mathfrak{C}$. All operations of $\mathbf{C}$ preserve D (viewed as a d-ary relation over $\mathfrak{C}$) since $\mathbf{D}$ is a subalgebra of $\mathbf{C}^d$. By Theorem 6.1.12, this implies that D has a primitive positive definition in $\mathfrak{C}$, which becomes the domain formula of the interpretation. Since h is an algebra homomorphism, the kernel K of h is a congruence of $\mathbf{D}$. It follows that K, viewed as a $2d$-ary relation over C, is preserved by all operations from $\mathbf{C}$. Theorem 6.1.12 implies that K has a primitive positive definition in $\mathfrak{C}$. This definition becomes the interpreting formula of the equality relation on B.

To see that h is a full interpretation, let $R \subseteq B^k$ be a relation of $\mathfrak{B}$, let τ be the signature of $\mathbf{C}$, and let $f \in \tau$ be arbitrary. By assumption, $f^{\mathbf{B}}$ preserves R. Therefore, $f^{\mathbf{C}}$ preserves $h^{-1}(R)$. Hence, all polymorphisms of $\mathfrak{C}$ preserve $h^{-1}(R)$, and because $\mathfrak{C}$ is ω-categorical, the relation $h^{-1}(R)$ has a primitive positive definition in $\mathfrak{C}$ (Theorem 6.1.12), which becomes the interpreting formula for $R(x_1, \ldots, x_k)$. We have verified that h is a primitive positive interpretation of $\mathfrak{B}$ in $\mathfrak{C}$. To see that h is a full interpretation, let $R \subseteq B^k$ be a relation such that $h^{-1}(R)$ is primitively positively definable in $\mathfrak{C}$. Then $h^{-1}(R)$ is preserved by $\mathrm{Clo}(\mathbf{C})$ and R is preserved by $\mathrm{Clo}(\mathbf{B})$ and therefore also by $\overline{\mathrm{Clo}(\mathbf{B})}$. By assumption $\overline{\mathrm{Clo}(\mathbf{B})} = \mathrm{Pol}(\mathfrak{B})$, and hence R is preserved by all polymorphisms of $\mathfrak{B}$ and primitively positively definable in $\mathfrak{B}$ by Theorem 6.1.12. ⊣

The proof of Theorem 6.3.7 above gives more information about the link between primitive positive interpretations in $\mathfrak{C}$ and the algebras in Exp $\text{HSP}^{\text{fin}}(\mathbf{C})$, and we state it explicitly.

THEOREM 6.3.8. *Let $\mathfrak{C}$ be a countable ω-categorical structure with polymorphism algebra $\mathbf{C}$, let $\mathfrak{B}$ be an arbitrary structure, let $d \in \mathbb{N}$, and let $h\colon D^d \to B$ be a partial surjection. Then the following are equivalent.*

1. *h is a primitive positive interpretation of $\mathfrak{B}$ in $\mathfrak{C}$;*
2. *h is a surjective homomorphism from an algebra $\mathbf{S} \in \mathrm{S}(\mathbf{C}^d)$ to an algebra $\mathbf{B}$ such that $\mathrm{Clo}(\mathbf{B}) \subseteq \mathrm{Pol}(\mathfrak{B})$.*

As in the case of primitive positive interpretations, we can also characterise primitive positive bi-interpretations in terms of pseudo-varieties of the respective polymorphism algebras.

PROPOSITION 6.3.9 (Proposition 25 in [91]). *Let $\mathfrak{A}$ and $\mathfrak{B}$ be countable ω-categorical structures. Then the following are equivalent.*

1. *$\mathfrak{A}$ has a polymorphism algebra $\mathbf{A}$ and $\mathfrak{B}$ has a polymorphism algebra $\mathbf{B}$ such that $\text{HSP}^{\text{fin}}(\mathbf{A}) = \text{HSP}^{\text{fin}}(\mathbf{B})$.*
2. *$\mathfrak{A}$ and $\mathfrak{B}$ are primitively positively bi-interpretable.*

PROOF. For the implication from (1) to (2), we assume that there is a $d_1 \geq 1$, a subalgebra $\mathbf{S}_1$ of $\mathbf{A}^{d_1}$, and a surjective homomorphism h_1 from $\mathbf{S}_1$ to $\mathbf{B}$. Moreover, we assume that there is a $d_2 \geq 1$, a subalgebra $\mathbf{S}_2$ of $\mathbf{B}^{d_2}$, and a surjective homomorphisms h_2 from $\mathbf{S}_2$ to $\mathbf{A}$. By Theorem 6.3.8, $I_1 := (d_1, S_1, h_1)$ is an interpretation of $\mathfrak{B}$ in $\mathfrak{B}$, and $I_2 := (d_2, S_2, h_2)$ is an interpretation of $\mathfrak{A}$ in $\mathfrak{B}$. Because the statement is symmetric it suffices to show that the (graph of the) function $h_1 \circ h_2\colon (S_2)^{d_1} \to B$ defined by

$$(y_{1,1}, \ldots, y_{1,d_2}, \ldots, y_{d_1,1}, \ldots, y_{d_1,d_2}) \mapsto \\ h_1(h_2(y_{1,1}, \ldots, y_{1,d_2}), \ldots, h_2(y_{d_1,1}, \ldots, y_{d_1,d_2}))$$

is primitively positively definable in $\mathfrak{B}$. Theorem 6.1.12 asserts that this is equivalent to showing that $h_1 \circ h_2$ is preserved by all operations $f^{\mathbf{B}}$ of $\mathbf{B}$. So let k be the arity of $f^{\mathbf{B}}$ and let $b^i = (b^i_1, \ldots, b^i_{d_1})$ be elements of $(S_2)^{d_1}$, for $1 \leq i \leq k$. Then indeed

$$\begin{aligned} & f^{\mathbf{B}}((h_1 \circ h_2)(b^1), \ldots, (h_1 \circ h_2)(b^k)) \\ &= h_1(f^{\mathbf{A}}(h_2(b^1_1), \ldots, h_2(b^k_1)), \ldots, f^{\mathbf{A}}(h_2(b^1_{d_1}), \ldots, h_2(b^k_{d_1}))) \\ &= (h_1 \circ h_2)(f^{\mathbf{B}}(b^1, \ldots, b^k)). \end{aligned}$$

For the implication from (2) to (1), suppose that $\mathfrak{A}$ and $\mathfrak{B}$ are primitive positive bi-interpretable via an interpretation $I_1 = (d_1, S_1, h_1)$ of $\mathfrak{B}$ in $\mathfrak{A}$ and an interpretation $I_2 = (d_2, S_2, h_2)$ of $\mathfrak{A}$ in $\mathfrak{B}$. Let $\mathbf{A}$ be a polymorphism algebra of $\mathfrak{A}$. Proposition 6.3.8 states that S_1 induces an algebra $\mathbf{S}_1$ in $\mathbf{A}^{d_1}$ and h_1 is a surjective homomorphism from $\mathbf{S}_1$ to an algebra $\mathbf{B}$ satisfying $\mathrm{Clo}(\mathbf{B}) = \mathrm{Pol}(\mathfrak{B})$.

Hence, $\mathbf{B}$ is a polymorphism algebra of $\mathfrak{B}$ that has the same signature τ as $\mathbf{A}$. Similarly, S_2 is the domain of a subalgebra $\mathbf{S}_2$ of $\mathbf{B}^{d_2}$ and h_2 is a homomorphism from $\mathbf{S}_2$ onto an algebra $\mathbf{A}'$ such that $\mathrm{Clo}(\mathbf{A}') = \mathrm{Pol}(\mathfrak{A})$.

We claim that $\mathrm{HSP}^{\mathrm{fin}}(\mathbf{A}) = \mathrm{HSP}^{\mathrm{fin}}(\mathbf{B})$. The inclusion '$\supseteq$' is clear since $\mathbf{B} \in \mathrm{HSP}^{\mathrm{fin}}(\mathbf{A})$. For the reverse inclusion it suffices to show that $\mathbf{A} = \mathbf{A}'$ since $\mathbf{A}' \in \mathrm{HSP}^{\mathrm{fin}}(\mathbf{B})$. Let $f \in \tau$ be k-ary; we show that $f^{\mathbf{A}} = f^{\mathbf{A}'}$. Let $a_1, \dots, a_k \in A$. Since $h_2 \circ h_1$ is surjective onto A, there are $c^i = (c^i_{1,1}, \dots, c^i_{d_1,d_2}) \in A^{d_1 d_2}$ such that $a_i = h_2 \circ h_1(c^i)$. Then

$$\begin{aligned}
f^{\mathbf{A}'}(a_1, \dots, a_k) &= f^{\mathbf{A}'}(h_2 \circ h_1(c^1), \dots, h_2 \circ h_1(c^k)) \\
&= h_2\big(f^{\mathbf{B}}(h_1(c^1_{1,1}, \dots, c^1_{d_1,1}), \dots, h_1(c^k_{1,1}, \dots, c^k_{d_1,1})), \dots, \\
&\qquad f^{\mathbf{B}}(h_1(c^1_{1,d_2}, \dots, c^1_{d_1,d_2}), \dots, h_1(c^k_{1,d_2}, \dots, c^k_{d_1,d_2}))\big) \\
&= h_2 \circ h_1(f^{\mathbf{A}}(c^1, \dots, c^k)) \\
&= f^{\mathbf{A}}(h_2 \circ h_1(c^1), \dots, h_2 \circ h_1(c^k)) \\
&= f^{\mathbf{A}}(a_1, \dots, a_k)
\end{aligned}$$

where the second and third equations hold since h_2 and h_1 are algebra homomorphisms, and the fourth equation holds because $f^{\mathbf{A}}$ preserves $h_2 \circ h_1$, because $I_2 \circ I_1$ is pp-homotopic to the identity. ⊣

The ω-categorical structures $\mathfrak{B}$ which primitively positively interpret every finite structure can be characterised algebraically as follows.

THEOREM 6.3.10. *Let $\mathfrak{B}$ be a countable ω-categorical structure and let $\mathbf{B}$ be a polymorphism algebra of $\mathfrak{B}$. Then the following are equivalent.*

1. $\mathrm{I}(\mathfrak{B})$ *contains all finite structures.*
2. $\mathrm{I}(\mathfrak{B})$ *contains K_n, for some $n \geq 3$.*
3. $\mathrm{I}(\mathfrak{B})$ *contains* $(\{0,1\}; \mathrm{NAE})$.
4. $\mathrm{I}(\mathfrak{B})$ *contains* $(\{0,1\}; \mathrm{1IN3})$.
5. $\mathrm{I}(\mathfrak{B})$ *contains for every finite set A a structure on A all of whose polymorphisms are projections.*
6. $\mathrm{HSP}^{\mathrm{fin}}(\mathbf{B})$ *contains for every finite set A an algebra on A all of whose operations are projections.*
7. $\mathrm{HSP}^{\mathrm{fin}}(\mathbf{B})$ *contains an algebra on a domain of size at least 2 all of whose operations are projections.*
8. $\mathrm{I}(\mathfrak{B})$ *contains a structure with at least two elements all of whose polymorphisms are projections.*
9. $\mathrm{I}(\mathfrak{B})$ *contains a structure with at least two elements where all first-order formulas are equivalent to primitive positive formulas.*

If these conditions apply then $\mathfrak{B}$ has a finite-signature reduct with an NP-hard CSP.

PROOF. We first show the cyclic sequence of implications

$$(1) \Rightarrow (5) \Rightarrow (6) \Rightarrow (7) \Rightarrow (8) \Rightarrow (9) \Rightarrow (1).$$

To show (1) ⇒ (5), let $\mathfrak{A}$ be the structure with domain A, the relations P_3^A, and for each $i \in A$ the unary relation $\{i\}$. By (1) there is a primitive positive interpretation of $\mathfrak{A}$ in $\mathfrak{B}$. All polymorphisms of $\mathfrak{A}$ are projections (Corollary 6.1.20), proving (5).

(5) ⇒ (6). For a finite set A, let $\mathfrak{A}$ be the structure with domain A from (5). Then Theorem 6.3.7 implies that there is an algebra $\mathbf{A} \in \mathrm{HSP}^{\mathrm{fin}}(\mathbf{B})$ such that all operations of $\mathbf{A}$ are polymorphisms of $\mathfrak{A}$.

The implication (6) ⇒ (7) is trivial. For the implication from (7) to (8), let $\mathfrak{A}$ be the algebra on a domain A with $|A| \geq 2$ such that all operations of $\mathfrak{A}$ are projections. Then all operations in $\mathfrak{A}$ preserve P_A and the unary relation $\{i\}$ for each $i \in \{1, \ldots, m\}$; Theorem 6.3.7 then implies (8). The implications (8) ⇒ (9) and (8) ⇒ (1) are by Theorem 3.2.2.

Clearly, (1) implies (2), (3), and (4). Proposition 6.1.43 shows that all first-order formulas are over K_n, for $n \geq 3$, equivalent to a primitive positive formula, so (2) implies (9). Similarly, Proposition 6.2.8 shows that (3) implies (9). Finally, every polymorphism of $(\{0,1\}; \mathrm{1IN3})$ is a projection (Corollary 6.2.2) so (4) implies (8). ⊣

6.4. Reflections

In Section 6.3.5 we have seen that the $\mathrm{HSP}^{\mathrm{fin}}$ operator is the algebraic counterpart to full primitive positive interpretations. This section treats a relatively new universal-algebraic operator, for forming *reflections* (introduced in [27]), which can be used to characterise the structure-building operator HI. Recall from Section 3.6 that HI is the operator that is most relevant for constraint satisfaction.

DEFINITION 6.4.1. Let $\mathbf{B}$ be a τ-algebra, let A be a set, and let $h \colon B \to A$ and $g \colon A \to B$ be two maps. Then the *reflection* of $\mathbf{B}$ with respect to h and g is the τ-algebra $\mathbf{A}$ with domain A where for all $x_1, \ldots, x_n \in A$ and $f \in \tau$ of arity n we define

$$f^{\mathbf{A}}(x_1, \ldots, x_n) := h(f^{\mathbf{B}}(g(x_1), \ldots, g(x_n))).$$

The class of reflections of a class of τ-algebras $\mathcal{C}$ is denoted by $\mathrm{Refl}(\mathcal{C})$.

As for the other operators on algebras, we write $\mathrm{Refl}(\mathbf{B})$ instead of $\mathrm{Refl}(\{\mathbf{B}\})$. The analog to the HSP-lemma (Lemma 6.3.5) is the following.

LEMMA 6.4.2 (from [27]). *Let $\mathcal{C}$ be a class of τ-algebras.*

- *The smallest class of τ-algebras that contains $\mathcal{C}$ and is closed under* Refl, H, S, *and* P *equals* Refl P($\mathcal{C}$).

- *The smallest class of* τ*-algebras that contains* $\mathcal{C}$ *and is closed under* Refl, H, S, *and* $\mathrm{P}^{\mathrm{fin}}$ *equals* $\mathrm{Refl}\,\mathrm{P}^{\mathrm{fin}}(\mathcal{C})$.

PROOF. For the first statement, it suffices to prove that $\mathrm{Refl}\,\mathrm{P}(\mathcal{C})$ is closed under Refl, H, S, P, and for the second that $\mathrm{Refl}\,\mathrm{P}^{\mathrm{fin}}(\mathcal{C})$ is closed under Refl, H, S, $\mathrm{P}^{\mathrm{fin}}$. For the operator Refl this follows from the simple fact that $\mathrm{Refl}\,\mathrm{Refl}(\mathcal{K}) = \mathrm{Refl}(\mathcal{K})$ for any class $\mathcal{K}$.

To prove that $\mathrm{Refl}\,\mathrm{P}(\mathcal{C})$ and $\mathrm{Refl}\,\mathrm{P}^{\mathrm{fin}}(\mathcal{C})$ are closed under H, we show that $\mathrm{H}(\mathcal{K}) \subseteq \mathrm{Refl}(\mathcal{K})$ for any class $\mathcal{K}$. Let $\mathbf{B} \in \mathcal{K}$ and $h\colon B \to A$ be a surjective homomorphism to an algebra $\mathbf{A}$. Pick any function g such that $h \circ g$ is the identity on A. Then h and g witness that $\mathbf{A}$ is a reflection of $\mathbf{B}$ since

$$
\begin{aligned}
h(f^{\mathbf{B}}(g(x_1),\dots,g(x_n))) &= f^{\mathbf{A}}(h \circ g(x_1),\dots,h \circ g(x_n)) \\
&\qquad\qquad \text{(since } h \text{ is a homomorphism)} \\
&= f^{\mathbf{A}}(x_1,\dots,x_n) \qquad \text{(by the choice of } g\text{)}.
\end{aligned}
$$

To prove that $\mathrm{Refl}\,\mathrm{P}(\mathcal{C})$ and $\mathrm{Refl}\,\mathrm{P}^{\mathrm{fin}}(\mathcal{C})$ are closed under S, we show that $\mathrm{S}(\mathcal{K}) \subseteq \mathrm{Refl}(\mathcal{K})$ for any class $\mathcal{K}$. Let $\mathbf{B} \in \mathcal{K}$ and suppose that $\mathbf{A}$ is a subalgebra of $\mathbf{B}$. Let $g\colon A \to B$ be the identity on A, and $h\colon B \to A$ be any extension of g to B. Then h and g show that $\mathbf{A}$ is a reflection of $\mathbf{B}$ since

$$h(f^{\mathbf{B}}(g(x_1),\dots,g(x_n))) = f^{\mathbf{B}}(x_1,\dots,x_n) = f^{\mathbf{A}}(x_1,\dots,x_n).$$

Let I be an arbitrary set, $(\mathbf{B}_i)_{i\in I}$ be algebras from $\mathrm{Refl}\,\mathrm{P}(\mathcal{C})$, and suppose that $\mathbf{A}_i$ is a reflection of $\mathbf{B}_i$ for every $i \in I$, witnessed by functions $h_i\colon B_i \to A_i$ and $g_i\colon A_i \to B_i$. Then the map $h\colon \prod_{i\in I} B_i \to \prod_{i\in I} A_i$ that sends $(b_i)_{i\in I}$ to $(h_i(b_i))_{i\in I}$ and the map $g\colon \prod_{i\in I} A_i \to \prod_{i\in I} B_i$ that sends $(a_i)_{i\in I}$ to $(g_i(a_i))_{i\in I}$ witness that $\prod_{i\in I} \mathbf{A}_i$ is a reflection of $\prod_{i\in I} \mathbf{B}_i$. This shows that $\mathrm{P}(\mathrm{Refl}\,\mathrm{P}(\mathcal{C})) \subseteq \mathrm{Refl}\,\mathrm{P}(\mathcal{C})$ and likewise that $\mathrm{P}^{\mathrm{fin}}(\mathrm{Refl}\,\mathrm{P}^{\mathrm{fin}}(\mathcal{C})) \subseteq \mathrm{Refl}\,\mathrm{P}^{\mathrm{fin}}(\mathcal{C})$. ⊣

THEOREM 6.4.3. *Let* $\mathfrak{B}, \mathfrak{C}$ *be at most countable* ω*-categorical relational structures and let* $\mathbf{C}$ *be a polymorphism algebra of* $\mathfrak{C}$. *Then*

1. $\mathfrak{B} \in \mathrm{H}\,\mathrm{Red}(\mathfrak{C})$ *if and only if there is an algebra* $\mathbf{B} \in \mathrm{Exp}\,\mathrm{Refl}(\mathbf{C})$ *such that* $\mathrm{Clo}(\mathbf{B}) = \mathrm{Pol}(\mathfrak{B})$.
2. $\mathfrak{B} \in \mathrm{HI}(\mathfrak{C})$ *if and only if there is an algebra* $\mathbf{B} \in \mathrm{Exp}\,\mathrm{Refl}\,\mathrm{P}^{\mathrm{fin}}(\mathbf{C})$ *such that* $\mathrm{Clo}(\mathbf{B}) = \mathrm{Pol}(\mathfrak{B})$.

PROOF. To show (1), first suppose that $\mathfrak{B} \in \mathrm{H}(\mathfrak{C}')$ and $\mathfrak{C}' \in \mathrm{Red}(\mathfrak{C})$; let $h\colon \mathfrak{C}' \to \mathfrak{B}$ and $g\colon \mathfrak{B} \to \mathfrak{C}'$ be homomorphisms witnessing homomorphic equivalence of $\mathfrak{B}$ and $\mathfrak{C}'$. Let $\mathbf{C}'$ be the reduct of $\mathbf{C}$ obtained by dropping all functions that are not polymorphisms of $\mathfrak{C}'$, so $\mathbf{C}'$ is a polymorphism algebra of $\mathfrak{C}'$. Let $\mathbf{B}'$ be the reflection of $\mathbf{C}'$ with respect to h and g. Every operation of $\mathrm{Clo}(\mathbf{B}')$ is obtained as a composition of homomorphisms, so preserves all the relations of $\mathfrak{B}$, so $\mathrm{Clo}(\mathbf{B}') \subseteq \mathrm{Pol}(\mathfrak{B})$. This shows the existence of an algebra

$$\mathbf{B} \in \mathrm{Exp}(\mathbf{B}') = \mathrm{Exp}\,\mathrm{Refl}(\mathbf{C}') = \mathrm{Exp}\,\mathrm{Refl}\,\mathrm{Exp}(\mathbf{C}) = \mathrm{Exp}\,\mathrm{Refl}(\mathbf{C})$$

such that $\mathrm{Clo}(\mathbf{B}') = \mathrm{Pol}(\mathfrak{B})$.

Now suppose that conversely the reflection $\mathbf{B}'$ of $\mathbf{C}$ at $h: C \to B$ and $g: B \to C$ is such that $\mathrm{Clo}(\mathbf{B}') \subseteq \mathrm{Pol}(\mathfrak{B})$. Let $\mathfrak{C}'$ be the structure with domain C and the same signature as $\mathfrak{B}$ which contains for every k-ary relation symbol R of $\mathfrak{B}$ the relation

$$R^{\mathfrak{C}'} := \{f(g(b_1), \dots, g(b_k)) \mid f \in \mathrm{Pol}(\mathfrak{C}), b_1, \dots, b_k \in R^{\mathfrak{B}}\}.$$

These relations are preserved by $\mathrm{Pol}(\mathfrak{C})$, so they are primitively positively definable in $\mathfrak{C}$ by Theorem 6.1.12, and hence $\mathfrak{C}' \in \mathrm{Red}(\mathfrak{C})$. Clearly, g is a homomorphism from $\mathfrak{B}$ to $\mathfrak{C}'$. We claim that h is a homomorphism from $\mathfrak{C}'$ to $\mathfrak{B}$. Indeed, if $(f(g(b_1), \dots, g(b_k))) \in R^{\mathfrak{C}'}$, then $h(f(g(b_1), \dots, g(b_k))) \in R^{\mathfrak{B}}$ because the operation $(x_1, \dots, x_k) \mapsto h(f(g(x_1), \dots, g(x_k)))$ is an operation of $\mathbf{B}' \in \mathrm{Refl}(\mathbf{C})$ and hence a polymorphism of $\mathfrak{B}$ since $\mathrm{Clo}(\mathbf{B}') \subseteq \mathrm{Pol}(\mathfrak{B})$. Thus, $\mathfrak{B} \in \mathrm{H}(\mathfrak{C}') = \mathrm{H}(\mathrm{Red}(\mathfrak{C}))$.

Item (2) is a straightforward combination of item (1) with Theorem 6.3.7: If $\mathfrak{B} \in \mathrm{HI}(\mathfrak{C})$ then there exists a structure $\mathfrak{D} \in \mathrm{I_{full}}(\mathfrak{C})$ such that $\mathfrak{B} \in \mathrm{H}\,\mathrm{Red}(\mathfrak{D})$. By Theorem 6.3.7 there is an algebra $\mathbf{D} \in \mathrm{HSP^{fin}}(\mathbf{C})$ such that $\mathrm{Clo}(\mathbf{D}) = \mathrm{Pol}(\mathfrak{D})$, and by item (1) there is an algebra $\mathbf{B} \in \mathrm{Exp}\,\mathrm{Refl}(\mathbf{D})$ such that $\mathrm{Clo}(\mathbf{B}) = \mathrm{Pol}(\mathfrak{B})$. This proves the statement since

$$\begin{aligned} \mathbf{B} \in \mathrm{Exp}\,\mathrm{Refl}(\mathbf{D}) &\subseteq \mathrm{Exp}\,\mathrm{Refl}\,\mathrm{HSP^{fin}}(\mathbf{C}) \\ &= \mathrm{Exp}\,\mathrm{Refl}\,\mathrm{P^{fin}}(\mathbf{C}) \qquad \text{(by Lemma 6.4.2).} \quad \dashv \end{aligned}$$

The following hardness condition for CSPs is stronger than the one presented in 6.3.10: there are situations where the conditions of the following corollary applies, but where the conditions of Theorem 6.3.10 do not.

COROLLARY 6.4.4. *Let $\mathfrak{B}$ be an at most countable ω-categorical structure and let $\mathbf{B}$ be a polymorphism algebra of $\mathfrak{B}$. Then the following are equivalent.*

1. $\mathrm{HI}(\mathfrak{B})$ *contains* K_3.
2. $\mathrm{HI}(\mathfrak{B})$ *contains all finite structures.*
3. $\mathrm{HI}(\mathfrak{B})$ *contains* $(\{0,1\}; \mathrm{1IN3})$.
4. $\mathrm{Refl}\,\mathrm{P^{fin}}(\mathbf{B})$ *contains an algebra of size at least* 2 *all of whose operations are projections.*
5. $\mathrm{Refl}\,\mathrm{P^{fin}}(\mathbf{B})$ *contains for every finite set A an algebra on A all of whose operations are projections.*

If these condition apply then $\mathfrak{B}$ has a finite-signature reduct with an NP-hard CSP.

PROOF. The implication from (1) to (2) follows from the fact that $\mathrm{I}(K_3)$ contains all finite structures (Corollary 3.2.1), and that $\mathrm{I}\,\mathrm{H}\,\mathrm{I}(\mathfrak{B}) \subseteq \mathrm{HI}(\mathfrak{B})$ by Theorem 3.6.2. The implication from (2) to (3) is trivial. The implication from (3) (4) follows from the fact that all polymorphisms of $(\{0,1\}; \mathrm{1IN3})$ are projections (Corollary 6.2.2), and Theorem 6.4.3. For the implication from (4) to (5), suppose that $\mathrm{Refl}\,\mathrm{P}^{\mathrm{fin}}(\mathbf{B})$ contains an algebra $\mathbf{A}$ of size at least 2 all of whose operations are projections. By Theorem 6.3.10, $\mathrm{HSP}^{\mathrm{fin}}(\mathbf{A})$ contains for every finite set S an algebra on S all of whose operations are projections. The statement follows since $\mathrm{HSP}^{\mathrm{fin}}(\mathbf{A}) \subseteq \mathrm{HSP}^{\mathrm{fin}}(\mathrm{HI}(\mathfrak{B})) \subseteq \mathrm{HI}(\mathfrak{B})$ by Theorem 3.6.2. To show that (5) implies (1), let $\mathbf{A} \in \mathrm{H}(\mathfrak{B})$ be such that $A = \{0,1,2\}$ and all operations of $\mathbf{A}$ are projections. Then $\mathrm{Pol}(\mathbf{A}) \subseteq \mathrm{Pol}(K_3)$ and hence $K_3 \in \mathrm{HI}(\mathfrak{B})$ by Theorem 6.4.3. The final statement follows from Corollary 3.7.1. ⊣

6.5. Varieties, abstract clones, and Birkhoff's theorem

Varieties (see Section 6.3.4) are a fascinatingly powerful concept to study classes of algebras. They are also central for the study of the complexity of CSPs: We will see that the complexity of $\mathrm{CSP}(\mathfrak{B})$ for finite $\mathfrak{B}$ only depends on the variety generated by a polymorphism algebra $\mathbf{B}$ of $\mathfrak{B}$. This comes from the fact that a finite algebra is in the variety generated by a finite algebra $\mathbf{B}$ if and only if it is in the pseudo-variety generated by $\mathbf{B}$; and the link between the pseudo-variety generated by the polymorphism algebra of $\mathfrak{B}$ and $\mathrm{CSP}(\mathfrak{B})$ has already been explained in Section 6.3.5.

The central theorem for the study of varieties is Birkhoff's HSP theorem (Section 6.5.1), which links varieties with equational classes. By Birkhoff's theorem, there is also a close relationship between varieties and the concept of an *abstract clone* (Section 6.5.2).

6.5.1. Birkhoff's theorem. Varieties have the advantage that they can be described by the identities (or *equations*) satisfied by its members. Let τ be a functional signature. An *identity (over τ)* is a τ-sentence of the form

$$\forall x_1, \dots, x_n \colon s(x_1, \dots, x_n) = t(x_1, \dots, x_n)$$

where s and t are τ-terms.

THEOREM 6.5.1 (Birkhoff [42]; see e.g. [206] or [119]). *Let* $\mathbf{A}$ *and* $\mathbf{B}$ *be algebras with the signature* τ. *Then the following are equivalent.*

1. *Every identity that holds in* $\mathbf{B}$ *also holds in* $\mathbf{A}$.
2. $\mathbf{A} \in \mathrm{HSP}(\mathbf{B})$.

Moreover, if A and B are finite then we can add the following to the list:

3. $\mathbf{A} \in \mathrm{HSP}^{\mathrm{fin}}(\mathbf{B})$.

PROOF. We first consider the case that A and B are finite, and prove that (1) implies (3): Let $a_1, \dots, a_k$ be the elements of $\mathbf{A}$, define $m := |B|^k$, and let C be B^A. Let $c^1, \dots, c^m$ be the elements of C; for $i \leq k$, define $c_i := (c^1(a_i), \dots, c^m(a_i)) \in B^m$. Let $\mathbf{S}$ be the smallest subalgebra of $\mathbf{B}^m$ that contains $c_1, \dots, c_k$; so the elements of $\mathbf{S}$ are precisely those of the form $t^{\mathbf{S}}(c_1, \dots, c_k)$, for a k-ary τ-term t. Define $\mu\colon S \to A$ by

$$\mu(t^{\mathbf{S}}(c_1,\dots,c_k)) := t^{\mathbf{A}}(a_1,\dots,a_k).$$

Claim 1. μ is well defined. Suppose that $t^{\mathbf{S}}(c_1,\dots,c_k) = s^{\mathbf{S}}(c_1,\dots,c_k)$. We first show that $t^{\mathbf{B}} = s^{\mathbf{B}}$. Let $b \in B^k$. Then there exists an $i \leq m$ such that $(c^i(a_1),\dots,c^i(a_k)) = b$. Thus,

$$\begin{aligned} t^{\mathbf{B}}(b) = t^{\mathbf{B}}(c^i(a_1),\dots,c^i(a_k)) &= (t^{\mathbf{S}}(c_1,\dots,c_k))_i \\ &= (s^{\mathbf{S}}(c_1,\dots,c_k))_i \\ &= s^{\mathbf{B}}(c^i(a_1),\dots,c^i(a_k)) = s^{\mathbf{B}}(b). \end{aligned}$$

Hence, $t^{\mathbf{B}} = s^{\mathbf{B}}$. By assumption, $t^{\mathbf{A}} = s^{\mathbf{A}}$ and in particular $t^{\mathbf{A}}(a_1,\dots,a_k) = s^{\mathbf{A}}(a_1,\dots,a_k)$.

Claim 2. μ is surjective. Let $i \leq k$, and let $t(x_1,\dots,x_k)$ be the τ-term x_i. Then

$$\mu(c_i) = \mu(t^{\mathbf{S}}(c_1,\dots,c_k)) = t^{\mathbf{A}}(a_1,\dots,a_k) = a_i.$$

Claim 3. μ is a homomorphism from $\mathbf{S}$ to $\mathbf{A}$. Let $f \in \tau$ be of arity n and let $s_1,\dots,s_m \in S$. For $i \leq n$, write $s_i = t_i^{\mathbf{S}}(c_1,\dots,c_k)$ for some τ-term t. Then

$$\begin{aligned} \mu(f^{\mathbf{S}}(s_1,\dots,s_n)) &= \mu(f^{\mathbf{S}}(t_1^{\mathbf{S}}(c_1,\dots,c_k),\dots,t_n^{\mathbf{S}}(c_1,\dots,c_k))) \\ &= \mu(f^{\mathbf{S}}(t_1^{\mathbf{S}},\dots,t_n^{\mathbf{S}})(c_1,\dots,c_k)) \\ &= \mu((f(t_1,\dots,t_n))^{\mathbf{S}}(c_1,\dots,c_k)) \\ &= \left(f(t_1,\dots,t_n)\right)^{\mathbf{A}}(a_1,\dots,a_k) \\ &= f^{\mathbf{A}}(t_1^{\mathbf{A}}(a_1,\dots,a_k),\dots,t_n^{\mathbf{A}}(a_1,\dots,a_k)) \\ &= f^{\mathbf{A}}(\mu(s_1),\dots,\mu(s_n)). \end{aligned}$$

Therefore, $\mathbf{A}$ is the homomorphic image of the subalgebra $\mathbf{S}$ of $\mathbf{B}^m$, and so $\mathbf{A} \in \mathrm{HSP}^{\mathrm{fin}}(\mathbf{B})$. The same proof shows that for general A we have $\mathbf{A} \in \mathrm{HSP}(\mathbf{B})$, and hence (1) implies (2).

To show that (2) implies (1) (and, likewise, (3) implies (1)), let

$$\phi := \forall x_1,\dots,x_n\colon s(x_1,\dots,x_n) = t(x_1,\dots,x_n)$$

be an identity that holds in $\mathbf{B}$. To see that ϕ is preserved in powers $\mathbf{A} = \mathbf{B}^I$ of $\mathbf{B}$ let $a_1,\dots,a_n \in A$ be arbitrary. By assumption $\mathbf{B} \models s(a_1[j],\dots,a_n[j]) = t(a_1[j],\dots,a_n[j])$ holds for all $j \in I$. Hence, $\mathbf{B}^I \models s(a_1,\dots,a_n) = t(a_1,\dots,a_n)$. Since $a_1,\dots,a_n$ were chosen arbitrarily, we conclude that $\mathbf{A} \models \phi$.

Moreover, ϕ is true in subalgebras of algebras that satisfy ϕ (this is true for universal sentences in general). Finally, suppose that $\mathbf{B}$ is an algebra that satisfies ϕ, and μ is a surjective homomorphism from $\mathbf{B}$ to some algebra $\mathbf{A}$.

Let $a_1, \ldots, a_n \in A$; by surjectivity of μ we can choose $b_1, \ldots, b_n$ such that $\mu(b_i) = a_i$ for all $i \leq n$. Then

$$\begin{aligned} s^{\mathbf{B}}(b_1, \ldots, b_n) = t^{\mathbf{B}}(b_1, \ldots, b_n) &\Rightarrow \mu\big(s^{\mathbf{B}}(b_1, \ldots, b_n)\big) = \mu\big(t^{\mathbf{B}}(b_1, \ldots, b_n)\big) \\ &\Rightarrow t^{\mathbf{A}}\big(\mu(b_1), \ldots, \mu(b_n)\big) = s^{\mathbf{A}}\big(\mu(b_1), \ldots, \mu(b_n)\big) \\ &\Rightarrow t^{\mathbf{A}}(a_1, \ldots, a_n) = s^{\mathbf{A}}(a_1, \ldots, a_n). \end{aligned}$$

Hence, $\mathbf{A} \models \phi$. ⊣

Birkhoff's theorem has an elegant reformulation based on the concept of *abstract clones*, which we discuss in the next section.

6.5.2. (Abstract) clones. Clones (in the literature often called *abstract clones*) have the same relation to operation clones as (abstract) groups do to permutation groups. The elements of a clone correspond to the functions of an operation clone, and the signature contains constant symbols for the projections, and composition symbols to code how operations compose. Since an operation clone contains operations of various arities, a clone will be formalised as a multi-sorted structure, with a sort for each arity. Abstract clones have also been formalised in category theoretic terms; this perspective is not needed here and we refer the interested reader to [336].

DEFINITION 6.5.2. A *clone* $\mathbf{C}$ is a multi-sorted structure with sorts $\{C^{(k)} \mid k \in \mathbb{N}^+\}$ and the signature $\{\mathrm{pr}_i^k \mid 1 \leq i \leq k\} \cup \{\mathrm{comp}_l^k \mid k, l \geq 1\}$. The elements of the sort $C^{(k)}$ will be called the *k-ary operations* of $\mathbf{C}$. We denote a clone by

$$\mathbf{C} = (C^{(1)}, C^{(2)}, \ldots; (\mathrm{pr}_i^k)_{1 \leq i \leq k}, (\mathrm{comp}_l^k)_{k,l \geq 1})$$

and require that pr_i^k is a constant in $C^{(k)}$, and that $\mathrm{comp}_l^k : C^{(k)} \times (C^{(l)})^k \to C^{(l)}$ is an operation of arity $k+1$. Moreover, we require that

$$\mathrm{comp}_k^k(f, \mathrm{pr}_1^k, \ldots, \mathrm{pr}_k^k) = f \tag{1}$$

$$\mathrm{comp}_l^k(\mathrm{pr}_i^k, f_1, \ldots, f_k) = f_i \tag{2}$$

$$\begin{aligned} &\mathrm{comp}_l^m(\mathrm{comp}_m^k(f, g_1, \ldots, g_k), h_1, \ldots, h_m) \\ &= \mathrm{comp}_l^k(f, \mathrm{comp}_l^m(g_1, h_1, \ldots, h_m), \ldots, \mathrm{comp}_l^m(g_k, h_1, \ldots, h_m)). \end{aligned} \tag{3}$$

We also write $f(g_1, \ldots, g_k)$ instead of $\mathrm{comp}_l^k(f, g_1, \ldots, g_k)$ when l is clear from the context. Equation (3) can then be phrased as

$$f(g_1, \ldots, g_k)(h_1, \ldots, h_m) = f\big(g_1(h_1, \ldots, h_m), \ldots, g_m(h_1, \ldots, h_m)\big)$$

and is a generalised form of associativity. Note that in this equation, the arguments $h_1, \ldots, h_m$ are duplicated (cloned), justifying (for me) the name *clone*. According to Cohn [135], the name is due to Philipp Hall. Michael Pinsker [302] suggests the origin "closed operation network".

In the following, we also use the term *abstract clone* when we want to stress that we are working with a clone and *not* with an operation clone. Every operation clone $\mathcal{C}$ gives rise to an abstract clone $\mathbf{C}$ in the obvious way: $\mathrm{pr}_i^k \in C^{(k)}$ denotes the projection $\pi_i^k \in \mathcal{C}$, and $\mathrm{comp}_l^k(f, g_1, \ldots, g_k) \in C^{(l)}$ denotes the composed function $(x_1, \ldots, x_l) \mapsto f(g_1(x_1, \ldots, x_l), \ldots, g_k(x_1, \ldots, x_l)) \in \mathcal{C}$. All the terminology that we have introduced for operation clones and that only depends on the associated abstract clone will also be used for abstract clones.

DEFINITION 6.5.3. Let $\mathbf{C}$ and $\mathbf{D}$ be clones. A function $\xi \colon C \to D$ is called a *(clone) homomorphism* iff

- ξ *preserves arities*, i.e., $\xi(C^{(i)}) \subseteq D^{(i)}$ for all $i \in \mathbb{N}^+$;
- ξ *preserves the projections*, i.e., $\xi\big((\mathrm{pr}_i^k)^{\mathbf{C}}\big) = (\mathrm{pr}_i^k)^{\mathbf{D}}$ for all $1 \leq i \leq k$;
- for all $n, m \geq 1$, $f \in C^{(n)}$, and $g_1, \ldots, g_n \in C^{(m)}$

$$\xi(f(g_1, \ldots, g_n)) = \xi(f)(\xi(g_1), \ldots, \xi(g_n)).$$

An *(clone) isomorphism* between $\mathbf{C}$ and $\mathbf{D}$ is a bijective homomorphism whose inverse is also a homomorphism.

In the following, we write pr_i^k for the element $(\mathrm{pr}_i^k)^{\mathbf{C}}$ of C when the reference to $\mathbf{C}$ is clear from the context. The following generalisation of Cayley's theorem for groups shows that for every clone $\mathbf{C}$ there exists an operation clone whose abstract clone is $\mathbf{C}$. In other words, every clone has an injective homomorphism into an operation clone. We present the proof here since it may serve as a motivation for the precise choice of the axioms (1), (2), and (3) for clones.

THEOREM 6.5.4 (Cayley's theorem for clones). *Every clone $\mathbf{C}$ is isomorphic to the clone of an operation clone.*

PROOF. Let $D := \prod_{n=1}^{\infty} C^{(n)}$. We define $\xi \colon C \to \mathcal{O}_D$ as follows. Let $f \in C$ be k-ary. For $d \in D$ and $i \in \mathbb{N}$, we write $d[i]$ for the i-th component of d (which is an element of $C^{(i)}$)). For $d_1, \ldots, d_k \in D$, let $d_0 \in D$ be such that $d_0[i] = \mathrm{comp}_i^k(f, d_1[i], \ldots, d_k[i])$ for all $i \in \mathbb{N}$. We then define $\xi(f)(d_1, \ldots, d_k) := d_0$.

We verify that ξ is a homomorphism from $\mathbf{C}$ to $\mathcal{O}_D$: for $d_1, \ldots, d_k \in D$ and $k, l, i \in \mathbb{N}$ we have

$$\begin{aligned} \xi(\mathrm{pr}_l^k)(d_1, \ldots, d_k)[i] &= \mathrm{comp}_i^k(\mathrm{pr}_l^k, d_1[i], \ldots, d_k[i]) && \text{(by definition of } \xi\text{)} \\ &= d_l[i] && \text{(by (2))} \end{aligned}$$

and therefore $\xi(\mathrm{pr}_l^k)(d_1, \ldots, d_k) = d_l$ and $\xi(\mathrm{pr}_l^k) = \pi_l^k$ as required.

For $f \in C^k, g_1, \ldots, g_k \in C^{(l)}, d_1, \ldots, d_l \in D, i \in \mathbb{N}$, using (3) we have that

$$\begin{aligned} &\xi(\mathrm{comp}_l^k(f, g_1, \ldots, g_k))(d_1, \ldots, d_l)[i] \\ &= \mathrm{comp}_i^l(\mathrm{comp}_l^k(f, g_1, \ldots, g_k), d_1[i], \ldots, d_l[i]) \\ &= \mathrm{comp}_i^k(f, \mathrm{comp}_i^l(g_1, d_1[i], \ldots, d_l[i]), \ldots, \mathrm{comp}_i^l(g_k, d_1[i], \ldots, d_l[i])) \end{aligned}$$

$$= \text{comp}_l^k(f, \xi(g_1)(d_1, \ldots, d_l)[i], \ldots, \xi(g_k)(d_1, \ldots, d_l)[i])$$
$$= \xi(f)(\xi(g_1)(d_1, \ldots, d_l), \ldots, \xi(g_k)(d_1, \ldots, d_l))[i]$$

and thus the desired

$$\xi(\text{comp}_l^k(f, g_1, \ldots, g_k)) = \xi(f)(\xi(g_1), \ldots, \xi(g_k)).$$

Moreover, ξ is injective: note that if $f \in C$ is k-ary, and $d_1, \ldots, d_k \in D$ are such that $d_j[k] = \text{pr}_j^k$ for all $j \in \{1, \ldots, k\}$, then

$$\begin{aligned} \xi(f)(d_1, \ldots, d_k)[i] &= \text{comp}_i^k(f, d_1[i], \ldots, d_k[i]) && \text{(by definition of } \xi) \\ &= \text{comp}_i^k(f, \text{pr}_1^k, \ldots, \text{pr}_k^k) && \text{(by the choice of } d_1, \ldots, d_k) \\ &= f && \text{(by (1)).} \end{aligned}$$

Hence, if $f \neq k$ then $\xi(f)(d_1, \ldots, d_k) \neq \xi(g)(d_1, \ldots, d_k)$ and hence $\xi(f) \neq \xi(g)$. Therefore, the image of ξ in $\mathcal{O}_D$ is the desired operation clone. ⊣

6.5.3. Clones and varieties. We are next going to describe the link between abstract clones and varieties. The first step is an (obvious, but formally cumbersome) translation between τ-terms over an algebra $\mathbf{A}$ and *clone terms*, i.e., terms over the signature of abstract clones. Each variable in such a term is equipped with a *rank* $k \in \mathbb{N}$. The *rank of a clone term* is defined as follows:

- the rank of variables is already defined;
- the rank of pr_l^k is defined to be k;
- the rank of a clone term of the form $\text{comp}_l^k(s_0, s_1, \ldots, s_k)$ is defined to be l.

Observe that every clone term is equivalent to a *normalised clone term*, i.e., a clone term such that the first argument of a comp symbol is always a variable; this can be achieved by applying Equation 2 and Equation 3.

DEFINITION 6.5.5. Let τ be a functional signature and let $f_1, \ldots, f_n \in \tau$. Then for every normalised clone term $r(z_1, \ldots, z_n)$ of rank m we write $r^*(f_1, \ldots, f_n)$ (or simply r^* if $f_1, \ldots, f_n$ are clear from the context) for the τ-term over the variables $x_1, \ldots, x_m$ inductively obtained as follows:

- if $r = \text{pr}_i^k$ then $r^* := x_i$;
- if $r = z_i$ then $r^* := f_i$;
- otherwise, $r = \text{comp}_l^k(z_i, s_1, \ldots, s_k)$ for normalised clone terms $s_1, \ldots, s_k$; in this case, define $r^* := f_i(s_1^*, \ldots, s_k^*)$.

A *clone formula* is a formula over the signature of clones. We may assume that all terms that appear in ϕ are normalised. If $\phi(z_1, \ldots, z_n)$ is a finite conjunction of atomic clone formulas, then we write ϕ^* for the conjunction of all identities

$$\forall x_1, \ldots, x_m : r^*(f_1, \ldots, f_n) = s^*(f_1, \ldots, f_n)$$

for each conjunct $r(z_1, \ldots, z_n) = s(z_1, \ldots, z_n)$ in ϕ where r and s are clone terms of rank m.

The central property of this translation is the following.

LEMMA 6.5.6. *Let* **A** *be a* τ*-algebra, let* $f_1, \dots, f_n \in \tau$, *and let* $\phi(z_1, \dots, z_n)$ *be a conjunction of atomic clone formulas. Then*

$$\mathrm{Clo}(\mathbf{A}) \models \phi(f_1^{\mathbf{A}}, \dots, f_n^{\mathbf{A}}) \text{ if and only if } \mathbf{A} \models \forall x_1, \dots, x_m \colon \phi^*(f_1, \dots, f_n)(x_1, \dots, x_m).$$

PROOF. Straightforward from the definitions. ⊣

We can also go in the other direction, translating τ-terms into clone terms as follows.

DEFINITION 6.5.7. For every τ-term $t(x_1, \dots, x_m)$ built from $f_1, \dots, f_n \in \tau$ we write $t^\dagger(z_1, \dots, z_n)$ for the (normalised) clone term inductively obtained as follows:

- if $t = x_i$ for some variable x_i then $t^\dagger := \mathrm{pr}_i^m$;
- if $t = f_i$ for some constant symbol $f_i \in \tau$ then $t^\dagger := z_i$;
- if $t = f_i(s_1, \dots, s_k)$ then $t^\dagger := \mathrm{comp}(z_i, s_1^\dagger, \dots, s_k^\dagger)$.

If $\psi(x_1, \dots, x_m)$ is a conjunction of atomic τ-formulas built from the function symbols $f_1, \dots, f_n$ then we write $\psi^\dagger(z_1, \dots, z_n)$ for the conjunction of $t^\dagger(f_1, \dots, f_n) = s^\dagger(f_1, \dots, f_n)$ for each conjunct $t = s$ in ψ.

Again, the central property of this translation is formulated in a lemma.

LEMMA 6.5.8. *Let* **A** *be a* τ*-algebra. Then for all conjunctions* $\psi(x_1, \dots, x_m)$ *of atomic* τ*-formulas built from* $f_1, \dots, f_n \in \tau$ *we have that*

$$\mathbf{A} \models \forall x_1, \dots, x_m \colon \psi(x_1, \dots, x_m)$$
$$\text{if and only if} \quad \mathrm{Clo}(\mathbf{A}) \models \psi^\dagger(f_1^{\mathbf{A}}, \dots, f_n^{\mathbf{A}})$$

PROOF. Straightforward from the definitions. ⊣

Example 6.5.9. For a τ-algebra **A** and $f \in \tau$ we have that

$$\mathrm{Clo}(\mathbf{A}) \models \mathrm{comp}_2^2(f^{\mathbf{A}}, p_1^2, p_2^2) = \mathrm{comp}_2^2(f^{\mathbf{A}}, \mathrm{pr}_2^2, \mathrm{pr}_1^2)$$

if and only if f is symmetric, i.e.,

$$\mathbf{A} \models \forall x_1, x_2 \colon f(x_1, x_2) = f(x_2, x_1).$$

In practice it can be more intuitive to manipulate τ-terms rather than clone terms, even if we work over a clone and not a τ-algebra; it is therefore standard to be sloppy with the distinction. We will refer to atomic clone formulas as identities, too.

If **A** and **B** are τ-algebras then the map $\xi \colon \mathrm{Clo}(\mathbf{B}) \to \mathrm{Clo}(\mathbf{A})$ defined by $f^{\mathbf{B}} \mapsto f^{\mathbf{A}}$ for all $f \in \tau$ is well defined if and only if for all τ-terms s, t

$$s^{\mathbf{B}} = t^{\mathbf{B}} \Rightarrow s^{\mathbf{A}} = t^{\mathbf{A}}.$$

In this case, ξ is in fact a surjective clone homomorphism, and we call it the *natural homomorphism* from $\mathrm{Clo}(\mathbf{B})$ to $\mathrm{Clo}(\mathbf{A})$. With this terminology, Birkhoff's theorem takes the following form.

THEOREM 6.5.10. *Let* $\mathbf{A}$ *and* $\mathbf{B}$ *be* τ*-algebras. The following are equivalent.*

1. *The natural homomorphism from* $\mathrm{Clo}(\mathbf{B})$ *onto* $\mathrm{Clo}(\mathbf{A})$ *exists.*
2. *All identities that hold in* $\mathbf{B}$ *also hold in* $\mathbf{A}$.
3. $\mathbf{A} \in \mathrm{HSP}(\mathbf{B})$.

When $\mathbf{A}$ *and* $\mathbf{B}$ *are finite, then we can add the following to the list:*

4. $\mathbf{A} \in \mathrm{HSP}^{\mathrm{fin}}(\mathbf{B})$

PROOF. The equivalence of (2), (3), and for finite $\mathbf{A}$ of (4) follows from Theorem 6.5.1.

(1) $\Rightarrow$ (2). Let $\xi\colon \mathrm{Clo}(\mathbf{B}) \to \mathrm{Clo}(\mathbf{A})$ be the natural homomorphism and let $\phi = \forall x_1, \dots, x_n \colon s = t$ be an identity that holds in $\mathbf{B}$. Then $s^{\mathbf{B}} = t^{\mathbf{B}}$ and hence $s^{\mathbf{A}} = \xi(s^{\mathbf{B}}) = \xi(t^{\mathbf{B}}) = t^{\mathbf{A}}$, which shows that $\mathbf{A} \models \forall x_1, \dots, x_n \colon s = t$.

(2) $\Rightarrow$ (1). Let $s(x_1, \dots, x_n), t(x_1, \dots, x_n)$ be τ-terms such that $s^{\mathbf{B}} = t^{\mathbf{B}}$. Then $\mathbf{B} \models \forall x_1, \dots, x_n \colon s(x_1, \dots, x_n) = t(x_1, \dots, x_n)$ and by assumption we have that $\mathbf{A} \models \forall x_1, \dots, x_n \colon s(x_1, \dots, x_n) = t(x_1, \dots, x_n)$ which shows that $s^{\mathbf{A}} = t^{\mathbf{A}}$, so the natural homomorphism from $\mathrm{Clo}(\mathbf{B})$ to $\mathrm{Clo}(\mathbf{A})$ exists. $\dashv$

The equivalence between (1) and (4) in Theorem 6.5.10 is relevant in the study of the complexity of CSPs since (4) is related to our most important tool to prove NP-hardness (see Theorem 6.3.10), and since (1) is the universal-algebraic property that will be used in the following (cf. Theorem 6.6.4 below). In Section 9.5.2, we will present a generalisation of the equivalence between (1) and (4) for algebras $\mathbf{B}$ with an infinite domain, replacing clone homomorphisms by uniformly continuous clone homomorphisms, or even by continuous clone homomorphisms if $\mathbf{B}$ is oligomorphic.

For easy reference we also spell out a consequence of this result for operation clones rather than algebras.

COROLLARY 6.5.11. *Let* $\mathcal{A}$ *and* $\mathcal{B}$ *be operation clones. Then there is a homomorphism from* $\mathcal{A}$ *onto* $\mathcal{B}$ *if and only if there are algebras* $\mathbf{A}$ *and* $\mathbf{B}$ *with the same signature such that* $\mathrm{Clo}(\mathbf{A}) = \mathcal{A}$ *and* $\mathrm{Clo}(\mathbf{B}) = \mathcal{B}$ *and* $\mathbf{A} \in \mathrm{HSP}(\mathbf{B})$. *If* $\mathbf{A}$ *and* $\mathbf{B}$ *are finite, then* $\mathbf{A} \in \mathrm{HSP}(\mathbf{B})$ *can be replaced by* $\mathbf{A} \in \mathrm{HSP}^{\mathrm{fin}}(\mathbf{B})$.

The setting of abstract clones is suitable for applying the compactness theorem, as demonstrated in the proof of the following lemma.

DEFINITION 6.5.12. An operation clone $\mathcal{C}$ *satisfies* a set Σ of identities over some signature τ if there exists a τ-algebra $\mathbf{A}$ such that $\mathbf{A} \models \Sigma$ and $\mathrm{Clo}(\mathbf{A}) \subseteq \mathcal{C}$.

LEMMA 6.5.13. *Let* $\mathcal{D}$ *be an operation clone over a finite domain and let* Σ *be a set of identities. Then* $\mathcal{D}$ *satisfies* Σ *if and only if it satisfies all finite subsets of* Σ.

PROOF. The forward implication is trivial. To prove the backwards implication, we introduce a constant symbol c_f of rank k for each operation symbol

f of arity k that appears in Σ. Let T be the first-order theory of $\mathfrak{D}$ in the signature of abstract clones. Let S be the set of all first-order sentences of the form $\psi^\dagger(c_{f_1},\ldots,c_{f_n})$ where ψ is an identity from Σ that contains the operation symbols $f_1,\ldots,f_n$ (recall Definition 6.5.7). If $\mathfrak{D}$ satisfies all finite subsets of Σ, then Lemma 6.5.8 implies that for all finite subsets F of S the theory $T \cup F$ is satisfiable. By the compactness theorem (Theorem 2.1.6) it follows that $T \cup S$ has a model $\mathbf{M}$. Consider the restriction of $\mathbf{M}$ to the elements that lie in $\bigcup_{i\in\mathbb{N}} M^{(i)}$, and consider the reduct of this restriction in the signature of abstract clones. Since $\mathbf{M}$ satisfies T and each $M^{(i)}$ is finite, we obtain a clone that is isomorphic to $\mathfrak{D}$. Let $\mathbf{A}$ be the algebra such that $f^{\mathbf{A}} = c_f^{\mathbf{M}}$ for all function symbols that appear in Σ. Then $\mathrm{Clo}(\mathbf{A}) \subseteq \mathfrak{D}$ and $\mathbf{M} \models S$ implies that $\mathfrak{D}$ satisfies Σ. ⊣

Clearly, if there is a clone homomorphism from $\mathbf{C}$ to $\mathbf{D}$, then every primitive positive clone sentence that holds in $\mathbf{C}$ also holds in $\mathbf{D}$.

COROLLARY 6.5.14. *Let* $\mathbf{C}$ *and* $\mathbf{D}$ *be clones. If* $\mathbf{D}$ *is the clone of a finite algebra, then there is a clone homomorphism from* $\mathbf{C}$ *to* $\mathbf{D}$ *if and only if every primitive positive sentence that holds in* $\mathbf{C}$ *also holds in* $\mathbf{D}$.

PROOF. We only have to prove the backwards direction. By Theorem 6.5.4 we may assume that there exists an algebra such that $\mathrm{Clo}(\mathbf{A}) = \mathbf{C}$. Let Σ be the set of all identities that hold in $\mathbf{A}$. By Lemma 6.5.8 there exists for every finite subset Δ of Σ a primitive positive clone sentence $\psi^\dagger(f_1,\ldots,f_n)$, where $f_1,\ldots,f_n$ are the function symbols that appear in Δ so that a τ-algebra $\mathbf{B}$ satisfies Δ if and only if $\mathrm{Clo}(\mathbf{B}) \models \psi^\dagger(f_1^{\mathbf{B}},\ldots,f_n^{\mathbf{B}})$. Hence, $\mathrm{Clo}(\mathbf{A}) \models \psi^\dagger(f_1^{\mathbf{A}},\ldots,f_n^{\mathbf{A}})$, and by assumption $\mathbf{D} \models \exists x_1,\ldots,x_n \colon \psi^\dagger(x_1,\ldots,x_n)$. This in turn implies that $\mathbf{D}$ satisfies Δ. Lemma 6.5.13 then shows that $\mathbf{D}$ satisfies Σ, i.e., there exists an algebra $\mathbf{B}$ such that $\mathbf{B} \models \Sigma$ and $\mathrm{Clo}(\mathbf{B}) = \mathbf{D}$. We may now define the homomorphism ξ from $\mathbf{C}$ to $\mathbf{D}$: for every $f \in C$ there exists a term t such that $t^{\mathbf{A}} = f$, and we define $\xi(f) := t^{\mathbf{B}}$. ⊣

We close this section with a simple lemma which can be useful when we want to verify that a given map is a clone homomorphism.

LEMMA 6.5.15. *Let* $\mathbf{C}$ *and* $\mathbf{D}$ *be clones and let* $\xi\colon \mathbf{C} \to \mathbf{D}$ *be a map which preserves arities, projections, and for all* $i, m, n \in \mathbb{N}$ *the formulas of the form*

$$f(\mathrm{pr}_1^n,\ldots,\mathrm{pr}_i^n, g(\mathrm{pr}_{i+1}^n,\ldots,\mathrm{pr}_m^n), \mathrm{pr}_{m+1}^n,\ldots,\mathrm{pr}_n^n) = h.$$

Then ξ *is a clone homomorphism.*

PROOF. We need to show that ξ also preserves formulas of the form

$$f(g_1,\ldots,g_k) = h.$$

Let n be the rank of $g_1,\ldots,g_k$ and h. We illustrate the idea of the proof for $k = 2$. Put

$$h_1 := f(\mathrm{pr}_1^{n+1}, g_2(\mathrm{pr}_2^{n+1},\ldots,\mathrm{pr}_{n+1}^{n+1}))$$

$$h_2 := h_1(g_1(\mathrm{pr}_1^{2n}, \dots, \mathrm{pr}_n^{2n}), \mathrm{pr}_{n+1}^{2n}, \dots, \mathrm{pr}_{2n}^{2n})$$

and note that

$$\begin{aligned}
&h_2(\mathrm{pr}_1^n, \dots, \mathrm{pr}_n^n, \mathrm{pr}_1^n, \dots, \mathrm{pr}_n^n)\\
&= h_1(g_1(\mathrm{pr}_1^{2n}, \dots, \mathrm{pr}_n^{2n}), \mathrm{pr}_{n+1}^{2n}, \dots, \mathrm{pr}_{2n}^{2n})(\mathrm{pr}_1^n, \dots, \mathrm{pr}_n^n, \mathrm{pr}_1^n, \dots, \mathrm{pr}_n^n)\\
&= h_1(g_1, \mathrm{pr}_1^n, \dots, \mathrm{pr}_n^n)\\
&= f(\mathrm{pr}_1^{n+1}, g_2(\mathrm{pr}_2^{n+1}, \dots, \mathrm{pr}_{n+1}^{n+1}))(g_1, \mathrm{pr}_1^n, \dots, \mathrm{pr}_n^n)\\
&= f(\mathrm{pr}_1^{n+1}(g_1, \mathrm{pr}_1^n, \dots, \mathrm{pr}_n^n), g_2(\mathrm{pr}_2^{n+1}, \dots, \mathrm{pr}_{n+1}^{n+1})(g_1, \mathrm{pr}_1^n, \dots, \mathrm{pr}_n^n))\\
&= f(g_1, g_2(\mathrm{pr}_2^{n+1}(g_1, \mathrm{pr}_1^n, \dots, \mathrm{pr}_n^n), \dots, \mathrm{pr}_{n+1}^{n+1}(g_1, \mathrm{pr}_1^n, \dots, \mathrm{pr}_n^n)))\\
&= f(g_1, g_2(\mathrm{pr}_1^n, \dots, \mathrm{pr}_n^n)) = f(g_1, g_2).
\end{aligned}$$

Hence, $f(g_1, g_2) = h$ is equivalent to

$$\begin{aligned}
\exists h_1, h_2 (h_1 &= f(\mathrm{pr}_1^{n+1}, g_2(\mathrm{pr}_2^{n+1}, \dots, \mathrm{pr}_{n+1}^{n+1}))\\
\wedge\, h_2 &= h_1(g_1(\mathrm{pr}_1^{2n}, \dots, \mathrm{pr}_n^{2n}), \mathrm{pr}_{n+1}^{2n}, \dots, \mathrm{pr}_{2n}^{2n})\\
\wedge\, h &= h_2(\mathrm{pr}_1^n, \dots, \mathrm{pr}_n^n, \mathrm{pr}_1^n, \dots, \mathrm{pr}_n^n))
\end{aligned}$$

and this formula is preserved by ξ by assumption. The general case for arbitrary k can be shown similarly. ⊣

6.5.4. Applications and examples. Birkhoff's theorem is important for the study of constraint satisfaction problems since it can be used to transform the 'negative' condition of not interpreting primitively positively certain finite structures into a 'positive' condition of having polymorphisms satisfying non-trivial identities, as we will see in the following corollary. Section 6.6 presents an application of this philosophy.

COROLLARY 6.5.16. *Let $\mathfrak{A}$ and $\mathfrak{B}$ be relational structures. Then* (1) ⇒ (2) ⇒ (3):

1. $\mathfrak{A} \in \mathrm{I}(\mathfrak{B})$.
2. *There is a clone homomorphism from* $\mathrm{Pol}(\mathfrak{B})$ *to* $\mathrm{Pol}(\mathfrak{A})$.
3. *Every primitive positive clone sentence which holds in* $\mathrm{Pol}(\mathfrak{B})$ *also holds in* $\mathrm{Pol}(\mathfrak{A})$.

If $\mathfrak{A}$ and $\mathfrak{B}$ are finite structures, then the converse implications hold as well. In particular, if $\mathfrak{A} \notin \mathrm{I}(\mathfrak{B})$, then there is a sentence in the signature of abstract clones which holds in $\mathrm{Pol}(\mathfrak{A})$ *but not in* $\mathrm{Pol}(\mathfrak{B})$.

PROOF. Let $\mathbf{B}$ be a polymorphism algebra of $\mathfrak{B}$. If $\mathfrak{A} \in \mathrm{I}(\mathfrak{B})$ then there exists an algebra $\mathbf{A} \in \mathrm{Exp}\,\mathrm{HSP}^{\mathrm{fin}}(\mathbf{B})$ such that $\mathrm{Clo}(\mathbf{A}) = \mathrm{Pol}(\mathfrak{A})$ (see Theorem 6.3.7). By Theorem 6.5.10, (3) ⇒ (1), there exists a clone homomorphism from $\mathrm{Clo}(\mathbf{B}) = \mathrm{Pol}(\mathfrak{B})$ to $\mathrm{Clo}(\mathbf{A}) = \mathrm{Pol}(\mathfrak{A})$. The implication (2) ⇒ (3) and its converse if $\mathfrak{A}$ is finite follows from Lemma 6.5.13.

Now suppose that $\mathfrak{A}$ and $\mathfrak{B}$ are finite structures, and suppose that every primitive positive sentence that holds in $\text{Pol}(\mathfrak{B})$ also holds in $\text{Pol}(\mathfrak{A})$. By the contrapositive of Lemma 6.5.13 from the previous section there is a clone homomorphism $\xi\colon \text{Pol}(\mathfrak{B}) \to \text{Pol}(\mathfrak{A})$. Let $\mathbf{A}$ be the algebra with domain A and the same signature τ as $\mathfrak{B}$ where $f \in \tau$ denotes $\xi(f^{\mathbf{B}})$. That is, ξ is the natural homomorphism from $\text{Clo}(\mathbf{B}) = \text{Pol}(\mathfrak{B})$ to $\text{Clo}(\mathbf{A}) \subseteq \text{Pol}(\mathfrak{A})$, and the implication from (1) to (4) in Theorem 6.5.10 implies that $\mathbf{A} \in \text{HSP}^{\text{fin}}(\mathbf{B})$. So there exists an algebra $\mathbf{A}' \in \text{Exp}\,\text{HSP}^{\text{fin}}(\mathbf{B})$ such that $\text{Clo}(\mathbf{A}') = \text{Pol}(\mathfrak{A})$ and so item (3) in Theorem 6.3.7 implies that $\mathfrak{A} \in \text{I}(\mathfrak{B})$. ⊣

The situation in which $\mathfrak{A}$ and $\mathfrak{B}$ are not finite, but countable and ω-categorical will be treated in Corollary 9.5.20. Already the easy first part of Corollary 6.5.16 has many applications, illustrated by the following example.

Example 6.5.17. The structure K_3 does not have a primitive positive interpretation in $\mathfrak{B} := (\mathbb{Q}; \{(x, y, z) \mid x > y \vee x > z\})$. The reason is that $\text{Pol}(\mathfrak{B})$ contains the symmetric operation $(x, y) \mapsto \min(x, y)$, but $\text{Pol}(K_3)$ does not contain a binary symmetric operation, as we have seen in Proposition 6.1.43.

The tractability conjecture for finite-domain constraint satisfaction can be formulated in terms of abstract clones; for this we need the following definition.

DEFINITION 6.5.18. We write **Proj** (pronounced *clone of projections*) for the abstract clone that just contains the projections and for which $\text{pr}^2_1 \neq \text{pr}^2_2$; this condition characterises **Proj** uniquely up to isomorphism.

Note that if $\mathfrak{C}$ is an operation clone that contains only projections then $\mathfrak{C}$ is isomorphic to **Proj** (as an abstract clone) if and only if its domain has size at least two. The following follows immediately from Corollary 6.2.2.

PROPOSITION 6.5.19. $\text{Pol}(\{0, 1\}; \text{1IN3})$ *is isomorphic to* **Proj**.

A set of identities is called *trivial* if it is satisfied by some algebra $\mathbf{A}$ whose clone is isomorphic to **Proj**, and *non-trivial* otherwise.

COROLLARY 6.5.20. *Let* $\mathbf{B}$ *be an algebra. Then the following are equivalent.*

1. *All identities satisfied by* $\mathbf{B}$ *are trivial;*
2. $\text{HSP}(\mathbf{B})$ *contains a 2-element algebra all of whose operations are projections;*
3. $\text{Clo}(\mathbf{B})$ *maps homomorphically to* **Proj**.

Moreover, if the polymorphism algebra of a finite structure $\mathfrak{B}$ *satisfies the above conditions, then* $\mathfrak{B}$ *has a finite-signature reduct* $\mathfrak{B}'$ *such that* $\text{CSP}(\mathfrak{B}')$ *is NP-hard.*

PROOF. (1) $\Rightarrow$ (3). Suppose that all identities satisfied by $\mathbf{B}$ are trivial, i.e., they are satisfied by some algebra $\mathbf{A}$ with $\text{Clo}(\mathbf{A}) = \mathbf{Proj}$. Then Theorem 6.5.10 implies that $\mathbf{A} \in \text{HSP}(\mathbf{B})$.

(2) $\Rightarrow$ (3). Let $\mathbf{A}$ be a 2-element algebra all of whose operations are projections; so $\text{Clo}(\mathbf{A})$ is isomorphic to **Proj**. Theorem 6.5.10 implies that if $\mathbf{A} \in \text{HSP}(\mathbf{B})$, then there exists a homomorphism from $\text{Clo}(\mathbf{B})$ to $\text{Clo}(\mathbf{A})$.

(3) ⇒ (1). If there is a homomorphism ξ: Clo(**B**) → **Proj**, then every identity that holds in **B** also holds in **Proj**, so every identity that holds in **B** is trivial.

The second condition is one of the equivalent conditions from Theorem 6.3.10, so the last statement follows from that theorem. ⊣

Corollary 6.5.20 shows that for finite structures $\mathfrak{B}$ the equivalent conditions from Theorem 6.3.10 (which imply that CSP($\mathfrak{B}$) is NP-hard) correspond to a property of the abstract polymorphism clone of $\mathfrak{B}$.

6.6. Idempotent algebras and Taylor terms

This section studies idempotent operation clones without a homomorphism to **Proj**, the clone of projections on an at least two-element set. A fundamental result about such clones is that they contain a *Taylor operation*. This has been improved later to the existence of operations that satisfy even stronger identities than the identities for Taylor operations, leading to several interesting equivalent characterisations of the border between polynomial-time tractable and NP-complete finite-domain CSPs. Many of these results have not yet found analogues over infinite domains. The classical result of Taylor also holds for operation clones on infinite sets. However, the assumption of idempotence limits the applicability of this result to oligomorphic clones, because an oligomorphic clone over an infinite set is certainly never idempotent (it may or may not contain interesting idempotent operations, though).

The results presented in this section are in fact about abstract clones; note that an operation clone is idempotent if and only if its abstract clone satisfies

$$\mathrm{comp}_1^k(f, \mathrm{pr}_1^1, \dots, \mathrm{pr}_1^1) = \mathrm{pr}_1^1$$

for all its k-ary operations f, for all $k \in \mathbb{N}$. However, we present the results using algebras and terms because this makes the notation more intuitive.

DEFINITION 6.6.1 (Taylor identities). A finite set of identities $\phi_1, \dots, \phi_n$, for $n \geq 2$, is called a set of *Taylor identities* if ϕ_i is of the form

$$\forall x, y \colon f(z_1, \dots, z_n) = f(z_1', \dots, z_n')$$

where $z_1, \dots, z_n, z_1', \dots, z_n'$ are variables from $\{x, y\}$ with $z_i \neq z_i'$. An operation $f: B^n \to B$, for $n \geq 2$, is called a *Taylor operation* if it satisfies some Taylor identities.

Examples of Taylor operations are constant operations, binary commutative operations, majority operations, and Maltsev operations. We do not insist on idempotence for Taylor operations. Note that an n-ary operation f is Taylor if and only if it satisfies a set of n equations that can be written as

$$f\begin{pmatrix} x & ? & ? & \cdots & ? \\ ? & x & ? & & \vdots \\ \vdots & ? & \ddots & \ddots & \vdots \\ \vdots & & \ddots & x & ? \\ ? & \cdots & \cdots & ? & x \end{pmatrix} = f\begin{pmatrix} y & ? & ? & \cdots & ? \\ ? & y & ? & & \vdots \\ \vdots & ? & \ddots & \ddots & \vdots \\ \vdots & & \ddots & y & ? \\ ? & \cdots & \cdots & ? & y \end{pmatrix}$$

where f is applied row-wise and ? stands for either x or y.

DEFINITION 6.6.2 (Taylor terms). Let **B** be a τ-algebra. A *Taylor term* of **B** is a τ-term $t(x_1, \ldots, x_n)$, for $n \geq 2$, such that $t^{\mathbf{B}}$ is a Taylor operation.

We make the same convention for other types of operations; e.g., a *majority term* over an algebra **B** is a term that denotes a majority operation in **B**. The following theorem goes back to Taylor (Corollary 5.3 in [335]); we give a slightly expanded presentation of the proof of Lemma 9.4 in [205]).

DEFINITION 6.6.3. For x of rank $l \in \mathbb{N}$ and y of rank $m \in \mathbb{N}$, the *star product* $x * y$ is defined as

$$\mathrm{comp}^l_{ml}(x, \mathrm{comp}^m_{ml}(y, \mathrm{pr}^{ml}_1, \ldots, \mathrm{pr}^{ml}_m), \ldots, \mathrm{comp}^m_{ml}(y, \mathrm{pr}^{ml}_{(l-1)m+1}, \ldots, \mathrm{pr}^{ml}_{ml})).$$

THEOREM 6.6.4. *Let* **B** *be an idempotent algebra with signature* τ. *Then there is no homomorphism from* Clo(**B**) *to* **Proj** *if and only if* **B** *has a Taylor term.*

PROOF. To show the easy backward implication, suppose that t is a Taylor term, and suppose for contradiction that $\xi\colon \mathrm{Clo}(\mathbf{B}) \to \mathbf{Proj}$ is a clone homomorphism. By the definition of **Proj** we have $\xi(t^{\mathbf{B}}) = \mathrm{pr}^n_l$ for some $l \leq n$. By assumption, **B** satisfies

$$\forall x, y\colon t(z_1, \ldots, z_n) = t(z'_1, \ldots, z'_n) \tag{4}$$

for $z_1, \ldots, z_n, z'_1, \ldots, z'_n \in \{x, y\}$ such that $z_l \neq z'_l$. Put differently, Clo(**B**) satisfies

$$\mathrm{comp}^n_2(t^{\mathbf{B}}, \mathrm{pr}^2_{i_1}, \ldots, \mathrm{pr}^2_{i_n}) = \mathrm{comp}^n_2(t^{\mathbf{B}}, \mathrm{pr}^2_{j_1}, \ldots, \mathrm{pr}^2_{j_n}) \tag{5}$$

for $i_1, \ldots, i_n, j_1, \ldots, j_n \in \{1, 2\}$ such that $i_l = 1$ if $z_l = x$ and $i_l = 2$ otherwise, $j_l = 1$ if $z'_l = x$ and $j_l = 2$ otherwise, and $i_l \neq j_l$. Since $\xi(t^{\mathbf{B}}) = \mathrm{pr}^n_l$ we therefore obtain that $\mathrm{pr}^2_1 = \mathrm{pr}^2_2$, which does not hold in **Proj**, a contradiction.

To show the forward implication, suppose that Clo(**B**) does not map homomorphically to **Proj**. Then Lemma 6.5.13 implies that there is a primitive positive sentence over the signature of clones that holds in Clo(**B**) but not in **Proj**. Note that by introducing new existentially quantified variables we can assume that this sentence is of the form $\exists u_1, \ldots, u_r\colon \phi$ where ϕ is a conjunction of atoms of the form $y = \mathrm{comp}^m_l(x_0, x_1, \ldots, x_m)$ or of the form $y = \mathrm{pr}^m_l$

for $y, x_0, x_1, \ldots, x_m \in \{u_1, \ldots, u_r\}$ and $l, m \in \mathbb{N}$. For example, the equation $\text{comp}_2^2(x_0, \text{pr}_1^2, \text{pr}_2^2) = \text{comp}_2^2(x_0, \text{pr}_2^2, \text{pr}_1^2)$ is equivalent to

$$\begin{aligned}\exists x_1, x_2, y\ (y &= \text{comp}_2^2(x_0, x_1, x_2)\\ &\wedge\ y = \text{comp}_2^2(x_0, x_2, x_1)\\ &\wedge\ x_1 = \text{pr}_1^2\\ &\wedge\ x_2 = \text{pr}_2^2).\end{aligned}$$

Note that $\text{Clo}(\mathbf{B})$ satisfies

$$(y = \text{comp}_l^{ml}(x * y, \text{pr}_1^{ml}, \ldots, \text{pr}_l^{ml}, \text{pr}_1^{ml}, \ldots, \text{pr}_l^{ml}, \ldots, \text{pr}_1^{ml}, \ldots, \text{pr}_l^{ml})) \quad (6)$$

and

$$(x = \text{comp}_m^{ml}(x * y, \text{pr}_1^{ml}, \ldots, \text{pr}_1^{ml}, \text{pr}_2^{ml}, \ldots, \text{pr}_2^{ml}, \ldots, \text{pr}_l^{ml}, \ldots, \text{pr}_l^{ml})) \quad (7)$$

since $\mathbf{B}$ is idempotent. Define

$$u := u_1 * (u_2 * (\cdots (u_{r-1} * u_r) \cdots)). \quad (8)$$

Observe that for each $i \in \{1, \ldots, r\}$ we can obtain u_i from u by composing u with projections. In order to formalise this, we need a compact notation for strings of arguments consisting of projection constants. In this notation, (6) reads as $y = \text{comp}_l^n(x * y, (1, \ldots, l)^m)$, and (7) reads as $x = \text{comp}_m^n(x * y, 1^m, \ldots, l^m)$. Similarly, if $k_i \in \mathbb{N}$ is the arity of u_i and $k := k_1 \cdots k_r$ we have

$$u_i = \text{comp}_{k_i}^k(u, \bar{p}^i)$$
$$\text{where } \bar{p}^i := (1^{k_1 \cdots k_{i-1}}, \ldots, k_i^{k_1 \cdots k_{i-1}})^{k_{i+1} \cdots k_n}) \in \{1, \ldots, k_i\}^k.$$

Let $n = k^2$. Then every term of the form u_j can be written as $\text{comp}_{k_j}^n(u * u, \bar{q}^{u_j})$ for $\bar{q}^{u_j} \in \{1, \ldots, k_j\}^n$ obtained by k times concatenating $\bar{p}^j$ with itself. Moreover, every term t of the form $\text{comp}_{k_j}^{k_i}(u_i, u_{i_1}, \ldots, u_{i_{k_i}})$ can be written as $\text{comp}_{k_j}^n(u * u, \bar{q}^t)$ for $\bar{q}^t \in \{1, \ldots, k_j\}^n$ obtained from $\bar{p}^i$ by replacing character $l \leq k_j$ by the string $\bar{p}^{i_l}$. In this way, every conjunct of ϕ of the form

$$u_j = \text{comp}_{k_j}^{k_i}(u_i, u_{i_1}, \ldots, u_{i_{k_i}})$$

can be written in the form

$$\text{comp}_l^n(u * u, \bar{q}^{u_j}) = \text{comp}_l^n(u * u, \bar{q}^t)$$

for appropriate $\bar{q}^{u_j}, \bar{q}^t \in \{1, \ldots, k_j\}^n$. Conjuncts of ϕ of the form $u_j = \text{pr}_l^{k_j}$ can be rewritten similarly. Let ψ be the conjunction of all these equations; note that ϕ implies ψ. Let θ be the formula obtained from ψ by replacing each occurrence of $u * u$ by a variable symbol f. Note that $\text{Clo}(\mathfrak{B}) \models \exists f : \theta$. It suffices to show that every $f \in \text{Clo}(\mathbf{B})$ that satisfies θ is a Taylor operation.

Suppose for contradiction that for $\ell \in \{1, \dots, n\}$ there are no $\bar{v}, \bar{v}' \in \{1,2\}^n$ with $\bar{v}_\ell \neq \bar{v}'_\ell$ such that $\text{comp}^n_2(f, \bar{v}) = \text{comp}^n_2(f, \bar{v}')$. It is easy to see that this implies that for all $m \leq n$

$$\text{if } \text{comp}^n_m(f, \bar{v}) = \text{comp}^n_m(f, \bar{v}') \text{ then } \bar{v}_\ell = \bar{v}'_\ell. \tag{9}$$

We claim that the assignment ρ that maps for every $s \in \{1, \dots, r\}$ the variable u_s to $\text{pr}^{k_s}_{o_s}$ for $o_s := \bar{q}^{u_s}_\ell$ satisfies all conjuncts of ϕ, contradicting our assumptions. First note that o_s equals the ℓ_1-st entry of $\bar{p}^s$ where $\ell_1 \in \{1, \dots, k\}$ is such that $\ell = (\ell_1 - 1)k + \ell_2$ for some $\ell_2 \in \{1, \dots, k\}$. To prove the claim, consider a conjunct of ϕ of the form

$$u_j = t \text{ for } t = \text{comp}^{k_i}_{k_j}(u_i, u_{i_1}, \dots, u_{i_{k_i}}).$$

By construction, θ contains the conjunct

$$\text{comp}(f, \bar{q}^{u_j}) = \text{comp}(f, \bar{q}^t).$$

Therefore, (9) implies that

$$\bar{q}^{u_j}_\ell = \bar{q}^t_\ell. \tag{10}$$

Then the assignment ρ satisfies the conjunct ϕ since

$$\begin{aligned}
\text{comp}^{k_i}_{k_j}(\rho(u_i), \rho(u_{i_1}), \dots)) &= \text{comp}^{k_i}_{k_j}(\text{pr}^{k_i}_{o_i}, \text{pr}^{k_{i_1}}_{o_{i_1}}, \dots) \\
&= \text{pr}^{k_j}_{\bar{q}^t_\ell} && \text{(by definition of } \bar{q}^t) \\
&= \text{pr}^{k_j}_{\bar{q}^{u_j}_\ell} && \text{(by 10))} \\
&= \rho(u_j).
\end{aligned}$$

The verification that ρ also satisfies conjuncts of ϕ of the form $u_j = \text{pr}^{k_j}_l$ is similar. ⊣

For idempotent algebras **A** there is yet another characterisation of the existence of clone homomorphisms to **Proj**, due to Bulatov and Jeavons [115].

THEOREM 6.6.5. *Let* **B** *be an idempotent algebra. Then* $\text{HSP}^{\text{fin}}(\mathbf{B})$ *contains an algebra with at least two elements all of whose operations are projections if and only if* $\text{HS}(\mathbf{B})$ *does.*

PROOF. Suppose that $\mathbf{C} \in S(\mathbf{B}^d)$ for some $d \in \mathbb{N}$ has a congruence K with at least two classes such that all operations of $\mathbf{A} := \mathbf{C}/K$ are projections.

Let $I \subseteq \{1, \dots, d\}$ be a maximal set such that for each $i \in I$ there exists $b_i \in B$ such that the set

$$C\big((b_i)_{i \in I}\big) := \{c \in C \mid c_i = b_i \text{ for all } i \in I\}$$

is not contained in a class of K. Note that such a set exists because for $I = \emptyset$ we have that $C\big((b_i)_{i \in I}\big) = C$ is not contained in one class of K since $\mathbf{C}/K$ has at least two elements.

Without loss of generality, we may assume that $\{1,\dots,d\} \setminus I = \{1,\dots,k\}$. Since $\mathbf{B}$ is idempotent, $C\big((b_i)_{i\in I}\big)$ is the domain of a subalgebra $\mathbf{C}'$ of $\mathbf{C}$. Let $K' := K \cap (C')^2$. All operations of $\mathbf{A}' := \mathbf{C}'/K'$ are restrictions of operations in $\mathbf{A}$ and hence projections. Let

$$B' := \{b \in B \mid (b_1,\dots,b_{k-1},b,b_{k+1},\dots,b_d) \in C'\}.$$

Then B' is the domain of a subalgebra $\mathbf{B}'$ of $\mathbf{B}$. The image

$$K'' := \{(c_k,e_k) \in (B')^2 \mid (c,e) \in K'\}$$

of K' under the k-th projection is a congruence of $\mathbf{B}'$, and it must have more than one equivalence class: otherwise there would be $b_1,\dots,b_d,b'_1,\dots,b'_{k-1} \in B$ such that the tuples $(b_1,\dots,b_{k-1},b_k,b_{k+1},\dots,b_d)$ and $(b'_1,\dots,b'_{k-1},b_k,b_{k+1},\dots,b_d)$ are in different K'-classes. But then $I \cup \{k\}$ is such that $C\big((b_i)_{i\in I\cup\{k\}}\big)$ is not contained in one class of K, contradicting the maximality of I. Therefore, $\mathbf{B}'/K'' \in \mathrm{HS}(\mathbf{B})$ is an algebra with at least two elements of all whose operations are projections. ⊣

Since the size of the algebras in $\mathrm{HS}(\mathbf{B})$ is bounded by the size of $\mathbf{B}$, this leads to an algorithm that decides whether a given finite structure $\mathfrak{B}$ satisfies the equivalent conditions in Theorem 6.6.4. Another effective condition can be found in Section 6.8.2.

It is not hard to see that Theorem 6.6.5 is false for oligomorphic $\mathbf{B}$, even in the case where the invertible elements of $\mathrm{Clo}(\mathbf{B})$ are dense in $\mathrm{Clo}(\mathbf{B})^{(1)}$. Consider for example a polymorphism algebra $\mathbf{B}$ of $(\mathbb{N}; P^3_{\mathbb{N}}, \neq)$ (Definition 6.1.16), which has no non-trivial subalgebras and homomorphic images, but where $\mathrm{HS}(\mathbf{B}^2)$ *does* contain a two-element algebra all of whose operations are projections. The following corollary summarises the consequences for finite idempotent algebras.

COROLLARY 6.6.6. *For a finite idempotent algebra* $\mathbf{B}$, *the following are equivalent.*

1. *Every 2-element algebra in* $\mathrm{HSP}(\mathbf{B})$ *contains an essential operation.*
2. *Every 2-element algebra in* $\mathrm{HSP}^{\mathrm{fin}}(\mathbf{B})$ *contains an essential operation.*
3. *Every algebra with at least two elements in* $\mathrm{HS}(\mathbf{B})$ *contains an essential operation.*
4. *There is no homomorphism from* $\mathrm{Clo}(\mathbf{B})$ *to* **Proj**.
5. $\mathbf{B}$ *satisfies a non-trivial identity.*
6. $\mathbf{B}$ *has a Taylor term.*

PROOF. (1) ⇔ (2): Theorem 6.5.1.

(2) ⇔ (3): Theorem 6.6.5.

(1) ⇔ (4) ⇔ (5): Corollary 6.5.20.

(5) ⇔ (6): Theorem 6.6.4. ⊣

DEFINITION 6.6.7. A sentence ϕ over the signature of abstract clones is called *trivial* if **Proj** $\models \phi$, and *non-trivial* otherwise.

For easy reference, we rephrase some of these results with structures and clones instead of algebras.

COROLLARY 6.6.8. *Let $\mathfrak{B}$ be a finite structure such that* $\mathrm{Pol}(\mathfrak{B})$ *is idempotent. Then the following are equivalent.*

1. *There exists a finite structure without primitive positive interpretation in $\mathfrak{B}$.*
2. $K_3 \notin \mathrm{I}(\mathfrak{B})$.
3. $(\{0,1\}; \mathrm{1IN3}) \notin \mathrm{I}(\mathfrak{B})$.
4. $\mathrm{I}(\mathfrak{B})$ *does not contain a structure with at least two elements all of whose polymorphisms are projections.*
5. *There is no homomorphism from* $\mathrm{Pol}(\mathfrak{B})$ *to* **Proj**.
6. $\mathrm{Pol}(\mathfrak{B})$ *satisfies a non-trivial sentence in the signature of abstract clones.*
7. $\mathrm{Pol}(\mathfrak{B})$ *contains a Taylor operation.*

PROOF. The equivalence of (1), (2), (3), and (4) has already been shown in Theorem 6.3.10. Let $\mathbf{B}$ be a polymorphism algebra of $\mathfrak{B}$. Theorem 6.3.10 also implies the equivalence to the statement that all two-element algebras in $\mathrm{HSP}^{\mathrm{fin}}(\mathbf{B})$ contain a two-element algebra all of whose operations are projections. This in turn is equivalent to the non-existence of a homomorphism from $\mathrm{Clo}(\mathbf{B}) = \mathrm{Pol}(\mathfrak{B})$ to **Proj**, to the existence of a non-trivial sentence that holds in $\mathrm{Clo}(\mathbf{B})$, and to the existence of a Taylor operation in $\mathrm{Clo}(\mathbf{B}) = \mathrm{Pol}(\mathfrak{B})$ (by Corollary 6.6.6) proving the equivalence of (4), (5), (6) and (7). ⊣

6.7. Minor-preserving maps and height-one identities

There are finite cores that do not interpret all finite structures primitively positively, but which do allow such interpretations modulo homomorphic equivalence (Example 6.7.1 below). This shows that clone homomorphisms from $\mathrm{Pol}(\mathfrak{B})$ to **Proj** are not a necessary condition for the NP-hardness of $\mathrm{CSP}(\mathfrak{B})$. In this section we present an algebraic condition on $\mathrm{Pol}(\mathfrak{B})$ which, for finite structures $\mathfrak{B}$, applies if and only if $\mathfrak{B}$ can interpret all finite structures up to homomorphic equivalence. A generalisation to countably infinite ω-categorical structures relies on concepts from topology and has to wait until Section 9.5.2.

Example 6.7.1 (from [29]). Consider the structure $\mathfrak{B} = (B; R, S)$ where

$$\begin{aligned} B &= \{1,2,3\} \times \{0,1\} \\ R &= \{((a,i),(b,j)) \mid i = j \wedge a \neq b\} \\ S &= \{((a,i),(b,j)) \mid i \neq j\}. \end{aligned}$$

The structure $\mathfrak{B}$ is a core since all maps from B to B that preserve R and S must be injective. Observe that for $c \in B$ chosen arbitrarily the substructure of $(B; R)$ induced on $\{x \in B \mid S(x, c)\}$ is isomorphic to K_3, and hence $(\mathfrak{B}, c)$ interprets K_3 primitively positively. It follows (Theorem 3.6.2) that $K_3 \in \mathrm{HI}(\mathfrak{B})$.

However, using the concepts from Section 6.5.4, one can show that $\mathfrak{B}$ does not interpret K_3 primitively positively. Indeed, note that the map defined by

$$\alpha((a, i)) := (a, 1 - i)$$

is an automorphism of $\mathfrak{B}$ and that

$$s((a,i),(b,j),(c,k)) := \begin{cases} (c,k) & \text{if } i = j \\ (a,i) & \text{if } i \neq j \end{cases}$$

is a ternary polymorphism of $\mathfrak{B}$. These maps satisfy the non-trivial identity

$$\forall x, y \colon s(x, x, y) = y = s(y, \alpha(y), x).$$

Therefore, the contraposition of (3) $\Rightarrow$ (1) in Corollary 6.5.20 shows that $\mathrm{Pol}(\mathfrak{B})$ has no homomorphism to **Proj**.

This example illustrates that we may want to consider from the algebraic perspective a reduction between CSPs that is more widely applicable than primitive positive interpretations; such a reduction is provided by the operator HI from Section 3.6. On the algebraic side, if $\mathbf{B}$ is the polymorphism algebra of $\mathfrak{B}$, then understanding $\mathrm{HI}(\mathfrak{B})$ amounts to understanding the class $\mathrm{Exp}\,\mathrm{Refl}\,\mathrm{P}^{\mathrm{fin}}(\mathbf{B})$ (Theorem 6.4.3 (2)) rather than $\mathrm{Exp}\,\mathrm{HSP}^{\mathrm{fin}}(\mathbf{B})$. It turns out that $\mathrm{Exp}\,\mathrm{Refl}\,\mathrm{P}^{\mathrm{fin}}(\mathbf{B})$ can be studied using $\mathrm{Clo}(\mathbf{B})$ and the existence of minor-preserving maps between clones in the manner of Section 6.5.

6.7.1. Minor-preserving maps. Let A, B be non-empty sets and $f : A^k \to B$ a k-ary function. A *minor of* f is an operation of the form $f(g_1, \ldots, g_n)$ where $g_1, \ldots, g_n$ are m-ary projections, for some $m \in \mathbb{N}$. A *minion on* (A, B) is a non-empty subset $\mathcal{M}$ of $\bigcup_{k \geq 1} A^k \to B$ which is closed under taking minors, i.e., which contains all minors of all operations in $\mathcal{M}$. Clearly, every operation clone is a minion (where $A = B$). Minions are not required to contain the projections.

DEFINITION 6.7.2. Let $\mathcal{M}_1, \mathcal{M}_2$ be minions. A map $\xi \colon \mathcal{M}_1 \to \mathcal{M}_2$ is called *minor-preserving* (or a *minion homomorphism*) if it maps functions from $\mathcal{M}_1$ to functions in $\mathcal{M}_2$ of the same arity, and if it preserves composition with projections, i.e., if

$$\xi(f(p_1, \ldots, p_n)) = \xi(f)(p_1, \ldots, p_n)$$

for all n-ary functions $f \in \mathcal{M}_1$ and projections $p_1, \ldots, p_n$ of equal arity.

We mention that minions and minor-preserving maps also play an important role in the emerging theory of *promise CSPs* [117]. Minor-preserving maps are not required to preserve the projections; we present an example of a minor-preserving map that does not preserve the projections.

Example 6.7.3. Let $\mathcal{C} := \mathrm{Pol}(\{0,1\}; \mathrm{NAE})$; note that $\mathcal{C}$ is the operation clone generated by $\neg\colon x \mapsto 1-x$ (Proposition 6.2.8). Then the map $\xi\colon \mathcal{C} \to \mathcal{C}$ that maps f to $\neg f$ is minor-preserving and does not preserve the projections.

However, in some cases minor-preserving maps must preserve projections.

PROPOSITION 6.7.4. *Let $\mathcal{M}$ be a minion and let $\mathcal{P}$ be an operation clone that just contains projections. Then every minor-preserving map $\xi\colon \mathcal{M} \to \mathcal{P}$ preserves all projections in $\mathcal{M}$.*

PROOF. Suppose that $\xi(\pi^k_i) = \pi^k_j$ for some $i \neq j$. Then

$$\begin{aligned}\pi^k_j = \xi(\pi^k_i) &= \xi(\pi^k_i(\pi^k_i, \dots, \pi^k_i)) \\ &= \xi(\pi^k_i)(\pi^k_i, \dots, \pi^k_i) = \pi^k_j(\pi^k_i, \dots, \pi^k_i) = \pi^k_i\end{aligned}$$

which is impossible unless $\mathcal{P}$ is an operation clone on a one-element set, where the statement holds trivially. ⊣

We present an important example from [20] of an ω-categorical structure $\mathfrak{B}$ such that $\mathrm{Pol}(\mathfrak{B})$ has a minor-preserving map to the clone of projections on $\{0,1\}$, but no clone homomorphism.

Example 6.7.5. Let $\mathfrak{A} = (\mathbb{A}; \wedge, \vee, -, 0, 1)$ be the countable atomless Boolean algebra (see Example 4.1.4 and Section 5.3). Note that $\mathfrak{A}$ is homomorphically equivalent to the two-element Boolean algebra, while its expansion $(\mathfrak{A}, \neq)$ is an ω-categorical model-complete core.

Also note that $\mathfrak{A}^2$ is again a countable atomless Boolean algebra, and hence there exists an isomorphism $e\colon \mathfrak{A}^2 \hookrightarrow \mathfrak{A}$ and an automorphism $\alpha \in \mathrm{Aut}(\mathfrak{A})$ such that $e(x,y) = \alpha e(y,x)$ for all $x, y \in \mathbb{A}$. Note that both e and α preserve $\neq$, and so $\mathrm{Pol}(\mathfrak{A}, \neq)$ satisfies

$$\exists e, \alpha\colon e = \mathrm{comp}(\alpha, \mathrm{comp}(e, \mathrm{pr}^2_2, \mathrm{pr}^2_1))$$

but **Proj** does not. Therefore, there is no clone homomorphism from $\mathrm{Pol}(\mathfrak{A}, \neq)$ to **Proj**. However, there exists a minor-preserving map from $\mathrm{Pol}(\mathfrak{A})$ (and in particular from $\mathrm{Pol}(\mathfrak{A}, \neq)$) to the clone of projections on a two-element set. To define this map, we pick an *ultrafilter U on* $\mathbb{A}$, i.e., a subset of $\mathbb{A}$ which is a maximal proper filter:

- $U \notin \{\emptyset, \mathbb{A}\}$,
- for $x, y \in U$ there is an element $z \in U$ such that $z \leq x$ and $z \leq y$,
- for every $x \in U$ and $y \in \mathbb{A}$ we have $y \in U$ whenever $x \leq y$, and
- U is maximal with these properties.

Observe that if $f \in \mathrm{Pol}(\mathfrak{A})$ is n-ary then exactly one of the elements

$$a_1 := f(1,0,\dots,0), a_2 := f(0,1,0,\dots,0), \dots, a_n := f(0,\dots,0,1)$$

is contained in U: this can be seen as follows.

- $a_i \wedge a_j = 0$ whenever $i \neq j$ since f preserves $\wedge$ and 0. Hence, if both a_i and a_j are in U, then $a_i \wedge a_j = 0$ must also be in U, and thus $U = \mathbb{A}$ in contradiction to the assumptions.
- $a_1 \vee \cdots \vee a_n = 1$ since f preserves $\vee$ and 1; hence, if none of the a_i is contained in U then U was not maximal.

Let $i \leq n$ be the unique index such that $a_i \in F$. We claim that $\xi(f) := \pi_i^n$ defines a minor-preserving map to **Proj**. Indeed, let $f \in \mathrm{Pol}(\mathfrak{A})$ be n-ary, let $p_1, \dots, p_n$ be projections of the same arity m, and let $i \in \{1, \dots, n\}$ be such that

$$a_i = f(b_i^n) \in U \quad \text{where } b_i^n := (\underbrace{0,\dots,0}_{i-1}, 1, \underbrace{0,\dots,0}_{n-i-1}).$$

Then $\xi(f) = \pi_i^n$. Let $k \in \{1, \dots, m\}$ be such that $p_i = \pi_k^m$. Then $\pi_k^m(b_k^m) = 1$ and hence

$$\begin{aligned} f(p_1,\dots,p_n)(b_k^m) &= f(p_1(b_k^m),\dots,p_n(b_k^m)) \\ &= f(p_1(b_k^m),\dots,p_{i-1}(b_k^m),1,p_{i+1}(b_k^m),\dots,p_n(b_k^m)) \in U \end{aligned}$$

showing that $\xi(f(p_1,\dots,p_n)) = \pi_k^m = p_i$. We conclude that

$$\xi(f)(p_1,\dots,p_n) = \pi_i^n(p_1,\dots,p_n) = p_i = \xi(f(p_1,\dots,p_n))$$

which shows that ξ is minor-preserving.

The following result from [29] provides a sufficient condition for a minor-preserving map to **Proj** to be a clone homomorphism (Theorem 6.7.7).

DEFINITION 6.7.6. Let **C** and **D** be clones, let $\xi\colon \mathbf{C} \to \mathbf{D}$ be a map, and let $S \subseteq C^{(1)}$ be a set of unary operations.

- *ξ preserves left composition with S* if for all $f \in C$ and $e \in S$
$$\xi(e \circ f) = \xi(e) \circ \xi(f).$$
- *ξ preserves right composition with S* if
$$\xi(f(e_1,\dots,e_n)) = \xi(f)(\xi(e_1),\dots,\xi(e_n))$$
for all $f \in C$ and $e_1, \dots, e_n \in S$.
- *ξ preserves composition with S* if it preserves left and right composition with S.

THEOREM 6.7.7 (Proposition 5.6 in [29]). *Let* **C** *be a clone and let* $\xi\colon \mathbf{C} \to$ **Proj** *be a minor-preserving map that preserves composition with* $C^{(1)}$. *Then* ξ *is a clone homomorphism.*

PROOF. First observe that if $f \in C$ depends only on its i-th argument, then f satisfies the identity

$$\forall x_1, \ldots, x_n, y_1, \ldots, y_n \colon f(x_1, \ldots, x_n) = f(y_1, \ldots, y_{i-1}, x_i, y_{i+1}, \ldots, y_n)$$

and the only projection satisfying this identity is pr_i^n. Since ξ is minor preserving, we deduce that $\xi(f) = \mathrm{pr}_i^n$.

Claim 1. ξ preserves identities of the form

$$\forall \bar{x} \colon f(g(x_{1,1}, \ldots, x_{1,m}), \ldots, g(x_{n,1}, \ldots, x_{n,m})) = h(x_{1,1}, \ldots, x_{n,m}).$$

Suppose that $\xi(f) = \mathrm{pr}_i^n$ and $\xi(g) = \mathrm{pr}_j^m$. We have to show that $\xi(h)$ satisfies the identity

$$\forall \bar{x} \colon \xi(h)(x_{1,1}, \ldots, x_{n,m}) = x_{i,j}.$$

Note that in $\mathbf{C}$ the following identity holds:

$$\begin{aligned}\forall x_1, \ldots, x_n \colon & h(x_1, \ldots, x_1, x_2, \ldots, x_2, \ldots, x_n, \ldots, x_n) \\ &= f(\hat{g}(x_1), \hat{g}(x_2), \ldots, \hat{g}(x_n)).\end{aligned}$$

Since ξ preserves right composition with $C^{(1)}$ and $\xi(\hat{g}) = \mathrm{pr}_1^1$, we obtain

$$\begin{aligned}\forall x_1, \ldots, x_n \colon & \xi(h)(x_1, \ldots, x_1, x_2, \ldots, x_2, \ldots, x_n, \ldots, x_n) \\ &= \xi(f)(x_1, x_2, \ldots, x_n) = x_i\end{aligned}$$

and it follows that for some $1 \leq l \leq m$

$$\forall \bar{x} \colon \xi(h)(x_{1,1}, \ldots, x_{n,m}) = x_{i,l}.$$

Similarly, from

$$\forall x_1, \ldots, x_m \colon h(x_1, \ldots, x_m, x_1, \ldots, x_m, \ldots, x_1, \ldots, x_m) = \hat{f}(g(x_1, \ldots, x_m))$$

and the assumption that ξ preserves left composition with $C^{(1)}$ we get that

$$\forall \bar{x} \colon \xi(h)(x_{1,1}, \ldots, x_{n,m}) = x_{k,j}$$

for some $1 \leq k \leq n$, proving Claim 1.

Claim 2. ξ preserves identities of the form

$$\forall \bar{x} \colon f(g(x_1, \ldots, x_m), x_{m+1}, \ldots, x_n) = h(x_1, \ldots, x_n).$$

Let $t(x_{1,1}, \ldots, x_{n,m}) := f(g(x_{1,1}, \ldots, x_{1,m}), \ldots, g(x_{n,1}, \ldots, x_{n,m}))$. We obtain

$$\begin{aligned}&\xi(f)(\xi(g)(x_1, \ldots, x_m), x_{m+1}, \ldots, x_n) \\ &= \xi(f)(\xi(g)(x_1, \ldots, x_m), \xi(g)(x_{m+1}, \ldots, x_{m+1}), \ldots, \xi(g)(x_n, \ldots, x_n)) \\ &= \xi(t)(x_1, \ldots, x_m, x_{m+1}, \ldots, x_{m+1}, \ldots, x_n, \ldots, x_n) \qquad \text{(Claim 1)} \\ &= \xi(h)(x_1, \ldots, x_n).\end{aligned}$$

The final equation holds because ξ preserves right composition with $C^{(1)}$ and hence preserves the identity

$$\forall \bar{x}\colon t(x_1,\dots,x_m,x_{m+1},\dots,x_{m+1},\dots,x_n,\dots,x_n)$$
$$= h(x_1,\dots,x_m,\hat{g}(x_m),\dots,\hat{g}(x_n)).$$

This concludes the proof of the claim. Analogously, ξ preserves identities of the form

$$\forall \bar{x}\colon f(x_1,\dots,x_k,g(x_{k+1},\dots,x_m),x_{m+1},\dots,x_n) = h(x_1,\dots,x_n).$$

The statement now follows from Lemma 6.5.15. ⊣

Minor-preserving maps that preserve right composition with invertible elements of $\mathcal{C}^{(1)}$ will play a role in Section 9.6, and minor-preserving maps that preserve left composition with $\mathcal{C}^{(1)}$ will be the topic of Section 10.1.

6.7.2. Birkhoff's theorem for height-one identities. Recall that an *identity* is a sentence of the form $\forall x_1,\dots,x_n\colon s = t$ where $s(x_1,\dots,x_n)$ and $t(x_1,\dots,x_n)$ are terms. A *height-one identity* is an identity where the involved terms have *height one*, i.e., each term involves exactly one function symbol. Some examples of properties that can be expressed as finite sets of height-one identities are

$$\forall x,y\colon f(x,y) = f(y,x) \qquad (f \text{ is commutative})$$
$$\forall x,y\colon f(x,x,y) = f(x,y,x) = f(y,x,x) = f(x,x,x) \qquad (f \text{ is a quasi majority})$$
$$\forall x,y\colon f(x,x,y) = f(y,x,x) = f(y,y,y) \qquad (f \text{ is quasi Maltsev})$$

and, more generally, all Taylor identities are finite sets of height-one identities. A non-example is furnished by the Maltsev identities $f(x,x,y) = f(y,x,x) = y$ because the term y involves no function symbol. Identities where each term involves *at most* one function symbol are called *linear*; so the Maltsev identities are an example of a set of linear identities. An example of a non-linear identity is the associativity law

$$\forall x,y,z\colon f(x,f(y,z)) = f(f(x,y),z).$$

If $\mathbf{A}$ is a τ-algebra, then we write $\mathrm{Minion}(\mathbf{A})$ for the smallest minion that contains $\{f^{\mathbf{A}} \mid f \in \tau\}$. If $\mathbf{A}$ and $\mathbf{B}$ are τ-algebras then there exists a minor-preserving map $\xi\colon \mathrm{Minion}(\mathbf{B}) \to \mathrm{Minion}(\mathbf{A})$ that maps $f^{\mathbf{B}}$ to $f^{\mathbf{A}}$ if and only if for all $f,g \in \tau$ of arity k and l and all m-ary projections $p_1,\dots,p_k,q_1,\dots,q_l$ we have that $f^{\mathbf{A}}(p_1,\dots,p_k) = g^{\mathbf{A}}(q_1,\dots,q_l)$ whenever $f^{\mathbf{B}}(p_1,\dots,p_k) = g^{\mathbf{B}}(q_1,\dots,q_k)$. If this map exists it must be surjective and we call it the *natural minor-preserving map from* $\mathrm{Minion}(\mathbf{B})$ *to* $\mathrm{Minion}(\mathbf{A})$. The following theorem is a variant of Birkhoff's theorem (Theorem 6.5.10) for height-one identities.

THEOREM 6.7.8 (cf. Proposition 5.3 of [27]). *Let* **A** *and* **B** *be* τ*-algebras such that* Minion(**A**) *and* Minion(**B**) *are operation clones. Then the following are equivalent.*

1. *The natural minor-preserving map from* Minion(**B**) *to* Minion(**A**) *exists.*
2. *All height-one identities that hold in* **B** *also hold in* **A**.
3. $\mathbf{A} \in \operatorname{Refl} \mathrm{P}(\mathbf{B})$.

Moreover, if A and B are finite then we can add the following to the list:

4. $\mathbf{A} \in \operatorname{Refl} \mathrm{P}^{\mathrm{fin}}(\mathbf{B})$.

PROOF. The equivalence of (1) and (2) is straightforward from the definitions, as in the proof of Theorem 6.5.10.

The proof that (2) implies (3) is similar to the proof of Theorem 6.5.1. For every $a \in A$, let $\pi_a^A \in C := B^{B^A}$ be the function that maps every tuple in B^A to its a-th entry. Let **S** be the subalgebra of $\mathbf{B}^{B^A}$ generated by $\{\pi_a^A \mid a \in A\}$. Define $h \colon S \to A$ as

$$h(f^{\mathbf{B}}(\pi_{a_1}^A, \ldots, \pi_{a_n}^A)) := f^{\mathbf{A}}(a_1, \ldots, a_n).$$

Similarly as in the proof of Theorem 6.5.1 one can show that h is well defined using that all height-one identities that hold in **B** also hold in **A**. Note that h is defined on all of S because Minion(**B**) is an operation clone.

Let $g \colon A \to S$ be the mapping which sends every $a \in A$ to π_a^A. Then h and g show that $\mathbf{A} \in \operatorname{Refl}(\mathbf{S}) \subseteq \operatorname{Refl} \mathrm{S}\,\mathrm{P}(\mathbf{B}) = \operatorname{Refl} \mathrm{P}(\mathbf{B})$: for all $a_1, \ldots, a_n \in A$

$$f^{\mathbf{A}}(a_1, \ldots, a_n) = g(f^{\mathbf{B}}(h(a_1), \ldots, h(a_n))).$$

If A and B are finite, then B^A is finite and hence $\mathbf{C} \in \mathrm{S}\,\mathrm{P}^{\mathrm{fin}}(\mathbf{B})$, so the proof implies that $\mathbf{A} \in \operatorname{Refl} \mathrm{P}^{\mathrm{fin}}(\mathbf{B})$.

(3) $\Rightarrow$ (2). If $\mathbf{A} \in \mathrm{P}(\mathbf{B})$ then the statement follows from Theorem 6.5.1. Now suppose that **A** is a reflection of **B** via the maps $h \colon B \to A$ and $g \colon A \to B$. Let ϕ be the identity $\forall x_1, \ldots, x_n \colon f_1(x_{i_1}, \ldots, x_{i_k}) = f_2(x_{j_1}, \ldots, x_{j_l})$ for $f_1, f_2 \in \tau$ and suppose that $\mathbf{A} \models \phi$. For all $b_1, \ldots, b_n \in B$ we have

$$\begin{aligned} f_1^{\mathbf{B}}(b_{i_1}, \ldots, b_{i_k}) &= g(f_1^{\mathbf{A}}(h(b_{i_1}), \ldots, h(b_{i_k}))) \\ &= g(f_2^{\mathbf{A}}(h(b_{j_1}), \ldots, h(b_{j_l}))) \\ &= f_2^{\mathbf{B}}(b_{j_1}, \ldots, b_{j_l}) \end{aligned}$$

Since $b_1, \ldots, b_n$ were chosen arbitrarily, we have that $\mathbf{B} \models \phi$. ⊣

A generalisation of this theorem to oligomorphic algebras will be presented in Section 9.6.

6.7.3. Minor conditions. If we apply our translation between identities and sentences in the signature of abstract clones (see Definition 6.5.5) to height-one identities $\forall x_1, \ldots, x_n \colon s = t$ we obtain clone formulas of a special form. To describe these sentences in a more readable form, we introduce a convenient notation. If $\sigma \colon \{1, \ldots, n\} \to \{1, \ldots, k\}$ then we also write f_σ instead of

$\text{comp}(f, \text{pr}^k_{\sigma(1)}, \dots, \text{pr}^k_{\sigma(n)})$. Correspondingly, if f is an n-ary operation over a set B (rather than a variable), then we also write f_σ for the k-ary operation $(x_1, \dots, x_k) \mapsto f(x_{\sigma(1)}, \dots, x_{\sigma(n)})$.

DEFINITION 6.7.9. A primitive positive clone sentence is called a *minor condition* if each of its conjuncts is of the form $f_\sigma = g_\rho$ for $\sigma \colon \{1, \dots, n_1\} \to \{1, \dots, k\}$ and $\rho \colon \{1, \dots, n_2\} \to \{1, \dots, k\}$.

Recall from Definition 6.6.7 that a clone sentence ϕ is called trivial if $\mathbf{Proj} \models \phi$, and *non-trivial* otherwise. A *loop condition* is a special case of a minor condition [294, 295, 305].

DEFINITION 6.7.10. A *loop condition* is a minor condition of the form

$$\exists f : f_\tau = f_\sigma$$

where f is n-ary, $\tau, \sigma \colon \{1, \dots, n\} \to \{1, \dots, k\}$, and $k \in \mathbb{N}$.

Every loop condition ϕ can be represented by a directed graph D_ϕ with vertex set $\{1, \dots, k\}$ and edge set $\{(\sigma(i), \tau(i)) \mid i \in \{1, \dots, n\}\}$. The name *loop condition* comes from the observation that if D_ϕ maps homomorphically to a digraph H, and H is preserved by an operation f satisfying $f_\tau = f_\sigma$, then H must have a loop: simply feed in the images of the n edges of D_ϕ into the arguments of f; the identity above then implies that H contains an edge of the form (x, x). An important loop condition is the loop condition for the digraph K_3 which will be treated in Section 6.8.2.

Note that if ϕ is a finite conjunction of height-one identities built from the function symbols $f_1, \dots, f_m$, then $\exists z_1, \dots, z_m \colon \phi^*(z_1, \dots, z_m)$ is a minor condition (recall Definition 6.5.5). Lemma 6.5.13 has the following variant for minions.

LEMMA 6.7.11. *Let $\mathfrak{M}_i$, for $i \in \{1, 2\}$, be a minion on (A_i, B_i). If there is a minor-preserving map from $\mathfrak{M}_1$ to $\mathfrak{M}_2$ then every minor condition that holds in $\mathfrak{M}_1$ also holds in $\mathfrak{M}_2$. If A_2 and B_2 are finite, then the converse holds as well.*

PROOF. Let $\xi \colon \text{Pol}(\mathfrak{B}) \to \text{Pol}(\mathfrak{A})$ be minor preserving, let ϕ be a minor condition, and suppose that $\text{Pol}(\mathfrak{B}) \models \phi$. Let $f_\sigma = g_\rho$ be a conjunct of ϕ. We use the same letters f and g to denote witnesses in $\text{Pol}(\mathfrak{B})$ of the variables f and g in ϕ. Then

$$\begin{aligned} \xi(f)_\sigma &= \xi(f_\sigma) && \text{(since } \xi \text{ is minor preserving)} \\ &= \xi(g_\rho) && \text{(by assumption)} \\ &= \xi(g)_\rho && \text{(since } \xi \text{ is minor preserving).} \end{aligned}$$

showing that ϕ holds in $\text{Pol}(\mathfrak{A})$. The converse can be shown by a compactness argument similarly as in Lemma 6.5.13. ⊣

Corollary 6.5.16 for clones has the following variant for minor-preserving maps.

COROLLARY 6.7.12. *Let* $\mathfrak{A}$ *and* $\mathfrak{B}$ *be relational structures. Then* (1) $\Rightarrow$ (2) $\Rightarrow$ (3):

1. $\mathfrak{A} \in \mathrm{HI}(\mathfrak{B})$.
2. *There exists a minor-preserving map from* $\mathrm{Pol}(\mathfrak{B})$ *to* $\mathrm{Pol}(\mathfrak{A})$.
3. *Every minor condition that holds in* $\mathrm{Pol}(\mathfrak{B})$ *also holds in* $\mathrm{Pol}(\mathfrak{A})$.

If $\mathfrak{A}$ *is finite then the implication from* (3) *to* (2) *holds as well. If additionally* $\mathfrak{B}$ *is finite, then the implication from* (2) *to* (1) *holds as well.*

PROOF. Let $\mathbf{B}$ be the polymorphism algebra of $\mathfrak{B}$.

(1) $\Rightarrow$ (2). If $\mathfrak{A} \in \mathrm{HI}(\mathfrak{B})$ then there exists an algebra $\mathbf{A} \in \mathrm{Exp}\,\mathrm{Refl}\,\mathrm{P}^{\mathrm{fin}}(\mathbf{B})$ such that $\mathrm{Clo}(\mathbf{A}) = \mathrm{Pol}(\mathfrak{A})$ (see Theorem 6.4.3). By Theorem 6.7.8, (3) $\Rightarrow$ (1), there exists a minor-preserving map from $\mathrm{Clo}(\mathbf{B}) = \mathrm{Pol}(\mathfrak{B})$ to $\mathrm{Clo}(\mathbf{A}) = \mathrm{Pol}(\mathfrak{A})$.

(2) $\Rightarrow$ (3). An immediate consequence of Lemma 6.7.11, which also shows the implication (3) $\Rightarrow$ (2) if $\mathfrak{A}$ is finite.

Let $\mathbf{A}$ be the algebra with domain A and the same signature τ as $\mathbf{B}$ where $f \in \tau$ denotes $\xi(f^{\mathbf{B}})$. That is, ξ is the natural minor-preserving map from $\mathrm{Clo}(\mathbf{B}) = \mathrm{Pol}(\mathfrak{B})$ to $\mathrm{Clo}(\mathbf{A}) \subseteq \mathrm{Pol}(\mathfrak{A})$. For finite $\mathbf{A}$ and $\mathbf{B}$ the implication from (1) to (4) in Theorem 6.7.8 implies that $\mathbf{A} \in \mathrm{Exp}\,\mathrm{Refl}\,\mathrm{P}^{\mathrm{fin}}(\mathbf{B})$. So there exists an algebra $\mathbf{A}' \in \mathrm{Exp}\,\mathrm{Refl}\,\mathrm{HSP}^{\mathrm{fin}}(\mathbf{B})$ such that $\mathrm{Clo}(\mathbf{A}') = \mathrm{Pol}(\mathfrak{A})$ and so item (2) in Theorem 6.4.3 implies that $\mathfrak{A} \in \mathrm{HI}(\mathfrak{B})$. ⊣

Corollary 6.7.13 below shows that for finite structures $\mathfrak{B}$ the condition from Conjecture 3.1 (item (1)) is equivalent to the condition from Conjecture 4.1 (item (7)). A proof for the larger class of reducts of homogeneous structures with finite relational signature can be found in Section 10.3.

COROLLARY 6.7.13. *Let* $\mathfrak{B}$ *be a finite structure. Then the following are equivalent.*

1. $K_3 \in \mathrm{HI}(\mathfrak{B})$.
2. $\mathrm{HI}(\mathfrak{B})$ *contains all finite structures.*
3. $\mathrm{HI}(\mathfrak{B})$ *contains* $(\{0,1\}; \mathrm{1IN3})$.
4. *There is a minor-preserving map from* $\mathrm{Pol}(\mathfrak{B})$ *to* $\mathbf{Proj}$.
5. $\mathrm{Pol}(\mathfrak{B})$ *satisfies no non-trivial minor condition.*
6. $\mathrm{Pol}(\mathfrak{B})$ *contains no Taylor operation.*
7. $K_3 \in \mathrm{ICH}(\mathfrak{B})$.

If these conditions hold, then $\mathfrak{B}$ *has a finite-signature reduct* $\mathfrak{B}'$ *such that* $\mathrm{CSP}(\mathfrak{B}')$ *is NP-hard.*

PROOF. The equivalence of (1), (2), and (3) has been stated in Corollary 6.4.4. Since $\mathbf{Proj}$ is isomorphic to $\mathrm{Pol}(\{0,1\}; \mathrm{1IN3})$ (Proposition 6.5.19) the implications from (3) to (4) and from (4) to (5) follow from Corollary 6.7.12. The implication from (5) to (6) is immediate since the existence of a Taylor operation is a non-trivial minor condition. To prove the contraposition of the implication from (6) to (7), suppose that $K_3 \notin \mathrm{ICH}(\mathfrak{B})$. Let $\mathfrak{C}$ be the core of $\mathfrak{B}$

with domain $C = \{c_1, \dots, c_n\}$, and note that $(\mathfrak{C}, c_1, \dots, c_n) \in \mathrm{CH}(\mathfrak{B})$. Hence, $K_3 \notin \mathrm{I}(\mathfrak{C}, c_1, \dots, c_n)$. Polymorphisms of $(\mathfrak{C}, c_1, \dots, c_n)$ are idempotent and therefore, by Corollary 6.6.8, $(\mathfrak{C}, c_1, \dots, c_n)$ and therefore also $\mathfrak{C}$ has a Taylor polymorphism. The implication from (7) to (1) follows from Theorem 3.6.2. ⊣

6.8. Siggers operations

In Section 1.1 the theorem of Hell and Nešetřil [195] was mentioned, which states that finite graphs $\mathfrak{B}$ exhibit a complexity dichotomy: CSP($\mathfrak{B}$) is in P or NP-complete. Long before the proofs of the Feder-Vardi conjecture a strengthened version of the theorem of Hell and Nešetřil was known, namely Theorem 6.8.1, which is of independent interest was known (in particular, the stronger statement remains interesting even if P = NP).

THEOREM 6.8.1. *Let $\mathfrak{B}$ be a finite loopless undirected graph. Then*

- $\mathfrak{B}$ *is* bipartite, *i.e., admits a homomorphism to* K_2, *or*
- $K_3 \in \mathrm{HI}(\mathfrak{B})$.

In the former case, the problem is to determine whether a given instance of the problem is bipartite. This problem can be solved easily in polynomial time by reducing to the connected case and attempting to compute the (unique possible) bipartition. This is an instance of the algorithmic methods described in Chapter 8. In the latter case, CSP($\mathfrak{B}$) is NP-complete by Corollary 3.7.1. Theorem 6.8.1 has a remarkable consequence in universal algebra, discovered by Siggers in 2010 (see Section 6.8.2), whose significance goes beyond the study of the complexity of CSPs.

6.8.1. Proof of the Hell-Nešetřil theorem. The graph $K_4 - \{0, 1\}$ (a clique where one edge is missing) is called a *diamond*. A graph is called *diamond-free* if it does not contain a copy of a diamond as a (not necessarily induced) subgraph.[18] Note that these are precisely those undirected graphs for which every edge is covered by at most one copy of a K_3. For every $l \in \mathbb{N}$, the graph $(K_3)^l$ is an example of diamond-free graph.

LEMMA 6.8.2. *Let $\mathfrak{B}$ be a finite loopless undirected graph which is not bipartite. Then* $\mathrm{HI}(\mathfrak{B})$ *contains a finite diamond-free core graph containing a triangle.*

PROOF. We may assume that

1. $\mathrm{HI}(\mathfrak{B})$ does not contain a non-bipartite loopless graph with fewer vertices than $\mathfrak{B}$, because otherwise we could replace $\mathfrak{B}$ by this graph.
2. $\mathfrak{B}$ contains a triangle: if the length of the shortest odd cycle is k, then $(B; E^{k-2})$, where E^{k-2} is defined via the usual composition of binary

[18]Note that when we view a graph as a relational structure $\mathfrak{B}$, then a substructure of $\mathfrak{B}$ in the sense of Definition 1.1.6 is what a graph theorist would call an *induced subgraph*; in contrast, a (*weak*) *subgraph* of a graph $(V; E)$ is a graph $(V'; E')$ where $V' \subseteq V$ and $E' \subseteq E \cap \binom{V'}{2}$.

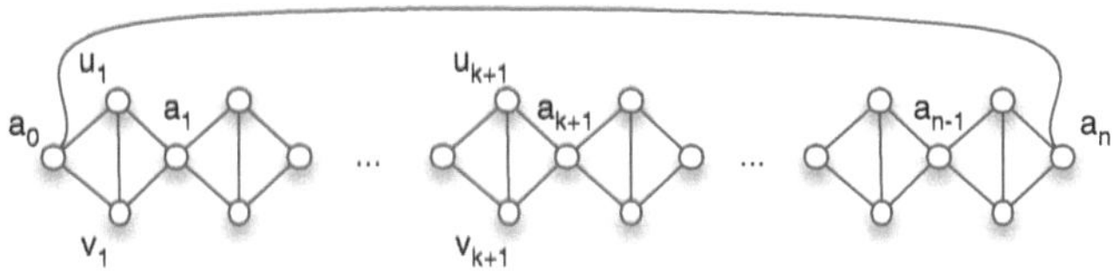

FIGURE 1. Diagram for the proof of Lemma 6.8.2.

relations and so primitively positively definable in $\mathfrak{B}$, is an undirected graph and contains a triangle, so it can replace $\mathfrak{B}$. Moreover, the new graph has the same number of vertices, so the assumption (1) is still satisfied.

Claim 1. $\mathfrak{B}$ is a core. If $\mathfrak{B}$ were not a core, then $H(\mathfrak{B})$ would contain a non-bipartite graph with fewer vertices, in contradiction to assumption (1) above.

Claim 2. Every vertex of $\mathfrak{B}$ is contained in a triangle. Otherwise, the subgraph of $\mathfrak{B}$ induced on the set of vertices defined by the primitive positive formula $\exists u, v\,(E(x,u) \wedge E(x,v) \wedge E(u,v))$ in $\mathfrak{B}$ still contains a triangle, and has fewer vertices than $\mathfrak{B}$, in contradiction to assumption (1).

Claim 3. $\mathfrak{B}$ does not contain a copy of K_4. Otherwise, if a is an element from a copy of K_4, then the subgraph of $\mathfrak{B}$ induced on $\{x \in B \mid E(a,x)\}$ is a non-bipartite graph $\mathfrak{A}$, which has strictly less vertices than $\mathfrak{B}$ because $a \notin A$. Moreover, $\mathfrak{A} \in \mathrm{HI}(\mathfrak{B})$ by Theorem 3.6.2, contrary to assumption (1).

Claim 4. $\mathfrak{B}$ is diamond-free. To see this, let R be the binary relation with the primitive positive definition

$$R(x,y) :\Leftrightarrow \exists u, v\,\big(E(x,u) \wedge E(x,v) \wedge E(u,v) \wedge E(u,y) \wedge E(v,y)\big) \qquad (11)$$

and let T be the transitive closure of R. The relation T is clearly symmetric, and since every vertex of $\mathfrak{B}$ is contained in a triangle, it is also reflexive, and hence an equivalence relation of $\mathfrak{B}$. If $(x,y) \in T$, we also say that x and y are *diamond connected*; see Figure 1. Since B is finite, for some n the formula $\delta_n(x,y)$

$$\exists a_1,\dots,a_{n-1}\,\big(R(x,a_1) \wedge R(a_1,a_2) \wedge \cdots \wedge R(a_{n-1},y)\big) \qquad (12)$$

defines T, showing that T is primitively positively definable in $\mathfrak{B}$.

We claim that the graph $\mathfrak{B}/T$ (see Example 3.1.2) is loopless. It suffices to show that $T \cap E = \emptyset$. Otherwise, let $(a,b) \in T \cap E$ be chosen so that $\delta_n(a,b)$ holds with n minimised. So there exists a sequence $a = a_0, a_1, \dots, a_n = b$ with $R(a_0,a_1)$, $R(a_1,a_2)$, … , $R(a_{n-1},a_n)$ in $\mathfrak{B}$; again, see Figure 1. This chain cannot have the form $R(a_0,a_1)$ because $\mathfrak{B}$ does not contain K_4 subgraphs. Suppose first that $n = 2k$ is even. Let the vertices u_1, v_1, u_{k+1} and v_{k+1} be as depicted in Figure 1. Let S be the set of elements $x \in B$ defined in $\mathfrak{B}$ by

$$\exists x_1, x_k \left(E(u_{k+1}, x_1) \wedge E(v_{k+1}, x_1) \wedge \delta_{k-1}(x_1, x_k) \wedge E(x_k, x)\right).$$

The vertices of the triangle (a_0, u_1, v_1) lie in S, so the subgraph induced on the primitively positively definable set S is non-bipartite. The vertex a_n does not lie in S, by the minimal choice of n. So $\mathrm{HI}(\mathfrak{B})$ contains a loopless non-bipartite graph with fewer vertices than $\mathfrak{B}$, in contradiction to the initial assumption.

If $n = 2k+1$ is odd, we can argue analogously with the set S defined by the formula $\exists y \left(\delta_k(a_{k+1}, y) \wedge E(y, x)\right)$ and again obtain a contradiction. Hence, $\mathfrak{B}/T$ does not contain loops. Since $\mathfrak{B}$ contains a triangle, say on $\{a, b, c\}$, it also follows that $\mathfrak{B}/T$ has a triangle on the classes of a, b, and c. The initial assumption for $\mathfrak{B}$ then implies that T must be the trivial equivalence relation, which in turn implies that $\mathfrak{B}$ does not contain any diamonds. ⊣

DEFINITION 6.8.3. For $I = \{i_1, \dots, i_m\} \subseteq \{1, \dots, k\}$ with $x_1 < \dots < i_m$ we write π^k_I for the function

$$(x_1, \dots, x_k) \mapsto (x_{i_1}, \dots, x_{i_m}).$$

LEMMA 6.8.4 (Bulatov [107]). *Let $\mathfrak{B}$ be a diamond-free loopless undirected graph and let $h\colon (K_3)^k \to \mathfrak{B}$ be a homomorphism. Then there exists $I \subseteq \{1, \dots, k\}$ such that h has the same kernel as π^k_I, and the image of h is isomorphic to $(K_3)^{|I|}$.*

PROOF. Let $I \subseteq \{1, \dots, k\}$ be maximal such that the kernel of h is contained in the kernel of π^k_I. Such a set exists, because the kernel of $\pi^k_\emptyset$ is the total relation. We claim that the kernel of h equals the kernel of π^k_I. We have to show that for every $j \in \{1, \dots, k\} \setminus I$ and for all $z_1, \dots, z_k, z'_j \in \{0, 1, 2\}$

$$h(z_1, \dots, z_j, \dots, z_k) = h(z_1, \dots, z_{j-1}, z'_j, z_{j+1}, \dots, z_k).$$

By the maximality of I, there are $x, y \in (K_3)^k$ such that $h(x) = h(y)$ and $x_j \neq y_j$. We may suppose that $z_j \neq x_j$ and $z'_j = x_j$. To simplify the notation we assume that $j = k$. It is easy to see that any two vertices in $(K_3)^k$ have a common neighbour.

- Let r be a common neighbour of x and $(z, z_k) := (z_1, \dots, z_k)$. Note that r and (z, z'_k) are adjacent, too.
- For all $i \neq k$ we choose an element s_i of K_3 that is distinct from both r_i and y_i. Since x_k is distinct from r_k and y_k we have that $(s, x_k) := (s_1, \dots, s_{k-1}, x_k)$ is a common neighbour of r and y.
- The tuple $(r, z_k) := (r_1, \dots, r_{k-1}, z_k)$ is a common neighbour of both x and (s, x_k).
- Finally, for $i \neq k$ choose t_i to be distinct from z_i and r_i, and choose t_k to be distinct from z_k and from z'_k. Then $t := (t_1, \dots, t_{k-1}, t_k)$ is a common neighbour of (z, z_k), of (z, z'_k), and of (r, z_k).

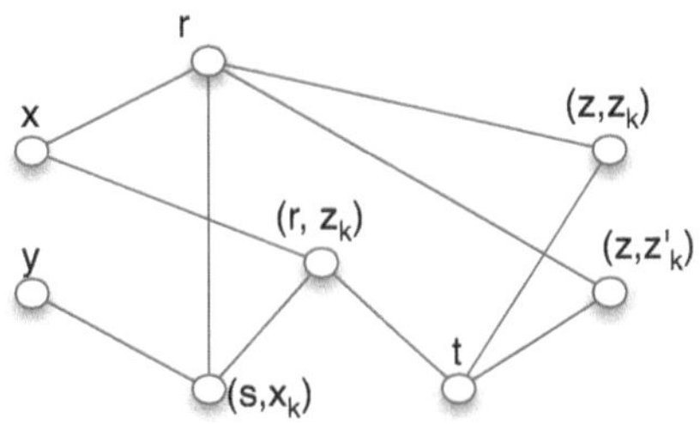

FIGURE 2. Diagram for the proof of Lemma 6.8.4.

The situation is illustrated in Figure 2. Since $\mathfrak{B}$ is diamond-free, $h(x) = h(y)$ implies that $h(r) = h(r, z_k)$ and for the same reason $h(z, z_k) = h(z, z'_k)$ which completes the proof of the claim.

We finally show that the image of h is isomorphic to $(K_3)^m$ where $m = |I|$. In fact, the map $\pi^k_I \circ h^{-1}$ provides an isomorphism:

- This map is well defined since h and π^k_I have the same kernel.
- If $((u_1, \dots, u_m), (v_1, \dots, v_m))$ is an edge in $(K_3)^m$, then let $a, b \in (K_3)^k$ be such that $\pi^k_I(a) = (u_1, \dots, u_m)$ and $\pi^k_I(b) = (v_1, \dots, v_m)$, and $a_i = u_1$, $b_i = v_1$ if $i \notin I$. Then (a, b) is an edge in $(K_3)^k$ and hence $(h(a), h(b))$ is an edge in $\mathfrak{B}$.
- if $(u_1, \dots, u_m)$ and $(v_1, \dots, v_m)$ are distinct but not adjacent then $u_i = v_i$ for some $i \in \{1, \dots, m\}$. We can therefore find two distinct elements p and q of $(K_3)^m$ that are adjacent to both $(u_1, \dots, u_m)$ and $(v_1, \dots, v_m)$. Similarly as in the previous item we can find a, b, c, d such that $\pi^k_I(a) = (u_1, \dots, u_m)$, $\pi^k_I(b) = (v_1, \dots, v_m)$, $\pi^k_I(c) = p$, $\pi^k_I(d) = q$, and such that c and d are adjacent to both a and b. Hence, the same holds for $h(c), h(d), h(a), h(b)$ as h is a homomorphism. Moreover, since h and π^k_I have the same kernel we have $h(a) \neq h(b)$ and $h(p) \neq h(q)$. Since $\mathfrak{B}$ is diamond-free we obtain that $h(a)$ and $h(b)$ are not adjacent in $\mathfrak{B}$.

This concludes the proof. ⊣

LEMMA 6.8.5 (Bulatov [107]). *If a finite diamond-free loopless undirected graph $\mathfrak{B}$ contains a copy of a K_3, then $\mathfrak{B}$ interprets $(K_3)^k$ primitively positively with parameters, for some $k \in \mathbb{N}$.*

PROOF. We construct a strictly increasing sequence of subgraphs $G_1 \subset G_2 \subset \cdots$ of $\mathfrak{B}$ such that G_i is isomorphic to $(K_3)^{k_i}$ for some $k_i \in \mathbb{N}$. Let G_1 be any triangle in $\mathfrak{B}$. Suppose now that G_i has already been constructed. If the domain of G_i is primitively positively definable in $\mathfrak{B}$ with constants, then we are done. Otherwise, there exists an idempotent polymorphism f of $\mathfrak{B}$ and $v_1, \dots, v_k \in G_i$ such that $f(v_1, \dots, v_k) \notin G_i$. The restriction of f

to G_i provides a homomorphism from $(K_3)^{k_i}$ to the diamond-free graph $\mathfrak{B}$. Lemma 6.8.4 shows that $G_{i+1} := f((G_i)^k)$ induces a copy of $(K_3)^{k_{i+1}}$ for some $k_{i+1} \leq k$. Since f is idempotent, we have that $G_i \subseteq G_{i+1}$, and by the choice of f the containment is strict. Since $\mathfrak{B}$ is finite, for some m the set G_m must have a primitive positive definition in $\mathfrak{B}$ with constants. ⊣

PROOF OF THEOREM 6.8.1. Let $\mathfrak{B}$ be a finite loopless undirected graph that is not bipartite. Lemma 6.8.2 states that there is a diamond-free core $\mathfrak{C}$ containing a triangle in $\mathrm{HI}(\mathfrak{B})$. Then Lemma 6.8.5 applied to $\mathfrak{C}$ implies that for some $k \in \mathbb{N}$ there is a primitive positive interpretation of $(K_3)^k$ with constants in $\mathfrak{C}$. Since $\mathfrak{C}$ is a core, and since $(K_3)^k$ is homomorphically equivalent to K_3, it follows that $K_3 \in \mathrm{HI}(\mathfrak{C})$. ⊣

6.8.2. Siggers' theorem. We present a strengthening of Taylor's theorem (Theorem 6.6.4). An operation $s\colon B^6 \to B$ is called a *Siggers operation* if

$$s(x, y, x, z, y, z) = s(y, x, z, x, z, y)$$

holds for all $x, y, z \in B$. Note that the existence of a Siggers operation can be formulated as a loop condition (Definition 6.7.10). Clearly, a Siggers operation is a Taylor operation. Also note that whether or not a clone contains a Siggers operation can be formulated as a (non-trivial) minor condition in the sense of the previous section.

THEOREM 6.8.6 (Siggers [329]). *Let $\mathfrak{B}$ be a finite structure. Then either $K_3 \in \mathrm{HI}(\mathfrak{B})$ or $\mathfrak{B}$ has a Siggers polymorphism and the two cases are mutually exclusive.*

PROOF. By Corollary 6.7.12, if $\mathfrak{A} \in \mathrm{HI}(\mathfrak{B})$ then every minor condition that holds in $\mathrm{Pol}(\mathfrak{B})$ also holds in $\mathrm{Pol}(\mathfrak{A})$. Since all polymorphisms of K_3 are essentially unary (Propoisition 6.1.43), $\mathrm{Pol}(K_3)$ does not satisfy any non-trivial minor condition. But having a Siggers polymorphism is a non-trivial minor condition. This shows that the two cases are mutually exclusive.

To show that one of the two cases applies, let $k \geq 1$ and $a, b, c \in B^k$ be such that $\{(a_i, b_i, c_i) \mid i \leq k\} = B^3$. Let R be the binary relation on B^k such that $(u, v) \in R$ iff there exists a 6-ary $s \in \mathrm{Pol}(\mathfrak{B})$ such that $u = s(a, b, a, c, b, c)$ and $v = s(b, a, c, a, c, b)$. We make the following series of observations.

- The vertices $a, b, c \in B^k$ induce in $(B^k; R)$ a copy of K_3: each of the six edges of K_3 is witnessed by one of the six 6-ary projections from $\mathrm{Pol}(\mathfrak{B})$.
- The relation R is symmetric: Suppose that $(u, v) \in R$ and let $s \in \mathrm{Pol}(\mathfrak{B})$ be such that $u = s(a, b, a, c, b, c)$ and $v = s(b, a, c, a, c, b)$. Define $s' \in \mathrm{Pol}(\mathfrak{B})$ by $s'(x_1, \ldots, x_6) := s(x_2, x_1, x_4, x_3, x_6, x_5)$; then

$$v = s(b, a, c, a, c, b) = s'(a, b, a, c, b, c)$$
$$u = s(a, b, a, c, b, c) = s'(b, a, c, a, c, b)$$

  and hence s' witnesses that $(v, u) \in R$.

- If the graph $(B^6; R)$ contains a loop $(w, w) \in R$, then there exists a 6-ary $s \in \mathrm{Pol}(\mathfrak{B})$ such that

$$s(a, b, a, c, b, c) = w = s(b, a, c, a, c, b).$$

The operation s is Siggers: for all $x, y, z \in B$ there exists an $i \leq k$ such that $(x, y, z) = (a_i, b_i, c_i)$, and the above implies that

$$s(a_i, b_i, a_i, c_i, b_i, c_i) = s(b_i, a_i, c_i, a_i, c_i, b_i)$$

and we are done in this case.

So we may assume in the following that $(B^k; R)$ is a undirected and loopless graph that contains a copy of K_3. The relation R (as a $2k$-ary relation over B) is preserved by $\mathrm{Pol}(\mathfrak{B})$, and hence $(B^k; R)$ has a primitive positive interpretation in $\mathfrak{B}$ (Theorem 6.1.12). By Theorem 6.8.1 applied to the undirected graph $(B^k; R)$, we have $K_3 \in \mathrm{HI}(B^k; R)$ and hence also $K_3 \in \mathrm{HI}(\mathfrak{B})$, and this concludes the proof. ⊣

In Section 10.2 we present a generalisation of this result for ω-categorical model-complete core structures, due to Barto and Pinsker [28].

Note that for an explicitly given finite structure $\mathfrak{B}$, the existence of a Siggers polymorphism can be decided, and it follows that the condition of the dichotomy for finite domain CSPs is decidable.

6.9. Weak near-unanimity operations

For finite structures, the existence of a Taylor polymorphism is not only equivalent to the existence of a Siggers polymorphism, but also to the existence of a *weak near-unanimity* polymorphism, to the existence of a *cyclic* polymorphism, and to the existence of a 4-*ary Siggers* polymorphism. The results in this section have not yet been generalised to ω-categorical structures, and we just give a survey to a small selection of recent results about finite algebras.

We also present some famous stronger systems of identities; such identities are a central topic in universal algebra and their discussion would fill an entire book. We focus on some of the strongest known systems (in Section 6.9.3 and Section 6.9.2) such that for ω-categorical structures $\mathfrak{B}$ with polymorphisms satisfying these identities the tractability of $\mathrm{CSP}(\mathfrak{B})$ is still open.

DEFINITION 6.9.1. An operation $f : B^n \to B$, for $n \geq 2$, is called

- *cyclic* if it satisfies

$$\forall x_1, \dots, x_n : f(x_1, \dots, x_n) = f(x_2, \dots, x_n, x_1);$$

- *a weak near-unanimity* if it satisfies

$$\forall x, y : f(x, \dots, x, y) = f(x, \dots, x, y, x) = \cdots = f(y, x, \dots, x).$$

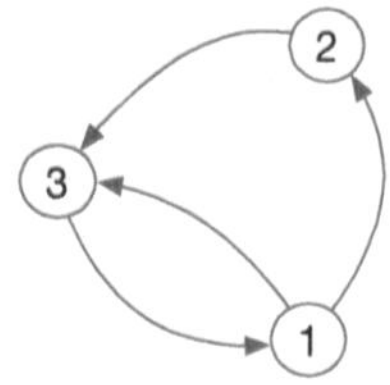

FIGURE 3. Operations satisfying the loop condition of this digraph are 4-ary Siggers operations.

- 4-*ary Siggers* if $n = 4$ and f satisfies

$$\forall a, e, r \colon f(r, a, r, e) = f(a, r, e, a).$$

Again, we do not require idempotence in this definition. Clearly, every cyclic operation is also a weak near-unanimity operation. Also note that Siggers terms and cyclic terms are (special) Taylor terms. The identity in the definition of 4-ary Siggers operations is the loop condition (Definition 6.7.10) of the digraph

$$(\{1, 2, 3\}; \{(1, 2), (2, 3), (3, 1), (1, 3)\});$$

see Figure 3. A full proof of Theorem 6.9.2 below is beyond the scope of this text; but we will provide exact pointers to the literature.

THEOREM 6.9.2. *Let* **B** *be a finite algebra. Then the following are equivalent.*

1. **B** *has a Taylor term.*
2. *for all prime numbers* $p > |B|$, *the algebra* **B** *has a* p*-ary cyclic term.*
3. **B** *has a cyclic term.*
4. **B** *has a weak near-unanimity term.*
5. **B** *has a* 4*-ary Siggers term.*

PROOF. The implications $(2) \Rightarrow (3) \Rightarrow (4) \Rightarrow (1)$ and $(5) \Rightarrow (1)$ are trivial.

$(1) \Rightarrow (2)$. The statement can be found for finite *idempotent* algebras in [23]. Since a structure has a Taylor polymorphism if and only if its core does (recall that in this text we do not require that Taylor operations are idempotent), and since a core has a Taylor polymorphism if and only if it has an idempotent Taylor polymorphism, the idempotent case implies the statement as given in the theorem.

It therefore suffices to show that $(2) \Rightarrow (5)$. Let $p = 3k + 2$ be some prime number larger than $|B|$ (which exists by Dirichlet's theorem) and let $c(x_1, \ldots, x_p)$ be a cyclic term of **B**. Define $s(x, y, z, w)$ to be the term

$$c(\underbrace{x, x, \ldots, x}_{k+1 \text{ times}}, \underbrace{y, \ldots, y}_{k \text{ times}}, w, \underbrace{z, \ldots, z}_{k \text{ times}}).$$

Then

$$\begin{aligned} s(x,y,z,y) &= c(\underbrace{x,x,\dots,x}_{k+1 \text{ times}},\underbrace{y,y,\dots,y,y}_{k+1 \text{ times}},\underbrace{z,\dots,z}_{k \text{ times}}) \\ &= c(\underbrace{y,y,\dots,y}_{k+1 \text{ times}},\underbrace{z,z,\dots,z}_{k \text{ times}},\underbrace{x,x,\dots,x}_{k+1 \text{ times}}) \quad \text{(by the cyclicity of } c\text{)} \\ &= s(y,z,x,x). \end{aligned}$$

Note that also this identity is the loop condition of the digraph shown in Figure 3, so a 4-ary Siggers operation can be obtained from s by permuting the arguments accordingly. ⊣

6.9.1. H-colouring revisited. As an application of Theorem 6.9.2, we show how to derive the complexity classification of the H-colouring problem for finite undirected simple graphs H, following [23], previously treated in Section 6.8.

THEOREM 6.9.3 (Hell and Nešetřil [195]). *Let H be a finite undirected graph. If H is bipartite then* CSP(H) *is in P. Otherwise,* CSP(H) *is NP-complete.*

PROOF. As remarked in Section 6.8, if H is bipartite then CSP(H) can be solved in polynomial time. Otherwise, G there exists a cycle $a_0, a_1, \dots, a_{2k}, a_0$ of odd length in H. If H has no Taylor polymorphism, then by Corollary 6.7.13, $K_3 \in \mathrm{HI}(\mathfrak{B})$ and CSP(H) is NP-hard.

Otherwise, if H has a Taylor polymorphism, then Theorem 6.9.2 asserts that there exists a p-ary cyclic polymorphism c of H where p is a prime number greater than $\max\{2k, |A|\}$. Since the edges in H are undirected, we can also find a cycle $a_0, a_1, \dots, a_{p-1}, a_0$ in H. Then $c(a_0, a_1, \dots, a_{p-1}) = c(a_1, \dots, a_{p-1}, a_0)$, which implies that H contains a loop, a contradiction to the assumption that the core of H has more than one element. ⊣

An alternative proof of the implication from (1) to (5) in Theorem 6.9.2, similar to the proof of Theorem 6.8.6, can be given using the following theorem. The *algebraic length* of a digraph $\mathfrak{B}$ is the greatest common divisor of the length of all directed cycles in $\mathfrak{B}$. A *sink* in a digraph is a vertex with no outgoing edges; similarly, a *source* in a digraph is a vertex with no incoming edges. Digraphs without sources and sinks are also called *smooth*.

THEOREM 6.9.4 (Barto, Kozik, Niven [25]). *Let $\mathfrak{B}$ be a finite smooth digraph of algebraic length* 1. *Then either $K_3 \in \mathrm{HI}(\mathfrak{B})$ or $\mathfrak{B}$ contains a loop.*

6.9.2. Edge Operations. In this section we briefly discuss another algebraic criterion for polynomial-time decidability of CSPs given in [212, 211], which has not yet been generalised to the ω-categorical setting (Question 34). This is the existence of an *edge polymorphism.*

DEFINITION 6.9.5. An operation $e\colon B^{k+1} \to B$, for $k \geq 2$, is called a *quasi k-edge operation* if it satisfies for all $x, y \in B$

$$e(y,y,x,x,x,\dots,x) = e(x,\dots,x)$$

$$\begin{aligned} e(y,x,y,x,x,\dots,x) &= e(x,\dots,x) \\ e(x,x,x,y,x,\dots,x) &= e(x,\dots,x) \\ \vdots \quad &= \quad \vdots \\ e(x,x,x,x,\dots,x,y) &= e(x,\dots,x). \end{aligned}$$

An idempotent quasi edge operation is called an *edge operation.* The equations of a k-edge operation from Definition 6.9.5 are perhaps more easily readable in matrix form as follows.

$$e\begin{pmatrix} y & y & x & x & \cdots & x \\ y & x & y & x & \cdots & x \\ x & x & x & y & & x \\ \vdots & \vdots & \vdots & & \ddots & \\ x & x & x & x & & y \end{pmatrix} = e\begin{pmatrix} x & \cdots & x \\ \vdots & \ddots & \vdots \\ x & \cdots & x \end{pmatrix}$$

Clearly, every quasi edge operation is a Taylor operation. Note that a quasi 2-edge operation is a quasi Maltsev operation (modulo interchanging the first and second argument). Bulatov and Dalmau [113] proved that if $\mathfrak{B}$ is a finite structure with finite relational signature and a Maltsev polymorphism, then CSP($\mathfrak{B}$) can be solved in polynomial time (the algorithm even works in polynomial time if the signature is infinite and the relations are represented by listing all tuples in the relation). The Bulatov-Dalmau algorithm has been generalised to all finite structures with an edge polymorphism [212].

The existence of an edge term in a finite algebra is an important property in universal algebra because it is equivalent to a number of equivalent fundamental properties: for example, a finite algebra $\mathbf{A}$ has an edge term if and only if $\mathbf{A}$ has the *few subpowers property* [38], i.e., the number of subalgebras of $\mathbf{A}^n$ is in $O(2^{n^k})$ for some $k \in \mathbb{N}$.

Example 6.9.6. The algebra $\mathbf{A} := (\{0,1\}; \min)$ does not have the few subpowers property: the number of subalgebras of $\mathbf{A}^n$ is at least the number of subalgebras of $\mathbf{A}^n$ which are generated by sets of the form $\{a \in \{0,1\}^n \mid \sum_i a_i = \lfloor n/2 \rfloor\}$. Any two distinct sets of this form generate different subalgebras, because applying min to distinct tuples in the set strictly decreases the number of entries with a 1. Thus, the number of subalgebras of $\mathbf{A}^n$ is at least $2^{\binom{n}{\lfloor n/2 \rfloor}}$, and $\binom{n}{\lfloor n/2 \rfloor}$ clearly grows faster than any polynomial. Hence, the facts mentioned above imply that $(\{0,1\}; \min)$ does not have an edge term.

Another result that should be mentioned in this context links the existence of an edge term with a fundamental property in universal algebra. A variety is called *congruence modular* if the congruence lattice of every algebra in the variety is modular; for such varieties, a strong structure theory is known (called commutator theory [228]). Barto [19] showed that a finite structure $\mathfrak{B}$

with finite relational signature has an edge polymorphism if and only if the polymorphism algebra of $\mathfrak{B}$ generates a congruence modular variety.

6.9.3. Jónsson chains. Jónsson chains were introduced by Bjarni Jónsson [218]; they provide an equivalent characterisation of congruence distributive varieties. If the variety is generated by the polymorphism clone of a finite structure $\mathfrak{B}$ with finite relational signature, this condition has drastic consequences for CSP($\mathfrak{B}$). Barto [18] proved that in this case $\mathfrak{B}$ must also have a near-unanimity polymorphism and hence can be solved in polynomial time by the methods that will be presented in Section 8.5. Unfortunately, a generalisation of Barto's theorem for polymorphism clones of ω-categorical structures $\mathfrak{B}$ with finite relational signature is not yet known (see Question 18).

There are several equivalent definitions of Jónsson chains; we present a variant that follows the terminology in [227].

DEFINITION 6.9.7. A sequence $j_1, \dots, j_{2n+1}$ of ternary operations on a set B is called a *chain of quasi Jónsson operations* if for all $x, y, z \in B$

$$\begin{aligned} j_1(x,x,y) &= j_1(x,x,x) && \\ j_i(x,y,x) &= j_i(x,x,x) && \text{for all } i \in \{1,\dots,2n+1\} \\ j_{2i-1}(x,y,y) &= j_{2i}(x,y,y) && \text{for all } i \in \{1,\dots,n\} \\ j_{2i}(x,x,y) &= j_{2i+1}(x,x,y) && \text{for all } i \in \{1,\dots,n\} \\ j_{2n+1}(x,y,y) &= j_{2n+1}(y,y,y). && \end{aligned}$$

A chain of quasi Jónsson operations for which the operations are idempotent is called a *chain of Jónsson operations* (or a *Jónsson chain*).

Note that if j_1, j_2, j_3 is a quasi Jónsson chain, then j_2 is a quasi majority operation. Clearly, the existence of a quasi Jónsson chain is a non-trivial minor condition (Definition 6.7.9), so any clone that contains a quasi Jónsson chain has no minor-preserving map to **Proj**. A variety is *congruence distributive* if the congruence lattice of every algebra in the variety is distributive. Jónsson [218] showed that this is the case if and only if there exists a chain of terms which denotes a Jonsson chain in all algebras of the variety.

We mention two further strengthenings of the existence of chains of quasi Jónsson operations whose impact on the complexity of CSP($\mathfrak{B}$) for ω-categorical $\mathfrak{B}$ is also unclear.

DEFINITION 6.9.8. A sequence $d_1, \dots, d_n$ of ternary operations on a set B is called a *chain of quasi directed Jónsson operations* if for all $x, y, z \in B$

$$\begin{aligned} d_1(x,x,y) &= d_1(x,x,x) && \\ d_i(x,y,x) &= d_i(x,x,x) && \text{for all } i \in \{1,\dots,n\} \\ d_i(x,y,y) &= d_{i+1}(x,x,y) && \text{for all } i \in \{1,\dots,n-1\} \\ d_n(x,y,y) &= d_n(y,y,y). && \end{aligned}$$

A chain of quasi directed Jónsson operations where the operations are idempotent is called a *chain of directed Jónsson operations*.

PROPOSITION 6.9.9. *If a clone contains a chain $d_1, \dots, d_n$ of quasi directed Jónsson operationsx then it also contains a chain $j_1, \dots, j_{2n-1}$ of quasi Jónsson operations.*

PROOF. Define $j_1(x, y, z) := d_1(x, y, z)$ and for $i \in \{1, \dots, n-1\}$

$$\begin{aligned} j_{2i}(x, y, z) &:= d_{i+1}(x, x, z) \\ j_{2i+1}(x, y, z) &:= d_{i+1}(x, y, z). \end{aligned}$$

Then

$$\begin{aligned} j_1(x, x, y) &= d_1(x, x, y) = d_1(x, x, x) = j_1(x, x, x), \\ j_{2n-1}(x, y, y) &= d_n(x, y, y) = d_n(y, y, y) = j_{2n-1}(y, y, y), \end{aligned}$$

and for $i \in \{1, \dots, n-1\}$:

$$\begin{aligned} j_{2i}(x, y, x) &= d_{i+1}(x, x, x) = j_{2i}(x, x, x) \\ j_{2i+1}(x, y, x) &= d_{i+1}(x, y, x) = d_{i+1}(x, x, x) = j_{2i+1}(x, x, x) \\ j_{2i-1}(x, y, y) &= d_i(x, y, y) = d_{i+1}(x, x, y) = j_{2i}(x, y, y) \\ j_{2i}(x, x, y) &= d_{i+1}(x, x, y) = j_{2i+1}(x, x, y). \end{aligned}$$

⊣

PROPOSITION 6.9.10. *If a clone contains an n-ary quasi near-unanimity operation f, then it also contains a chain $d_1, \dots, d_n$ of quasi directed Jónsson operations.*

PROOF. For $i \in \{1, \dots, n\}$ define $d_i(x, y, z) := f(x, \dots, x, y, z, \dots, z)$ where the argument y is at position $n - i + 1$. ⊣

It has been shown by Kazda, Kozik, McKenzie, and Moore that a clone contains a chain of Jónsson operations if and only if it contains a chain of directed Jónsson operations [227]. Whether the same is true for quasi Jónsson operations and quasi directed Jónsson operations is not clear to the author (see Question 19). A slight variation of the previous definition from [227] yields a much stronger condition.

DEFINITION 6.9.11. A sequence $p_1, \dots, p_n$ of ternary operations on a set B is called a *chain of quasi Pixley operations* if for all $x, y \in B$

$$\begin{aligned} p_1(x, y, y) &= p_1(x, x, x) & \\ p_i(x, y, x) &= p_i(x, x, x) & \text{for all } i \in \{1, \dots, n\} \\ p_i(x, x, y) &= p_{i+1}(x, y, y) & \text{for all } i \in \{1, \dots, n-1\} \\ p_n(x, x, y) &= p_n(y, y, y). & \end{aligned}$$

A chain of quasi Pixley operations whose operations are idempotent is called a *chain of Pixley operations*.

PROPOSITION 6.9.12. *If a clone contains a chain $p_1, \dots, p_n$ of quasi Pixley operations, then it also contains a chain $j_1, \dots, j_{2n+1}$ of quasi Jónsson operations.*

PROOF. Define

$$\begin{aligned} j_1(x,y,z) &:= p_1(x,x,x) \\ j_{2i}(x,y,z) &:= p_i(x,y,z) && \text{for } i \in \{1,\dots,n\} \\ j_{2i+1}(x,y,z) &:= p_{i+1}(x,z,z) && \text{for } i \in \{1,\dots,n-1\} \\ j_{2n+1}(x,y,z) &:= p_n(z,z,z). \end{aligned}$$

Then

$$j_1(x,x,y) = p_1(x,x,x) = j_1(x,x,x),$$
$$j_{2n+1}(x,y,y) = p_n(x,y,y) = j_{2n+1}(y,y,y),$$

and for $i \in \{1,\dots,n\}$:

$$\begin{aligned} j_{2i}(x,y,x) &= p_i(x,y,x) = p_i(x,x,x) = j_{2i}(x,x,x) \\ j_{2i+1}(x,y,x) &= p_{i+1}(x,x,x) = j_{2i+1}(x,x,x) \\ j_{2i-1}(x,y,y) &= p_i(x,y,y) = j_{2i}(x,y,y) \\ j_{2i}(x,x,y) &= p_i(x,x,y) = p_{i+1}(x,y,y) = j_{2i+1}(x,x,y). \end{aligned}$$ ⊣

Note that if p_1, p_2, p_3 is a chain of quasi Pixley operations of length $n = 3$, then p_2 is a quasi Maltsev operation. We have already see that an ω-categorical model-complete core never has a quasi Maltsev polymorphism (Proposition 6.1.44). However, there are ω-categorical model-complete cores with essentially infinite signature and a chain of quasi Pixley polymorphisms of length $n = 4$ (Proposition 8.5.18).

Chapter 7

EQUALITY CONSTRAINT SATISFACTION PROBLEMS

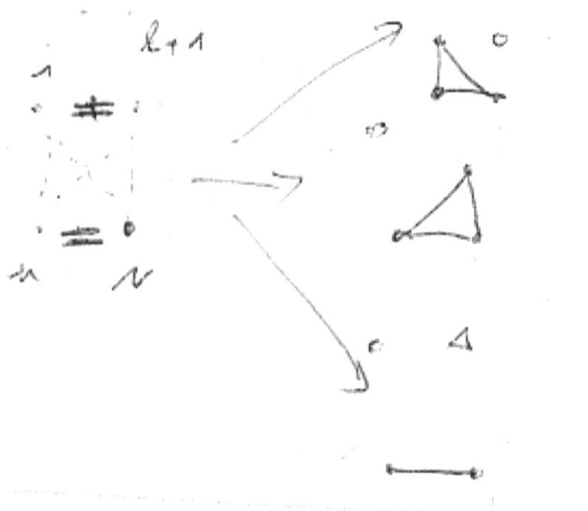

Jan Kára, 2005

This section is about first-order reducts of $(\mathbb{N}; =)$; From a model theory perspective, such structures appear to be altogether trivial. However, the set of all such structures, ordered by primitive positive definability, is a quite complicated object; there are actually 2^{ω} many first-order reducts of $(\mathbb{N}; =)$ up to primitive positive interdefinability [54].

First-order reducts of $(\mathbb{N}; =)$ will be called *equality constraint languages*, in particularly when used as templates for CSPs. It is easy to show that a structure $\mathfrak{B}$ is isomorphic to an equality constraint language if and only if $\mathfrak{B}$ is preserved by all permutations of its domain (cf. Section 4.2). Therefore, by the connection presented in Section 6.1 the results in this chapter concern locally closed clones that contain all permutations of the domain. Clones on infinite sets that contain all permutations are of independent interest in universal algebra [194, 303, 269, 304]. On a *finite* domain, such clones have been completely described in [191]; it turns out that the number of clones that contain all permutations of a fixed finite domain is finite. On *infinite* domains, local closure is a strong additional assumption, which allows a good understanding of the lattice of all locally closed clones that contain all permutations [54].

The CSP for a finite equality constraint language is called an *equality constraint satisfaction problem*. Equality CSPs are of fundamental importance in infinite-domain constraint satisfaction; we mention some reasons.

- NP-hard equality CSPs are useful for establishing hardness results of other infinite-domain CSPs. For instance, it follows from the results presented in this section that every structure which admits a primitive positive definition of the relation $\{(x, y, z) \mid (x = y \neq z) \vee (x \neq y = z)\}$ has an NP-hard CSP.
- When analysing an ω-categorical structure $\mathfrak{B}$ via the universal-algebraic approach, the question which equality constraint languages can be primitively positively defined in $\mathfrak{B}$ is of crucial importance, as we will see for instance in Chapter 12. For example, if the relation $\{(x, y, u, v) \mid x = y \Leftrightarrow u = v\}$ is primitively positively definable in $\mathfrak{B}$, then every polymorphism of $\mathfrak{B}$ that depends on all its arguments must be injective (Lemma 7.5.1).

The complexity of equality CSPs has been completely classified [68]; they are in P or NP-complete. In this chapter we present a new proof of this result. We show that a structure $\mathfrak{B}$ that is preserved by all permutations either has a binary injective polymorphism, in which case $\mathfrak{B}$ has a quantifier-free Horn definition in $(\mathbb{N}; =)$, and CSP($\mathfrak{B}$) is in P, or else allows primitive positive interpretations of all finite structures, in which case CSP($\mathfrak{B}$) is NP-complete. This implies that the tractability conjecture for reducts of finitely bounded homogeneous structures (Conjecture 3.1; also see Theorem 6.3.10) is true for equality CSPs.

The proof given here has the advantage that it divides the argument into several steps, some of which generalise to much larger classes of structures. Indeed, the central argument (Theorem 7.2.1) applies to all structures with a 2-transitive automorphism group, and several other results of this chapter also turn out to be useful in later classification arguments.

The fact that satisfiability of quantifier-free Horn clauses over $(\mathbb{N}; =)$ can be decided in polynomial time was observed in [224]. Here, we derive an algorithm from more general principles that will also be important for algorithmic results in Chapter 12.

7.1. Independence of disequality

We will now discuss a useful notion of *independence* of a particular relation from a template[19]. This notion has been discovered by a number of authors

[19]The notion of independence in constraint satisfaction should not be confused with independence in the sense of model theory, e.g., as in [5].

independently, beginning with [259, 260]. Here we focus on the case where the relation in question is disequality, $\neq$. A general definition of independence has been worked out in [132].

Independence has been applied in the study of temporal reasoning [243, 219] and qualitative reasoning calculi [104, 105], and is related to a notion called *convexity* in the literature on combining decision procedures [289, 16].

Definition 7.1.1 (Independence of disequality). Let $\mathfrak{B}$ be a structure with relational signature τ. Then we say that $\neq$ *is independent from* $\mathfrak{B}$ if for every primitive positive τ-formula ϕ with variables $x_1, y_1, \dots, x_n, y_n$, if for all $i \leq n$ the formula $\phi \wedge x_i \neq y_i$ is satisfiable over $\mathfrak{B}$, then $\phi \wedge \bigwedge_{i \leq n} x_i \neq y_i$ is satisfiable over $\mathfrak{B}$ as well.

In this section we prove that for arbitrary ω-categorical structures, independence of disequality is equivalent to the existence of a binary injective polymorphism. The following definition comes from [68].

Definition 7.1.2. A relation $R \subseteq B^k$ is called *intersection-closed* if for all k-tuples $(u_1, \dots, u_k), (v_1, \dots, v_k) \in R$ there is a tuple $(w_1, \dots, w_k) \in R$ such that for all $1 \leq i, j \leq k$ we have $w_i \neq w_j$ whenever $u_i \neq u_j$ or $v_i \neq v_j$.

The following lemma appears in [90]; see also [65].

Lemma 7.1.3. *Let $\mathfrak{B}$ be a countable ω-categorical structure. Then the following are equivalent.*

1. *Disequality $\neq$ is independent from $\mathfrak{B}$.*
2. *Every finite substructure of $\mathfrak{B}^2$ admits an injective homomorphism into $\mathfrak{B}$.*
3. *$\mathfrak{B}$ has a binary injective polymorphism.*
4. *Every primitively positively definable relation in $\mathfrak{B}$ is intersection-closed.*

Proof. We show the implications in cyclic order. Throughout the proof, let $b_1, b_2, \dots$ be an enumeration of the domain B of $\mathfrak{B}$.

(1) $\Rightarrow$ (2). Let $\mathfrak{A}$ be a finite induced substructure of $\mathfrak{B}^2$. Then the domain of $\mathfrak{A}$ is contained in $\{b_1, \dots, b_n\}^2$, for sufficiently large n. It clearly suffices to show that the substructure of $\mathfrak{B}^2$ induced on $\{b_1, \dots, b_n\}^2$ homomorphically and injectively maps to $\mathfrak{B}$, so let us assume without loss of generality that the domain of $\mathfrak{A}$ is $\{b_1, \dots, b_n\}^2$.

Consider the formula ϕ whose variables $x_1, \dots, x_{n^2}$ are the elements of $\mathfrak{A}$,

$$\begin{aligned} x_1 := (b_1, b_1), \dots, x_n := (b_1, b_n), \dots, x_{n^2-n+1} \\ := (b_n, b_1), \dots, x_{n^2} \\ := (b_n, b_n), \end{aligned}$$

and which is the conjunction over all literals $R((b_{i_1}, b_{j_1}), \dots, (b_{i_k}, b_{j_k}))$ such that $R(b_{i_1}, \dots, b_{i_k})$ and $R(b_{j_1}, \dots, b_{j_k})$ hold in $\mathfrak{B}$; so ϕ states precisely which relations hold in $\mathfrak{A}$ (by the ω-categoricity of $\mathfrak{B}$, the conjunction can be chosen to be finite). We claim that the formula $\phi \wedge \bigwedge_{1 \leq k \leq m} x_{i_k} \neq x_{j_k}$ with the property that $i_k \neq j_k$ for all $1 \leq k \leq m$ is satisfiable over $\mathfrak{B}$. This implies that there exists an injective homomorphism from $\mathfrak{A}$ into $\mathfrak{B}$. To prove the claim, let $i := i_k$ and $j := j_k$ for $k \leq m$ be distinct and let r, s be the n^2-tuples defined as follows.

$$r := (b_1, \dots, b_1, b_2, \dots, b_2, \dots, b_n, \dots, b_n);$$
$$s := (b_1, b_2, \dots, b_n, b_1, b_2, \dots, b_n, \dots, b_1, b_2, \dots, b_n).$$

These two tuples satisfy ϕ, because the projections to the first and second coordinate, respectively, are homomorphisms from $\mathfrak{A}$ to $\mathfrak{B}$. Now r or s satisfies $x_i \neq x_j$, proving that $\phi \wedge x_i \neq x_j$ is satisfiable in $\mathfrak{B}$. The claim now follows from the independence of $\neq$ from $\mathfrak{B}$.

The implication (2) $\Rightarrow$ (3) follows from Lemma 4.1.10, because the property that a function is injective can be described by the universal first-order sentence $\forall \bar{x}, \bar{y}\, (\bar{x} \neq \bar{y} \Rightarrow f(\bar{x}) \neq f(\bar{y}))$.

(3) $\Rightarrow$ (4). If f is a binary injective polymorphism of $\mathfrak{B}$, then it also preserves all relations that have a primitive positive definition in $\mathfrak{B}$. Clearly, every relation preserved by an injective function is intersection-closed.

The implication (4) $\Rightarrow$ (1) is straightforward as well. ⊣

Under the condition that the relation $\neq$ is primitively positively definable in $\mathfrak{B}$, we have a further characterisation of independence of disequality, which appeared in [90]; see also [65]. The condition given in this characterisation is in many situations easier to verify.

DEFINITION 7.1.4 (Local Independence of Disequality). Let $\mathfrak{B}$ be a structure with relational signature τ. Then we say that *$\neq$ is locally independent from $\mathfrak{B}$* if for all primitive positive τ-formula ϕ with variables x_1, y_1, x_2, y_2, if each of the formulas $\phi \wedge x_1 \neq y_2$ and $\phi \wedge x_2 \neq y_2$ is satisfiable over $\mathfrak{B}$, then $\phi \wedge x_1 \neq y_1 \wedge x_2 \neq y_2$ is satisfiable over $\mathfrak{B}$ as well.

LEMMA 7.1.5. *Let $\mathfrak{B}$ be a countable ω-categorical structure where $\neq$ is primitively positively definable. Then the following are equivalent.*

1. *$\neq$ is independent from $\mathfrak{B}$.*
2. *$\neq$ is locally independent from $\mathfrak{B}$.*

PROOF. Clearly, (1) implies (2). To prove that (2) implies (1) we use the criterion of Lemma 7.1.3 (2). Arguing as in the proof of Lemma 7.1.3, we let $\mathfrak{A}$ be the substructure of $\mathfrak{B}^2$ induced on a finite subset of the form $\{b_1, \dots, b_n\}^2$. Let ϕ be defined from $\mathfrak{A}$ as in that proof. Using induction over the number of disequalities, we will now show that for any conjunction $\sigma := \bigwedge_{1 \leq k \leq m} x_{i_k} \neq x_{j_k}$ with the property that $i_k \neq j_k$ for all $1 \leq k \leq m$, the formula $\phi \wedge \sigma$ is satisfiable over $\mathfrak{B}$. This implies that there exists an n^2-tuple t in $\mathfrak{B}$ with pairwise distinct entries which satisfies ϕ; the assignment that sends every x_i to t_i is an injective homomorphism from $\mathfrak{A}$ into $\mathfrak{B}$.

For the base of the induction, let $x_i \neq x_j$ be any disequality. Let r, s be the n^2-tuples defined as follows.

$$r := (b_1, \ldots, b_1, b_2, \ldots, b_2, \ldots, b_n, \ldots, b_n);$$
$$s := (b_1, b_2, \ldots, b_n, b_1, b_2, \ldots, b_n, \ldots, b_1, b_2, \ldots, b_n).$$

These two tuples satisfy ϕ, because the projections to the first and second coordinate, respectively, are homomorphisms from $\mathfrak{A}$ to $\mathfrak{B}$. Now either r or s satisfies $x_i \neq x_j$, proving that $\phi \wedge x_i \neq x_j$ is satisfiable in $\mathfrak{B}$.

In the induction step, let a conjunction $\sigma := \bigwedge_{1 \leq k \leq m} x_{i_k} \neq x_{j_k}$ be given, where $m \geq 2$. Set $\sigma' := \bigwedge_{3 \leq k \leq m} x_{i_k} \neq x_{j_k}$, and $\phi' := \phi \wedge \sigma'$. Observe that ϕ' has a primitive positive definition in $\mathfrak{B}$, as ϕ and $\neq$ have such definitions. By induction hypothesis, both $\phi' \wedge x_{i_1} \neq x_{j_1}$ and $\phi' \wedge x_{i_2} \neq x_{j_2}$ are satisfiable in $\mathfrak{B}$. But then $\phi' \wedge x_{i_1} \neq x_{j_1} \wedge x_{i_2} \neq x_{j_2} = \phi \wedge \sigma$ is satisfiable over $\mathfrak{B}$ as well by (2), concluding the proof. ⊣

We close with an application to CSPs. When $\mathfrak{A}$ is an instance of CSP($\mathfrak{B}$), then an injective homomorphism from $\mathfrak{A}$ to $\mathfrak{B}$ is also called an *injective solution* for $\mathfrak{A}$.

PROPOSITION 7.1.6. *Suppose that $\mathfrak{B}$ has a binary injective polymorphism. Then every satisfiable instance $\mathfrak{A}$ of* CSP($\mathfrak{B}$) *either has an injective solution, or $\mathfrak{A}$ has two distinct elements a, a' such that $s(a) = s(a')$ in all solutions s for $\mathfrak{A}$.*

PROOF. Suppose that $\mathfrak{A}$ has a solution, but no injective solution. Let f be a solution such that the cardinality of f is maximal. Since there is no injective solution, there are two elements a, a' of $\mathfrak{A}$ such that $f(a) = f(a')$. We claim that $s(a) = s(a')$ in *all* solutions s of $\mathfrak{A}$. Otherwise, if $s(a) \neq s(a')$ for some solution s, then by the choice of f there must be another pair b, b' such that $s(b) \neq s(b')$ but $f(b) \neq f(b')$. Let h be the binary injective polymorphism of $\mathfrak{B}$. Then the mapping $x \mapsto h(f(x), s(x))$ is also a solution to $\mathfrak{A}$, but has a strictly larger image than f, a contradiction. ⊣

7.2. Two-transitive templates

We show that every structure with a 2-transitive automorphism group (in particular, every equality constraint language) with an essential polymorphism but without constant polymorphisms also has a binary injective polymorphism. Here we use Corollary 6.1.31 about the existence of binary essential polymorphisms, and Lemma 7.1.5 about the existence of binary injective polymorphisms.

THEOREM 7.2.1. *Let $\mathfrak{B}$ be a structure with a 2-transitive automorphism group such that* Pol($\mathfrak{B}$) *contains an essential operation but no constant operation. Then* Pol($\mathfrak{B}$) *also contains a binary injective operation.*

PROOF. Corollary 6.1.31 implies that Pol($\mathfrak{B}$) contains a binary essential operation f. Since $\mathfrak{B}$ has no constant polymorphism and is 2-set transitive, Corollary 6.1.27 implies that all polymorphisms of $\mathfrak{B}$ preserve $\neq$, and hence $\neq$ is primitively positively definable in $\mathfrak{B}$. So we can apply Lemma 7.1.5, and have to show that for every primitive positive formula ϕ the formula $\phi \wedge x \neq y \wedge u \neq v$ is satisfiable over $\mathfrak{B}$ whenever $\phi \wedge x \neq y$ and $\phi \wedge u \neq v$ are satisfiable over $\mathfrak{B}$.

Let V be the variables of ϕ, and let $s\colon V \to B$ be a satisfying assignment for $\phi \wedge x \neq y$, and $t\colon V \to B$ be a satisfying assignment for $\phi \wedge u \neq v$. We can assume that $s(u) = s(v)$ and $t(x) = t(y)$, as otherwise we are done. Let k be the cardinality of the set $\{s(x), s(y), s(u), t(u), t(v), t(x)\}$; note that $4 \leq k \leq 6$. Suppose that $k = 6$, the other cases are simpler. Since f is essential, it does not preserve the relation P^3_B (Lemma 6.1.17).

Since f preserves $\neq$, we can therefore assume that there are tuples (a, a, b) for $a \neq b$ and (c, d, d) for $c \neq d$ such that $f(a, c) \neq f(a, d)$ and $f(a, d) \neq f(b, d)$. By 2-transitivity of Aut($\mathfrak{B}$), there are $\alpha, \beta \in$ Aut($\mathfrak{B}$) such that $\alpha(s(u), s(x)) = (a, b)$, and $\beta(t(u), t(x)) = (c, d)$. Since $s(x) \neq s(y)$ and $t(v) \neq t(u)$, and f preserves $\neq$, we have

$$f(\alpha(s(x)), \beta(t(v))) \neq f(\alpha(s(y)), \beta(t(u))).$$

This implies that

$$f(\alpha(s(x)), \beta(t(v))) = f(\alpha(s(y)), \beta(t(v)))$$

and

$$f(\alpha(s(y)), \beta(t(v))) = f(\alpha(s(y)), \beta(t(u)))$$

cannot both be true. By 2-transitivity of Aut($\mathfrak{B}$), there exist α', β' such that $\alpha'(a, \alpha(u)) = (\alpha(u), a)$, and $\beta'(d, \beta(v)) = (\beta(v), d)$.

If $f(\alpha(s(x)), \beta(t(v))) \neq f(\alpha(s(y)), \beta(t(v)))$, then $z \mapsto f(\alpha s(z), \beta' t(z))$ is a satisfying assignment for $\phi \wedge x \neq y \wedge u \neq v$. If $f(\alpha(s(y)), \beta(t(v))) \neq f(\alpha(s(y)), \beta(t(u)))$, then $z \mapsto f(\alpha' s(z), \beta t(z))$ is a satisfying assignment for $\phi \wedge x \neq y \wedge u \neq v$. ⊣

7.3. Existential Horn formulas and square embeddings

In this section we show that $\mathfrak{B}$ has certain binary injective polymorphisms if and only if all relations in $\mathfrak{B}$ have a quantifier-free Horn definition over a 'base' structure $\mathfrak{C}$; this is useful to design algorithms for CSP($\mathfrak{B}$). We then apply these results to equality CSPs.

The following is a simple, but very useful definition for syntactic purposes.

DEFINITION 7.3.1. A quantifier-free first-order formula ϕ in conjunctive normal form is called *reduced* (over a structure $\mathfrak{B}$) if every formula obtained from ϕ by removing a literal is not equivalent to ϕ (over $\mathfrak{B}$).

Clearly, every quantifier-free formula is equivalent to a reduced formula over $\mathfrak{B}$, because we can find one by successively removing literals from ϕ. The following theorem is from [53] and [61] (stated there for quantifier-free Horn formulas only).

THEOREM 7.3.2. *Let $\mathfrak{B}$ be a structure with an embedding $e \colon \mathfrak{B}^2 \hookrightarrow \mathfrak{B}$. Then a relation R with an existential definition (quantifier-free definition) in $\mathfrak{B}$ has an existential Horn definition (or quantifier-free Horn definition, respectively) in $\mathfrak{B}$ if and only if R is preserved by e.*

PROOF. Forwards (necessity). Let δ be an existential Horn definition of R over $\mathfrak{B}$, written in prenex conjunctive normal form. It suffices to demonstrate that e preserves each clause in δ. Note that a Horn clause ψ of δ can always be written in the form $(\phi_1 \wedge \cdots \wedge \phi_l) \rightarrow \phi_0$, for atomic τ-formulas $\phi_0, \ldots, \phi_l$. Let V be the variables of ψ, and let $s_1, s_2 \colon V \rightarrow \mathbb{N}$ be two assignments that satisfy the clause. We claim that $s_3 \colon V \rightarrow \mathbb{N}$ defined by $s_3(x, y) = e(s_1(x), s_2(y))$ satisfies ψ. There are two cases to consider. Either there is an $i \leq l$ such that s_1 or s_2 does not satisfy ϕ_i. In this case, since $e \colon \mathfrak{B}^2 \hookrightarrow \mathfrak{B}$ is an embedding, s_3 does not satisfy ϕ_i, and therefore satisfies ψ. Or, if for all $i \leq l$ both s_1 and s_2 satisfy ϕ_i, then they also satisfy ϕ_0. Since e is a polymorphism of $\mathfrak{B}$, it follows that s_3 satisfies ϕ_0, and therefore also ψ.

Backwards (sufficiency). Consider an existential definition δ of R in $\mathfrak{B}$ such that δ is in prenex normal form, and that the quantifier-free part η of δ is a reduced formula in conjunctive normal form over $\mathfrak{B}$. Assume for contradiction that δ is not existential Horn, that is, η has a clause $\psi = \phi_1 \vee \phi_2 \vee \phi_3 \vee \cdots \vee \phi_l$ where ϕ_1, ϕ_2 are positive literals, and $\phi_3, \ldots, \phi_l$ are positive or negative literals. Let V be the variables of η. Since η is reduced, it has a satisfying assignment $s_1 \colon V \rightarrow \mathbb{N}$ such that ϕ_i is false for all $i \leq l$ except for $i = 1$. Similarly, η has a satisfying assignment $s_2 \colon V \rightarrow \mathbb{N}$ such that ϕ_i is false for all $i \leq l$ except for $i = 2$. Then $s_3 \colon V \rightarrow \mathbb{N}$ defined by $s_3(x, y) = e(s_1(x), s_2(y))$ does not satisfy ψ, a contradiction.

It is clear that the same proof works in the special case that the relation R has a *quantifier-free* definition over $\mathfrak{B}$: the proof in fact shows that every quantifier-free reduced formula preserved by e must be quantifier-free Horn. ⊣

Example 7.3.3. The structure $\mathfrak{B} := (\mathbb{N}; =)$ is an obvious example with an embedding from $\mathfrak{B}^2$ into $\mathfrak{B}$.

Example 7.3.4. Let $V_{\mathbb{F}}$ be a countable vector space over a finite field $\mathbb{F}$. Then $(V_{\mathbb{F}})^2$ is again a countable vector space over $\mathbb{F}$, again of countably infinite dimension, and hence isomorphic to $V_{\mathbb{F}}$.

Example 7.3.5. Let $\mathbb{A}$ be the countable atomless Boolean algebra (Example 4.1.4). Then $\mathbb{A}^2$ is a countable and atomless Boolean algebra, and thus isomorphic to $\mathbb{A}$.

For the structures $\mathfrak{C}$ from the previous examples we even had an isomorphism between $\mathfrak{C}^2$ to $\mathfrak{C}$. The following is an example where $\mathfrak{C}^2$ embeds into $\mathfrak{C}$, but not surjectively.

Example 7.3.6. Let $(V; E)$ be the countable random graph (defined in Section 4.1.1). By the universality of $(V; E)$, the graph $(V; E)^2$ embeds into $(V; E)$. Note that $(V; E)^2$ is not isomorphic to $(V; E)$: When $E(x_1, x_2)$ and $E(y_1, y_2)$, then there is no point in $(V; E)^2$ that is adjacent to (x_1, y_1) and (x_2, y_2) but not adjacent to (x_1, y_2) and (x_2, y_1), violating the extension property of $(V; E)$.

For a relational structure $\mathfrak{B}$ let $\mathfrak{B}^\neg$ be the expansion of $\mathfrak{B}$ by all relations that are the complement of a relation from $\mathfrak{B}$. The following formulation is from [53], but similar statements were proved earlier [134].

THEOREM 7.3.7. *Let $\mathfrak{C}$ be a structure with an embedding $e \colon \mathfrak{C}^2 \hookrightarrow \mathfrak{C}$. Let $\mathfrak{B}$ be a relational structure with finite signature σ that is preserved by e and has an existential definition in $\mathfrak{C}$. Then there is a polynomial-time Turing reduction from* CSP($\mathfrak{B}$) *to* CSP($\mathfrak{C}^\neg$).

```
// Input: An instance φ of CSP(𝔅)
// Assumption: 𝔅 has an existential Horn definition in a τ-structure ℭ.
Replace each constraint R(x_1, ..., x_n) from φ by δ(x_1, ..., x_n),
   where δ is an existential Horn definition of R in ℭ.
Let ψ be the resulting τ-sentence, written in prenex conjunctive normal
form.
Do
   Let Ψ be the set of all singleton clauses in ψ.
   If Ψ is unsatisfiable over ℭ then reject.
   For each negative literal η of ψ do
     If Ψ ∪ {η}, considered as an instance of CSP(ℭ¬), is unsatisfiable
       Remove η from its clause in ψ
Loop until no further changes are made
Accept
```

FIGURE 1. A polynomial-time Turing reduction from CSP($\mathfrak{B}$) to CSP($\mathfrak{C}^\neg$) if $\mathfrak{B}$ is preserved by an embedding $\mathfrak{C}^2 \hookrightarrow \mathfrak{C}$.

PROOF. We use the algorithm shown in Figure 1. By Theorem 7.3.2, every relation of $\mathfrak{B}$ has an existential Horn definition in $\mathfrak{C}$. Let ϕ be an input instance of CSP($\mathfrak{B}$), and let ψ be the sentence over the signature of $\mathfrak{C}$ obtained from ϕ as described in the algorithm. Since σ is finite and fixed, and does not depend on the input, only a linear number of literals can be deleted from ψ in the course of the algorithm. It is thus clear that the algorithm works in polynomial time.

To show that the algorithm is correct, observe that ϕ is false in $\mathfrak{B}$ if and only if ψ is false in $\mathfrak{C}$. We first show that if the algorithm rejects, then ψ is false in $\mathfrak{B}$. The reason is that whenever a negative literal η is removed from a clause of ψ, then in fact $\neg\eta$ is implied by the other clauses in ψ, and therefore removing η from ψ leads to an equivalent formula.

Finally, we show that if the algorithm accepts, then ψ is true in $\mathfrak{C}$. Let B be the domain of $\mathfrak{B}$ and $\mathfrak{C}$, and let V be the set of variables of ψ. Consider the negative literals $\eta_1, \ldots, \eta_m$ that are in clauses of ψ at the final stage of the algorithm. For all $i \leq m$, let $t_i \colon V \to B$ be an assignment that satisfies all clauses of ψ without negative literals, and which also satisfies η_i. Such an assignment must exist, since otherwise η_i would have been false in all solutions, and our algorithm would have removed η_i in the inner loop of the algorithm. We claim that $s \colon V \to B$ given by

$$s(x) = e(t_1(x), e(t_2(x), \ldots e(t_{m-1}(x), t_m(x)) \ldots))$$

satisfies all clauses of ψ. Negative literals η_k are satisfied because t_k satisfies η_k and $e \colon \mathfrak{C}^2 \hookrightarrow \mathfrak{C}$ is an embedding. Positive literals from ψ are satisfied by s because they are satisfied by all the t_i, and since e is a polymorphism of $\mathfrak{C}$. $\dashv$

Note that in Theorem 7.3.7 we did not assume the ω-categoricity of $\mathfrak{B}$ or $\mathfrak{C}$.

7.4. Classification

In this section we prove a complexity classification for CSP($\mathfrak{B}$) where $\mathfrak{B}$ is an equality constraint language; the classification confirms the tractability conjecture (Conjecture 3.1). The structure $(\mathbb{N}; =)$ has quantifier elimination: this follows from Corollary 4.3.3 by the observation that every bijection between finite subsets of $\mathbb{N}$ can be extended to a permutation of $\mathbb{N}$. In fact, by the same argument every equality constraint language has quantifier elimination.

THEOREM 7.4.1. *Let $\mathfrak{B}$ be an equality constraint language. Then one of the following cases applies.*

1. *$\mathfrak{B}$ has a constant polymorphism. In this case, for every reduct $\mathfrak{B}'$ of $\mathfrak{B}$ with finite signature* CSP($\mathfrak{B}'$) *can be solved in polynomial time.*
2. *$\mathfrak{B}$ has a binary injective polymorphism. In this case, for every reduct $\mathfrak{B}'$ of $\mathfrak{B}$ with finite signature* CSP($\mathfrak{B}'$) *can be solved in polynomial time.*
3. *All polymorphisms of $\mathfrak{B}$ are essentially unary and preserve $\neq$; equivalently, every first-order formula is over $\mathfrak{B}$ equivalent to a primitive positive formula; equivalently, the relation $\neq$ and the relation P^4_B (Definition* 6.1.16*) are primitively positively definable in $\mathfrak{B}$. In this case, there exists a reduct $\mathfrak{B}'$ of $\mathfrak{B}$ with finite signature such that* CSP($\mathfrak{B}'$) *is NP-hard.*

PROOF. If $\mathfrak{B}$ has a constant endomorphism, then the claim for finite reducts of $\mathfrak{B}$ follows from Proposition 1.1.12. So suppose in the following that $\mathfrak{B}$ does not have a constant polymorphism. Since equality constraint languages have 2-transitive automorphism groups, we can use the contrapositive of Corollary 6.1.27 to derive that all polymorphisms of $\mathfrak{B}$ must preserve $\neq$. The endomorphisms of $\mathfrak{B}$ are therefore injective, and locally generated by the automorphisms of $\mathfrak{B}$. If $\mathfrak{B}$ does not have essential polymorphisms, then Corollary 6.1.20 shows that all relations that are first-order definable in $\mathfrak{B}$ are also primitively positively definable in $\mathfrak{B}$, and we are in case (3); for the equivalent characterisations of this case, see Proposition 6.1.19. If $\mathfrak{B}$ has an essential polymorphism, then $\mathfrak{B}$ has a binary injective polymorphism by Theorem 7.2.1. Since every relation of $\mathfrak{B}$ has a quantifier-free definition over $(\mathbb{N}; =)$, Theorem 7.3.2 shows that every relation of $\mathfrak{B}$ even has a quantifier-free Horn definition over $\mathfrak{B}$. By Theorem 7.3.7, the CSP for every finite signature reduct of $\mathfrak{B}$ can be reduced to CSP$(\mathbb{N}; =, \neq)$ in polynomial time. Tractability of CSP$(\mathbb{N}; =, \neq)$ has been shown in Section 1.1. ⊣

We can now give the complexity classification for equality constraint languages. The classification confirms the tractability conjecture (Conjecture 3.1) in the special case of equality CSPs (note that every equality CSP is the CSP for a reduct of a finitely bounded homogeneous structure).

THEOREM 7.4.2. *Let $\mathfrak{B}$ be an equality constraint language. Then exactly one of the following cases applies.*

- *$\mathfrak{B}$ has a polymorphism f and an automorphism α such that*
$$f(x, y) = \alpha f(y, x)$$
for all elements x and y of $\mathfrak{B}$. In this case, for every finite reduct $\mathfrak{B}'$ of $\mathfrak{B}$ the problem CSP$(\mathfrak{B}')$ *can be solved in polynomial time.*
- *All finite structures have a primitive positive interpretation in $\mathfrak{B}$. In this case, there is a finite reduct $\mathfrak{B}'$ of $\mathfrak{B}$ such that* CSP$(\mathfrak{B}')$ *is NP-complete.*

PROOF. If $\mathfrak{B}$ has a constant polymorphism, then clearly there are f and α such that $f(x, y) = \alpha f(y, x)$ for all $x, y \in B$. Now suppose that $\mathfrak{B}$ has a binary injective polymorphism f. Then there exists a permutation α of B such that $f(x, y) = \alpha f(y, x)$ for all $x, y \in \mathbb{N}$. By Theorem 7.4.1, the only remaining case is that over $\mathfrak{B}$ all first-order formulas are equivalent to primitive positive formulas. In this case the claim follows from Theorem 3.2.2. That the two cases are disjoint follows from Corollary 6.6.8. ⊣

7.5. Essential injectivity

We have seen in Section 6.1.4 that clones of operations that are essentially unary can be characterised using relations definable from equality. Also the situation where all operations are *essentially injective* can be characterised using such relations. An operation $f: B^k \to B$ is called *essentially injective* if there exist $i_1, \dots, i_l$ and an injective function $g: B^k \to B$ such that for all $x_1, \dots, x_n \in B$ we have

$$f(x_1, \dots, x_k) = g(x_{i_1}, \dots, x_{i_l}).$$

We first prove the following intermediate results.

LEMMA 7.5.1. *Let f be an operation from B^k to B that depends on all arguments. Then the following are equivalent.*

1. *f is injective.*
2. *f preserves the relation defined by $x = y \Leftrightarrow u = v$.*
3. *f preserves the relation I_4 defined by $x = y \Rightarrow u = v$.*

PROOF. For the implication from (1) to (2), suppose that f is injective. We check that f preserves $x = y \Leftrightarrow u = v$. Let a, b, c, d be elements of B^k such that $a_i = b_i \Leftrightarrow c_i = d_i$ for all $i \leq k$, and let t be the tuple $(f(a), f(b), f(c), f(d))$. If $a = b$, we thus have that $c_i = d_i$ for all $i \leq k$, and so $c = d$. In this case, t satisfies $t_1 = t_2$ and $t_3 = t_4$, and we are done. Similarly, if $c = d$ then $a = b$ and we are done. Otherwise, $a \neq b$ and $c \neq d$, and by injectivity of f we have $t_1 \neq t_2$ and $t_3 \neq t_4$. So we have in all cases that $t_1 = t_2$ if and only if $t_3 = t_4$.

For the implication from (2) to (3), note that $x = y \Rightarrow u = v$ is equivalent to a primitive positive formula over $(B; R)$ where $R = \{(a, b, c, d) \mid a = b \Leftrightarrow c = d\})$. The primitive positive formula is

$$\exists w \, \big(R(x, y, u, w) \wedge R(x, y, w, v)\big).$$

Finally, for the implication from (3) to (1), suppose that there are distinct $a, b \in B^k$ such that $f(a) = f(b)$. We want to prove that f does not preserve $x = y \Rightarrow u = v$. Let I be the set of all $i \in \{1, \dots, k\}$ such that $a_i \neq b_i$. Since a and b are distinct, I is non-empty; let $i \in I$ be arbitrary. Since f depends on the i-th argument, there are $c, d \in D^k$ with $c_j = d_j$ for all $j \neq i$, and $c_i \neq d_i$. We claim that (a, b, c, d) shows that f does not preserve $x = y \Rightarrow u = v$. First, note that for all $j \in \{1, \dots, k\} \setminus I$, we have that $a_j = b_j$ and $c_j = d_j$. Next, note that for all $j \in I$ we have that $a_j \neq b_j$. We conclude that for all $j \in \{1, \dots, k\}$ we have that $a_i = b_i$ implies $c_i = d_i$. However, $f(a) = f(b)$ and $f(c) \neq f(d)$. ⊣

We obtain several equivalent descriptions of a particularly important oligomorphic clone, the *Horn clone* $\mathcal{H}$. In general, there are four natural ways to specify an operation clone over a domain B:

- Give a preferably finite set of operations that generates $\mathcal{C}$.
- Give an explicit description of all operations in the clone.
- Give a preferably finite set of relations $R_1, R_2, \dots$ on B such that $\mathcal{C} = \mathrm{Pol}(B; R_1, R_2, \dots)$.

- Give an explicit description of all relations in $\text{Inv}(\mathcal{C})$.

For most oligomorphic clones, not all of these four potential descriptions of the clone are known. For the Horn clone, however, we have all four descriptions, and we present them in the following theorem. The proof follows easily from the facts developed in this chapter and is left to the reader.

THEOREM 7.5.2. *The following four definitions are equivalent.*

- $\mathcal{H} := \overline{\langle \{i\} \cup \text{Sym}(\mathbb{N})\rangle}$ *where* $i\colon \mathbb{N}^2 \to \mathbb{N}$ *is any injective map.*
- $\mathcal{H}$ *is the set of all essentially injective operations.*
- $\mathcal{H} := \text{Pol}(\mathbb{N}; \neq, I_4)$.
- $\mathcal{H}$ *is the set of all operations such that* $\text{Inv}(\mathcal{H})$ *is precisely the set of all relations with a quantifier-free Horn definition over* $(\mathbb{N}; =)$.

The clone thus defined is called the *Horn clone*.

7.6. Injectivity in one direction

In this section we present another locally closed clone, introduced in [54], that contains $\text{Sym}(\mathbb{N})$ and that provides examples and counterexamples for many basic questions that can be asked about oligomorphic clones.

DEFINITION 7.6.1. Let B be a set and $n \geq 1$. An operation $f\colon B^n \to B$ is called *injective in the i-th direction* if for all $a, b \in B^n$

$$a_i \neq b_i \Rightarrow f(a) \neq f(b).$$

We say that f is *injective in one direction* if there exists an $i \in \{1, \ldots, n\}$ such that f is injective in the i-th direction. Let $\mathcal{R}$ be the set of all operations on $B = \mathbb{N}$ that are injective in one direction.

It is easy to verify that $\mathcal{R}$ is indeed a clone.

DEFINITION 7.6.2. A quantifier-free formula ϕ over the empty signature is called *negative* if it is in conjunctive normal form and if each clause of ϕ either

- consists of a single positive literal (which must be of the form $x = y$ for variables x and y), or
- does not contain positive literals (and so must be of the form $x_1 \neq y_1 \vee \cdots \vee x_n \neq y_n$ for variables $x_1, y_1, \ldots, x_n, y_n$).

The following theorem of Bodirsky, Chen, and Pinsker [54] gives equivalent characterisations of the clone $\mathcal{R}$.

THEOREM 7.6.3. *Let* $R \subseteq \mathbb{N}^k$ *be a relation that is preserved by all permutations of* $\mathbb{N}$ *(i.e.,* R *is first-order definable over the empty signature). Then the following are equivalent.*

1. *R is preserved by $\mathcal{R}$.*

2. *In* $(\mathbb{N}; R, \neq)$ *there is no primitive positive definition of the relation*

$$\{(a, b, c) \in \mathbb{N}^3 \mid a = b = c \vee |\{a, b, c\}| = 3\}.$$

3. *R has a negative definition.*

The clone $\mathcal{R}$ appears as a fundamental example for instance in Section 8.5.5.

Chapter 8

DATALOG

In Chapter 6 we presented algebraic conditions for finite or ω-categorical structures $\mathfrak{B}$ which imply that CSP($\mathfrak{B}$) is NP-hard. In this chapter we present an important technique to show that CSP($\mathfrak{B}$) is in P.

The most important and prominent family of algorithmic approaches to constraint satisfaction is based on *establishing local consistency*. There are numerous definitions and variations of what *local consistency* can mean. A convenient framework to formulate and study many of these variants is *Datalog*, which is a heavily studied formalism in database theory. Datalog can be viewed from a number of different perspectives, for example as

- an extension of *conjunctive queries* (see Section 1.2) by a mechanism for recursion;
- the fragment of the logic programming language *Prolog* [266] where function symbols are not allowed;

- the existential positive fragment of *least fixed-point logic* (Section 8.7);
- a fragment of monotone SNP (Section 1.4.2), namely *existential second-order Horn logic* [182].

Typical CSPs that can be solved by a Datalog program are

- CSP($\{0, 1\}; \{1\}, \{(x, y, z) \mid x \wedge y \Rightarrow z\}$), which is closely related to Horn-SAT; see Section 1.6.7 and Theorem 6.2.7;
- CSP($\{0, 1\}; \{(x, y) \mid x \Rightarrow y\}, \{(x, y) \mid x \vee y\}, \{(x, y) \mid \neg x \vee \neg y\}$), which is essentially 2-SAT; see Theorem 6.2.7;
- CSP($\mathbb{Q}; <$); see Example 1.1.2;
- CSP($\mathbb{Q}; \leq, \neq$); see Example 1.5.2;
- CSP($\mathbb{N}; \neq, \{(x, y, u, v) \mid x = y \Rightarrow u = v\}$), treated in Section 7.3;
- CSP($\mathbb{Q}; \neq, \{(x, y, u, v) \mid x = y \Rightarrow u \leq v\}$); this is closely related to the Ord-Horn constraints from Section 1.6.9.

The finite-domain structures $\mathfrak{B}$ such that CSP($\mathfrak{B}$) can be solved by Datalog have been characterised by Barto and Kozik, resolving the so-called (*Larose-Zadori*) *bounded width conjecture* [257], which was anticipated by Feder and Vardi [170] (see [256] for a discussion). Informally, the result states that CSP($\mathfrak{B}$) can be solved by Datalog unless CSP($\mathfrak{B}$) can 'simulate' linear equation systems over a finite field (Section 8.8). The bounded width conjecture is complemented by a result of Atserias, Bulatov, and Dawar [14] which states that if CSP($\mathfrak{B}$) *can* simulate systems of linear equations over a finite field, then CSP($\mathfrak{B}$) it not even expressible in the counting extension of least fixed-point logic, LFP+C. We also mention that Zhuk's algorithm [348] confirms the (informal) conjecture of Feder and Vardi that if CSP($\mathfrak{B}$) can be solved in polynomial time, then this is because CSP($\mathfrak{B}$) can be solved either by Datalog, by Gaussian elimination, or by (clever) combinations of these two approaches.

For countable ω-categorical structures $\mathfrak{B}$ a universal-algebraic characterisation of solvability of CSP($\mathfrak{B}$) in Datalog is not known. However, many of the fundamental definitions and results about Datalog and constraint satisfaction apply to all countable ω-categorical structures.

8.1. Introducing Datalog

There are (at least) two possibilities to formally introduce the semantics of Datalog programs. In this text, we choose the more concise approach via least fixed points. From this it will become obvious that every Datalog program is equivalent to a monotone SNP sentence.

Another approach is 'operational' in the sense that we explicitly specify how to evaluate a Datalog program on a given finite structure; this has the advantage that it is then easy to see that Datalog programs can be evaluated in polynomial time. The proof that the two approaches are equivalent can be found in Section 8.1.3.

8.1.1. Syntax of Datalog. A Datalog program consists of a finite set of *rules*, i.e., expressions that are traditionally written in the following form

$$\psi :- \phi_1, \ldots, \phi_m$$

where $m \geq 0$ and $\psi, \phi_1, \ldots, \phi_m$ are atomic formulas over some relational signature. The formula ψ is called the *head* of the rule, and $\phi_1, \ldots, \phi_m$ is called the *body*. We also require that all variables in the head also occur in the body.

The relation symbols occurring in the head of some clause are called *intensional database predicates* (or *IDBs*), and all other relation symbols in the clauses are called *extensional database predicates* (or *EDBs*). A Datalog program has *width* (ℓ, k) if all IDBs are at most ℓ-ary, and if all rules have at most k distinct variables.

Example 8.1.1. Before we give formal definitions of the semantics of Datalog, consider the following Datalog program Π, which has width $(2, 3)$:

$$tc(x, y) :- x < y$$
$$tc(x, y) :- tc(x, u), tc(u, y)$$
$$\mathit{false} :- tc(x, x)$$

Here, $<$ is a binary EDB (written in infix notation), tc is a binary IDB, and *false* is an IDB of arity 0. The program Π can be used to solve CSP$(\mathbb{Q}; <)$: the idea is that on a given instance $\mathfrak{A}$ of CSP$(\mathbb{Q}; <)$ the program Π computes the transitive closure of the relation $<$ in $\mathfrak{A}$, and that Π derives *false* if and only if it finds that (x, x) is in the transitive closure, i.e, if and only if $\mathfrak{A}$ contains a directed cycle. This will be made more precise in the following section.

8.1.2. Declarative semantics of Datalog. Let Π be a Datalog program with EDBs τ and IDBs σ and let $\mathfrak{A}$ be a τ-structure. An expansion $\mathfrak{A}'$ with signature $\tau \cup \sigma$ is called a *fixed point of* Π *on* $\mathfrak{A}$ if $\mathfrak{A}'$ satisfies the sentence $\forall \bar{x}(\psi \vee \neg\phi_1 \vee \cdots \vee \neg\phi_m)$ for each rule $\psi :- \phi_1, \ldots, \phi_m$ of Π. Recall from Section 1.1 that for two $(\tau \cup \sigma)$-structures $\mathfrak{A}_1$ and $\mathfrak{A}_2$ with the same domain, the structure $\mathfrak{A}_1 \cap \mathfrak{A}_2$ is the $(\tau \cup \sigma)$-structure where the relations are defined to be the intersections of the respective relations in $\mathfrak{A}_1$ and $\mathfrak{A}_2$. The following is easy to see from the definitions.

LEMMA 8.1.2. *Let* $\mathfrak{A}_1$ *and* $\mathfrak{A}_2$ *be two fixed points of* Π *on* $\mathfrak{A}$; *then* $\mathfrak{A}_1 \cap \mathfrak{A}_2$ *is a fixed point of* Π *on* $\mathfrak{A}$ *as well.*

So we can compare fixed-points by declaring $\mathfrak{A}_1 \leq \mathfrak{A}_2$ if $\mathfrak{A}_1 \cap \mathfrak{A}_2 = \mathfrak{A}_1$. The lemma above implies that for every finite structure $\mathfrak{A}$, the program Π has a unique smallest fixed-point on $\mathfrak{A}$, which we denote by $\Pi(\mathfrak{A})$.

Datalog programs can be used to describe classes of finite structures (and in particular CSPs). We require that there exists a distinguished IDB *false* of arity 0. We say that

- Π *derives false on* $\mathfrak{A}$ if the empty tuple is in $false^{\Pi(\mathfrak{A})}$;
- Π is *sound* for CSP($\mathfrak{B}$) if $\mathfrak{A}$ does not map homomorphically to $\mathfrak{B}$ whenever Π derives *false* on $\mathfrak{A}$;
- Π *solves* CSP($\mathfrak{B}$) if Π derives *false* on precisely those finite structures $\mathfrak{A}$ that do not map homomorphically to $\mathfrak{B}$;
- CSP($\mathfrak{B}$) has *(Datalog-) width* (ℓ, k) if there exists a Datalog program of width (ℓ, k) that solves CSP($\mathfrak{B}$);
- CSP($\mathfrak{B}$) has *(Datalog-) width* ℓ if it has width (ℓ, k) for some $k \in \mathbb{N}$;
- CSP($\mathfrak{B}$) *is in Datalog* if there exists a Datalog program that solves CSP($\mathfrak{B}$).

Clearly, CSP($\mathfrak{B}$) is in Datalog if and only it has *bounded width*, i.e., if CSP($\mathfrak{B}$) has Datalog width ℓ for some $\ell \in \mathbb{N}$.

Example 8.1.3. A constraint satisfaction problem satisfies the equivalent conditions given in Theorem 5.6.2 if and only if it has Datalog width 0: if CSP($\mathfrak{B}$) $=$ Forb$^{\text{hom}}(\mathcal{F})$ for a finite set of finite connected structures $\mathcal{F}$, then we introduce for each $\mathfrak{F} \in \mathcal{F}$ a rule with head *false* whose body is the canonical query of $\mathfrak{F}$. Conversely, if CSP($\mathfrak{B}$) has width 0, consider the set $\mathcal{F}$ of canonical databases of the bodies of the rules whose head is *false*. Then CSP($\mathfrak{B}$) $=$ Forb$^{\text{hom}}(\mathcal{F})$.

PROPOSITION 8.1.4. *If* CSP($\mathfrak{B}$) *is in Datalog, then* CSP($\mathfrak{B}$) *can be expressed in monotone SNP.*

PROOF. Let Π be a Datalog program that solves CSP($\mathfrak{B}$). We may assume that each of the rules of Π does not involve atomic formulas of the form $x = y$ in the rule body, because such rules can be equivalently replaced by rules where each occurrence of y is replaced by x and the atomic formula has been removed. We use the following monotone SNP sentence Φ (Section 1.4):

- the IDBs of Π are the existentially quantified relations of Φ;
- each rule ψ :– $\phi_1, \dots, \phi_m$ of Π corresponds to a conjunct $\psi \vee \neg\phi_1 \vee \dots \vee \neg\phi_m$ of Φ;
- we additionally add the conjunct $\neg$*false* to Φ.

Note that if Π does not derive false on a finite τ-structure $\mathfrak{A}$, then the least fixed point of Π on $\mathfrak{A}$ satisfies the first-order part of Φ. Conversely, if there exists a $(\tau \cup \sigma)$-expansion $\mathfrak{A}'$ of $\mathfrak{A}$ wich satisfies the first-order part of Φ, then also the least fixed point does. Hence, Π does not derive *false* on $\mathfrak{A}$ if and only if $\mathfrak{A} \models \Phi$. ⊣

Example 8.1.5. In the proof of Proposition 5.8.1 we presented an SNP sentence Φ that describes CSP($\mathbb{Z}$; Succ). It is easy to see that there exists a

Datalog program Π that describes CSP($\mathbb{Z}$; Succ): each conjunct of Φ gives rise to a rule of Π.

While there are NP-hard CSPs that can be expressed in monotone SNP, there is a polynomial-time algorithm to evaluate whether a Datalog program Π derives *false* on a given finite structure $\mathfrak{A}$; we prove this in the next section.

8.1.3. Operational semantics of Datalog. In this section we interpret Datalog programs algorithmically; that is, we describe a constructive approach to the construction of the least fixed point.

Let Π be a Datalog program with EDBs τ and IDBs σ, and let $\mathfrak{A}$ be a finite τ-structure. An *evaluation* of Π on $\mathfrak{A}$ proceeds in steps $i = 0, 1, \dots$, constructing a chain $S^0 \subseteq S^1 \subseteq \cdots$ of sets of atomic $(\sigma \cup \tau)$-formulas with parameters from A. Each clause of Π is understood as a rule that may derive a new atomic $(\sigma \cup \tau)$-formula from the formulas in S^i. We define S^0 to be the canonical query of $\mathfrak{A}$ (see Section 1.2.1). Now suppose that $R_1(a_1^1, \dots, a_{k_1}^1), \dots, R_l(a_1^m, \dots, a_{k_m}^m) \in S^i$ for some $a_1^1, \dots, a_{k_1}^1, \dots, a_1^m, \dots, a_{k_m}^m \in A$ and that Π contains the rule

$$R_0(x_1^0, \dots, x_{k_0}^0) :- R_1(x_1^1, \dots, x_{k_1}^1), \dots, R_l(x_1^m, \dots, x_{k_m}^m)$$

where $a_j^i = a_{j'}^{i'}$ if $x_j^i = x_{j'}^{i'}$. Then S^{i+1} additionally contains $R_0(a_1^0, \dots, a_m^0)$ where $a_j^0 = a_{j'}^i$ if and only if $x_j^0 = x_{j'}^i$.

Since $\mathfrak{A}$ is finite, there exists an $r \in \mathbb{N}$ such that $S^i = S^r$ for all $i \geq r$. Moreover, S^r can be computed in polynomial time for a given Π and a given finite structure $\mathfrak{A}$. If $\psi \in S^r$, we also say that Π *derives* ψ *on* $\mathfrak{A}$.

THEOREM 8.1.6. *Let* Π *be a Datalog program with EDBs* τ *and IDBs* σ. *Let* $\mathfrak{A}$ *be a* τ*-structure. Then the canonical database of the set of all* $(\tau \cup \rho)$*-formulas derived by* Π *on* $\mathfrak{A}$ *equals* $\Pi(\mathfrak{A})$.

PROOF. Since S^r contains the canonical query of $\mathfrak{A}$, the canonical database $\mathfrak{A}^*$ of S^r is an expansion of $\mathfrak{A}$. It is straightforward to show by induction over the steps in the definition of S^r that all formulas added to S^r must hold in every fixed point of Π on $\mathfrak{A}$, so in particular in $\Pi(\mathfrak{A})$. Since $S^r = S^{r+1}$, the canonical database of S^r itself is a fixed point, which implies that $\mathfrak{A}^* = \Pi(\mathfrak{A})$. ⊣

COROLLARY 8.1.7. *If* CSP($\mathfrak{B}$) *is in Datalog then* CSP($\mathfrak{B}$) *is in P.*

8.2. The expressive power of Datalog

In this section we introduce important concepts used to analyse the expressive power of Datalog.

8.2.1. The canonical Datalog program. For every ω-categorical structure $\mathfrak{B}$ and all $\ell, k \in \mathbb{N}$ there exists a Datalog program $\Pi_{\ell,k}$ with the property that if *some* Datalog program of width (ℓ, k) solves CSP($\mathfrak{B}$) then $\Pi_{\ell,k}$ solves CSP($\mathfrak{B}$) (Theorem 8.2.8). The Datalog program $\Pi_{\ell,k}$ that we construct will also be called the *canonical Datalog program of width* (ℓ, k). For *finite* τ-structures $\mathfrak{B}$ the canonical Datalog program was defined by Feder and Vardi [170]. Our definition (from [56]) and generalises this definition to ω-categorical structures $\mathfrak{B}$.

DEFINITION 8.2.1 (Canonical Datalog program). Let $\mathfrak{B}$ be an ω-categorical τ-structure and let $\mathfrak{B}'$ be the expansion of $\mathfrak{B}$ by all primitively positively definable relations of arity at most ℓ; let σ be the (finite) signature of $\mathfrak{B}'$. Then the *canonical* (ℓ, k)*-Datalog program* for $\mathfrak{B}$ has IDBs σ and EDBs τ. The empty 0-ary relation serves as *false*. Theorem 4.1.6 asserts that over $\mathfrak{B}'$ there is a finite number of inequivalent formulas $\Psi(\overline{x}, \overline{y})$ of the form

$$(\psi_1(\overline{x}, \overline{y}) \wedge \cdots \wedge \psi_j(\overline{x}, \overline{y})) \Rightarrow R(\overline{x})$$

where $\overline{x}$ is a tuple of at most ℓ variables, $(\overline{x}, \overline{y})$ is a tuple of at most k variables, $\psi_1, \ldots, \psi_j$ are atomic $(\tau \cup \sigma)$-formulas, and $R \in \sigma$. For each of these formulas $\Psi(\overline{x}, \overline{y})$ such that $\mathfrak{B}' \models \forall \overline{x} \exists \overline{y} \colon \Psi(\overline{x}, \overline{y})$ we introduce a rule

$$R(\overline{x}) :- \psi_1(\overline{x}, \overline{y}), \ldots, \psi_j(\overline{x}, \overline{y})$$

into the canonical Datalog program.

The canonical (ℓ, k)-Datalog program is often called the (ℓ, k)*-consistency procedure* (more on the concept of (ℓ, k)*-consistency* can be found in Section 8.2.2). We would like to mention that for binary homogeneous structures $\mathfrak{B}$, executing the canonical $(2, 3)$-Datalog program on an instance of CSP($\mathfrak{B}$) has been called *establishing (strong) path consistency* in the temporal and spatial reasoning community. The following is easy to see.

PROPOSITION 8.2.2. *Let $\mathfrak{B}$ be an ω-categorical structure with finite relational signature τ. Then the canonical (ℓ, k)-Datalog program for $\mathfrak{B}$ is sound for* CSP($\mathfrak{B}$).

PROOF. We have to show that if the canonical (ℓ, k)-Datalog Π program derives *false* on a given τ-structure $\mathfrak{A}$, then there is no homomorphism from $\mathfrak{A}$ to $\mathfrak{B}$. More generally, we claim that if Π derives $R(\bar{c})$ for some tuple $\bar{c} = (c_1, \ldots, c_d) \in A^d$, and the IDB R was introduced for the relation with the primitive positive formula $\phi(x_1, \ldots, x_d)$ over $\mathfrak{B}$, then for all homomorphisms f from $\mathfrak{A}$ to $\mathfrak{B}$ we have that $\mathfrak{B}$ satisfies $\phi(f(c_1), \ldots, f(c_d))$. This follows by a straightforward induction over the evaluation of canonical Datalog programs, using the fact that the rules of the canonical Datalog program have been introduced for valid implications in the expansion $\mathfrak{B}'$ of $\mathfrak{B}$ by all at most ℓ-ary primitively positively definable relations in $\mathfrak{B}$. ⊣

We will see later that if the canonical Datalog program for CSP($\mathfrak{B}$) does not derive *false*, then no other sound Datalog program for CSP($\mathfrak{B}$) does (Theorem 8.2.8).

8.2.2. The existential pebble game. The existential (ℓ, k)-pebble game is used in the context of constraint satisfaction to characterise the expressive power of Datalog [238, 146, 170]. In particular, it can be used to show that certain constraint satisfaction problems *cannot* be solved by Datalog programs.

The game is played by the players Spoiler and Duplicator on (possibly infinite) structures $\mathfrak{A}$ and $\mathfrak{B}$ with the same relational signature. Let $k, \ell \in \mathbb{N}$ be such that $\ell \leq k$. Each player has k pebbles, $p_1, \dots, p_k$ for Spoiler and $q_1, \dots, q_k$ for Duplicator. Spoiler places his pebbles on elements of $\mathfrak{A}$, Duplicator her pebbles on elements of $\mathfrak{B}$. Initially, no pebbles are placed. In each round of the game Spoiler picks $k - \ell$ pebbles. If some of these pebbles are already placed on $\mathfrak{A}$, then Spoiler removes them from $\mathfrak{A}$, and Duplicator responds by removing the corresponding pebbles from B. Spoiler places the $k - \ell$ pebbles on elements of $\mathfrak{A}$, and Duplicator responds by placing the corresponding pebbles on elements of B. Let $i_1, \dots, i_m$ be the indices of the pebbles that are placed on $\mathfrak{A}$ (and thus on $\mathfrak{B}$) after the r-th round. Let $a_{i_1}, \dots, a_{i_m}$ $(b_{i_1}, \dots, b_{i_m})$ be the elements of $\mathfrak{A}$ ($\mathfrak{B}$, respectively) pebbled with the pebbles $p_{i_1}, \dots, p_{i_m}$ $(q_{i_1}, \dots, q_{i_m})$ after the r-th round. If for some r the partial mapping h_r from $\mathfrak{A}$ to $\mathfrak{B}$ defined by $h_r(a_{i_j}) := b_{i_j}$, for $1 \leq j \leq m$, is not a homomorphism from $\mathfrak{A}[\{a_{i_1}, \dots, a_{i_m}\}]$ to $\mathfrak{B}$, then Spoiler wins, and the game ends at that stage. Duplicator wins if the game continues forever.

We are interested in the situations where Duplicator can win the game no matter how Spoiler plays, that is, where Duplicator has a winning strategy.[20] Our description of a winning strategy above is rather informal; we do not formalise it, but rather give a formal definition of winning strategy that is easier to work with (and that is equivalent to the notion of a winning strategy hinted at above, once it is properly formalised).

DEFINITION 8.2.3 (of [237]). A *winning strategy for Duplicator* for the existential (ℓ, k)-pebble game on $(\mathfrak{A}, \mathfrak{B})$ is a non-empty set $\mathcal{H}$ of partial homomorphisms from $\mathfrak{A}$ to $\mathfrak{B}$ such that

- $\mathcal{H}$ is closed under restrictions of its members, and
- for all functions h in $\mathcal{H}$ with domain size $d \leq \ell$ and for all $a_1, \dots, a_{k-d} \in A$ there is an extension $h' \in \mathcal{H}$ of h such that h' is also defined on $a_1, \dots, a_{k-d}$.

The second item is often called the (ℓ, k)-*extension property*. Clearly, if there exists a homomorphism $h\colon \mathfrak{A} \to \mathfrak{B}$ then Duplicator has a winning strategy, namely the set of all restrictions of h to subsets of A of size at most k. The full link between the existential pebble game and the expressivity of Datalog can be found in Section 8.2.5.

[20] It turns out that in this case Duplicator also has a *positional* strategy in the sense that the decisions of Duplicator are only based on the current position of the pebbles, and not the previous decisions in the game.

8.2.3. Bounded treewidth duality. Let $0 \leq \ell < k$ be positive integers. An (ℓ, k)-*tree* is defined inductively as follows:

- A k-clique is an (ℓ, k)-tree;
- For every (ℓ, k)-tree G and for every ℓ-clique induced on the nodes $v_1, \dots, v_\ell$ in G, the graph G' obtained by adding $k - \ell$ new nodes $v_{l+1}, \dots, v_k$ to G and adding edges so that a k-clique is induced on $\{v_1, \dots, v_k\}$ is also an (ℓ, k)-tree.

A *partial* (ℓ, k)-*tree* is a (not necessarily induced) subgraph of an (ℓ, k)-tree.

DEFINITION 8.2.4. Let $0 \leq \ell < k$ and let τ be a relational signature. We say that a τ-structure $\mathfrak{A}$ has *treewidth at most* (ℓ, k) if the Gaifman graph of $\mathfrak{A}$ (Definition 2.1.5) is a partial (ℓ, k)-tree.

If a structure has treewidth at most $(k, k+1)$ we also say that it has *treewidth at most* k, and it is not difficult to see that these structures are precisely the structures of treewidth at most k in the sense of [238]. It is also possible to define partial (ℓ, k)-trees by using tree decompositions.

DEFINITION 8.2.5. A *tree decomposition* of a graph $(V; E)$ is a set $\{X_t \mid t \in T\}$ of subsets of V together with a tree with vertex set T such that

1. $\bigcup_{t \in T} X_t = V$;
2. for each $e \in E$ there exists $t \in T$ with $e \subseteq X_t$;
3. if $v \in X_r \cap X_s$ then $v \in X_t$ for every t on the unique path from r to s in T.

A tree decomposition $\{X_t \mid t \in T\}$ has *width* (ℓ, k) if $|X_t| \leq k$ for all $t \in T$ and $|X_r \cap X_s| \leq \ell$ for distinct $r, s \in T$.

PROPOSITION 8.2.6. *A graph is a partial* (ℓ, k)*-tree if and only if it has a tree decomposition of width* (ℓ, k).

PROOF. Let G be a partial (ℓ, k)-tree. Any tree decomposition of G is also a tree decomposition of graphs obtained by removing edges from G, so we can assume without loss of generality that G is an (ℓ, k)-tree. We construct a tree decomposition of width (ℓ, k) inductively. The base case is that G is a k-clique, in which case there exists a tree decomposition with $T = \{t\}$ and X_t containing this clique, which clearly has width (ℓ, k). For the inductive step, suppose that G is obtained from an (ℓ, k)-tree G' which contains an ℓ-clique $v_1, \dots, v_\ell$ by adding the new vertices $v_{l+1}, \dots, v_k$ and adding edges so that $v_1, \dots, v_k$ induce a k-clique. By induction, there exists a tree decomposition $\{X_t \mid t \in T'\}$ of G' of width (ℓ, k). As $v_1, \dots, v_\ell$ induce a clique, there exists $t \in T'$ such that $\{v_1, \dots, v_\ell\} \subseteq X_t$. Let $T := T' \cup \{s\}$ for a new element s, define $X_s := \{v_1, \dots, v_k\}$, and link s to t in T. Clearly, T is a tree decomposition of G of width (ℓ, k).

Conversely, suppose that G has a tree decomposition T of width (ℓ, k). We prove by induction on the number of vertices of T that G is a subgraph of an

(ℓ, k)-tree H such that every subset of X_t of size at most ℓ for every $t \in T$ is contained in some ℓ-clique in H. Let r be a leaf of T, let T' be the tree decomposition obtained from T by removing r, and let G' be the subgraph of G induced on $\bigcup_{s \in T'} X_s$. By induction, G' is a subgraph of an (ℓ, k)-tree H' with the desired properties. Let t be the unique node of T that is adjacent to r. By assumption, $|X_r \cap X_t| \leq \ell$ and hence $X_r \cap X_t$ is contained in some ℓ-clique Y in H'. Note that the elements of $X_r \setminus X_t$ cannot appear in X_s for any $s \neq r$ since then they would then have to also appear in X_t. Add the elements of $X_r \setminus X_t$ to H' and add edges so that $Y \cup (X_r \setminus X_t)$ forms a clique in the resulting graph H, which satisfies all the requirements. ⊣

We say that a relational structure $\mathfrak{B}$ has *(ℓ, k)-treewidth duality* if there is a set $\mathcal{N}$ of finite structures of treewidth at most (ℓ, k) such that every finite τ-structure $\mathfrak{A}$ is homomorphic to $\mathfrak{B}$ if and only if no structure in $\mathcal{N}$ is homomorphic to $\mathfrak{A}$. We say that $\mathfrak{B}$ has *bounded treewidth duality* if there exist $\ell, k \in \mathbb{N}$ such that $\mathfrak{B}$ has (ℓ, k)-treewidth duality. The link between bounded treewidth duality and the expressivity of Datalog can be found in Section 8.2.5.

8.2.4. Finite variable logics. Finite variable logics have been introduced in the context of constraint satisfaction by Kolaitis and Vardi as a tool to study the expressive power of Datalog [237, 238]. Our presentation is based on [56].

Let $0 \leq \ell < k$ be positive integers. A conjunction $\bigwedge \Psi$ is called *ℓ-bounded* if Ψ is a collection of formulas ψ that are quantifier-free or have at most ℓ free variables. We denote by $\mathcal{L}^{\ell,k}$ the logic where the formulas have at most k variables and are obtained inductively from atomic formulas using ℓ-bounded conjunctions and existential quantification; this logic will be called *ℓ-bounded existential positive k-variable logic* [56].

LEMMA 8.2.7. *Let $\mathfrak{A}$ be a finite structure of treewidth at most (ℓ, k). Then the canonical query of $\mathfrak{A}$ is logically equivalent to a sentence from $\mathcal{L}^{\ell,k}$.*

PROOF. Let G be the Gaifman graph of $\mathfrak{A}$ and let $\{X_t \mid t \in T\}$ be a tree decomposition of G. Pick a leaf $t \in T$ and let $X_t = \{a_1, \dots, a_{k'}\}$. We show by induction on the size of T that there exists a formula $\phi_{\mathfrak{A}}(y_1, \dots, y_{k'})$ in $\mathcal{L}^{\ell,k}$ such that for every structure $\mathfrak{B}$ and elements $b_1, \dots, b_{k'} \in B$ the following two statements are equivalent:

(1) The partial mapping from $\mathfrak{A}$ to $\mathfrak{B}$ that maps a_i to b_i for $1 \leq i \leq k'$ can be extended to a homomorphism from $\mathfrak{A}$ to $\mathfrak{B}$;
(2) $\mathfrak{B} \models \phi_{\mathfrak{A}}(b_1, \dots, b_{k'})$.

The base case is that the tree contains only one node t. In this case $\phi_{\mathfrak{A}}$ is the canonical conjunctive query of $\mathfrak{A}$. For the inductive step, let $t_1, \dots, t_m$ be neighbours of t in T. Consider the m subtrees $T_1, \dots, T_m$ of T obtained by removing t, and let $\mathfrak{A}_i$ be the substructure of $\mathfrak{A}$ induced on $\{X_s \mid s \in T_i\}$. Then T_i is a tree decomposition of $\mathfrak{A}_i$ and the induction hypothesis provides a formula $\phi_{\mathfrak{A}_i}$ for which (1) and (2) are equivalent. Let $\phi_{\mathfrak{A}}(y_1, \ldots, y_{k'})$ be the conjunction of the following formulas:

(a) for each $i \in \{1, \ldots, m\}$, existentially quantify all free variables y_j in $\phi_{\mathfrak{A}_i}$ where $a_j \notin X_t$. Note that the resulting formula has at most ℓ free variables;

(b) the canonical query of the substructure of $\mathfrak{A}$ induced on X_t (as in the base case).

To show that (1) and (2) are equivalent, let $\mathfrak{B}$ be an arbitrary structure. By the properties of the tree decomposition we know that h is a homomorphism from $\mathfrak{A}$ to $\mathfrak{B}$ that maps a_i to b_i if and only if for every $i \in \{1, \ldots, m\}$ the restriction of h to $\mathfrak{A}_i$ is a homomorphism from $\mathfrak{A}_i$ to $\mathfrak{B}$ and the restriction of h to the elements of X_t is a partial homomorphism from $\mathfrak{A}$ to $\mathfrak{B}$ as well. The former condition is equivalent to the fact that the assignment $y_{a_i} \mapsto b_i$ satisfies every formula of $\phi_{\mathfrak{A}}(y_1, \ldots, y_{k'})$ included in (a). The latter condition is equivalent to the fact that the same assignment satisfies the formula included in (b). ⊣

8.2.5. Characterising the expressive power of Datalog. The following theorem is the promised link between Datalog, the existential pebble game, finite variable logics, and treewidth duality for ω-categorical structures. We present it here in its most general form with both parameters ℓ and k.

THEOREM 8.2.8. *Let τ be a finite relational signature. Let $\mathfrak{B}$ be an ω-categorical τ-structure and $\mathfrak{A}$ be a finite τ-structure. Then for all ℓ, k with $\ell \leq k$ the following statements are equivalent.*

1. *Every sound (ℓ, k)-Datalog program for* CSP($\mathfrak{B}$) *does not derive false on $\mathfrak{A}$.*
2. *The canonical (ℓ, k)-Datalog program for $\mathfrak{B}$ does not derive false on $\mathfrak{A}$.*
3. *Duplicator wins the existential (ℓ, k)-pebble game on $\mathfrak{A}$ and $\mathfrak{B}$.*
4. *All sentences in $\mathcal{L}^{\ell,k}$ that hold in $\mathfrak{A}$ also hold in $\mathfrak{B}$.*
5. *Every finite τ-structure with a core of treewidth at most (ℓ, k) that maps homomorphically to $\mathfrak{A}$ also maps homomorphically to $\mathfrak{B}$.*

PROOF. The implication from (1) to (2) follows from Proposition 8.2.2.

To show that (2) implies (3), we define a winning strategy for Duplicator as follows. Let $\mathfrak{B}'$ be the expansion of $\mathfrak{B}$ by all at most ℓ-ary primitively positively definable relations, let σ be the signature of $\mathfrak{B}'$, and let $\mathfrak{A}'$ be the σ-structure computed by the canonical (ℓ, k)-Datalog program for $\mathfrak{B}$ on input $\mathfrak{A}$. Let $\mathcal{H}$ be the strategy for Duplicator that consists of all partial homomorphisms $f : A' \to B'$ with domain of size at most k. By construction, $\mathcal{H}$ is closed under restrictions and is non-empty (since *false* is not derived, $\mathcal{H}$ contains the partial mapping with the empty domain). We claim that $\mathcal{H}$ has the (ℓ, k)-extension property. Let h be a function with domain $v_1, \ldots, v_{\ell'}$ of size at most ℓ and let $D = \{v_1, \ldots, v_{\ell'}, v_{\ell'+1}, \ldots, v_{k'}\} \subseteq A$ be a superset of $\{v_1, \ldots, v_{\ell'}\}$ of size at most k. Consider the following rule with variables D of the canonical Datalog program. The body of the rule is the canonical query ϕ of $\mathfrak{A}'[D]$. The head of the rule is $S(v_1, \ldots, v_{\ell'})$ where $S^{\mathfrak{B}'}$ is the projection of the relation defined by ϕ in $\mathfrak{B}'$ to the first ℓ' arguments. This rule shows that the canonical Datalog program for $\mathfrak{A}$ derived $S(v_1, \ldots, v_{\ell'})$ in $\mathfrak{A}'$, and, by the definition of $\mathcal{H}$, $(h(v_1), \ldots, h(v_{\ell'}))$ belongs to $S^{\mathfrak{B}'}$. By the definition of $S^{\mathfrak{B}'}$, there exist $b_{\ell'+1}, \ldots, b_{k'}$ such that $(h(v_1), \ldots, h(v_{\ell'}), b_{\ell'+1}, \ldots, b_{k'})$ satisfies ϕ in $\mathfrak{B}'$. Hence, if we extend h by $v_i \mapsto b_i$ for $i \in \{l'+1, \ldots, k'\}$ we obtain the desired function.

Next, we show the implication from (3) to (4). Suppose Duplicator has a winning strategy $\mathcal{H}$ for the existential (ℓ, k)-pebble game on $\mathfrak{A}$ and $\mathfrak{B}$. Let ϕ be a τ-sentence from $\mathcal{L}^{\ell,k}$ that holds in $\mathfrak{A}$. We have to show that ϕ also holds in $\mathfrak{B}$. For that, we prove by induction on the syntactic structure of $\mathcal{L}^{\ell,k}$ formulas that if $\psi(v_1, \ldots, v_m)$ is an ℓ-bounded conjunction or has at most ℓ free variables (i.e., $m \leq \ell$), then for all $h \in \mathcal{H}$ and all elements $a_1, \ldots, a_m$ from the domain of h, if $\mathfrak{A} \models \psi(a_1, \ldots, a_m)$, then $\mathfrak{B} \models \psi(h(a_1), \ldots, h(a_m))$. Clearly, for $m = 0$ this implies that ϕ holds in $\mathfrak{B}$. The base case of the induction is obvious since atomic formulas are preserved under homomorphisms. Next, suppose that $\psi(v_1, \ldots, v_m)$ is an ℓ-bounded conjunction of a set of formulas Ψ. Then each formula in Ψ has at most ℓ free variables or is quantifier-free. In both cases we can use the inductive hypothesis and the inductive step follows directly. Assume that the formula $\psi(v_1, \ldots, v_m)$ is of the form

$$\exists u_1, \ldots, u_n \colon \chi(v_1, \ldots, v_m, u_1, \ldots, u_n)$$

where χ is an ℓ-bounded conjunction or an atomic formula. We will use the inductive hypothesis for the formula $\chi(v_1, \ldots, v_m, u_1, \ldots, u_n)$. Let h be a homomorphism in $\mathcal{H}$. We have to show that if $a_1, \ldots, a_m$ are arbitrary elements from the domain of h such that $\mathfrak{A} \models \psi(a_1, \ldots, a_m)$, then $\mathfrak{B} \models \psi(h(a_1), \ldots, h(a_m))$. Since $\mathfrak{A} \models \exists u_1, \ldots, u_n \colon \chi(a_1, \ldots, a_m)$, there exist $a_{m+1}, \ldots, a_{m+n}$ such that

$$\mathfrak{A} \models \chi(a_1, \ldots, a_m, a_{m+1}, \ldots, a_{m+n}).$$

Consider the restriction h^* of h to the subset $\{a_1, \ldots, a_m\}$ of the domain of h. Because of the first property of winning strategies $\mathcal{H}$, the homomorphism h^* is in $\mathcal{H}$. Since $m \leq l$, we can apply the extension property of $\mathcal{H}$ to h^* and $a_{m+1}, \ldots, a_{m+n}$, and there are $b_1, \ldots, b_n$ such that the extension h' of h^* with domain $\{a_1, \ldots, a_{m+n}\}$ that maps a_{m+i} to b_i is in $\mathcal{H}$. By applying the induction hypothesis to $\chi(v_1, \ldots, v_m, u_1, \ldots, u_n)$ and to h', we infer that $\mathfrak{B} \models \chi(h'(a_1), \ldots, h'(a_{m+n}))$, and hence $\mathfrak{B} \models \psi(h(a_1), \ldots, h(a_m))$.

(4) $\Rightarrow$ (5). Let $\mathfrak{T}$ be a finite τ-structure whose core $\mathfrak{T}'$ has treewidth at most (ℓ, k) such that $\mathfrak{T}$ maps homomorphically to $\mathfrak{A}$. By Lemma 8.2.7 there exists an $\mathcal{L}^{\ell,k}$-sentence ϕ such that ϕ is equivalent to the canonical query of $\mathfrak{B}$, and hence ϕ holds in a structure if and only if $\mathfrak{T}'$ maps homomorphically to that structure. In particular, ϕ must hold in $\mathfrak{A}$. Then (4) implies that ϕ holds in $\mathfrak{B}$, and therefore $\mathfrak{T}'$ maps homomorphically to $\mathfrak{B}$. But then we can compose the

homomorphism from $\mathfrak{T}$ to $\mathfrak{T}'$ and the homomorphism from $\mathfrak{T}'$ to $\mathfrak{B}$ to obtain the desired homomorphism from $\mathfrak{T}$ to $\mathfrak{B}$.

We finally show that (5) implies (1). Suppose that there is a sound (ℓ, k)-Datalog program Π for $\mathfrak{B}$ that derives *false* on $\mathfrak{A}$. The idea is to use the 'derivation tree of *false*' to construct a τ-structure $\mathfrak{S}$ of treewidth at most (ℓ, k) that maps homomorphically to $\mathfrak{A}$, but not to $\mathfrak{B}$. The construction proceeds by induction over the evaluation of Π on $\mathfrak{A}$. Suppose that $R_0(y_1^0, \dots, y_{k_0}^0)$ is an atomic formula derived by Π on $\mathfrak{A}$ from previously derived atomic formulas $R_1(y_1^1, \dots, y_{k_1}^1), \dots, R_s(y_1^s, \dots, y_{k_s}^s)$. We will prove that there exists a structure $\mathfrak{S}_0$ with distinguished vertices $v_1^0, \dots, v_{k_0}^0$ and an (ℓ, k)-tree G_0 such that

1. the Gaifman graph of S_0 is a (not necessarily induced) subgraph of G_0,
2. $v_1^0, \dots, v_{k_0}^0$ induce a clique in G_0,
3. there is a homomorphism from $\mathfrak{S}_0$ to $\mathfrak{A}$ that maps v_i^0 to y_i^0 for every $1 \leq i \leq k_0$, and
4. the program Π derives $R_0(v_1^0, \dots, v_{k_0}^0)$ on $\mathfrak{S}_0$.

Let $i \in \{1, \dots, s\}$. If R_i is an IDB of Π, then let S_i, $v_1^i, \dots, v_{k_i}^i$, and G_i be given by the inductive hypothesis. If R_i is an EDB, we create fresh vertices $v_1^i, \dots, v_{k_i}^i$, and define $\mathfrak{S}_i$ to be the following structure with vertices $v_1^i, \dots, v_{k_i}^i$. The relation R_i in $\mathfrak{S}_i$ equals $\{(v_1^i, \dots, v_{k_i}^i)\}$, and all other relations in $\mathfrak{S}_i$ are empty. Clearly, $\{v_1^i, \dots, v_{k_i}^i\}$ induces a clique in the Gaifman graph of $\mathfrak{S}_i$, and the Gaifman graph of $\mathfrak{S}_i$ is a partial (ℓ, k)-tree.

Now, the structure $\mathfrak{S}_0$ has the distinguished vertices $v_1^0, \dots, v_{k_0}^0$, and is obtained from the τ-structures $\mathfrak{S}_1, \dots, \mathfrak{S}_s$ as follows. We start from the disjoint union of $\mathfrak{S}_1, \dots, \mathfrak{S}_s$. When $y_j^i = y_s^r$ for $i, r \in \{0, \dots, s\}$, $j \in \{1, \dots, k_i\}$, and $s \in \{1, \dots, k_r\}$, then we identify v_j^i and v_s^r. To define G_0 we form a disjoint union of $G_1, \dots, G_s$ and the isolated nodes $v_1^0, \dots, v_{k_0}^0$, and do the same node identifications as before. We finally add an edge for every pair of distinct vertices in $v_1^0, \dots, v_{k_0}^0$. The resulting graph, G_0, satisfies the requirements of the claim. Observe that since Π derives $R_1(v_1^1, \dots, v_{k_1}^1), \dots, R_s(v_1^s, \dots, v_{k_s}^s)$ on $\mathfrak{S}_0$ by the inductive assumption, it also derives $R_0(v_1^0, \dots, v_{k_0}^0)$ on $\mathfrak{S}_0$.

In this fashion we proceed for all inference steps of the Datalog program. Let $\mathfrak{S}$ be the resulting structure for the final derivation of false. It has treewidth at most (ℓ, k), and maps to $\mathfrak{A}$, but does not map to $\mathfrak{B}$, since Π (which is sound) derives also *false* on $\mathfrak{S}$. ⊣

COROLLARY 8.2.9. *Let $\mathfrak{B}$ be an ω-categorical structure and let $\ell, k \in \mathbb{N}$. Then every instance of* CSP($\mathfrak{B}$) *whose core has treewidth at most (ℓ, k) can be solved in polynomial time by the canonical (ℓ, k)-Datalog program.*

PROOF. It is clear that the canonical (ℓ, k)-Datalog program can be evaluated on a (finite) instance $\mathfrak{A}$ of CSP($\mathfrak{B}$) in polynomial time. If the canonical (ℓ, k)-Datalog program derives *false* on $\mathfrak{A}$, then, because the canonical Datalog program is always sound, there is no homomorphism from $\mathfrak{A}$ to $\mathfrak{B}$. Now, suppose that the canonical Datalog program does not derive *false* on a finite structure $\mathfrak{A}$ whose core has treewidth at most (ℓ, k). Then, by Theorem 8.2.8, every τ-structure whose core has treewidth at most (ℓ, k) that maps homomorphically to $\mathfrak{A}$ also maps homomorphically to $\mathfrak{B}$. This holds in particular for $\mathfrak{A}$ itself, and hence $\mathfrak{A}$ is homomorphic to $\mathfrak{B}$. ⊣

The following direct consequence of Theorem 8.2.8 yields other characterisations of bounded Datalog width.

THEOREM 8.2.10. *Let $\mathfrak{B}$ be an ω-categorical structure with a finite relational signature τ. Then for all $\ell, k \in \mathbb{N}$ with $\ell \leq k$ the following statements are equivalent.*

1. *There is an (ℓ, k)-Datalog program that solves* CSP($\mathfrak{B}$).
2. *The canonical (ℓ, k)-Datalog program solves* CSP($\mathfrak{B}$).
3. *For every finite τ-structures $\mathfrak{A}$, if Duplicator has a winning strategy for the existential (ℓ, k)-pebble game on $\mathfrak{A}$ and $\mathfrak{B}$, then $\mathfrak{A}$ maps homomorphically to $\mathfrak{B}$.*
4. *For every finite τ-structures $\mathfrak{A}$, if all sentences in $\mathcal{L}^{\ell,k}$ that hold in $\mathfrak{A}$ also hold in $\mathfrak{B}$, then $\mathfrak{A}$ maps homomorphically to $\mathfrak{B}$.*
5. *for every finite τ-structures $\mathfrak{A}$, if every finite τ-structure $\mathfrak{S}$ of treewidth at most (ℓ, k) that maps homomorphically to $\mathfrak{A}$ also maps homomorphically to $\mathfrak{B}$, then $\mathfrak{A}$ maps homomorphically to $\mathfrak{B}$.*
6. *$\mathfrak{B}$ has (ℓ, k)-*treewidth duality, *i.e., there is a set $\mathcal{N}$ of finite structures of treewidth at most (ℓ, k) such that every finite τ-structure $\mathfrak{A}$ is homomorphic to $\mathfrak{B}$ if and only if no structure in $\mathcal{N}$ is homomorphic to $\mathfrak{A}$.*

PROOF. To prove the implication from (1) to (2), suppose that an (ℓ, k)-Datalog program Π solves CSP($\mathfrak{B}$), and let $\mathfrak{A}$ be an instance of CSP($\mathfrak{B}$). If the canonical (ℓ, k)-Datalog program derives *false* on $\mathfrak{A}$, then by Proposition 8.2.2 the structure $\mathfrak{A}$ is not homomorphic to $\mathfrak{B}$. Otherwise, since Π is sound, the implication from (2) to (1) in Theorem 8.2.8 shows that the canonical (ℓ, k)-Datalog program does not derive *false* on $\mathfrak{A}$ as well. Hence, the canonical Datalog program solves CSP($\mathfrak{B}$).

The implications (2) $\Rightarrow$ (3) $\Rightarrow$ (4) $\Rightarrow$ (5) $\Rightarrow$ (1) are straightforward consequences of Theorem 8.2.8.

To show that (5) implies (6), let $\mathcal{N}$ be the set of all those structures of treewidth at most (ℓ, k) that does not map homomorphically to $\mathfrak{B}$. Let $\mathfrak{A}$ be a finite τ-structure. If $\mathfrak{A}$ maps homomorphically to $\mathfrak{B}$, then clearly there is no structure $\mathfrak{S}$ in $\mathcal{N}$ that maps to $\mathfrak{A}$, because then $\mathfrak{S}$ would also map to $\mathfrak{B}$, a contradiction to the definition of $\mathcal{N}$. Conversely, suppose that no structure in $\mathcal{N}$ maps homomorphically to $\mathfrak{A}$. In other words, every structure that homomorphically maps to $\mathfrak{A}$ also maps to $\mathfrak{B}$. Using 5, this implies that $\mathfrak{A}$ homomorphically maps to $\mathfrak{B}$.

Finally, to show that (6) implies (5) let $\mathcal{N}$ be as in item (6). It follows that all structures in $\mathcal{N}$ do not map homomorphically to $\mathfrak{B}$. Let $\mathfrak{A}$ be a finite τ-structure such that every finite τ-structure $\mathfrak{S}$ of treewidth at most (ℓ, k) that maps homomorphically to $\mathfrak{A}$ also maps homomorphically to $\mathfrak{B}$. In particular, no structure in $\mathcal{N}$ maps homomorphically to $\mathfrak{A}$. Therefore, $\mathfrak{A}$ maps homomorphically to $\mathfrak{B}$. ⊣

8.3. Datalog and primitive positive interpretations

Primitive positive interpretations not only preserve polynomial-time tractability of CSPs, but they also preserve bounded Datalog width. This has been proved by Larose and Zadori [257], but only stated for finite domains and with slightly weaker bounds on the Datalog width than those given below. An alternative approach, with worse bounds on the width, can be found in [14].

We first treat primitive positive definitions rather than primitive positive interpretations; here it is convenient to view instances of CSP($\mathfrak{B}$) as primitive positive sentences ϕ rather than structures; the relation between the two points of view is discussed in Section 1.2. Thus, we write $\Pi(\phi)$ for the set of $(\tau \cup \rho)$-formulas derived by Π on input ϕ. Recall that one of the differences between formulas and structures in the input is that formulas might contains conjuncts using the equality symbol even if the signature does not contain a symbol that denotes the equality relation. However, the following proposition shows that Datalog is able to deal with expansions by a binary relation symbol that denotes the equality relation.

LEMMA 8.3.1. *Let $\mathfrak{B}$ be an ω-categorical structure such that* CSP($\mathfrak{B}$) *has Datalog width (ℓ, k) and let $\mathfrak{B}'$ be the expansion of $\mathfrak{B}$ with a new relation symbol $\equiv$ which denotes the equality relation. Then* CSP($\mathfrak{B}'$) *has Datalog width (ℓ', k') where $\ell' := \max(2, l)$ and $k' := \max(3, 2k)$.*

PROOF. Let ϕ be an instance of CSP($\mathfrak{B}'$). If ϕ contains a conjunct of the form $x \equiv y$, replace all occurrences of y by x, and repeat this step until the resulting formula ψ has no more conjuncts of the form $x \equiv y$. For later use, we define E to be the equivalence relation on the variables of ϕ that relates two variables if they have been replaced by the same variable in ψ. Clearly, ψ is satisfiable in $\mathfrak{B}$ if and only if ϕ is satisfiable in $\mathfrak{B}'$.

If ϕ is satisfiable then by Proposition 8.2.2 the canonical (ℓ', k')-Datalog program Π' for $\mathfrak{B}'$ does not derive false on ϕ. So let us assume that ϕ is unsatisfiable. In this case ψ is unsatisfiable, too, and by assumption the canonical (ℓ, k)-Datalog program Π derives *false* on ψ. We use the derivation of *false* by Π on ψ to inductively construct a derivation of *false* on ϕ by Π'.

Note that since $\ell' \geq 2$ and $k' \geq 3$, the program Π' has a rule that has been introduced for transitivity of equality. Hence, if $E(x_1, x_2)$ holds for two variables x_1 and x_2 of ϕ, then Π' will derive $E'(x_1, x_2)$ where E' is the IDB that has been introduced for the primitive positive formula $x \equiv y$. Finally, suppose that Π contains the rule

$$R_0(\bar{x}) :- R_1(y_1^1, \dots, y_{i_1}^1), \dots, R_s(y_1^s, \dots, y_{i_s}^s).$$

Then Π' must contain the rule

$$R_0(\bar{x}) :- E'(y_1^1, z_1^1), \dots, E'(y_{j_1}^1, z_{j_1}^1), \dots, E'(y_1^s, z_1^s), \dots, E'(y_{j_s}^s, z_{j_s}^s),$$
$$R_1(z_1^1, \dots, z_{i_1}^1), \dots, R_s(z_1^s, \dots, z_{i_s}^s).$$

With these rules it is straightforward to translate a derivation of *false* of Π on ψ into a derivation of *false* of Π' on ϕ, which concludes the proof. ⊣

LEMMA 8.3.2. *Let $\mathfrak{B}$ be an ω-categorical structure that contains the equality relation and such that* CSP($\mathfrak{B}$) *has Datalog width* (ℓ, k). *Let $\mathfrak{B}'$ be the expansion of $\mathfrak{B}$ by all primitively positively definable relations of arity at most m. Then* CSP($\mathfrak{B}'$) *has Datalog width (ℓ', k') for $\ell' := m\ell$ and $k' := mk$.*

PROOF. To simplify the presentation, we prove the lemma for an expansion of $\mathfrak{B}$ by a single m-ary relation R with a primitive positive definition θ in $\mathfrak{B}$; the general case can be shown similarly. Let ϕ be an instance of CSP($\mathfrak{B}'$). Let τ be the signature of $\mathfrak{B}$, and let ψ be the τ-formula obtained from ϕ as follows (this is as in the proof of Lemma 1.2.6): replace each conjunct of ϕ of the form $R(\bar{x})$ by $\theta(\bar{x})$. Rewrite the formula into prenex normal form, and replace equalities in the primitive positive definition by the symbol for equality in the signature τ. Clearly, the resulting τ-formula ψ is satisfiable in $\mathfrak{B}$ if and only if ϕ is satisfiable in $\mathfrak{B}'$. If the canonical (ℓ', k')-Datalog program Π' for $\mathfrak{B}'$ derives *false* on ϕ then ϕ is unsatisfiable by Proposition 8.2.2 and there is nothing to be shown. Suppose that the canonical (ℓ', k')-Datalog program Π' for $\mathfrak{B}'$ does not derive *false* on ψ. By Theorem 8.2.8 this means that Duplicator has a winning strategy $\mathcal{H}$ for the existential (ℓ', k')-pebble game on ψ and $\mathfrak{B}'$.

We claim that Duplicator also has a winning strategy $\mathcal{J}$ for the existential (ℓ, k)-pebble game on ψ and $\mathfrak{B}$. Let $\bar{x}$ be a tuple of at most l variables from ψ. If an entry $\bar{x}_i$ is not a variable from ϕ, then it must be an existentially quantified variable of some primitive positive formula $\theta(\bar{y}^i)$ that replaced a conjunct $R(\bar{y}^i)$ of ϕ. Let $T(\bar{x})$ be the set of variables in $\bar{x}$ together with all the variables in $\theta(\bar{y}^i)$ for every i such that $\bar{x}_i$ in not a variable from ϕ. Since $\mathcal{H}$ satisfies the $(0, k')$-extension property, there exists a partial homomorphism $h \in \mathcal{H}$ from ϕ to $\mathfrak{B}'$ which is defined on all the variables from $T(\bar{x})$ that lie in ϕ (there are at most $k' = mk$ such variables). Let $R(\bar{z})$ be a conjunct of ϕ where the entries of $\bar{z}$ are in the domain of h. Then h has an extension to the existential quantifiers that appear in $\theta(\bar{z})$. So we may assume that h has an extension g to all the variables in $T(\bar{x})$; we then include into our strategy $\mathcal{J}$ the restriction of g to $\bar{x}$. This map will clearly be a partial homomorphism from ψ to $\mathfrak{B}$.

To show that $\mathcal{J}$ has the (ℓ, k)-extension property, let $\bar{x}$ be a tuple of at most l variables from ψ and let $h \in \mathcal{J}$ be defined on all entries from x. Let $\bar{y}$ be some tuples of variables from ψ such that $(\bar{x}, \bar{y})$ contains at most k variables. We have to find an extension of h to $(\bar{x}, \bar{y})$ that also lies in $\mathcal{J}$. The map h was defined as the restriction of some partial homomorphism g from ψ to $\mathfrak{B}$ that was defined on all of the variables from $T(\bar{x})$. The restriction of this map to the variables in $\bar{z}$ that also lie in ϕ has at most $\ell' = m\ell$ variables, and thus can be extended by the (ℓ', k')-extension property of $\mathcal{H}$ to a homomorphism defined on all variables of $T(\bar{x}, \bar{y})$ that lie in ϕ, which are at most $k' = mk$ variables. This map in turn has an extension g to all variables in $T(\bar{x}, \bar{y})$. The restriction of g to $(\bar{x}, \bar{y})$ defines a partial map from $\mathcal{J}$ which extends h, concluding the proof. ⊣

LEMMA 8.3.3. *Let $\mathfrak{B}$ be an ω-categorical structure such that* CSP($\mathfrak{B}$) *has Datalog width (ℓ, k). Suppose that $\mathfrak{D}$ has maximal arity m and is homomorphically equivalent to a structure with a d-dimensional primitive positive interpretation in $\mathfrak{B}$. Then* CSP($\mathfrak{D}$) *has Datalog width (ℓ', k') for $\ell' := \max(2, dm\ell)$ and $k' := \max(3, dmk)$.*

PROOF. Clearly, if two structures $\mathfrak{B}$ and $\mathfrak{D}$ are homomorphically equivalent, then a Datalog program solves CSP($\mathfrak{B}$) if and only if it solves CSP($\mathfrak{D}$). It is also clear that if $\mathfrak{D}$ is a reduct of $\mathfrak{B}$ then a Datalog program for CSP($\mathfrak{B}$) also solves CSP($\mathfrak{D}$). So to show the statement for $\mathfrak{D} \in \mathrm{HI}(\mathfrak{B}) = \mathrm{H\,Red\,P}^{\mathrm{fin}}_{\mathrm{full}}(\mathfrak{B})$ we can suppose without loss of generality that $\mathfrak{D}$ is the reduct of the full d-th power of $\mathfrak{B}$ containing all at most m-ary relations (Theorem 3.6.2). We also assume that $\mathfrak{B}$ contains the equality relation; otherwise, the statement can be shown by combining the argument below with the proof of Lemma 8.3.1 (this is why we need that $\ell' \geq 2$ and $k' \geq 3$).

Let $\mathfrak{C}$ be the expansion of $\mathfrak{B}$ by all primitively positively definable relations of arity dm. Lemma 8.3.2 shows that CSP($\mathfrak{B}$) has Datalog width (ℓ', k'). We show that CSP($\mathfrak{D}$) has Datalog width (ℓ', k'), too. Let ϕ be an instance of CSP($\mathfrak{D}$) with variable set $X = \{x_1, \dots, x_n\}$. From ϕ we construct an instance ψ of CSP($\mathfrak{C}$) as follows. Note that for every primitive positive formula over the signature of $\mathfrak{B}$ with md free variables the structure $\mathfrak{D}$ has an m-ary relation R', and the structure $\mathfrak{C}$ has an md-ary relation R. Let $Y := \{y^i_j \mid 1 \leq i \leq d, 1 \leq j \leq n\}$ be fresh and pairwise distinct variables. The formula ψ contains a conjunct of the form $R(y^1_{i_1}, \dots, y^d_{i_1}, \dots, y^1_{i_k}, \dots, y^d_{i_k})$ for every conjunct $R'(x_{i_1}, \dots, x_{i_k})$ of ϕ.

If the canonical (ℓ', k')-Datalog program for $\mathfrak{D}$ derives *false* on ϕ then by Proposition 8.2.2 ϕ is unsatisfiable and there is nothing to be shown. Otherwise, we have to show that ϕ is satisfiable. By Theorem 8.2.8 Duplicator has a winning strategy for the existential (ℓ', k')-pebble game $\mathcal{H}$ on ϕ and $\mathfrak{D}$. We claim that the Duplicator also has a winning strategy $\mathcal{J}$ for the existential (ℓ', k')-pebble game on ψ and $\mathfrak{C}$.

Let $(y^{i_1}_{j_1}, \ldots, y^{i_s}_{i_s})$ be a tuple of variables of ψ of length $s \leq k'$. By the $(0, k')$-extension property the strategy $\mathcal{H}$ contains a partial homomorphism h from ϕ to $\mathfrak{D}$ defined on $x_{j_1}, \ldots, x_{j_s}$. We then include in to $\mathcal{J}$ the map which sends $(y^{i_1}_{j_1}, \ldots, y^{i_s}_{j_s})$ to $(h(x_{j_1})_{i_1}, \ldots, h(x_{j_s})_{i_s})$. To show that $\mathcal{J}$ has the (ℓ', k')-extension property, let $(y^{i_1}_{j_1}, \ldots, y^{i_s}_{j_s})$ be a tuple of variables of ψ of length $s \leq \ell'$ and let $g \in \mathcal{J}$ a partial homomorphism from ψ to $\mathfrak{C}$ defined on all of these variables. Let $\bar{y} = (y^{i_1}_{j_1}, \ldots, y^{i_r}_{j_r})$ be an extension of the tuple above of length $r \leq k'$. We have to show that $\mathcal{J}$ contains an extension of h defined on all of $\bar{y}$. Then g was included into $\mathcal{J}$ for some $h \in \mathcal{H}$ defined on $x_{j_1}, \ldots, x_{j_s}$. By the (ℓ', k')-extension property h has an extension in $\mathcal{H}$ defined an all of $x_{j_1}, \ldots, x_{j_r})$. Then the map which sends $(y^{i_1}_{j_1}, \ldots, y^{i_r}_{j_r})$ to $(h(x_{j_1})_{i_1}, \ldots, h(x_{j_r})_{i_r}))$ is an extension of g with the desired properties.

This shows via Theorem 8.2.8 that the canonical (ℓ', k')-Datalog program does not derive *false* on ψ and $\mathfrak{C}$, and hence by assumption there exists a homomorphism f from ψ to $\mathfrak{C}$. This in turn implies that there exists a homomorphism from ϕ to $\mathfrak{D}$: the mapping from $X \to D$ that sends x_i to $(f(y^1_i), \ldots, f(y^d_i))$ satisfies all conjuncts of ϕ in $\mathfrak{D}$. ⊣

COROLLARY 8.3.4. *Let $\mathfrak{B}$ be an ω-categorical structure with finite signature such that* CSP($\mathfrak{B}$) *is in Datalog. Let $\mathfrak{D}$ be a structure with finite signature such that $\mathfrak{D}$ is homomorphically equivalent to a structure with a primitive positive interpretation in $\mathfrak{B}$. Then* CSP($\mathfrak{D}$) *is in Datalog as well.*

Remark 8.3.5. Corollary 8.3.4 also holds without the assumption of ω-categoricity; this has been stated for finite structures in Theorem 18 in [14], in a formulation that is equivalent to our formulation of Corollary 8.3.4 if the domain is finite. Their proof, however, can be used to prove our formulation of the statement for arbitrary domains; for the details, we refer to [87].

8.4. Arc consistency

The *arc consistency procedure* (AC) is an algorithm for constraint satisfaction problems that is intensively studied in Artificial Intelligence.[21] If m is the maximal arity of the input relations, then AC can be described as the subset of the canonical $(1, m)$-Datalog program that consists of all rules whose body contains at most one non-IDB. For finite templates $\mathfrak{B}$ it is known that AC solves CSP($\mathfrak{B}$) if and only if $\mathfrak{B}$ has Datalog width one [170]. For infinite structures, this is no longer true, as the following example shows.

[21] In the literature, the name *arc consistency* is reserved for binary CSPs, and the general case with constraints of arity larger than two is then called *hyper-arc consistency* or *generalised arc consistency*.

Example 8.4.1. The problem Triangle-Freeness from Figure 2 has Datalog width 0 (cf. Example 8.1.3). On the other hand, it is easy to see that this CSP cannot be solved by AC: to see this, let $\mathfrak{B}$ be the countably triangle-free (K_3-free) Henson graph (Example 2.3.10). Since $\mathfrak{B}$ is homogeneous and has no loops, the only primitive positive definable unary relations are the empty and the full relation, and since each of the rules involves at most one edge, none of the rules can detect the existence of a triangle in the input graph.

In the following sections we present general conditions that imply that CSP($\mathfrak{B}$) can be solved by AC.

8.4.1. The definable subset structure. Let $\mathfrak{B}$ be a countable ω-categorical structure with finite relational signature τ. Since $\mathfrak{B}$ is ω-categorical there is only a finite number of primitive positive definable non-empty sets $S_1, \ldots, S_n$.

DEFINITION 8.4.2. The *definable subset structure* $P(\mathfrak{B})$ of $\mathfrak{B}$ is the finite relational τ-structure with domain $\{S_1, \ldots, S_n\}$ where a k-ary relation R from τ holds on $S_{i_1}, \ldots, S_{i_k}$ if for every $j \in \{1, \ldots, k\}$ and every vertex v_j in the orbit S_{i_j} there are vertices $v_1, \ldots, v_{j-1}, v_{j+1}, \ldots, v_k$ from $S_{i_1}, \ldots, S_{i_{j-1}}, S_{i_{j+1}}, \ldots, S_{i_k}$, respectively, such that $(v_1, \ldots, v_k) \in R^{\mathfrak{B}}$.

LEMMA 8.4.3. *Let $\mathfrak{B}$ be an ω-categorical structure with finite relational signature τ. For every finite τ-structure $\mathfrak{A}$ the following two statements are equivalent*:

1. *The arc consistency procedure Π for $\mathfrak{B}$ does not derive* false *on $\mathfrak{A}$.*
2. *$\mathfrak{A}$ is homomorphic to $P(\mathfrak{B})$.*

PROOF. Let $\mathfrak{B}'$ be the expansion of $\mathfrak{B}$ by all unary primitive positive definable relations.

(1) $\Rightarrow$ (2). If Π derives $R(a)$ on $\mathfrak{A}$ then R is an IDB and has been introduced for a relation that is primitive positive definable over $\mathfrak{B}$, and hence $R^{\mathfrak{B}'} \in \{S_1, \ldots, S_n\}$. For every $a \in A$, let T^a be the set of all $R^{\mathfrak{B}'} \in \{S_1, \ldots, S_n\}$ such that Π derives $R(a)$ on $\mathfrak{A}$. By the definition of the rules of Π, the set T^a is closed under intersection. Hence, T^a contains a smallest element with respect to set inclusion, which will be denoted by $\cap T^a$. Define h to be the mapping from A to $\{S_1, \ldots, S_n\}$ that maps $a \in A$ to $\cap T^a$. We shall show that h is a homomorphism from $\mathfrak{A}$ to $P(\mathfrak{B})$. Let $R \in \tau$, and let $(a_1, \ldots, a_k)$ be a tuple of $R^{\mathfrak{A}}$. Then $(\cap T^{a_1}, \ldots, \cap T^{a_k})$ is the image of this tuple under h. Fix any $j \in \{1, \ldots, k\}$ and let S be the set containing all those v_j such that there are vertices $b_1, \ldots, b_{j-1}, b_{j+1}, \ldots, b_k$ from $\cap T^{a_1}, \ldots, \cap T^{a_{j-1}}, \cap T^{a_{j+1}}, \ldots, \cap T^{a_k}$, respectively, such that $(b_1, \ldots, b_k) \in R^{\mathfrak{B}}$. Then S is primitive positive definable in $\mathfrak{B}$, and Π contains an IDB U such that $U^{\mathfrak{B}'} = S$ and the rule

$$U(x_j) :- R(x_1, \ldots, x_k), \cap T^{a_1}(x_1), \ldots, \cap T^{a_{j-1}}(x_{j-1}), \cap T^{a_{j+1}}(x_{j+1}), \ldots, \cap T^{a_k}(x_k)$$

which allows to derive $U(a_j)$. As $\cap T^{a_j} \subseteq S = U^{\mathfrak{B}'}$ we conclude that

$$(\cap T^{a_1}, \dots, \cap T^{a_k}) \in R^{P(\mathfrak{B})}.$$

(2) $\Rightarrow$ (1). Let h be a homomorphism from $\mathfrak{A}$ to $P(\mathfrak{B})$. It is easy to prove by induction on the evaluation of Π on $\mathfrak{A}$ that $h(a) \subseteq R^{\mathfrak{B}'}$ for every $R(a)$ derived by Π on $\mathfrak{A}$ (as described in Section 8.1.3). Hence, *false* cannot be derived by Π on $\mathfrak{A}$. ⊣

The next theorem is from Feder and Vardi [170], formulated for CSPs for finite templates; we follow the presentation in [56] for ω-categorical templates.

THEOREM 8.4.4. *Let $\mathfrak{B}$ be an ω-categorical structure with finite relational signature. Then AC solves* CSP($\mathfrak{B}$) *if and only if $P(\mathfrak{B})$ is homomorphic to $\mathfrak{B}$.*

PROOF. Since $P(\mathfrak{B})$ maps homomorphically to $P(\mathfrak{B})$, Lemma 8.4.3 shows that AC does not derive *false* on $P(\mathfrak{B})$. Hence, if the arc consistency procedure solves CSP($\mathfrak{B}$) then $P(\mathfrak{B})$ maps homomorphically to $\mathfrak{B}$.

Conversely, suppose that there is a homomorphism h from $P(\mathfrak{B})$ to $\mathfrak{B}$. To show that Π solves CSP($\mathfrak{B}$), it suffices to show that a finite τ-structure $\mathfrak{A}$ such that Π does not derive *false* on $\mathfrak{A}$ maps homomorphically to $\mathfrak{B}$. By Lemma 8.4.3 there is a homomorphism g from $\mathfrak{A}$ to $P(\mathfrak{B})$. Composing g and h yields the desired homomorphism from $\mathfrak{A}$ to $\mathfrak{B}$. ⊣

Example 8.4.5. Let $\mathfrak{B}$ be a countable ω-categorical digraph with a directed cycle but without loop. Then CSP($\mathfrak{B}$) cannot be solved by AC. Indeed, suppose that $\mathfrak{B}$ contains a directed cycle of length k. The subset S of all vertices of $\mathfrak{B}$ that lie in a cycle of $\mathfrak{B}$ whose length divides k is primitively positively definable. Hence, S is a vertex in $P(\mathfrak{B})$, and (S, S) is an edge in $P(\mathfrak{B})$. Since $\mathfrak{B}$ has no loop, there is no homomorphism from $P(\mathfrak{B})$ to $\mathfrak{B}$, and thus Theorem 8.4.4 implies the statement. If $\mathfrak{B}$ is a finite structure then the converse holds as well: if $P(\mathfrak{B})$ contains a loop, then $\mathfrak{B}$ must contain a directed cycle. The structure $\mathfrak{B} = (\mathbb{Q}; <)$ shows that this is no longer true for ω-categorical structures in general, since $P(\mathfrak{B})$ has the loop $(\mathbb{Q}, \mathbb{Q})$ but $\mathfrak{B}$ has no directed cycle.

COROLLARY 8.4.6. *Let $\mathfrak{B}$ be an ω-categorical structure with finite relational signature. If* CSP($\mathfrak{B}$) *is solved by the arc consistency algorithm, then $\mathfrak{B}$ is homomorphically equivalent to a finite structure.*

PROOF. The definable subset structure $\mathfrak{C}$ of $\mathfrak{B}$ is finite, and by Theorem 8.4.4, it has a homomorphism to $\mathfrak{B}$. Thus, it suffices to prove that $\mathfrak{B}$ has a homomorphism to $\mathfrak{C}$. Let $\mathfrak{A}$ be a finite substructure of $\mathfrak{B}$. Since substructures of $\mathfrak{B}$ are satisfiable instances of CSP($\mathfrak{B}$), the program Π does not derive *false* on such a structure $\mathfrak{A}$. So by Lemma 8.4.3, there exists a homomorphism from $\mathfrak{A}$ to $\mathfrak{C}$. Lemma 4.1.7 then implies that also $\mathfrak{B}$ maps homomorphically to $\mathfrak{C}$. ⊣

In spite of Corollary 8.4.6, the arc consistency procedure is also important for infinite-domain CSPs, because many infinite-domain CSPs can be reduced in polynomial time to finite-domain CSPs (cf. Section 10.5.3) that can then be solved by the arc consistency procedure (a simple example is CSP($\mathbb{N}; \neq, I_4$) from Section 7.5).

8.4.2. Totally symmetric polymorphisms. There is also a characterisation of solvability by AC which is based on polymorphisms, due to [147].

DEFINITION 8.4.7. A function $f : B^k \to B$ is called *totally symmetric* if

$$f(x_1, \ldots, x_m) = f(y_1, \ldots, y_m) \text{ whenever } \{x_1, \ldots, x_m\} = \{y_1, \ldots, y_m\}.$$

Note that the existence of a totally symmetric polymorphism is a non-trivial minor condition (Definition 6.7.9).

Example 8.4.8. The operation $(x_1, \ldots, x_n) \mapsto \min(x_1, \ldots, x_n)$ is a totally symmetric polymorphism of the structure $(\{0, 1\}; \{(x, y, z) \mid (x \wedge y) \Rightarrow z, \{0\}, \{1\})$ (cf. Section 1.6.7 and Proposition 6.2.5).

More generally, if a structure has a semilattice operation (Example 2.1.2), then it has totally symmetric polymorphisms of all arities.

THEOREM 8.4.9. *Let $\mathfrak{B}$ be a finite structure and maximal arity m. Then the following are equivalent.*

1. *$P(\mathfrak{B})$ maps homomorphically to $\mathfrak{B}$.*
2. *$\mathfrak{B}$ has totally symmetric polymorphisms of all arities.*
3. *$\mathfrak{B}$ has a totally symmetric polymorphism of arity $m|B|$.*

PROOF. (1) $\Rightarrow$ (2). Suppose that g is a homomorphism from $P(\mathfrak{B})$ to $\mathfrak{B}$, and let $k \in \mathbb{N}$ be arbitrary. Let f be defined by $f(x_1, \ldots, x_k) := g(S(\{x_1, \ldots, x_k\}))$ where $S(\{x_1, \ldots, x_k\})$ is the smallest primitive positive definable subset of $\mathfrak{B}$ that contains $\{x_1, \ldots, x_k\}$. Clearly, f is totally symmetric. If $R \in \tau$ is ℓ-ary and $t^1, \ldots, t^k \in R^{\mathfrak{B}}$, then Corollary 6.1.14 (2) implies that

$$(S(\{t_1^1, \ldots, t_1^k\}), \ldots, S(\{t_\ell^1, \ldots, t_\ell^k\})) \in R^{P(\mathfrak{B})}$$

and hence $(f(t^1), \ldots, f(t^\ell)) \in R^{\mathfrak{B}}$. Therefore, f is a polymorphism of $\mathfrak{B}$.

The implication (2) $\Rightarrow$ (3) is trivial. To prove (3) $\Rightarrow$ (1), suppose that f is a totally symmetric polymorphism of $\mathfrak{B}$ of arity $m|B|$. Let $g: P(\mathfrak{B}) \to B$ be defined by

$$g(\{x_1, \ldots, x_n\}) := f(x_1, \ldots, x_{n-1}, x_n, x_n, \ldots, x_n)$$

which is well defined because f is totally symmetric. Let $R \in \tau$ be of arity $\ell \leq m$, let $(U_1, \ldots, U_\ell) \in R^{P(\mathfrak{B})}$, and for each $i \leq \ell$ let $u_1^i, \ldots, u_{p_i}^i$ be an enumeration of the elements of U_i. The properties of $P(\mathfrak{B})$ imply that for each $i \leq \ell$ and $q \leq p_i$ there are $v_{i,q}^1 \in U_1, \ldots, v_{i,q}^{i-1} \in U_{i-1}, v_{i,q}^{i+1} \in U_{i+1}, \ldots, v_{i,q}^\ell \in U_\ell$ such that

$$(v_{i,q}^1, \ldots, v_{i,q}^{i-1}, u_q^i, v_{i,q}^{i+1}, \ldots, v_{i,q}^\ell) \in R^{\mathfrak{B}}.$$

Then

$$
\begin{aligned}
g(U_1) &= g(\{u_1^1, \dots, u_{p_1}^1\}) \\
&= f(u_1^1, \dots, u_{p_1}^1, v_{2,1}^1, \dots, v_{2,p_2}^1, \dots, v_{\ell,1}^1, \dots, v_{\ell,p_\ell}^1, u_1^1, \dots, u_1^1) \\
&\vdots \\
g(U_\ell) &= g(\{u_1^\ell, \dots, u_{p_\ell}^\ell\}) \\
&= f(v_{1,1}^\ell, \dots, v_{1,p_1}^\ell, v_{2,1}^\ell, \dots, v_{2,p_2}^\ell, \dots, u_1^\ell, \dots, u_{p_\ell}^\ell, u_1^\ell, \dots, u_1^\ell)
\end{aligned}
$$

and $(g(U_1), \dots, g(U_\ell)) \in R^{\mathfrak{B}}$ since f preserves R. ⊣

Note that the assumption that $\mathfrak{B}$ is finite is necessary, as the following example shows.

Example 8.4.10. Let $\mathfrak{B} := (\mathbb{Q}; <)$. Clearly, $\mathfrak{B}$ has totally symmetric polymorphisms of all arities, namely the operations $(x_1, \dots, x_n) \mapsto \min(x_1, \dots, x_n)$. On the other hand, $P(\mathfrak{B})$ contains a loop, and hence does not map homomorphically to $\mathfrak{B}$. And indeed, AC does not solve CSP($\mathfrak{B}$).

8.5. Strict width and quasi near-unanimity operations

The notion of *strict width* has been introduced for finite-domain CSPs by Feder and Vardi [170]. Bounded strict width of CSP($\mathfrak{B}$) is a strong form of bounded width of CSP($\mathfrak{B}$), and can be defined for arbitrary countable ω-categorical structures $\mathfrak{B}$. For countable ω-categorical structures $\mathfrak{B}$ has bounded strict width has particularly elegant universal-algebraic characterisations in terms of Pol($\mathfrak{B}$) and *(quasi-) near-unanimity polymorphisms* (Section 8.5.2). Bounded strict width and quasi near-unanimity polymorphisms are also tightly linked to the concept of decomposability, defined in Section 8.5.3. The link between the three concepts is presented in Section 8.5.4. In Section 8.5.5 we revisit the clone from Section 7.6 which provides an example of a structure with an infinite signature and no quasi near-unanimity polymorphism, but where every finite reduct of the structure has such a polymorphism.

8.5.1. Bounded strict width. Feder and Vardi [170] defined strong width for finite-domain CSPs in terms of the canonical Datalog program. Based on our notion of canonical Datalog programs, we study the analogous concept for countable ω-categorical structures. In the terminology of the constraint satisfaction literature in Artificial Intelligence, strict width ℓ is equivalent to the property that *'strong ℓ-consistency implies global consistency'*.

Recall that a Datalog program over the signature τ receives as input a τ-structure $\mathfrak{A}$ and returns an expansion $\Pi(\mathfrak{A})$ of $\mathfrak{A}$ in the signature that contains τ as well as a symbol for every IDB of Π. If Π is the canonical

(ℓ, k)-Datalog program for CSP($\mathfrak{B}$), then $\Pi(\mathfrak{A})$ can be viewed as an instance of CSP($\mathfrak{B}'$) where $\mathfrak{B}'$ is the expansion of $\mathfrak{B}$ by all at most ℓ-ary primitively positively definable relations. The instance $\Pi(\mathfrak{A})$ is called *globally consistent* if every partial homomorphism, i.e, every homomorphism from an induced substructure of $\Pi(\mathfrak{A})$ to $\mathfrak{B}'$, can be extended to a homomorphism from $\Pi(\mathfrak{A})$ to $\mathfrak{B}'$.

DEFINITION 8.5.1. Let $\ell \geq 2$ and $k \geq \ell$. Then $\mathfrak{B}$ has *strict width* (ℓ, k) if all instances of CSP($\mathfrak{B}'$) that are computed by the canonical (ℓ, k)-Datalog program Π are globally consistent. We say that $\mathfrak{B}$ has *strict width* ℓ if it has strict width (ℓ, k) for some k.

Note that if $\mathfrak{B}$ has strict width ℓ then CSP($\mathfrak{B}$) has Datalog width ℓ. To state some of our later results in the strongest possible form, we define strict width also for ω-categorical structures $\mathfrak{B}$ with an *infinite* relational signature: such a structure has strict width ℓ if every reduct of $\mathfrak{B}$ with a finite signature has strict width ℓ.

Also note that if Π derives *false* on input $\mathfrak{A}$, then $\Pi(\mathfrak{A})$ does not have any partial homomorphisms to $\mathfrak{B}'$, and hence $\Pi(\mathfrak{A})$ is in this case by definition globally consistent. If the reader feels uneasy about calling unsatisfiable instances globally consistent, one may also define global consistence only for satisfiable instances; for strict width ℓ we would then require that the instances computed by the canonical (ℓ, k)-program that do not contain the predicate *false* are globally consistent. With our definition we follow what is standard in the literature.

Example 8.5.2. The infinite clique $\mathfrak{B} := (\mathbb{N}; \neq)$ has strict width $(0, 2)$. Indeed, let $\mathfrak{A}$ be a graph such that the canonical $(0, 2)$-Datalog program for $\mathfrak{B}$ does not derive *false* on $\mathfrak{A}$. Then $\mathfrak{A}$ does not contain loops, i.e., unsatisfiable constraints of the form $x \neq x$, and hence every partial mapping from $\mathfrak{A}$ to $(\mathbb{N}; \neq)$ can be extended to a homomorphism from $\mathfrak{A}$ to $(\mathbb{N}; \neq)$ by always picking new elements in $\mathbb{N}$.

Example 8.5.3. The structure $\mathfrak{B} := (\mathbb{Q}; <)$ has strict width $(2, 3)$. Indeed, let $\mathfrak{A}$ be a finite digraph such that the canonical $(2, 3)$-Datalog program Π for $\mathfrak{B}$ does not derive *false* on $\mathfrak{A}$. Let $\mathfrak{B}'$ be the expansion of $\mathfrak{B}$ by all at most binary relations with a primitive positive definition in $\mathfrak{B}$; let T be the binary relation symbol of $\mathfrak{B}'$ introduced for the primitive positive formula $x < y$. Then $T^{\Pi(\mathfrak{A})}$ equals the transitive closure of $<^{\mathfrak{A}}$. Moreover, it is easy to see that any partial homomorphism from $\Pi(\mathfrak{A})$ to $\mathfrak{B}'$ can be extended to all of $\Pi(\mathfrak{A})$ (as in the step of '*going forth*' in the proof of Proposition 4.1.1)), so $\Pi(\mathfrak{A})$ is globally consistent.

To study the strict width of an ω-categorical structure $\mathfrak{B}$, we may always assume that $\mathfrak{B}$ contains the equality relation, thanks to the following lemma [56]

LEMMA 8.5.4. *Let $\mathfrak{B}$ be an ω-categorical structure of strict width ℓ and let $\mathfrak{B}'$ be the expansion of $\mathfrak{B}$ by the equality relation. Then $\mathfrak{B}'$ has strict width ℓ as well.*

PROOF. Let $\mathfrak{C}$ be a reduct of $\mathfrak{B}$ with finite signature. Let k be such that CSP($\mathfrak{C}$) has strict width (ℓ, k). Let $\equiv$ the binary symbol that denotes the equality relation in $\mathfrak{B}'$; it suffices to show that CSP($\mathfrak{C}; \equiv$) has strict width (ℓ, k). Let Π' be the canonical (ℓ, k)-program for CSP($\mathfrak{C}, \equiv$). Let $\mathfrak{A}$ be an instance of CSP($\mathfrak{C}, \equiv$). Let E be the finest equivalence relation on A that contains $\equiv^{\mathfrak{A}}$. Let $\mathfrak{A}/E$ be the τ-reduct of $\mathfrak{A}$ obtained by factoring $\mathfrak{A}$ by the equivalence relation E. The definition is analogous to the construction in Example 3.1.2: the universe of $\mathfrak{A}/E$ are the equivalence classes of E and for every $R \in \tau$, say r-ary, $R^{\mathfrak{A}/E} = \{(a_1/E, \dots, a_r/E) \mid (a_1, \dots, a_r) \in R^{\mathfrak{A}}\}$. We now consider $\mathfrak{A}/E$ as an instance of CSP($\mathfrak{B}$). Let Π be the canonical (ℓ, k)-program of $\mathfrak{B}$. It is easy to prove by induction on the evaluation of Π on $\mathfrak{A}/E$ that if R is an IDB, say r-ary, and $R(a_1/E, \dots, a_r/E)$ is derived by Π on $\mathfrak{A}/E$, then $R(a_1, \dots, a_r)$ is derived by Π' on $\mathfrak{A}$. We have to show that $\Pi(\mathfrak{A})$ is globally consistent. So suppose that there is a partial homomorphism h from $\Pi(\mathfrak{A})$ to an expansion of $(\mathfrak{C}; \equiv)$ by all ℓ-ary primitively positively definable relations. Since $\ell \geq 2$ and $k \geq 3$, the Datalog program Π' will be able to derive that all elements in the same equivalence class of E have to get the same value and hence, if h is a partial homomorphism then this implies that $h(a) = h(b)$ for all elements a, b in the domain of h such that $(a, b) \in E$. Define h/E to be the partial mapping that maps a/E to $h(a)$ for every a in the domain of h. Then h/E is a partial homomorphism from $\mathfrak{A}/E$ to $\mathfrak{B}$. Hence h/E can be extended to a full homomorphism h' from $\mathfrak{A}/E$ to $\mathfrak{B}$. Finally, the mapping defined by $a \mapsto (h/E)(a/E)$ is a homomorphism from $\mathfrak{A}$ to $(\mathfrak{C}, \equiv)$. ⊣

8.5.2. Quasi near-unanimity polymorphisms. In Section 8.5.4 we will show that for ω-categorical model-complete cores, $\mathfrak{B}$ bounded strict width if and only if $\mathfrak{B}$ has a quasi near-unanimity polymorphism. Quasi near-unanimity operations have already been introduced in Section 6.1.8 in the context of minimal clones. Recall that an operation f is a *quasi near-unanimity operation* if it satisfies the identities

$$\forall x, y \colon f(x, \dots, x, y) = f(x, \dots, x, y, x) = \dots = f(y, x, \dots, x) = f(x, \dots, x).$$

A *near-unanimity operation* is additionally idempotent. If $A \subseteq B$ we say that $f \colon B^k \to B$ is a *near-unanimity on A* if the near-unanimity identities are satisfied for all $x, y \in A$. A ternary near-unanimity operation is called a *majority operation*, and a ternary quasi near-unanimity operation is called a *quasi majority operation*. We present some examples with majority or quasi majority polymorphisms.

Example 8.5.5. Let $(D; <)$ be any linearly ordered set. Then the ternary *median operation* defined as

$$\begin{aligned}(x, y, z) \mapsto \min(&\max(x, y), \max(y, z), \max(z, x)) \\ &= \max(\min(x, y), \min(y, z), \min(z, x))\end{aligned}$$

is a majority operation, which also preserves $\leq$ (increasing all arguments of the operation can only increase the function value). Hence, the structure $(\mathbb{Q}; <, \leq)$ has a majority polymorphism. If D is finite, the median operation also preserves the successor relation. The definition of the median can be generalised to any lattice defined on D, with meet in place of min and join in place of max.

Example 8.5.6. *Directed cycles* $\mathfrak{C}_n$ of length $n \geq 1$, defined as

$$\mathfrak{C}_n := (\{0, \dots, n-1\}; E) \text{ where } E := \{(x, y) \mid y = x + 1 \mod n\},$$

have a majority polymorphism: Let f be the ternary operation that maps $(u, v, w) \in \{0, \dots, n-1\}^3$ to u if u, v, w are pairwise distinct, and otherwise acts as a majority operation. We claim that f is a polymorphism of $\mathfrak{C}_n$. Let $(u, u'), (v, v'), (w, w') \in E$ be arcs. If u, v, w are all distinct, then u', v', w' are clearly all distinct as well, and hence $(f(u, v, w), f(u', v', w')) = (u, u') \in E$. Otherwise, if two elements of u, v, w are equal, say $u = v$, then u' and v' must be equal as well, and hence $(f(u, v, w), f(u', v', w')) = (u, u') \in E$.

Example 8.5.7. The infinite clique $(\mathbb{N}; \neq)$ has no near-unanimity operation: if $f : \mathbb{N}^k \to \mathbb{N}$ is a near-unanimity, we distinguish two cases:

- $a := f(1, 2, \dots, k) \notin \{1, \dots, k\}$. In this case we have $a \neq 1, \dots, a \neq k$, and $f(a, \dots, a) = a = f(1, \dots, k)$, so f does not preserve $\neq$.
- $a := f(1, 2, \dots, k) \in \{1, \dots, k\}$; say $a = 1$. Then $2 \neq 1, 1 \neq 2, \dots, 1 \neq k$ and $f(2, 1, \dots, 1) = 1 = f(1, 2, \dots, k)$, so again f does not preserve $\neq$.

However, $(\mathbb{N}; \neq)$ does have a quasi majority: to see this, let E be the smallest equivalence relation on $\mathbb{N}^3$ that contains all pairs of the form $((x, x, y), (x, x, x))$, $((x, y, x), (x, x, x))$, and $((y, x, x), (x, x, x))$. We claim that $(\mathbb{N}; \neq)^3/E$ (Definition given in Example 3.1.2) does not contain loops: the only edges in $(\mathbb{N}; \neq)^3/E$ come from edges $((x_1, x_2, x_3), (y_1, y_2, y_3))$ in $(\mathbb{N}; \neq)^3$ where $x_1 \neq y_1$, $x_2 \neq y_2$, and $x_3 \neq y_3$. All triples that are E-equivalent to (x_1, x_2, x_3) must have two equal entries, and the same holds for all triples that are E-equivalent to (y_1, y_2, y_3). This shows that (x_1, x_2, x_3) cannot be E-equivalent to (y_1, y_2, y_3) and proves the claim. Hence, every injection from $\mathbb{N}/E$ to $\mathbb{N}$ is a homomorphism h from $(\mathbb{N}; \neq)^3/E$ to $(\mathbb{N}; \neq)$, and $(x_1, x_2, x_3) \mapsto h((x_1, x_2, x_3)/E)$ is the desired quasi majority operation.

Quasi near-unanimity identities have height one (cf. Section 6.7.2), so if $\mathfrak{B}$ has a quasi near-unanimity polymorphism, then so have all structures that are homomorphically equivalent to $\mathfrak{B}$ (see Corollary 6.7.12). This holds in particular for the model-complete core of an ω-categorical structure $\mathfrak{B}$. The following lemma shows that the existence of a quasi near-unanimity polymorphism follows from the local existence of quasi near-unanimity polymorphisms.

PROPOSITION 8.5.8. *Let $\mathfrak{B}$ be countable ω-categorical and let $k \geq 2$. Then $\mathfrak{B}$ has a k-ary quasi near-unanimity polymorphism if for every finite $F \subseteq B$ there exists a polymorphism of $\mathfrak{B}$ of arity k which is a quasi near-unanimity operation on F. Conversely, if $\mathfrak{B}$ is an ω-categorical model-complete core and has a quasi near-unanimity polymorphism of arity k, then for all finite $F \subseteq B$ there is a k-ary polymorphism of $\mathfrak{B}$ whose restriction to $\mathfrak{A}$ is a near-unanimity.*

PROOF. The statement can be shown using an argument based on König's tree lemma as in the proof of Lemma 4.1.7. Alternatively, we can apply Lemma 4.1.10, because whether f is a k-ary quasi near-unanimity operation can be expressed as a universal first-order sentence in a two-sorted structure with the sorts B^k and B.

Now suppose that $\mathfrak{B}$ is an ω-categorical model-complete core. Let f be a polymorphism of $\mathfrak{B}$ which is a quasi near-unanimity operation and let $b_1, \dots, b_n \in B$. The map $\hat{f}$ is an endomorphism of $\mathfrak{B}$, and hence $\hat{f} \in \overline{\mathrm{Aut}(\mathfrak{B})}$ since $\mathfrak{B}$ is a model-complete core (Theorem 4.5.1). So there exists an $\alpha \in \mathrm{Aut}(\mathfrak{B})$ such that $\hat{f}(b_i) = \alpha(b_i)$ for every $i \leq n$. Then the map $\alpha^{-1}\hat{f} \in \mathrm{Pol}(\mathfrak{B})$ is a near-unanimity on $\{b_1, \dots, b_n\}$. ⊣

The following example shows that the assumption that $\mathfrak{B}$ is a core is necessary for the converse implication in Proposition 8.5.8.

Example 8.5.9. The core of the structure $\mathfrak{B} := (\mathbb{N}; P^3_{\mathbb{N}})$ (Definition 6.1.16) has just one element, and $\mathfrak{B}$ has a constant polymorphism of arity 2, which is a quasi near-unanimity operation. However, by Proposition 6.1.19 all polymorphisms of $(\mathbb{N}; P^3_{\mathbb{N}})$ are essentially unary, so for any finite $F \subset \mathbb{N}$ with at least two elements there does not exist a near-unanimity operation on F.

8.5.3. Decomposability. Bounded strict width and the existence of quasi near-unanimity polymorphisms of ω-categorical model-complete cores can also be characterised relationally. The corresponding result for finite structures and (idempotent) near-unanimity polymorphisms is known as the theorem of Baker and Pixley, and is a special case of the result presented in Theorem 8.5.12.

DEFINITION 8.5.10. Let $l \geq 1$. We say that a relation $R \subseteq B^k$ is *ℓ-decomposable* if it consists of all tuples $(b_1, \dots, b_k)$ such that for all $i_1, \dots, i_l \leq k$ there exists $(c_1, \dots, c_k) \in R$ such that $b_{i_1} = c_{i_1}, \dots, b_{i_k} = c_{i_k}$.

We say that a structure $\mathfrak{B}$ is *ℓ-decomposable* if every relation with a primitive positive definition in $\mathfrak{B}$ is ℓ-decomposable. The following shows that ℓ-decomposability can be seen as a special form of primitive positive quantifier elimination; the proof is straightforward from the definitions and ω-categoricity is not needed.

LEMMA 8.5.11. *Let $l \geq 2$. A structure is ℓ-decomposable if and only if every primitive positive formula is equivalent over $\mathfrak{B}$ to a conjunction of primitive positive formulas each having at most ℓ free variables.*

8.5.4. Characterising strict width. We finally state and prove the connection between strict width, near-unanimity polymorphisms, and decomposability. The proof of this result combines a result from [56] and from [50] and is based on ideas from [170] and [216] for finite structures $\mathfrak{B}$. It is convenient to treat instances of CSPs as primitive positive sentences rather than structures.

THEOREM 8.5.12. *Let $\mathfrak{B}$ be an ω-categorical relational structure and let $l \geq 2$. Then the following are equivalent.*

1. *$\mathfrak{B}$ has strict width ℓ.*
2. *$\mathfrak{B}$ is ℓ-decomposable.*
3. *For every finite subset F of B there is an $(\ell + 1)$-ary polymorphism of $\mathfrak{B}$ whose restriction to F is a near-unanimity.*

PROOF. Let τ be the signature of $\mathfrak{B}$. We first show that (1) implies (2). Without loss of generality, $\mathfrak{B}$ contains a binary symbol $\equiv$ that denotes the equality relation on B (see Proposition 8.5.4). Let $\mathfrak{B}'$ be the expansion of $\mathfrak{B}$ by all ℓ-ary primitively positively definable relations, and let τ' be the signature of $\mathfrak{B}'$. Let $\phi(x_1, \dots, x_n)$ be a primitive positive τ-formula. We have to show that ϕ is equivalent to a conjunction of atomic τ'-formulas.

Let $\mathfrak{C}$ be the reduct of $\mathfrak{B}$ that only contains the relations that appear in ϕ. Let $k \in \mathbb{N}$ be such that $\mathfrak{C}$ has strict width (ℓ, k), and let Π be the canonical (ℓ, k)-Datalog program. Add the new variables $y_1, \dots, y_n$ and the conjuncts $x_1 \equiv y_1, \dots, x_n \equiv y_n$ to ϕ, and then run Π on the resulting formula. Let $\psi(y_1, \dots, y_n)$ be the set of atomic σ-formulas derived by Π on $y_1, \dots, y_n$. Clearly, all of them have arity at most ℓ, and $\phi(y_1, \dots, y_n)$ implies $\psi(y_1, \dots, y_n)$. The converse also holds since $\Pi(\phi)$ is globally consistent.

(2) $\Rightarrow$ (3). Let $F \subseteq B$ be finite. We have to prove that there is a polymorphism of $\mathfrak{B}$ that is an $(\ell + 1)$-ary near-unanimity on F. Let τ_F be the superset of τ that additionally contains a unary relation symbol R_a for each $a \in F$. Let $\mathfrak{B}_F$ be the τ_F-expansion of $\mathfrak{B}$ where $R_a = \{a\}$ for every $a \in F$. Consider the set G of tuples $(a_0, \dots, a_l)$ in F^{l+1} that have identical entries $a_i = a$ except for possibly one position. Let $\mathfrak{C}$ be the τ_F-expansion of $\mathfrak{B}^{l+1}$ where R_a denotes the set of all tuples in G where at most one entry is not a. Every homomorphism from $\mathfrak{C}$ to $\mathfrak{B}_F$ is by construction a polymorphism of $\mathfrak{B}$ that is a near-unanimity on F. To show that such a homomorphism from $\mathfrak{C}$ to $\mathfrak{B}_F$ indeed exists, is suffices to show that every finite substructure of $\mathfrak{C}$ maps homomorphically to $\mathfrak{B}_F$ (Lemma 4.1.7; here we use ω-categoricity).

Let $\mathfrak{S}$ be a finite substructure of $\mathfrak{C}$. Let ψ be the canonical query of the τ-reduct of $\mathfrak{S}$. Let ϕ be the primitive positive formula obtained by existentially quantifying all variables in ψ except for the variables from G. We claim that the map h that sends $(a,\dots,a,b,a,\dots,a)$ to a satisfies ϕ, which shows that $\mathfrak{S}$ maps homomorphically to $\mathfrak{B}_F$. By assumption, ϕ is equivalent to a conjunction of primitive positive formulas ψ with at most $s \leq l$ free variables. Let R be the relation of arity s that has been introduced in $\mathfrak{B}'$ for such a formula ψ and suppose that $\mathfrak{S} \models R(t^1,\dots,t^s)$. For every $j \in \{1,\dots,s\}$ the tuple t^j is of the from $(a_j,\dots,a_j,b_j,a_j,\dots,a_j)$. Since $s \leq l$, the pigeon-hole principle guarantees that there exists an index $i \in \{1,\dots,l+1\}$ such that for every $1 \leq j \leq l$ the i-th entry of t^j equals a_j. Hence, $R(a_1,\dots,a_s)$ holds and h preserves ϕ. We conclude that $\mathfrak{S}$ maps homomorphically to $\mathfrak{B}_F$.

(3) $\Rightarrow$ (1). Let $\mathfrak{C}$ be a reduct of $\mathfrak{B}$ with finite signature $\sigma \subseteq \tau$. Let k be the maximum of $l+1$ and the maximal arity of the relations in $\mathfrak{C}$. Let Π be the canonical (ℓ,k)-program for $\mathfrak{B}$ and let $\mathfrak{C}'$ be the expansion of $\mathfrak{C}$ by all at most ℓ-ary primitively positively definable relations. Let $\mathfrak{A}$ be a finite σ-structure. We have to prove that every partial homomorphism h from $\Pi(\mathfrak{A})$ to $\mathfrak{C}'$ with domain $\{v_1,\dots,v_i\}$, for $i < |A|$, has an extension to any other element v of $\Pi(\mathfrak{A})$ such that the extension is still a partial homomorphism from $\Pi(\mathfrak{A})$ to $\mathfrak{C}'$. We prove this by induction on i.

For the case that $i \leq l$, let Ψ be the set of all atomic formulas that hold in $\Pi(\mathfrak{A})$ on variables from $\{v_1,\dots,v_i,v\}$, and let R be the IDB associated to the primitive positive formula $\exists v \bigwedge \Psi$ with free variables $v_1,\dots,v_i$. Since each formula in Ψ is derived by Π on $\mathfrak{A}$, the conjunct $R(v_1,\dots,v_i)$ is also derived by Π on $\mathfrak{A}$. Since h preserves R, we have that $(h(v_1),\dots,h(v_i))$ satisfies $\exists v \bigwedge \Psi$; hence, there exists an extension of h to v which is a partial homomorphism from $\Pi(\mathfrak{A})$ to $\mathfrak{C}'$.

For the induction step where $i \geq l+1$, let h_j be the restriction of h where v_j is undefined, for $j \in \{1,\dots,l+1\}$. By induction, h_j can be extended to a homomorphism h'_j from the substructure of $\Pi(\mathfrak{A})$ induced on $\{v_1,\dots,v_i,v\} \setminus \{v_j\}$ to $\mathfrak{C}'$.

Claim. For each $(u_1,\dots,u_r) \in R^{\Pi(\mathfrak{A})}$, there exists a tuple $(b^j_1,\dots,b^j_r) \in R^{\mathfrak{C}'}$ such that $h'_j(u_i) = b^j_i$ for all u_i where h'_j is defined.

Let $i_1,\dots,i_s$ be a list of the indices $i \in \{1,\dots,r\}$ such that h'_j is defined on u_i, and let $i'_1,\dots,i'_t$ be a list of the other indices in $\{1,\dots,r\}$ (so we have $s+t=r$). We prove the statement by induction on s. For $s \leq l$, let R' be the IDB associated to the $\exists u_{i'_1},\dots,u_{i'_t} \colon R(u_1,\dots,u_r)$ with free variables $u_{i_1},\dots,u_{i_s}$. Since $R'(u_{i_1},\dots,u_{i_s}) \mathrel{:-} R(u_1,\dots,u_r)$ is a rule in Π, we have $(h'_j(u_{i_1}),\dots,h'_j(u_{i_s})) \in R'^{\mathfrak{C}'}$. Then the witnesses for the existentially quantified variables $u_{i'_1},\dots,u_{i'_t}$ in $\mathfrak{C}'$ along with $h'_j(u_{i_1}),\dots,h'_j(u_{i_s})$ determine the tuple $(d_1,\dots,d_r) \in R^{\mathfrak{C}'}$ with the desired property. For $s \geq l+1$, consider for all

$i \in \{i_1, \ldots, i_{l+1}\}$ the tuple $b^i = (b^i_1, \ldots, b^i_r) \in R^{\mathfrak{C}'}$ given inductively for the restriction of h'_j to $S \setminus \{u_i\}$. Let g be an $(\ell + 1)$-ary polymorphism which is a near-unanimity on the set containing all elements in all tuples b^i. Then $(g(b^1_1, \ldots, b^{l+1}_1), \ldots, g(b^1_r, \ldots, b^{l+1}_r))$ has the desired properties, concluding the proof of the claim.

Let $F \subset B$ be the finite set that contains all elements b^j_i of $\mathfrak{C}'$, for all tuples $(u_1, \ldots, u_r)$ in all relations R of $\Pi(\mathfrak{A})$. Let g be an $(\ell + 1)$-ary polymorphism of $\mathfrak{C}'$ that is a near-unanimity on F (observe that $\mathfrak{B}$ and $\mathfrak{C}'$ have the same polymorphisms). We claim that the extension h' of h mapping v to

$$b := g(h'_1(v), \ldots, h'_{l+1}(v))$$

is a homomorphism from the substructure of $\Pi(\mathfrak{A})$ induced on $\{v_1, \ldots, v_i, v\}$ to $\mathfrak{C}'$. Arbitrarily choose $(u_1, \ldots, u_r) \in R^{\Pi(\mathfrak{A})}$; we want to show that $(h'(u_1), \ldots, h'(u_r)) \in R^{\Pi(\mathfrak{A})}$. Recall that $(b^j_1, \ldots, b^j_r) \in R^{\mathfrak{C}'}$ is such that $h'_j(u_i) = b^j_i$ for all u_i where h'_j is defined. Then $(g(b^1_1, \ldots, b^{l+1}_1), \ldots, g(b^1_r, \ldots, b^{l+1}_r)) \in R^{\mathfrak{C}'}$. Moreover, we claim that $g(b^1_s, \ldots, b^{l+1}_s) = h'(u_s)$ for every $s \leq r$: if $u_s \in \{v_1, \ldots, v_i\}$, note that for all but at most one j from $\{1, \ldots, l+1\}$ we have that $b^j_s = h'_j(u_s) = h(u_s)$, and since g is a near-unanimity on the entries of the tuples b^j we obtain that

$$g(b^1_s, \ldots, b^{l+1}_s) = h'(u_s) = h(u_s).$$

Otherwise, if $u_s = v$, then

$$g(b^1_s, \ldots, b^{l+1}_s) = g(h'_1(v), \ldots, h'_{l+1}(v)) = b = h'(v)$$

by the definition of h'. We conclude that $(h'(u_1), \ldots, h'(u_r)) \in R^{\Pi(\mathfrak{A})}$. ⊣

COROLLARY 8.5.13. *If $\mathfrak{B}$ is an ω-categorical structure with finite relational signature and a quasi near-unanimity polymorphism. Then* CSP($\mathfrak{B}$) *is in Datalog.*

PROOF. Let $\mathfrak{C}$ be the model-complete core of $\mathfrak{B}$, which also has a quasi near-unanimity polymorphism. By Proposition 8.5.8 for every finite $F \subseteq C$ the structure $\mathfrak{C}$ has a polymorphism of arity $l + 1$ which is a near-unanimity operation on F. Hence, $\mathfrak{C}$ has bounded strict width by Theorem 8.5.12 and hence is in Datalog. This proves the statement since $\mathfrak{C}$ and $\mathfrak{B}$ have the same CSP. ⊣

COROLLARY 8.5.14. *Let $\mathfrak{B}$ be an ω-categorical model-complete core. Then for every $l \geq 2$ the following are equivalent.*

- *$\mathfrak{B}$ has an $(\ell + 1)$-ary quasi near-unanimity polymorphism.*
- *Every primitive positive formula is over $\mathfrak{B}$ equivalent to a conjunction of primitive positive formulas each with at most ℓ free variables.*
- *$\mathfrak{B}$ is ℓ-decomposable.*
- *$\mathfrak{B}$ has strict width ℓ.*

PROOF. The equivalence of (4), (3), and (2) holds without the assumption of $\mathfrak{B}$ being a model-complete core (Theorem 8.5.12 and Lemma 8.5.11). The equivalence with (1) holds by Proposition 8.5.8. ⊣

Example 8.5.15. Let $\mathcal{C}$ be the smallest locally closed operation clone that contains $\mathrm{Aut}(\mathbb{Q}; <)$ and the median operation on $\mathbb{Q}$ (Example 8.5.5). We claim that

$$\mathcal{C} = \mathrm{Pol}(\mathbb{Q}; <, \leq).$$

To see this, first recall from Example 8.5.5 that median preserves both $<$ and $\leq$. Conversely, we have to show that every polymorphism of $(\mathbb{Q}; <, \leq)$ is locally generated by the median and automorphisms of $(\mathbb{Q}; <)$. Equivalently (Theorem 6.1.13) one can show that every relation that is first-order definable over $(\mathbb{Q}; <)$ and preserved by the median operation is primitively positively definable over $(\mathbb{Q}; <)$. Since the structure $\mathfrak{B}$ with domain $\mathbb{Q}$ that contains all these relations is an ω-categorical model-complete core and median is a majority operation, Theorem 8.5.14 implies that $\mathfrak{B}$ is 2-decomposable. However, there are only a few binary relations in $\mathfrak{B}$: by the homogeneity of $(\mathbb{Q}; <)$ and $\mathfrak{B}$, these are $\leq, <, \geq, >, =$, the full relation and the empty relation (the disequality relation $\neq$ is not preserved as we have seen in Example 8.5.7). All of these relations are primitively positively definable in $(\mathbb{Q}; <, \leq)$ which concludes the proof.

Example 8.5.16. the structure $(\mathbb{Q}; \leq, \neq)$ has a quasi near-unanimity polymorphism of arity five, but not of arity four [51].

8.5.5. The strict width hierarchy. For every $\ell \geq 1$ and every set B of cardinality at least two, there exists a structure with domain B which has strict width ℓ, but not strict width $\ell - 1$. Define the relation

$$D_\ell := \{(x_1, y_1, \dots, x_\ell, y_\ell) \in B^{2\ell} \mid x_1 \neq y_1 \vee \dots \vee x_\ell \neq y_\ell\}.$$

Then the structure $(B; D_\ell)$ does not have a quasi near-unanimity polymorphism: if $a, b \in B$ are distinct, then

$$\begin{aligned} t^1 &:= (a, b, a, a, \dots, a, a) \in D_\ell \\ t^2 &:= (a, a, a, b, \dots, a, a) \in D_\ell \\ &\vdots \\ t^\ell &:= (a, a, a, a, \dots, a, b) \in D_\ell \end{aligned}$$

but $f(t^1, \dots, t^\ell) = (\hat{f}(a), \dots, \hat{f}(a)) \notin D_\ell$. However, the structure $(B; D_\ell)$ does have a quasi near-unanimity operation of arity $\ell + 1$. Pick any function from $B^{\ell+1}$ to B whose kernel contains precisely the pairs of $\ell + 1$-tuples that must be mapped to the same value under any quasi near-unanimity operation. We claim that f preserves D_ℓ: suppose that $t^1, \dots, t^{\ell+1}$ are such that $(f(t_1^1, \dots, t_1^{\ell+1}), \dots, f(t_{2\ell}^1, \dots, t_{2\ell}^{\ell+1})) \notin D_\ell$, i.e.,

$$f(t_1^1, \dots, t_1^{\ell+1}) = f(t_2^1, \dots, t_2^{\ell}), \dots, f(t_{2\ell-1}^1, \dots, t_{2\ell-1}^{\ell+1}) = f(t_{2\ell}^1, \dots, t_{2\ell}^{\ell+1}).$$

By the pigeon-hole principle there must exist an $i \in \{1, \dots, \ell+1\}$ such that $t_1^i = t_2^i, \dots, t_{2\ell-1}^i = t_{2\ell}^i$, and hence $t^i \notin D_\ell$, proving the claim.

COROLLARY 8.5.17. *For any set B of cardinality at least two, the structure $\mathfrak{B} := (B; D_1, D_2, \dots)$ has no quasi near-unanimity polymorphism and an essentially infinite signature (Definition 3.3.5).*

PROOF. Suppose that $\mathfrak{B}'$ is a structure with domain B and finite signature such that all relations of $\mathfrak{B}'$ are primitively positively definable in $\mathfrak{B}$. Then there is a maximal ℓ such that D_ℓ is used in the definitions of the relations of $\mathfrak{B}'$ in $\mathfrak{B}$. Let f be the function defined above. It preserves all relations of $\mathfrak{B}'$, but it does not preserve D_ℓ, as we have seen above. This shows that $\mathfrak{B}$ is not primitively interdefinable with $\mathfrak{B}'$ and proves the statement. ⊣

Since the structure $\mathfrak{B} := (B; D_1, D_2, \dots)$ does not have quasi near-unanimity polymorphisms, one might wonder whether $\mathrm{Pol}(\mathfrak{B})$ satisfies some other nontrivial minor conditions. The following example shows that $\mathrm{Pol}(\mathfrak{B})$ contains a chain of quasi Pixley operations (and hence also a chain of quasi Jónsson operations; see Section 6.9.3) and is taken from the master thesis of Sergej Scheck [320].

PROPOSITION 8.5.18. *The clone $\mathrm{Pol}(\mathbb{N}; D_1, D_2, \dots)$ contains a chain of quasi Pixley operations of length 4.*

PROOF. First note that $\mathrm{Pol}(B; D_1, D_2, \dots) = \mathcal{R}$ (cf. Section 7.6): for every $\ell \in \mathbb{N}$, the relation D_ℓ has a negative definition (Definition 7.6.2), and conversely, every relation with a negative definition has a (quantifier-free) primitive positive definition in $(B; D_1, D_2, \dots)$. Let $f : \mathbb{N}^2 \to \mathbb{N}$ be an injection. Define

$$p_1(x, y, z) = f(x, x)$$

$$p_2(x, y, z) = \begin{cases} f(x, z) & \text{if } x = y, \\ f(x, x) & \text{otherwise;} \end{cases}$$

$$p_3(x, y, z) = \begin{cases} f(x, z) & \text{if } y = z, \\ f(z, z) & \text{otherwise;} \end{cases}$$

$$p_4(x, y, z) = f(z, z).$$

Clearly, p_1 and p_2 are injective in the first and p_3 and p_4 are injective in the third direction, so all of these operations are contained in $\mathcal{R}$. Moreover, the operations satisfy the quasi Pixley identities:

$$p_1(x, y, z) = f(x, x) = p_1(x, x, x)$$
$$p_1(x, x, z) = f(x, x) = p_2(x, z, z)$$
$$p_2(x, y, x) = f(x, x) = p_2(x, x, x)$$

$$p_2(x,x,z) = f(x,z) = p_3(x,z,z)$$
$$p_3(x,y,x) = f(x,x) = p_3(x,x,x)$$
$$p_3(x,x,z) = f(z,z) = p_4(z,z,z)$$
$$p_4(x,y,z) = f(z,z) = p_4(z,z,z). \qquad \dashv$$

Chains of quasi Pixley operations are of particular interest to the theory of infinite-domain constraint satisfaction, because the existence of polymorphisms of $\mathfrak{B}$ that form such a chain is among the strongest universal-algebraic conditions which is not yet known to imply polynomial-time tractability of CSP($\mathfrak{B}$), even if $\mathfrak{B}$ is a reduct of a finitely bounded homogeneous structure (Question 35).

8.6. Datalog inexpressibility results

In this section we show that certain CSPs of ω-categorical structures *cannot* be solved by Datalog programs. We first prove that Datalog becomes strictly more powerful by increasing the width. This holds even for CSPs of ω-categorical digraphs (Section 8.6.1), but not for finite-domain CSPs, where the hierarchy stops at width two (Theorem 8.8.2). We then present applications of the existential pebble game and Theorem 8.2.8 to show that the and/or scheduling problem, the rooted triple satisfiability problem, and solving linear equations over a non-trivial abelian group are not in Datalog (Section 8.6.2).

8.6.1. The width hierarchy. The following has been shown by Grohe [187] (Corollary 6.7) using finite graphs with the so-called *Hrushovski property* [201]. Two Datalog programs with a distinguished predicate *false* are called *equivalent* if they derive *false* for the same class of finite structures.

THEOREM 8.6.1 (Grohe [187]). *The Datalog width hierarchy is strict over graphs, that is, for every $l \geq 2$ there exists a Datalog program of width l which is not equivalent to a Datalog program of width $l-1$.*

We will give an alternative simple proof of the analogous result for *directed* graphs, Theorem 8.6.3. Let $n \geq 1$ be a fixed integer. We write E for the binary arc relation. Consider the following $(2n, 3n)$-Datalog program; we denote it by Π_n.

$$\begin{aligned}
R(x_1,\dots,x_n,y_1,\dots,y_n) &:- \{E(y_i,y_j), E(x_i,x_j) \mid i,j \in \{1,\dots,n\}, i \neq j\},\\
&\qquad \{E(x_i,y_j) \mid i,j \in \{1,\dots,n\}\}\\
R(x_1,\dots,x_n,y_1,\dots,y_n) &:- R(x_1,\dots,x_n,z_1,\dots,z_n),\\
&\qquad \{E(z_i,y_j) \mid i,j \in \{1,\dots,n\}\},\\
&\qquad \{E(y_i,y_j) \mid i,j \in \{1,\dots,n\}, i \neq j\}\\
\mathit{false} &:- R(x_1,\dots,x_n,x_1,\dots,x_n)
\end{aligned}$$

We will show that there is no Datalog program of width $2n-1$ that computes the same query as Π_n. Define a *bad sequence* in an $\{E\}$-structure $\mathfrak{C}$ to be a sequence $x^0, \ldots, x^m$ of elements $x^i = (x_1^i, \ldots, x_n^i)$ of C^n such that

- $E(x_p^i, x_q^i)$ for all $p, q \in \{1, \ldots, n\}$, $p \neq q$, and $i \in \{0, \ldots, m\}$;
- $E(x_p^i, x_q^{i+1})$ for all $p, q \in \{1, \ldots, n\}$ and $i \in \{0, \ldots, m-1\}$.

A *bad cycle* is a bad sequence $x^0, x^1, \ldots, x^m$ where $x^m = x^0$. Observe that if Π_n derives $R(x_1, \ldots, x_n, y_1, \ldots, y_n)$ on $\mathfrak{C}$ for some $x_1, \ldots, x_n, y_1, \ldots, y_n \in C$, then this means that $\mathfrak{C}$ contains a bad sequence $x^0, \ldots, x^m$ with $x^0 = (x_1, \ldots, x_n)$ and $x^m = (y_1, \ldots, y_n)$. Moreover, if Π_n derives *false* then $\mathfrak{C}$ must contain a bad cycle.

Let $\mathcal{K}_n'$ be the class of all structures with signature $\{E, R, \mathit{false}\}$ that are computed by runs of Π_n on $\{E\}$-structures that do not derive *false*.

LEMMA 8.6.2. *The class $\mathcal{K}_n$ of all substructures of the class $\mathcal{K}_n'$ defined above is an amalgamation class.*

PROOF. The verification of the amalgamation property is the only interesting part of the proof. Let $\mathfrak{B}_1$ and $\mathfrak{B}_2$ be structures in $\mathcal{K}_n$, and suppose that $\mathfrak{A}$ is a common substructure of $\mathfrak{B}_1$ and $\mathfrak{B}_2$. Let $\mathfrak{B}_i$ be a substructure of $\mathfrak{B}_i' \in \mathcal{K}_n'$, for $i \in \{1, 2\}$. Suppose without loss of generality that $B_1' \cap B_2' = A$. Let $\mathfrak{C}'$ be the free amalgam of the $\{E\}$-reduct of $\mathfrak{B}_1'$ and $\mathfrak{B}_2'$ over the $\{E\}$-reduct of $\mathfrak{A}$. Execute Π_n on $\mathfrak{C}'$. It suffices to show that Π_n does not derive new atomic formulas on variables from B_1', nor on variables from B_2' (and in particular does not derive *false*) because then the desired amalgam $\mathfrak{C} \in \mathcal{K}_n$ of $\mathfrak{B}_1$ and $\mathfrak{B}_2$ over $\mathfrak{A}$ can be obtained as a substructure of the $\{E, R, \mathit{false}\}$-expansion computed by Π_n on $\mathfrak{C}'$.

Suppose for contradiction that $\mathfrak{C}'$ contains a bad sequence $x^0, x^1, \ldots, x^m$ such that $x^0, x^m \in (B_1')^n$ and $\mathfrak{B}_1'$ does not satisfy $R(x^0, x^m)$. Choose this sequence such that the number of elements $x_j^i \in B_2'$, for $i \in \{0, \ldots, m\}$ and $j \in \{1, \ldots, n\}$, is minimal. This number must be at least one, because otherwise Π_n would have derived $R(x^0, x^m)$ on $\mathfrak{B}_1'$ alone. Choose $u \leq m$ minimal such that $\{x_1^u, \ldots, x_n^u\} \subsetneq B_1'$. Then $x_1^{u-1}, \ldots, x_n^{u-1}$ must all lie in A since every vertex of $x_1^{u-1}, \ldots, x_n^{u-1}$ is connected with every vertex of $x_1^u, \ldots, x_n^u$, but all edges between B_1' and B_2' must involve vertices from A. Choose $v \in \{u+1, \ldots, m\}$ minimal with the property that $\{x_1^v, \ldots, x_n^v\} \subseteq B_1'$; such a v must exist because $x_1^m, \ldots, x_n^m \subseteq B_1'$. Again, note that $\{x_1^v, \ldots, x_n^v\} \subseteq A$. But then $R(x^{u-1}, x^v)$ holds in $\mathfrak{B}_2'$, and hence also in $\mathfrak{A}$, and hence also in $\mathfrak{B}_1'$, which means that in $\mathfrak{B}_1'$ there exists a bad sequence from x^{u-1} to x^v. So we may replace the subsequence $x^{u-1}, x^u, \ldots, x^v$ by this sequence in $\mathfrak{B}_1'$, obtaining a new bad sequence with fewer vertices in B_2', a contradiction. ⊣

Write $\mathfrak{C}_n$ for the Fraïssé-limit of $\mathcal{K}_n$, expanded by the binary relation N with the quantifier-free definition $\neg E(x, y) \wedge x \neq y$, which is ω-categorical since it is homogeneous and has a finite relational signature. We claim that $\mathrm{CSP}(\mathfrak{C}_n)$

has width $(2n, 3n)$. To see this, it suffices to add to Π_n the rules

$$false :- N(x,x)$$
$$false :- E(x,y), N(x,y).$$

THEOREM 8.6.3. *The Datalog width hierarchy is strict over directed graphs.*

PROOF. By Theorem 8.2.10, it suffices to show that for every $k \in \mathbb{N}$ there exists an unsatisfiable instance $\mathfrak{A}$ of CSP($\mathfrak{C}_n$) such that the canonical $(2n-1,k)$-Datalog program Π for $\mathfrak{C}_n$ does not derive *false* on $\mathfrak{A}$. Let $k \in \mathbb{N}$ be given. Choose $\mathfrak{A}$ to be an $\{E, N, false\}$-structure; the relation $E^{\mathfrak{A}}$ consists of the edges of a bad cycle $x^0, x^1, \dots, x^{k+1}$ with disjoint vertices, and $N^{\mathfrak{A}}$ is empty. We claim that Π does not derive false on $\mathfrak{A}$. Let P be a relation of arity $2n-1$ with a primitive positive definition in $\mathfrak{B}_n$. By the homogeneity of $\mathfrak{C}_n$, the relation P is definable by a quantifier-free formula ϕ; Since the relation R only contains tuples of pairwise distinct elements of $\mathfrak{B}$, and is of arity $2n$, we can even assume that ϕ is a Boolean combination of atomic formulas over the signature $\{E, N, false\}$. Moreover, we may assume that ϕ is written in disjunctive normal form and does not contain occurrences of *false*. We may then eliminate formulas of the form $\neg E(x,y)$ by $N(x,y) \vee x = y$, formulas of the form $\neg N(x,y)$ by $E(x,y) \vee x = y$, and formulas of the form $x \neq y$ by $E(x,y) \vee N(x,y)$. Let $\theta_1, \dots, \theta_q$ be the disjuncts of the resulting positive formula. Now, every rule of Π of the form

$$S(\bar{x}) :- \beta(\bar{y}), P(\bar{z})$$

can be replaced by the rules

$$S(\bar{x}) :- \beta(\bar{y}), \theta_1(\bar{z}),$$
$$\vdots$$
$$S(\bar{x}) :- \beta(\bar{z}), \theta_q(\bar{z}).$$

In this way we can eliminate all IDBs from the bodies of the rules of Π and obtain an equivalent Datalog program. Since $N^{\mathfrak{A}}$ is empty and every proper subgraph (not necessarily induced) of the graph $(A; E^{\mathfrak{A}})$ does not contain a bad cycle, Π does not derive *false* on $\mathfrak{A}$. ⊣

8.6.2. High girth. In this section we present a general lemma that allows us to prove that various CSPs cannot be solved by Datalog. The proof of the following lemma is taken from [83][22]. The *girth* of an undirected graph G is the length of a shortest cycle in G. A graph is called k-*regular* if every vertex has precisely k neighbours. A 3-regular graph is also called *cubic*. A great deal is known about the existence of finite graphs of high girth.

FACT 8.6.4. *For every $k \in \mathbb{N}$ there exists*

[22]The statement of the corresponding lemma in [83] is false; here we present a corrected version.

- *a finite four-regular graph of girth at least k (it is even known that there are such graphs of size exponential in k* [214]);
- *a cubic graph of girth at least k with a Hamiltonian cycle (see the comments after the proof of Theorem* 3.2 *in* [41]).

DEFINITION 8.6.5. Let τ be a relational structure. Then the *incidence graph* $G(\mathfrak{A})$ of $\mathfrak{A}$ is the graph whose set of vertices is the disjoint union of A and the set of tuples in relations from $\mathfrak{A}$, and where an element from A is connected to a tuple if and only if the element appears in that tuple.

Note that $G(\mathfrak{A})$ is bipartite. We say that $\mathfrak{A}$ has *girth k* if all tuples in relations from $\mathfrak{A}$ have pairwise distinct entries and the shortest cycle of $G(\mathfrak{A})$ has $2k$ edges.

LEMMA 8.6.6. *Let $\mathfrak{B}$ be a countable ω-categorical structure whose signature τ consists of finitely many relation symbols of arity at least* 3 *such that for every relation R of $\mathfrak{B}$ of arity r, every $(r-1)$-ary projection of R equals the full relation B^{r-1}. Suppose that for every k there exists an unsatisfiable instance $\mathfrak{A}_k$ of girth at least* $2k+1$. *Then* CSP($\mathfrak{B}$) *cannot be solved by Datalog.*

Note that the assumption about minimum arity three in Lemma 8.6.6 is necessary, because otherwise the structure $(\mathbb{Q}; <)$ would satisfy the assumptions, but we have already seen a Datalog program that solves CSP$(\mathbb{Q}; <)$. To prove the lemma, we use a notion of *controlled sets* to specify winning strategies for Duplicator for the existential k-pebble game.

DEFINITION 8.6.7. Let $k \geq 2$ and $\mathfrak{A}$ an instance of CSP($\mathfrak{B}$) of girth at least $2k+1$ where k elements are pebbled. A non-empty set $S \subseteq A$ is called *controlled* if it satisfies the following conditions.

1. The incidence graph $G[S] := G(\mathfrak{A}[S])$ is a tree.
2. All but at most one of the elements that are leaves of $G[S]$ are pebbled.

An example of a controlled set S can be found in Figure 1.

PROOF OF LEMMA 8.6.6. Due to Theorem 8.2.8 it suffices to prove that Duplicator wins for some $k \geq 1$ the existential $(k-1, k)$-pebble game on $\mathfrak{A}_k$ and $\mathfrak{B}$. Suppose that in the course of the game, $a_{\ell+1}$ is an unpebbled leaf of a controlled set S with pebbled leaves $a_1, \dots, a_\ell$, for $\ell \leq k-1$, and let $b_1, \dots, b_\ell$ be the corresponding responses of Duplicator. Duplicator always maintains the following condition:

(∗) whenever Spoiler places a pebble on $a_{\ell+1}$, Duplicator can play a value $b_{\ell+1}$ from B such that the mapping h that assigns a_i to b_i for $1 \leq i \leq \ell+1$ has an extension to a homomorphism from $\mathfrak{A}_k[S]$ to $\mathfrak{B}$.

Condition (∗) clearly holds at the beginning of the game. Suppose that at move i Spoiler pebbles a variable a. Let $S_1, \dots, S_m$ be the controlled sets S at stage $i-1$ that contain a. Note that $m \geq 1$ since the set $\{a\}$ is a controlled set. Let $T_1, \dots, T_q$ be the controlled sets at stage i which contain a such that each T_j

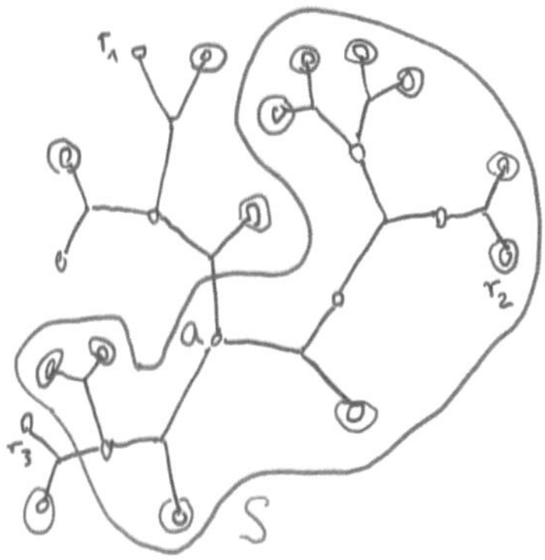

FIGURE 1. A situation in the proof of Lemma 8.6.6. The encircled vertices are already pebbled, Spoiler is about to pebble vertex a. The encircled sets of vertices are controlled sets before the vertex a has been pebbled.

has some unpebbled leaf r_j. We have to show that under the assumption that Duplicator in her previous move has maintained condition $(*)$, she will be able to make a move that again satisfies condition $(*)$.

Note that $S := S_1 \cup \cdots \cup S_m$ is a controlled set, too: indeed, $G[S]$ is clearly connected since all trees share the vertex a. To prove that $G[S]$ has no cycles, let $j \in \{1, \ldots, m\}$ and let p be the number of pebbled vertices in S_j. If $p = 0$ then $|S| = 1$ and the statement is clear. Otherwise, $|S_j| \leq 2p$ because of the assumption that $G[S_j]$ is a tree, that all but one of the leaves of $G[S]$ are pebbled, and that the minimum arity of the relation symbols in τ is three. Therefore, $|S| \leq 2k$ and S cannot contain a cycle because $\mathfrak{A}_k$ has girth at least $2k + 1$. By assumption, the map that sends the pebbled vertices in S to the responses of Duplicator can be extended to a homomorphism h from $\mathfrak{A}_k[S]$ to $\mathfrak{B}$. Duplicator plays $b := h(a)$. For an illustration, see Figure 1.

For each $j \in \{1, \ldots, q\}$, we have to prove that h can be extended to T_j. We prove this by induction on the number of elements where h is already defined. Here we will make use of the assumption on the relations of $\mathfrak{B}$ in the statement of the lemma. If h is defined on all of T_j, then we are done, so let us suppose that this is not the case. If $a_1, \ldots, a_r \in T_j$ are such that $(a_1, a_2, \ldots, a_r) \in R^{\mathfrak{A}_k[T_j]}$, let $P_1, P_2, \ldots, P_r$ be the connected components of the structure obtained from $\mathfrak{A}_k[T_j]$ by removing this tuple from $R^{\mathfrak{A}_k[T_j]}$. At most one of these components, say P_ℓ, can contain r_j. Also note that all the elements where h is not yet defined lie on the path from a to r_j in $\mathfrak{A}_k[T_j]$. Pick $(a_1, a_2, \ldots, a_r) \in R^{\mathfrak{A}_k[T_j]}$ such that h is undefined on a_ℓ and h has already been defined for all other elements; such a choice must exist because of our assumption that h is not yet defined on all of T_j. By the assumption on the relations of $\mathfrak{B}$, the map h can be extended to a homomorphism that is also defined on a_ℓ, concluding the inductive step. ⊣

8.6.3. Applications. For our first application of Lemma 8.6.6, let $R^{\min}$ be the ternary relation $\{(x, y, z) \in \mathbb{Q}^3 \mid x > y \vee x > z\}$ from Section 1.6.8. The CSP for $(\mathbb{Q}; R^{\min})$ is one of the simplest computational problems that cannot be solved by Datalog (Theorem 8.6.8 below). We will later see (in Section 8.7) that this problem can be solved in polynomial (even linear) time. Another relatively easy computational problem that can be solved in polynomial time but does not lie in Datalog is the rooted triple satisfiability problem from Section 1.6.2 (Theorem 8.6.10). As a third application of Lemma 8.6.6, we prove that Datalog programs cannot be used to solve satisfiability of systems of linear equations over non-trivial abelian groups (Theorem 8.6.11).

THEOREM 8.6.8 (Theorem 5.2 in [70]). *There is no Datalog program that solves* CSP$(\mathbb{Q}; R^{\min})$.

PROOF. Note that the relation $R^{\min}$ satisfies the requirement from the statement of Lemma 8.6.6. To apply Lemma 8.6.6 we only have to construct an unsatisfiable instance $\mathfrak{A}$ of CSP$(\mathbb{Q}; R^{\min})$ of girth k for every $k \in \mathbb{N}$. For this, let $\mathfrak{G}$ be a 4-regular graph of girth at least k (Fact 8.6.4). Since $\mathfrak{G}$ is 4-regular, there exists an Euler tour for $\mathfrak{G}$ (see e.g. [150]). Orient the edges in $\mathfrak{G}$ along this Euler tour such that there are exactly two outgoing and two incoming edges for each vertex in $\mathfrak{G}$. Now we can define our instance $\mathfrak{A}$ of CSP$(\mathbb{Q}; R^{\min})$ as follows. The domain of $\mathfrak{A}$ is the vertex set of $\mathfrak{G}$, and $R^{\min}(w, u, v)$ holds in $\mathfrak{A}$ iff uw and vw are the two incoming edges at vertex w. If $G(\mathfrak{A})$ contains a cycle $a_1, t_1, a_2, t_2, \ldots, a_n, t_n, a_1$ of length $2n$ then $\mathfrak{G}$ has the cycle $a_1, a_2, \ldots, a_n, a_1$ of length n. Since $\mathfrak{G}$ has girth at least k, we conclude that $\mathfrak{A}$ has girth at least k.

We claim that $\mathfrak{A}$ does not have a solution: if there is a homomorphism s from $\mathfrak{A}$ to $(\mathbb{Q}; R^{\min})$ then for some $w \in A$ the value $s(w)$ is minimal. But for every $w \in A$ there is a constraint $R^{\min}(w, u, v)$ in $\mathfrak{A}$, and this constraint is not satisfied by s since either $s(u)$ or $s(v)$ must be strictly smaller than $s(w)$. ⊣

We now present a second application of Lemma 8.6.6 to prove the following result from [83]. We need the following sufficient condition that a rooted triple satisfiability problem has no solution, which goes back to Aho, Sagiv, Szymanski, and Ullman [9].

LEMMA 8.6.9. *Let $\mathfrak{A}$ be an instance of the rooted triple satisfiability problem. Let $\mathfrak{G}$ be the graph whose vertices are the elements of $\mathfrak{A}$ and which has an edge $\{a, b\}$ if $\mathfrak{A} \models ab|c$ for some $c \in A$. Suppose that $\mathfrak{G}$ is connected. Then $\mathfrak{A}$ is unsatisfiable.*

PROOF. Suppose for contradiction that there exists a rooted tree $\mathfrak{T}$ with leaves L and a solution $s \colon A \to L$. Let r be the $yca(A)$ (Definition 5.1.1). It cannot be that all the vertices of $s(A)$ lie below the same child of r in $\mathfrak{T}$, by the definition of $yca(A)$. Since $\mathfrak{G}$ is connected, there is an edge $\{a, b\}$ in $\mathfrak{G}$ such that $s(a)$ and $s(b)$ lie below different children of r in T. Hence, there is an element $c \in A$ such that $ab|c$ holds in $\mathfrak{A}$. By assumption, $yca(s(a), s(b)) = r$ lies strictly below $yca(s(a), s(c))$, a contradiction to the choice of r. ⊣

THEOREM 8.6.10. *There is no Datalog program that solves the rooted triple satisfiability problem* (*Section* 1.6.2).

PROOF. Let $\mathfrak{B}$ be the template constructed for the rooted triple satisfiability problem constructed in Section 5.1. It is clear that the rooted triple relation satisfies the assumptions on R from Lemma 8.6.6.

To construct an unsatisfiable instance $\mathfrak{A}$ for CSP($\mathfrak{B}$) of girth at least k, let $\mathfrak{G}$ be a cubic graph of girth at least k that has a Hamiltonian cycle (Fact 8.6.4). Note that $\mathfrak{G}$ must have an even number of vertices. Let $H = (v_1, v_2, \dots, v_n)$ be the Hamiltonian cycle of $\mathfrak{G}$. For any vertex a of $\mathfrak{G}$, let $r(a)$ be the vertex that precedes a on H, let $s(a)$ the vertex that follows a on H, and let $t(a)$ the third remaining neighbour of a in $\mathfrak{G}$.

We now define $\mathfrak{A}$. The domain A of $\mathfrak{A}$ are the vertices of $\mathfrak{G}$, and

$$|^{\mathfrak{A}} := \{(r(a), s(a), t(a)) \mid a \in A\}.$$

Note that the graph from the statement of Lemma 8.6.9 equals the (connected) graph $\mathfrak{G}$. Therefore, Lemma 8.6.9 implies that $\mathfrak{A}$ is an unsatisfiable instance of CSP($\mathfrak{B}$).

The only remaining point for the application of Lemma 8.6.6 is the verification that $G(\mathfrak{A})$ has girth at least k. But this is obvious since any cycle of length $2l < 2k$ in the incidence graph $G(\mathfrak{A})$ would give rise to a cycle of length $l < k$ in $\mathfrak{G}$, in contradiction to $\mathfrak{G}$ having girth at least k. ⊣

Our third application of Lemma 8.6.6 is a simple proof that solving linear equations over non-trivial abelian groups is not in Datalog. Let $\mathfrak{G}$ be an abelian group and $c \in G$. We define

$$R_c^k := \{(x_1, \dots, x_k) \in G^k \mid x_1 + \cdots + x_k = c\}.$$

THEOREM 8.6.11 (Feder and Vardi [170]). *Let $\mathfrak{G}$ be an abelian group and $a \in G \setminus \{0\}$. Then the problem* CSP($G; R_0^3, R_a^3$) *is not in Datalog.*

PROOF. Let S_a^i be the relation of arity $3 + 2i$ defined by

$$x_1 + x_2 + x_3 = a \wedge \bigwedge_{s \in \{1,\dots,i\}} x_{2s+2} + x_{2s+3} = 0.$$

Note that for every $i \in \mathbb{N}$, the relation S_a^i has a primitive positive definition in CSP($G; R_0^3, R_a^3$). This is easy to see from the observations that

- $x = 0$ if and only if $R_0^3(x, x, x) \wedge \exists y \big(R_0^3(x, x, y) \wedge R_0^3(x, y, y) \wedge R^3(y, y, y)\big)$;
- $x + y = 0$ if and only if $R_0^3(x, y, 0)$.

Let $\mathfrak{B}$ be the structure $(G; R_0^0, S_0^1, S_0^2, S_0^3, R_a^0, S_a^1, S_a^2, S_a^3)$. By Lemma 8.3.2 it suffices to show that CSP($\mathfrak{B}$) is not in Datalog. Note that all the relations of $\mathfrak{B}$ satisfy the requirements from Lemma 8.6.6. To construct an unsatisfiable instance $\mathfrak{A}$ of CSP($\mathfrak{B}$) of girth at least k, let $(V; E)$ be a cubic graph of girth at least k (Fact 8.6.4). Orient the edges E arbitrarily. The domain of $\mathfrak{A}$ is $V \times E$. For each $v \in V$ with three incoming edges e_1, e_2, e_3 we add the constraint $R_0^3((v, e_1), (v, e_2), (v, e_3))$ to $\mathfrak{A}$. For each $v \in V$ with two incoming edges e_2, e_3 and one outgoing edge $e_1 = (v, w)$ we add the constraint $S_0^1((v, e_1), (v, e_2), (v, e_3), (v, e_1), (w, e_1))$. Likewise, for each $v \in V$ with one or no incoming edges and two or three outgoing edges we add a constraint involving the relation S_0^2 or S_0^3. Finally, we move exactly one of the tuples from S_0^i in $\mathfrak{A}$ to the relation S_a^i. Suppose for contradiction that $s: A \to G$ is a solution for $\mathfrak{A}$. Sum over all constraints. Since each element of A appears once in a two-variable constraint and once in a three-variable constraint, we obtain $2\sum_{e \in E, u \in e} s(u, e)$ on the left-hand side. Since s satisfies $s(u, e) + s(v, e) = 0$ for every edge $e = \{u, v\} \in E$, the left-hand side can be rewritten as $2\sum_{\{u,v\} \in E} \sum_{u \in e} s(u, e) = \sum_{\{u,v\} \in E}(s(u, e) + s(v, e)) = 0$. On the right-hand side we obtain a since we have precisely one tuple in S_a^i in $\mathfrak{A}$. Hence, s cannot be a homomorphism. ⊣

8.7. Fixed-point logic

Fixed-point logics are powerful logics for expressing computational problems that are in P; they properly extend Datalog. There are various fixed-point logics, for instance *least fixed-point logics* (*LFP*), *inflationary fixed-point logic* (*IFP*), and extensions by *counting quantifiers*; they are treated in finite model theory textbooks such as [263, 213, 162]. It is known that least fixed-point logic and inflationary fixed-point logic have the same expressive power (even over infinite structures [247]); since inflationary fixed-point logic has a simpler definition and is all that will be needed in our examples, we only introduce IFP. In this section we show that satisfiability of and/or precedence constraints and the rooted triple satisfiability problem can be expressed in IFP (but not in Datalog, as we have seen in the previous section). On the other hand, IFP has the same expressive power as Datalog for finite-domain CSPs (Theorem 8.8.2 below).

8.7.1. Inflationary fixed points. Let A be a set. We write $\mathcal{P}(A)$ for the set of all subsets of A. An operator $F: \mathcal{P}(A) \to \mathcal{P}(A)$ is called *inflationary* if $X \subseteq F(X)$ for every $X \in \mathcal{P}(A)$. A *fixed point* of F is an element $X \in \mathcal{P}(A)$ such that $X = F(X)$. Clearly, if A if finite and F is inflationary then the sequence $(X^i)_{i \in \mathbb{N}}$ given by

$$X^0 := \emptyset, \quad X^{i+1} := F(X^i)$$

will eventually be constant, and

$$X^\infty := \bigcup_{i=1}^{\infty} X^i$$

is a fixed point. If $G: \mathcal{P}(A) \to \mathcal{P}(A)$ is an arbitrary operator, we associate to G the inflationary operator $F(X) := X \cup G(X)$, and the least fixed point X^∞ of F defined above will be called the *inflationary fixed point of* G and denoted by $\text{ifp}(G)$.

8.7.2. Inflationary fixed-point logic. Let τ be a relational signature and let ϕ be a $(\tau \cup \{R\})$-formula for some k-ary relation symbol $R \notin \tau$. If $\mathfrak{A}$ is a τ-structure and $X \subseteq A^k$, then we write $(\mathfrak{A}, X)$ for the expansion of $\mathfrak{A}$ with the signature $\tau \cup \{R\}$. We define the operator $[\![\mathfrak{A}, \phi]\!]: \mathcal{P}(A^k) \to \mathcal{P}(A^k)$ by

$$X \mapsto \{a \in A^k \mid (\mathfrak{A}, X) \models \phi(a)\}.$$

DEFINITION 8.7.1 (IFP). *Inflationary fixed-point formulas* over a relational signature τ (short, *IFP τ-formulas*) are defined inductively as follows. Every atomic τ-formula is an IFP τ-formula, and formulas built from IFP τ-formulas by the usual first-order constructors are again IFP τ-formulas. Finally, if $\phi(x_1, \dots, x_k, \bar{y})$ is an IFP $(\tau \cup \{R\})$-formula where R has arity k then $\text{ifp}_R\, \phi$ is an IFP τ-formula with the free variables $x_1, \dots, x_k, \bar{y}$.

The semantics of inflationary fixed-point logic is defined similarly as for first-order logic; we just discuss how to interpret the inflationary fixed-point constructor. Let $\mathfrak{A}$ be a τ-structure and let R be a k-ary relation symbol not from τ. Let $a_1, \dots, a_k, b_1, \dots, b_l \in A$ and let $\phi(x_1, \dots, x_k, y_1, \dots, y_l)$ be an IFP $(\tau \cup \{R\})$-formula. Then

$$\mathfrak{A} \models (\text{ifp}_R\, \phi)(a_1, \dots, a_k, b_1, \dots, b_l) :\Leftrightarrow \\ (a_1, \dots, a_k) \in \text{ifp}([\![\mathfrak{A}, \phi(x_1, \dots, x_k, b_1, \dots, b_l)]\!]).$$

As we have mentioned above, it has been shown that least fixed-point logic and inflationary fixed-point logic have the same expressive power, so following an established convention in the literature we refer from now on to both logics as *fixed-point logic* (*FP*). We present a series of examples of problems expressible in FP of increasing difficulty.

Example 8.7.2 (Connectivity). Let $(V; E)$ be a finite graph and $a, b \in V$. The property that a and b are connected in the graph $(V; E)$ can be expressed by a fixed-point formula as follows. Let C be a new unary relation symbol, and let $\phi(x, y)$ be the formula $x = y \vee \exists z(E(x, z) \wedge C(z))$. Then $\text{ifp}_C\, \phi(x, a)(b)$ holds if and only if a and b are connected in the graph $(V; E)$. This fixed-point formula will be useful later, in Example 8.7.5. The following notation is convenient for a flexible usage: for a formula $\psi(x, z)$ (which we imagine as a formula defining the edge relation E) we write $\text{conn}(a, b, \psi)$ for the fixed-point formula above where we replace the occurrence of $E(x, z)$ by $\psi(x, z)$.

Generalising Example 8.7.2, it is easy to see that for every Datalog program Π there exists an FP-formula ϕ such that for all structures $\mathfrak{A}$, the program Π derives *false* on $\mathfrak{A}$ if and only if $\mathfrak{A} \models \phi$ (cf. [213, 263, 162]). We now present examples of properties that can be expressed in fixed-point logic, but not in Datalog.

Example 8.7.3 (Precedence Constraints). Let $R^{\min}$ be the relation

$$\{(a, b, c) \in \mathbb{Q}^3 \mid a > b \vee a > c\}$$

from Section 1.6.8 on and/or precedence constraints. We will show that $\text{CSP}(\mathbb{Q}; R^{\min})$ can be expressed in FP (recall that it is not in Datalog; Theorem 8.6.8). We first describe the algorithmic idea. Let $\mathfrak{A}$ be an instance of this problem, and suppose that for every $a \in A$ there exist b_1, b_2 such that $\mathfrak{A} \models R^{\min}(a, b_1, b_2)$. Then $\mathfrak{A}$ has no homomorphism to $(\mathbb{Q}; R^{\min})$ because no element from A can take a minimal value in a solution. On the other hand, let F be the set of all $a \in A$ that satisfy $\forall b_1, b_2 : \neg R^{\min}(a, b_1, b_2)$. If there is a homomorphism h from $\mathfrak{A}[A \setminus F]$ to $(\mathbb{Q}; R^{\min})$ then it can be extended to a homomorphism from $\mathfrak{A}$ to $(\mathbb{Q}; R^{\min})$ by setting $h(a)$, for all $a \in F$, to some value from $\mathbb{Q}$ that is smaller than all other values in the image of h. This gives rise to a recursive polynomial-time algorithm for $\text{CSP}(\mathbb{Q}; R^{\min})$. To formulate the problem in FP, let P be a new unary relation symbol and let $\phi(x)$ be the formula

$$\forall y, z \left(R^{\min}(x, y, z) \Rightarrow (P(y) \vee P(z))\right).$$

We claim that $\mathfrak{A}$ maps homomorphically to $(\mathbb{Q}; R^{\min})$ if and only if

$$\mathfrak{A} \models \forall x (\text{ifp}_P \, \phi)(x).$$

To prove the claim, define $P^0 := \emptyset$, and for $i \geq 1$ define $P^i := P^{i-1} \cup [\![\mathfrak{A}, \phi]\!](P^{i-1})$. Then $P^\infty = \text{ifp}([\![\mathfrak{A}, \phi]\!])$. We need to show that $P^\infty = A$ if and only if there exists a homomorphism $h \colon \mathfrak{A} \to (\mathbb{Q}; R^{\min})$.

First suppose that there does exist such a homomorphism. Let $i \in \mathbb{N}$; if $P^i = A$ then we are done, so suppose that $P^i \neq A$. Let

$$F_i := \{a \in A \setminus P^i \mid f(a) \leq f(b) \text{ for all } b \in A\}.$$

Note that $F_i \neq \emptyset$. Also note that $F_i \subseteq [\![\mathfrak{A}, \phi]\!](P^i)$: if $x \in A$ is such that $(\mathfrak{A}, P^i) \models \neg\phi(x)$, then there exist $y, z \in A \setminus P^i$ such that $R^{\min}(x, y, z)$. But this in turn means that either $h(x) > h(y)$ or $h(x) > h(z)$, and hence $x \notin F_i$. We conclude that P^{i+1} is strictly larger than P^i. Since A is finite, we conclude that $P^\infty = A$.

Conversely, suppose that $P^\infty = A$. Define $h \colon A \to \mathbb{Q}$ by mapping a to the smallest $i \in \mathbb{N}$ such that $a \in P^i$. We claim that h is a homomorphism from $\mathfrak{A}$ to

$(\mathbb{Q}; R^{\min})$. Let $a, b, c \in A$ be such that $\mathfrak{A} \models R^{\min}(a, b, c)$. Let $i \in \mathbb{N}$ be smallest so that $\{a, b, c\} \cap P^i$ is non-empty (hence, $i \geq 1$). Then $(\mathfrak{A}, P^{i-1}) \models \neg\phi(a)$, and hence $a \notin P^i$, which shows that $R^{\min}(h(a), h(b), h(c))$ is true.

Example 8.7.4. Let $R^{\min}_{\leq}$ be the relation $\{(a, b, c) \in \mathbb{Q}^3 \mid a \geq b \vee a \geq c\}$. Then $\mathrm{CSP}(\mathbb{Q}; R^{\min}_{\leq}, <)$ is a more expressive variant of the CSP in the previous example; it will play an important role in Chapter 12. It can be shown that this CSP can be formulated in IFP [87], using the algorithm presented in Section 12.8.3.

Example 8.7.5. The rooted triple satisfiability problem (Section 1.6.2) can be formulated in FP (recall that it is not in Datalog; Theorem 8.6.10); this has been observed by Stefan Mengel [283]. The fixed-point formula that we present below is inspired by a polynomial-time algorithm due to Aho, Sagiv, Szymanski, and Ullman [9], which is based on the unsatisfiability criterion from Lemma 8.6.9, and proceeds inductively if the criterion does not apply. The criterion states that a certain graph associated to the instance is connected; note that connectivity can be tested by a fixed-point computation as we have seen in Example 8.7.2.

Let D be a new binary relation symbol. Informally, we use D to store all pairs (a, b) of elements of an input instance $\mathfrak{A}$ such that (a, b) is not an edge in the graph from Lemma 8.6.9, or such that (a, b) is an edge that has been discarded in later stages of the inductive algorithm. Formally, let $\phi(u, v)$ be the formula

$$\forall w \big(uv|w \Rightarrow \neg\, \mathrm{conn}(u, w, \neg D(x, z))\big).$$

We claim that $\mathfrak{A}$ is a satisfiable instance of the rooted triple satisfiability problem if and only if

$$\mathfrak{A} \models \forall a, b (\mathrm{ifp}_D\, \phi)(a, b).$$

To prove the claim, define $D^0 := \emptyset$, and for $i \geq 1$ define $D^i := D^{i-1} \cup [\![\mathfrak{A}, \phi]\!](D^{i-1})$. Then $D^\infty = \mathrm{ifp}([\![\mathfrak{A}, \phi]\!]) = D^\ell$ for some $\ell \in \mathbb{N}$. We need to show that $D^\infty = A^2$ if and only if $\mathfrak{A}$ is satisfiable.

First suppose that $\mathfrak{A}$ is satisfiable. Let $i \in \mathbb{N}$; if $D^i = A^2$ then there is nothing to be shown, so suppose that there exists $(u, v) \in A^2 \setminus D^i$ and $i \geq 1$. By the definition of D^i there exists $w \in A$ such that $uv|w$ and u, v, w are in the same connected component C in the graph $(A; A^2 \setminus D^{i-1})$. Then there must be a bipartition of C into non-empty subsets C_1 and C_2 such that $h(C_1)|h(C_2)$ (Definition 5.1.4). Since C is a connected component in $(A; A^2 \setminus D^{i-1})$, there exists a pair $(x, y) \in A^2 \setminus D^{i-1}$ which is contained in D^i, which shows that D^i is strictly larger than D^{i-1}. Since A^2 is finite, we conclude that $D^\infty = A^2$.

Conversely, suppose that $D^\infty = A^2$. Create a rooted tree $\mathfrak{T}$ in stages $i = 0, 1, 2, \ldots, \ell$ as follows. Create a new root vertex and add one child vertex for each connected component of the graph $(A; A^2 \setminus D^1)$. At state i, create a new vertex for each connected component of the graph $(A; A^2 \setminus D^i)$. For $i = 0$, there is a single connected component which becomes the root of $\mathfrak{T}$. For $i \geq i$, each connected component C is a subset of a connected component C' of $(A; A^2 \setminus D^{i-1})$. We then add the vertex that has been created for C as a child below C'. At stage $i = \ell$, we know that the graph $(A; A^2 \setminus D^i)$ does not have any edges, so we may identify the newly created vertices with the elements of A; they become the leaves of $\mathfrak{T}$.

We verify that whenever $ab|c$ holds in $\mathfrak{A}$, then $ab|c$ holds in $\mathfrak{T}$. Note that $(a,b) \in A^2 \setminus D^1$. Let $i \geq 2$ be smallest so that $(a,b) \in D^i$; such an i exists because $D^\ell = A^2$. Then a and c lie in the same connected component in $(A^2 \setminus D^{i-2})$: otherwise $\phi(a,b)$ holds in $(\mathfrak{A}; D^{i-2})$ and hence $(a,b) \in [\![\mathfrak{A},\phi]\!](D^{i-1}) = D^{i-1}$, contrary to the assumption. On the other hand, a and c lie in different connected components in $(A^2 \setminus D^{i-1})$, because $(a,b) \in D^i = [\![\mathfrak{A},\phi]\!](D^{i-1})$ implies that $\phi(a,b)$ holds in $(\mathfrak{A}; D^{i-1})$, so in particular a and c are not connected in $(A; A^2 \setminus D^{i-1})$. This means that $ab|c$ holds in $\mathfrak{T}$, as required.

8.7.3. Fixed-point logic with counting. In this section we mention two fundamental facts about CSPs in the context of fixed-point logic and its most important extension, *fixed-point logic with counting* (*FP+C*). In the following we do not need a formal definition of this logic and instead refer to Libkin's textbook [263]. FP+C is important since it can express most problems that are known to be solvable in polynomial time (not just CSPs). A remarkable exception is satisfiability of systems of linear equations over abelian groups. Theorem 8.6.11, concerning the inexpressibility of this problem in Datalog, can be strengthened as follows.

THEOREM 8.7.6 (Atserias, Bulatov, Dawar [14]). *Let A be a finite abelian group with a nonzero element $a \in A$. Then* CSP$(A; R_0^2, R_0^3, R_a^3)$ *(defined in Theorem* 8.6.11*) is not in FP+C.*

The importance of this theorem for the descriptive complexity of CSPs stems from the following result whose proof, given in [14] for the case of finite domains, is valid in general.

THEOREM 8.7.7. *Let $\mathfrak{B}$ be a structure with finite relational signature such that* CSP$(\mathfrak{B})$ *is in FP+C (FP, Datalog), and let $\mathfrak{D} \in$* HI$(\mathfrak{B})$ *with finite relational signature. Then* CSP$(\mathfrak{D})$ *is in FP+C (FP, Datalog) as well.*

Hence, if HI$(\mathfrak{B})$ contains the structure $(A; R_0^2, R_0^3, R_a^3)$ from Theorem 8.7.6, then CSP$(\mathfrak{B})$ is not in FP+C.

8.8. Datalog for finite templates

The combination of Theorem 8.7.6 with Theorem 8.7.7 from the previous section provides a powerful condition implying that CSP$(\mathfrak{B})$ is not in FP+C, and in particular not in Datalog. Barto and Kozik [21] proved that if this condition does not apply, the CSP($\mathfrak{B}$) is in Datalog. This has been conjectured by Feder and Vardi [170] and later, in a different formulation, by Larose and Zadori [257] (see [256] for the equivalence).

In this section we present several other characterisations of finite structures $\mathfrak{B}$ such that CSP($\mathfrak{B}$) is in Datalog. In fact, each of the characterisation we present has its own advantages and many of them are needed to conveniently derive their equivalence from the proofs that can be found in the literature. Recall from Section 6.9 that a *weak near-unanimity* is an operation of arity at least two that satisfies the height-one identities

$$\forall x, y\colon w(x, \dots, x, y) = w(x, \dots, y, x) = \cdots = w(y, x, \dots, x).$$

We write WNU(k) for the k-ary weak near-unanimity operations. Again, we warn the reader that some authors have additionally assumed that weak near-unanimity operations are idempotent; we do not make this assumption (like in many other recent articles, e.g., in the survey [26]).

Example 8.8.1. For $n \in \mathbb{N} \setminus \{0\}$ consider the algebra $\mathbf{A}_n := (\mathbb{Z}_n; m)$ where $\mathbb{Z}_n := \{0, \dots, n-1\}$ and $m(x, y, z) := x - y + z$. Then $\mathrm{Clo}(\mathbf{A}_n)$ consists of precisely the operations defined as

$$g(x_1, \dots, x_k) := \sum_i a_i x_i$$

where $a_1, \dots, a_k \in \mathbb{Z}_n$ with $\sum_i a_i = 1$. We claim that $\mathbf{A}_n$ has a WNU(k) term if and only if $gcd(k, n) = 1$:

- If $gcd(k, n) = 1$ then there is an $a \in \mathbb{Z}_n$ such that $ak = 1$. Then the operation $\sum_i ax_i$ is in WNU(k), and as $\sum_i a = ka = 1$ it also belongs to $\mathrm{Clo}(\mathbf{A}_n)$.
- Conversely, let $g \in$ WNU(k). In particular, we have

$$\begin{aligned} g(1, 0, \dots, 0) &= a_1 \\ = g(0, 1, \dots, 0) &= a_2 \\ = \cdots = g(0, \dots, 0, 1) &= a_k \end{aligned}$$

  and it follows that $a := a_1 = \cdots = a_k$. But $1 = \sum_i a_i = ka$, which implies that n and m are pairwise prime.

For example, $\mathrm{Clo}(\mathfrak{A}_6)$ has a WNU(5) term, but no WNU(k) term for $k \leq 4$.

THEOREM 8.8.2. *Let $\mathfrak{B}$ be a finite structure with finite relational signature. Then the following are equivalent.*

1. CSP($\mathfrak{B}$) *has width $(2, k)$ where k is the maximal arity of $\mathfrak{B}$.*
2. CSP($\mathfrak{B}$) *has width (l, k) for some $l, k \in \mathbb{N}$.*

3. $\mathrm{HI}(\mathfrak{B})$ *does not contain* $(A; R_0^2, R_0^3, R_a^3)$ *where* A *is an abelian group with an element* $a \in A \setminus \{0\}$ *and*
$$R_c^k := \{(x_1, \dots, x_k) \in A^k \mid x_1 + \dots + x_k = c\} \quad \text{for } c \in A.$$
4. *If* $\mathfrak{C}$ *is the core of* $\mathfrak{B}$ *and* $\mathbf{C}$ *is an algebra whose operations are all idempotent polymorphisms of* $\mathfrak{C}$, *then* $\mathrm{HS}(\mathbf{C})$ *does not contain an algebra with domain* $\mathbb{Z}_n$, *for some* $n \geq 2$, *whose operations are generated by* $(x, y, z) \mapsto x - y + z$.
5. $\mathfrak{B}$ *has for all but finitely many* $n \in \mathbb{N}$ *a polymorphism in* $\mathrm{WNU}(n)$.
6. $\mathfrak{B}$ *has polymorphisms* $f \in \mathrm{WNU}(3)$ *and* $g \in \mathrm{WNU}(4)$ *satisfying*
$$\forall x, y \colon f(y, x, x) = g(y, x, x, x).$$
7. $\mathfrak{B}$ *has a binary polymorphism* f_2 *and polymorphisms* $f_n \in \mathrm{WNU}(n)$ *for every* $n \geq 3$ *such that*
$$\forall x, y \colon f_n(x, y, \dots, y) = f_2(x, y).$$
8. $\mathfrak{B}$ *has ternary polymorphisms* p, q *such that* $p \in \mathrm{WNU}(3)$ *and*
$$\forall x, y \ \big(p(x, x, y) = q(x, y, x) \wedge q(x, x, y) = q(x, y, y)\big).$$

We do not give a complete proof of this important result, but only show some of the easy implications and we explain how to deduce the remaining implications from statements that can be found explicitly in the literature.

PROOF. We first show implications between (1), (2), (3), (4), (5) in cyclic order and then prove the remaining equivalences.

(1) ⇒ (2) is trivial.

(2) ⇒ (3). By Lemma 8.3.2, it suffices to show that $\mathrm{CSP}(A; R_0^2, R_0^3, R_a^3)$ does not have bounded width, which is Theorem 8.6.11.

(3) ⇒ (4). Let $\bar{c}$ be a tuple enumerating all the elements of $\mathbf{C}$. Suppose that $\mathrm{HS}(\mathbf{C})$ contains an algebra $\mathbf{A}$ with domain $\mathbb{Z}_n$ all of whose operations are generated by $(x, y, z) \mapsto x - y + z$. Let $\mathfrak{A}$ be a structure with domain $\mathbb{Z}_n$ such that $\mathrm{Pol}(\mathfrak{A}) = \mathrm{Clo}(\mathbf{A})$. Then $\mathfrak{A}$ has a primitive positive interpretation in $\mathfrak{C}, \bar{c})$ (Theorem 6.3.7). Moreover, Theorem 3.6.2 implies that $\mathbf{A} \in \mathrm{HI}(\mathfrak{B})$. Note that the relations R_0^2, R_0^3, R_1^3 are preserved by $(x, y, z) \mapsto x - y + z$, and hence $(A; R_0^2, R_0^3, R_1^3) \in \mathrm{HI}(\mathfrak{B})$.

The implication (4) ⇒ (5) was proved by Maroti and McKenzie [279].

The implication (5) ⇒ (1) was proved by Barto and Kozik [21].

A self-contained proof of the implication (2) ⇒ (6) was given by Kozik, Krokhin, Valeriote, and Willard [244]; the implication follows from a combination of Theorem 2.7 and Theorem 2.8 in [244].

The proof from [244] has been generalised by Jovanović, Marković, McKenzie, and Moore [222] to prove the implication (2) ⇒ (8) (Corollary 3.3) and the implication (2) ⇒ (7) (Proposition 4.1). They work in the setting of idempotent algebras; but since a structure has polymorphisms satisfying a given minor condition if and only if its core does, and since the polymorphisms of a core satisfy a given minor condition if and only if the idempotent polymorphisms do, the idempotent case implies the statement as given in the theorem.

The implication (7) $\Rightarrow$ (5) is trivial.

For the implication (8) $\Rightarrow$ (4), suppose that $\mathrm{HS}(\mathbf{C})$ contains an algebra $\mathbf{A}$ with domain $\mathbb{Z}_n$, for $n \geq 2$, whose operations are generated by the operation $(x, y, z) \mapsto x - y + z$. Note that if p and q are ternary term operation of $\mathbf{A}$, then there are a_1, a_2, a_3 with $a_1 + a_2 + a_3 = 1$ and b_1, b_2, b_3 with $a_1 + a_2 + a_3 = 1$ such that $p(x, y, z) = a_1 x + a_2 y + a_3 z$ and $q(x, y, z) = b_1 x + b_2 y + b_3 z$. The identities from (8) imply that $p(0, 0, y) = p(0, y, 0) = p(0, 0, y)$ and hence $a := a_1 = a_2 = a_3$. Moreover, $p(0, 0, y) = q(0, y, 0)$ and hence $b_2 = a$, and $q(0, 0, y) = q(0, y, y)$ and hence $b_3 = b_2 + b_3$. Therefore, $a = b_2 = 0$ which implies that $A = \mathbb{Z}_n = \{0\}$, in contradiction to $n \geq 2$.

The implication (6) $\Rightarrow$ (4) can be shown similarly. ⊣

Remark 8.8.3. The equivalent conditions in Theorem 8.8.2 are also equivalent to an important congruence lattice condition: namely that all algebras in the variety generated by the idempotent polymorphisms of the core of $\mathfrak{B}$ have a *congruence meet semidistributive* congruence lattice [205]. This property is in turn equivalent to a fundamental condition from tame congruence theory [205], namely *omitting types* 1 *and* 2. This terminology is omnipresent in the literature cited in the proof of Theorem 8.8.2, but as we demonstrated above these more advanced universal-algebraic concepts are not needed in the proof of Theorem 8.8.2.

Chapter 9

TOPOLOGY

In this chapter we will see that the computational complexity of CSP($\mathfrak{B}$) for a countable ω-categorical structure $\mathfrak{B}$ only depends on the polymorphism clone of $\mathfrak{B}$, viewed as a *topological clone*. The definition of *topological clone* is analogous to the definition of *topological group*: a topological clone is an abstract clone together with a topology on the set of operations such that composition is continuous. Every operation clone is a topological clone with respect to the topology of pointwise convergence. In fact, for operation clones on countable domains, this topology is induced by a natural metric.

In this chapter we present a general correspondence between the pseudo-variety generated by a algebra **B** with a countable domain, and Clo(**B**) viewed as an abstract clone together with the mentioned metric.[23] The next step is

[23]Alternatively, we could work with *uniformities*; in this text, we focus on countable domains and it will suffice to work with metrics rather than uniformities.

a correspondence between the pseudo-variety generated by an oligomorphic algebra $\mathbf{B}$ and the topological clone $\text{Clo}(\mathbf{B})$; this result can be seen as a topological variant of Birkhoff's theorem from Section 6.5. We also present a modification of this result that captures the class $\text{Exp Refl P}^{\text{fin}}(\mathbf{B})$ topologically. If $\mathbf{B}$ is the polymorphism algebra of an ω-categorical structure $\mathfrak{B}$, then this class can be used to study the complexity of constraint satisfaction problems, because it corresponds to the class $\text{HI}(\mathfrak{B})$ of structures that have a primitive positive interpretation in $\mathfrak{B}$ modulo homomorphic equivalence.

Several other important properties of ω-categorical structures only depend on the topological polymorphism clone or even the topological automorphism group. In particular, this is the case for certain Ramsey properties that become important in the next chapter. Section 9.1 is introductory and can be skipped by readers familiar with topology.

9.1. Topological spaces

A *topological space* is a set S together with a collection of subsets of S, called the *open* sets of S, such that

1. the empty set and S are open;
2. arbitrary unions of open sets are open;
3. the intersection of two open sets is open.

The collection of open sets is also called a *topology* on S. If $U \subseteq S$ is open, then the complement $S \setminus U$ of U in S is called *closed*. For $E \subseteq S$, the *closure* of E is the set of all points x such that every open set in S that contains x also contains a point from E. Clearly, the closure of E is a closed set. A subset E of S is called *dense (in S)* if its closure is the full space S. The *subspace of S induced on E* is the topological space on E where the open sets are exactly the intersections of E with the open sets of S. A *basis* of S is a collection of open subsets of S such that every open set in S is the union of sets from the collection. For $s \in S$, a collection of open subsets of S is called a *basis at s* if each set from the collection contains s, and every open set containing s also contains an open set from the collection. A topological space S is called

- *discrete* if every subset of S is open (and hence also closed);
- *Hausdorff* if for any two distinct points u, v of S there are disjoint open sets U and V that contain u and v, respectively;
- *separable* if it contains a countable dense set;
- *first-countable* if for all $s \in S$ there exists a countable basis at s;
- *second-countable* if it has a countable basis.

Note that if S is second-countable, it is also first-countable and separable.

9.1.1. Convergence and continuity. A function between topological spaces is called *continuous* if the pre-images of open sets are open, and *open* if images of open sets are open. A bijective open and continuous map is called a *homeomorphism*.

There are equivalent characterisations of continuity of maps from a first-countable space S to a topological space T that are often easier to work with and which we recall in Proposition 9.1.1. For a sequence $(s_n)_{n\geq 1}$ of elements of S we say that s_n *converges against* s if for every open set U of S that contains s there exists an n_0 such that $s_n \in U$ for all $n \geq n_0$. Note that if T is Hausdorff, then s is unique, and called *the limit* of $(s_n)_{n\geq 1}$, and we write $\lim_{n\to\infty} s_n$ for s. For $x \in S$, we say that $\xi\colon S \to T$ is *continuous at* x if for every open $V \subseteq T$ containing $\xi(x)$ there is an open $U \subseteq S$ containing x whose image $\xi(U)$ is contained in V.

PROPOSITION 9.1.1. *Let S be a first-countable space and T an arbitrary topological space. Then for every $\xi\colon S \to T$ the following are equivalent.*

1. *ξ is continuous.*
2. *For all s_n, if s_n converges against s then $\xi(s_n)$ converges against $\xi(s)$.*
3. *ξ is continuous at every $x \in S$.*

PROOF. The implication from (1) to (2) is true even without the assumption that S is first-countable. Let $(s_n)_{n\geq 1}$ be a sequence that converges against s, and let V be open so that $\xi(s) \in V$. Then $U := \xi^{-1}(V)$ is open, and $s \in U$. So there exists an n_0 such that $s_n \in U$ for all $n \geq n_0$. For then $\xi(s_n) \in V$ for all $n \geq n_0$. So $\xi(s_n)$ converges against $\xi(s)$.

For the implication from (2) to (3), we show the contraposition. Suppose that ξ is not continuous at some $s \in S$. That is, there exists an open set V containing $\xi(s)$ such that all open sets U that contain x have an image that is not contained in V. Since S is first-countable, there exists a countable collection U_n of open sets containing x so that any open V that contains x also contains some U_n. Replacing U_n by $\cap_{k=1}^{n} U_k$ where necessary, we may assume that $U_1 \supset U_2 \supset \cdots$. If $U_n \subseteq \xi^{-1}(V)$, then $\xi(U_n) \subseteq V$, in contradiction to our assumption; so we can pick an $x_n \in U_n \setminus \xi^{-1}(V)$ for all n, and obtain a sequence that converges to x. But $s_n \notin \xi^{-1}(V)$ for all n, and so $\xi(s_n)$ does not converge to $\xi(s) \in V$.

Finally, the implication from (3) to (1) again holds in arbitrary topological spaces. Let $V \subseteq T$ be open. We want to show that $U := \xi^{-1}(V)$ is open. If s is a point from U, then because ξ is continuous at s, and V contains $\xi(s)$ and is open, there is an open set $U_s \subseteq S$ containing s whose image $\xi(U_s)$ is contained in V. Then $\bigcup_{s\in U} U_s = U$ is open as a union of open sets. ⊣

9.1.2. Sum and quotient spaces. Let $(S_i)_{i\in I}$ be a family of topological spaces. If the S_i are pairwise disjoint, then the *sum space* $\oplus_{i\in I} S_i$ is the topological space on $\bigcup_{i\in I} S_i$ where the topology is given by the basis

$$\{U \subseteq \bigcup_{i\in I} S_i \mid U \in S_i \text{ for some } i \in I\}$$

Hence, $U \subseteq \bigcup_{i \in I}$ is open if and only if $U \cap S_i$ is open in S_i for every $i \in I$.

Let S be a topological space and let E be an equivalence relation on S. We write S/E for the set of equivalence classes of E and for $s \in S$ we write s/E for the equivalence class of s with respect to E. Then S/E can be equipped with the following topology, called *quotient topology*: First define $p: S \to S/E$ by setting $p(s) = s/E$ (called the *projection map*). Define $U \subseteq S/E$ to be open if and only if $p^{-1}(U)$ is open in S. So the topology is smallest possible so that the projection map is continuous.

9.1.3. Product spaces. The *product* $T = \prod_{i \in I} S_i$ of a family of topological spaces $(S_i)_{i \in I}$ is the topological space on the cartesian product $\times_{i \in I} S_i$ where the open sets are unions of sets of the form $\times_{i \in I} U_i$ where U_i is open in S_i for all $i \in I$, and $U_i = S_i$ for all but finitely many $i \in I$. For $i \in I$, the i-th *projection map* $\pi_i : T \to S_i$ is the function defined by $\pi_i(u) := u(i)$. Note that the product topology on T is the smallest topology such that each projection map π_i is continuous.

LEMMA 9.1.2. *Let S and S_i, for $i \in I$, be topological spaces and let $f : S \to \prod_{i \in I} S_i$ be any map. Then f is continuous if and only if for every $i \in I$ the composition $\pi_i \circ f : S \to S_i$ is continuous.*

PROOF. Clearly, if f is continuous, then so is $\pi_i \circ f$ since the composition of continuous functions is continuous. To prove the converse implication, it suffices to prove that the preimage under f of every basic open set $V = \times_{i \in I} V_i$ is open in S. By the definition of the product topology, there exists a finite set $F \subseteq I$ such that $V_i = S_i$ for all $I \setminus F$. By assumption, $\pi_i \circ f$ is continuous for all $i \in I$, and hence $f^{-1}(\pi_i^{-1}(V_i))$ is open, for all $i \in I$. We have that

$$f^{-1}(V) = \bigcap_{i \in F} f^{-1}(\pi_i^{-1}(V_i))$$

which is open as a finite intersection of open sets. ⊣

If I has just two elements, say 1 and 2, we also write $S_1 \times S_2$ for the product; this operation is clearly associative. We denote by S^k for the k-th power $S \times \cdots \times S$ of S. We also write S^I to a $|I|$-th power of S, where the factors are indexed by the elements of I. In this case, we can view each element of $T := S^I$ as a function from I to S in the obvious way. The product topology on T is also called the *topology of pointwise convergence*, due to the following.

PROPOSITION 9.1.3. *Let S be a topological Hausdorff space and let I be a set. Let $(\xi_n)_{n \in \mathbb{N}}$ be a sequence of elements of the product space $T := S^I$. Then T is Hausdorff and $\lim_{n \to \infty} \xi_n = \xi$ if and only if $\lim_{n \to \infty} \xi_n(j) = \xi(j)$ in S for all $j \in I$.*

PROOF. Let $u, v \in T$ be distinct. Then there exists $n \in \mathbb{N}$ such that $u(n) \neq v(n)$ are distinct elements of the Hausdorff space S. It follows that there are disjoint open sets $U, V \subseteq S$ such that $u(n) \in U$ and $v(n) \in V$. By the definition of the product topology, the sets $\pi_n^{-1}(U)$ and $\pi_n^{-1}(V)$ are open. Clearly, $u \in \pi_n^{-1}(U)$, $v \in \pi_n^{-1}(V)$, and $\pi_n^{-1}(U)$ and $\pi_n^{-1}(V)$ are disjoint.

To prove the second statement, suppose first that $\lim_{n\to\infty} \xi_n = \xi$ in T. Let $j \in I$ be arbitrary and let V be an open set that contains $\xi(j)$. Then the set $U := \prod_{i\in I} T_i$ where $T_i = V$ if $i = j$, and $T_i = S$ otherwise, is open in T and contains ξ, so there is an n_0 such that $\xi_n \in U$ for all $n \geq n_0$. But then $\xi_n(j) \in V$ for all $n \geq n_0$, and so $\lim_{n\to\infty} \xi_n(j) = \xi(j)$.

Now suppose that $\lim_{n\to\infty} \xi_n(j) = \xi(j)$ in S for all $j \in I$, and let V be an open set of T that contains ξ. Then there exists a finite $J \subseteq I$ and open subsets $(V_j)_{j\in J}$ of S such that $f \in \prod_{i\in I} T_i$ where $T_i = V_i$ if $i \in J$ and $T_i = S$ otherwise. For each $j \in J$ there exists an n_j so that $\xi_n(j) \in V_j$ for all $n \geq n_j$. Then $\xi_n \in V$ for all $n \geq \max_{j\in J} n_j$, and hence $\lim_{n\to\infty} \xi_n = \xi$. $\dashv$

Example 9.1.4. If we equip $\{0, 1\}$ with the discrete topology, then $\{0, 1\}^{\mathbb{N}}$ with the product topology is called the *Cantor space*.

Example 9.1.5. If we equip the natural numbers $\mathbb{N}$ with the discrete topology, then $\mathbb{N}^{\mathbb{N}}$ with the product topology is called the *Baire space*. The open sets are exactly the unions of sets of the form $\{g \in \mathbb{N} \to \mathbb{N} \mid g(\bar{a}) = \bar{b}\}$ for some $\bar{a}, \bar{b} \in \mathbb{N}^k$, $k \in \mathbb{N}$.

In Section 9.2 we primarily work with the following topological space.

Example 9.1.6. For any base set B, we view $\mathrm{Sym}(B)$ as a subspace of $B \to B$ with the product topology, where B is taken to be discrete, and also refer to this topology as the topology of pointwise convergence on $\mathrm{Sym}(B)$.

In Section 9.4 we primarily work with the topological space introduced in Example 9.1.7. Let B be a set, and let $f \in B^k \to B$ for some $k \geq 1$. If $A \in B^{m\times k}$ for some $m \geq 1$, and A is viewed as a matrix with entries in B, then we write $f(A)$ for the m-tuple obtained by applying f to each row of the matrix A.

Example 9.1.7. For any countable base set B, we equip the set

$$\mathcal{O}_B := \bigcup_{n\geq 1} \mathcal{O}_B^{(n)}$$

with the sum topology where for each $n \in \mathbb{N}$ the set $\mathcal{O}_B^{(n)} = B^{B^n}$ is equipped with the product topology. Note that the open sets are exactly the unions of sets of the form

$$T_{A,b} := \{g \in \mathcal{O}_B^{(k)} \mid g(A) = \bar{b}\}$$

for some $n, k \in \mathbb{N}$, $A \in B^{n\times k}$, and $b \in B^n$, so the topology is second-countable.

9.1.4. Metrics. Important examples of topologies come from metric spaces. A *metric space* is a pair (M, d) where M is a set and d is a *metric* on M, i.e., a function

$$d\colon M \times M \to \mathbb{R}$$

such that for any $x, y, z \in M$, the following holds:

1. $d(x, y) \geq 0$ (*non-negativity*);
2. $d(x, y) = 0 \Leftrightarrow x = y$ (*indiscernibility*);
3. $d(x, y) = d(y, x)$ (*symmetry*);
4. $d(x, z) \leq d(x, y) + d(y, z)$ (*subadditivity* or *triangle inequality*).

When $M' \subseteq M$ then the restriction of d to M' is clearly a metric, too. Every metric on M gives rise to a topology on M, namely the topology with the basis

$$\{\{y \in M \mid d(x, y) < \varepsilon\} \mid 0 < \varepsilon \in \mathbb{R}, x \in M\}.$$

A topological space S is *metrisable* if there exists a metric d on S which is *compatible* with the topology, i.e., the topology equals the topology that arises from the metric as described above. Clearly, metrisable spaces are Hausdorff.

DEFINITION 9.1.8. A metric d is called an *ultrametric* if it satisfies $d(x, z) \leq max(d(x, y), d(y, z))$ for all x, y, z.

Example 9.1.9. The Baire space (Example 9.1.5) has the following compatible ultrametric d. For elements $f, g \in \mathcal{G}$ we define $d(f, g) = 0$ if $f = g$, and otherwise $d(f, g) = 1/2^{n+1}$ where n is the least natural number such that $f(n) \neq g(n)$.

Example 9.1.10. The *box metric* on $\mathbb{N}^k \to \mathbb{N}$ is the metric defined by

$$d(f, g) := 2^{-\min\{n \in \mathbb{N} \mid \text{there is } s \in \{1,\dots,n\}^k \text{ such that } f(s) \neq g(s)\}}.$$

This metric is compatible with the product topology of $\mathbb{N}^k \to \mathbb{N}$ if $\mathbb{N}$ is taken to be discrete (Example 9.2.1), and it is an ultrametric.

Metric spaces have the advantage that we can use Cauchy sequences to talk about points that are not really there.

DEFINITION 9.1.11. Let (M, d) be a metric space. A sequence $(x_n)_{n \in \mathbb{N}}$ of elements in M is called *Cauchy* if for any $\varepsilon > 0$ there exists an n_0 such that for all $m, n > n_0$ we have $d(x_m, x_n) < \varepsilon$. The metric space (M, d) is called *complete* if every Cauchy sequence has a limit in M.

Let (X, d_X) and (Y, d_Y) be metric spaces. A map $f\colon (X, d_X) \to (Y, d_Y)$ is called *Cauchy continuous* if for every Cauchy sequence $(x_n)_{n \in \mathbb{N}}$ in (X, d_X) the sequence $(f(x_n))_{n \in \mathbb{N}}$ is Cauchy in (Y, d_Y).

Example 9.1.12. The standard distance metric on $\mathbb{R}$ is complete. The same metric on $\mathbb{Q}$ is not complete.

Example 9.1.13. The ultrametric d from Example 9.1.9, restricted to Sym($\mathbb{N}$), is compatible with the topology of pointwise convergence. This metric space is not complete: to see this, let f be an arbitrary injective non-surjective mapping from $\mathbb{N} \to \mathbb{N}$. For each n, there exists a permutation h_n of $\mathbb{N}$ such that $h_n(i) = f(i)$ for all $i \leq n$. We claim that the sequence $(h_n)_{n\geq 1}$ in Sym($\mathbb{N}$) is Cauchy. Let $\varepsilon > 0$. By the definition of the metric d, there exists an $\ell \in \mathbb{N}$ such that for all $p, q \in \mathbb{N}$, if $h_p(i) = h_q(i)$ for all $i \leq \ell$, then $d(h_p, h_q) < \varepsilon$. By the definition of the sequence $(h_n)_{n\geq 1}$, it follows that for all $n, m \geq \ell$ we have that $h_n(i) = h_m(i)$ for all $i \leq \ell$. Hence, $d(h_n, h_m) < \varepsilon$, proving that $(h_n)_{n\geq 1}$ is Cauchy. On the other hand, $(h_n)_{n\geq 1}$ does not converge to a permutation.

A topological space S is called *completely metrisable* if it has a compatible complete metric. It is called *Polish* if S is separable and completely metrisable.

Example 9.1.14. The subspace Sym($\mathbb{N}$) of the Baire space is completely metrisable. While the ultrametric d is not complete (Example 9.1.13), one can define a compatible complete metric d' on Sym($\mathbb{N}$) as follows. We define $d'(f, g) := 0$ if $f = g$, and otherwise $d'(f, g) := 1/2^{n+1}$ where n is the least natural number such that $f(n) \neq g(n)$ or $f^{-1}(n) \neq g^{-1}(n)$. It is again easy to verify that d' is an ultrametric. To see that it is complete, let $(f_n)_{n\in\mathbb{N}}$ be a Cauchy sequence in Sym($\mathbb{N}$). We define $f \in$ Sym($\mathbb{N}$) as follows. For $n \in \mathbb{N}$, choose $\varepsilon > 0$ such that for all $p, q \in \mathbb{N}$, if $d(f_p, f_q) < \varepsilon$ then $f_p(i) = f_q(i)$ and $f_p^{-1}(i) = f_q^{-1}(i)$ for all $i \leq n$. Since $(f_n)_{n\in\mathbb{N}}$ is Cauchy, there exists an $n_0 \in \mathbb{N}$ such that for all $p, q \geq n_0$ we have $d(f_p, f_q) < \varepsilon$. Define $f(n) := f_{n_0}(n)$. Then it is straightforward to verify that f is a permutation and that $(f_n)_{n\in\mathbb{N}}$ converges against f. Since Sym($\mathbb{N}$) is also separable, we have that Sym($\mathbb{N}$) is Polish.

9.1.5. Uniform continuity. Given metric spaces (X, d_1) and (Y, d_2), a function $f : X \to Y$ is called *uniformly continuous* if

$$\forall \varepsilon > 0 \, \exists \delta > 0 \, \forall x, y \in X \, (d_1(x, y) < \delta \Rightarrow d_2(f(x), f(y)) < \varepsilon).$$

For comparison: continuity of f with respect to the topologies induced by d_1 and d_2 only requires that

$$\forall \varepsilon > 0, x \in X \, \exists \delta > 0 \, \forall y \in X \, (d_1(x, y) < \delta \Rightarrow d_2(f(x), f(y)) < \varepsilon).$$

Example 9.1.15. An endomorphism ξ of the Baire space with the metric d defined in Example 9.1.9 is uniformly continuous if for every finite $F \subseteq \mathbb{N}$ there exists a finite $G \subseteq \mathbb{N}$ such that for all $f, g \in \mathbb{N}^{\mathbb{N}}$ if $f|_G = g|_G$ then $\xi(f)|_F = \xi(g)|_F$.

For comparison: an endomorphism of the Baire space is continuous if and only if for every finite $F \subseteq \mathbb{N}$ and every $f \in \mathbb{N}^{\mathbb{N}}$ there exists a finite $G \subseteq \mathbb{N}$ such that if $g \in \mathbb{N}^{\mathbb{N}}$ is such that $f|_G = g|_G$ then $\xi(f)|_F = \xi(g)|_F$.

PROPOSITION 9.1.16. *A uniformly continuous map f between metric spaces maps Cauchy sequences to Cauchy sequences.*

PROOF. Let $(s_n)_{n\in\mathbb{N}}$ be a Cauchy sequences, and let $\varepsilon > 0$. By uniform continuity of f there exists $\delta > 0$ such that $d(f(x)-f(y)) < \varepsilon$ for $d(x-y) < \delta$. Since s_n is Cauchy, there exists an $n_0 > 0$ such that $d(s_n - s_m) < \delta$ for all $n, m > n_0$. Hence, $d(f(s_n) - f(s_m)) < \varepsilon$ for all $n, m > n_0$. Therefore, $(f(s_n))_{n\in\mathbb{N}}$ is Cauchy. ⊣

9.1.6. Compactness. A *cover* of a space S is a family $\mathcal{C}$ of subsets of S whose union equals S; we say that $\mathcal{C}$ *covers* S. A topological space S is called *compact* if for any cover $\mathcal{C}$ of S consisting of open subsets of S (also called an *open cover of* S) there is a finite subset of $\mathcal{C}$ that covers S (also called a *subcover* of $\mathcal{C}$). Equivalently, if $\{V_i\}_{i\in A}$ is a collection of closed sets such that $\bigcap_{i\in B} V_i \neq \emptyset$ for every finite subset B of A, then $\bigcap_{i\in A} V_i \neq \emptyset$. The Baire space is clearly not compact, but the Cantor space is. More generally, Tychonoff's theorem states that products of compact spaces are compact; for a proof, see for example [344].

THEOREM 9.1.17 (Tychonoff). *Products of compact spaces are compact.*

PROPOSITION 9.1.18. *Let T be a compact space. Then closed subspaces of T and quotients of T are compact as well.*

PROOF. Let C be a closed subspace of T. Let $\mathcal{U}$ be an open cover of C. By assumption, $T \setminus C$ is open in T. Hence, $\mathcal{U} \cup \{T \setminus C\}$ is an open cover of T. As T is compact, there is a finite subcover of $\mathcal{U}$, say $\{U_1, U_2, \dots, U_r\}$. This also covers C by the fact that it covers T. If $T \setminus C$ is among $U_1, U_2, \dots, U_r$, then it can be removed and the remaining sets still cover C. Thus we have found a finite subcover of $\mathcal{U}$ which covers C, and hence C is compact.

Let E be an equivalence relation and let $\mathcal{C}$ be an open cover of T/E. Then $\{\{s/E \mid s \in U\} \mid U \in \mathcal{C}\}$ is an open cover of T. Since T is compact there is a finite subcover of $\{\{s/E \mid s \in U_1\}, \dots, \{s/E \mid s \in U_n\}\}$. But then $\{U_1, \dots, U_n\}$ is a finite subcover of $\mathcal{C}$, showing compactness of T/E. ⊣

Let T be a topological space and $S \subseteq T$. Then $x \in T$ is called a *limit point of* S if $x \in \overline{S \setminus \{x\}}$. In other words, $x \in T$ is not a limit point of S if there is an open $U \subseteq T$ such that $U \cap S = \{x\}$.

PROPOSITION 9.1.19. *Let T be compact. Then every infinite $S \subseteq T$ has a limit point in T.*

PROOF. Let $S \subseteq T$ be without limit points. Then S is closed and hence compact (Proposition 9.1.18). For any $s \in S$ there is an open U_s such that $U_s \cap S = \{s\}$, because s is not a limit point of S. Then $\{U_s \mid s \in S\}$ is an open cover of S, which must have a finite subcover by compactness. It follows that S is finite. ⊣

9.2. Topological groups

A *topological group* is an (abstract) group $\mathbf{G}$ together with a topology on the elements G of $\mathbf{G}$ such that the function $(x, y) \mapsto xy^{-1}$ from G^2 to G is continuous. In other words, we require that the binary group operation and the inverse function are continuous. Note that every open subgroup $\mathbf{H}$ of a topological group $\mathbf{G}$ is closed, since the complement of H in G is the open set given by the union of open sets gH for $g \in G \setminus H$. A *topological isomorphism* of topological groups is a group isomorphism which is a homeomorphism between the respective topologies.

Example 9.2.1. The group $\mathrm{Sym}(\mathbb{N})$ with the pointwise convergence topology induced by the Baire space is a topological group: if $U \subseteq G$ is a basic open set of the form $S_{a,c} := \{f \in G \mid f(a) = c\}$ for some $a, c \in \mathbb{N}^k$, then the preimage of U under the composition operation $\circ\colon G^2 \to G$ is

$$\begin{aligned}\{(f, g) \in G^2 \mid f \circ g \in S_{a,c}\} &= \{(f, g) \in G^2 \mid \exists b \in \mathbb{N}^k \text{ s.t. } g \in S_{a,b} \\ &\qquad\qquad\qquad\qquad\qquad \text{and } f \in S_{b,c}\} \\ &= \bigcup_{b \in \mathbb{N}^k} (S_{b,c} \times S_{a,b})\end{aligned}$$

which is open as it is a union of open sets.

As in Example 9.2.1, every permutation group on a set B gives rise to a topological group with respect to the product topology on $\mathrm{Sym}(B)$ where B is taken to be discrete; then the closed subsets of $\mathrm{Sym}(B)$ are precisely as described in Definition 4.2.1.

A topological group is *Hausdorff* (*first-countable, metrisable, Polish*) if its topology is Hausdorff (first-countable, metrisable, Polish, respectively). Note that **G** is first-countable if and only if **G** it has a countable basis at the identity: if $\mathcal{B}$ is a basis of open sets at the identity, and $g \in G$, then $\{g^{-1}U \mid U \in \mathcal{B}\}$ is a basis at g.

The continuity of a homomorphism between two topological groups can be checked at the identity.

LEMMA 9.2.2. *Let* **G** *and* **H** *be topological groups and* $h\colon G \to H$ *a homomorphism from* **G** *to* **H**. *Then h is continuous if and only if it is continuous at* 1.

PROOF. Continuity implies continuity at 1. To prove the converse direction, let U be an open set of **H**, and let $g \in U$. Since g^{-1} is continuous we have that $S := g^{-1}U$ is open. $\xi^{-1}(U) = \xi^{-1}(gS) = \xi^{-1}(g)\xi^{-1}(S)$ because ξ is a homomorphism. Since $1 \in S$, the assumption implies that $\xi^{-1}(S)$ is open, and since multiplication by $\xi^{-1}(g)$ is continuous, $\xi^{-1}(g)\xi^{-1}(S) = \xi^{-1}(U)$ is open, which establishes continuity of ξ. ⊣

9.2.1. Continuous actions. An action of a topological group $\mathbf{G}$ on a topological space S is *continuous* if it is continuous as a function from $G \times S$ into S. If S is a topological space, then $\mathrm{Homeo}(S) \subseteq \mathrm{Sym}(S)$ denotes the set of all homeomorphisms of S. We will view $\mathrm{Homeo}(S)$ as a topological space with the subspace topology inherited from S^S which carries the product topology[24].

PROPOSITION 9.2.3. *Every continuous action of a topological group* $\mathbf{G}$ *on a topological space* S *is a continuous homomorphism from* $\mathbf{G}$ *into* $\mathrm{Homeo}(S)$.

PROOF. Suppose that $\xi\colon \mathbf{G} \to \mathrm{Sym}(S)$ is a continuous action of $\mathbf{G}$ on S, so the map $\chi(g, s) := \xi(g)(s)$ is continuous from $G \times S$ to S. For every $g \in G$, the map t_g defined by $s \mapsto \chi(g, s)$ is continuous. The inverse of t_g is $s \mapsto \chi(g^{-1}, s)$, which is also continuous. Hence, t_g is a homeomorphism.

To show that ξ is continuous, let U be a basic open subset of $\mathrm{Homeo}(S)$, i.e., $U = \prod_{s\in S} U_s$ where U_s is open in S for all $s \in S$, and there exists a finite set F such that $U_s = S$ for all $s \in S \setminus F$. Note that for fixed s, the map $t_s\colon G \to S$ given by $g \mapsto \xi(g)(s)$ is continuous, and hence for all $s \in F$ the set $\{g \in G \mid \xi(g)(s) \in U_s\}$ is open. Therefore,

$$\begin{aligned}\xi^{-1}(U) &= \{g \in G \mid \xi(g)(s) \in U_s \text{ for all } s \in F\}\\ &= \bigcap_{s\in F}\{g \in G \mid \xi(g)(s) \in U_s\}\end{aligned}$$

is a finite intersection of open sets and hence open. ⊣

If S carries the discrete topology (in which case $\mathrm{Homeo}(S) = \mathrm{Sym}(S)$), the statement of Proposition 9.2.3 can be strengthened to obtain an equivalent characterisation of continuity of actions.

LEMMA 9.2.4. *Let* $\mathbf{G}$ *be a topological group and* ξ *an action of* $\mathbf{G}$ *on a set* S *equipped with the discrete topology. Then the action* ξ *is continuous if and only if* ξ *is continuous as a map from* $\mathbf{G}$ *to* $\mathrm{Sym}(S)$.

PROOF. The forward implication follows from Proposition 9.2.3. For the converse implication, we have to show that the function $\chi\colon G \times S \to S$ given by $(g, s) \mapsto \xi(g)(s)$ is continuous. Let $S' \subseteq S$ and $s' \in S'$; it suffices to show that there exists an open $U \subseteq G$ and an open $T \subseteq S'$ such that $\chi(U, T)$ contains s'. Since S is discrete, in particular $T := \{s'\}$ is open. Let $U := \xi^{-1}(\mathrm{Sym}(S)_{s'})$ which is by assumption an open subset of G. Then $\chi(U, T)$ contains s'. ⊣

An important example of a continuous action of a topological group $\mathbf{G}$ is the action of $\mathbf{G}$ on the coset space of an open subgroup by left translation (Example 9.2.6).

DEFINITION 9.2.5. A *left coset* of a subgroup $\mathbf{V}$ of $\mathbf{G}$ is a set of the form $\{hg \mid g \in V\}$ for $h \in G$, also written hV. Clearly, the set of all left cosets of

[24]Note that it is not clear (and depends on S) whether $\mathrm{Homeo}(S)$ with this topology is a topological group.

$\mathbf{G}$ partitions G, and is denoted by $\mathbf{G}/V$. The cardinality of $\mathbf{G}/V$ is called the *index* of $\mathbf{V}$ in $\mathbf{G}$. The space $V\backslash\mathbf{G}$ of all *right cosets* Vh is defined analogously.

We also view $\mathbf{G}/V$ as a topological space with the quotient topology, i.e., a set of left-cosets is open if their union is an open subset of G.

Example 9.2.6. We define a continuous action of $\mathbf{G}$ on $\mathbf{G}/V$ by setting

$$g \cdot hV := ghV.$$

This action is also called the *action of* $\mathbf{G}$ *on* $\mathbf{G}/V$ *by left translation*. Suppose now that V is open. Then the action by left translation is continuous: to see this, let $S \subseteq \mathbf{G}/V$ be open, and let $gV \in S$. It suffices to show that there are open subsets $U \subseteq G$ and $T \subseteq G/V$ such that $gV \in \{\xi(u)(t) \mid u \in U, t \in T\} \subseteq S$. By the definition of the quotient topology $p^{-1}(S)$ is open in G. Since composition in G is continuous the set $\{(g_1, g_2) \in G^2 \mid g_1 g_2 \in p^{-1}(S)\}$ is open in G^2. This set contains $(1, g)$ since $p(1g) = gV \in S$. So there exists an open $U \subseteq G$ containing 1 and an open $H \subseteq G$ containing g such that $\{uv \mid u \in U, v \in H\} \subseteq p^{-1}(S)$. Then $T := p(H) = \{vV \mid v \in H\}$ is open in $\mathbf{G}/H$, and

$$gV \in \{\xi(u)(t) \mid u \in U, t \in T\} = \{uvV \mid u \in U, v \in H\} \subseteq S.$$

We present an example of a discontinuous group action of an oligomorphic permutation group.

Example 9.2.7. The structure $\mathfrak{A}$ presented in this example is due to Cherlin and Hrushovski (see also [258]). Let $\mathcal{K}$ be the class of all finite structures $(A; E_1, E_2, \dots)$ where E_i denotes an equivalence relation on i-tuples with pairwise distinct entries from A with at most two equivalence classes. Clearly, $\mathcal{K}$ is closed under substructures and isomorphism. It is easy to verify that it also has the amalgamation property (Section 2.3). Let $\mathfrak{A}$ be the Fraïssé-limit of $\mathcal{K}$. Then $\mathrm{Aut}(\mathfrak{A})$ has a homomorphism ξ_1 to $(\mathbb{Z}_2)^{\mathbb{N}}$ (which is equipped with the product topology): for $\alpha \in \mathrm{Aut}(\mathfrak{A})$ we define $\xi_1(\alpha) := (\alpha_i)_{i \in \mathbb{N}}$ where $\alpha_i := 0$ if α fixes the equivalence classes of E_{i+1} and $\alpha_i := 1$ otherwise. This map is clearly a group homomorphism.

To construct a discontinuous group homomorphism, let $\mathcal{U}$ be an ultrafilter on $\mathbb{N}$, and let $\xi_2 \colon (\mathbb{Z}_2)^{\mathbb{N}} \to \mathbb{Z}_2$ be the function that maps $(\alpha_i)_{i \in \mathbb{N}}$ to 0 if $\{i \mid \alpha_i = 0\} \in \mathcal{U}$, and to 1 otherwise. Again, it is straightforward to verify that ξ_2 is a group homomorphism. It is continuous if and only if $\mathcal{U}$ is principal. For a non-principal ultrafilter $\mathcal{U}$ the map $\xi_2 \circ \xi_1$ is a discontinuous group homomorphism from an oligomorphic permutation group to $\mathbb{Z}_2$.

We often consider continuous actions on product spaces; in this context, the following basic fact is useful.

PROPOSITION 9.2.8. *Let X be a topological Hausdorff space, and Y be any set. Let $\mathbf{G}$ be a topological group with an action ξ on X^Y. Then ξ is continuous if and* *only if for every $y \in Y$, the map $f_y \colon G \times X^Y \to X$ given by $f_y(g, \xi) := (g \cdot \xi)(y)$ is continuous.*

PROOF. Suppose that $\lim_{n\to\infty}(g_n, \xi_n) = (g, \xi)$. Then by Proposition 9.1.3 we have $\lim_{n\to\infty} g_n = g$ and $\lim_{n\to\infty} \xi_n(y) = \xi(y)$ for all $y \in Y$. Since f_y is continuous and by Proposition 9.1.1

$$\lim_{n\to\infty} (g_n \cdot \xi_n)(y) = \lim_{n\to\infty} f_y(g_n, \xi_n) = f_y(g, \xi) = (g \cdot \xi)(y)$$

for all $y \in Y$. We again apply Proposition 9.1.3 and obtain that $\lim_{n\to\infty}(g_n \cdot x_n) = g \cdot \xi$, which implies continuity of the action of $\mathbf{G}$, again using Proposition 9.1.1. The proof of the converse implication is straightforward, too. ⊣

There are closed oligomorphic subgroups of $\mathrm{Sym}(\mathbb{N})$ with a continuous action on $\mathbb{N}$ such that the image of the action is not closed in $\mathrm{Sym}(\mathbb{N})$. The basic idea of this example is due to Dugald Macpherson and can be found in Hodges' *Model Theory* [206] on page 354.

Example 9.2.9. Using the technique presented in Section 2.3.6, it is easy to construct a homogeneous structure $\mathfrak{Q} := (\mathbb{Q}; <, P)$ where

- $<$ is the usual strict order of the rational numbers, and
- $P \subseteq \mathbb{Q}$ is such that both P and $O := \mathbb{Q} \setminus P$ are dense in $(\mathbb{Q}; <)$.

Let $\mathfrak{P}$ be the substructure of $\mathfrak{Q}$ induced on P. It is easy to see (and also follows from Theorem 9.5.25) that the mapping which sends $f \in \mathrm{Aut}(\mathfrak{Q})$ to $f|_P$ induces a continuous homomorphism μ from $\mathrm{Aut}(\mathfrak{Q})$ to $\mathrm{Aut}(\mathfrak{P})$ whose image is dense in $\mathrm{Aut}(\mathfrak{P})$. We claim that μ is not surjective. To prove this, we consider *Dedekind cuts* (S, T) *of* $\mathfrak{P}$, that is, partitions of P into subsets S, T with the property that for all $s \in S$ and $t \in T$ we have $s < t$. Note that for each element $o \in O$ we obtain a Dedekind cut (S, T) with $S := \{a \in P \mid a < o\}$ and $T := \{a \in P \mid a > o\}$. But since there are uncountably many Dedekind cuts and only countably many elements of O, there also exists a Dedekind cut (S', T') which is not of this form. By a standard back-and-forth argument, there exists an $\alpha \in \mathrm{Aut}(P, <)$ that maps S to S' and T to T'. Suppose for contradiction that there is $\beta \in \mathrm{Aut}(\mathfrak{Q})$ with $\beta|_P = \alpha$. Then $s < \beta(o) < t$ for all $s \in S'$, $t \in T'$, in contradiction to the assumptions on (S', T').

On the other hand, for some natural actions the image will be closed.

Example 9.2.10. Let $\mathbf{G}$ be a topological group and $\mathbf{H}$ be an open subgroup of $\mathbf{G}$. Then the action of $\mathbf{G}$ on $\mathbf{G}/H$ by left translation (Example 9.2.6) has a closed image in $\mathrm{Sym}(\mathbf{G}/H)$. This follows from the fact that H is open in G if and only if $\mathbf{G}/H$ is discrete.

9.2.2. Metrics on topological groups. Let $\mathbf{G}$ be a closed subgroup of $\mathrm{Sym}(\mathbb{N})$. Then the ultrametric d inherited from the Baire space (Example 9.1.9) is compatible with the pointwise convergence topology on G, and, in addition, is *left-invariant*, i.e., $d(gf_1, gf_2) = d(f_1, f_2)$ for all $g, f_1, f_2 \in G$.

PROPOSITION 9.2.11. *Let* $\xi \colon \mathbf{G} \to \mathbf{H}$ *be a continuous homomorphism between topological groups with compatible left-invariant metrics* d_1 *and* d_2. *Then* f *is uniformly continuous.*

PROOF. Let $\varepsilon > 0$. Since ξ is continuous, there exists a $\delta > 0$ such that for all $g \in G$ with $d_1(1^{\mathbf{G}}, g) < \delta$ we have $d_2(1^{\mathbf{H}}, \xi(g)) < \varepsilon$. Let $g_1, g_2 \in G$ be such that $d_1(g_1, g_2) < \delta$. Then $d_1(1^{\mathbf{G}}, g_1^{-1}g_2) < \delta$, and hence

$$d_2(\xi(g_1), \xi(g_2)) = d_2(1^{\mathbf{H}}, \xi(g_1)^{-1}\xi(g_2)) = d_2(1^{\mathbf{H}}, \xi(g_1^{-1}g_2)) < \varepsilon$$

which shows uniform continuity of ξ. ⊣

9.2.3. Closed subgroups. In this text, we are mostly interested in topological groups that arise as automorphism groups of countable structures. These groups can be characterised in topological terms. We have already seen in Section 9.2.2 that every closed subgroup of Sym($\mathbb{N}$) is Polish. But the group $(\mathbb{R}; +)$ with the usual topology on $\mathbb{R}$ is also a Polish group and certainly not a closed subgroup of Sym($\mathbb{N}$).

A topological group is called *non-archimedean* if it has a basis at the identity consisting of open subgroups. It is clear that Sym($\mathbb{N}$) is non-archimedean.

THEOREM 9.2.12. (Section 1.5 in [33]; also see Theorem 2.4.1 and Theorem 2.4.4 in [175]). *Let* $\mathbf{G}$ *be a topological group. Then the following are equivalent.*

1. $\mathbf{G}$ *is topologically isomorphic to the automorphism group of a countable relational structure.*
2. $\mathbf{G}$ *is topologically isomorphic to a closed subgroup of* Sym($\mathbb{N}$).
3. $\mathbf{G}$ *is Polish and admits a compatible left-invariant ultrametric.*
4. $\mathbf{G}$ *is Polish and non-archimedean.*
5. $\mathbf{G}$ *is Polish and has a countable basis* closed under left multiplication, *i.e., a countable basis* $\mathcal{B}$ *of* G *so that for any* $U \in \mathcal{B}$ *and* $g \in G$ *we have* $gU \in \mathcal{B}$.

PROOF. The equivalence of (1) and (2) has been shown in Proposition 4.2.2. For the implication from (2) to (3), we have already discussed the left-invariant ultrametric. Note that G is separable: for all finite tuples $\bar{a}, \bar{b}$ that lie in the same orbit we fix an element of G that maps $\bar{a}$ to $\bar{b}$; the (countable) set of all the selected elements of $\mathbf{G}$ is clearly dense in G. We have seen in Example 9.1.14 that G is also completely metrisable.

For the implication from (3) to (4), let d be a left-invariant ultrametric on G. Let $U_n = \{x \in G \mid d(x, 1) < 2^{-n}\}$, for $n \in \mathbb{N}$. We claim that the set of all those U_n forms a basis at the identity consisting of open subgroups. Since d is a left-invariant ultrametric, for $x, y \in G$ we have

$$d(x^{-1}y, 1) = d(y, x) \leq \max(d(y, 1), d(1, x))$$

and thus U_n is a indeed a subgroup.

For the implication from (4) to (5), assume (4). Let $\{U_1, U_2, \dots\}$ be an at most countable basis at the identity (which exists since G is metrizable). Each U_i has an open subset V_i which is a subgroup, since $\mathbf{G}$ has a basis at the identity consisting of open subgroups. Then $\{V_1, V_2, \dots\}$ is a countable basis of the identity consisting of open subgroups. Each V_i has at most countably many cosets since G is separable. So the set of all cosets of those groups gives an at most countable basis that is closed under left multiplication.

Finally, we show that (5) implies (2). Let $\mathcal{B} = \{U_1, U_2, \dots\}$ be a countable basis closed under left multiplication. We define the map $\xi \colon G \to \mathrm{Sym}(\mathbb{N})$ by setting

$$\xi(g)(n) = m \Leftrightarrow gU_n = U_m.$$

(If $|\mathcal{B}| = n_0$ is finite, we define the map $\xi \colon G \to \mathrm{Sym}(\mathbb{N})$ similarly, but set $\xi(g)(n) = n$ for all $n > n_0$.) It is straightforward to verify that $\xi(fg) = \xi(f)\xi(g)$. The mapping ξ is injective: when f and g are distinct, then there are disjoint open subsets U and V with $f \in U$ and $g \in V$, because the topology is Hausdorff; since $\mathcal{B}$ is a basis, we can assume that $U = U_{n_1}$ and $V = U_{n_2}$, for some $n_1, n_2 \geq 1$. If $fU_{n_1} = gU_{n_1}$, then $g \in U_{n_1} = U$ since $f \in U_{n_1}$, contradicting the assumption that U and V are disjoint. Hence, $\xi(f)(n_1) \neq \xi(g)(n_1)$, and so $\xi(f) \neq \xi(g)$. Since bijective algebra homomorphisms are isomorphisms, ξ is an isomorphism between $\mathbf{G}$ and a subgroup of $\mathrm{Sym}(\mathbb{N})$. To verify that ξ is continuous, let $g \in G$ be arbitrary, and let $V \subseteq \mathrm{Sym}(\mathbb{N})$ be an open set containing $\xi(g)$. Then V is a union of basic open sets of the form $V_{\bar{a},\bar{b}} := \{f \in \mathrm{Sym}(\mathbb{N}) \mid f(\bar{a}) = \bar{b}\}$ for some $\bar{a}, \bar{b} \in \mathbb{N}^n$. The preimage of $V_{\bar{a},\bar{b}}$ under ξ is $\{g \in G \mid gU_{a_1} = U_{b_1} \wedge \cdots \wedge gU_{a_n} = U_{b_n}\}$. Since multiplication in G is continuous, this set is open. Hence, the preimage of V is a union of open sets and therefore open as well, which concludes the proof that ξ is continuous.

It can also be verified that ξ is open, and that the image of ξ is closed; for the details of this last step, we refer to [175] (Theorem 2.4.4). ⊣

9.2.4. Open subgroups. Let $\mathbf{G}$ be a subgroup of $\mathrm{Sym}(B)$. If $n \in \mathbb{N}$ and $a \in B^n$, then G_a denotes the set of all elements of G that fix a; they form a subgroup of $\mathbf{G}$. Similarly, if $F \subseteq B$ then we write $G_{(F)}$ for the set of all elements of G that fix every $a \in F$. These subgroups are called *point stabilisers of* $\mathbf{G}$ (*at* a and *at* F, respectively).

LEMMA 9.2.13. *Let* $\mathbf{G}$ *be a subgroup of* $\mathrm{Sym}(\mathbb{N})$ *and* $\mathbf{U}$ *a subgroup of* $\mathbf{G}$. *Then the following are equivalent.*

1. *U is open in G;*
2. *U contains the point stabiliser of G at some finite subset of $\mathbb{N}$;*
3. *U contains an open subgroup of* $\mathbf{G}$.

PROOF. (1) ⇒ (2). Since U is open in G it must contain $S(a, b) \cap G$ for some $a, b \in \mathbb{N}^n$. Every element of G_a can be written as $\alpha\beta$ with $\alpha \in G \cap S(b, a) \subseteq U$ and $\beta \in G \cap S(a, b) \subseteq U$. Hence, U contains G_a.

(2) ⇒ (3). Trivial.

(3) ⇒ (1). Let **H** ben an open subgroup of **U**. Then $U = \bigcup_{\alpha \in U} \alpha H$. Since H and αH are open, it follows that U is open, too. ⊣

For a subset A of B, the *set stabiliser* G_A of **G** is the set of all $\alpha \in G$ that fix A setwise, that is, $\alpha A = A$.

LEMMA 9.2.14. *Let* **G** *be a subgroup of* Sym(B). *Then* **U** *is an open subgroup of* **G** *if and only if* $U = G_S$ *is the set stabiliser of a block* S *of the componentwise action of* **G** *on* B^n *for some* $n \in \mathbb{N}$.

PROOF. Let $S \subseteq B^n$ be a block of the componentwise action of **G** on B^n (viewed as a permutation group on B^n); let C be a congruence of this permutation group such that S is a congruence class of C. We first prove that G_S is open in **G**. Arbitrarily pick an $s \in S$. Let $\alpha \in G_s$ and $t \in S$. Then $(s, t) \in C$ and hence $(\alpha(s), \alpha(t)) \in C$. Since $\alpha(s) = s \in S$ we conclude that $\alpha(t) \in S$. So $\alpha \in G_S$ and $G_s \subseteq G_S$. Therefore, G_S contains an open subgroup and Lemma 9.2.13 implies that G_S is open.

Conversely, let **U** be an open subgroup of **G**. Then U must contain G_t for some $n \in \mathbb{N}$ and some $t \in B^n$, again using Lemma 9.2.13. We claim that $S := \{h(t) \mid h \in U\}$ is a block of the componentwise action of **G** on B^n. By Lemma 4.2.11 it suffices to verify that $g(S) = S$ or $g(S) \cap S = \emptyset$ for all $g \in \mathbf{G}$. Suppose that $g(S) \cap S \neq \emptyset$. Then there are $h_1, h_2 \in U$ such that $h_1(t) = g(h_2(t))$. Thus, $(h_1^{-1} \circ g \circ h_2) \in G_t$, and $g \in h_1 G_t h_2^{-1} \subseteq U$. But then $g(S) = S$, which concludes the proof that S is a block.

Now we verify that $U = G_S$. Let $g \in U$ and $s \in S$. Then $s = h(t)$ for some $h \in U$. Hence, $g(s) = gh(t) \in S$ by the definition of S, because $gh \in U$. Therefore, $U \subseteq G_S$. Now suppose that $g \in G_S$. Since g preserves S, we have $g(t) \in S$, and thus there exists an $h \in U$ with $hg(t) = t$. So $hg \in G_t \subseteq U$, and thus $g \in U$. ⊣

Lemma 9.2.14 has the following immediate consequence.

COROLLARY 9.2.15. *Every permutation group on a countable set has countably many open subgroups.*

9.2.5. Closed normal subgroups. Every open subgroup of a topological group is closed, but the converse is of course false. However, if we additionally require that the subgroup is *normal*, more can be said. A subgroup **N** of a group **G** is called *normal* if $gN = Ng$ for every $g \in G$. Recall the following equivalent characterisations of normality of subgroups, which can be seen as a refinement of Proposition 6.3.2 for the case of groups.

PROPOSITION 9.2.16. *Let* **G** *be a group, and* **N** *be a subgroup of* **G**. *Then the following are equivalent.*

1. **N** *is normal.*
2. **G** *has the congruence* $E = \{(a, b) \mid ab^{-1} \in N\}$.
3. *There is a homomorphism* h *from* **G** *to some group such that* $N = h^{-1}(0)$.
4. *For every* $g \in G$ *and every* $v \in N$ *we have* $gvg^{-1} \in N$.

The notion of a *quotient algebra* (Definition 6.3.3) of course also applies to groups. If E is a congruence of a group $\mathbf{G}$, then $\mathbf{G}/E$ is also called a *quotient group*. Proposition 9.2.16 shows that in the case of groups, every congruence E arises from a normal subgroup $\mathbf{N}$, and we then also use the notation $\mathbf{G}/N$ instead of $\mathbf{G}/E$; note that this is compatible with Definition 9.2.5, since the elements of the quotient group are the left cosets of $\mathbf{N}$ (which are the same as the right cosets of $\mathbf{N}$ since $\mathbf{N}$ is a normal subgroup). It can be shown that the quotient topology on $\mathbf{G}/N$ turns the quotient group into a topological group.

Congruences of $\mathbf{G}$ should not be confused with congruences of actions of $\mathbf{G}$; but if $\mathbf{G}$ is a closed subgroup of $\mathrm{Sym}(B)$, we can use the latter to understand the former.

PROPOSITION 9.2.17. *Let* $\mathbf{G}$ *be a closed subgroup of* $\mathrm{Sym}(B)$.

- *Let* C *be a congruence of the componentwise action of* $\mathbf{G}$ *on* B^n, *for some* n. *Then* $\bigcap_{S \in B^n/C} G_S$ *is a closed normal subgroup of* $\mathbf{G}$.
- *The closed normal subgroups of* $\mathbf{G}$ *are precisely the countable intersections of closed normal subgroups of the above form.*

PROOF. We first show that $U := \bigcap_{S \in B^n/C} G_S$ is a closed normal subgroup of $\mathbf{G}$. Let $\alpha \in G$ and $\beta \in U$, let $S \in \mathcal{S}$ and $s \in S$. Note that $(\alpha^{-1}s, \beta\alpha^{-1}s) \in C$ since β fixes the equivalence classes of C. Hence, $(s, \alpha\beta\alpha^{-1}s)$ are in the same equivalence classes since α preserves C. It follows that $\alpha \circ \beta \circ \alpha^{-1}$ preserves each equivalence class of C, and thus is in U. Normality of U then follows from Proposition 9.2.16. Also note that U is closed as an intersection of closed sets.

For the second statement, let $\mathbf{N}$ be a closed normal subgroup of $\mathbf{B}$ and for each $n \in \mathbb{N}$, consider the relation

$$R_n := \{(x, y) \mid x, y \in B^n \text{ and there is } \beta \in N \text{ such that } \beta x = y\}.$$

This relation is obviously an equivalence relation, and it is preserved by all permutations in G. For this, we have to show that for all $\alpha \in G$ and all $(x, y) \in R_n$ we have that $(\alpha x, \alpha y) \in R_n$. So suppose that $x, y \in B^n$ such that $\beta x = y$ for some $\beta \in N$. Then

$$\alpha y = \alpha\beta x \in (\alpha N)x = (N\alpha)x$$

by the normality of N. Hence there exists a $\beta' \in N$ such that $\beta'\alpha x = \alpha y$, which shows that $(\alpha x, \alpha y) \in R_n$.

For $n \geq 1$, let $\mathcal{S}_n$ be the set of equivalence classes of R_n and define

$$U := \bigcap_{n \geq 1} \bigcap_{S \in \mathcal{S}_n} G_S;$$

we have to show that $U = N$. For every $n \leq 1$, every $\beta \in N$ preserves every $S \in \mathcal{S}_n$, and hence $N \subseteq \bigcap_{S \in \mathcal{S}_n} G_S$. For the converse inclusions, let $\alpha \in U$, and let x, y be from B^n so that $\alpha x = y$. Since α preserves the equivalence classes of R_n, there exists an $\beta \in N$ such that $\beta x = y$. Hence, $\alpha \in \overline{N}$, which implies that $g \in N$ since N is closed by assumption. ⊣

We illustrate Proposition 9.2.17 with an example.

Example 9.2.18. Let $\text{Betw} = \{(x, y, z) \in \mathbb{Q}^3 \mid (x < y < z) \vee (z < y < x)\}$ be the Betweenness Relation on $\mathbb{Q}$. Then $\text{Aut}(\mathbb{Q}; \text{Betw})$ is 2-transitive and therefore primitive. However, the relation

$$\{((x_1, x_2), (y_1, y_2)) \mid (x_1 < x_2 \wedge y_1 < y_2) \vee (x_1 > x_2 \wedge y_1 > y_2) \vee (x_1 = x_2 \wedge y_1 = y_2)\}$$

is an invariant equivalence relation on $\mathbb{Q}^2$. And indeed, $\text{Aut}(\mathbb{Q}; \text{Betw})$ has the closed normal subgroup $\text{Aut}(\mathbb{Q}; <)$, and $\text{Aut}(\mathbb{Q}; \text{Betw}) / \text{Aut}(\mathbb{Q}; <)$ has two elements, corresponding to the automorphisms that reverse the order $<$ and the automorphisms that preserve the order.

9.2.6. Reconstruction of topology. A surprising amount of information about the topology of the automorphism group of a countable structure $\mathfrak{B}$ may be coded into $\text{Aut}(\mathfrak{B})$ viewed as an abstract group. For example, $\text{Sym}(B)$ has only two separable group topologies, namely the trivial one and the Polish topology from Example 9.2.1 (Kechris and Rosendal [230], Theorem 6.26). In this section we discuss the question whether we can reconstruct the topology of closed subgroups of $\text{Sym}(\mathbb{N})$ from the abstract group. Section 9.4.2 discusses the analogous problem for polymorphism clones instead of automorphism groups, which is relevant for the complexity of constraint satisfaction problems with ω-categorical templates.

DEFINITION 9.2.19. Let $\mathbf{G}$ be a closed subgroup of $\text{Sym}(\mathbb{N})$. We say that

- $\mathbf{G}$ has *automatic continuity* iff every homomorphism from $\mathbf{G}$ to $\text{Sym}(\mathbb{N})$ is continuous;
- $\mathbf{G}$ has *automatic homeomorphicity* iff every group isomorphism between $\mathbf{G}$ and a closed subgroup of $\text{Sym}(\mathbb{N})$ is a homeomorphism;
- $\mathbf{G}$ is *reconstructible* (or $\mathbf{G}$ has *reconstruction*) iff for every other closed subgroup $\mathbf{H}$ of $\text{Sym}(\mathbb{N})$, if there exists an isomorphism between $\mathbf{H}$ and $\mathbf{G}$, then there also exists a group isomorphism between $\mathbf{H}$ and $\mathbf{G}$ which is a homeomorphism.

Automatic continuity implies automatic homeomorphicity (Corollary 2.8 in [258]), and clearly automatic homeomorphicity implies reconstruction. There are two dominant methods for proving that a group is reconstructible. The first method is via showing the *small index property* and the second is based on Mati Rubin's *forall-exists interpretations*. We have seen an example of a closed oligomorphic permutation group without automatic continuity in Example 9.2.7. This example has still reconstruction; this can be shown using the results of Rubin (see Remark 5.4.3 in [270]). A more involved example of a closed oligomorphic permutation group without reconstruction has been found by Evans and Hewitt [167].

Recall from Lemma 9.2.13 that a subgroup $\mathbf{G}$ of $\mathrm{Sym}(\mathbb{N})$ is open if it contains the point stabiliser $G_{(A)}$ for some finite $A \in \mathbb{N}$. Clearly, these groups have countable index, so all open subgroups of $\mathrm{Sym}(\mathbb{N})$ have countable index. The situation that the converse holds as well deserves a name.

DEFINITION 9.2.20. A topological group $\mathbf{G}$ has the *small index property* if every subgroup of $\mathbf{G}$ of at most countable index is open.

Some authors define the small index property slightly differently: they require that every subgroup of $\mathbf{G}$ of cardinality less than 2^{ω} is open. There is no example known of closed oligomorphic permutation group where the two definitions differ [270]. We have chosen our formulation essentially because of the following proposition.

PROPOSITION 9.2.21 (Folklore). *Let* $\mathbf{G}$ *be a closed subgroup of* $\mathrm{Sym}(\mathbb{N})$. *Then* $\mathbf{G}$ *has automatic continuity if and only if it has the small index property.*

PROOF. Suppose that $\mathbf{G}$ has automatic continuity and let $\mathbf{U}$ be a subgroup of $\mathbf{G}$ of at most countable index. We have to show that $\mathbf{U}$ is open. Let $\xi \colon \mathbf{G} \to \mathbf{G}/U$ be the action of $\mathbf{G}$ on the left cosets of $\mathbf{U}$ in $\mathbf{G}$ by left translation (Example 9.2.6), where $\mathbf{G}/U$ is equipped with the discrete topology. By automatic continuity, ξ is continuous. In particular, the pre-image of the open set $\{\alpha \in \mathrm{Sym}(\mathbf{G}/U) \mid \alpha(U) = U\}$ is open. But this pre-image is precisely U, which proves the small index property.

Now suppose that $\mathbf{G}$ has the small index property, and let ξ be an isomorphism between $\mathbf{G}$ and another closed subgroup of $\mathbf{H}$ of $\mathrm{Sym}(\mathbb{N})$. By Lemma 9.2.2, it suffices to prove continuity at 1. Note that the basic open subsets of $\mathbf{H}$ that contain 1 are of the form $\mathbf{H}_a$ for $a \in \mathbb{N}^n$, $n \in \mathbb{N}$. The subgroup $\mathbf{H}_a$ of $\mathbf{H}$ has countable index, and therefore $\xi^{-1}(\mathbf{H}_a)$ is a subgroup of $\mathbf{G}$ of countable index, too, and hence open by assumption. This establishes continuity of ξ. ⊣

The small index property has been verified for the following groups:

1. $\mathrm{Sym}(\mathbb{N})$ [311, 328, 151];
2. the automorphism groups of countable vector spaces over finite fields [166];
3. all automorphism groups of ω-categorical ω-stable structures [208];
4. $\mathrm{Aut}(\mathbb{Q};<)$ [341];
5. the automorphism group of the atomless Boolean algebra [341];
6. the automorphism group of the ω-categorical dense semilinear order giving rise to a meet-semilattice [157];
7. the automorphism group of the random graph [208];
8. the automorphism groups of the Henson graphs [201].

On the other hand, the small index property is not known for the automorphism group of the countable universal homogeneous tournament , the countable universal poset, or the random permutation (see [270]). But it is known to be consistent with Zermelo-Fraenkel set theory (even in the presence of the axiom of dependent choice) that every closed subgroup of $\mathrm{Sym}(\mathbb{N})$ has automatic continuity [258].

9.3. Oligomorphic groups

In the previous section we have seen conditions that describe when a topological group is topologically isomorphic to a closed subgroup of $\mathrm{Sym}(\mathbb{N})$. In this section, we give conditions that characterise the topological groups which are topologically isomorphic to an *oligomorphic* closed subgroup of $\mathrm{Sym}(\mathbb{N})$.

A topological group $\mathbf{G}$ is called *Roelcke precompact* if for every open set $U \subseteq G$ that contains the identity there exists a finite set $F \subseteq G$ such that $G = UFU$. The following theorem is essentially from Tsankov [342]; there, the focus has been a characterisation of Roelcke precompact groups in terms of oligomorphic permutation groups. Here, on the other hand, the focus will be the characterisation of oligomorphic permutation groups in terms of Roelcke precompact ones, and this motivates the following formulation of Tsankov's theorem[25].

THEOREM 9.3.1. *Let* $\mathbf{G}$ *be topologically isomorphic to a closed subgroup of* $\mathrm{Sym}(\mathbb{N})$. *Then the following are equivalent.*

1. $\mathbf{G}$ *is the automorphism group of a countably infinite* ω*-categorical structure.*
2. $\mathbf{G}$ *is Roelcke precompact, and* $\mathbf{G}$ *has an open subgroup* V *of countably infinite index such that for all open subgroups* U *of* $\mathbf{G}$ *there are* $g_1, \dots, g_n \in G$ *such that* $\bigcap_{i \leq n} g_i V g_i^{-1} \subseteq U$.
3. $\mathbf{G}$ *has a faithful transitive continuous action on* $\mathbb{N}$, *equipped with the discrete topology, whose image is closed in* $\mathrm{Sym}(\mathbb{N})$, *and every such action of* $\mathbf{G}$ *is oligomorphic.*
4. $\mathbf{G}$ *is the automorphism group of a countably infinite* ω*-categorical structure with only one orbit.*

PROOF. The implication from (4) to (1) is trivial, and we prove (1) $\Rightarrow$ (2) $\Rightarrow$ (3) $\Rightarrow$ (4). For the implication from (1) to (2), suppose that $\mathbf{G}$ is the automorphism group of an ω-categorical structure $\mathfrak{B}$, and let G be the domain of $\mathbf{G}$, which is a set of permutations of the domain B of $\mathfrak{B}$. Since $\mathfrak{B}$ is ω-categorical, $\mathrm{Aut}(\mathfrak{B})$ has a finite number k of orbits by Theorem 4.1.6; choose orbit representatives $b_1, \dots, b_k \in B$, and write $\bar{b}$ for $(b_1, \dots, b_k)$. Then the stabiliser $V := G_{\bar{b}}$ is an open subgroup of $\mathbf{G}$ of countably infinite index. Let U be an arbitrary open subgroup of $\mathbf{G}$. Then U contains $G_{\bar{a}}$ for some $\bar{a} \in B^n$. For $j \leq n$, let $g_j \in G$ be such that $g_j(a_j) = b$ where $b \in \{b_1, \dots, b_k\}$

[25] I am grateful to Todor Tsankov for his help with the presented reformulation of his result.

is from the same orbit as a_j. We claim that $K := \bigcap_{j \leq n} g_j^{-1} V g_j \subseteq U$. To see this, let $h \in K$ be arbitrary. Since $h \in g_j^{-1} V g_j$ we find that $h(a_j) = a_j$. Hence, $h \in G_{\bar{a}} \subseteq U$.

To show that $\mathbf{G}$ is Roelcke precompact, let $U \subseteq G$ be open with $1 \in U$. Then there exists an n such that U contains the stabiliser $G_{\bar{a}}$ for an n-tuple $\bar{a}$ of elements of B. It suffices to show the existence of a finite number of elements $g_1, \ldots, g_k$ of G such that $G = \bigcup_{i \leq k} G_{\bar{a}} g_i G_{\bar{a}}$. By Theorem 4.1.6, G has finitely many orbits of $2n$-tuples; so let $(\bar{a}, g_1 \cdot \bar{a}), \ldots, (\bar{a}, g_k \cdot \bar{a})$ be a complete list of representatives for those orbits of $2n$-tuples that are contained in $G \cdot \bar{a} \times G \cdot \bar{a}$. We claim that $G_{\bar{a}} g_1 G_{\bar{a}} \cup \cdots \cup G_{\bar{a}} g_k G_{\bar{a}} = G$. Let $f \in G$ be arbitrary. Let $i \leq k$ be such that $(\bar{a}, f \cdot \bar{a})$ and $(\bar{a}, g_i \cdot \bar{a})$ lie in the same orbit of n-tuples under $G_{\bar{a}}$. So there exists an $h \in G_{\bar{a}}$ such that $f \cdot \bar{a} = h g_i \cdot \bar{a}$. Then $f^{-1} \circ h \circ g_i$ lies in $G_{\bar{a}}$, so $f \in G_{\bar{a}} g_i G_{\bar{a}}$ as required.

(2) $\Rightarrow$ (3). Since V is open, $\mathbf{G}/V$ has the discrete topology, and B is countably infinite by assumption. The action of $\mathbf{G}$ on $\mathbf{G}/V$ by left translation is continuous and transitive (Example 9.2.6) and its image is closed in $\mathrm{Sym}(\mathbf{G}/V)$ (Example 9.2.10). We show by induction that this action has only finitely many orbits of n-tuples for all $n \in \mathbb{N}$. Since the action is transitive, this is true for $n = 1$. For the induction step, fix $\bar{a} = (a_1, \ldots, a_n) \in (\mathbf{G}/V)^n$, and let c be an arbitrary element from $(\mathbf{G}/V) \setminus \{a_1, \ldots, a_n\}$. Since $\mathbf{G}$ is Roelcke precompact, there exists a finite set $\{f_1, \ldots, f_k\} \subseteq G$ such that $G = G_{\bar{a}c} f_1 G_{\bar{a}c} \cup \cdots \cup G_{\bar{a}c} f_k G_{\bar{a}c}$. Let $B(\bar{a})$ be $\{f_1 \cdot c, \ldots, f_k \cdot c\}$.

Claim 1. For every $d \in (\mathbf{G}/V) \setminus \{a_1, \ldots, a_n\}$ there is an $h \in G_{\bar{a}}$ and $b \in B(\bar{a})$ such that $d = h \cdot b$. By transitivity of G, there is a $g \in G$ so that $d = g \cdot c$, for arbitrary $d \in (\mathbf{G}/V) \setminus \{a_1, \ldots, a_n\}$. Let i, h_1, h_2 be such that $h_1, h_2 \in G_{\bar{a}c}$ and $g = h_1 f_i h_2$. Then $d = gc = h_1 f_i h_2 \cdot c = h_1 f_i \cdot c$, proving Claim 1.

Claim 2. When $\{\bar{a}_1, \ldots, \bar{a}_s\}$ is a complete set of representatives for the orbits of n-tuples under the permutation group G, then

$$\{(\bar{a}_i, b) \mid i \in [s], b \in B(\bar{a}_i)\}$$

is a complete set of representatives for the orbits of $(n+1)$-tuples. Let $(\bar{c}, d) \in (\mathbf{G}/V)^{n+1}$. By assumption there exists $g \in G$ such that $g \cdot \bar{a}_i = \bar{c}$. Find $h \in G_{\bar{a}_i}$ and $b \in B(\bar{a}_i)$ such that $g^{-1} \cdot d = h \cdot b$. Then

$$gh \cdot (\bar{a}_i, b) = g \cdot (\bar{a}_i, h \cdot b) = (\bar{c}, d)$$

shows that $\mathbf{G}$ has finitely many orbits of $(n+1)$-tuples and concludes the induction step.

The implication from (3) to (4) follows from Corollary 4.2.10. Suppose that ξ is a continuous transitive action of $\mathbf{G}$ on $\mathbb{N}$ whose image is closed in $\mathrm{Sym}(\mathbb{N})$. Then the image of ξ is a closed oligomorphic subgroup of $\mathrm{Sym}(\mathbb{N})$, and hence the automorphism group of an ω-categorical relational structure with domain $\mathbb{N}$. ⊣

Note that closed oligomorphic permutation groups on countably infinite sets must always have continuum cardinality; this follows from the following theorem.

THEOREM 9.3.2 (cf. Corollary 4.1.5 in [207]). *Let G be a closed subgroup of* $\mathrm{Sym}(\mathbb{N})$. *Then the following are equivalent.*

1. *There is a finite $A \subseteq \mathbb{N}$ such that $|G_{(A)}| = 1$;*
2. $|G| \leq \omega$;
3. $|G| < 2^{\omega}$.

SKETCH OF PROOF. The implications from (1) to (2) and from (2) to (3) are trivial. For the implication from (3) to (1), suppose that $\neg(1)$. Construct a binary branching tree with the levels indexed by $\mathbb{N}$; since G is closed in $\mathrm{Sym}(\mathbb{N})$ this can be done in such a way that the infinite branches of the tree correspond to pairwise distinct elements of G; we deduce that $\neg(3)$ since there are uncountably many infinite branches of the tree. ⊣

Note that if G is an oligomorphic subgroup of $\mathrm{Sym}(\mathbb{N})$, then for every finite $A \subseteq \mathbb{N}$ the point stabiliser $G_{(A)}$ is oligomorphic (Lemma 4.7.3, Theorem 4.1.6), and hence item (1) in Theorem 9.3.2 does not apply. Hence, for every oligomorphic permutation group on a countably infinite set there is a bijection between the group elements and $\mathrm{Sym}(\mathbb{N})$. This bijection can be chosen to be a homeomorphism; we prove something even stronger in Proposition 9.3.3. A bijection ξ between two metric spaces is called a *uniform homeomorphism* if ξ and ξ^{-1} are uniformly continuous.

PROPOSITION 9.3.3. *Every closed oligomorphic subgroup G of* $\mathrm{Sym}(\mathbb{N})$ *is uniformly homeomorphic to* $\mathrm{Sym}(\mathbb{N})$ (*both spaces are equipped with the metric d inherited from the Baire space*).

PROOF. Let $\mathbb{N}^*$ be the set of words over the alphabet $\mathbb{N}$, i.e., the set of finite tuples of natural numbers. If $k \in \mathbb{N}$ and $w \in \mathbb{N}^k \subseteq \mathbb{N}^*$ then we write $|w|$ for k and ε for the unique element $w \in \mathbb{N}^*$ with $|w| = 0$. A word $w' \in \mathbb{N}^*$ is a *prefix* of $w \in \mathbb{N}^*$ if $w'(i) = w(i)$ for all $i \in \{1, \ldots, |w'|\}$. Let

$$S := \{w \in \mathbb{N}^* \mid \text{there exists } \alpha \in G \text{ such that } \alpha(0, \ldots, |w| - 1) = w\}.$$

We first construct an injection $f : \mathbb{N}^* \to S$ with the property that if w' is a prefix of w, then $f(w')$ is a prefix of $f(w)$. The uniform homeomorphism $\xi \colon \mathrm{Sym}(\mathbb{N}) \to G$ that we are going to construct afterwards will then have the property that for every $k \in \mathbb{N} \setminus \{0\}$

$$\xi(\alpha)(0, \ldots, |f(0 \cdots k)| - 1) = f\big(\alpha(0) \cdots \alpha(k-1)\big). \tag{1}$$

Since G is oligomorphic some elements of $\mathbb{N}$ must lie in infinite orbits under G. Let $n_\varepsilon \in \mathbb{N}$ be smallest such that the orbit of n_ε under G is infinite. Pick a bijection f_ε between $\mathbb{N}$ and the (infinite) orbit of $(0, \dots, n_\varepsilon)$ under G. We define $f(\varepsilon) := \varepsilon$ and for every $\ell \in \mathbb{N}$ we define $f(\ell) := f_\varepsilon(\ell)$.

Now suppose that inductively we have already defined f for $w \in \mathbb{N}^*$; we want to define f for words of the form $w\ell$ for $\ell \in \mathbb{N}$. Since $f(w) \in S$ there exists $\alpha_w \in G$ such that $\alpha_w(0, \dots, |f(w)| - 1) = f(w)$. The permutation group $G_{(0,\dots,|f(w)|)}$ is oligomorphic, and hence it must have elements in infinite orbits. Let $n_w \in \mathbb{N}$ be smallest so that the orbit of n_w under $G_{(0,\dots,|f(w)|)}$ is infinite; note that $n_w \geq |w|$. Fix a bijection f_w between $\mathbb{N}$ and the (infinite) orbit of $(0, 1, \dots, n_w)$ under $G_{(0,\dots,|f(w)|)}$. Define $f(w\ell) := \alpha_w f_w(\ell)$. We prove that $f(w)$ is a prefix of $f(w\ell)$. First note that

$$(0, \dots, |f(w)|) = (f_w(\ell)_0, \dots, f_w(\ell)_{|f(w)|}) \tag{2}$$

since $(0, \dots, n_w)$ and $f_w(\ell)$ lie in the same orbit under $G_{(0,\dots,|f(w)|)}$. For $i \leq |f(w)|$, we have

$$\begin{aligned}
f(w\ell)_i &= \big(\alpha_w f_w(\ell)\big)_i && \text{(by the definition of } f(w\ell))\\
&= \alpha_w(f_w(\ell)_i)\\
&= \alpha_w\big((0, \dots, |f(w)|)_i\big) && \text{(by (2))}\\
&= f(w)_i && \text{(by the definition of } \alpha_w).
\end{aligned}$$

It is straightforward to verify that f is injective.

To define $\xi\colon \mathrm{Sym}(\mathbb{N}) \to G$, let $\beta \in \mathrm{Sym}(\mathbb{N})$. Since G is a closed subset of $\mathrm{Sym}(\mathbb{N})$, the observations above imply that the sequence $(\alpha_{(\beta(0),\dots,\beta(n))})_{n\in\mathbb{N}}$ converges to an element $\alpha \in G$, and we define $\xi(\beta) := \alpha$.

Claim 1. ξ is injective. If $\beta_1, \beta_2 \in \mathrm{Sym}(\mathbb{N})$ are distinct, then there exists a smallest $n \in \mathbb{N}$ such that $\beta_1(n) \neq \beta_2(n)$. Then $f(\beta_1(0), \dots, \beta_1(n)) \neq f(\beta_2(0), \dots, \beta_2(n))$, which in turn implies that $\alpha_{(\beta_1(0),\dots,\beta_1(n))} \neq \alpha_{(\beta_2(0),\dots,\beta_2(n))}$ and thus $\xi(\beta_1) \neq \xi(\beta_2)$.

Claim 2. ξ is surjective. Let $\alpha \in G$. We claim that for every $k \in \mathbb{N} \setminus \{0\}$ there exists $w^k \in \mathbb{N}^k$ such that

- $f(w^k) = \alpha(0) \cdots \alpha(|f(w^k)| - 1)$, and
- if $k \geq 2$ then w^{k-1} is a prefix of w^k.

We prove this by induction on k. For $k = 1$, there exists an $\ell \in \mathbb{N}$ such that $f(\ell) = f_\varepsilon(\ell) = \alpha(0) \cdots \alpha(n_\varepsilon)$ since f_ε is a bijection between $\mathbb{N}$ and the orbit of $(0, \dots, n_\varepsilon)$ under G. For the inductive step we assume that there exists $w^k \in \mathbb{N}^k$ such that $f(w^k) = \alpha(0) \cdots \alpha(|f(w^k)| - 1)$. Note that $\alpha_{w^k}^{-1}\alpha(i) = i$ for all $i \in \{0, \dots, |f(w^k)| - 1\}$. Since f_{w^k} is a bijection between $\mathbb{N}$ and the orbit of $(0, \dots, n_{w^k})$ under $G_{(0,\dots,|f(w^k)|)}$, there exists an ℓ so that $f_{w^k}(\ell) = \alpha_{w^k}^{-1}\alpha(0) \cdots \alpha_{w^k}^{-1}\alpha(n_{w^k})$. Then

$$f(w^k\ell) = \alpha_{w^k} f_{w^k}(\ell) = \alpha(0) \cdots \alpha(n_{w^k})$$

so $w^{k+1} := w^k\ell$ satisfies the requirement for the inductive step.

Observe that $(w^k)_{k \in \mathbb{N}}$ converges to an element $\beta \in \mathrm{Sym}(\mathbb{N})$ and that $(\alpha_{w^k})_{k \in \mathbb{N}}$ converges to α, so $\xi(\beta) = \alpha$, proving the surjectivity of ξ.

Claim 3. ξ is uniformly continuous. Let $F \subseteq \mathbb{N}$ be finite. Let $k := \max(F)$ and suppose that $\beta_1, \beta_2 \in \mathrm{Sym}(\mathbb{N})$ are such that $\beta_1(0, \dots, k) = \beta_2(0, \dots, k) =: t$. Then $\xi(\beta_1)(0, \dots, |f(t)|) = \xi(\beta_2)(0, \dots, |f(t)|)$, proving uniform continuity of ξ.

Claim 4. ξ^{-1} is uniformly continuous. If $F \subseteq \mathbb{N}$ is finite, and $\beta_1, \beta_2 \in \mathrm{Sym}(\mathbb{N})$ are such that $\xi(\beta_1)$ and $\xi(\beta_2)$ agree on the entries of $f(0 \cdots \max(F))$, then β_1 and β_2 agree on F. ⊣

9.4. Topological clones

DEFINITION 9.4.1. A *topological clone* is a clone

$$\mathbf{C} = (C^{(1)}, C^{(2)}, \dots; (\mathrm{pr}_i^k)_{1 \le i \le k}, (\mathrm{comp}_l^k)_{1 \le k,l})$$

together with a topology on $C = \bigcup_i C^{(i)}$ such that each $C^{(n)}$ is closed and open in C and such that the composition operations are continuous.

A *topological isomorphism* between topological clones is an isomorphism between the respective abstract clones which is also a homeomorphism.

Example 9.4.2. Let $\mathcal{C} \subseteq \mathcal{O}_B$ be an operation clone over the domain B. Then $\mathcal{C}$ is a topological clone with respect to the topology on $\mathcal{C}$ inherited from the topology on $\mathcal{O}_B$ introduced in Example 9.1.7 (similarly as for permutation groups, see Example 9.2.1).

In this text, when we view operation clones as topological clones, this will always be meant with respect to the topology in Example 9.4.2.

9.4.1. Metrics on clones. Let $\mathbf{C}$ be a topological clone and d a compatible metric on C. We say that d is *left non-expansive* if for all $f \in C^{(k)}$ and $g_1, h_1, \dots, g_k, h_k \in C^{(l)}$ we have that

$$d(\mathrm{comp}_l^k(f, g_1, \dots, g_k), \mathrm{comp}_l^k(f, h_1, \dots, h_k)) \le \max(d(g_1, h_1), \dots, d(g_k, h_k)).$$

We say that d is *projection right invariant* if for all $k \in \mathbb{N}$, $f, g \in C^{(k)}$, and pairwise distinct $i_1, \dots, i_k \in \{1, \dots, l\}$ we have

$$d(f, g) = d(\mathrm{comp}_l^k(f, \mathrm{pr}_{i_1}^l, \dots, \mathrm{pr}_{i_k}^l), \mathrm{comp}_l^k(g, \mathrm{pr}_{i_1}^l, \dots, \mathrm{pr}_{i_k}^l)).$$

The box metric from on $\mathcal{O}_{\mathbb{N}}^{(k)} = (\mathbb{N}^k \to \mathbb{N})$, defined in Example 9.1.10 can be extended to all of $\mathcal{O}_{\mathbb{N}}$ by defining $d(f, g) := 1$ if $f, g \in \mathcal{O}_{\mathbb{N}}$ have distinct arities. It is straightforward to verify that this metric is compatible with the topology of pointwise convergence on $\mathcal{O}_{\mathbb{N}}$ (Example 9.2.1), and that it is complete, left non-expansive, and projection right invariant. In the following, when we work with operation clones $\mathcal{C}$ on countable base sets and refer to concepts the require a metric, such as uniform continuity, we refer to box metric after identifying the countable base set with $\mathbb{N}$.

Theorem 9.2.12 for topological groups has the following topological clone analog.

THEOREM 9.4.3 (Theorem 5.4 in [97]). *Let* $\mathbf{C}$ *be a topological clone. The following are equivalent.*

1. $\mathbf{C}$ *is isomorphic to the polymorphism clone of a countable (homogeneous and relational) structure.*
2. $\mathbf{C}$ *is isomorphic to a closed subclone of* $\mathcal{O}_{\mathbb{N}}$.
3. $\mathbf{C}$ *is separable and admits a compatible complete left non-expansive and projection right invariant ultrametric.*

PROOF. The equivalence between (1) and (2) has already been shown in Corollary 6.1.6. For the implication from (2) to (3), separability of $\mathcal{O}_{\mathbb{N}}$, and hence also of the topology on C has already been noted in Example 9.1.7. The box metric from Example 9.1.10 is a compatible complete left non-expansive and projection right invariant ultrametric on $\mathcal{O}_{\mathbb{N}}$.

(3) $\Rightarrow$ (4). Let d be a compatible left non-expansive and projection right invariant ultrametric of $\mathbf{C}$. Note that for every $n \in \mathbb{N}$ the relation

$$U_n := \{(x, y) \in C^2 \mid d(x, y) < 1/n\}$$

is an equivalence relation. Let $N_n := C/U_n$ and let

$$N := \bigcup_{n \in \mathbb{N}} (N_n \times \{n\}).$$

Define $\xi \colon C \to \mathcal{O}_N$ as follows. Let $k \in \mathbb{N}$, $f \in C^{(k)}$, $e_1, \dots, e_k \in N$, and for $i \in \{1, \dots, k\}$ let $l_i \in \mathbb{N}$ and $g_i \in C^{(l_i)}$ be such that $e_i = (g_i/U_{n_i}, n_i)$ for $n_1, \dots, n_k \in \mathbb{N}$. Let $n := \min(n_1, \dots, n_k)$ and $l := \max(l_1, \dots, l_k)$. Define

$$\xi(f)(e_1, \dots, e_k) := \big(f(g_1(\mathrm{pr}^l_1, \dots, \mathrm{pr}^l_{l_1}), \dots, g_k(\mathrm{pr}^l_1, \dots, \mathrm{pr}^l_{l_k}))/U_n, n\big).$$

This is well-defined: for $i \in \{1, \dots, k\}$ let $h_i \in C$ be such that $(g_i, h_i) \in U_{n_i}$. Then $(g_i(\mathrm{pr}^l_1, \dots, \mathrm{pr}^l_{l_i}), h_i(\mathrm{pr}^l_1, \dots, \mathrm{pr}^l_{l_i})) \in U_{n_i} \subseteq U_n$ since d is projection right invariant. Since U_n is left invariant is contains the pair

$$\begin{aligned}(f(g_1(\mathrm{pr}^l_1, \dots, \mathrm{pr}^l_{l_1}), \dots, g_k(\mathrm{pr}^l_1, \dots, \mathrm{pr}^l_{l_k})),\\ f(h_1(\mathrm{pr}^l_1, \dots, \mathrm{pr}^l_{l_1}), \dots, h_k(p^l_1, \dots, \mathrm{pr}^l_{l_k})).\end{aligned}$$

Claim 1. The map ξ is a homomorphism. Let $f \in C^{(k)}$ and $g_1, \dots, g_k \in C^{(l)}$. Let $n_1, \dots, n_l \in \mathbb{N}$ and $h_1 \in C^{(m_1)}, \dots, h_l \in C^{(m_l)}$. For $i \in \{1, \dots, l\}$ define $e_i := (h_i/U_{n_i}, n_i) \in N$. Let $m := \max(m_1, \dots, m_l)$ and let $n :=$

$\min(n_1, \ldots, n_l)$. We write h_i' for $h_i(\mathrm{pr}_1^m, \ldots, \mathrm{pr}_{m_i}^m)$. Then

$$\begin{aligned}&\xi(f(g_1, \ldots, g_k))(e_1, \ldots, e_l)\\&= (f(g_1, \ldots, g_n)(h_1', \ldots, h_l')/U_n, n)\\&= (f(g_1(h_1', \ldots, h_l'), \ldots, g_k(h_1', \ldots, h_l')))/U_n, n)\\&= \xi(f)((g_1(h_1', \ldots, h_l')/U_n, n), \ldots, (g_k(h_1', \ldots, h_l')/U_n, n))\\&= \xi(f)(\xi(g_1)(e_1, \ldots, e_n), \ldots, \xi(g_l)(e_1, \ldots, e_n))\end{aligned}$$

and

$$\xi(\mathrm{pr}_i^l)(e_1, \ldots, e_l) = (\mathrm{pr}_i^l(h_1, \ldots, h_l)/U_n, n) = (h_i/U_n, n) = e_i$$

and thus, $\xi(\mathrm{pr}_i^l)$ is mapped to the i-th l-ary projection in $\mathcal{O}_N$.

Claim 2. The map ξ is injective. Let $f_1, f_2 \in C$ be distinct. As the topology on C is Hausdorff there exists an $n \in \mathbb{N}$ such that $(f_1, f_2) \notin U_n$. Therefore,

$$\begin{aligned}\xi(f_1)((g_1/U_n, n), \ldots, (g_k/U_n, n)) &= f_1(g_1, \ldots, g_k)/U_n\\&\neq f_2(g_1, \ldots, g_k)/U_n = \xi(f_2)((g_1/U_n, n), \ldots, (g_k/U_n, n)).\end{aligned}$$

Claim 3. The map ξ is continuous. Let $f \in C^{(k)}$ be arbitrary and let $E \subseteq N$ be finite. Define $W_{f,E} := \{g \in \xi(C) \mid g(e) = \xi(f)(e) \text{ for all } e \in E^k\}$, and observe that the sets of the form $W_{f,E}$ form a basis for the topology on $\xi(C)$ induced by $\mathcal{O}_N$. Then

$$\begin{aligned}\xi^{-1}(W_{f,E}) &= \{g \in C \mid \xi(g)(e) = \xi(f)(e) \text{ for all } e \in E^k\}\\&= \bigcap_{e_1,\ldots,e_k \in E} \{g \in C \mid \xi(g)(e_1, \ldots, e_k) = \xi(f)(e_1, \ldots, e_k)\}.\end{aligned}$$

Let $e_1, \ldots, e_k \in E$ be arbitrary and let $h_i \in C^{(l_i)}$ be such that $e_i = (h_i/U_{n_i}, n_i)$. Let $n := \min(n_1, \ldots, n_k)$ and $l := \max(l_1, \ldots, l_k)$. Let $\rho_{h_1,\ldots,h_k} : C^{(k)} \to C^{(l)}$ be defined by

$$\rho_{h_1,\ldots,h_k}(f) := f(h_1(\mathrm{pr}_1^l, \ldots, \mathrm{pr}_{l_1}^l), \ldots, h_k(\mathrm{pr}_1^l, \ldots, \mathrm{pr}_{l_1}^l)).$$

Now,

$$\begin{aligned}S &:= \{g \in C \mid \xi(g)(e_1, \ldots, e_k) = \xi(f)(e_1, \ldots, e_k)\}\\&= \{g \in C^{(k)} \mid \rho_{h_1,\ldots,h_k}(g) \in \rho_{h_1,\ldots,h_k}(f)/U_n\}\\&= \rho_{h_1,\ldots,h_k}^{-1}(f/U_n).\end{aligned}$$

The set f/U_n is open because d is a compatible metric. As composition in topological clones is continuous, ρ is continuous, and therefore S is open, too. Hence, $\xi^{-1}(W_{f,E})$ is a finite intersection of open sets, and open. We have thus shown that for every open subset V of $\xi(C)$ containing $\xi(f)$ there is an open subset of C that contains f and whose image is contained in V, that is, ξ is continuous at f. The claim follows from Proposition 9.1.1.

Claim 4. The map ξ^{-1} is uniformly continuous. We write d' for the box metric on $\mathcal{O}_N$ (where we identify N with $\mathbb{N}$ in an arbitrary way). We have to show that for every $n \in \mathbb{N}$ there exists an $m \in \mathbb{N}$ such that for all $(f,g) \in C$ if $d'(\xi(f),\xi(g)) < 1/m$ then $d(f,g) \leq 1/n$. We may assume that f and g have the same arity k. By the definition of the box metric we may choose m large enough so that if $d'(\xi(f),\xi(g)) < 1/m$ then

$$\xi(f)\big((\mathrm{pr}_1^k/U_n,n),\dots,(\mathrm{pr}_k^k/U_n,n)\big) = \xi(g)\big((\mathrm{pr}_1^k/U_n,n),\dots,(\mathrm{pr}_k^k/U_n,n)\big).$$

Note that this is the case if and only if $(f(\mathrm{pr}_1^k,\dots,\mathrm{pr}_k^k), g(\mathrm{pr}_1^k,\dots,\mathrm{pr}_k^k)) \in U_n$, i.e., $d(f,g) < 1/n$.

Claim 5. The image of C under ξ is closed in $\mathcal{O}_N$. Let $(f_i)_{i\in\mathbb{N}}$ be a sequence of elements of C such that $(\xi(f_i))_{i\in\mathbb{N}}$ converges to $g \in \mathcal{O}_N$. Then $(\xi(f_i))_{i\in\mathbb{N}}$ is a Cauchy sequence. Since ξ^{-1} is uniformly continuous, $(f_i)_{i\in\mathbb{N}}$ is a Cauchy sequence, too (Proposition 9.1.16). Since d is complete, $(f_i)_{i\in\mathbb{N}}$ converges to an element $f \in C$. Then $\xi(f) = g$ by the continuity of ξ. ⊣

9.4.2. Reconstruction of topology. As in the case of groups, one may ask how much information about the topology of a closed subclone $\mathcal{C}$ of $\mathcal{O}_B$ is coded in $\mathcal{C}$ viewed as an abstract clone. This is an important question for complexity classifications for CSPs of ω-categorical structures $\mathfrak{B}$ because we will see in Section 9.5.3 that the computational complexity of CSP($\mathfrak{B}$) is determined by Pol($\mathfrak{B}$) viewed as a topological clone. Indeed, if $\mathcal{C}$ is the Horn clone (Theorem 7.5.2), the polymorphism clone of the random graph, or $\mathcal{O}_B$ itself it can be shown that any (abstract) clone isomorphism between $\mathcal{C}$ and a closed subclone of $\mathcal{O}_{\mathbb{N}}$ is a homeomorphism [94]; we call this property *automatic homeomorphicity*, analogously to the definition of automatic homeomorphicity for topological groups (Definition 9.2.19). Further results in this direction can be found in [301, 34]. An ω-categorical structure $\mathfrak{A}$ with finite relational signature whose polymorphism clone does not have automatic homeomorphicity has been described in [57]. The structure $\mathfrak{A}$ even has the property that there exist other ω-categorical structures such that Pol($\mathfrak{A}$) and Pol($\mathfrak{B}$) are isomorphic as abstract clones, but no isomorphism exists which is additionally a homeomorphism. The construction used requires the axiom of choice and the structure $\mathfrak{A}$ is not homogeneous in a finite relational signature.

9.5. The topological Birkhoff theorem

In this section we present a result which can be seen as a topological variant of Birkhoff's theorem (Theorem 6.5.1); a strengthened form of the statement for oligomorphic algebras appeared first in [91] (see Theorem 9.5.15 below). Strictly more general variants of the result can be found in [177, 323].

9.5.1. Uniformly continuous Birkhoff. Let $\mathcal{C}_1$ and $\mathcal{C}_2$ be operation clones over countable sets C_1 and C_2. A function $\xi\colon \mathcal{C}_1 \to \mathcal{C}_2$ is uniformly continuous if and only if for every finite $G \subseteq C_2$ there is a finite set $F \subseteq C_1$ such that for every $n \in \mathbb{N}$ and all $f, g \in \mathcal{C}_1^{(n)}$ if $f|_{F^n} = g|_{F^n}$ then $\xi(f)|_{G^n} = \xi(g)|_{G^n}$.

THEOREM 9.5.1. *Let* $\mathbf{A}$ *and* $\mathbf{B}$ *be* τ*-algebras such that* B *is countable and* $\mathbf{A}$ *is generated by a finite set* G*. Then the following are equivalent.*

1. $\mathbf{A} \in \mathrm{HSP}^{\mathrm{fin}}(\mathbf{B})$;
2. *there exists a finite* $F \subseteq B$ *such that if* $t_1(x_1, \dots, x_n)$ *and* $t_2(x_1, \dots, x_n)$ *are* τ*-terms with* $t_1^{\mathbf{B}}|_{F^n} = t_2^{\mathbf{B}}|_{F^n}$*, then* $t_1^{\mathbf{A}}|_{G^n} = t_2^{\mathbf{A}}|_{G^n}$;
3. *the natural homomorphism* ξ *from* $\mathrm{Clo}(\mathbf{B})$ *onto* $\mathrm{Clo}(\mathbf{A})$ *exists and is uniformly continuous.*

PROOF. For the implication from (1) to (2), suppose that μ is a surjective homomorphism from a subalgebra $\mathbf{S}$ of $\mathbf{B}^k$ to $\mathbf{A}$. For each $a \in G$ pick an $s \in S$ such that $\mu(s) = a$; let F be the (finite) set of all entries of all the picked tuples s. Now let $t_1(x_1, \dots, x_n)$ and $t_2(x_1, \dots, x_n)$ be τ-terms such that $t_1^{\mathbf{B}}|_{F^n} = t_2^{\mathbf{B}}|_{F^n}$. Since $G \subseteq \mu(F^n)$ it follows that $t_1^{\mathbf{A}}|_{G^n} = t_2^{\mathbf{A}}|_{G^n}$, which proves the statement.

(2) $\Rightarrow$ (3). Immediate from the definition of uniform continuity.

(3) $\Rightarrow$ (1). Let $G = \{a_1, \dots, a_k\}$. Since ξ is uniformly continuous, there exists a finite $F \subseteq B$ such that for all $f, g \in$ if $f|_{F^k} = g_{F^k}$ then $\xi(f)|_{G^k} = \xi(g)_{G^k}$. Let $C := F^G$, and let $m := |C| = |F|^k$. Let $c^1, \dots, c^m$ be an enumeration of C, and for $j \leq k$ define $c_j := (c_j^1, \dots, c_j^m) \in F^m$. Let $\mathbf{S}$ be the subalgebra of $\mathbf{B}^m$ generated by $c_1, \dots, c_k$; so the elements of $\mathbf{S}$ are precisely those of the form $t^{\mathbf{B}^m}(c_1, \dots, c_k)$, for a k-ary τ-term t. Define a function $\mu\colon S \to A$ by setting

$$\mu(t^{\mathbf{S}}(c_1, \dots, c_k)) := t^{\mathbf{A}}(a_1, \dots, a_k).$$

Claim 1. μ is well defined. Suppose that $t^{\mathbf{S}}(c_1, \dots, c_k) = s^{\mathbf{S}}(c_1, \dots, c_k)$. We first show that $t^{\mathbf{B}}|_{G^k} = s^{\mathbf{B}}|_{G^k}$. Let $b \in F^k$. Note that there is some $i \leq m$ such that $c^i(a_j) = b_j$ for all $j \leq k$. Hence,

$$\begin{aligned} t^{\mathbf{B}}(b) = t^{\mathbf{B}}(c^i(a_1), \dots, c^i(a_k)) &= t^{\mathbf{B}}((c_1)_i, \dots, (c_k)_i) \\ &= s^{\mathbf{B}}((c_1)_i, \dots, (c_k)_i) \\ &= s^{\mathbf{B}}(c^i(a_1), \dots, c^i(a_k)) = s^{\mathbf{B}}(b). \end{aligned}$$

So $t^{\mathbf{B}}|_{F^k} = s^{\mathbf{B}}|_{F^k}$ and by the property of F it follows that $t^{\mathbf{A}}|_{G^k} = s^{\mathbf{A}}|_{G^k}$, and in particular that $t^{\mathbf{A}}(a_1, \dots, a_k) = s^{\mathbf{A}}(a_1, \dots, a_k)$.

Claim 2. μ is surjective. This follows immediately from the assumption that $\mathbf{A}$ is generated by G.

Claim 3. μ is a homomorphism. Let $f \in \tau$ be an n-ary function symbol and let $s_1, \dots, s_n \in S$. Since $\mathbf{S}$ is generated by $c_1, \dots, c_k$, there exist τ-terms $t_1, \dots, t_n$ such that $s_i = t_i^{\mathbf{S}}(c_1, \dots, c_k)$ for all $i \in \{1, \dots, n\}$. Writing $\bar{c}$ for

$(c_1, \ldots, c_k)$ we obtain

$$\begin{aligned}\mu(f^{\mathbf{S}}(s_1, \ldots, s_n)) &= \mu(f^{\mathbf{S}}(t_1^{\mathbf{S}}(\bar{c}), \ldots, t_n^{\mathbf{S}}(\bar{c}))) = \mu(f^{\mathbf{S}}(t_1^{\mathbf{S}}, \ldots, t_n^{\mathbf{S}})(\bar{c}))\\ &= \mu((f(t_1, \ldots, t_n))^{\mathbf{S}}(\bar{c})) = (f(t_1, \ldots, t_n))^{\mathbf{A}}(a_1, \ldots, a_k)\\ &= f^{\mathbf{A}}(t_1^{\mathbf{A}}(a_1, \ldots, a_k), \ldots, t_n^{\mathbf{A}}(a_1, \ldots, a_k))\\ &= f^{\mathbf{A}}(\mu(s_1), \ldots, \mu(s_n)).\end{aligned}$$

It follows that $\mathbf{A}$ is the homomorphic image of the subalgebra $\mathbf{S}$ of $\mathbf{B}^m$, and so $\mathbf{A} \in \mathrm{HSP}^{\mathrm{fin}}(\mathbf{B})$. ⊣

Note that Theorem 9.5.1 has a version for permutation groups instead of polymorphism clones (Theorem 9.5.3 below). Because of Proposition 9.2.11, the assumptions can be relaxed in this case to continuity instead of uniform continuity.

DEFINITION 9.5.2. Let $\mathcal{G}$ be a permutation group on a set B. Then a $\mathcal{G}$-*set* is an algebra $\mathbf{B}$ with unary functions only such that the term functions for one-variable terms over $\mathbf{B}$ are precisely the elements of G, i.e., $\mathrm{Clo}(\mathbf{B})^{(1)} = \mathcal{G}$.

THEOREM 9.5.3. *Let $\mathcal{G}$ be a closed subgroup of* $\mathrm{Sym}(B)$ *and let ξ be a homomorphism from $\mathcal{G}$ to* $\mathrm{Sym}(A)$. *Then the following are equivalent.*

- *ξ is continuous and the permutation group $\xi(\mathcal{G})$ has finitely many orbits.*
- *There exists a $\mathcal{G}$-set* $\mathbf{B}$ *such that* $\mathrm{HSP}^{\mathrm{fin}}(\mathbf{B})$ *contains an algebra* $\mathbf{A}$ *such that* $\mathrm{Clo}(\mathbf{A})^{(1)} = \xi(\mathcal{G})$.

If ξ is a map from $\mathcal{C}_1$ to $\mathcal{C}_2$ and $\mathcal{C}_2$ is a clone over a finite domain, then uniform continuity has an equivalent description.

PROPOSITION 9.5.4. *Let $\mathcal{C}_1$ be an operation clone over a countable domain B and let $\mathcal{C}_2$ be an operation clone over a finite domain A. Then a minor-preserving map $\xi \colon \mathcal{C}_1 \to \mathcal{C}_2$ is uniformly continuous if and only if for every $n \in \mathbb{N}$ there is a finite $F \subseteq B$ such that and all $f_1, f_2 \in \mathcal{C}_1^{(n)}$ if $f_1|_{F^n} = f_2|_{F^n}$ then $\xi(f_1) = \xi(f_2)$.*

PROOF. The forwards implication is trivial. For the backwards implication, let $A = \{a_1, \ldots, a_k\}$. By assumption there exists a finite $F \subseteq B$ such that for all $g_1, g_2 \in \mathcal{C}_1{}^{(k)}$, if $g_1|_{F^k} = g_2|_{F^k}$, then $\xi(g_1) = \xi(g_1)|_{A^k} = \xi(f_2)_{A^k} = \xi(g_2)$. Let $n \in \mathbb{N}$ and $f_1, f_2 \in \mathcal{C}_1^{(n)}$ be such that $f_1|_{F^n} = f_2|_{F^n}$. Let $c_1, \ldots, c_n \in A$. Let $p \colon \{1, \ldots, n\} \to \{1, \ldots, k\}$ be such that $c_i = a_{p(i)}$. For $i \in \{1, 2\}$, let $g_i \in \mathcal{C}_1$ be the minor $(f_i)_p$ (for the notation, see Section 6.7.3). We then have $g_1|_{F^k} = g_2|_{F^k}$, and hence $\xi(g_1) = \xi(g_2)$. Since ξ is minor-preserving we get $\xi(f_1)(c_1, \ldots, c_n) = \xi(g_1)(a_1, \ldots, a_k) = \xi(g_2)(a_1, \ldots, a_k) = \xi(f_2)(c_1, \ldots, c_n)$. Since $c_1, \ldots, c_n \in A$ were chosen arbitrarily, we conclude that $\xi(f) = \xi(g)$. ⊣

The existence of uniformly continuous clone homomorphisms to clones of finite algebras can also be characterised locally using the following definition.

DEFINITION 9.5.5. Let $\mathbf{B}$ be a τ-algebra and let ϕ be an identity over τ of the form $\forall x_1, \dots, x_n : r(x_1, \dots, x_n) = s(x_1, \dots, x_n)$. Then we say that $\mathbf{B}$ *satisfies* ϕ *on* F if for all $a_1, \dots, a_n \in F$ we have $r^{\mathbf{B}}(a_1, \dots, a_n) = s^{\mathbf{B}}(a_1, \dots, a_n)$.

PROPOSITION 9.5.6. *Let* $\mathbf{B}$ *be a* τ*-algebra and let* $\mathcal{C}$ *be an operation clone over a finite domain* A. *Then the following are equivalent.*

1. *There is no uniformly continuous clone homomorphism from* $\mathrm{Clo}(\mathbf{B})$ *to* $\mathcal{C}$.
2. *For every finite* $F \subseteq B$ *there exists a finite set of identities that are satisfied by* $\mathbf{B}$ *on* F *but not in* $\mathcal{C}$ (*Definition* 6.5.12).

PROOF. Suppose that there is a uniformly continuous clone homomorphism from $\mathrm{Clo}(\mathbf{B})$ to $\mathcal{C}$. so that there exists a finite $F \subseteq B$ such that if $f_1, f_2 \in \mathrm{Clo}(\mathbf{B})^{(n)}$ are such that $f_1|_{F^n} = f_2|_{F^n}$, then $\xi(f_1) = \xi(f_2)$. Hence, all identities that are satisfied by $\mathbf{B}$ on F (Definition 9.5.5) are also satisfied in $\mathcal{C}$ (Definition 6.5.12).

Now suppose that there exists a finite $F \subseteq B$ such that all finite sets of identities that are satisfied by $\mathbf{B}$ on F are also satisfied in $\mathcal{C}$. Then by Lemma 6.5.13 all the identities Σ that are satisfied by $\mathbf{B}$ on F are also satisfied in $\mathcal{C}$. That is, there exists an algebra $\mathbf{A}$ such that $\mathbf{A} \models \Sigma$ and $\mathrm{Clo}(\mathbf{A}) \subseteq \mathcal{C}$. The map that sends $f^{\mathbf{B}}$, for $f \in \tau$, to $f^{\mathbf{A}}$ is a uniformly continuous clone homomorphism from $\mathrm{Clo}(\mathbf{B})$ to $\mathrm{Clo}(\mathbf{A})$. ⊣

The previous lemma can also be formulated in the language of clones.

DEFINITION 9.5.7. Let $\mathcal{C}$ be an operation clone and let $f_1, \dots, f_n \in \mathcal{C}$. Let $\phi(x_1, \dots, x_n)$ be a primitive positive clone formula. Then $\mathcal{C}$ *satisfies* $\phi(f_1, \dots, f_n)$ *on* F if there are operations in $\mathcal{C}$ for the existentially quantified variables in ϕ such that for each conjunct $r = s$ in ϕ, where r and s have rank k, we have that $r(x_1, \dots, x_k) = s(x_1, \dots, x_k)$ for all $x_1, \dots, x_k \in F$.

The following is essentially the same statement as in Proposition 9.5.6, but formulated in the language of clones; we state it here for easy reference.

COROLLARY 9.5.8. *Let* $\mathcal{C}$ *be an operation clone over a countable domain and let* $\mathcal{D}$ *be an operation clone over a finite domain. Then the following are equivalent.*

- *There exists a uniformly continuous minor-preserving map from* $\mathcal{C}$ *to* $\mathcal{D}$.
- *There is a finite subset* F *of the domain of* $\mathcal{C}$ *such that every minor condition* ϕ *that is satisfied by* $\mathcal{C}$ *on* F *is also satisfied by* $\mathcal{D}$.

9.5.2. The oligomorphic case. The uniformly continuous Birkhoff theorem (Theorem 9.5.1) refers to the metrics of the operation clones under consideration. For oligomorphic algebras, this reference can be eliminated by a compactness argument that we present here. More precisely, we show that every continuous clone homomorphism from a closed oligomorphic subclone of $\mathcal{O}_{\mathbb{N}}$ to $\mathcal{O}_{\mathbb{N}}$ is uniformly continuous (with respect to the box metric). We start with a simple observation.

PROPOSITION 9.5.9. *Every oligomorphic algebra is finitely generated.*

PROOF. The permutation group of invertible unary term operations in an oligomorphic algebra **B** has finitely many orbits; picking a representative from each orbit one obtains a generating set for **B**. ⊣

DEFINITION 9.5.10. Let **G** be a topological group with a continuous action on a topological space X. Let $\sim$ be the orbit equivalence relation on X where $x \sim y$ if there exists $g \in G$ such that $x = gy$. We write $X/\mathbf{G}$ for the quotient space $X/_{\sim}$ with the quotient topology.

The following statement is taken from [91] and closely related to the compactness theorems that we presented in Section 4.1.2.

PROPOSITION 9.5.11. *Let X, Y be countably infinite sets and let $\mathcal{G}$ be a permutation group on Y. Equip Y with the discrete topology and Y^X with the product topology. Then $Y^X/\mathcal{G}$ is compact if and only if $\mathcal{G}$ is oligomorphic.*

PROOF. Suppose that $X = \mathbb{N}$. We first prove that if $\mathcal{G}$ is oligomorphic, then $Y^X/\mathcal{G}$ is compact. Let $\mathcal{U} := \{U_i \mid i \in A\}$ be a family of open subsets of $Y^X/\mathcal{G}$ such that no finite subset of $\mathcal{U}$ covers $Y^X/\mathcal{G}$. For $n \in \mathbb{N}$, let $\sim_n$ be the equivalence relation on Y^X where $f \sim_n g$ if there exists an $\alpha \in \mathcal{G}$ such that $f(x) = \alpha g(x)$ for all $x \in \{0, \dots, n-1\}$. Note that each equivalence class of $\sim_n$ is a union of elements of $Y^X/\mathcal{G}$, and that oligomorphicity implies that $\sim_n$ has finitely many classes for each $n \in \mathbb{N}$. If each of the finitely many equivalence classes of $\sim_n$ were contained in the complement of U_i for some $i \in A$, then we would have found a finite subset of $\mathcal{U}$ that covers $Y^X/\mathcal{G}$, contrary to our assumptions. So for each n there exists a $\sim_n$-equivalence class which is not contained in $\bigcup_{i \in A} U_i$.

Consider the following tree: the vertices of the tree are the equivalence classes of $\sim_n$, for all $n \in \mathbb{N}$, that are not contained $\bigcup_{i \in A} U_i$. Let the equivalence class of $f: \{0, \dots, n-1\} \to \mathbb{N}$ be adjacent to the equivalence class of $g: \{0, \dots, n\} \to \mathbb{N}$ if f is the restriction of g. Clearly, the resulting tree is finitely branching and by Kőnigs tree lemma contains an infinite path. From this infinite path $F_1, F_2, \dots$ one can construct a function $f \in Y^X$ inductively as follows. Initially, pick any function f_1 from F_1. By the definition of edges in the tree there exists an $\alpha \in G$ such that αf_1 is the restriction of some $g_2 \in F_2$. We define f_2 to be $\alpha^{-1} g_2$ which is an extension of f_1 and in F_2. We continue with f_2 instead of f_1, and iterate to obtain an infinite sequence of functions $f_1, f_2, \dots$ which converges against some $f \in Y^X$. Note that $f/_{\sim}$ is not contained in $\bigcup_{i \in A} U_i$ which finishes the proof that $Y^X/_{\sim}$ is compact.

For the other direction, assume that $\mathcal{G}$ is not oligomorphic. Pick an $n \geq 1$ such that the componentwise action of $\mathcal{G}$ on Y^n has infinitely orbits, and enumerate these orbits by $(O_i)_{i \in \omega}$. For each $i \in \omega$ let U_i consist of all classes $f/_{\sim}$ in $Y^X/\mathcal{G}$ with the property that $f|_{\{1,\dots,n\}}$ belongs to O_i; this is well defined since for all $f, g \in Y^X$ with $f \sim g$ we have that $f|_{\{1,\dots,n\}}$ belongs to O_i if and only if $g|_{\{1,\dots,n\}}$ belongs to O_i. Then $Y^X/\mathcal{G}$ is the disjoint union of the U_i. But each U_i is open, and hence $Y^X/\mathcal{G}$ is not compact. ⊣

We mention that the countability of X is necessary in the statement of Proposition 9.5.11: in fact with Y countable and X uncountable, even $Y^X/\operatorname{Sym}(Y)$ is non-compact (Example 4.5 in [323]).

If $\mathcal{G}$ is an oligomorphic permutation group on a countable set Y, then the space $Y^Y/\mathcal{G}$ is not Hausdorff, as the following example shows.

Remark 9.5.12. Consider any function f in Y^Y which lies in the closure of $\mathcal{G}$ but not in $\mathcal{G}$; Proposition 4.4.7 shows that if $\mathcal{G}$ is oligomorphic, such functions must exist. Then f is inequivalent to every element of $\mathcal{G}$, but $f/_\sim$ cannot be separated from $\mathrm{id}_Y/_\sim$ by open sets: if U is an open subset of $Y^Y/_\sim$ that contains $f/_\sim$, then $\bigcup U$ is open in Y^Y and hence must contain a basic open set $T_{a,b}$ where $a, b \in Y^n$ for some $n \in \mathbb{N}$ and $f(a) = b$. Since f is in the closure of $\mathcal{G}$ there also exists an $\alpha \in \mathcal{G}$ with $\alpha a = b$, and $\alpha \sim \mathrm{id}_Y$. So every open set that contains $f/_\sim$ also contains $\mathrm{id}_Y/_\sim$.

COROLLARY 9.5.13. *Let $\mathfrak{B}$ be an ω-categorical structure and let $k \geq 1$. Then* $\operatorname{Pol}(\mathfrak{B})^{(k)}/\operatorname{Aut}(\mathfrak{B})$ *is compact.*

PROOF. $\operatorname{Pol}(\mathfrak{B})^{(k)}$ is a closed subset of B^{B^k} which is preserved by $\operatorname{Aut}(\mathfrak{B})$. Since $\mathfrak{B}$ is ω-categorical, $\operatorname{Aut}(\mathfrak{B})$ is an oligomorphic permutation group by the theorem of Engeler, Svenonius, and Ryll-Nardzewski (Theorem 4.1.6). Proposition 9.5.11 implies that $B^{B^k}/\operatorname{Aut}(\mathfrak{B})$ is compact. Note that $\operatorname{Pol}(\mathfrak{B})^{(k)}/\mathcal{G}$ is a closed subspace of $B^{B^k}/\operatorname{Aut}(\mathfrak{B})$, so the statement follows from Proposition 9.1.18 ⊣

Note that $\operatorname{Pol}(\mathfrak{B})/\operatorname{Aut}(\mathfrak{B})$ is never compact since it is the disjoint union of the spaces $\operatorname{Pol}(\mathfrak{B})^{(k)}/\operatorname{Aut}(\mathfrak{B})$.

Let $\mathbf{C}$ and $\mathbf{D}$ be clones and $\xi\colon C \to D$ a map. Recall from Definition 6.7.6 that ξ *preserves left composition with invertibles* if for all $f \in C$ and $\alpha \in C^{(1)}$ invertible in $C^{(1)}$ we have that

$$\xi(\alpha \circ f) = \xi(\alpha) \circ \xi(f).$$

LEMMA 9.5.14. *Let $\mathcal{C}$ be a closed oligomorphic subclone of $\mathcal{O}_{\mathbb{N}}$ and let $\xi\colon \mathcal{C} \to \mathcal{O}_{\mathbb{N}}$ be a continuous function that preserves left composition with invertible elements. Then ξ is uniformly continuous. In particular, every continuous clone homomorphism ξ is uniformly continuous.*

PROOF. Let $F \subseteq \mathbb{N}$ be finite and $k \in \mathbb{N}$. It suffices to prove the existence of a matrix $A \in \mathbb{N}^{m \times k}$ for some $m \in \mathbb{N}$ such that for all $f, g \in \mathcal{C}^{(k)}$ we have that $f(A) = g(A)$ implies $\xi(f)|_F = \xi(g)|_F$.

Let $\mathcal{G}$ be the oligomorphic permutation group formed by the invertible operations in $\mathcal{C}^{(1)}$, and let $\mathcal{D}$ be the subclone $\xi(\mathcal{C})$ of $\mathcal{O}_{\mathbb{N}}$. First note that for each $f \in \mathcal{C}^{(k)}$ the set

$$\{h \in \mathcal{D}^{(k)} \mid h|_F = \xi(f)|_F\}$$

is open in $\mathcal{D}^{(k)}$. Since ξ is continuous, this means that

$$U_f := \{g \in \mathcal{C}^{(k)} \mid \xi(g)|_F = \xi(f)|_F\}$$

is open in $\mathcal{C}^{(k)}$. By the definition of the topology on $\mathcal{C}$ there exist $l_f \in \mathbb{N}$, a matrix $A_f \in \mathbb{N}^{l_f \times k}$, and a vector $b_f \in \mathbb{N}^{l_f}$ such that the basic open set

$$T_{A_f,b_f} = \{g \in \mathcal{O}_{\mathbb{N}}^{(k)} \mid g(A_f) = b_f\}$$

is contained in U_f and contains f. Hence,

$$\mathcal{C}^{(k)} = \bigcup_{f \in \mathcal{C}^{(k)}} T_{A_f,b_f}.$$

Also note that $\mathcal{G} \cdot T_{A_f,b_f} = \{\alpha f \mid \alpha \in \mathcal{G}, f \in T_{A_f,b_f}\}$ is open and preserved by $\mathcal{G}$. So there exists an open $V_f \subseteq \mathcal{C}^{(k)}/\mathcal{G}$ such that $\bigcup V_f = \mathcal{G} \cdot T_{A_f,b_f}$. Note that

$$\mathcal{C}^{(k)}/\mathcal{G} = \bigcup_{f \in \mathcal{C}^{(k)}} V_f.$$

By compactness of $\mathcal{C}^{(k)}/\mathcal{G}$ (Proposition 9.5.11) there exists an $n \in \mathbb{N}$ and $f_1, \dots, f_n \in \mathcal{C}^{(k)}$ such that

$$\mathcal{C}^{(k)}/\mathcal{G} = \bigcup_{i \in \{1,\dots,n\}} V_{f_i}.$$

Set $m := l_{f_1} + \cdots + l_{f_n}$. Let $A \in B^{m \times k}$ be the matrix obtained by superposing the matrices $A_{f_1}, \dots, A_{f_n}$.

To see that A satisfies the desired property, let $f, g \in \mathcal{C}^{(k)}$. Assume without loss of generality that $f \in \mathcal{G} \cdot T_{A_{f_1},b_{f_1}}$; then there exists $\alpha \in \mathcal{G}$ such that $f(A_{f_1}) = \alpha(b_{f_1})$. Since $f(A) = g(A)$ we have $f(A_{f_1}) = g(A_{f_1})$ and so also $g(A_{f_1}) = \alpha b_{f_1}$. Hence, $\alpha^{-1} f$ and $\alpha^{-1} g$ are in $T_{A_{f_1},b_{f_1}}$, implying that $\xi(\alpha^{-1} f)|_F = \xi(\alpha^{-1} g)|_F$. Thus, $\xi(f)|_F = \xi(g)|_F$ since ξ is a clone homomorphism. ⊣

We can now readily combine Theorem 9.5.1, Proposition 9.5.9, and Lemma 9.5.14 and obtain a result which we call 'topological Birkhoff theorem'.

THEOREM 9.5.15. *Let* **A** *and* **B** *be countable oligomorphic algebras with the same signature such that* Clo(**B**) *is closed in* $\mathcal{O}_B$. *Then the following three statements are equivalent.*

1. *The natural homomorphism from* Clo(**B**) *onto* Clo(**A**) *exists and is continuous.*
2. *The natural homomorphism from* Clo(**B**) *onto* Clo(**A**) *exists and is uniformly continuous.*
3. $\mathbf{A} \in \mathrm{HSP}^{\mathrm{fin}}(\mathbf{B})$.

PROOF. The implication from (2) to (1) is trivial, and the implication from (1) to (2) follows from Lemma 9.5.14 applied to the closed oligomorphic clone $\mathfrak{C} := \mathrm{Clo}(\mathbf{B})$, identifying the countable domain with $\mathbb{N}$. For the equivalence of (2) and (3), first recall that $\mathbf{A}$ is finitely generated (Proposition 9.5.9). So the equivalence follows from Theorem 9.5.1. ⊣

Theorem 6.5.1 for finite algebras $\mathbf{A}$ and $\mathbf{B}$ is a special case of Theorem 9.5.15, since the topology of any operation clone on a finite set is discrete, and hence the natural homomorphism from the operation clone of a finite algebra to that of any other algebra is continuous. For convenient reference, we also present a straightforward consequence of this result which departs from operation clones rather than algebras.

COROLLARY 9.5.16. *Let $\mathfrak{C}$ and $\mathfrak{D}$ be oligomorphic operation clones on countable domains, and suppose that $\mathfrak{C}$ is locally closed. Then the following three statements are equivalent.*

1. *There is a uniformly continuous clone homomorphism from $\mathfrak{C}$ onto $\mathfrak{D}$.*
2. *There is a continuous clone homomorphism from $\mathfrak{C}$ onto $\mathfrak{D}$.*
3. *there are τ-algebras $\mathbf{C}$ and $\mathbf{D}$ such that $\mathrm{Clo}(\mathbf{C}) = \mathfrak{C}$, $\mathrm{Clo}(\mathbf{D}) = \mathfrak{D}$, and $\mathbf{D} \in \mathrm{HSP}^{\mathrm{fin}}(\mathbf{C})$.*

Applying Theorem 9.5.15 in both directions we also obtain the following corollary.

COROLLARY 9.5.17. *Let $\mathfrak{C}$ and $\mathfrak{D}$ be oligomorphic operation clones on countable domains. Then $\mathfrak{C}$ and $\mathfrak{D}$ are topologically isomorphic if and only if there are τ-algebras $\mathbf{C}$ and $\mathbf{D}$ such that $\mathrm{Clo}(\mathbf{C}) = \mathfrak{C}$, $\mathrm{Clo}(\mathbf{D}) = \mathfrak{D}$, and $\mathrm{HSP}^{\mathrm{fin}}(\mathbf{A}) = \mathrm{HSP}^{\mathrm{fin}}(\mathbf{B})$.*

9.5.3. Continuous homomorphisms and interpretations. The results from the previous section can be combined with the results from Section 6.3.5 about pseudo-varieties and primitive positive interpretations. We start with an elegant characterisation of topological isomorphism of polymorphism clones of ω-categorical structures.

COROLLARY 9.5.18. *Let $\mathfrak{A}$ and $\mathfrak{B}$ be countable ω-categorical structures. Then $\mathrm{Pol}(\mathfrak{A})$ and $\mathrm{Pol}(\mathfrak{B})$ are isomorphic as topological clones if and only if $\mathfrak{A}$ and $\mathfrak{B}$ are primitively positively bi-interpretable.*

PROOF. By Proposition 6.3.9 $\mathfrak{A}$ and $\mathfrak{B}$ are primitively positive bi-interpretable if and only if $\mathfrak{A}$ has a polymorphism algebra $\mathbf{A}$ and $\mathfrak{B}$ has a polymorphism algebra $\mathbf{B}$ such that $\mathrm{HSP}^{\mathrm{fin}}(\mathbf{A}) = \mathrm{HSP}^{\mathrm{fin}}(\mathbf{B})$. This in turn is the case if and only if $\mathrm{Clo}(\mathbf{A}) = \mathrm{Pol}(\mathfrak{A})$ and $\mathrm{Clo}(\mathbf{B}) = \mathrm{Pol}(\mathfrak{B})$ are topologically isomorphic (Corollary 9.5.17). ⊣

Corollary 9.5.18 shows that the complexity of the CSP of a finite-signature ω-categorical structure $\mathfrak{B}$ only depends on $\mathrm{Pol}(\mathfrak{B})$ as a topological clone. The same is true for many other properties of an ω-categorical structure $\mathfrak{B}$. We present another example of such a property.

PROPOSITION 9.5.19. *Whether an oligomorphic clone $\mathfrak{C}$ is finitely related (Definition 6.1.2) depends only on $\mathfrak{C}$ viewed as a topological clone.*

PROOF. The statement is an immediate consequence of Proposition 3.3.6 via Corollary 9.5.18. ⊣

Recall that $\mathfrak{A}$ is primitive positive interpretable in $\mathfrak{B}$ if and only if $\mathfrak{A}$ is the reduct of a structure $\mathfrak{A}'$ with a full primitive positive interpretation in $\mathfrak{B}$. We now present a topological characterisation of the structures that have a full primitive positive interpretation in an ω-categorical structure.

COROLLARY 9.5.20. *Let $\mathfrak{B}$ be a countable ω-categorical structure and $\mathfrak{A}$ an arbitrary structure. Then $\mathfrak{A}$ has a full primitive positive interpretation in $\mathfrak{B}$ if and only if $\mathfrak{A}$ is countable ω-categorical and there exists a continuous clone homomorphism from* $\text{Pol}(\mathfrak{B})$ *to* $\text{Pol}(\mathfrak{A})$ *whose image is dense in* $\text{Pol}(\mathfrak{A})$.

PROOF. By Theorem 6.3.7 we have that $\mathfrak{A} \in \mathrm{I}_{\text{full}}(\mathfrak{B})$ if and only if there exists $\mathbf{A} \in \text{HSP}^{\text{fin}}(\mathbf{B})$ such that $\overline{\text{Clo}(\mathbf{A})} = \text{Pol}(\mathfrak{A})$. This in turn is the case if and only if the natural homomorphism from $\text{Clo}(\mathbf{B}) = \text{Pol}(\mathfrak{B})$ to $\text{Clo}(\mathbf{A})$ exists, by Theorem 9.5.15. ⊣

9.5.4. Continuous homomorphisms to the clone of projections. The topological Birkhoff theorem in the form of Corollary 9.5.20 can also be combined with the results from Section 6.3.5 to add yet another equivalent item to Theorem 6.3.10.

COROLLARY 9.5.21. *Let $\mathfrak{B}$ be a countable ω-categorical structure. Then the following are equivalent.*

1. *$I(\mathfrak{B})$ contains all finite structures.*
2. *$I(\mathfrak{B})$ contains K_3.*
3. *$I(\mathfrak{B})$ contains* $(\{0,1\}; \text{1IN3})$.
4. $\text{Pol}(\mathfrak{B})$ *has a continuous homomorphism to* **Proj**.
5. $\text{Pol}(\mathfrak{B})$ *has a uniformly continuous homomorphism to* **Proj**.
6. *There is a finite $F \subseteq B$ such that every primitive positive clone sentence that is satisfied by* $\text{Pol}(\mathfrak{B})$ *on F is trivial.*

PROOF. The equivalence of (1), (2) and (3) (among many other equivalences) has already been shown in Theorem 6.3.10.

(3) ⇒ (4). Suppose that $(\{0,1\}; \text{1IN3})$ has an expansion $\mathfrak{A}$ with a full primitive positive interpretation in $\mathfrak{B}$. Recall that $\text{Pol}(\{0,1\}; \text{1IN3})$ is isomorphic to **Proj** (Proposition 6.5.19), hence $\text{Pol}(\mathfrak{A}) = \text{Pol}(\{0,1\}; \text{1IN3})$. Corollary 9.5.20 asserts the existence of a continuous clone homomorphism from $\text{Pol}(\mathfrak{B})$ to $\text{Pol}(\mathfrak{A})$, and hence to **Proj**.

(4) ⇒ (3). Let ξ be the continuous homomorphism from $\text{Pol}(\mathfrak{B})$ to **Proj**. Since ξ is surjective, its image is dense, and we apply Corollary 9.5.20 in the other direction to obtain a (full) primitive positive interpretation of $(\{0,1\}; \text{1IN3})$ in $\mathfrak{B}$.

The equivalence of (4) and (5) follows from Corollary 9.5.16. The equivalence of (5) and (6) follows from Corollary 9.5.8 (for this equivalence we do not need the ω-categoricity assumption). ⊣

Theorem 9.5.15 can also be used to reformulate the hardness condition from Corollary 4.5.4 and the infinite-domain tractability conjecture in its original form (Conjecture 4.1).

COROLLARY 9.5.22. *Let $\mathfrak{B}$ be an ω-categorical structure, and let $\mathfrak{C}$ be its model-complete core. If $\mathfrak{C}$ has an expansion by finitely many constants $c_1, \ldots, c_n$ such that* $\mathrm{Pol}(\mathfrak{C}, c_1, \ldots, c_n)$ *has a continuous homomorphism to* **Proj** *then $\mathfrak{B}$ has a finite-signature reduct with an NP-hard* CSP. *Otherwise, Conjecture* 4.1 *states that* $\mathrm{CSP}(\mathfrak{B}')$ *is in P for every finite-signature reduct $\mathfrak{B}'$ of $\mathfrak{B}$.*

The following lemma can be useful to verify that a given clone homomorphism to the projections is continuous.

LEMMA 9.5.23. *Let* **C** *be a topological clone and let* $\xi\colon \mathbf{C} \to \mathbf{Proj}$ *be a homomorphism. Then ξ is continuous if and only if its restriction to $C^{(2)}$ is.*

PROOF. For $1 \leq i \leq n$ and $f \in C^{(n)}$ define

$$f_i := \mathrm{comp}(f, \mathrm{pr}^2_1, \ldots, \mathrm{pr}^2_1, \mathrm{pr}^2_2, \mathrm{pr}^2_1, \ldots, \mathrm{pr}^2_1)$$

i.e., the composition of f with the first binary projection except at the i-th argument, where we choose the second projection. Note that if $\xi(f) = \mathrm{pr}^n_j$ with $j \leq n$, then $\xi(f_i) = \mathrm{pr}^2_2$ if $i = j$ and $\xi(f_i) = \mathrm{pr}^2_1$ otherwise. Hence,

$$\xi^{-1}(\mathrm{pr}^n_i) = \{f \in C^{(n)} \mid f_i \in \xi^{-1}(\mathrm{pr}^2_2)\}.$$

Since composition in **C** is continuous the map that sends $f \in C^{(n)}$ to f_i is continuous. As $\xi^{-1}(\mathrm{pr}^2_2)$ is open it follows that $\xi^{-1}(\mathrm{pr}^n_i)$ is open, too. ⊣

The following example shows that there are closed oligomorphic operation clones on a countable domain with a discontinuous homomorphism to **Proj**.

Example 9.5.24. The following example of a closed oligomorphic operation clone on a countable set has been given in [95]; it is the polymorphism clone of a first-order reduct of the structure presented in Example 9.2.7. Let τ be the signature consisting of a relation symbol R_n of arity $2n$ for each $n \geq 1$. The class of all finite τ-structures where each R_n is interpreted as an equivalence relation on n-tuples of distinct entries with two equivalence classes is a Fraïssé class. Let $\mathfrak{D}$ be its Fraïssé limit, with domain D; it is ω-categorical since it is homogeneous and has for all $n \geq 1$ only finitely many inequivalent atomic formulas with n variables.

Let $\mathfrak{B}$ be the structure with domain D that has for all $n \geq 1$ the relation R_n, as well as the $3n$-ary relation

$$S_n := \big\{(x, y, z) \in D^{3n} \mid \neg\big(R_n(x, y) \wedge R_n(y, z) \wedge R_n(z, x)\big)\big\}.$$

Then $\mathfrak{B}$ is first-order definable over $\mathfrak{D}$ and therefore also ω-categorical. Since the elements of $\mathrm{Pol}(\mathfrak{B})$ preserve R_n for each $n \geq 1$, the operation clone $\mathrm{Pol}(\mathfrak{B})$, viewed as a topological clone, acts naturally on the equivalence classes of R_n. Write ξ_n for the mapping which sends every $f \in \mathrm{Pol}(\mathfrak{B})$ to its corresponding function on the equivalence classes of R_n. Then ξ_n is a continuous clone homomorphism, and its image is an operation clone on a domain with two elements, which we will denote by 0 and 1 in the following (independently of n, since the name of the elements of the base set is irrelevant). We claim that for every $f \in \mathrm{Pol}(\mathfrak{B})$, the operation $\xi_n(f)$ depends on one of its arguments only. To see this, observe that $\xi_n(f)$ preserves the Boolean relation $\{0,1\}^3 \setminus \{(0,0,0),(1,1,1)\}$ because f preserves S_n; the claim therefore follows from Proposition 6.2.8.

Let $\mathcal{U}$ be a non-principal ultrafilter on ω. Let $\xi\colon \mathrm{Pol}(\mathfrak{B}) \to \mathbf{Proj}$ be the mapping which sends every k-ary $f \in \mathrm{Pol}(\mathfrak{B})$ to the projection $\pi_i^k \in \mathbf{Proj}$ if and only if

$$\{n \geq 1 \mid \xi_n(f) \text{ depends on the } i\text{-th argument}\} \in \mathcal{U}.$$

It is easy to verify that ξ is a clone homomorphism. Moreover, ξ is not continuous; by Lemma 9.5.23, it suffices to verify this for the binary polymorphisms of $\mathfrak{B}$. Observe that for any $S \subseteq \omega$ there exists a binary $f \in \mathrm{Pol}(\mathfrak{B})$ such that $\xi_n(f)$ depends on the first argument if and only if $n \in S$. This function f can be constructed by defining a structure on D^2 with the same signature as $\mathfrak{D}$ in which for each $t \in D^2$ the membership in R_n depends only on the membership in R_n of the projection of t onto its first coordinate when $n \in S$, and onto its second coordinate when $n \notin S$. Choosing f as any embedding of this structure into $\mathfrak{D}$ using universality, we obtain a polymorphism of $\mathfrak{B}$ with the desired property. But since $\mathcal{U}$ is non-principal, the membership in $\mathcal{U}$ cannot be determined on any finite subset of S, and the discontinuity of ξ follows.

We mention that the clone $\mathrm{Pol}(\mathfrak{B})$ also has continuous homomorphisms to **Proj**: for example, each single ξ_n is continuous, and the image of $\mathrm{Pol}(\mathfrak{B})$ under ξ_n has a homomorphism to **Proj** which is necessarily continuous since the topology on the image is discrete.

QUESTION 9.1. *Does there exist a reduct of a finitely bounded homogeneous structure $\mathfrak{B}$ such that* $\mathrm{Pol}(\mathfrak{B})$ *has a homomorphism to* **Proj**, *but no continuous one?*

9.5.5. Consequences for first-order interpretability. The results in the previous sections can be specialised to first-order interpretability and topological automorphism groups; in particular, we obtain an older result which Hodges credits to Ahlbrandt and Ziegler [8]; Ahlbrandt and Ziegler cite it as an unpublished result of Coquand.

THEOREM 9.5.25 (Theorem 5.3.5 and 7.3.7 in [206]). *Let* $\mathfrak{B}$ *be a countable* ω*-categorical structure. Then a structure* $\mathfrak{A}$ *has a full first-order interpretation in* $\mathfrak{B}$ *if and only if there is a continuous group homomorphism* $\xi\colon \operatorname{Aut}(\mathfrak{B}) \to \operatorname{Aut}(\mathfrak{A})$ *such that the image of* ξ *is dense in* $\operatorname{Aut}(\mathfrak{A})$ *and* $\operatorname{Aut}(\mathfrak{A})$ *has finitely many orbits.*

PROOF. Let $\mathfrak{A}'$ and $\mathfrak{B}'$ be the expansions of $\mathfrak{A}$ and $\mathfrak{B}$ by all first-order definable relations. Then $\mathfrak{A}$ has a full first-order interpretation in $\mathfrak{B}$ if and only if $\mathfrak{A}'$ has a full primitive positive interpretation in $\mathfrak{B}'$. By Corollary 9.5.20, this is the case if and only if there exists a continuous clone homomorphism ξ from $\operatorname{Pol}(\mathfrak{B}')$ to $\operatorname{Pol}(\mathfrak{A}')$ whose image is dense in $\operatorname{Pol}(\mathfrak{A}')$ and $\mathfrak{A}'$ is ω-categorical. Note that $\operatorname{Aut}(\mathfrak{B}) = \operatorname{Aut}(\mathfrak{B}')$ lies dense in $\operatorname{End}(\mathfrak{B}')$ and $\operatorname{Aut}(\mathfrak{A}) = \operatorname{Aut}(\mathfrak{A}')$ lies dense in $\operatorname{End}(\mathfrak{A}')$, and every polymorphism of $\mathfrak{A}'$ and of $\mathfrak{B}'$ is essentially unary, hence ξ has a continuous extension to a clone homomorphism from $\operatorname{Pol}(\mathfrak{B}')$ to $\operatorname{Pol}(\mathfrak{A}')$. Moreover, the ω-categoricity of $\mathfrak{A}'$ implies that $\operatorname{Aut}(\mathfrak{A}') = \operatorname{Aut}(\mathfrak{A})$ has finitely many orbits. Conversely, if $\operatorname{Aut}(\mathfrak{A})$ has finitely many orbits then $\mathfrak{A}$ must be ω-categorical (see the proof of the implication (3) implies (4) in Theorem 9.3.1). Hence, there exists a continuous clone homomorphism from $\operatorname{Pol}(\mathfrak{B}')$ to $\operatorname{Pol}(\mathfrak{A}')$ whose image is dense in $\operatorname{Pol}(\mathfrak{A}')$ and $\mathfrak{A}'$ is ω-categorical if and only if there exists a continuous group homomorphism from $\operatorname{Aut}(\mathfrak{B})$ to $\operatorname{Aut}(\mathfrak{A})$ whose image is dense in $\operatorname{Aut}(\mathfrak{A})$ and $\operatorname{Aut}(\mathfrak{A})$ has finitely many orbits. ⊣

See Example 9.2.9 for an ω-categorical structure $\mathfrak{B}$ and a group homomorphism $\xi\colon \operatorname{Aut}(\mathfrak{B}) \to \operatorname{Sym}(\mathbb{N})$ where the image of ξ is dense but not surjective. As in the primitive positive case from Section 9.5.3, the situation that two structures are first-order bi-interpretable has an even more elegant topological characterisation.

COROLLARY 9.5.26 (Corollary 1.4 in [8]). *Two* ω*-categorical structures* $\mathfrak{A}$ *and* $\mathfrak{B}$ *are first-order bi-interpretable if and only if* $\operatorname{Aut}(\mathfrak{A})$ *and* $\operatorname{Aut}(\mathfrak{B})$ *are isomorphic as topological groups.*

PROOF. Similarly as in the proof of Theorem 9.5.25, the statement is an immediate consequence of Corollary 9.5.18. ⊣

Example 9.5.27. The structures $\mathfrak{C} := (\mathbb{N}^2; \{(x,y),(u,v) \mid x = u\})$ and $\mathfrak{D} := (\mathbb{N}; =)$ are mutually primitive positive interpretable, but *not* bi-interpretable. To see this, observe that $\operatorname{Aut}(\mathfrak{C})$ has a proper non-trivial closed normal subgroup $\mathbf{N}$ such that $\operatorname{Aut}(\mathfrak{C})/\mathbf{N}$ is isomorphic to $\operatorname{Aut}(\mathfrak{D})$ (see Proposition 9.2.17), whereas $\operatorname{Aut}(\mathfrak{D})$, the symmetric permutation group of a countably infinite set, has no proper non-trivial closed normal subgroups (it has exactly three proper non-trivial normal subgroups [325], none of which is closed).

Corollary 9.5.26 has many consequences. For instance, it shows in combination with Theorem 9.3.1 that every ω-categorical structure is bi-interpretable with an ω-categorical structure whose automorphism group has only one orbit.

Several fundamental properties of ω-categorical structures $\mathfrak{B}$ are preserved by bi-interpretability, and therefore, by Corollary 9.5.26, only depend on the topological automorphism group of $\mathfrak{B}$. As we will see in Chapter 11, this is for instance the case for the property whether an ω-categorical structure has the Ramsey property.

9.6. Uniformly continuous minor-preserving maps

There is also a topological variant for Birkhoff's theorem for height-one identities from Section 6.7.2. This variant can be used to study the class $\mathrm{HI}(\mathfrak{B})$ from Section 3.6 for ω-categorical structures $\mathfrak{B}$; the results presented here are essentially from [27]. Note that for minor-preserving maps, unlike for clone homomorphisms, one cannot simply replace uniform continuity by continuity; the compactness argument from Section 9.5.2 involves left compositions with invertible elements which need not be preserved by a minor-preserving map.

THEOREM 9.6.1. *Let $\mathcal{C}$ and $\mathcal{D}$ be closed oligomorphic subclones of $\mathcal{O}_{\mathbb{N}}$ and let $\xi\colon \mathcal{C} \to \mathcal{D}$ be a surjective minor-preserving map. Then* $(1) \Rightarrow (2) \Rightarrow (3) \Rightarrow (4)$.

1. *ξ is continuous and preserves composition with invertibles (Definition 6.7.6);*
2. *ξ is uniformly continuous and preserves right composition with invertibles;*
3. *there are τ-algebras $\mathbf{C}, \mathbf{D}$ such that $\mathrm{Clo}(\mathbf{C}) = \mathcal{C}$, $\mathrm{Clo}(\mathbf{D}) = \mathcal{D}$, $\xi(f^{\mathbf{C}}) = f^{\mathbf{D}}$, and $\mathbf{D} \in \mathrm{Refl}\, \mathrm{P}^{\mathrm{fin}}(\mathbf{C})$;*
4. *ξ is uniformly continuous.*

PROOF. $(1) \Rightarrow (2)$. Lemma 9.5.14 states that if ξ preserves left composition with invertibles and is continuous, then it is uniformly continuous.

$(2) \Rightarrow (3)$. Let $\mathbf{C}$ be any τ-algebra such that $\mathrm{Clo}(\mathbf{C}) = \mathcal{C}$, and let $\mathbf{D}$ be the τ-algebra on D given by $f^{\mathbf{D}} := \xi(f^{\mathbf{C}})$. Let $d_1, \dots, d_k$ be representatives for each of the orbits of $\mathcal{D}$ and let $F := \{d_1, \dots, d_k\}$. By uniform continuity there exists a finite subset G of the domain of $\mathcal{C}$ such that $f|_{G^k} = g|_{G^k}$ implies $\xi(f)|_{F^k} = \xi(f)|_{F^k}$ for all $f, g \in \mathcal{C}^{(k)}$. Let $A := G^F$ and let $m := |A| = |G|^k$. Let $a^1, \dots, a^m$ be an enumeration of A, and for $j \leq k$ define $a_j := (a_j^1, \dots, a_j^m) \in G^m$. Let $\mathbf{S}$ be the subalgebra of $\mathbf{C}^m$ generated by $a_1, \dots, a_k$; so the elements of $\mathbf{S}$ are precisely those of the form $f^{\mathbf{C}^m}(a_1, \dots, a_k)$, for some k-ary $f \in \tau$. Define a function $\mu\colon S \to D$ by setting

$$\mu(f^{\mathbf{S}}(a_1, \dots, a_k)) := f^{\mathbf{D}}(d_1, \dots, d_k).$$

The function μ is well defined: the proof is the same as the proof of Claim 1 in the proof of Theorem 9.5.1.

Conversely, to define a function $\nu\colon D \to S$ for an element $e \in D$, note that there exists a $u \in \tau$ and an $i \in \{1, \dots, k\}$ such that $e = u^{\mathbf{D}}(d_i)$. Then we define

$$\nu(e) = \nu(u^{\mathbf{D}}(d_i)) := u^{\mathbf{S}}(a_i).$$

To prove that $\mathbf{D}$ is the reflection of $\mathbf{S}$ at μ and ν, let $f \in \tau$ be an n-ary function symbol and let $e_1, \dots, e_n \in S$. For every $i \in \{1, \dots, n\}$ there is a $j_i \in \{1, \dots, k\}$ and a unary $u_i \in \tau$ such that $u_i^{\mathbf{D}}(d_{j_i}) = e_i$. Since $\mathfrak{C}$ is a clone there is some $g \in \tau$ such that $g^{\mathbf{C}} = f^{\mathbf{C}}(u_1^{\mathbf{C}}, \dots, u_n^{\mathbf{C}})$. By assumption, ξ preserves right composition with invertibles, and hence

$$\begin{aligned} g^{\mathbf{D}} = \xi(g^{\mathbf{C}}) = \xi(f^{\mathbf{C}}(a_1^{\mathbf{C}}, \dots, a_n^{\mathbf{C}})) &= \xi(f^{\mathbf{C}})\xi(a_1^{\mathbf{C}}, \dots, a_n^{\mathbf{C}}) \\ &= f^{\mathbf{D}}(u_1^{\mathbf{D}}, \dots, u_n^{\mathbf{D}}). \end{aligned}$$

Hence,

$$\begin{aligned} f^{\mathbf{D}}(e_1, \dots, e_n) &= f^{\mathbf{D}}(u_1^{\mathbf{D}}(d_{j_1}), \dots, u_n^{\mathbf{D}}(d_{j_n})) \\ &= g^{\mathbf{D}}(d_{j_1}, \dots, d_{j_n}) \\ &= \mu(g^{\mathbf{S}}(a_{j_1}, \dots, a_{j_n})) \\ &= \mu(f^{\mathbf{S}}(u_1^{\mathbf{S}}(a_{j_1}), \dots, u_n^{\mathbf{S}}(a_{j_n})) \\ &= \mu(f^{\mathbf{S}}(\nu(e_1), \dots, \nu(e_n))). \end{aligned}$$

It follows that $\mathbf{D}$ is a reflection of a subalgebra $\mathbf{S}$ of $\mathbf{C}^m$, and so $\mathbf{D} \in \operatorname{Refl} \mathrm{P}^{\mathrm{fin}}(\mathbf{C})$.

(3) $\Rightarrow$ (4). We have already seen in Theorem 9.5.1 that the map $f^{\mathbf{C}} \mapsto f^{\mathbf{B}}$ is uniformly continuous if $\mathbf{B} \in \mathrm{P}^{\mathrm{fin}}(\mathbf{C})$; so it suffices to show that the map $f^{\mathbf{B}} \mapsto f^{\mathbf{D}}$ is uniformly continuous if $\mathbf{D} \in \operatorname{Refl}(\mathbf{B})$. Suppose that $\mathbf{D}$ is a reflection of $\mathbf{B}$ via the maps $h \colon B \to D$ and $i \colon D \to B$. Let $F \subseteq D$ be finite. Choose $G := i(F)$, and suppose that f, g are k-ary and satisfy $f^{\mathbf{B}}|_{G^k} = g^{\mathbf{B}}|_{G^k}$. Then for every k-ary f, g and $d \in F^k$ we have

$$f^{\mathbf{D}}(d) = h(f^{\mathbf{B}}(i(d))) = h(g^{\mathbf{B}}(i(d))) = g^{\mathbf{D}}(d)$$

which proves uniform continuity of $f^{\mathbf{B}} \mapsto f^{\mathbf{D}}$. ⊣

The implication (4) $\Rightarrow$ (3) in Theorem 9.6.1 is false in general as the following example shows (the author is grateful to Michael Pinsker for pointing this out).

Example 9.6.2. Let $\mathfrak{B}$ be a homogeneous digraph with an undecidable constraint satisfaction problem (Example 2.3.12). Then all polymorphisms of $(\mathfrak{B}, P_B^3)$ are essentially unary (Lemma 6.1.17). We claim that there exists a uniformly continuous homomorphism from $\operatorname{Pol}(B; P_B^3, \neq)$ to $\operatorname{Pol}(\mathfrak{B}, P_B^3, \neq)$. Pick a uniform homeomorphism ξ from $\operatorname{Aut}(B; P_B^3)$ to $\operatorname{Aut}(\mathfrak{B}, P_B^3)$ which exists by Proposition 9.3.3. Since $\operatorname{Aut}(B; P_B^3)$ and $\operatorname{Aut}(\mathfrak{B}, P_B^3)$ are dense in $\operatorname{End}(B; P_B^3, \neq)$ and in $\operatorname{End}(\mathfrak{B}, P_B^3, \neq)$, respectively, the map ξ has a unique extension to a uniform homeomorphism from $\operatorname{End}(B; P_B^3, \neq)$ to $\operatorname{End}(\mathfrak{B}, P_B^3, \neq)$. Finally, ξ has a minor-preserving uniformly continuous extension to $\operatorname{Pol}(B; P_B^3, \neq)$ to $\operatorname{Pol}(\mathfrak{B}, P_B^3, \neq)$ because all operations in f are essentially unary.

Since $\operatorname{CSP}(B; P_B^3)$ is clearly in NP, but $\operatorname{CSP}(\mathfrak{B}, P_B^3)$ is undecidable, the structure $(\mathfrak{B}; P_B^3)$ cannot have a primitive positive interpretation in $(B; P_B^3)$

(Theorem 3.1.4). Theorem 6.3.7 therefore implies that item (3) in Theorem 9.6.1 does not apply.

In Example 9.6.5 below we show that the implication (3) $\Rightarrow$ (2) in Theorem 9.6.1 is false even if $\mathcal{D} = \mathbf{Proj}$ and $\mathcal{C}$ is a clone on a finite domain. An interesting reduct $\mathfrak{C}$ of a finitely bounded homogeneous structure such that $\mathrm{Pol}(\mathfrak{C})$ has a uniformly continuous homomorphism to **Proj**, but no clone homomorphism to **Proj**, is presented in the following example.

Example 9.6.3. Let $\mathfrak{C}$ be the structure $(\mathbb{Q}; T_3)$ where

$$T_3 := \{(x, y, z) \mid x = y < z \vee x > y = z\})$$

(Definition 3.1.8). In Proposition 3.1.9 we have proved the NP-hardness of $\mathrm{CSP}(\mathfrak{C})$ via a primitive positive interpretation of the structure $(\{0, 1\}; \mathrm{1IN3})$ in an expansion of $\mathfrak{C}$ by to constants. Hence, if $\mathbf{C}$ is a polymorphism algebra of $\mathfrak{C}$, Theorem 6.4.3 implies that $\mathrm{Exp}\,\mathrm{Refl}\,\mathrm{P}^{\mathrm{fin}}(\mathbf{C})$ contains a two-element algebra all of whose operations are projections. The implication (3) $\Rightarrow$ (1) in Theorem 6.7.8 therefore gives us a minor-preserving map from $\mathrm{Pol}(\mathfrak{C})$ to **Proj**, and the implication (3) $\Rightarrow$ (4) in Theorem 6.4.3 shows that $\mathrm{Pol}(\mathfrak{C})$ is a uniformly continuous.

In Section 12.5 we prove that $\mathfrak{C}$ has endomorphisms $\alpha_1, \alpha_2, \beta_1, \beta_2$ and a binary polymorphism pp satisfying

$$\forall x, y \; (\mathrm{pp}(\alpha_1(x), y) = \beta_1(y) \wedge \mathrm{pp}(\alpha_2(x), y) = \beta_2(x)).$$

Every binary operation over a 2-element set that satisfies this sentence is either essential or constant. So $\mathrm{Pol}(\mathfrak{C})$ cannot have a homomorphism to **Proj**.

Note that the existence of such operations is preserved by minor-preserving maps that preserve right composition with unary operations. It follows that there is no minor-preserving map to **Proj** that preserves right composition with $\mathrm{End}(\mathfrak{C})$. This example shows that if we replace 'right composition with invertibles' by 'right composition with unaries' in item (2) of Theorem 9.6.1, then the implication (3) $\Rightarrow$ (2) is false.

9.6.1. Uniformly continuous minor-preserving maps to Proj. For *finite* structures $\mathfrak{A}$ we have a necessary and sufficient condition for containment of $\mathfrak{A}$ in $\mathrm{HI}(\mathfrak{B})$ in terms of uniformly continuous minor-preserving maps.

THEOREM 9.6.4 (essentially from [27]). *Let $\mathfrak{A}$ be a finite structure and $\mathfrak{B}$ an at most countable ω-categorical structure. Then the following are equivalent.*

1. *There is a uniformly continuous minor-preserving map* $\xi\colon \mathrm{Pol}(\mathfrak{B}) \to \mathrm{Pol}(\mathfrak{A})$.
2. *If* $\mathbf{B}$ *is a polymorphism algebra of* $\mathfrak{B}$ *then there exists* $\mathbf{A} \in \mathrm{Refl}\,\mathrm{P}^{\mathrm{fin}}(\mathbf{B})$ *such that* $\mathrm{Clo}(\mathbf{A}) \subseteq \mathrm{Pol}(\mathfrak{A})$.
3. $\mathfrak{A} \in \mathrm{HI}(\mathfrak{B})$.

PROOF. The equivalence between (2) and (3) is Theorem 6.4.3. For the implication from (2) to (1), it follows from Theorem 6.7.8 that the natural minor-preserving map from $\text{Pol}(\mathfrak{B})$ to $\text{Pol}(\mathfrak{A})$ exists, and it is uniformly continuous by Theorem 9.6.1 (3) $\Rightarrow$ (4).

The implication from (1) to (2) has essentially been shown in the proof of Theorem 6.7.8; we only need a similar modification as in the proof of Theorem 9.5.1. Alternatively, one may modify the proof of Theorem 9.6.1 and enumerate all elements of A instead of orbit representatives. ⊣

Example 9.6.5. In Example 6.7.1 we have seen a finite structure $\mathfrak{B}$ such that $K_3 \notin I(\mathfrak{B})$ and hence $\text{Pol}(\mathfrak{B})$ has no clone homomorphism to **Proj**; the identity presented there shows that $\text{Pol}(\mathfrak{B})$ has no minor-preserving map to **Proj** that preserves composition with $\text{End}(\mathfrak{B})$; since $\mathfrak{B}$ is a core, this shows that item (2) in Theorem 9.6.1 does not apply.

On the other hand, $K_3 \in \text{HI}(\mathfrak{B})$ and hence, by Theorem 9.6.4 above, item (3) in Theorem 9.6.1 applies for $\mathcal{C} = \text{Pol}(\mathfrak{B})$ and $\mathcal{D} = \textbf{Proj}$.

Theorem 9.6.4 provides another equivalent formulation of the condition from the infinite-domain tractability conjecture (Conjecture 3.1).

COROLLARY 9.6.6. *Let $\mathfrak{B}$ be an at most countable ω-categorical structure. Then the following are equivalent.*

1. $\text{HI}(\mathfrak{B})$ *contains* $(\{0,1\}; \text{1IN3})$ *(and the other equivalent conditions in Corollary* 6.4.4*).*
2. *There is a uniformly continuous minor-preserving map* $\text{Pol}(\mathfrak{B}) \to \textbf{Proj}$.

If these conditions apply, then $\mathfrak{B}$ has a finite-signature reduct with an NP-hard CSP.

PROOF. Since **Proj** is isomorphic to $\text{Pol}(\{0,1\}; \text{1IN3})$ (Proposition 6.5.19) the equivalence of (3) and (4) follows from Theorem 9.6.4. The final statement follows from Corollary 6.4.4. ⊣

Example 9.6.7. We consider again the atomless Boolean algebra $\mathfrak{A}$ from Example 6.7.5, where we described a minor-preserving map from $\text{Pol}(\mathfrak{A})$ to **Proj**. This map is uniformly continuous: for every $n \in \mathbb{N}$ and $f, g \in \text{Pol}(\mathfrak{A})^{(n)}$, if $f|_{\{\mathbf{0},\mathbf{1}\}} = g|_{\{\mathbf{0},\mathbf{1}\}}$ then $\xi(f) = \xi(g)$. Corollary 9.6.6 implies that $(\{0,1\}; \text{1IN3}) \in \text{HI}(\mathfrak{A})$. Indeed, $(\{0,1\}; \text{1IN3})$ embeds into

$$\mathfrak{B} := \big(\mathbb{A}; \{(x,y,z) \mid (c(x) \cap y \cap z) \cup (x \cap c(y) \cap z) \cup (x \cap y \cap c(z)) = \mathbf{1}\}\big)$$

via the map that sends 0 to $\mathbf{0}$ and 1 to $\mathbf{1}$, and conversely there is a homomorphism h from $\mathfrak{B}$ to $(\{0,1\}; \text{1IN3})$: as in the construction of the minor-preserving map from $\text{Pol}(\mathfrak{A})$ to **Proj** in Example 6.7.5, we pick an ultrafilter $U \subset \mathbb{A}$ that contains $\mathbf{0}$. Then h maps all elements in U to $\mathbf{0}$ and all other elements to $\mathbf{1}$.

Remark 9.6.8. An example of an ω-categorical structure $\mathfrak{B}$ such that $\mathrm{Pol}(\mathfrak{B})$ has a minor-preserving map to **Proj**, but no uniformly continuous minor-preserving map to **Proj** has been constructed recently [82]; the structure $\mathfrak{B}$ can be chosen to have a finite relational signature [179].

9.6.2. Local satisfaction of minor conditions. The existence of uniformly continuous minor-preserving maps to an operation clone over a finite domain can be characterised in terms of *local satisfaction of minor conditions* (Proposition 9.6.11).

Let ϕ be a formula in the language of clones. Clearly, if a clone $\mathcal{C}$ satisfies $\phi(f_1, \dots, f_n)$ then $\mathcal{C}$ satisfies $\phi(f_1, \dots, f_n)$ on every finite subset of the domain of $\mathcal{C}$ (Definition 9.5.7). The converse is false, as the following example illustrates.

Example 9.6.9. Consider the Horn clone $\mathcal{H}$ (see Theorem 7.5.2); it locally satisfies the clone sentence

$$\exists a, b, f \; \big(f(x,x) = x \wedge a(f(x,y)) = b(f(y,x))\big) \tag{3}$$

since for every finite $F \subseteq \mathbb{N}$ we can pick a binary injection $f \colon \mathbb{N}^2 \to \mathbb{N}$ and unary injections a, b satisfying (3) for elements from F. However, if f satisfies (3) globally then it must be idempotent because of the first conjunct, and essential because of the second conjunct. Let $x, y \in \mathbb{N}$ be distinct, and let $z := f(x,y)$. Then $f(z,z) = z = f(x,y)$ contradicting the fact that in the Horn clone all functions that depend on all their arguments are injective.

If $\mathcal{C}$ is a locally closed oligomorphic clone, and if ϕ is a fixed minor condition (i.e., all variables are quantified), then local and global satisfaction of ϕ are again equivalent; this is a direct consequence of Lemma 4.1.10.

LEMMA 9.6.10. *Let $\mathcal{C}$ be a locally closed oligomorphic operation clone and let ϕ be a minor condition. Then $\mathcal{C}$ satisfies ϕ if and only if $\mathcal{C}$ satisfies ϕ locally.*

In particular, a locally closed oligomorphic operation clone has a minor-preserving map to **Proj** if and only if all minor conditions ϕ that are satisfied in $\mathcal{C}$ on all finite subsets of the domain of $\mathcal{C}$ are trivial. Also the existence of *uniformly continuous* minor-preserving maps to **Proj** can be characterised locally, but in a different sense; this is a consequence of the following more general proposition (a variant of Corollary 9.5.8).

PROPOSITION 9.6.11. *Let $\mathcal{C}$ be an operation clone and let $\mathcal{D}$ be an operation clone over a finite domain. Then the following are equivalent.*

- *There exists a uniformly continuous minor-preserving map from $\mathcal{C}$ to $\mathcal{D}$.*
- *There is a finite subset F of the domain of $\mathcal{C}$ such that every minor condition ϕ that is satisfied by $\mathcal{C}$ on F is also satisfied on $\mathcal{D}$.*

PROOF. If ξ is a uniformly continuous minor-preserving map from $\mathcal{C}$ to $\mathcal{D}$, then there is a finite subset F of the domain of $\mathcal{C}$ such that for all $n \in \mathbb{N}$ and

$f, g \in \mathcal{C}^{(n)}$, if $f|_{F^n} = g|_{F^n}$ the $\xi(f) = \xi(g)$. Hence, every minor condition that holds on F also holds in $\mathcal{D}$.

Conversely, suppose that there exists a finite subset F of the domain of $\mathcal{C}$ such that every minor condition that holds on F also holds in $\mathcal{D}$. Consider the minion of all restrictions of operations in $\mathcal{C}$ to F. Lemma 6.7.11 shows that there exists a minor-preserving map ξ from this minion to $\mathcal{D}$. Then the map $\zeta \colon \mathcal{C} \to \mathcal{D}$ which is defined for an operation $f \in \mathcal{C}$ of arity n by $\zeta(f) := \xi(f|_{F^n})$ is minor preserving, and uniformly continuous, because if $f|_{F^n} = g|_{F^n}$ then $\zeta(f) = \zeta(g)$. ⊣

Chapter 10

OLIGOMORPHIC CLONES

We have seen in Sections 6.6, 6.8.2, and 6.9 that polynomial-time tractability of finite-domain CSPs can be expressed in terms of the existence of polymorphisms that satisfy certain minor conditions, such as the existence of a Taylor polymorphism, a Siggers polymorphism, or a cyclic polymorphism. For ω-categorical structures, and even for reducts of finitely bounded homogeneous structures, the same minor conditions cannot be used to characterise polynomial-time tractability; in fact, failing these minor conditions no longer implies NP-hardness.

For example, the tractability condition for equality constraint languages from Theorem 7.4.2 requires the existence of a polymorphism f and an automorphism α such that

$$f(x, y) = \alpha f(y, x)$$

holds for all $x, y \in B$, which is not a minor condition. The Horn clone $\mathcal{H}$ introduced in Section 7.5 is an example of a clone that satisfies this condition but does not have any Taylor operation. This motivates our definition of *pseudo-minor conditions*. In Section 10.1 we present some general facts about pseudo-minor conditions and how to construct operations that satisfy these conditions.

In Section 10.2 we present a break-through result of Barto and Pinsker [28] which states that every ω-categorical model-complete core that satisfies the condition from the infinite-domain tractability conjecture in its original formulation (Conjecture 3.1) has a pseudo-Siggers polymorphism. In Section 10.3 we prove that the two versions of the infinite-domain tractability conjecture are indeed equivalent; this is based on results from [20]. Some of the results that we present in this chapter are under the assumption that the ω-categorical structure under consideration is a model-complete core, and for some results we do not need this assumption; this will be discussed in Section 10.4.

We dedicate the final section to the important concept of *canonical functions*, covering material from [81, 95, 92]. If $\mathcal{C}$ is the polymorphism clone of a structure $\mathfrak{C}$ which is homogeneous in a finite relational signature, and all polymorphisms of $\mathfrak{C}$ are canonical, then $\mathcal{C}$ is determined by a clone on a finite set, and CSP($\mathfrak{C}$) is polynomial-time equivalent to a finite-domain CSP; all this follows from the results in Section 10.5 and is based on material from [81]. The assumption that all polymorphisms of a structure $\mathfrak{B}$ are canonical might appear to be rather strong. But it turns out that very often, if $\mathfrak{B}$ is an ω-categorical structure such that CSP($\mathfrak{B}$) is not NP-hard, then $\mathfrak{B}$ has an expansion $\mathfrak{C}$ such that Pol($\mathfrak{C}$) is canonical and CSP($\mathfrak{C}$) (and hence CSP($\mathfrak{B}$)) can be solved in polynomial time. A partial explanation as to why this happens so often can be found in the next chapter about Ramsey theory.

10.1. Pseudo minor conditions

Let $\mathbf{C}$ be a clone. A *pseudo minor condition* is a primitive positive clone sentence whose conjuncts are of the form (using the notation from Section 6.7.3)

$$a \circ f_\sigma = b \circ g_\tau$$

where a and b are variables of rank one, f and g are variables of rank n_1 and n_2, respectively, and $\sigma \colon \{1, \dots, n_1\} \to \{1, \dots, k\}$ and $\tau \colon \{1, \dots, n_2\} \to \{1, \dots, k\}$ are functions. If there exist $a, b \in S \subseteq C^{(1)}$ such that $a \circ f_\sigma = b \circ g_\tau$ holds in $\mathbf{C}$ then we also say that *$f_\sigma = g_\tau$ holds modulo S*.

Let $\mathbf{C}$ and $\mathbf{D}$ be clones. A function $\xi \colon C \to D$ is called a *pseudo minor-preserving map* if it is minor-preserving and additionally preserves left composition with $C^{(1)}$. Using the terminology from Section 6.7.3 we thus have that if

$f \in C^{(n)}$, $a \in C^{(1)}$, and $\sigma \colon \{1, \dots, n\} \to \{1, \dots, k\}$, then

$$\xi(a \circ f_\sigma) = \xi(a) \circ \xi(f)_\sigma.$$

Clearly, pseudo minor-preserving maps preserve pseudo minor conditions.

Example 10.1.1. Recall that a Siggers operation (Section 6.8.2) is an operation s satisfying $\forall x, y, z \colon s(x, y, x, z, y, z) = s(y, x, z, x, z, y)$. We say that s is *Siggers modulo* $S \subseteq C^{(1)}$ if there are unary operations $p, q \in S$ such that

$$\forall x, y, z \colon p(s(x, y, x, z, y, z)) = q(s(y, x, z, x, z, y)).$$

If s is Siggers modulo $C^{(1)}$ we also say that s is a *pseudo-Siggers operation of* $\mathbf{C}$. To express that s is a pseudo-Siggers operation, we can use the clone formula $\phi_{\text{Sig}}(s)$ defined as

$$\begin{aligned}\exists p, q \colon\ & p \circ \text{comp}(s, \text{pr}^3_1, \text{pr}^3_2, \text{pr}^3_1, \text{pr}^3_3, \text{pr}^3_2, \text{pr}^3_3) \\ &= q \circ \text{comp}(s, \text{pr}^3_2, \text{pr}^3_1, \text{pr}^3_3, \text{pr}^3_1, \text{pr}^3_3, \text{pr}^3_2).\end{aligned} \tag{1}$$

The clone formula $\exists s \colon \phi_{\text{Sig}}(s)$, stating the existence of a pseudo-Siggers operation, is an example of a pseudo minor condition. Section 10.2 presents a proof, due to Pinsker and Barto [28], that an ω-categorical model-complete core satisfies the conditions from the infinite-domain tractability conjecture (Conjecture 4.1) if and only if it has a pseudo-Siggers polymorphism.

Example 10.1.2. An operation $f \colon B^2 \to B$ is called *symmetric modulo* $S \subseteq C^{(1)}$ if there are $p, q \in S$ such that for all $x, y \in B$

$$p(f(x, y)) = q(f(y, x)),$$

and is called a *pseudo-symmetric operation in* $\mathcal{C}$ if it is symmetric modulo $C^{(1)}$. Note that every binary injective operation from $B^2 \to B$ is pseudo-symmetric with respect to the set of all injective maps from $B \to B$, and so is every binary constant operation. Hence, by Theorem 7.4.1 the existence of a pseudo-symmetric polymorphism characterises polynomial-time tractability of the CSP of equality constraint languages.

10.1.1. Pseudo versus quasi. Clearly, oligomorphic operation clones over infinite sets cannot be idempotent. Some important operations on infinite sets, such as the binary minimum operation over $(\mathbb{Q}; <)$ or the projections, are idempotent, but this is rather rare (see for example the clones that appear in Chapter 12). One way to relax the assumption of idempotence, for instance for majority operations f, is to replace the identities

$$\forall x, y \colon f(x, x, y) = f(x, y, x) = f(y, x, x) = x$$

by their "quasi" form (cf. Definition 6.1.39, Section 8.5.2)

$$\forall x, y \colon f(x, x, y) = f(x, y, x) = f(y, x, x) = f(x, x, x).$$

Here we present an observation, due to Kozik and Wrona, that implies that certain pseudo minor conditions imply the corresponding 'quasi identities'.

LEMMA 10.1.3. *Let $\mathcal{C}$ be an operation clone whose unary operations are injective and let $f \in \mathcal{C}$ and $a_1, a_2 \in \mathcal{C}^{(1)}$ be such that for some $i_0, i_1, \dots, i_k \leq n$*

$$\forall x_1, \dots, x_n \colon a_1(f(x_{i_1}, \dots, x_{i_k})) = a_2(f(x_{i_0}, \dots, x_{i_0})).$$

Then

$$\forall x_1, \dots, x_n \colon f(x_{i_1}, \dots, x_{i_k}) = f(x_{i_0}, \dots, x_{i_0}).$$

PROOF. The assumption asserts in particular that for all x_{i_0}

$$a_1(f(x_{i_0}, \dots, x_{i_0})) = a_2(f(x_{i_0}, \dots, x_{i_0})).$$

Since $a_2(f(x_{i_0}, \dots, x_{i_0})) = a_1(f(x_{i_1}, \dots, x_{i_k}))$ we therefore have

$$a_1(f(x_{i_0}, \dots, x_{i_0})) = a_1(f(x_{i_1}, \dots, x_{i_k})).$$

By assumption a_1 injective and hence $f(x_{i_0}, \dots, x_{i_0}) = f(x_{i_1}, \dots, x_{i_k})$. ⊣

COROLLARY 10.1.4. *Let $\mathcal{C}$ be an operation clone such that all operations in $\mathcal{C}^{(1)}$ are injective, and suppose that there are $a_0, a_1, \dots, a_k \in \mathcal{C}^{(1)}$ and $f \in \mathcal{C}^{(k)}$ such that*

$$\begin{aligned}\forall x, y \colon a_1(f(y, x, x, \dots, x)) &= a_2(f(x, y, \dots, x)) \\ &= \dots = a_k(f(x, \dots, x, y)) = a_0(f(x, \dots, x)).\end{aligned}$$

Then f is a quasi near-unanimity operation.

10.1.2. The lift lemma. In order to show that an oligomorphic operation clone satisfies a pseudo minor condition it suffices to verify the condition locally (recall Definition 9.5.7). While this is in parallel to the case of minor conditions (see Lemma 9.6.10), it requires a different compactness proof. We actually prove an even stronger statement where we can fix the operations that satisfy the given identities modulo some unary operations; we refer to this result as the *lift lemma*. The lift lemma is essentially from [95], but there are many variations and different presentations of the result (some of which appeared e.g. in [92, 28, 64]). The result even holds for infinite systems of identities, but since the infinite version is not needed in this text and requires additional technical effort, we do not prove it here, and refer the interested reader to [92] instead. On the other hand, we also state an addition that first appeared in [20] which allows us to prove that a given clone satisfies a system of pseudo-identities where some of the unary maps are required to be equal. Moreover, our formulation of the lift lemma can also be used for ω-categorical structures that are not necessarily model-complete cores (like the one in [28], which concerns the particular case of the pseudo-Siggers identity).

LEMMA 10.1.5 (Lift lemma). *Let $\mathfrak{B}$ be a countable ω-categorical structure and for $i \in \{1,\dots,s\}$ let $f_i, g_i \colon B^{k_i} \to B$ be operations such that for every finite $F \subseteq B$ there are $a_{i,F}, b_{i,F} \in \mathrm{End}(\mathfrak{B})$ such that $a_{i,F} \circ f_i|_F = b_{i,F} \circ g_i|_F$. Then there are $d_1,\dots,d_s,e_1,\dots,e_s \in \mathrm{End}(\mathfrak{B})$ such that $d_i \circ f_i = e_i \circ g_i$ for every $i \in \{1,\dots,s\}$. Moreover, if $i,j \in \{1,\dots,s\}$ are such that $a_{i,F} = a_{j,F}$ for all finite $F \subseteq B$, then $d_i = d_j$, and likewise if $b_{i,F} = b_{j,F}$ for all finite $F \subseteq B$, then $e_i = e_j$.*

PROOF. Suppose without loss of generality that $B = \mathbb{N}$. For $e_1, e_2 \in \mathrm{End}(\mathfrak{B})$ we write $e_1 =_n e_2$ if $e_1|_{\{0,\dots,n-1\}} = e_2|_{\{0,\dots,n-1\}}$. For $i \in \{1,\dots,s\}$ and $n \in \mathbb{N}$ let $P_{i,n}$ be the set of all pairs $(a_{i,n}, b_{i,n}) \in \mathrm{End}(\mathfrak{B})^2$ such that $a_{i,n} \circ f_i =_n b_{i,n} \circ g_i$. Note that $P_{i,n}$ is non-empty by assumption. Define $P_n := P_{1,n} \times \cdots \times P_{s,n}$, and define the equivalence relation $\sim$ on P_n as follows: $((a_1,b_1),\dots,(a_s,b_s)) \sim ((a'_1,b'_1),\dots,(a'_s,b'_s))$ if there exists $\beta \in \mathrm{Aut}(\mathfrak{B})$ such that for all $i \in \{1,\dots,s\}$

$$a_i =_n \beta \circ a'_i \quad \text{and} \quad b_i =_n \beta \circ b'_i.$$

Note that for each n, the relation $\sim$ has finitely many equivalence classes on P_n, because their number is bounded by the number of orbits of sn-tuples under $\mathrm{Aut}(\mathfrak{B})$, which is finite by the ω-categoricity of $\mathfrak{B}$. We now construct a rooted tree as follows. Each vertex of the tree lies on some level $n \in \mathbb{N}$. The vertices of the tree on level n are precisely the equivalence classes of $\sim$ on P_n. We define adjacency in the tree as follows: if E is a vertex on level n, and E' is a vertex on level $n+1$, and $((a_1,b_1),\dots,(a_s,b_s)) \in E$ and $((a'_1,b'_1),\dots,(a'_s,b'_s)) \in E'$ such that for all $i \in \{1,\dots,s\}$

$$a_i =_n a'_i \quad \text{and} \quad b_i =_n b'_i$$

then we make E and E' adjacent in the tree. Note that the resulting tree is finitely branching and contains vertices on all levels. Hence, by Kőnig's tree lemma there exists an infinite path $E_0, E_1, E_2, \dots$ in the tree. We construct an infinite sequence $(p_n)_{n\in\mathbb{N}}$ with

$$p_n = ((a_{1,n}, b_{1,n}),\dots,(a_{s,n}, b_{s,n})) \in E_n$$

such that for all $i \in \{1,\dots,s\}$ and $n \in \mathbb{N}$

$$a_{i,n-1} =_n a_{i,n} \quad \text{and} \quad b_{i,n-1} =_n b_{i,n}.$$

Initially, p_0 is defined to be any element of E_0. Suppose we have already defined p_n. Since E_n and E_{n+1} are adjacent, there are $((a_1,b_1),\dots,(a_s,b_s)) \in E_n$ and $((a'_1,b'_1),\dots,(a'_s,b'_s)) \in E_{n+1}$ such that $a_i =_n a'_i$ and $b_i =_n b'_i$ for all $i \in \{1,\dots,s\}$. By the definition of $\sim$ on P_n there exists $\beta \in \mathrm{End}(\mathfrak{B})$ such that for all $i \in \{1,\dots,s\}$

$$a_{i,n} =_n \beta \circ a_i \quad \text{and} \quad b_{i,n} =_n \beta \circ b_i.$$

Define $a_{i,n+1} := \beta \circ a'_i$ and $b_{i,n+1} := \beta \circ b'_i$. We verify that this definition has the required properties. First, observe that

$$a_{i,n+1} = \beta \circ a'_i =_n \beta \circ a_i = a_{i,n}.$$

Moreover,

$$p_{n+1} := ((a_{1,n+1}, b_{1,n+1}), \dots, (a_{s,n+1}, b_{s,n+1})) \in E_{n+1}$$

since $p_{n+1} \sim ((a'_1, b'_1), \dots, (a'_s, b'_s)) \in E_{n+1}$. For $i \in \{1, \dots, s\}$ and $n \in \mathbb{N}$ define $d_i(n) := a_{s,n+1}(n)$ and $e_i(n) := b_{s,n+1}(n)$; these maps have the properties that are required in the statement of the lemma. $\dashv$

In many situations, in particular if the structure $\mathfrak{B}$ from Lemma 10.1.5 is a model-complete core, the following permutation-group formulation of the lift lemma is sufficient. We find it instructive to give two proofs: the first as a direct consequence of Lemma 10.1.5, and the second using the terminology from topology.

COROLLARY 10.1.6. *Let $\mathcal{G}$ be an oligomorphic permutation group on a countable base set B and for $i \in \{1, \dots, s\}$ let $f_i, g_i \colon B^{k_i} \to B$ be operations such that for every finite $F \subseteq B$ there is an $\alpha_{i,F} \in \mathcal{G}$ such that $\alpha_{i,F} f_i|_F = g_i|_F$. Then there are $e_0, e_1, \dots, e_s \in \overline{\mathcal{G}}$ such that $e_i \circ f_i = e_0 \circ g_i$ for every $i \in \{1, \dots, s\}$. Moreover, if $i, j \in \{1, \dots, s\}$ are such that $\alpha_{i,F} = \alpha_{j,F}$ for all finite $F \subseteq B$, then $e_i = e_j$.*

PROOF USING LEMMA 10.1.5. Take $\mathfrak{B}$ any structure with domain B such that $\overline{\mathcal{G}} = \operatorname{End}(\mathfrak{B})$ (Proposition 4.4.2). Then we apply Lemma 10.1.5 with $a_{i,F} := \alpha_{i,F}$ and $b_{i,F} := \mathrm{id}$ for all $i \in \{1, \dots, s\}$. $\dashv$

For our second proof of Corollary 10.1.6, based on the presentation in [92], we use a certain compact Hausdorff space that will be important later again in Section 11.4.2. Recall that the space $B^B/\mathcal{G}$ from Definition 9.5.10 is compact (Proposition 9.5.11) but not Hausdorff (Remark 9.5.12). We consider the following quotient space which is still compact and additionally Hausdorff.

DEFINITION 10.1.7. Let $\mathcal{G}$ be a permutation group on a set B and let A be any set. For $f, g \in B^A$, define $f \sim g$ if $f \in \overline{\mathcal{G}g} := \overline{\{\alpha \circ g \mid \alpha \in \mathcal{G}\}}$. Then we also write $B^A /\!\!/ \mathcal{G}$ instead of $B^A/{\sim}$.

PROPOSITION 10.1.8. *Let $\mathcal{G}$ be an oligomorphic permutation group on a set B, and let A be countable. Then $B^A /\!\!/ \mathcal{G}$ is a compact Hausdorff space.*

PROOF. Since $B^A /\!\!/ \mathcal{G}$ is a quotient of $B^A/\mathcal{G}$, and since $B^A/\mathcal{G}$ is compact (Proposition 9.5.11), the compactness of $B^A /\!\!/ \mathcal{G}$ follows from Proposition 9.1.18. To prove that $B^A /\!\!/ \mathcal{G}$ is Hausdorff, let $s_1/{\sim}, s_2/{\sim}$ be elements of $B^A /\!\!/ \mathcal{G}$. If these two elements are distinct, there exists $t \in A^n$ such that $s_1(t), s_2(t) \in B^n$ lie in different orbits of n-tuples under $\mathcal{G}$. Then $s_1 \in U_1 := \{u \in B^A \mid u(t) = s_1(t)\}$ and $s_2 \in U_2 := \{u \in B^A \mid u(t) = s_2(t)\}$, and U_1 and U_2 are open and disjoint. $\dashv$

SECOND PROOF OF COROLLARY 10.1.6. We may assume that $B = \mathbb{N}$, and we write $\alpha_{i,j}$ instead of $\alpha_{i,\{0,\dots,j-1\}}$. By adding fictitious arguments to the f_i and g_i we may assume that $k_i = m$ for every $i \in \{1,\dots,s\}$. Let $\mathcal{H}$ be the componentwise action of $\mathcal{G}$ on B^{s+1}. Then $(B^{s+1})^{B^{s+1}} /\!\!/ \mathcal{H}$ is compact by Proposition 10.1.8, and hence its closed subspace $\overline{\mathcal{G}}^{s+1} /\!\!/ \mathcal{H}$ is compact as well (Proposition 9.1.18). Let S be the set of equivalence classes of elements of $\overline{\mathcal{G}}^{s+1}$ of the form $(\mathrm{id}, \alpha_{1,j}, \dots, \alpha_{s,j})$ for $j \in \mathbb{N}$. Note that $\lim_{j\to\infty}(\alpha_{i,j} f_i) = g_i$. Hence, if S is finite then there exists $\delta \in \mathcal{G}$ and $j \in \mathbb{N}$ such that for every $i \in \{1,\dots,s\}$ we have $\delta\alpha_{i,j} f_i = g_i$, and we are done. Otherwise, by Proposition 9.1.19 the set S has a limit point $(b, a_1, \dots, a_s)/\!\sim$. Hence, there are $\delta_j \in \mathcal{G}$ for $j \in \mathbb{N}$ such that $\lim_{j\to\infty}(\delta_j, \delta_j\alpha_{1,j}, \dots, \delta_j\alpha_{s,j}) = (b, a_1, \dots, a_s)$. We obtain that for every $i \in \{1,\dots,s\}$

$$\begin{aligned} a_i \circ f_i &= \left(\lim_{j\to\infty} \delta_j \alpha_{i,j}\right) \circ f_i \\ &= \lim_{j\to\infty} \delta_j \circ \lim_{j\to\infty} (\alpha_{i,j} f_i) = b \circ g_i. \end{aligned}$$

For the second part of the statement it suffices to observe that if there are $i, i' \in \{1,\dots,s\}$ such that $\alpha_{i,j} = \alpha_{i',j}$ for all $j \in \mathbb{N}$, then $a_i = a_{i'}$ for the representative $(b, a_1, \dots, a_s)$ of the limit point. ⊣

For illustration we present an application of the lift lemma for pseudo-Siggers operations (recall Example 10.1.1).

COROLLARY 10.1.9. *Let $\mathcal{C}$ be an oligomorphic operation clone such that the invertible elements in $\mathcal{C}^{(1)}$ are dense in $\mathcal{C}^{(1)}$, and let $s \in \mathcal{C}^{(6)}$. Then s is a pseudo-Siggers operation in $\mathcal{C}$ if and only if $\mathcal{C}$ locally satisfies the pseudo-Siggers clone formula $\phi_{\mathrm{Sig}}(s)$ given in* (1).

PROOF. The statement is an immediate consequence of the lift lemma in the form of Corollary 10.1.6 applied to $f_1(x, y, z) := s(x, y, x, z, y, z)$ and $g_1(x, y, z) := s(y, x, z, x, z, y)$. ⊣

Note that the corollary does not immediately imply that $\mathcal{C}$ has a pseudo-Siggers polymorphism if and only if $\mathcal{C}$ satisfies the pseudo-Siggers condition $\exists s : \phi_{\mathrm{Sig}}(s)$ locally. To prove this we have to combine the lift lemma with yet another compactness argument. The argument does not only apply to the pseudo-Siggers condition, but shows the following.

LEMMA 10.1.10. *Let $\mathfrak{B}$ be a countable ω-categorical structure and ϕ a pseudo minor condition. Then ϕ holds in* Pol($\mathfrak{B}$) *if and only if ϕ holds in* Pol($\mathfrak{B}$) *locally.*

PROOF. Clearly, if ϕ holds in Pol($\mathfrak{B}$) then it also holds locally in Pol($\mathfrak{B}$). For the converse, let $a_1 \circ (f_1)_{\sigma_1} = b_1 \circ (g_1)_{\tau_1}, \dots, a_s \circ (f_s)_{\sigma_s} = b_s \circ (g_s)_{\tau_s}$ be the conjuncts of the pseudo minor condition. The existence of functions $f_1, g_1, \dots, f_s, g_s$ such that for every finite subset $F \subseteq B$ there are

$a'_1, \dots, a'_s, b'_1, \dots, b'_s \in \mathrm{End}(\mathfrak{B})$ such that

$$a'_i \circ (f_i)_{\sigma_i}|_F = b'_i \circ (g_i)_{\tau_i}|_F$$

for every $i \in \{1, \dots, s\}$ follows from Lemma 4.1.10. The statement then follows from Lemma 10.1.5. ⊣

10.1.3. Stable pseudo minor conditions. In this section we consider a certain class of pseudo minor conditions that are preserved by *taking point stabilisers* and by *diagonal interpolation*. A pseudo minor condition is called *stable* if each rank 1 clone variable in the pseudo minor condition always appears in front of the same clone variable. Formally, a clone formula $\phi(f_1, \dots, f_n)$ is called *stable* if it is of the form

$$\exists e_{1,1}, \dots, e_{1,k_1}, \dots, e_{n,1}, \dots, e_{n,k_n} : \psi$$

where $e_{1,1}, \dots, e_{1,k_1}, \dots, e_{n,1}, \dots, e_{n,k_n}$ are unary and ψ is a conjunction of atomic clone formulas of the form

$$e_{i,j} \circ (f_i)_\sigma = e_{p,q} \circ (f_p)_\tau$$

for $i, p \leq n$, $j \leq k_i$, $q \leq k_p$,

$$\sigma\colon \{1, \dots, \mathrm{ar}(f_i)\} \to \{1, \dots, k\} \text{ and } \tau\colon \{1, \dots, \mathrm{ar}(f_p)\} \to \{1, \dots, k\}.$$

Correspondingly, a pseudo minor condition $\exists f_1, \dots, f_n \colon \phi(f_1, \dots, f_n)$ is called *stable* if ϕ is stable.

Example 10.1.11. *Pseudo-loop conditions* [305] are defined analogously to loop conditions (Definition 6.7.10): they are pseudo minor conditions of the form

$$\exists f, a, b \colon a \circ f_\tau = b \circ f_\sigma$$

where a, b are unary, f is n-ary, and $\tau, \sigma\colon \{1, \dots, n\} \to \{1, \dots, k\}$. The existence of a pseudo-Siggers operation is an example of a pseudo-loop condition. Note that each pseudo-loop condition ϕ can be represented by a directed graph D_ϕ, as in the case of loop conditions. Observe that every pseudo-loop condition is stable.

Example 10.1.12. An example of a pseudo minor condition that is not stable is

$$\exists f_1, f_2, e_1, e_2 \big(e_1 \circ f(x, y) = e_2 \circ g(x, y) \wedge e_2 \circ f(x, y) = e_1 \circ g(y, x)\big).$$

The following proposition states that if the polymorphism clone of an ω-categorical model-complete core structure satisfies a stable pseudo minor condition, then this condition also holds in the polymorphism clone of the expansion of the structure by finitely many elements.

PROPOSITION 10.1.13. *Let $\mathcal{C}$ be an operation clone on a base set B such that the group $\mathcal{G}$ of invertible elements of $\mathcal{C}^{(1)}$ is dense in $\mathcal{C}^{(1)}$. If $\mathcal{C}$ satisfies a stable pseudo minor condition ϕ then the stabiliser clone $\mathcal{C}_t$, for some $t \in B^n$ and $n \in \mathbb{N}$, also satisfies ϕ.*

PROOF. Let $f_1, \dots, f_m \in \mathcal{C}$ be operations of arity $k_1, \dots, k_m$, respectively, and let $e_{1,1}, \dots, e_{1,k_1}, \dots, e_{m,1}, \dots, e_{m,k_m} \in \mathcal{C}^{(1)}$ be witnesses for the variables of ϕ showing that $\mathcal{C} \models \phi$. Then for each $i \in \{1, \dots, m\}$ the operation $\hat{f}_i \in \mathcal{C}^{(1)}$ given by $x \mapsto f_i(x, \dots, x)$ is in $\overline{\mathcal{G}}$. In particular, there exists $\beta_i \in \mathcal{G}$ such that $\beta_i \hat{f}_i(t) = t$. The operation $f'_i := \beta_i f_i \in \mathcal{C}$ preserves t, and hence $f'_i \in \mathcal{C}_t$.

Let $e_{i,j} \circ (f_i)_\sigma = e_{p,q} \circ (f_p)_\rho$ be a conjunct of ϕ, where $\sigma \colon \{1, \dots, k_i\} \to \{1, \dots, k\}$ and $\rho \colon \{1, \dots, k_p\} \to \{1, \dots, k\}$. Since $\mathcal{C}^{(1)} = \overline{\mathcal{G}}$ and $e_{i,j}(\hat{f}_i(t)) = e_{p,q}(\hat{f}_p(t))$ there exists $\alpha \in \mathcal{G}$ such that $\alpha e_{i,j}(\hat{f}_i(t)) = t = \alpha e_{p,q}(\hat{f}_p(t))$. Define

$$\begin{aligned} e'_{i,j} &:= \alpha e_{i,j} \beta_i^{-1} \\ \text{and } e'_{p,q} &:= \alpha e_{p,q} \beta_i^{-1}. \end{aligned}$$

We will show that $e'_{i,j}(t) = t = e'_{p,q}(t)$. We have

$$e'_{i,j}(t) = e'_{i,j}(f'_i(t, \dots, t)) = \alpha e_{i,j} \beta_i^{-1} \beta_i \hat{f}_i(t) = \alpha e_{i,j} \hat{f}_i(t) = t.$$

Therefore, $e'_{i,j} \in \mathcal{C}_t^{(1)}$. The proof that $e'_{p,q} \in \mathcal{C}_t^{(1)}$ is analogous. We finally verify that $e'_{i,j} \circ (f'_i)_\sigma = e'_{p,q} \circ (f'_p)_\rho$.

$$\begin{aligned} e'_{i,j} \circ (f'_i)_\sigma &= \alpha e_{i,j} \beta_i^{-1} \beta_i (f_i)_\sigma \\ &= \alpha e_{i,j} (f_i)_\sigma \\ &= \alpha e_{p,q} (f_p)_\rho \\ &= \alpha e_{p,q} \beta_i^{-1} \beta_i (f_p)_\sigma \\ &= e'_{p,q} \circ (f'_p)_\rho. \end{aligned}$$

This concludes the proof that $\mathcal{C}_t$ satisfies ϕ. ⊣

Let $\mathcal{G}$ be a permutation group on a set B and $f : B^k \to B$. Then the *diagonal interpolation space*

$$\overline{\{(x_1, \dots, x_k) \mapsto \gamma f(\beta(x_1), \dots, \beta(x_k)) \mid \beta, \gamma \in \mathcal{G}\}}$$

becomes important in Section 11.6 for the task of algorithmically deciding whether a reduct of a finitely bounded homogeneous structure has a pseudo-Siggers polymorphism. The following lemma implies for instance that if f is a pseudo-Siggers polymorphism of an ω-categorical model-complete core $\mathfrak{C}$ then every operation in the diagonal interpolation space is a pseudo-Siggers polymorphism of $\mathfrak{C}$, too.

LEMMA 10.1.14. *Let $\mathcal{C}$ be an oligomorphic operation clone on a countable set B such that the set $\mathcal{G}$ of invertible elements of $\mathcal{C}^{(1)}$ is dense in $\mathcal{C}^{(1)}$. Let $f_1, \dots, f_n \in \mathcal{C}$ and let $\phi(x_1, \dots, x_n)$ be a stable clone formula such that $\mathcal{C}$ satisfies $\phi(f_1, \dots, f_n)$. For every $i \in \{1, \dots, n\}$, let*

$$g_i \in \overline{\{(x_1, \dots, x_{\text{ar}(f_i)}) \mapsto \gamma f_i(\beta(x_1), \dots, \beta(x_{\text{ar}(f_i)})) \mid \beta, \gamma \in \mathcal{G}\}}.$$

Then $\mathcal{C}$ also satisfies $\phi(g_1, \dots, g_n)$.

PROOF. We first show the statement for the formula

$$\exists a, b \colon a \circ f_\sigma = b \circ f_\tau$$

where $f = f_i$ is k-ary and a, b unary. Let $g := g_i$ be as in the statement. By Lemma 10.1.5 it suffices to show that for every finite $F \subset B^k$ there is an $\eta \in \mathcal{G}$ such that $g_\sigma|_F = \eta g_\tau|_F$. Let $F \subseteq B^k$ be finite. Then there are $\gamma, \delta \in \mathcal{G}$ such that $g(t_1, \dots, t_k) = \gamma f(\delta(t_1), \dots, \delta(t_k))$ for every $(t_1, \dots, t_k) \in F$. Since $a \in \overline{\mathcal{G}}$ there exists an $\alpha \in \mathcal{G}$ such that $\alpha a(t) = t$ for every $t \in f(F)$. Moreover, since $b \in \overline{\mathcal{G}}$ there exists a $\beta \in \mathcal{G}$ such that $\beta(t) = b(t)$ for every $t \in f(F)$. Let $\eta := \gamma\alpha\beta\gamma^{-1}$. Then for every $t \in F$

$$\begin{aligned} g_\sigma(t) &= \gamma f_\sigma(\delta t) \\ &= \gamma\alpha a(f_\sigma(\delta t)) \\ &= \gamma\alpha b(f_\tau(\delta t)) \\ &= \eta\gamma f_\tau(\delta t)) = \eta g_\tau(t) \end{aligned}$$

and thus $\eta \in \mathcal{G}$ has the desired properties. The proof for arbitrary stable pseudo minor conditions is similar. ⊣

10.2. Pseudo-Siggers operations

Barto and Pinsker [28] proved the following remarkable result; the title of the journal version starts with the words *'topology is irrelevant'* because item (1) involves topology whereas condition (3) is purely algebraic: it only refers to $\text{Pol}(\mathfrak{B})$ as an abstract clone.

THEOREM 10.2.1 (Pseudo-Siggers theorem). *Let $\mathfrak{B}$ be an ω-categorical model-complete core. Then the following are equivalent.*

1. *For any $n \geq 1$ and $c_1, \dots, c_n \in B$ there is no continuous clone homomorphism from* $\text{Pol}(\mathfrak{B}, c_1, \dots, c_n)$ *to* **Proj**.
2. *There is no primitive positive interpretation of K_3 in $\mathfrak{B}$ with parameters.*
3. $\text{Pol}(\mathfrak{B})$ *contains a pseudo-Siggers operation.*
4. $\text{Pol}(\mathfrak{B})$ *satisfies a non-trivial stable pseudo-minor condition.*
5. *For any $n \geq 1$ and $c_1, \dots, c_n \in B$ there is no clone homomorphism from* $\text{Pol}(\mathfrak{B}, c_1, \dots, c_n)$ *to* **Proj**.

All the implications in cyclic order are either trivial or follow straightforwardly from previous observations, except for the implication from (2) to (3). The proof of this implication is similar to the proof of Theorem 6.8.6; however, instead of Theorem 6.8.1, we use a lemma that has been called the *pseudo-loop lemma* and that will be presented in Section 10.2.1. The proof of Theorem 10.2.1 can be found in Section 10.2.2.

10.2.1. The pseudo-loop lemma. Before we state the pseudo-loop lemma, we first state the loop lemma for undirected graphs. Using Corollary 6.7.13, Theorem 6.8.1 can be reformulated as follows.

LEMMA 10.2.2 (Loop lemma). *Let $(V; E)$ be a finite undirected graph which is not bipartite. If $K_3 \notin \mathrm{ICH}(V; E)$, then E contains a loop.*

Let $\mathcal{G}$ be a permutation group on a set B and let $E \subseteq B^2$ be a binary relation. An edge in E is called *pseudo-loop with respect to* $\mathcal{G}$ if it is of the form $(v, \alpha(v))$ for some $v \in B$ and $\alpha \in \mathcal{G}$.

LEMMA 10.2.3 (Pseudo-loop lemma). *Let $\mathcal{G}$ be an oligomorphic permutation group on a countable set B and suppose that there exists a binary symmetric relation $E \in \mathrm{Inv}(\mathcal{G})$ such that the graph $(B; E)$ embeds K_3. If $K_3 \notin \mathrm{IC}(\mathrm{Orb}(\mathcal{G}), E)$ (Definition 6.1.7), then $(B; E)$ contains a pseudo-loop with respect to $\mathcal{G}$.*

Let $\mathcal{G}$ be an oligomorphic permutation group on a countable set B and let E be a symmetric binary relation on B. We call the graph $(B; E)$ *expansive with respect to* $\mathcal{G}$ if

- $E \in \mathrm{Inv}(\mathcal{G})$,
- $(B; E)$ embeds K_3, and
- $(B; E)$ has no pseudo-loops with respect to $\mathcal{G}$.

In the proof of the pseudo-loop lemma we need the following definition from [28].

DEFINITION 10.2.4. Let $\mathcal{G}$ be an oligomorphic permutation group on a countable set B and let $(B; E)$ be expansive with respect to $\mathcal{G}$. Then $(B; E)$ is called *minimal with respect to* $\mathcal{G}$ if the following two properties hold.

1. If $B' \subseteq B$ is primitively positively definable in $(\mathrm{Orb}(\mathcal{G}), E)$ and E' is a binary symmetric relation on B' with a primitive positive definition in $(\mathrm{Orb}(\mathcal{G}), E)$ such that $(B'; E')$ is expansive with respect to $\mathcal{G}|_{B'}$, then $B' = B$.
2. If $\sim$ is an equivalence relation on B which is primitively positively definable in $(\mathrm{Orb}(\mathcal{G}), E)$ and E' is a binary symmetric relation on B with a primitive positive definition in $(\mathrm{Orb}(\mathcal{G}), E)$ such that the graph $(B; E')/{\sim}$ (see Example 3.1.2) is expansive with respect to $\mathcal{G}/{\sim}$, then $\sim$ is the equality relation.

When proving Lemma 10.2.3, we may assume without loss of generality that E is minimal with respect to $\mathcal{G}$. Otherwise, suppose that B' is a proper subset of B with a primitive positive definition in $(\mathrm{Orb}(\mathcal{G}), E)$ and E' is a primitively positively definable relation on B' such that $(B'; E')$ has no pseudo-loops with respect to $\mathcal{G}' := \mathcal{G}|_{B'}$ and $(B'; E')$ embeds K_3. Note that $\mathcal{G}'$ is again oligomorphic and has strictly fewer orbits than $\mathcal{G}$ (since the number of orbits of $\mathcal{G}$ equals the number of orbits of $\mathcal{G}'$ plus the number of orbits of $\mathcal{G}'' := \mathcal{G}|_{B \setminus B'}$, which is at least one). Then we may replace $\mathcal{G}$ by $\mathcal{G}'$ and E by E'; this will be called a *shrinking step*.

Now suppose that $(B; E)$ is not minimal with respect to $\mathcal{G}$ because there exists a non-trivial equivalence relation $\sim$ on B which is primitively positively definable in $(\mathrm{Orb}(\mathcal{G}), E)$ and E' is a primitively positively definable binary relation on B such that the graph $(B; E)/{\sim}$ embeds K_3 and does not have pseudo-loops with respect to $\mathcal{G}/{\sim}$. In this case we may again replace $\mathcal{G}$ by $\mathcal{G}/{\sim}$ and E by E'; this will be called a *factoring step*.

We claim that after a finite number of such replacement steps we reach an oligomorphic permutation group $\mathcal{G}$ on a countable set B and a graph $(B; E)$ such that E is minimal with respect to $\mathcal{G}$. The reason is that in the second type of replacement, the number of orbits does not increase. So in total, there can only be a finite number of shrinking steps, and from some point on, only factoring steps can be made. But note that we may combine two factoring steps into one factoring step with respect to a coarser primitively positively definable equivalence relation. Since there are only finitely many primitively positively definable equivalence relations in $(\mathrm{Orb}(\mathcal{G}), E)$, the factoring step can also only be performed a finite number of times, which proves the claim.

LEMMA 10.2.5. *Let $\mathcal{G}$ be an oligomorphic permutation group on a countable set B and let $(B; E)$ be minimal with respect to $\mathcal{G}$ and such that every edge of E is contained in a copy of K_3 in $(B; E)$. Then $(B; E)$ is diamond-free (see Section 6.8).*

PROOF. Let R be the binary symmetric relation with the primitive positive definition over $(B; E)$ given in (11) expressing that x and y are linked by a diamond, and for $n \in \mathbb{N}$ let $\delta_n(x, y)$ be the formula (12) expressing that x and y are linked by a chain of at most $n + 1$ diamonds. Let T be the transitive closure of R, which is the equivalence relation of being *diamond connected* as in the proof of Lemma 6.8.2; reflexivity follows from the assumption that every edge of E is contained in a copy of K_3 in $(B; E)$. The oligomorphicity of $\mathcal{G}$ implies that there are finitely many relations with a primitive positive definition in $(B; E)$, so there must exist an n such that δ_n is equivalent to $\delta_{n'}$ for all $n' \geq n$. Therefore, the formula δ_n provides a primitive positive definition of T in (B, E).

Claim. If $(x, y) \in T$ and y' is in the same orbit as y, then x is not adjacent to y'. The proof is similar to the one given in the proof of Lemma 6.8.2. Suppose otherwise, and choose x, y, y' with a shortest possible sequence $(a_0, \ldots, a_m)$ in B such that $a_0 = x$, $a_m = y$, and $R(a_0, a_1), \ldots, R(a_{m-1}, a_m)$. First suppose

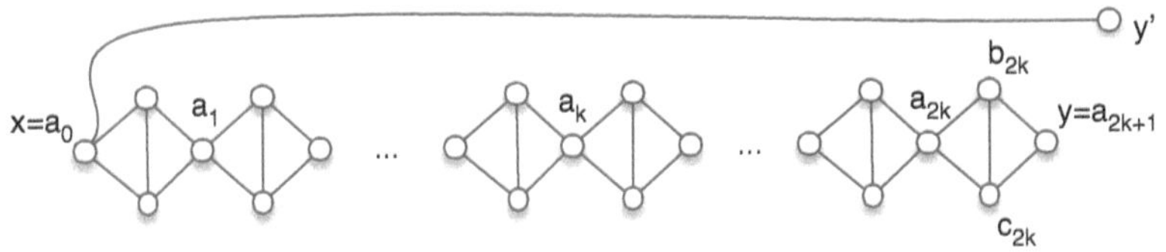

FIGURE 1. Diagram for the proof of Lemma 10.2.5.

that $m = 2k + 1$ is odd; see Figure 1. Consider the set S consisting of all $w \in B$ satisfying

$$\exists u, v \left(u \in \mathcal{G}a_k \wedge \delta_k(u, v) \wedge E(v, w)\right).$$

Let b_{2k}, c_{2k} be the two witnesses showing that $R(a_{2k}, a_{2k+1})$ holds. Then $b_{2k}, c_{2k} \in S$. We also have $y' \in S$, and therefore $y \in S$ since y' and y lie in the same orbit and S has a first-order definition. Hence, S contains the triangle formed by b_{2k}, c_{2k}, and $y = a_{2k+1}$ in $(B; E)$. By the minimality of $(B; E)$, we must have that $S = B$. Let $u, v \in B$ be as in the definition of S witnessing that $S(x)$ holds. Then $\delta_k(u, v)$, but also $\delta_k(u, x')$ for some $x' \in \mathcal{G}x$ because $\delta_k(a_k, x)$, $u \in \mathcal{G}a_k$, and δ_k is preserved by $\mathcal{G}$. Therefore, $\delta_{2k}(v, x')$, which together with $E(v, x)$ contradicts the choice of x, y, y' such that m is minimal if $m \geq 3$. The case $m = 1$ is impossible, since in this case $\delta_0(v, x')$ implies that $v = x'$, so $E(x, x')$ in contradiction to the assumption that $(B; E)$ does not have pseudo-loops. The case where m is even can be shown similarly.

The claim shows that $(B; E)/T$ does not have pseudo-loops. Moreover, it is easy to see that every edge in $(B; E)/T$ is contained in a copy of a K_3. By minimality, T must be the equality relation, which means that $(B; E)$ is diamond-free. ⊣

An important example of a diamond-free graph is $(K_3)^k$. Clearly, $K_3 \in \mathrm{H}((K_3)^k)$; but we even have the following.

LEMMA 10.2.6. *For every $k \in \mathbb{N}$, the graph K_3 has a primitive positive interpretation in $(K_3)^k$ with parameters.*

PROOF. Consider the following equivalence relation

$$C := \big\{((u_1, \ldots, u_k), (v_1, \ldots, v_k)) \mid (u_1, \ldots, u_k), (v_1, \ldots, v_k) \in (K_3)^k \text{ and } u_1 = v_1\big\};$$

we claim that C is preserved by every idempotent $f \in \mathrm{Pol}((K_3)^k)$. Let l be the arity of f. We may view f as a homomorphism from $(K_3)^{kl}$ to the diamond-free graph $(K_3)^k$. Hence, Lemma 6.8.4 implies that there exists $I \subseteq \{1, \ldots, kl\}$ such that the kernel of f equals the kernel of π_I^{kl} and the image of f is isomorphic to $(K_3)^{|I|}$. It follows that $|I| \leq k$ and by the idempotence of f we have that $|I| = k$. So there are $\{(i_1, j_1), \ldots, (i_k, j_k)\} \subseteq \{(1, 1), \ldots, (l, k)\}$ and an automorphism

α of $(K_3)^k$ such that $f(x_{1,1},\dots,x_{1,k},\dots,x_{l,1},\dots,x_{l,k}) = \alpha(x_{i_1,j_1},\dots,x_{i_k,j_k})$. The idempotence of f implies that

$$f(x_{i,1},\dots,x_{i,k},x_{i,1},\dots,x_{i,k},\dots,x_{i,1},\dots,x_{i,k}) = (x_{i,1},\dots,x_{i,k})$$

and hence $j_1 = 1,\dots,j_k = k$. Note that for every automorphism α of K_3^k there exists a permutation q of $\{1,\dots,k\}$ and permutations $g_1,\dots,g_k$ of $\{1,2,3\}$ such that for all $u_1,\dots,u_k \in \{1,2,3\}$

$$\alpha(u_1,\dots,u_k) = (g_1(u_{q(1)}),\dots,g_k(u_{q(k)})).$$

Now, if $(\bar{u}_1,\bar{v}_1),\dots,(\bar{u}_l,\bar{v}_l) \in C$ then

$$\begin{aligned} f(\bar{u}_1,\dots,\bar{u}_l)_1 &= \alpha(\bar{u}_{i_1,1},\dots,\bar{u}_{i_k,k})_1 \\ &= g_1(\bar{u}_{1,q(1)}) \\ &= g_1(\bar{v}_{1,q(1)}) \qquad \text{(by assumption)} \\ &= f(\bar{v}_1,\dots,\bar{v}_l)_1 \end{aligned}$$

and hence $(f(\bar{u}_1,\dots,\bar{u}_l), f(\bar{v}_1,\dots,\bar{v}_l)) \in C$. Analogously one can show that

$$E := \{((u_1,\dots,u_k),(v_1,\dots,v_k)) \mid (u_1,\dots,u_k),(v_1,\dots,v_k) \in (K_3)^k \text{ and } u_1 \neq v_1\}$$

is preserved by every idempotent polymorphism of $\mathrm{Pol}((K_3)^k)$. Hence, C and E are primitively positively definable in the expansion of $(K_3)^k$ by a constant for each element (Theorem 6.1.12). This shows that the first projection is a 1-dimensional primitive positive interpretation of K_3 in $(K_3)^k$ with parameters. ⊣

LEMMA 10.2.7 (Lemma 3.5 in [28]). *Let $\mathcal{G}$ be an oligomorphic permutation group on a countable set B and let E be a binary symmetric relation on B such that*

- *$(B; E)$ is minimal with respect to $\mathcal{G}$;*
- *every edge from E is contained in a copy of K_3 in $(B; E)$;*
- *$(B; E)$ contains an induced subgraph H isomorphic to $(K_3)^k$ for some $k \geq 1$ such that $|(K_3)^k| = 3^k$ is larger than the number of orbits under $\mathcal{G}$.*

Then $(\mathrm{Orb}(\mathcal{G}), E)$ defines primitively positively a symmetric relation E' such that

- *E' properly contains E,*
- *E' has no pseudo-loops with respect to $\mathcal{G}$, and*
- *every edge from E' is contained in a copy of K_3 in $(B; E')$.*

PROOF. Let $\{1,2,3\}^k$ be the vertices of H. By assumption, H has two distinct vertices o and o' that lie in the same orbit under $\mathcal{G}$. Let p and q be vertices of H such that $\{o,p,q\}$ induces a copy of K_3. Let O, P, Q be the orbits of o, p, q under $\mathcal{G}$, respectively. Since E does not contain pseudo-loops, O,

P, and Q are pairwise distinct. Let $S(u, v)$ be the binary symmetric relation defined by the following primitive positive formula in $(\text{Orb}(\mathcal{G}), E)$.

$$\begin{aligned} \exists x, y, z, n_o, n_p, n_q \big(E(u, n_o) \wedge E(v, n_o) \wedge E(n_o, x) \wedge O(x) \\ \wedge E(u, n_p) \wedge E(v, n_p) \wedge E(n_p, y) \wedge P(y) \\ \wedge E(u, n_q) \wedge E(v, n_q) \wedge E(n_q, z) \wedge Q(z)\big) \end{aligned}$$

The set N of neighbours of neighbours of elements in O is primitively positively definable in $(\text{Orb}(\mathcal{G}); E)$. Note that N contains $O \cup P \cup Q$. Indeed, if $x \in P$, pick $\alpha \in \mathcal{G}$ such that $\alpha p = x$. Then $\alpha o \in O$, and since $(o, p) \in E$ we have $(\alpha o, x) \in E$ and hence x is a neighbour of an element in O. We can reason analogously if $x \in Q$. This shows that $P \cup Q$ belongs to the neighbours of elements in O. Repeating this argument one more time, we obtain that $O \cup P \cup Q$ belongs to the neighbours of the neighbours of elements in O. The restriction of E to N still contains a copy of K_3 (namely $\{o, p, q\}$). Hence, the minimality of $(B; E)$ with respect to $\mathcal{G}$ implies that every element of B is at distance at most two from an element in O. The same applies to P, and to Q. This implies that the relation S is reflexive.

The set $M(u, v)$ of elements of B adjacent to a common neighbour of an element of $\mathcal{G}u$ and an element of $\mathcal{G}v$ is primitively positively definable in $(\text{Orb}(\mathcal{G}); E)$. Note that $M(u, v)$ contains $O \cup P \cup Q$, and again minimality of $(B; E)$ implies that $M = B$. Let $T(u, v)$ be the binary relation defined by the primitive positive formula

$$\exists s, t \big(E(u, s) \wedge S(s, v) \wedge S(u, t) \wedge E(t, v)\big).$$

Note that $E \subseteq T$: since S is reflexive, setting $s = v$ and $t = u$ in the above definition shows that $E(u, v)$ implies $T(u, v)$. Moreover, T is symmetric by definition. Let E' consist of those edges of T that are contained in a copy of a K_3 in $(B; T)$. We still have that $E \subseteq E'$.

We now show that $(B; T)$, and hence also $(B; E')$, does not contain pseudo-loops with respect to $\mathcal{G}$. Suppose for contradiction that $T(u, \alpha u)$ for some $\alpha \in \mathcal{G}$. Then there exists s such that $E(u, s) \wedge S(s, \alpha u)$, and since E is covered by copies of K_3 in $(B; E)$ there exists some $w \in B$ such that $\{u, s, w\}$ induces a copy of K_3. Since $w \in M(s, u)$, it has a neighbour z which is also adjacent to an element of $\mathcal{G}s$ and an element of $\mathcal{G}u$. The set of vertices with a neighbour in $\mathcal{G}z$

- is primitively positively definable in $(\text{Orb}(\mathcal{G}); E)$;
- contains $\mathcal{G}u$, $\mathcal{G}s$, and $\mathcal{G}w$, and hence contains $\{u, s, w\}$ which induces a K_3 in $(B; E)$;
- is a proper subset of B: it does not contain $\mathcal{G}z$ because $(B; E)$ does not have pseudo-loops with respect to $\mathcal{G}$.

This contradicts the minimality of $(B; E)$ with respect to $\mathcal{G}$.

We finally show that E is *properly* contained in E'. It suffices to show that $E'(o', p)$ and $E'(o', q)$, because if a vertex of $(K_3)^k$ is adjacent to both p and q, then the vertex must be equal to o, but o and o' are distinct by assumption. We only show that $E'(o', p)$ since $E'(o', q)$ can be shown analogously. We may assume without loss of generality that $o = (1, \dots, 1)$, $p = (2, \dots, 2)$, and $q = (3, \dots, 3)$. Let j be the number of entries of o' that are distinct from 2; we may assume that $o'_i \neq 2$ for all $i \in \{1, \dots, j\}$ and $o'_i = 2$ for all $i \in \{j+1, \dots, k\}$. Since $o' \neq p$ we have $j \geq 1$. Observe that whenever $u, v \in \{1, 2, 3\}^k$ are of the form

$$u = (w_1, \dots, w_j, 2, \dots, 2) \text{ and } v = (w_1, \dots, w_j, 3, \dots, 3)$$

then $S(u, v)$: this is witnessed by

- their common neighbour $(x_1, \dots, x_j, 1, \dots, 1)$, where $x_i \notin \{o'_i, w_i\}$ for all $i \in \{1, \dots, j\}$, which is E-related to $o' \in O$,
- their common neighbour $(y_1, \dots, y_j, 1, \dots, 1)$ with $y_i \notin \{2, w_i\}$ for all $i \in \{1, \dots, j\}$, which is E-related to $p \in P$, and
- their common neighbour $(z_1, \dots, z_j, 1, \dots, 1)$ with $z_i \notin \{3, w_i\}$ for all $i \in \{1, \dots, j\}$, which is E-related to $q \in Q$.

Then $T(o', p)$ holds: setting

$$t := (o'_1, \dots, o'_j, 3, \dots, 3)$$

we have that $S(o', t)$ by the above observation, and clearly $E(t, p)$. Setting

$$s := (\underbrace{2, \dots, 2}_{j \text{ times}}, 3, \dots, 3)$$

we have that $E(o', s)$ and that $S(s, p)$ again by the above observation. We can then conclude that $E'(o', p)$ holds, because any two elements of $\{1, 2, 3\}^k$, and in particular o' and p, have a common neighbour with respect to E, and hence also with respect to T, showing that $(o', p) \in T$ is covered by a copy of a K_3 in $(B; T)$. ⊣

PROOF OF LEMMA 10.2.3. Suppose that the graph $(B; E)$ has no pseudo-loop with respect to $\mathcal{G}$. We must show that $K_3 \in \text{IC}(\text{Orb}(\mathcal{G}), E)$. We may assume that $(B; E)$ is minimal with respect to $\mathcal{G}$ as explained above. Moreover, we may also assume that every edge of E is contained in a copy of K_3 in $(B; E)$: consider the relation E' with the primitive positive definition

$$E'(x, y) :\Leftrightarrow \exists z \, \big(E(x, y) \wedge E(y, z) \wedge E(z, x)\big).$$

Note that $(B; E')$ must be minimal with respect to $\mathcal{G}$, and we may replace E by E' to obtain the desired property.

It then follows from Lemma 10.2.5 that $(B; E)$ is diamond-free. If $(B; E)$ contains an induced subgraph isomorphic to $(K_3)^k$ such that $|(K_3)^k|$ is larger than the number of orbits under $\mathcal{G}$, then by Lemma 10.2.7, there is a symmetric relation E' that properly contains E, is primitively positively definable in

$(\text{Orb}(\mathcal{G}), E)$, and whose edges are covered by copies of K_3 in $(B; E')$. Then $(B; E')$ is minimal with respect to $\mathcal{G}$ (and in particular diamond-free), and we may therefore replace E by E'. Since there are finitely many binary relations in $\text{Inv}(\mathcal{G})$, after a finite number of steps we arrive at a graph $(B; E)$ which is minimal with respect to $\mathcal{G}$ and every subgraph isomorphic to $(K_3)^k$ has fewer vertices than the number of orbits under $\mathcal{G}$. Let $k \in \mathbb{N}$ be the maximal k such that $(B; E)$ has an isomorphic copy of $(K_3)^k$ induced on $A := \{a_1, \dots, a_\ell\}$. We show that A has a primitive positive definition in $(\text{Orb}(\mathcal{G}); E)$ with parameters $a_1, \dots, a_\ell$. By Theorem 6.1.12, it suffices to show that every $f \in \text{Pol}(\text{Orb}(\mathcal{G}); E, a_1, \dots, a_\ell)$ preserves A. The restriction f' of f to A is a homomorphism from $(K_3)^k$ to the diamond-free graph $(B; E)$. The image of f' contains A because f' preserves $a_1, \dots, a_\ell$. Hence, if f does not preserve A, then the image of f is strictly larger than A, and by Lemma 6.8.4 the image of f' induces a copy of $(K_3)^m$ for some $m > k$, contradicting the maximality of k. We therefore get $(K_3)^k \in \text{IC}(\text{Orb}(\mathcal{G}), E)$. Lemma 10.2.6 states that K_3 has an interpretation with parameters in $(K_3)^k$. By Corollary 4.5.7 we get that $K_3 \in \text{IC}(\text{Orb}(\mathcal{G}), E)$. ⊣

10.2.2. Constructing the pseudo-Siggers polymorphism. In this section we prove the pseudo-Siggers theorem (Theorem 10.2.1). The important step is proving that if an ω-categorical model-complete core $\mathfrak{B}$ does not have a pseudo-Siggers polymorphism, then $K_3 \in \text{IC}(\mathfrak{B})$.

PROOF OF THEOREM 10.2.1. The equivalence of (1) and (2) is a consequence of Corollary 9.5.21.

(2) ⇒ (3). We use the lift lemma (Lemma 10.1.5) and verify the existence of a pseudo-Siggers polymorphism locally. Let $F \subseteq B$ be finite. As in the proof of Theorem 6.8.6 where we constructed a Siggers polymorphism, let $k \geq 1$ and $a, b, c \in F^k$ be such that $\{(a_i, b_i, c_i) \mid i \leq k\} = F^3$. Let R be the binary relation on B^k such that $(u, v) \in R$ iff there exists a 6-ary $s \in \text{Pol}(\mathfrak{B})$ such that $u = s(a, b, a, c, b, c)$ and $v = s(b, a, c, a, c, b)$.

- The vertices $a, b, c \in B^k$ induce in $(B^k; R)$ a copy of K_3: each of the six edges of K_3 is witnessed by one of the six 6-ary projections from $\text{Pol}(\mathfrak{B})$.
- The relation R is symmetric; this is as in the proof of Theorem 6.8.6.
- The relation R (as a $2k$-ary relation over B) is preserved by $\text{Pol}(\mathfrak{B})$, and hence $(B^k; R) \in \text{I}(\mathfrak{B})$ (Theorem 6.1.12). Thus, if $K_3 \in \text{IC}(B^k; R)$ then $K_3 \in \text{IC}(\mathfrak{B})$ by Corollary 4.5.7, contrary to our assumptions.

Note that $\text{Orb}(\text{Aut}(\mathfrak{B}))$ is primitively positively definable in $\mathfrak{B}$ since $\mathfrak{B}$ is an ω-categorical model-complete core, so Lemma 10.2.3 implies that the graph $(B^k; R)$ has a pseudo-loop $(w, \alpha w) \in R$ with respect to $\text{Aut}(\mathfrak{B})$. Hence, there exists a 6-ary $s \in \text{Pol}(\mathfrak{B})$ such that

$$w = s(a, b, a, c, b, c) \text{ and } \alpha w = s(b, a, c, a, c, b).$$

We then have that $s(x, y, x, z, y, z)$ and $s(y, x, z, x, z, y)$ lie in the same orbit for all $x, y, z \in F$, because there exists an $i \leq k$ such that $(x, y, z) = (a_i, b_i, c_i)$. Therefore, Lemma 10.1.5 implies that $\mathfrak{B}$ has a pseudo-Siggers polymorphism.

(3) $\Rightarrow$ (4). Clearly, the existence of a pseudo-Siggers operation is a pseudo-minor condition which is stable and non-trivial.

(4) $\Rightarrow$ (5). This is Proposition 10.1.13.

(5) $\Rightarrow$ (1) is trivial. ⊣

In Section 11.6 we will present an algorithm that tests for a large class of ω-categorical structures with a suitable effective representation whether they have a pseudo-Siggers polymorphism.

10.3. Equivalence of two conjectures

Let $\mathfrak{B}$ be an ω-categorical model-complete core. By the results of the previous section we know that if $\mathfrak{B}$ has no pseudo-Siggers polymorphism, then $\mathrm{HI}(\mathfrak{B})$ contains all finite structures. What about the converse? Is it possible that $\mathrm{HI}(\mathfrak{B})$ contains all finite structures and $\mathfrak{B}$ has a pseudo-Siggers polymorphism? Indeed, this is possible and we have already seen an example, which we revisit here.

Example 10.3.1. Let $\mathfrak{A}$ be the countable atomless Boolean algebra (see Example 4.1.4). The structure $(\mathfrak{A}, \neq)$

- is an ω-categorical model-complete core;
- has a pseudo-Siggers polymorphism: $\mathfrak{A}^6$ has an embedding e into $\mathfrak{A}$ (see Example 6.7.5), which is a 6-ary polymorphism of $(\mathfrak{A}, \neq)$. Then $e(x, x, y, y, z, z)$ and $e(y, z, x, z, x, y)$ are embeddings of $\mathfrak{A}^3$ into $\mathfrak{A}$, so we can find endomorphisms u_1 and u_2 of $(\mathfrak{A}, \neq)$ using the lift lemma (Lemma 10.1.5);
- $(\{0, 1\}; \mathrm{1IN3}) \in \mathrm{HI}(\mathfrak{A})$ (Example 9.6.7).

The expansion of $(\mathfrak{A}; \neq)$ with finitely many constants still has a pseudo-Siggers polymorphism (Proposition 10.1.13). It follows that the structure $(\{0, 1\}; \mathrm{1IN3})$ cannot be interpreted in $(\mathfrak{A}; \neq)$ with finitely many parameters (Theorem 10.2.1 and Theorem 6.3.10). Note that this is *not* a counterexample to Conjecture 4.1, since this conjecture only applies to reducts of finitely bounded homogeneous structures. The structure $(\mathfrak{A}; \neq)$ is not even a reduct of a homogeneous structure with finite relational signature, because it has doubly exponential orbit growth (recall the argument given in the proof of Proposition 5.7.3).

In this section we present the result from [20] that if $\mathfrak{B}$ is an ω-categorical model-complete core such that

- the condition given in Conjecture 4.1 holds, i.e., $K_3 \notin \mathrm{IC}(\mathfrak{B})$,

- the condition given in Conjecture 3.1 fails, i.e., $K_3 \in \mathrm{HI}(\mathfrak{B})$,

then the structure $\mathfrak{B}$ must have doubly exponential orbit growth (as in the example that we have just seen), and therefore does not fall into the scope of the two conjectures. It follows that Conjecture 3.1 and 4.1 are equivalent.

10.3.1. Ambiguity degree. Let $\mathcal{C}$ be an operation clone over an infinite set C and let $S \subseteq C$ be of size two. A function $r \colon C \to S$ is called a *projectivity witness*[26] *for* $\mathcal{C}$ if $r \circ f|_S$ is a projection for every $f \in \mathcal{C}$.

DEFINITION 10.3.2. Let $f \in \mathcal{C}$ be k-ary and let $S \subseteq C$ be of size two. The *ambiguity degree of f on S* is the number of indices $i \in \{1, \ldots, k\}$ such that there exists a projectivity witness $r \colon C \to S$ such that $r \circ f|_S = \pi^k_i$. The *ambiguity degree of* $\mathcal{C}$ is the supremum of the ambiguity degrees over all operations $f \in \mathcal{C}$ and all two-element subsets S of C.

LEMMA 10.3.3 (Lemma 3.2 in [20]). *Let $\mathcal{C}$ be an operation clone of infinite ambiguity degree and let $\mathcal{G}$ be the permutation group of invertible elements of $\mathcal{C}^{(1)}$. Then for every $n \in \mathbb{N}$ the componentwise action of $\mathcal{G}$ on C^n has at least $2^{2^n} - 1$ orbits.*

PROOF. Let $n \in \mathbb{N}$ and let $f \in \mathcal{C}$ be of ambiguity degree at least 2^n with respect to a two-element set $S \subseteq C$. By identifying arguments of f we may assume that the arity of f is exactly 2^n. For every non-empty $R \subseteq S^n$ arbitrarily choose $q^R_1, \ldots, q^R_{2^n} \in R$ such that $R = \{q^R_1, \ldots, q^R_{2^n}\}$.

Claim. If $R, T \subseteq S^n$ are non-empty and distinct then the n-tuples $f(q^R_1, \ldots, q^R_{2^n})$ and $f(q^T_1, \ldots, q^T_{2^n})$ lie in distinct orbits under $\mathcal{G}$. Suppose otherwise that there is $\alpha \in \mathcal{G}$ such that $f(q^R_1, \ldots, q^R_{2^n}) = \alpha f(q^T_1, \ldots, q^T_{2^n})$. It suffices to show that for every $i \leq 2^n$ we have $q^R_i \in T$, because then $R \subseteq T$, and similarly $T \subseteq R$. Choose a projectivity witness $r \colon C \to S$ for $\mathcal{C}$ such that $r \circ f|_S$ is the i-th projection. Then $q^R_i = r \circ f(q^R_1, \ldots, q^R_{2^n}) = r \circ \alpha \circ f(q^T_1, \ldots, q^T_{2^n}) \in T$ since $r \circ \alpha \circ f|_S$ is a projection.

The statement follows from the claim because there are $2^{2^n} - 1$ non-empty binary relations over the domain S^n. ⊣

Recall the definition of pseudo-loop conditions from Example 10.1.11.

LEMMA 10.3.4. *Let $\mathcal{C}$ be an operation clone such that the set $\mathcal{G}$ of invertible elements of $\mathcal{C}^{(1)}$ is dense in $\mathcal{C}^{(1)}$. If*

- *$\mathcal{C}$ satisfies a non-trivial pseudo-loop condition, and*
- *$\mathcal{C}$ has a projectivity witness,*

then $\mathcal{C}$ has infinite ambiguity degree, and hence $\mathcal{G}$ has doubly exponential orbit growth.

PROOF. We show the statement by proving that for every $f \in \mathcal{C}$ of ambiguity degree $n \geq 1$ there exists an $f' \in \mathcal{C}$ of ambiguity degree $2n$. We may assume that f has arity n by identifying arguments; so there exists a two-element set

[26] In the original publication [20] projectivity witnesses were called *retractional witnesses*.

$S \subseteq C$ and projectivity witnesses $r_1, \ldots, r_n \colon C \to S$ such that $r_i \circ f|_S = \pi_i^n$. Set $F := \{f(s) \mid s \in S^n\}$. Because $\mathcal{G}$ is dense in $\mathfrak{C}^{(1)}$, the pointwise stabiliser $\mathfrak{C}_{(F)}$ satisfies a nontrivial pseudo-loop condition, too (Proposition 10.1.13). Let $g \in \mathfrak{C}_{(F)}$ be a function that witnesses this, i.e, there are $u, v \in \mathfrak{C}_{(F)}^{(1)}$ such that g, u, v satisfy the nontrivial identity $u \circ g(y_1, \ldots, y_m) = v \circ g(z_1, \ldots, z_m)$ for variables $y_1, \ldots, y_m, z_1, \ldots, z_m$ which are not necessarily distinct. We verify that the star product $g * f$ (Definition 6.6.3) has ambiguity degree at least $2n$.

We already know that $r_i \circ (g * f)|_S$ must be a projection. Writing $\pi_{k,l}^{m,n}$ for $\pi_{(k-1)n+l}^{mn}$, where $k \leq m, l \leq n$, we claim that $r_i \circ (g * f)|_S = \pi_{j,i}^{m,n}$ for some $j \leq m$. To see this, note that for all $x_1, \ldots, x_n \in S$ we have

$$
\begin{aligned}
& r_i \circ (g * f)(x_1, \ldots, x_n, \ldots, x_1, \ldots, x_n) \\
&= r_i \circ f(x_1, \ldots, x_n) && \text{(since } g \text{ fixes } F = f(S^n) \text{ pointwise)} \\
&= \pi_i^n(x_1, \ldots, x_n) && \text{(by the choice of } r_i\text{)}.
\end{aligned}
$$

Similarly, we can verify that $r_i \circ u \circ (g * f)|_S = \pi_{j_1,i}^{m,n}$ for some $j_1 \leq m$, and that $r_i \circ v \circ (g * f)|_S = \pi_{j_2,i}^{m,n}$ for some $j_2 \leq m$. We claim that $j_1 \neq j_2$. Otherwise, if $j_1 = j_2$ we have for all $y_1, \ldots, y_m \in S$

$$
\begin{aligned}
y_{j_1} &= \pi_{j_1,i}^{m,n}(y_1, \ldots, y_1, \ldots, y_m, \ldots, y_m) \\
&= r_i \circ u \circ (g * f)(y_1, \ldots, y_1, \ldots, y_m, \ldots, y_m) \\
&= r_i \circ u \circ g(f(y_1, \ldots, y_1), \ldots, f(y_m, \ldots, y_m)) \\
&= r_i \circ v \circ g(f(z_1, \ldots, z_1), \ldots, f(z_m, \ldots, z_m)) \\
&= r_i \circ v \circ (g * f)(z_1, \ldots, z_1, \ldots, z_m, \ldots, z_m) \\
&= \pi_{j_2,i}^{m,n}(z_1, \ldots, z_1, \ldots, z_m, \ldots, z_m) = y_{j_2}
\end{aligned}
$$

which contradicts the assumption that S has two elements. Hence, the operations $r_1 \circ u, \ldots, r_n \circ u, r_1 \circ v, \ldots, r_n \circ v$ witness that $g * f$ has ambiguity degree at least $2n$. The final part of the statement follows from Lemma 10.3.3. ⊣

THEOREM 10.3.5. *Let $\mathfrak{C}$ be a model-complete core with less than doubly exponential orbit growth. Then the following are equivalent.*

1. $\mathrm{Pol}(\mathfrak{C})$ *satisfies no non-trivial pseudo-loop condition.*
2. $\mathrm{Pol}(\mathfrak{C})$ *has no pseudo-Siggers operation.*
3. $K_3 \in \mathrm{IC}(\mathfrak{C})$.
4. $K_3 \in \mathrm{HI}(\mathfrak{C})$.
5. $\mathrm{Pol}(\mathfrak{C})$ *has a uniformly continuous minor-preserving map to* **Proj**.
6. *If* $\mathbf{C}$ *is a polymorphism algebra of* $\mathfrak{C}$ *then there exists* $\mathbf{A} \in \mathrm{Refl}\, \mathrm{P}^{\mathrm{fin}}(\mathbf{C})$ *such that* $\mathrm{Clo}(\mathbf{A})$ *is isomorphic to* **Proj**.

PROOF. The implication (1) ⇒ (2) is trivial, and the equivalence (2) ⇔ (3) holds for all ω-categorical model-complete cores by Theorem 10.2.1. The implication (3) $\Rightarrow$ (4) holds in general by Theorem 3.6.2. The equivalence (4) $\Leftrightarrow$ (5) holds by Corollary 9.6.6, and the equivalence (5) $\Leftrightarrow$ (6) follows from Theorem 9.6.4. To show (6) $\Rightarrow$ (1), suppose that $\mathbf{A}$ is a reflection of $\mathbf{C}^k$ for some $k \in \mathbb{N}$. This means that $\text{Clo}(\mathbf{C}^k)$ has a projectivity witness. Since $\text{Aut}(\mathfrak{C})$ does not have at least double exponential orbit growth neither has $\text{Aut}(\mathfrak{C})^k$. So Lemma 10.3.4 implies that $\text{Clo}(\mathbf{C}^k)$ and therefore also $\text{Clo}(\mathbf{C}) = \text{Pol}(\mathfrak{C})$ satisfies no non-trivial pseudo-loop condition. ⊣

In particular, we obtain that the condition from Conjecture 3.1 and the one from Conjecture 4.1 are equivalent, even though Conjecture 3.1 does not mention model-complete cores.

COROLLARY 10.3.6. *Let $\mathfrak{B}$ be a first-order reduct of a finitely bounded homogeneous structure. Then the following are equivalent.*

- $K_3 \in \text{IC}(\mathfrak{C})$ *for the model-complete core $\mathfrak{C}$ of $\mathfrak{B}$;*
- $K_3 \in \text{HI}(\mathfrak{B})$.

PROOF. Since $\mathfrak{B}$ is a first-order reduct of a finitely bounded homogeneous structure, $\text{Aut}(\mathfrak{B})$ does not have at least double exponential orbit growth by Proposition 5.7.1, and the same applies to the model-complete core $\mathfrak{C}$ of $\mathfrak{B}$ by Proposition 4.7.7. The equivalence of (3) and (4) in Theorem 10.3.5 states that $K_3 \in \text{IC}(\mathfrak{C})$ if and only if $K_3 \in \text{HI}(\mathfrak{C})$. Note that $\text{HI}(\mathfrak{C}) = \text{HI}(\text{H}(\mathfrak{C})) = \text{HI}(\text{H}(\mathfrak{B})) = \text{HI}(\mathfrak{B})$ (Theorem 3.6.2) which implies the statement. ⊣

10.4. The model-complete core assumption

The conditions from Theorem 10.3.5 were equivalent under the assumption that $\mathfrak{C}$ is a model-complete core. For which implications in Theorem 10.3.5 is this assumption necessary? Our proof shows that some parts of the statement hold in general, for example the equivalences (4) $\Leftrightarrow$ (5) $\Leftrightarrow$ (6) and the implications (3) $\Rightarrow$ (1) $\Rightarrow$ (2) and (3) $\Rightarrow$ (4). However, the implication (4) $\Rightarrow$ (2) no longer applies (and hence neither do the implications (4) $\Rightarrow$ (1) and (4) $\Rightarrow$ (3)), as the following example shows, which was found by Trung Van Pham (personal communication).

PROPOSITION 10.4.1. *There is a first-order reduct $\mathfrak{B}$ of a finitely bounded homogeneous structure which has a pseudo-symmetric polymorphism and a pseudo-Siggers polymorphism, but $K_3 \in \text{HI}(\mathfrak{B})$. In particular, the model-complete core of $\mathfrak{B}$ has no pseudo-Siggers polymorphism.*

PROOF. We write S for the ternary relation

$$\{(x, y, z) \in \mathbb{N}^3 \mid (x = y \lor y = z) \land x \neq z\}.$$

Let $\mathfrak{B}$ be the structure with domain $\mathbb{N}^2 \cup \mathbb{N}$ and a single ternary relation

$$R := \{((x_1, y_1), (x_2, y_2), (x_3, y_3)) \mid S(x_1, x_2, x_3) \land S(y_1, y_2, y_3)\} \cup S.$$

The expansion of this structure by a unary relation P for $\mathbb{N}$ and two binary relations

$$\{((x, y), (x, z)) \mid x, y, z \in \mathbb{N}\}$$
$$\{((x, y), (z, y)) \mid x, y, z \in \mathbb{N}\}$$

is clearly homogeneous, and it is finitely bounded since a finite structure does not embed into $\mathfrak{B}$ if and only if it contains a three-element structure that does not embed into $\mathfrak{B}$.

For $i \in \{1, 2\}$, define $e_i \colon B \to B$ by $e_i(x) := x$ if $x \in P$ and $e_i(x_1, x_2) := x_i$ otherwise; note that e_i is an endomorphism of $\mathfrak{B}$. We claim that the substructure $\mathfrak{C}$ of $\mathfrak{B}$ induced on P is the model-complete core of $\mathfrak{B}$. Note that $\mathfrak{C} = (\mathbb{N}; S)$ is an equality constraint language, so it is easy to see that $\mathfrak{C}$ is a model-complete core. Moreover, the range of the endomorphism e_i lies in P. Also note that S is not preserved by a constant operation and not preserved by binary injective operations. Theorem 7.4.1 therefore implies that $K_3 \in \mathrm{I}(\mathfrak{C})$, and hence $K_3 \in \mathrm{HI}(\mathfrak{B})$. Let $f : B^2 \to B$ be defined by

$$f(x, y) := \begin{cases} (x_1, y_1) & \text{if } x = (x_1, x_2), y = (y_1, y_2) \in \mathbb{N}^2, \\ (x, y_1) & \text{if } x \in \mathbb{N}, y = (y_1, y_2) \in \mathbb{N}^2, \\ (x_1, y) & \text{if } x = (x_1, x_2) \in \mathbb{N}^2, y \in \mathbb{N}, \\ (x, y) & \text{if } x, y \in \mathbb{N}. \end{cases}$$

It is easy to see that f preserves S, and therefore also preserves R. We claim that for all $x, y \in B$

$$e_1(f(x, y)) = e_2(f(y, x)).$$

We have

$$\begin{aligned} e_1(f((x_1, x_2), (y_1, y_2))) &= e_1(x_1, y_1) = x_1 \\ &= e_2(y_1, x_1) = e_2(f((y_1, y_2), (x_1, x_2))) \end{aligned}$$

and

$$\begin{aligned} e_1(f((x_1, x_2), y)) &= e_1(x_1, y) = x_1 \\ &= e_2(y, x_1) = e_2(f(y, (x_1, x_2))) \end{aligned}$$

and likewise we compute

$$e_1(f(x, y)) = e_2(f(y, x))$$
$$\text{and } e_1(f(x, (y_1, y_2))) = e_2(f((y_1, y_2), x)).$$

We conclude that f is a pseudo-symmetric polymorphism of $\mathfrak{B}$. A pseudo-Siggers polymorphism can be constructed similarly. The final statement follows from Theorem 10.3.5. ⊣

Note that we see here a behaviour of pseudo-minor conditions that is very different from minor conditions: for example, if $\mathfrak{B}$ has a Taylor polymorphism f, and $\mathfrak{C}$ is homomorphically equivalent to $\mathfrak{B}$ via homomorphisms $h \colon \mathfrak{B} \to \mathfrak{C}$ and $i \colon \mathfrak{C} \to \mathfrak{B}$, then $(x_1, \dots, x_n) \mapsto h(f(i(x_1), \dots, i(x_n)))$ is a Taylor polymorphism of $\mathfrak{C}$.

It turns out that the implication (2) $\Rightarrow$ (4) in Theorem 10.3.5 also holds for general ω-categorical structures; this is again an observation of Trung Van Pham and first appeared in [28]. We state the argument here in more generality for pseudo-minor conditions rather than the special case of the pseudo-Siggers condition.

THEOREM 10.4.2. *Let $\mathfrak{C}$ be the model-complete core of an ω-categorical structure $\mathfrak{B}$ and suppose that* $\mathrm{Pol}(\mathfrak{C})$ *satisfies a pseudo-minor condition ϕ. Then* $\mathrm{Pol}(\mathfrak{B})$ *satisfies ϕ as well.*

PROOF. Let h be a homomorphism from $\mathfrak{B}$ to $\mathfrak{C}$ and let i be a homomorphism from $\mathfrak{C}$ to $\mathfrak{B}$. Let $f_1, \dots, f_n$ be the operations witnessing that $\mathrm{Pol}(\mathfrak{C})$ satisfies all pseudo-minor identities in ϕ. For each $f \in \{f_1, \dots, f_n\}$ of arity n, define

$$f'(x_1, \dots, x_n) := i(f(h(x_1), \dots, h(x_n))).$$

We use the lift lemma (Lemma 10.1.5) to show that $f'_1, \dots, f'_n$ witness that $\mathrm{Pol}(\mathfrak{B})$ satisfies ϕ, too.

Let $a \circ f_\sigma = b \circ g_\tau$ be a pseudo-identity of ϕ where $\sigma \colon \{1, \dots, n_1\} \to \{1, \dots, k\}$ and $\tau \colon \{1, \dots, n_2\} \to \{1, \dots, k\}$. Let $m \in \mathbb{N}$ and $t_1, \dots, t_k \in B^m$. Then

$$s_1 := f(h(t_{\sigma(1)}), \dots, h(t_{\sigma(n_1)}))$$
$$s_2 := g(h(t_{\tau(1)}), \dots, h(t_{\tau(n_2)}))$$

lie in the same orbit under $\mathrm{Aut}(\mathfrak{C})$ because $a \circ f_\sigma = b \circ g_\tau$ holds in $\mathrm{Pol}(\mathfrak{C})$ and $\mathfrak{C}$ is a model-complete core. By Proposition 4.7.7 (2) there is an endomorphism of $\mathfrak{B}$ that maps $i(s_1) = f'(t_{\sigma(1)}, \dots, t_{\sigma(n_1)})$ to $i(s_2) = g'(t_{\tau(1)}, \dots, t_{\tau(n_2)})$, and an endomorphism that maps $i(s_2)$ to $i(s_1)$. The lift lemma (Lemma 10.1.5) then implies that there are operations in $\mathrm{End}(\mathfrak{B})$ that together with $f'_1, \dots, f'_n$ show that $\mathrm{Pol}(\mathfrak{B}) \models \phi$. ⊣

10.5. Clones of canonical operations

Canonical functions are a very fruitful concept with many applications, for instance for

- classifying first-order reducts with respect to first-order interdefinability [6, 307, 297, 93, 63, 7, 265, 89],
- classifying the computational complexity of constraint satisfaction problems [80, 100, 90, 64, 241]

- deciding various definability questions over homogeneous finitely bounded structures [96],
- lifting algorithmic results from finite-domain CSPs to problems about the existence of homomorphisms from definable infinite structures to finite structures [234], and
- decidability questions about automata with infinite state space in [235].

Operations that are canonical with respect to an oligomorphic permutation group resemble operations on a finite set, and many results about operation clones with a finite domain extend to operation clones on infinite domains that consist of canonical operations.

10.5.1. Canonical functions. Let $\mathcal{G}$ be a permutation group on a set A and let $\mathcal{H}$ be a permutation group on a set B. A function $f : A \to B$ is called *canonical with respect to* $(\mathcal{G}, \mathcal{H})$ if for every $n \in \mathbb{N}$, $t \in A^n$, and $\alpha \in \mathcal{G}$ there exists $\beta \in \mathcal{H}$ such that $f(\alpha(t)) = \beta(f(t))$. In other words, the orbit of the image of t under f only depends on the orbit of t. Therefore, functions that are canonical with respect to $(\mathcal{G}, \mathcal{H})$ induce for each integer $k \geq 1$ a function from the orbits of the componentwise action of $\mathcal{G}$ of A^k to the orbits of the componentwise action of $\mathcal{H}$ on B^k. For oligomorphic permutation groups we have the following equivalent characterisations of canonicity.

PROPOSITION 10.5.1. *Let $\mathcal{G}$ be a permutation group on a countable set A and let $\mathcal{H}$ be an oligomorphic permutation group on a countable set B. Then for any function $f : A \to B$ the following are equivalent.*

1. *f is canonical with respect to $(\mathcal{G}, \mathcal{H})$;*
2. *for every $\alpha \in \mathcal{G}$ we have $f \circ \alpha \in \overline{\mathcal{H}f} := \overline{\{\beta \circ f \mid \beta \in \mathcal{H}\}}$;*
3. *for every $\alpha \in \mathcal{G}$ there are $e_1, e_2 \in \overline{\mathcal{H}}$ such that $e_1 \circ f \circ \alpha = e_2 \circ f$.*

PROOF. The implications (3) $\Rightarrow$ (1) $\Rightarrow$ (2) follow straightforwardly from the definitions. The implication (2) $\Rightarrow$ (3) is a direct consequence of the lift lemma (Lemma 10.1.5). ⊣

A stronger condition than canonicity would be to require that for every $\alpha \in \mathcal{G}$ there is an $e \in \overline{\mathcal{H}}$ such that

$$f \circ \alpha = e \circ f. \tag{2}$$

To illustrate that this is strictly stronger, already when $\mathcal{G} = \mathcal{H}$, we give an explicit example.

Example 10.5.2 (thanks to Trung Van Pham). Let $\mathcal{G} := \mathcal{H} := \text{Aut}(\mathbb{Q}; <)$. Note that $(\mathbb{Q}; <)$ and $(\mathbb{Q} \setminus \{0\}; <)$ are isomorphic, and let f be such an isomorphism. Then f, viewed as a function from $\mathbb{Q} \to \mathbb{Q}$, is clearly canonical with respect to $(\mathcal{G}, \mathcal{H})$. But f does not satisfy the stronger condition (2). To see this, choose $a \in \mathbb{Q}$ such that $f(a) < 0$, and pick $\alpha \in \mathcal{G}$ such that $f(\alpha(a)) > 0$. Since the image of $f \circ \alpha$ equals the image of f, any $e \in \overline{\mathcal{H}}$ such that $f \circ \alpha = e \circ f$ must fix 0. Since e must also preserve $<$, it cannot map $f(a) < 0$ to $f(\alpha(a)) > 0$. Hence, there is no $e \in \overline{\mathcal{H}}$ such that $f \circ \alpha = e \circ f$.

A function $f: A \to B$ is called *m-canonical* with respect to $(\mathcal{G}, \mathcal{H})$ if for every m-tuple t, the orbit of $f(t)$ under $\mathcal{H}$ only depends on the orbits of t under $\mathcal{G}$. Hence, f is canonical if it is m-canonical for all finite m. The following is a straightforward consequence of homogeneity.

LEMMA 10.5.3. *Let $\mathfrak{B}$ be a homogeneous relational structure whose relations have maximal arity m (in fact, it suffices that $\mathfrak{B}$ is first-order interdefinable with such a structure). Then m-canonicity implies canonicity.*

10.5.2. Clones of canonical operations. Let $\mathcal{G}$ be a permutation group on a set B. Then an operation $f: B^k \to B$ is called *m-canonical with respect to* $\mathcal{G}$ if it is m-canonical with respect to $(\mathcal{G}^k, \mathcal{G})$. In other words, if $t_1, \ldots, t_k \in B^m$ then the orbit of $f(t_1, \ldots, t_k)$ under $\mathcal{G}$ only depends on the orbits of $t_1, \ldots, t_k$ under $\mathcal{G}$. It is called *canonical with respect to* $\mathcal{G}$ if it is m-canonical for all $m \geq 1$.

Example 10.5.4. Let lex be a binary operation on $\mathbb{Q}$ such that $\text{lex}(a, b) < \text{lex}(a', b')$ if either $a < a'$, or $a = a'$ and $b < b'$. Clearly, such an operation exists. Note that lex is injective, that it preserves $<$, and that it is canonical as a binary polymorphism of $(\mathbb{Q}; <)$.

If $\mathcal{B}$ is a clone whose operations are canonical with respect to $\mathcal{G}$ then we say that $\mathcal{B}$ *is canonical with respect to* $\mathcal{G}$. If $\mathcal{B}$ is an operation clone that contains $\mathcal{G}$, then the set $\mathcal{C}$ of operations of $\mathcal{B}$ that are canonical with respect to $\mathcal{G}$ is again an operation clone that contains $\mathcal{G}$. We also call the clone $\mathcal{C}$ the *canonical subclone of* $\mathcal{B}$ *with respect to* $\mathcal{G}$.

Example 10.5.5. Every elementary clone (Section 6.1.5) is canonical with respect to its automorphism group. The Horn clone $\mathcal{H}$ (see Theorem 7.5.2) is another example of a clone which is canonical with respect to its invertible unary operations (i.e., with respect to $\text{Sym}(\mathbb{N})$).

For every $m \in \mathbb{N}$, every operation $f \in \mathcal{C}$ induces a function on orbits of $\mathcal{G}$ on C^m, which we denote by $\xi_m^{\mathcal{G}}(f)$. The set of operations

$$\mathcal{C}_m^{\mathcal{G}} := \{\xi_m^{\mathcal{G}}(f) \mid f \in \mathcal{C}\}$$

is a clone; if $\mathcal{G}$ is oligomorphic then $\mathcal{C}_m^{\mathcal{G}}$ has a finite domain. The map $\xi_m^{\mathcal{G}}: \mathcal{C} \to \mathcal{C}_m^{\mathcal{G}}$ is easily seen to be a uniformly continuous clone homomorphism.

LEMMA 10.5.6. *Let $\mathfrak{B}$ be an ω-categorical model-complete core, and let $\mathcal{C}$ be the canonical subclone of* $\text{Pol}(\mathfrak{B})$ *with respect to* $\mathcal{G} := \text{Aut}(\mathfrak{B})$ *and let* $m \in \mathbb{N}$. *Then $\mathcal{C}_m^{\mathcal{G}}$ is an idempotent operation clone on a finite domain.*

PROOF. Let $f \in \text{Pol}(\mathfrak{B})$ and let A be an element of the domain of $\mathcal{C}_m^{\mathcal{G}}$ (which is finite by the ω-categoricity of $\mathfrak{B}$), i.e., an orbit of m-tuples under $\text{Aut}(\mathfrak{B})$. Since $\mathfrak{B}$ is an ω-categorical model-complete core, A is preserved by the endomorphisms of $\mathfrak{B}$, and in particular it is preserved by $\hat{f}$ (Definition 6.1.40). Hence, $\xi^{\mathcal{G}}_m(\hat{f}) \in \mathcal{C}^{\mathcal{G}}_m$ preserves A, showing the idempotence of $\xi^{\mathcal{G}}_m(\hat{f})$. ⊣

The following lemma shows that we can transfer information from $\mathcal{C}^{\mathcal{G}}_m$ to $\text{Pol}(\mathfrak{B})$.

LEMMA 10.5.7 (from [95]). *Let $\mathfrak{B}$ be a homogeneous structure with finite relational signature of maximal arity m, let $\mathfrak{A}$ be a first-order reduct of $\mathfrak{B}$, and let $\mathcal{C}$ be the polymorphisms of $\mathfrak{A}$ that are canonical with respect to $\mathcal{G} := \text{Aut}(\mathfrak{B})$. If $\mathcal{C}^{\mathcal{G}}_m$ satisfies a minor condition, then $\text{Pol}(\mathfrak{A})$ satisfies this condition modulo $\mathcal{G}$.*

PROOF. Fix operations from $\mathcal{C}^{\mathcal{G}}_m$ that witness that the given condition holds in $\mathcal{C}^{\mathcal{G}}_m$. We claim that the preimages of these operations witness that $\text{Pol}(\mathfrak{A})$ satisfies the condition modulo $\mathcal{G}$, and verify this using the lift lemma (Lemma 10.1.5). Let $F \subseteq A = B$ be finite and suppose that $f_\sigma = g_\tau$ is a conjunct of the condition. For all finite tuples $t_1, \dots, t_k$ over F, by the assumption we know that the two tuples $(\xi^{\mathcal{G}}_m)^{-1}(f)_\sigma(t_1, \dots, t_k)$ and $(\xi^{\mathcal{G}}_m)^{-1}(g)_\tau(t_1, \dots, t_k)$ lie in the same orbit under $\mathcal{G}$. So there exists an automorphism of $\mathfrak{B}$ mapping one tuple to the other and the statement follows from the lift lemma. ⊣

PROPOSITION 10.5.8. *Let $\mathfrak{B}$ be a homogeneous structure $\mathfrak{B}$ with a finite relational signature of maximal arity m, and let $\mathfrak{C}$ be a first-order reduct of $\mathfrak{B}$ which is a model-complete core. Let $\mathcal{C}$ be the clone of all polymorphisms of $\mathfrak{C}$ that are canonical with respect to $\mathcal{G} := \text{Aut}(\mathfrak{B})$. Then the following are equivalent.*

1. *$\mathcal{C}$ contains for every prime p that is larger than the number of orbits of m-tuples under $\mathcal{G}$ a pseudo-cyclic operation of arity p.*
2. *$\mathcal{C}$ contains a pseudo-cyclic operation of some arity $n \geq 2$.*
3. *$\mathcal{C}$ contains a pseudo weak near-unanimity of some arity $n \geq 2$.*
4. *$\mathcal{C}$ contains a pseudo Taylor operation.*
5. *$\mathcal{C}^{\mathcal{G}}_m$ contains a Taylor operation.*
6. *$\mathcal{C}^{\mathcal{G}}_m$ contains for every prime p that is larger than the number of orbits of m-tuples under $\mathcal{G}$ a cyclic operation of arity p.*
7. *$\mathcal{C}^{\mathcal{G}}_m$ contains a 4-ary (or a 6-ary) Siggers operation.*
8. *$\mathcal{C}$ contains a 4-ary (or a 6-ary) pseudo-Siggers operation.*
9. *$\mathcal{C}$ has no uniformly continuous minor-preserving map to* **Proj** *(and the other equivalent conditions in Theorem 10.3.5).*

PROOF. The implications (1) ⇒ (2) ⇒ (3) ⇒ (4) ⇒ (5) are trivial. The equivalence of (5), (6), and (7) follows from Theorem 6.9.2.

For the implication (7) ⇒ (8), let $g \in \text{Pol}(\mathfrak{C})$ be such that $\xi^{\mathcal{G}}_m(g)$ equals the 4-ary Siggers operation. Then g satisfies the 4-ary pseudo-Siggers condition by Lemma 10.5.7.

The implication (6) ⇒ (1) can be shown analogously.

For the equivalence of (8) and (9), let $\mathfrak{D}$ be any structure such that $\text{Pol}(\mathfrak{D})$ equals the set of all polymorphisms of $\mathfrak{C}$ that are canonical with respect to $\text{Aut}(\mathfrak{B})$. Recall that $\mathfrak{B}$ and hence also $\mathfrak{C}$ and $\mathfrak{D}$ have less than doubly exponential orbit growth, because $\mathfrak{B}$ is homogeneous in a finite relational language. So the implication follows from Theorem 10.3.5. ⊣

10.5.3. Reduction to the finite. In this section we present a polynomial-time reduction from a large class of infinite-domain CSPs to certain finite-domain CSPs. This reduction can be used to derive many tractability results for infinite-domain CSPs from algorithms for finite-domain CSPs; examples will be given in Section 10.5.5. In Section 10.5.4 we will use the concept of canonical functions to prove that in certain situations, the finite-domain CSP is even polynomial-time equivalent to the infinite-domain CSP we started from.

DEFINITION 10.5.9. Let $\mathfrak{B}$ be a finitely bounded structure, and let $\mathfrak{A}$ be a relational structure with the same domain as $\mathfrak{B}$ such that every relation of $\mathfrak{A}$ has a quantifier-free definition in $\mathfrak{B}$. Let m be a positive integer. Then the *associated type structure* $T_{\mathfrak{B},m}(\mathfrak{A})$ is defined to be the relational structure whose domain is the set of maximal quantifier-free (qf) m-types of $\mathfrak{B}$ and whose relations are as follows.

- For each relation R of $\mathfrak{A}$ of arity r, let $\chi(z_1, \ldots, z_r)$ be a definition of R in $\mathfrak{B}$. For $i\colon [r] \to [m]$ we write $\langle \chi(z_{i(1)}, \ldots, z_{i(r)}) \rangle$ for the unary relation that consists of all the maximal qf-types that contain $\chi(z_{i(1)}, \ldots, z_{i(r)})$, and add all such relations to $T_{\mathfrak{B},m}(\mathfrak{A})$.
- For each $r \in [m]$ and $i, j\colon [r] \to [m]$, the *compatibility relation* $\mathrm{Comp}_{i,j}$ is defined to be the binary relation that contains all the pairs (p, q) of maximal qf m-types such that for every quantifier-free formula $\chi(z_1, \ldots, z_s)$ of $\mathfrak{B}$ and $t\colon [s] \to [r]$, the formula $\chi(z_{it(1)}, \ldots, z_{it(s)})$ is in p if and only if $\chi(z_{jt(1)}, \ldots, z_{jt(s)})$ is in q.

Note that if $(a_1, \ldots, a_m)$ has type p and $(b_1, \ldots, b_m)$ has type q, then $\mathrm{Comp}_{i,j}(p, q)$ holds if and only if $(a_{i(1)}, \ldots, a_{i(r)})$ and $(b_{j(1)}, \ldots, b_{j(r)})$ have the same type in $\mathfrak{B}$. Also note that if $i\colon [m] \to [m]$ is the identity map, then $\mathrm{Comp}_{i,i}$ denotes the equality relation on the domain of $T_{\mathfrak{B},m}(\mathfrak{A})$.

Theorem 10.5.10 below holds for arbitrary finitely bounded structures $\mathfrak{B}$. In the next section we present a sufficient condition for the existence of a polynomial-time reduction in the other direction, from $\mathrm{CSP}(T_{\mathfrak{B},m}(\mathfrak{A}))$ to $\mathrm{CSP}(\mathfrak{A})$.

THEOREM 10.5.10 (Theorem 3.1 in [81]). *Let $\mathfrak{A}$ be a quantifier-free reduct of a finitely bounded structure $\mathfrak{B}$, and suppose that $\mathfrak{A}$ has a finite signature. Let m_a be the maximal arity of all relations in $\mathfrak{A}$ or $\mathfrak{B}$, and m_b be the maximal size of a bound for $\mathfrak{B}$. Let m be at least* $\max(m_a + 1, m_b, 3)$. *Then there is a polynomial-time reduction from* $\mathrm{CSP}(\mathfrak{A})$ *to* $\mathrm{CSP}(T_{\mathfrak{B},m}(\mathfrak{A}))$.

PROOF. Let ψ be an instance of $\mathrm{CSP}(\mathfrak{A})$, and let $V = \{x_1, \ldots, x_n\}$ be the variables of ψ. Assume without loss of generality that $n \geq m$. We build an instance ϕ of $\mathrm{CSP}(T_{\mathfrak{B},m}(\mathfrak{A}))$ as follows.

- The variable set of ϕ is the set $\mathcal{I}$ of increasing functions[27] from $[m] \to V$ (where the variables are endowed with an arbitrary linear order). The idea of the reduction is that the variable $v \in \mathcal{I}$ of ϕ represents the maximal qf-type of $(h(v(1)), \ldots, h(v(m)))$ in a satisfying assignment h for ψ.
- For each conjunct of the form $R(j(1), \ldots, j(r))$ of ψ, where $j \colon [r] \to V$, we add unary constraints to ϕ as follows. By assumption, R has a qf-definition $\chi(z_1, \ldots, z_r)$ over $\mathfrak{B}$. Let $v \in \mathcal{I}$ be such that its image $\mathrm{Im}(v)$ contains the image of j. Let U be the relation symbol of $T_{\mathfrak{B},m}(\mathfrak{A})$ that denotes the unary relation $\langle \chi(z_{v^{-1}j(1)}, \ldots, z_{v^{-1}j(r)}) \rangle$. We then add $U(v)$ to Φ.
- Finally, for all $u, v \in \mathcal{I}$ let $k \colon [r] \to \mathrm{Im}(u) \cap \mathrm{Im}(v)$ be a bijection. We then add the constraint $\mathrm{Comp}_{u^{-1}k, v^{-1}k}(u, v)$.

Before proving that the given reduction works, we illustrate it with an example.

Example 10.5.11. Let $\mathfrak{A}$ be $(\mathbb{N}; =, \neq)$ and consider the following instance of $\mathrm{CSP}(\mathfrak{A})$.

$$x_1 = x_2 \wedge x_2 = x_3 \wedge x_3 = x_4 \wedge x_1 \neq x_4$$

The structure $(\mathbb{N}; =, \neq)$ is a reduct of the homogeneous structure $\mathfrak{B}$ with domain $\mathbb{N}$ and the empty signature, which has no bounds. We have in this example $m = 3$. The maximal qf m-types of $\mathfrak{B}$ can be viewed as partitions of the variables where two variables z_i, z_j are in the same part if the type implies that $z_i = z_j$.

The structure $T_{\mathfrak{B},3}(\mathfrak{A})$ has a domain of size five, where each element corresponds to a partition of $\{z_1, z_2, z_3\}$. It has a unary relation U_1 for $\langle z_2 = z_3 \rangle$, containing all partitions in which z_2 and z_3 belong to the same part. Similarly, it has a relation U_2 for $\langle z_1 = z_3 \rangle$, U_3 for $\langle z_1 = z_2 \rangle$, V_1 for $\langle z_2 \neq z_3 \rangle$, V_2 for $\langle z_1 \neq z_3 \rangle$, and V_3 for $\langle z_1 \neq z_2 \rangle$. The instance ϕ of $\mathrm{CSP}(T_{\mathfrak{B},3}(\mathfrak{A}))$ created by our reduction has four variables, for the four order-preserving injections from $[3] \to \{x_1, x_2, x_3, x_4\}$ (where we order $x_1, \ldots, x_4$ according to their index). These four variables will be called v_1, v_2, v_3, v_4 and $\mathrm{Im}(v_j) = \{x_1, \ldots, x_4\} \setminus \{x_j\}$. We then have the following constraints in ϕ:

- $U_3(v_3)$ and $U_3(v_4)$ for the constraint $x_1 = x_2$ in ψ;
- $U_1(v_4)$ and $U_3(v_1)$ for the constraint $x_2 = x_3$ in ψ;
- $U_1(v_2)$ and $U_1(v_1)$ for the constraint $x_3 = x_4$ in ψ;
- $V_2(v_2)$ and $V_2(v_3)$ for the constraint $x_1 \neq x_4$ in ψ.

For the compatibility constraints we only give an example. Let $k, k' \colon [2] \to [4]$ be such that $k(1,2) = (1,3)$ and $k'(1,2) = (1,2)$. Then $\mathrm{Comp}_{k,k'}(v_4, v_2)$ and $\mathrm{Comp}_{k',k'}(v_4, v_3)$ are in ϕ.

---

[27]One could take $\mathcal{I}$ to be the set of all functions $[m] \to V$ without any change to the reduction. We choose here to only take increasing functions so that the presentation of the example below is more concise.

We now prove that the reduction is correct. Let $h\colon V \to B$ be an assignment of the variables to the domain of $\mathfrak{B}$. Let $\chi(z_1,\dots,z_r)$ be a qf-formula over the signature of $\mathfrak{B}$, let $j\colon [r] \to V$, and let $v \in \mathcal{I}$ be such that $\mathrm{Im}(j) \subseteq \mathrm{Im}(v)$. First note that

$$\mathfrak{B} \models \chi(h(j(1)),\dots,h(j(r))) \text{ iff } \\ (h(v(1)),\dots,h(v(m))) \text{ satisfies } \chi(z_{v^{-1}j(1)},\dots,z_{v^{-1}j(r)}) \text{ in } \mathfrak{B}. \quad (\ddagger)$$

The property (‡) holds since in the qf-type of the tuple $(h(v(1)),\dots,h(v(m)))$, the variable z_i represents the element $h(v(i))$, and therefore $z_{v^{-1}j(i)}$ represents $h(j(i))$.

Claim 1. ψ satisfiable implies ϕ satisfiable. Suppose that $h\colon V \to B$ satisfies ψ in $\mathfrak{A}$. To show that ϕ is satisfiable in $T_{\mathfrak{B},m}(\mathfrak{A})$ define $g\colon \mathcal{I} \to T_{\mathfrak{B},m}(\mathfrak{A})$ by setting $g(v)$ to be the qf-type of $(h(v(1)),\dots,h(v(m)))$ in $\mathfrak{B}$, for every $v \in \mathcal{I}$. To see that all the constraints of ϕ are satisfied by g, let $U(v)$ be a constraint in ϕ that has been introduced for a conjunct of the form $R(j(1),\dots,j(r))$ in ψ, where $j\colon [r] \to V$. Let $\chi(z_1,\dots,z_r)$ be a qf-formula that defines R in $\mathfrak{B}$. Then

$$\begin{aligned} \mathfrak{A} &\models R(h(j(1)),\dots,h(j(r))) \\ &\Rightarrow \mathfrak{B} \models \chi(h(j(1)),\dots,h(j(r))) \\ &\Rightarrow \chi(z_{v^{-1}j(1)},\dots,z_{v^{-1}j(m)}) \in g(v) \qquad \text{(because of (‡))} \\ &\Rightarrow T_{\mathfrak{B},m}(\mathfrak{A}) \models U(g(v)). \end{aligned}$$

Next, consider a constraint of the form $\mathrm{Comp}_{u^{-1}k,v^{-1}k}(u,v)$ in Φ, and let $r := |\mathrm{Im}(k)|$. Let $\chi(z_1,\dots,z_s)$ be a qf-formula over the signature of $\mathfrak{B}$ and let $t\colon [s] \to [r]$. Suppose that $\chi(z_{u^{-1}kt(1)},\dots,z_{u^{-1}kt(s)})$ is in $g(u)$. From (‡) we obtain that $\mathfrak{B} \models \chi(h(kt(1)),\dots,h(kt(s)))$. Again by (‡) we get that $\chi(z_{v^{-1}kt(1)},\dots,z_{v^{-1}kt(s)})$ is in $g(v)$. Hence, $T_{\mathfrak{B},m}(\mathfrak{A}) \models \mathrm{Comp}_{u^{-1}k,v^{-1}k}(g(u),g(v))$.

Claim 2. ϕ satisfiable implies ψ satisfiable. Suppose that ϕ is satisfiable in $T_{\mathfrak{B},m}(\mathfrak{A})$, that is, there exists a map h from I to the qf m-types of $\mathfrak{B}$ that satisfies all conjuncts of ϕ. We show how to obtain an assignment $\{x_1,\dots,x_n\} \to A$ that satisfies ψ in $\mathfrak{A}$. Define an equivalence relation $\sim$ on V as follows. Let $x, y \in V$. Let $u \in \mathcal{I}$ be such that there are $p, q \in [m]$ such that $u(p) = x$ and $u(q) = y$. We define $x \sim y$ if, and only if, $h(u)$ contains the formula $z_p = z_q$. Note that the choice of u is not important: if u', p', q' are such that $u'(p') = x$ and $u'(q') = y$, the intersection of $\mathrm{Im}(u)$ and $\mathrm{Im}(u')$ contains $\{x, y\}$. Let $k\colon [r] \to \mathrm{Im}(u) \cap \mathrm{Im}(u')$ be a bijection. By construction, the constraint $\mathrm{Comp}_{u^{-1}k,u'^{-1}k}(u,u')$ is satisfied by h, which by definition of the relation means that $h(u)$ contains $z_p = z_q$ iff $h(u')$ contains $z_{p'} = z_{q'}$.

We prove that $\sim$ is an equivalence relation. Reflexivity and symmetry are clear from the definition. Assume that $x \sim y$ and $y \sim z$. Let $w \in I, p, q, r$ be such that $w(p) = x$, $w(q) = y$, and $w(r) = z$, which is possible since $m \geq 3$.

Since $x \sim y$, the previous paragraph implies that $h(w)$ contains the formula $z_p = z_q$. Similarly, since $y \sim z$, the formula $z_q = z_r$ is in $h(w)$. Since $h(w)$ is a type, transitivity of equality implies that $z_p = z_r$ is in $h(w)$, so that $x \sim z$.

Define a structure $\mathfrak{C}$ on $V/_\sim$ as follows. For every k-ary relation symbol R of $\mathfrak{B}$ and k elements $y_1/_\sim, \dots, y_k/_\sim$ of $V/_\sim$, let $w \in \mathcal{I}, p_1, \dots, p_k \in [m]$ be such that $w(p_i) = y_i$ (such a w exists because $m \geq k$). Add the tuple $(y_1/_\sim, \dots, y_k/_\sim)$ to $R^{\mathfrak{C}}$ if and only if $h(w)$ contains the formula $R(z_{p_1}, \dots, z_{p_k})$. As in the paragraph above, this definition does not depend on the choice of the representatives $y_1, \dots, y_k$ or on the choice of w. Proving that the definition does not depend on w is straightforward. Suppose now that $y_1 \sim y_1'$, and let $w \in \mathcal{I}$ be such that $(w(p_1), \dots, w(p_k)) = (y_1, \dots, y_k)$ and such that $h(w)$ contains $R(z_{p_1}, \dots, z_{p_k})$. Let $w' \in \mathcal{I}$ be such that $(w'(q), w'(p_1'), \dots, w'(p_k')) = (y_1', y_1, y_2, \dots, y_k)$, which is possible since $m \geq k+1$. We prove that $h(w')$ contains $R(z_q, z_{p_2'}, \dots, z_{p_k'})$. Since $y \sim y'$, we have that $h(w')$ contains $z_q = z_{p_1'}$. Moreover, $\mathrm{Im}(w') \cap \mathrm{Im}(w) = \{y_1, \dots, y_k\}$, and since h satisfies the compatibility constraints we obtain that $h(w')$ contains $R(z_{p_1'}, \dots, z_{p_k'})$. It follows that $h(w')$ contains $R(z_q, z_{p_2'}, \dots, z_{p_k'})$. Therefore, the definition of R in $\mathfrak{C}$ does not depend on the choice of the representative for the first entry of the tuple. By iterating this argument for each coordinate, we obtain that $R^{\mathfrak{C}}$ is well defined.

We claim that $\mathfrak{C}$ embeds into $\mathfrak{B}$. Otherwise, there would exist a bound $\mathfrak{D}$ of size $k \leq m$ for $\mathfrak{B}$ such that $\mathfrak{D}$ embeds into $\mathfrak{C}$. Let $y_1/_\sim, \dots, y_k/_\sim$ be the elements of the image of $\mathfrak{D}$ under this embedding. Since $k \leq m$, there exist $w \in \mathcal{I}, p_1, \dots, p_k$ such that $(w(p_1), \dots, w(p_k)) = (y_1, \dots, y_k)$. The qf-type of $(y_1/_\sim, \dots, y_k/_\sim)$ in $\mathfrak{C}$ is in $h(w)$, by the previous paragraph. It follows that if $(a_1, \dots, a_m) \in B^m$ is a tuple whose qf-type is $h(w)$, then there is an embedding of $\mathfrak{D}$ into the substructure of $\mathfrak{B}$ induced on $\{a_1, \dots, a_m\}$. This contradicts the assumption that $\mathfrak{D}$ is a bound of $\mathfrak{B}$ and hence does not embed into $\mathfrak{B}$.

Let e be an embedding $\mathfrak{C} \hookrightarrow \mathfrak{B}$. We claim that $f\colon \{x_1, \dots, x_n\} \to A$ given by $f(x) := e(x/_\sim)$ is a valid assignment for ψ. Let $R(j(1), \dots, j(r))$ be a constraint from ψ, where $j\colon [r] \to V$. Let $v \in \mathcal{I}$ be such that $\mathrm{Im}(j) \subseteq \mathrm{Im}(v)$, and such that the constraint $\langle \chi(z_{v^{-1}j(1)}, \dots, z_{v^{-1}j(r)}) \rangle (v)$ is in ϕ. Since h satisfies this constraint, $h(v)$ contains $\chi(z_{v^{-1}j(1)}, \dots, z_{v^{-1}j(r)})$. It follows that $\mathfrak{C} \models \chi(j(1)/_\sim, \dots, j(r)/_\sim)$. Since e embeds $\mathfrak{C}$ into $\mathfrak{B}$, we obtain that $\mathfrak{B} \models \chi(f(j(1)), \dots, f(j(r)))$, whence $\mathfrak{A} \models R(f(j(1)), \dots, f(j(r)))$, as required.

The given reduction can be performed in polynomial time: the number of variables in the new instance is in $O(n^m)$, and if c is the number of constraints in ψ, then the number of constraints in ϕ is in $O(cn^m + n^{2m})$. Each of the new constraints can be constructed in constant time. ⊣

We mention that the reduction is in fact a *first-order reduction* (see [14] for a definition). We also note that Theorem 10.5.10 applies to all CSPs that can be described in SNP (see Section 1.4). Finally, note that this is one of the rare situations of a complexity reduction between CSPs which is *not* based on a pp-construction. For example, it is easy to see that there cannot be a pp-construction of

$$\mathfrak{B} := (\mathbb{N}; \neq, \{(x, y, u, v) \mid x = y \Rightarrow u = v\})$$

in any finite structure, and in particular not in $T_{\mathfrak{B},m}(\mathfrak{B})$ even though the reduction from Theorem 10.5.10 can be applied in this case.

10.5.4. Complexity classification for canonical clones. This section connects the canonical polymorphisms of a first-order reduct $\mathfrak{A}$ of a homogeneous structure $\mathfrak{B}$ with finite relational signature of maximal arity m with the polymorphism clone of the associated type structure $T_{\mathfrak{B},m}(\mathfrak{A})$ from the previous section. This will lead to a complete complexity classification for reducts $\mathfrak{A}$ of finitely bounded homogeneous structures $\mathfrak{B}$ whose polymorphisms are canonical with respect to $\mathfrak{B}$, based on the complexity classification for finite-domain CSPs.

LEMMA 10.5.12. *Let $\mathfrak{A}$ be a first-order reduct of a homogeneous relational structure $\mathfrak{B}$ and let $\mathfrak{C}$ be the canonical subclone of* Pol($\mathfrak{A}$) *with respect to* $\mathcal{G} :=$ Aut($\mathfrak{B}$). *Then*

$$\mathfrak{C}_m^{\mathcal{G}} \subseteq \mathrm{Pol}(T_{\mathfrak{B},m}(\mathfrak{A}))$$

for every $m \geq 1$. If m is larger than the size of every bound of $\mathfrak{B}$ and strictly larger than the maximal arity of $\mathfrak{A}$ and $\mathfrak{B}$, and at least 3, *then* $\mathfrak{C}_m^{\mathcal{G}} = \mathrm{Pol}(T_{\mathfrak{B},m}(\mathfrak{A}))$.

PROOF. Let $f \in \mathfrak{C}$. We have to show that $\xi_m^{\mathcal{G}}(f) \in \mathrm{Pol}(T_{\mathfrak{B},m}(\mathfrak{A}))$. Let k be the arity of f. Let $\chi(z_1, \dots, z_r)$ be a quantifier-free definition of a relation of $\mathfrak{A}$. Let $i \colon [r] \to [m]$ and let $p_1, \dots, p_k$ be from the relation $\langle \chi(z_{i(1)}, \dots, z_{i(r)}) \rangle$ of $T_{\mathfrak{B},m}(\mathfrak{A})$. Let $\bar{a}^1, \dots, \bar{a}^k$ be m-tuples whose qf-types are $p_1, \dots, p_k$ respectively. Since $\mathfrak{B}$ is homogeneous there is a one-to-one correspondence between the orbits of m-tuples under Aut($\mathfrak{B}$), the maximal qf-types of $\mathfrak{B}$, and the maximal types of $\mathfrak{B}$. Hence, $\xi_m^{\mathcal{G}}(f)$ can be seen as an operation on maximal qf m-types; $\xi_m^{\mathcal{G}}(f)(p_1, \dots, p_k)$ is the qf-type of $f(\bar{a}^1, \dots, \bar{a}^k)$ in $\mathfrak{B}$. Since f preserves the relation defined by $\chi(z_{i(1)}, \dots, z_{i(r)})$, it follows that $f(\bar{a}^1, \dots, \bar{a}^k)$ satisfies $\chi(z_{i(1)}, \dots, z_{i(r)})$, which means that $\chi(z_{i(1)}, \dots, z_{i(r)})$ is contained in the qf-type of this tuple. Therefore, $\xi_m^{\mathcal{G}}(f)$ preserves the relations of $T_{\mathfrak{B},m}(\mathfrak{A})$ from the first bullet in Definition 10.5.9.

We now prove that $\xi_m^{\mathcal{G}}(f)$ also preserves the compatibility relations in $T_{\mathfrak{B},m}(\mathfrak{A})$. Indeed, let $(p_1, q_1), \dots, (p_k, q_k)$ be pairs of qf-types in $\mathrm{Comp}_{i,j}$. Let $(\bar{a}^1, \bar{b}^1), \dots, (\bar{a}^k, \bar{b}^k)$ be pairs of m-tuples such that $\mathrm{tp}(\bar{a}^l) = p_l$ and $\mathrm{tp}(\bar{b}^l) = q_l$ for all $l \in [k]$. As noted above, the definition of $\mathrm{Comp}_{i,j}$ implies that the tuples $(a^l_{i(1)}, \dots, a^l_{i(r)})$ and $(b^l_{j(1)}, \dots, b^l_{j(r)})$ have the same type in $\mathfrak{B}$ for all $l \in [k]$.

Since f is canonical, we have that

$$(f(a^1_{i(1)},\dots,a^k_{i(1)}),\dots,f(a^1_{i(r)},\dots,a^k_{i(r)}))$$

has the same type as

$$(f(b^1_{j(1)},\dots,b^k_{j(1)}),\dots,f(b^1_{j(r)},\dots,b^k_{j(r)}))$$

in $\mathfrak{B}$. This implies that

$$\mathrm{Comp}_{i,j}(\xi^{\mathfrak{g}}_m(f)(p_1,\dots,p_k),\xi^{\mathfrak{g}}_m(f)(q_1,\dots,q_k))$$

holds in $T_{\mathfrak{B},m}(\mathfrak{A})$, which concludes the proof.

For the reverse inclusion, we prove that for every $g \in \mathrm{Pol}(T_{\mathfrak{B},m}(\mathfrak{A}))$ there exists an $f \in \mathfrak{C}$ such that $\xi^{\mathfrak{g}}_m(f) = g$. Let k be the arity of g. We prove that for every subset F of A there exists a function h from $F^k \to A$ such that for all $\bar{a}^1,\dots,\bar{a}^k \in F^m$ whose types are $p_1,\dots,p_k$, respectively, $h(\bar{a}^1,\dots,\bar{a}^k)$ has type $g(p_1,\dots,p_k)$. A standard compactness argument then shows the existence of a function $f\colon A^k \to A$ such that for all $\bar{a}^1,\dots,\bar{a}^k \in A^m$ whose types are $p_1,\dots,p_k$, respectively, $f(\bar{a}^1,\dots,\bar{a}^k)$ has type $g(p_1,\dots,p_k)$, and such a function must satisfy $\xi^{\mathfrak{g}}_m(f) = g$.

Note that we can assume without loss of generality that $\mathfrak{B}$ has for each relation symbol R also a relation symbol for the complement of $R^{\mathfrak{B}}$. This does not change $\mathfrak{C}^{\mathfrak{g}}_m$ or $T_{\mathfrak{B},m}(\mathfrak{A})$. The existence of a function h with the properties as stated above can then be expressed as an instance Ψ of CSP($\mathfrak{B}$) where the variable set is F^k and where we impose constraints from $\mathfrak{B}$ on $\bar{a}^1,\dots,\bar{a}^k$ to enforce that in any solution h to this instance the tuple $h(\bar{a}^1,\dots,\bar{a}^k)$ satisfies $g(p_1,\dots,p_k)$. Let Φ be the instance of CSP($T_{\mathfrak{B},m}(\mathfrak{B})$) obtained from Ψ under the reduction from CSP($\mathfrak{B}$) to CSP($T_{\mathfrak{B},m}(\mathfrak{B})$) described in the proof of Theorem 10.5.10. The variables of Φ are the order-preserving injections from $[m]$ to F^k. For $v\colon [m] \to F^k$ and $i \le k$, let p_i be the type of $(v(1)_i,\dots,v(m)_i)$ in $\mathfrak{B}$. Then the mapping h that sends v to $g(p_1,\dots,p_k)$, for all variables v of Φ, is a solution to Φ:

- the constraints of Φ of the form $\langle\chi(\dots)\rangle(v)$ have been introduced to translate constraints of Ψ, and it is easy to see that they are satisfied by the choice of these constraints of Ψ and by the choice of h.
- The other constraints of Φ are of the form $\mathrm{Comp}_{i,j}(u,v)$ where u, v are order-preserving injections from $[m]$ to F^k. Since g is a k-ary polymorphism of $\mathrm{Pol}(T_{\mathfrak{B},m}(\mathfrak{B}))$ and hence preserves the relations $\mathrm{Comp}_{i,j}$ of $T_{\mathfrak{B},m}(\mathfrak{B})$, it follows that h satisfies these constraints, too.

By Theorem 10.5.10, the instance Ψ of CSP($\mathfrak{B}$) is satisfiable as well. $\dashv$

We now derive a complexity classification for canonical clones from the complexity classification of finite-domain CSPs.

COROLLARY 10.5.13. *Let $\mathfrak{B}$ be a finitely bounded homogeneous structure in a finite relational signature and let $\mathfrak{A}$ be a reduct of $\mathfrak{B}$. Let m be larger than* 3, *larger than the maximal arity of* $\mathfrak{A}$, *and larger than the maximal bound of* $\mathfrak{B}$. *Suppose that all the polymorphisms of* $\mathfrak{A}$ *are canonical with respect to* $\mathrm{Aut}(\mathfrak{B})$. *Then* $\mathrm{CSP}(\mathfrak{A})$ *and* $\mathrm{CSP}(T_{\mathfrak{B},m}(\mathfrak{A}))$ *are polynomial-time equivalent. In particular,* $\mathfrak{A}$ *is either in P or NP-complete.*

PROOF. Lemma 10.5.12 implies that $\xi_m^{\mathcal{G}}$ is a surjective continuous clone homomorphism from $\mathrm{Pol}(\mathfrak{A})$ to $\mathrm{Pol}(T_{\mathfrak{B},m}(\mathfrak{A}))$. Therefore, Corollary 9.5.20 implies that there is a primitive positive interpretation of $T_{\mathfrak{B},m}(\mathfrak{A})$ in $\mathfrak{A}$, and hence there is a polynomial-time reduction from $\mathrm{CSP}(T_{\mathfrak{B},m}(\mathfrak{A}))$ to $\mathrm{CSP}(\mathfrak{A})$ by Theorem 3.1.4.

Lemma 10.5.10 shows that there is a polynomial-time reduction in the reverse direction. The statement now follows from the finite-domain complexity dichotomy theorem (Theorem 3.7.2). ⊣

10.5.5. Tractability conditions. The situation that *all* polymorphisms of a structure are canonical is quite rare. But for polynomial-time tractability it suffices that there are *some* canonical polymorphisms that guarantee tractability, and we will see that this applies in many situations. In this section we use Lemma 10.5.12 to derive new tractability conditions for reducts $\mathfrak{A}$ of finitely bounded homogeneous structures $\mathfrak{B}$.

Remark 10.5.14. Interestingly, if the structure $\mathfrak{A}$ under consideration is a model-complete core, then the tractability conditions in this section are purely *algebraic* in the sense that they are phrased in terms of the existence of polymorphisms satisfying certain identities (without any reference to the action or the topology of the clone). The reason is that if the invertible unary operations are dense in the unary operations, then canonicity of operations is an algebraic property (item (3) in Proposition 10.5.1).

We start with a powerful result for containment in Datalog (see Chapter 8).

THEOREM 10.5.15. *Let* $\mathfrak{A}$ *be a reduct of a finitely bounded homogeneous structure* $\mathfrak{B}$. *Suppose that* $\mathrm{Pol}(\mathfrak{A})$ *contains an operation* f *or arity three and an operation* g *of arity four such that*

- f *and* g *are canonical with respect to* $\mathcal{G} := \mathrm{Aut}(\mathfrak{B})$,
- f *and* g *are weak near-unanimity operations modulo* $\overline{\mathrm{Aut}(\mathfrak{B})}$, *and*
- $f(y,x,x) = g(y,x,x,x)$ *for all* $x, y \in B$.

Then $\mathrm{CSP}(\mathfrak{A})$ *is in Datalog.*

PROOF. Let m be as in the statement of Theorem 10.5.10. By Lemma 10.5.12, $f' := \xi_m^{\mathcal{G}}(f)$ and $g' := \xi_m^{\mathcal{G}}(g)$ are polymorphisms of $T_{\mathfrak{B},m}(\mathfrak{A})$. Moreover, f' and g' must be weak near-unanimity operations, and they satisfy $f'(y,x,x) = g'(y,x,x,x)$. It follows from Theorem 8.8.2 that $\mathrm{CSP}(T_{\mathfrak{B},m}(\mathfrak{A}))$ has width (ℓ, k) for some $\ell, k \in \mathbb{N}$, i.e., can be solved by a Datalog program. Since the reduction from $\mathrm{CSP}(\mathfrak{A})$ to $\mathrm{CSP}(T_{\mathfrak{B},m}(\mathfrak{A})$ presented in Section 10.5.3 is a first-order reduction, it is in particular a Datalog reduction. Theorem 10.5.10

and the transitivity of Datalog reductions implies that CSP($\mathfrak{A}$) is in Datalog, too. ⊣

This result generalises many tractability results from the literature, for instance

- polynomial-time tractable equality constraints [68] (see Chapter 7),
- the polynomial-time algorithms for partially-ordered time from [104, 241],
- the polynomial-time tractable fragments of RCC5 [220], and
- all polynomial-time tractable equivalence CSPs [100].

In all cases, the respective structures $\mathfrak{A}$ have a polymorphism f such that $\xi_2^{\mathfrak{g}}(f)$ is a semilattice operation(Example 2.1.2). Clearly, finite structures with a semilattice polymorphism also have weak near-unanimity polymorphisms f' and g' that satisfy $f'(y, x, x, x) = g'(y, x, x)$, and hence $\mathfrak{A}$ satisfies the assumptions of Theorem 10.5.15. The fact that all of these problems can be solved by Datalog can be seen without using Theorem 8.8.2 by combining Theorem 10.5.10 with the results presented in Section 8.4.2.

Also the finite-domain CSP tractability theorem (Theorem 3.7.2) can be lifted to a tractability condition for reducts of finitely bounded homogeneous structures.

THEOREM 10.5.16. *Let $\mathfrak{A}$ be a reduct of a finitely bounded homogeneous structure $\mathfrak{B}$. Suppose that $\mathfrak{A}$ has a polymorphism that is canonical with respect to* Aut($\mathfrak{B}$) *and that satisfies the Siggers identity modulo* $\overline{\text{Aut}(\mathfrak{B})}$ (*or any other of the equivalent conditions from Proposition* 10.5.8), *then* CSP($\mathfrak{A}$) *is in P.*

PROOF. Let m be as in the statement of Theorem 10.5.10. By Lemma 10.5.12, $\xi_m^{\mathfrak{g}}(f)$ is a polymorphism of $T_{\mathfrak{B},m}(\mathfrak{A})$. Since $\xi_m^{\mathfrak{g}}(f)$ is a Siggers operation, Theorem 3.7.2 implies that CSP($T_{\mathfrak{B},m}(\mathfrak{A})$) is in P. Then Theorem 10.5.10 implies that CSP($\mathfrak{A}$) is in P. ⊣

We mention that all the polynomial-time tractable cases in the classification for Graph-SAT problems [90] satisfy the assumptions of Theorem 10.5.16. Finally, we mention that the non-trivial polynomial-time tractable cases for first-order reducts of ($\mathbb{Q}$; <) (see Chapter 12) provide examples that cannot be lifted from finite-domain tractability results in this way, since the respective languages do not have non-trivial canonical polymorphisms.

Chapter 11

RAMSEY THEORY

The application of Ramsey theory to study the expressive power of constraint languages via polymorphisms is one of the central contributions of this text. The idea is that polymorphisms of Ramsey structures must behave canonically on large parts of their domain. This also leads to decidability results for several meta-questions about the expressive power of constraint languages. The same idea can be used to show statements of the type 'every polymorphism that does not preserve R must locally generate g', for certain relations R and certain operations g with good properties. Such statements will be crucial for instance in the CSP complexity classification for first-order reducts of $(\mathbb{Q}; <)$ presented in Chapter 12. The same Ramsey theoretic technique has been applied to classify the complexity of the CSP for all first-order reducts of the following structures:

- the Rado graph [90];
- the equivalence relation with infinitely many infinite classes [100];
- the homogeneous binary branching C-relation [64];
- the countable homogeneous universal poset [241];

- the Henson graphs [80];
- all structures where the number of orbits of n-tuples is bounded by an exponential function [81, 48].

In this chapter we first revisit classical concepts and results from structural Ramsey theory (Section 11.1). In order to apply Ramsey theory to analyse polymorphism clones, we need the product Ramsey theorem and other fundamental facts from Ramsey theory. Many of these general facts can be derived from a fundamental connection between Ramsey theory and topological dynamics due to Kechris, Pestov, and Todorčević [229]. This connection allows a more systematic understanding of Ramsey-theoretic principles, and we present it in Section 11.2. The way in which we apply Ramsey theory to polymorphisms is based on the concept of *canonical functions* as described in Section 11.4. We close with an application of the technique in Section 11.6, and prove the decidability of various *meta-problems* concerning constraint satisfaction problems, that is, problems where the input is a description of a template $\mathfrak{B}$, and where the question is, for instance, whether certain relations are primitively positively definable in $\mathfrak{B}$, or whether $\mathfrak{B}$ has polymorphisms satisfying certain stable clone formulas, for example the pseudo-Siggers condition.

Some of the results presented here have been published in [96]; there is also a survey article [88] that additionally covers the applications of our technique for the classification of the first-order reducts of a given homogeneous structure $\mathfrak{C}$ up to first-order interdefinability.

11.1. Ramsey classes

This section is concerned with classes $\mathcal{C}$ of finite structures that satisfy the following *Ramsey-type* property: for all $\mathfrak{A}, \mathfrak{B} \in \mathcal{C}$ and $c \in \mathbb{N}$ there exists a $\mathfrak{C} \in \mathcal{C}$ such that for every c-*colouring* of the substructures of $\mathfrak{C}$ that are isomorphic to $\mathfrak{A}$ there exists a *monochromatic* copy of $\mathfrak{B}$ in $\mathfrak{C}$. Before we formalise this in detail, we give the classical result of Ramsey, which provides a prototype of a class with the Ramsey property.

In this section we write $[n]$ for the set $\{1, \dots, n\}$. Subsets of cardinality s will be called s-*subsets*. Let $\binom{C}{s}$ denote the set of all s-subsets of the set C. We also refer to a mapping $f \colon \binom{C}{s} \to [c]$ as a *colouring* of C (with the *colours* $[c]$). If there exists a subset B of C such that f is constant on $\binom{B}{s}$ then we say that B is *monochromatic*.

THEOREM 11.1.1 (Ramsey). *Let C be a countably infinite set and let $s, c \in \mathbb{N}$. Then C has for every colouring $\chi \colon \binom{C}{s} \to [c]$ an infinite monochromatic subset.*

A proof of Theorem 11.1.1 can be found in [207] (Theorem 5.6.1); for a broader introduction to Ramsey theory see [185]. It is easy to derive the following finite version of Ramsey's theorem from Theorem 11.1.1 via a compactness argument.

THEOREM 11.1.2 (Finite version of Ramsey's theorem). *For all* $s, m, c \in \mathbb{N}$ *there exists a positive integer* l *such that for every* $\chi\colon \binom{[l]}{s} \to [c]$ *there exists a monochromatic* $S \in \binom{[l]}{m}$.

PROOF. Suppose that there are $s, m, c \in \mathbb{N}$ such that for every $l \in \mathbb{N}$ there is a $\chi\colon \binom{[l]}{s} \to [c]$ such that for all $S \in \binom{[l]}{m}$ the mapping χ is not constant on $\binom{S}{s}$. Since the property that for all k-subsets S of $[l]$ the mapping χ is not constant on $\binom{S}{m}$ is universal first-order, and since the image of χ is finite, we can apply Lemma 4.1.10 and get the existence of a mapping χ with the same property but defined on all integers. This contradicts Theorem 11.1.1. ⊣

The letters are chosen deliberately: we imagine s as a small number, m as a medium size number, and l as a large number. The *Ramsey number* $r(s, m, c)$ is the *smallest* number l whose existence is asserted by Theorem 11.1.2.

To discuss structural generalisations of Theorem 11.1.2, we introduce some general notation. If $\mathfrak{A}$ and $\mathfrak{B}$ are τ-structures, we write $\binom{\mathfrak{B}}{\mathfrak{A}}$ for the set of all induced substructures of $\mathfrak{B}$ that are isomorphic to $\mathfrak{A}$. For τ-structures $\mathfrak{S}, \mathfrak{M}, \mathfrak{L}$ and $c \in \mathbb{N}$ we write

$$\mathfrak{L} \to (\mathfrak{M})^{\mathfrak{S}}_c$$

if for all $\chi\colon \binom{\mathfrak{L}}{\mathfrak{S}} \to [c]$ there exists $\mathfrak{M}' \in \binom{\mathfrak{L}}{\mathfrak{M}}$ such that χ is constant on $\binom{\mathfrak{M}'}{\mathfrak{S}}$.

DEFINITION 11.1.3. A class of relational structures that is closed under isomorphisms and induced substructures is called *Ramsey*, or is said to have the *Ramsey property*, if for every $\mathfrak{S}, \mathfrak{M} \in C$ and for every $c \in \mathbb{N}$ there exists an $\mathfrak{L} \in C$ such that $\mathfrak{L} \to (\mathfrak{M})^{\mathfrak{S}}_c$.

Example 11.1.4. The class of all finite structures over the empty signature is Ramsey; this is an immediate consequence of Theorem 11.1.2.

Example 11.1.5. The class of all finite linear orders is a Ramsey class. This is again a direct consequence of Theorem 11.1.2, since whether or not a linear order with m elements is a substructure of a linear order with n elements only depends on n and m.

We will now present further examples of Ramsey classes; the proofs are non-trivial and fall outside the scope of this text, but we provide references.

Example 11.1.6. The class of all finite Boolean algebras $\mathfrak{B} = (B; \sqcup, \sqcap, c, \mathit{0}, \mathit{1})$ has the Ramsey property. This is explicitly stated in [229], page 147, line 3ff (see also page 112, line 9ff), where it is observed that this follows from a result of Graham and Rothschild [184].

It may also be instructive to see an example of a class of structures that is *not* Ramsey. Typical examples come from classes that contain structures with non-trivial automorphism groups, as in the following.

Example 11.1.7. The class of all finite graphs is not a Ramsey class. To see this, let $\mathfrak{S}$ be the (undirected) graph $(\{0,1,2\};\{(0,1),(1,0),(1,2),(2,1)\})$ with three vertices and two edges, and let $\mathfrak{M}$ be C_4, that is, the undirected four-cycle

$$(\{0,1,2,3\};\{(0,1),(1,2),(2,3),(3,0),(1,0),(2,1),(3,2),(0,3)\}).$$

Let $\mathfrak{L}$ be an arbitrary graph. We want to show that there is a way to colour the copies of $\mathfrak{S}$ in $\mathfrak{L}$ without producing a monochromatic copy of $\mathfrak{M}$. For that, fix an arbitrary linear order $<$ on the vertices of $\mathfrak{L}$. We define a colouring $\chi\colon \binom{\mathfrak{L}}{\mathfrak{S}} \to [2]$ as follows. If there is an embedding h of $\mathfrak{S}$ into $\mathfrak{L}$ such that $h(0) < h(1) < h(2)$, then we colour the copy of $\mathfrak{S}$ on $h(S)$ in $\mathfrak{L}$ by the colour 1; all other copies of $\mathfrak{S}$ in $\mathfrak{L}$ are coloured by the colour 2. We claim that any copy of $\mathfrak{M}$ in $\mathfrak{L}$ contains a copy of $\mathfrak{S}$ that is coloured by 2, and one that is coloured by 1. The reason is that for any ordering of the vertices of $\mathfrak{M}$ there is an embedding h' of $\mathfrak{S}$ into $\mathfrak{M}$ such that $h'(0) < h'(1) < h'(2)$, and an embedding h'' of $\mathfrak{S}$ into $\mathfrak{M}$ such that not $h''(0) < h''(1) < h''(2)$. Hence, $\mathfrak{L} \not\to (\mathfrak{M})^{\mathfrak{S}}_2$.

Frequently, a class without the Ramsey property can be made Ramsey by expanding its members appropriately with a linear ordering. We will see several examples of this.

Example 11.1.8. Nešetřil and Rödl [292] and independently Abramson and Harrington [1] showed that for any relational signature τ, the class C of all finite *ordered* τ-structures is a Ramsey class. That is, the members of C are finite structures $\mathfrak{A} = (A; <, R_1, R_2, \dots)$ for some fixed signature $\tau = \{<, R_1, R_2, \dots\}$, where $<$ is a linear order of A. A shorter and simpler proof of this substantial result can be found in [293] and [290] (see also [46]); the proof there is based on the *partite method*, which uses amalgamation to reduce the statement to proving the so-called *partite lemma*. The proof of the partite lemma relies on the Hales-Jewett theorem from Ramsey theory (see [185]).

Example 11.1.9. Recall from Section 5.1 the homogeneous structure $\mathfrak{B} = (B; |)$ carrying a C-relation. We consider the expansion of $\mathfrak{B}$ by a linear order, defined as follows. It is easy to see that for every finite tree $\mathfrak{T}$ there is an ordering $<$ on the leaves L of $\mathfrak{T}$ such that for all $u, v, w \in L$ with $u < v < w$ we have either $u|vw$ or $uv|w$ (recall Definition 5.1.4). This can be seen from the obvious existence of an embedding of $\mathfrak{T}$ on the plane so that all leaves lie on a line and none of the edges cross, and take the linear order induced by the line. We call such an ordering of L *compatible* with the tree $\mathfrak{T}$. The class of all finite substructures $\mathfrak{C}$ of $\mathfrak{B}$ expanded by a compatible ordering of the underlying tree of $\mathfrak{C}$ is a Ramsey class [261]; this also follows from more general results by Milliken (Theorem 4.3 in [284], building on work in [149]); see [46] for a direct proof.

Example 11.1.10. The Ramsey classes from Example 11.1.8 have been further generalised by Nešetřil and Rödl as follows [292]. Suppose that $\mathcal{F}$ is a (not necessarily finite) class of structures with a finite relational signature τ whose Gaifman graph (Definition 2.1.5) is a clique – such structures have been called *irreducible* in the Ramsey theory literature. It can be readily verified that $C := \text{Forb}^{\text{emb}}(\mathcal{F})$ is an amalgamation class. Then the class of all expansions of the structures in C by a linear order is a Ramsey class; again, this can be shown by the partite method [293]. This example is indeed a generalisation of Example 11.1.8 since we obtain the previous result by taking $\mathcal{F} = \emptyset$.

The following theorem indicates that the fact that each of the above Ramsey classes is the age of a homogeneous structure is not a coincidence.

THEOREM 11.1.11 (Nešetřil [291]). *Let τ be a relational signature, and let C be a class of finite ordered τ-structures that is closed under induced substructures, isomorphism, and has the joint embedding property (see Section 2.3). If C is Ramsey, then it has the amalgamation property.*

PROOF. Let $\mathfrak{A}, \mathfrak{B}_1, \mathfrak{B}_2$ be members of C such that $\mathfrak{A}$ is an induced substructure of both $\mathfrak{B}_1$ and $\mathfrak{B}_2$. Since C has the joint embedding property, there exists a structure $\mathfrak{C} \in C$ with embeddings e_1, e_2 of $\mathfrak{B}_1$ and $\mathfrak{B}_2$ into $\mathfrak{C}$. If e_1, e_2 have the same restriction to $\mathfrak{A}$, then we are done, so assume otherwise.

Let $\mathfrak{D}$ be such that $\mathfrak{D} \to (\mathfrak{C})^{\mathfrak{A}}_2$. Define a colouring $\chi \colon \binom{\mathfrak{D}}{\mathfrak{A}} \to \{1,2\}$ as follows. When $\mathfrak{A}' \in \binom{\mathfrak{D}}{\mathfrak{A}}$, and $f : \mathfrak{A} \to \mathfrak{A}'$ is an isomorphism, then $\chi(\mathfrak{A}') = 1$ if and only if there is an embedding $h \colon \mathfrak{C} \hookrightarrow \mathfrak{D}$ such that $f = h \circ e_1$. Since $\mathfrak{D} \to (\mathfrak{C})^{\mathfrak{A}}_2$, there exists $\mathfrak{C}' \in \binom{\mathfrak{D}}{\mathfrak{C}}$, witnessed by an embedding $h' \colon \mathfrak{C} \hookrightarrow \mathfrak{D}$ such that χ is constant on $\binom{\mathfrak{C}'}{\mathfrak{A}}$. Now any copy of $\mathfrak{C}$ in $\mathfrak{D}$ contains a copy $\mathfrak{A}'$ of $\mathfrak{A}$ with $\chi(\mathfrak{A}') = 1$. Thus χ is constant 1 on $\binom{\mathfrak{C}'}{\mathfrak{A}}$. In particular, the image of the embedding $h' \circ e_2 \colon \mathfrak{A} \hookrightarrow \mathfrak{D}$ is coloured 1. Thus there exists an embedding $h'' \colon \mathfrak{C} \hookrightarrow \mathfrak{D}$ such that $f = h'' \circ e_1 = h' \circ e_2$ (here we use the assumption that the structure $\mathfrak{A}$ is ordered). This shows that $\mathfrak{D}$ together with the embeddings $h'' \circ e_1 \colon B_1 \hookrightarrow D$ and $h' \circ e_2 \colon B_2 \hookrightarrow D$ is the amalgam of $\mathfrak{B}_1$ and $\mathfrak{B}_2$ over $\mathfrak{A}$. ⊣

It is often convenient to work with the Fraïssé-limit of a Ramsey class rather than the class itself.

DEFINITION 11.1.12. A homogeneous structure $\mathfrak{C}$ is called *Ramsey* if $\mathfrak{C} \to (\mathfrak{B})^{\mathfrak{A}}_c$ holds for all $\mathfrak{A}, \mathfrak{B} \in C$ and c.

PROPOSITION 11.1.13. *Let C be an amalgamation class and let $\mathfrak{C}$ be the Fraïssé-limit of C. Then $\mathfrak{C}$ is Ramsey if and only if C is a Ramsey class.*

PROOF. Let $\mathfrak{A}, \mathfrak{B} \in C$ and $c \geq 2$. If C has the Ramsey property then there exists a $\mathfrak{C}' \in C$ such that $\mathfrak{C}' \to (\mathfrak{B})^{\mathfrak{A}}_c$, and since $\mathfrak{C}'$ embeds into $\mathfrak{C}$ we also have that $\mathfrak{C} \to (\mathfrak{B})^{\mathfrak{A}}_c$.

Conversely, suppose that $\mathfrak{C} \to (\mathfrak{B})^{\mathfrak{A}}_c$. Let $k := \left|\binom{\mathfrak{B}}{\mathfrak{A}}\right|$. For any structure $\mathfrak{D}$ the fact that $\mathfrak{D} \to (\mathfrak{B})^{\mathfrak{A}}_c$ can be equivalently expressed in terms of c-colourability of a certain k-uniform hypergraph, defined as follows. Let $\mathfrak{G} = (V; E)$ be the structure whose vertex set V is $\binom{\mathfrak{D}}{\mathfrak{A}}$, and where $(\mathfrak{A}_1, \dots, \mathfrak{A}_k) \in E$ if there exists a $\mathfrak{B}' \in \binom{\mathfrak{D}}{\mathfrak{B}}$ such that $\binom{\mathfrak{B}'}{\mathfrak{A}} = \{\mathfrak{A}_1, \dots, \mathfrak{A}_k\}$. Let $\mathfrak{H} = ([c]; E)$ be the structure where E contains all tuples except the tuples $(1, \dots, 1), \dots, (c, \dots, c)$. Then $\mathfrak{D} \to (\mathfrak{B})^{\mathfrak{A}}_c$ if and only if $\mathfrak{G}$ maps homomorphically to $\mathfrak{H}$. By Lemma 4.1.7, this is the case if and only if all finite substructures of $\mathfrak{G}$ map homomorphically to $\mathfrak{H}$. Thus, $\mathfrak{C} \to (\mathfrak{B})^{\mathfrak{A}}_c$ if and only if $\mathfrak{C}' \to (\mathfrak{B})^{\mathfrak{A}}_c$ for all finite substructure $\mathfrak{C}'$ of $\mathfrak{C}$, and we conclude that C is Ramsey. ⊣

If $\mathfrak{C}$ is a finitely bounded homogeneous Ramsey structure, then the Ramsey property is useful for studying which relations are primitively positively definable in $\mathfrak{C}$, as we will see for instance in Section 11.6. In fact, for these applications of Ramsey theory it suffices to suppose that $\mathfrak{C}$ can be expanded to a finitely bounded homogeneous Ramsey structure. We make the following conjecture.

CONJECTURE 11.1. *Let $\mathfrak{C}$ be a homogeneous finitely bounded structure. Then $\mathfrak{C}$ has a homogeneous finitely bounded ordered Ramsey expansion.*

We mention that a related question has been answered in the negative.

THEOREM 11.1.14 (Evans, Hubička, Nešetřil [168]). *There exists an ω-categorical structure which has no ω-categorical Ramsey expansion.*

11.2. Extremely amenable groups

This section presents a link between the Ramsey property and the concept of *extreme amenability* in topological dynamics. The link rests on the theorem of Kechris, Pestov, and Todorčević that characterises those homogeneous ordered structures that are Ramsey in terms of their topological automorphism group. This theorem will be presented in Section 11.2.1. In Section 11.2.2, 4.2.5, and 11.2.4 we derive some general transfer principles for the Ramsey property based on this correspondence.

11.2.1. Extreme amenability. The Ramsey property for homogeneous ordered structures $\mathfrak{B}$ has an elegant characterisation in terms of the topological automorphism group of $\mathfrak{B}$: the age of $\mathfrak{B}$ is Ramsey if and only if the automorphism group of $\mathfrak{B}$ is extremely amenable. Extreme amenability is a concept from the theory of topological groups which has been studied since the 1960s [186].

DEFINITION 11.2.1. A topological group **G** is *extremely amenable* if every continuous action ξ of **G** on a compact Hausdorff space S has a fixed point, i.e., an $s \in S$ such that $\xi(g)(s) = s$ for every $g \in G$.

The following is a combination of Proposition 4.2, Proposition 4.3, Theorem 4.5, and Theorem 4.7 from [229]. Structures for which the identity is the only automorphism are called *rigid*.

THEOREM 11.2.2 (Kechris, Pestov, Todorčević [229]). *Let $\mathfrak{B}$ be a countable homogeneous relational structure. Then the following are equivalent.*

1. *$\mathfrak{B}$ is Ramsey and* $\mathrm{Aut}(\mathfrak{B})$ *preserves a linear order.*
2. *The age of $\mathfrak{B}$ has the Ramsey property and only contains rigid structures.*
3. $\mathrm{Aut}(\mathfrak{B})$ *is extremely amenable.*

PROOF. Clearly, if $\mathcal{G} := \mathrm{Aut}(\mathfrak{B})$ preserves a linear order, then by the homogeneity of $\mathfrak{B}$ all finite substructures of $\mathfrak{B}$ must be rigid; hence the implication (1) $\Rightarrow$ (2) follows from Proposition 11.1.13. For the implication (2) $\Rightarrow$ (3) we refer to [229]. For the implication from (3) to (1), consider the topological space 2^{B^2} consisting of all binary relations on B equipped with the topology of pointwise convergence, which is a compact space (Theorem 9.1.17) and Hausdorff (Proposition 9.1.3). The space $\mathcal{LO}$ of all linear orders on B is a closed subspace since the axioms of linear orders are universal first-order properties. Hence, $\mathcal{LO}$ is compact, too (Proposition 9.1.18). We define an action of $\mathcal{G}$ on $\mathcal{LO}$ by defining

$$g \cdot R := \{(ga, gb) \mid (a,b) \in R\}$$

for $R \in \mathcal{LO}$. This action is continuous: a basic open set in $\mathcal{LO}$ has the form

$$U_{F,R} := \{S \in \mathcal{LO} \mid S \cap F = R\}$$

for some finite $F, R \subseteq B^2$ with $R \subseteq F$. The preimage of $U_{F,R}$ is

$$\bigcup_{\alpha \in \mathcal{G}} S_{\alpha F,F} \times U_{\alpha F, \alpha R}$$

which is open. Since $\mathcal{G}$ is extremely amenable, the action has a fixed point R_0. So for every $g \in \mathcal{G}$ we have $g \cdot R_0 = R_0$ and thus $(a,b) \in R_0$ if and only if $(\alpha a, \alpha b) \in R_0$. This shows that $\mathcal{G} = \mathrm{Aut}(\mathfrak{B})$ preserves R_0.

To show that $\mathfrak{B}$ is Ramsey, let $c \in \mathbb{N}$, let $\mathfrak{S}, \mathfrak{M}$ be finite substructures of $\mathfrak{B}$, and let $\chi \colon \binom{\mathfrak{B}}{\mathfrak{S}} \to [c]$ be a coloring. Consider the compact topological space $X := [c]^{\binom{\mathfrak{B}}{\mathfrak{S}}}$ of all such colorings. Then $\mathcal{G}$ acts naturally on X by

$$g \cdot \theta := [\mathfrak{S}' \mapsto \theta(g^{-1}(\mathfrak{S}'))].$$

This action is again continuous. The closure of the orbit $\mathcal{G} \cdot \chi$ of χ is a compact Hausdorff subspace of X. Since $\mathcal{G}$ is extremely amenable there is a fixed point $\chi' \in \overline{\mathcal{G} \cdot \chi}$ for this action. We claim that any such fixed point is constant. For all $h \in \mathcal{G}$

$$\chi'(h(\mathfrak{S}')) = (h \cdot \chi')(\mathfrak{S}') = \chi'(\mathfrak{S}').$$

Thus, with respect to the colouring χ', all of $\mathfrak{B}$ is monochromatic. But χ' lies in the closure of the orbit of χ. Thus, there is some $g \in \mathcal{G}$ such that $g^{-1} \cdot \chi$ coincides with χ' on the finite set $\binom{\mathfrak{M}}{\mathfrak{S}}$. Hence, χ is constant on $\binom{g(\mathfrak{M})}{\mathfrak{S}}$. $\dashv$

For structures $\mathfrak{B}$ that are ω-categorical but not necessarily homogeneous, the equivalence between (3) and (4) remains valid, since every ω-categorical structure has a homogeneous expansion by first-order definable relations – and such an expansion has the same automorphism group as $\mathfrak{B}$. We can then apply Theorem 11.2.2 to the expansion. This justifies the following definition, which is compatible with Definition 11.1.12.

DEFINITION 11.2.3. An ω-categorical structure $\mathfrak{B}$ is *Ramsey* if the age of the expansion of $\mathfrak{B}$ by all first-order definable relations is Ramsey.

For an ω-categorical structure $\mathfrak{B}$ and a finite substructure $\mathfrak{A}$ of $\mathfrak{B}$ we write $\binom{\mathfrak{B}}{\mathfrak{A}}$ for the set of all substructures $\mathfrak{A}'$ of $\mathfrak{B}$ such that there exists an automorphism of $\mathfrak{B}$ that maps $\mathfrak{A}$ to $\mathfrak{A}'$. Then $\mathfrak{B}$ is Ramsey if and only if for all $\mathfrak{A}, \mathfrak{M} \in \text{Age}(\mathfrak{B})$ and $\chi \colon \binom{\mathfrak{B}}{\mathfrak{A}} \to [c]$ there exists $\mathfrak{M}' \in \binom{\mathfrak{B}}{\mathfrak{M}}$ such that χ is constant on $\binom{\mathfrak{M}'}{\mathfrak{A}}$. Whether an ω-categorical structure $\mathfrak{B}$ is Ramsey only depends on the automorphism group of $\mathfrak{B}$, viewed as a topological group. This fact has the following consequence.

COROLLARY 11.2.4. *Let $\mathfrak{B}$ be ω-categorical and $d \in \mathbb{N}$. Then $\mathfrak{B}$ is Ramsey if and only if $\mathfrak{B}^{[d]}$ (Definition 3.5.3) is Ramsey.*

PROOF. Since $\mathfrak{B}^{[d]}$ is a full power of $\mathfrak{B}$ (Proposition 3.5.4), the structures $\mathfrak{B}$ and $\mathfrak{B}^{[d]}$ are primitively positively bi-interpretable (Proposition 3.5.2), and hence their automorphism groups are isomorphic as topological groups (Corollary 9.5.18), so the result follows from the comments above. ⊣

We point out another consequence of Theorem 11.2.2 in combination with Theorem 4.2.9.

COROLLARY 11.2.5. *Let $\mathfrak{B}$ be an ω-categorical Ramsey structure. If all finite induced substructures of $\mathfrak{B}$ are rigid, then a linear order is first-order definable in $\mathfrak{B}$.*

Note that in Corollary 11.2.5, the assumption that $\mathfrak{B}$ is Ramsey is necessary, as the following example shows.

Example 11.2.6. Peter Cameron found the following example of a homogeneous structure $\mathfrak{B}$ such that

- all structures in the age of $\mathfrak{B}$ are rigid, and
- no linear order is first-order definable in $\mathfrak{B}$.

Let $(\mathbb{L}; |)$ be the binary branching homogeneous C-relation from Section 5.1.2 and let $(\mathbb{T}; E)$ be the countable homogeneous universal tournament, i.e., the Fraïssé-limit of the class of all finite tournaments. Since both structures have no algebraicity (they are Fraïssé-limits of strong amalgamation classes) there exists a generic superposition $\mathfrak{B} := (\mathbb{L}; |) * (\mathbb{T}; E)$. Then the second property listed above is straightforward to verify from homogeneity: there are only finitely many binary first-order definable relations in $\mathfrak{B}$, and none of them is a linear order. To prove the first property, choose a finite substructure $\mathfrak{A}$ of $\mathfrak{B}$ with a non-trivial automorphism $\alpha \in \mathrm{Aut}(\mathfrak{A})$, i.e., α has a cycle $(a_0, a_2, \dots, a_{n-1})$ for $n \geq 2$. Since $(\mathbb{L}; |)$ is binary branching there exists a non-empty proper subset I of $\{0, \dots, n-1\}$ such that $a_i a_j | a_k$ if $i, j \in I$ and $k \in \{0, \dots, n-1\} \setminus I$, or if $i, j \in \{0, \dots, n-1\} \setminus I$ and $k \in I$. Moreover, since $(a_0, a_2, \dots, a_{n-1})$ is a cycle of an automorphism, $|I| = |\{0, \dots, n-1\} \setminus I|$ and n must be even. Note that $\alpha^{n/2}(a_0) = a_{n/2}$ and $\alpha^{n/2}(a_{n/2}) = a_0$. Since we have either $(a_0, a_{n/2}) \in E$ or $(a_{n/2}, a_0) \in E$, but not both, $\alpha^{n/2} \in \mathrm{Aut}(\mathfrak{A})$ does not preserve the tournament edge relation E, a contradiction.

We already mentioned that many ω-categorical structures $\mathfrak{B}$ that are not Ramsey can be turned into Ramsey structures by expanding $\mathfrak{B}$ with an appropriate linear order (cf. Example 11.1.7 and Example 11.1.8). To give another example, consider again the countable atomless Boolean algebra. In this case an order expansion with an extremely amenable automorphism group has been specified in [229], and can be found below.

Example 11.2.7. Let $\mathfrak{B} = (B; \sqcup, \sqcap, c, \mathit{0}, \mathit{1})$ be a finite Boolean algebra and A its set of atoms (see Example 11.1.6 in Section 5.3). Then every ordering $a_1 < \cdots < a_n$ of A gives an ordering of B as follows (we follow [229]). For $x, y \in B$, we set $x < y$ if there exists an $i_0 \in \{1, \dots, n\}$ such that

- for all $i \in \{1, \dots, i_0 - 1\}$ we have that $a_i \sqcap x = a_i \sqcap y$, and
- $x \sqcap a_{i_0} = \mathit{0}$ and $y \sqcap a_{i_0} \neq \mathit{0}$.

Such an ordering of the elements of $\mathfrak{B}$ is called a *natural ordering*. It can be shown that the class $\mathcal{C}$ of all finite naturally ordered atomless Boolean algebras has the Ramsey property (see the comments preceding Theorem 6.14 in [229], and Proposition 5.6 in [229]). By Theorem 11.1.11, $\mathcal{C}$ is an amalgamation class. The reduct of the Fraïssé-limit of $\mathcal{C}$ with signature $\{\sqcup, \sqcap, c, \mathit{0}, \mathit{1}\}$ is the atomless Boolean algebra (Propositions 5.2 and 6.13 in [229]), so we have indeed found an extremely amenable order expansion of the atomless Boolean algebra.

Using Theorem 11.2.2, the negative result from Theorem 11.1.14 can also be phrased as follows.

THEOREM 11.2.8 (Evans, Hubička, Nešetřil [168]). *There exists an oligomorphic permutation group with no closed extremely amenable oligomorphic subgroup.*

The main focus of the article by Kechris, Pestov, and Todorčević [229] is the application of Theorem 11.2.2 to prove that certain groups are extremely amenable, using known and deep Ramsey results. Here, on the other hand, our interest lies in the opposite direction: we apply Theorem 11.2.2 in the following sections to obtain a more systematic understanding of which classes of structures have the Ramsey property. Concerning Conjecture 11.1 we also mention that Zucker [349] showed that the existence of an ω-categorical Ramsey expansion of an ω-categorical structure $\mathfrak{C}$ is equivalent to the metrisability of the *universal minimal flow* of $\mathrm{Aut}(\mathfrak{C})$; for the definition of the concept of the universal minimal flow from topological dynamics we refer to [229] and [349].

11.2.2. Surjective continuous homomorphisms. As we have seen, whether an ω-categorical structure $\mathfrak{B}$ is Ramsey only depends on the automorphism group of $\mathfrak{B}$ viewed as a *topological* group. More generally, we have the following.

PROPOSITION 11.2.9. *Let* $\mathbf{G}$, $\mathbf{H}$ *be topological groups. If there is a surjective continuous homomorphism* $f: \mathbf{G} \to \mathbf{H}$ *and* $\mathbf{G}$ *is extremely amenable, then so is* $\mathbf{H}$.

PROOF. Let $\xi: H \times S \to S$ be a continuous action of $\mathbf{H}$ on a compact Hausdorff space S. Then $\xi': G \times S \to S$ given by $(g, s) \mapsto \xi(f(g), s)$ is a continuous action of $\mathbf{G}$ on S. Since $\mathbf{G}$ is extremely amenable, ξ' has a fixed point s_0.

Now, let $h \in H$ be arbitrary. By surjectivity, there is a $g \in G$ such that $f(g) = h$. Then we have

$$s_0 = \xi'(g, s_0) = \xi(f(g), s_0) = \xi(h, s_0)$$

and hence s_0 is also a fixed point under ξ. ⊣

As a consequence of this and Theorem 9.5.25 we obtain the following.

COROLLARY 11.2.10. *Let* $\mathfrak{A}$ *be a structure with a full first-order interpretation in an* ω*-categorical ordered Ramsey structure* $\mathfrak{B}$. *If* $\mathfrak{B}$ *is Ramsey, then so is* $\mathfrak{A}$.

Example 11.2.11. Recall Allen's Interval Algebra (Section 1.6.1; Example 2.4.2) which is an ω-categorical structure $\mathfrak{A}$ with binary relations and a two-dimensional first-order interpretation over $(\mathbb{Q}; <)$. In Example 3.3.3 we have shown that $\mathfrak{A}$ is first-order bi-interpretable with $(\mathbb{Q}; <)$, and hence $\mathfrak{A}$ has a full interpretation in $(\mathbb{Q}; <)$ by Lemma 2.4.8. Since $(\mathbb{Q}; <)$ is Ramsey (Example 11.1.5), Corollary 11.2.10 implies that $\mathfrak{A}$ is Ramsey as well.

11.2.3. Products. In this section we present an important tool for building new extremely amenable groups from old ones, namely Theorem 11.2.12 below. Recall the definition of direct products of two groups from Section 4.2.5. The direct product of topological groups is the direct product of the respective abstract groups together with the product topology on the group elements. In the following theorem, but also at later occasions, we make the assumption that the topological group $\mathbf{G}$ is first-countable, because this covers all the situations of interest in this text, and because this simplifies the presentation of continuity proofs; however, the theorem holds without this assumption as well; cf. Proposition 6.7 in [229].

THEOREM 11.2.12. *Let* $\mathbf{G}$ *be a topological group.*

1. *If* $\mathbf{G}$ *has extremely amenable subgroups* $\mathbf{H}_1, \mathbf{H}_2, \dots$ *such that* $H_1 \subseteq H_2 \subseteq \cdots$ *and* $\bigcup_{i \in \mathbb{N}} H_i$ *is dense in* G, *then* $\mathbf{G}$ *is extremely amenable.*
2. *Let* $\mathbf{N}$ *be a closed normal subgroup of* $\mathbf{G}$. *If both* $\mathbf{N}$ *and the quotient group* $\mathbf{G}/\mathbf{N}$ *are extremely amenable, then so is* $\mathbf{G}$.
3. *If* $\mathbf{H}_1, \mathbf{H}_2, \dots$ *are extremely amenable topological groups, then so is* $\prod_{i \in \mathbb{N}} \mathbf{H}_i$.

PROOF. Suppose that G acts continuously on a compact Hausdorff space S. To show (1), let

$$S_i := \{s \in S \mid \text{for all } h \in H_i : h \cdot s = s\}.$$

Since $\mathbf{H}_i$ is extremely amenable, S_i is non-empty. Moreover, S_i is closed. Since for every finite subset $F \subseteq \mathbb{N}$ we have $\bigcap_{i \in F} S_i \neq \emptyset$, the compactness of S implies that $\bigcap_{i \in \mathbb{N}} S_i$ contains an element s_0. Then $h \cdot s_0 = s_0$ for every $h \in H := \bigcup_{i \in \mathbb{N}} H_i$, and since H is dense in G we have that s_0 is a fixed point of the action of $\mathbf{G}$ on S.

To prove (2), let $S_{\mathbf{N}}$ be the closed subspace of S induced on

$$\{s \in S \mid \text{for all } h \in N : h \cdot s = s\}$$

which is Hausdorff and compact since closed subsets of compact spaces are compact. As $\mathbf{N}$ is extremely amenable, $S_{\mathbf{N}}$ is non-empty. The set $S_{\mathbf{N}}$ is preserved by the action of $\mathbf{G}$ on S: if $x \in S_{\mathbf{N}}$ and $g \in G$, then for any $h \in N$

$$h \cdot (g \cdot x) = hg \cdot x = g \cdot (g^{-1}hg) \cdot x = g \cdot x$$

where the last equality is by the normality of N (Proposition 9.2.16). Thus, $g \cdot x \in S_{\mathbf{N}}$. Now, consider the action of $\mathbf{G}/\mathbf{N}$ on $S_{\mathbf{N}}$ defined by $(gN) \cdot x := g \cdot x$, which is clearly well defined. It is continuous, because the action of $\mathbf{G}$ on S is continuous. By the extreme amenability of $\mathbf{G}/\mathbf{N}$ there is a point $p \in S_{\mathbf{N}}$ such that $f \cdot p = p$ for all $f \in \mathbf{G}/\mathbf{N}$. But then p is also a fixed point for the action of $\boldsymbol{G}$ on S, since $g \cdot p = (gN) \cdot p = p$ for any $g \in G$.

(3) follows from (1) and (2): suppose that $\mathbf{G} = \mathbf{H}_1 \times \mathbf{H}_2$, and that $\mathbf{H}_1$ and $\mathbf{H}_2$ are extremely amenable. Then $\mathbf{H}_1$ is a normal subgroup of $\boldsymbol{G}$, and $\mathbf{G}/\mathbf{H}_1$ is isomorphic to $\mathbf{H}_2$, so $\mathbf{G}$ is extremely amenable by (2). The statement for $\mathbf{G} = \mathbf{H}_1 \times \cdots \times \mathbf{H}_n$ follows by induction on n. If $\mathbf{G} = \prod_{i \in \mathbb{N}} \mathbf{G}_i$, define $\mathbf{G}_i := \mathbf{H}_1 \times \cdots \times \mathbf{H}_i \times \{\text{id}\} \times \{\text{id}\} \times \cdots$. Then $\bigcup G_i$ is dense in $\prod_{i \in \mathbb{N}} \mathbf{H}_i$ and the statement follows from (1). ⊣

For every ordered Ramsey class there is a corresponding product Ramsey theorem which can be shown either directly or by applying the general results from topological dynamics. The underlying idea is best explained by the product Ramsey theorem for the class of all finite linear orders (Theorem 11.2.13, which will be used extensively for $d = m = 2$ in Chapter 12). We will use the following terminology in this case. If $S_1, \ldots, S_d$ are sets, we call a set of the form $S_1 \times \cdots \times S_d$ a *grid*, and we also write S^d for a product of the form $S \times \cdots \times S$ with d factors. A $[k]^d$*-subgrid* of a grid $S_1 \times \cdots \times S_d$ is a subset of $S_1 \times \cdots \times S_d$ of the form $S_1' \times \cdots \times S_d'$, where S_i' is a k-element subset of S_i.

THEOREM 11.2.13 (Product Ramsey Theorem). *For all positive integers d, c, s, and $m \geq m$, there is a positive integer ℓ such that for every colouring of the $[s]^d$-subgrids of $[\ell]^d$ with c colours there exists a* monochromatic *$[m]^d$-subgrid G of $[\ell]^d$, i.e., all the $[s]^d$-subgrids of G have the same colour.*

PROOF. Let d, c, s, and $m \geq m$ be positive integers. We claim that we can choose ℓ to be the Ramsey number $r(c, ds, dm)$. To verify this, let χ be a colouring of the $[s]^d$-subgrids of $[\ell]^d$ with c colours. We have to find a monochromatic $[m]^d$-subgrid of $[\ell]^d$. The colouring χ can be used to define a c-colouring ξ of the ds-subsets of $[\ell]$ as follows. Let $S = \{u_1, u_2, \ldots, u_{ds}\}$ be a ds-subset of $[\ell]$, with $u_1 < u_2 < \cdots < u_{ds}$. Then define

$$\xi(S) := \chi(\{u_1, \ldots, u_s\} \times \cdots \times \{u_{s(d-1)+1}, \ldots, u_{ds}\}).$$

By Theorem 11.1.2, there is a dm-subset $\{v_1, v_2, \ldots, v_{dm}\}$ of $[\ell]$ such that ξ is constant on the ds-element subsets of $\{v_1, v_2, \ldots, v_{dm}\}$. Suppose that $v_1 < v_2 < \cdots < v_{dm}$. Then $G = \{v_1, \ldots, v_m\} \times \cdots \times \{v_{m(d-1)+1}, \ldots, v_{dm}\}$ is a subgrid of $[\ell]^d$ that is monochromatic with respect to χ. ⊣

We next present our formulation of the product Ramsey theorem for arbitrary ω-categorical ordered Ramsey structures. The proof uses topological methods, namely, Theorem 11.2.2 and Theorem 11.2.12.

THEOREM 11.2.14. *Let $\mathfrak{B}_1, \ldots, \mathfrak{B}_d$ be ω-categorical ordered Ramsey structures. Then the algebraic product structure $\mathfrak{P} := \mathfrak{B}_1 \boxtimes \cdots \boxtimes \mathfrak{B}_d$ (Definition 4.2.18) is Ramsey.*

PROOF. By Theorem 11.2.2, the automorphism groups $\mathbf{G}_1, \ldots, \mathbf{G}_d$ of $\mathfrak{B}_1, \ldots, \mathfrak{B}_d$ are extremely amenable, and it suffices to show that the automorphism group $\boldsymbol{G}$ of $\mathfrak{P}$ is extremely amenable. The group $\boldsymbol{G}$ is given by the product action of $\mathbf{G}_1 \times \cdots \times \mathbf{G}_d$ on $B_1, \ldots, B_d$ (see Section 4.2.5.2). Hence, the extreme amenability of $\boldsymbol{G}$ follows from Theorem 11.2.12. ⊣

Theorem 11.2.14 indeed generalizes Theorem 11.2.13, which can be seen as follows. Let r, d, m, k be positive integers. We consider the ordered ω-categorical Ramsey structure $(\mathbb{Q}; <)$, and apply Theorem 11.2.14 where d in Theorem 11.2.14 equals the d given above. Let $\mathfrak{A}$ be the structure induced in $\mathfrak{P} := (\mathbb{Q}; <)^{(d)}$ by some (equivalently, every) $[s]^d$-subgrid of $\mathbb{Q}^d$, and let $\mathfrak{B}$ be the structure induced in $\mathfrak{P}$ by some (equivalently, every) $[m]^d$-subgrid of $\mathbb{Q}^d$. Since $\mathfrak{C}$ is Ramsey, there exists an induced substructure $\mathfrak{C}$ of $\mathfrak{P}$ such that $\mathfrak{C} \rightarrow (\mathfrak{B})^{\mathfrak{A}}_r$. If $\mathfrak{C}$ is not induced on an $[\ell]^d$-subgrid of $\mathbb{Q}^d$, for some large enough ℓ, we can clearly choose a larger substructure $\mathfrak{C}$ with this property, such that still $\mathfrak{C} \rightarrow (\mathfrak{B})^{\mathfrak{A}}_r$. The occurrences of $\mathfrak{A}$ in $\mathfrak{C}$ correspond precisely to the $[s]^d$-subgrids of $[\ell]^d$, which proves the claim.

11.2.4. Open subgroups. In this section we show that open subgoups of extremely amenable groups are again extremely amenable. This fact will be important in Section 11.4.5 and 11.6 when it comes to applications for the analysis of polymorphisms of first-order reducts of Ramsey structures.

PROPOSITION 11.2.15 (from [96]). *Let* $\mathbf{G}$ *be an extremely amenable group, and let* $\mathbf{H}$ *be an open subgroup of* $\mathbf{G}$. *Then* $\mathbf{H}$ *is also extremely amenable.*

PROOF. Let $\mathbf{H}$ act continuously on a compact space X; we have to show that this action has a fixed point. Define a continuous action of $\mathbf{G}$ on $X^{\mathbf{G}/H}$ as follows: for $g, k \in G$ and $\xi \in X^{\mathbf{G}/H}$, set

$$(g\xi)(kH) := \xi(g^{-1}kH).$$

This defines indeed a group action (Definition 4.2.13), because

$$\begin{aligned}((gh)\xi)(kH) &= \xi((gh)^{-1}kH) = \xi(h^{-1}g^{-1}kH)\\ &= (h\xi)(g^{-1}kH) = (g(h\xi))(kH).\end{aligned}$$

The second property of actions is similarly straightforward to verify. To prove that the action is continuous, let $\xi \in X^{\mathbf{G}/H}$, $g \in G$, and let $V \subseteq X^{\mathbf{G}/H}$ be a basic open set that contains $g\xi$. By the definition of the topology of $X^{\mathbf{G}/H}$, the basic open set V is of the form $\times_{fH \in \mathbf{G}/H} V_{fH}$ where V_{fH} is an open subset of X and $V_{fH} = X$ unless $f \in F$ for some finite $F \subseteq G$. For every $f \in F$ there exists an open subset $O_f \subseteq G$ containing 1 such that for every $e \in O_f$ we have $e^{-1}f \in fH$, because multiplication in G is continuous and fH is open and contains f. Set $O := \bigcap_{f \in F} O_f$. If $e \in O$ and $\xi' \in V$, then for all $f \in F$

$$(e\xi')(fH) = \xi'(e^{-1}fH) = \xi'(fH) \in V_{fH}.$$

Thus, $e\xi' \in V$ and we have proven that the action maps all elements of $O \times V$ into V. The set $g^{-1}V$ is open and contains ξ. Analogously as above one can show that there exists an open set $O \subseteq G$ containing 1 such that the action maps all elements of $O \times g^{-1}V$ into $g^{-1}V$. The set

$$P := gO \times g^{-1}V \subseteq G \times X^{\mathbf{G}/H}$$

is open and contains (g, ξ). Let $e \in O$ and $\xi' \in V$ be arbitrary. Then

$$(ge)(g^{-1}\xi') = g(eg^{-1}\xi') \in g(g^{-1}V) = V.$$

Therefore, P is mapped into V by the action; this concludes the proof that the action is continuous.

Denote by $\pi\colon \mathbf{G} \to \mathbf{G}/H$ the quotient map and let $s\colon \mathbf{G}/H \to \mathbf{G}$ be a *section* for π, i.e., a mapping satisfying $\pi \circ s = \mathrm{id}$. We additionally suppose that $s(H) = 1$. Consider the product space $X^{\mathbf{G}}$ which is Hausdorff by Proposition 9.1.3 and compact by Theorem 9.1.17. Let S be the subspace of $X^{\mathbf{G}}$ consisting of all $\chi \in X^{\mathbf{G}}$ such that $\chi(gh) = h\chi(g)$ for all $h \in H$ and $g \in s(\mathbf{G}/H)$. Any function from $s(\mathbf{G}/H)$ to X has a unique extension in S. Let $\eta\colon X^{\mathbf{G}/H} \to S$ be the function defined by $\eta(\xi)(gh) := h\xi(gH)$ for all $h \in H$ and $g \in s(\mathbf{G}/H)$ and let $\sigma\colon S \to X^{\mathbf{G}/H}$ be defined by $\sigma(\chi)(gH) := \chi(s(gH))$ for all $g \in G$. Then $\eta \circ \sigma = \mathrm{id}$ and $\sigma \circ \eta = \mathrm{id}$:

$$\begin{aligned}\eta(\sigma(\chi))(gh) &= h(\sigma(\chi))(gH) = h\chi(s(gH) = h\chi(g) = \chi(gh)\\ \sigma(\eta(\xi))(gH) &= (\eta(\xi))(s(gH)) = \xi(s(gH)H) = \xi(gH).\end{aligned}$$

Both η and σ are continuous. To verify this for σ, by Lemma 9.1.2 is suffices to verify the continuity of $\pi_g \circ \sigma$ for all $g \in G$. So let $U \subseteq X$ be open. Then

$$\begin{aligned}(p_g \circ \sigma)^{-1}(U) = \{\xi \mid p_g(\sigma(\xi)) \in U\} &= \{\xi \mid \sigma(\xi)(g) \in U\} \\ &= \{\xi \mid \xi(gH) \in U\}\end{aligned}$$

which is a basic open set in $X^{\mathbf{G}/H}$. Continuity of $\eta = \sigma^{-1}$ can be verified similarly: For $gH \in \mathbf{G}/H$, we compute

$$\begin{aligned}(p_{gH} \circ \eta)^{-1}(U) = \{\chi \mid p_{gH}(\eta(\chi)) \in U\} &= \{\chi \mid \eta(\chi)(gH) \in H\} \\ &= \{\chi \mid \chi(s(gH)) \in U\}\end{aligned}$$

which is a basic open set in S. So it follows that σ is a homeomorphism.

The function $g(\xi) := \sigma^{-1}(g(\sigma(\xi)))$ defines a continuous action of $\mathbf{G}$ on S. Since $\mathbf{G}$ is extremely amenable, this action has a fixed point $\xi \in S$, that is, $g(\xi) = \xi$ for all $g \in G$. In particular, $h(\xi)(1) = \xi(1)$ for all $h \in H$. Now

$$\begin{aligned}
h(\xi)(1) &= \sigma^{-1}(h(\sigma(\xi)))(1) && \text{(by the definition of the action of } \mathbf{G} \text{ on } S) \\
&= h\sigma(\xi)(H) && \text{(by the definition of } \sigma) \\
&= \sigma(\xi)(h^{-1}H) && \text{(by the definition of the action of } \mathbf{G} \text{ on } X^{\mathbf{G}/H}) \\
&= \xi(s(h^{-1}H)) && \text{(by the definition of } \sigma) \\
&= \xi(h^{-1}) && \text{(since } s(H) = 1) \\
&= h^{-1}(\xi(1)) && \text{(by the definition of } S).
\end{aligned}$$

Putting this together, we get that $h^{-1}(\xi(1)) = \xi(1)$ for all $h \in H$. Thus, $\xi(1)$ is a fixed point of X for the action of $\mathbf{H}$ on X. ⊣

Proposition 11.2.15 can be applied to provide a purely topological proof of the following Ramsey transfer principle.

COROLLARY 11.2.16 (from [96]). *Let $\mathfrak{B}$ be homogeneous ordered Ramsey, and let $c_1, \ldots, c_n \in B$. Then $(\mathfrak{B}, c_1, \ldots, c_n)$ is homogeneous ordered Ramsey as well.*

PROOF. It is easy to see that the expansion of any homogeneous structure $\mathfrak{B}$ by constants is again homogeneous. The automorphism group of $\mathfrak{B}$ is extremely amenable because $\mathfrak{B}$ is ordered Ramsey. The automorphism group of $(\mathfrak{B}, c_1, \ldots, c_n)$ is an open subgroup of $\mathrm{Aut}(\mathfrak{B})$. The statement thus follows directly from Proposition 11.2.15 and Theorem 11.2.2. ⊣

A combinatorial proof of this Ramsey transfer principle, due to Miodrag Sokic, can be found in [46].

11.3. Transfer principles for the Ramsey property

The previous section already gave one Ramsey transfer principle, derived from the topological characterisation of the Ramsey property from Theorem 11.2.2. In this section we present further Ramsey transfer principles that seem to be difficult to describe on the level of automorphism groups.

11.3.1. Taking model-complete cores. The following is from [45]. We mention that in [46], the assumption that $\mathfrak{B}$ is homogeneous has been removed even though it is necessary–the author thanks Michael Pinsker for pointing out this mistake.

THEOREM 11.3.1. *Let $\mathfrak{B}$ be homogeneous, ω-categorical, and Ramsey and let $\mathfrak{C}$ be the model-complete core of $\mathfrak{B}$. Then $\mathfrak{C}$ is also Ramsey.*

PROOF. Let h be a homomorphism from $\mathfrak{B}$ to $\mathfrak{C}$, and let i be an embedding from $\mathfrak{C}$ into $\mathfrak{B}$; see Remark 4.7.5. Let $\mathfrak{S}$ and $\mathfrak{M}$ be finite substructures of $\mathfrak{C}$ and $r \in \mathbb{N}$, and let $\chi\colon \binom{\mathfrak{C}}{\mathfrak{S}} \to [r]$ be arbitrary. Let $\mathfrak{T}$ be the substructure of $\mathfrak{B}$ induced on $i(S)$. We define $\chi'\colon \binom{\mathfrak{B}}{\mathfrak{T}} \to [r]$ as follows. Let $\mathfrak{T}' \in \binom{\mathfrak{B}}{\mathfrak{T}}$ and let $\mathfrak{S}'$ be the substructure of $\mathfrak{C}$ induced on $h(T')$. The assumption that $\mathfrak{C}$ is a model-complete core implies that $\mathfrak{S}' \in \binom{\mathfrak{C}}{\mathfrak{S}}$. Define $\chi'(\mathfrak{T}') := \chi(\mathfrak{S}')$. Let $\mathfrak{N}$ be the substructure of $\mathfrak{B}$ induced on $i(M)$. Since $\mathfrak{B}$ is Ramsey, there exists $\mathfrak{N}' \in \binom{\mathfrak{B}}{\mathfrak{N}}$ and $c \in [r]$ such that for all $\mathfrak{T}' \in \binom{\mathfrak{N}'}{\mathfrak{T}}$ we have $\chi'(\mathfrak{T}') = c$.

Let $\mathfrak{M}'$ be the substructure of $\mathfrak{C}$ induced on $h(N')$, and let $\mathfrak{S}' \in \binom{\mathfrak{M}'}{\mathfrak{S}}$. We claim that $\chi(\mathfrak{S}') = c$. Since $\mathfrak{N}' \in \binom{\mathfrak{B}}{\mathfrak{N}}$ there exists an automorphism $\alpha \in \mathrm{Aut}(\mathfrak{B})$ that maps $\mathfrak{N}$ to $\mathfrak{N}'$. Since $h \circ \alpha \circ i \in \mathrm{End}(\mathfrak{C})$ and $\mathfrak{C}$ is a model-complete core, there exists $\beta \in \mathrm{Aut}(\mathfrak{C})$ such that $\beta(h \circ \beta \circ i(x)) = x$ for all $x \in M$. Let $T' := \alpha \circ i \circ \beta(S')$ and note that $h(T') = S'$. Note that T' induces a copy of $\mathfrak{T}$ in $\mathfrak{B}$ by the homogeneity of $\mathfrak{B}$. Therefore, $\chi'(\mathfrak{T}') = c$, which by the definition of χ' means that $\chi(\mathfrak{S}') = c$. Hence, χ is constant on $\binom{\mathfrak{M}'}{\mathfrak{S}}$ and thus $\mathfrak{C}$ is Ramsey. ⊣

11.3.2. Generic superpositions. Generic superpositions have been introduced for Fraïssé-limits of strong amalgamation classes in Section 2.3.6, and more generally for ω-categorical structures without algebraicity in Section 4.7.1. Our proof of the Ramsey transfer result for generic superpositions (Theorem 11.3.2) is from [45] and uses Theorem 11.3.1 about model-complete cores.

THEOREM 11.3.2 (Theorem 1.5 in [45]). *Let $\mathfrak{B}_1$ and $\mathfrak{B}_2$ be ω-categorical homogeneous ordered structures with disjoint signatures and no algebraicity such that both $\mathfrak{B}_1$ and $\mathfrak{B}_2$ are Ramsey. Then $\mathfrak{B}_1 * \mathfrak{B}_2$ is Ramsey.*

PROOF. By Lemma 4.7.1 we can make the assumption that $\mathfrak{B}_1$ and $\mathfrak{B}_2$ are reducts of $\mathfrak{B}_1 * \mathfrak{B}_2$ to their signatures, and, in particular, all have the same domain B. Theorem 11.2.14 implies that the algebraic product $\mathfrak{B}_1 \boxtimes \mathfrak{B}_2$ is

Ramsey. Theorem 11.3.1 implies that the model-complete core $\mathfrak{C}$ of $\mathfrak{B}_1 \boxtimes \mathfrak{B}_2$ is Ramsey. It therefore suffices to show that $\mathfrak{C}$ is isomorphic to $\mathfrak{B}_1 * \mathfrak{B}_2$. Since $\mathfrak{B}_1 * \mathfrak{B}_2$ is a model-complete core (Lemma 4.7.2) it suffices to show that $\mathfrak{B}_1 \boxtimes \mathfrak{B}_2$ and $\mathfrak{C}$ are homomorphically equivalent. First note that $d \mapsto (d, d)$ is a homomorphism from $\mathfrak{C}$ to $\mathfrak{B}_1 \boxtimes \mathfrak{B}_2$. For the other direction, we verify that every finite substructure $\mathfrak{F}$ of $\mathfrak{B}_1 \boxtimes \mathfrak{B}_2$ maps homomorphically to $\mathfrak{C}$ (Lemma 4.1.7). By Lemma 4.3.6 there is an injective homomorphism h_i from the τ_i-reduct of $\mathfrak{F}$ to $\mathfrak{B}_i$. The superposition of the substructure of $\mathfrak{B}_1$ induced on the image of h_1 with the substructure of $\mathfrak{B}_2$ induced on the image of h_2 is a substructure of $\mathfrak{B}_1 \boxtimes \mathfrak{B}_2$, and therefore maps homomorphically to $\mathfrak{C}$. ⊣

In the following we present four examples of generic superpositions in the context of the Ramsey property. Example 11.3.3 and Example 11.3.5 show that the assumption in Theorem 11.3.2 that both structures $\mathfrak{B}_1$ and $\mathfrak{B}_2$ are ordered is necessary. Examples 11.3.4 and 11.3.6 are applications of Theorem 11.3.2 to obtain Ramsey expansions of the previous two examples, confirming Conjecture 11.1 in these cases.

Example 11.3.3. Let $\mathcal{E}$ be the class of all finite $\{E\}$-structures where E denotes an equivalence relation and let $\mathcal{LO}$ the class of all finite $\{<\}$-structures where $<$ denotes a linear order. It is easy to verify that $\mathcal{E} * \mathcal{LO}$ has the amalgamation property. Moreover, all automorphisms of structures in $\mathcal{E} * \mathcal{LO}$ have to preserve the linear order and hence must be the identity. But $\mathcal{E} * \mathcal{LO}$ does not have the Ramsey property: let $\mathfrak{A}$ be the structure with domain $\{u, v\}$ such that $<^{\mathfrak{A}} = \{(u, v)\}$, and such that u and v are not E-equivalent. Let $\mathfrak{B}$ be the structure with domain $\{a, b, c, d\}$ such that $a <^{\mathfrak{B}} b <^{\mathfrak{B}} c <^{\mathfrak{B}} d$ and such that $\{a, c\}$ and $\{b, d\}$ are the equivalence classes of $E^{\mathfrak{B}}$. There are four copies of $\mathfrak{A}$ in $\mathfrak{B}$.

Suppose for contradiction that there is $\mathfrak{C} \in \mathcal{E} * \mathcal{LO}$ such that $\mathfrak{C} \to (\mathfrak{B})^{\mathfrak{A}}_2$. Let $\prec$ be a *convex* linear ordering of the elements of C, that is, a linear ordering such that $E(x, z)$ and $x < y < z$ implies that $E(x, y)$ and $E(y, z)$. Let $\mathfrak{A}' \in \binom{\mathfrak{C}}{\mathfrak{A}}$. Define $\chi(\mathfrak{A}') = 1$ if $\prec \cap (A')^2 = <^{\mathfrak{A}'}$, and $\chi(g) = 2$ otherwise. For any $\mathfrak{B}' \in \binom{\mathfrak{C}}{\mathfrak{B}}$, the convex linear order induces a convex linear order on $\mathfrak{B}'$ in which one of the two equivalence classes precedes the other. It follows by inspection that both colours are realised in $\binom{\mathfrak{B}'}{\mathfrak{A}}$.

Let $\mathfrak{B}$ be the Fraïssé limit of the class $\mathcal{E} * \mathcal{LO}$ from the previous example. In the light of Conjecture 11.1 we look for a finitely bounded homogeneous Ramsey expansion of $\mathfrak{B}$.

Example 11.3.4. Let $\mathcal{C}$ be the class of all finite structures $(V; E, \prec)$ where E is an equivalence relation and $\prec$ is a linear order that is convex with respect to E. Let $\mathfrak{C}$ be the Fraïssé-limit of $\mathcal{C}$. Note that $\mathfrak{C}$ is a reduct of Allen's Interval Algebra $\mathfrak{A}$ (Example 2.4.2) which is homogeneous (Example 5.5.5), Ramsey (Example 11.2.11), and without algebraicity. By Theorem 11.3.2 the generic superposition $\mathfrak{A} * (\mathbb{Q}; <)$ is Ramsey. Then $\mathfrak{A} * (\mathbb{Q}; <)$ is isomorphic to a Ramsey expansion of the structure $\mathfrak{B}$ from Example 11.3.3.

Example 11.3.5. The class of finite ordered binary branching C-relations is an amalgamation class, but does not have the Ramsey property. To see how the Ramsey property fails, consider the structure $\mathfrak{B} \in \mathcal{C}$ with domain $\{a, b, c, d\}$ where $a < c < b < d$ such that $C(a; c, d), C(b; c, d), C(d; a, b), C(c; a, b)$, and the structure $\mathfrak{A} \in \mathcal{C}$ with domain $\{u, v\}$ where $u < v$. Now let $\mathfrak{C} \in \mathcal{C}$ be arbitrary. Let $\prec$ be a *convex* ordering of $\mathfrak{C}$, that is, a linear ordering such that for all $u, v, w \in L$, if $C(u; v, w)$ and $v \prec w$, then either $u \prec v \prec w$ or $v \prec w \prec u$. Define $\chi \colon \binom{\mathfrak{C}}{\mathfrak{A}} \to [2]$ as follows. For $\mathfrak{A}' \in \binom{\mathfrak{C}}{\mathfrak{A}}$ define $\chi(\mathfrak{A}') = 1$ if $\prec \cap (A')^2 = <^{\mathfrak{A}'}$, and $\chi(\mathfrak{A}') = 2$ otherwise. Note that for every convex ordering $\prec$ of B there exists an $\mathfrak{A}_1 \in \binom{\mathfrak{B}}{\mathfrak{A}}$ such that $\prec \cap (A_1)^2 = <^{\mathfrak{A}_1}$, and an $\mathfrak{A}_2 \in \binom{\mathfrak{B}}{\mathfrak{A}}$ such that $\prec \cap (A_2)^2 = <^{\mathfrak{A}_2}$. Hence, for every $\mathfrak{B}' \in \binom{\mathfrak{C}}{\mathfrak{B}}$ the colouring χ is not monochromatic on $\mathfrak{B}'$.

Again, we want to confirm Conjecture 11.1 for the Fraïssé-limit of the class from Example 11.3.5 and show how to find a homogeneous Ramsey expansion, again using Theorem 11.3.2.

Example 11.3.6. The class $\mathcal{C}$ of all finite structures $(L; C, <, \prec)$, where $<$ is an arbitrary linear order, and $\prec$ is convex with respect to C, is a Ramsey class. The class $\mathcal{C}$ can be described as the superposition of the Ramsey class $\mathcal{LO}$ with the class of all convexly ordered C-relations, which is Ramsey (Example 11.1.9), and hence is Ramsey by Theorem 11.3.2.

11.4. Canonisation

In this section we apply Ramsey theory to analyse polymorphisms of Ramsey structures $\mathfrak{B}$. We first consider the more general case of functions between two possibly distinct structures, and introduce a refinement of the notion of canonicity from Section 10.5.1.

DEFINITION 11.4.1. Let $\mathfrak{A}, \mathfrak{B}$ be structures and $S \subseteq A$. We say that a function $f : A \to B$ is *canonical on S with respect to* $(\mathfrak{A}, \mathfrak{B})$ if for every $n \in \mathbb{N}$ and every $t \in S^n$ the type of $f(t)$ in $\mathfrak{B}$ only depends on the type of t in $\mathfrak{A}$. We say that f is *canonical* (*with respect to* $(\mathfrak{A}, \mathfrak{B})$) if it is canonical on all of A.

Note that for ω-categorical structures $\mathfrak{A}$ and $\mathfrak{B}$ these definitions coincide with the corresponding definitions of canonicity for $\mathcal{G} = \mathrm{Aut}(\mathfrak{A})$ and $\mathcal{H} = \mathrm{Aut}(\mathfrak{B})$ from Section 10.5.1. Ramsey theory can be applied to the analysis of functions via the following lemma.

LEMMA 11.4.2. *Let $\mathfrak{A}$ be an ω-categorical ordered Ramsey structure, let $\mathfrak{B}$ be an ω-categorical structure, and let $f : A \to B$ be a function. Then for every finite*

$F \subseteq A$ there exists $\alpha \in \mathrm{Aut}(\mathfrak{A})$ so that $f \circ \alpha$ is canonical on F with respect to $(\mathfrak{A}, \mathfrak{B})$.

PROOF. Let $<$ be a linear order in the signature of $\mathfrak{A}$. Suppose without loss of generality that $\mathfrak{A}$ contains all relations that are first-order definable in $\mathfrak{A}$, so that $\mathfrak{A}$ is homogeneous and the age of $\mathfrak{A}$ is a Ramsey class. Let $F \subseteq A$ be finite let $\mathfrak{C}_1, \ldots, \mathfrak{C}_k$ be a list of all non-isomorphic substructures of $\mathfrak{A}[F]$, and let r be the number of m-types in $\mathfrak{B}$, which is finite by the ω-categoricity of $\mathfrak{B}$. Since $\mathfrak{A}$ is Ramsey, there is a substructure $\mathfrak{A}_1$ of $\mathfrak{A}$ such that $\mathfrak{A}_1 \to (\mathfrak{F})^{\mathfrak{C}_1}_r$. Further, there is a finite substructure $\mathfrak{A}_2$ of $\mathfrak{A}'$ such that $\mathfrak{A}_2 \to (\mathfrak{A}_1)^{\mathfrak{C}_2}_r$. We iterate this k times, arriving at a structure $\mathfrak{A}_k$. For each $i \leq k$, the operation f defines a colouring χ_i of $\binom{\mathfrak{A}_k}{\mathfrak{C}_i}$ with finitely many colours as follows: let $a_1, \ldots, a_m$ be the elements of A_i such that $a_1 < \cdots < a_m$. Then the colour of a copy $\mathfrak{C}$ of $\mathfrak{C}_i$ is just the type of $(f(a_1), \ldots, f(a_m))$ in $\mathfrak{B}$. By a downward induction following the construction of $\mathfrak{A}_k$ in reverse, we find copies $\mathfrak{A}'_i$ of $\mathfrak{A}_i$ monochromatic with the colourings associated with $\mathfrak{C}_j$ for $j > i$, and at the end a copy of $\mathfrak{A}[F]$ on which they are all monochromatic. Since $\mathfrak{A}$ is homogeneous, there exists an automorphism α of $\mathfrak{A}$ that sends F to this copy. Then $f \circ \alpha$ is canonical on F as a map from $\mathfrak{A}$ to $\mathfrak{B}$. ⊣

Note that the assumption in Lemma 11.4.2 that $\mathfrak{A}$ is ordered cannot simply be dropped: for instance, if $\mathfrak{A}$ is the structure $(A;=)$ and $f : A \to \mathbb{Q}$ is injective, then f is not canonical on any two-element subset of A with respect to $\big((A;=), (\mathbb{Q};<)\big)$.

11.4.1. Functions of higher arity. In this section we consider canonicity for functions of higher arity; this generalises the terminology from Section 10.5.2.

Let $\mathfrak{A}$ and $\mathfrak{B}$ be structures, $f : A^d \to B$ a function, and S a subset of A^d. Recall that $\mathfrak{A}^{(d)}$ denotes the algebraic product structure introduced in Definition 4.2.20. Note that f is canonical on S with respect to $(\mathfrak{A}^{(d)}, \mathfrak{B})$ if for all n and all n-tuples $t^1, \ldots, t^d$ where $(t^1_i, \ldots, t^d_i) \in S$ for all $i \leq n$ the n-type of $f(t^1, \ldots, t^d)$ in $\mathfrak{B}$ only depends on the n-types of $t^1, \ldots, t^d$ in $\mathfrak{A}$. If $\mathfrak{B} = \mathfrak{C}$, then we say that f *is canonical* (*on* S) *with respect to* $\mathfrak{B}$ if it is canonical (on S) with respect to $(\mathfrak{B}^{(d)}, \mathfrak{B})$.

Example 11.4.3. The function $(x, y) \mapsto \min(x, y)$ is not canonical with respect to $(\mathbb{Q};<)$. However, if $S \subseteq \mathbb{Q}$ is finite, and α is an automorphism such that $\alpha(\max(S)) < \min(S)$, then for all $x, y \in S$ we have $\min(\alpha(x), y) = \alpha(x)$, so the operation $(x, y) \mapsto \min(\alpha(x), y)$ is canonical on S with respect to $(\mathbb{Q};<)$.

In the proof of the following we use the product Ramsey theorem, Theorem 11.2.14.

PROPOSITION 11.4.4. *Let $\mathfrak{B}$ be an ω-categorical ordered Ramsey structure and let $f : B^d \to B$ be any operation. Then for all finite subsets $S_1, \ldots, S_d$*

of B there are $\alpha_1, \ldots, \alpha_d \in \mathrm{Aut}(\mathfrak{B})$ so that the operation $(x_1, \ldots, x_d) \mapsto f(\alpha_1 x_1, \ldots, \alpha_d x_d)$ is canonical on $S_1 \times \cdots \times S_d$ with respect to $\mathfrak{B}$.

PROOF. By Theorem 11.2.14, the structure $\mathfrak{B}^{(d)}$ is Ramsey. Hence, Lemma 11.4.2 shows the existence of $\alpha \in \mathrm{Aut}(\mathfrak{B}^{(d)})$ such that $x \mapsto f(\alpha x)$ is canonical on $S_1 \times \cdots \times S_d$ with respect to $(\mathfrak{B}^{(d)}, \mathfrak{B})$. By Proposition 4.2.19, there are $\alpha_1, \ldots, \alpha_d \in \mathrm{Aut}(\mathfrak{B})$ so that $\alpha(x_1, \ldots, x_d)) = (\alpha_1 x_1, \ldots, \alpha_d x_d)$. Now clearly the function $(x_1, \ldots, x_d) \mapsto f(\alpha_1 x_1, \ldots, \alpha_d x_d)$ is canonical on $S_1 \times \cdots \times S_d$ with respect to $\mathfrak{B}$. ⊣

11.4.2. Interpolation modulo automorphisms. One of the central problems when analysing a polymorphism of a structure $\mathfrak{B}$ is to determine what kind of operations it generates locally (these operations will also be polymorphisms of $\mathfrak{B}$, cf. Proposition 6.1.5). Proposition 11.4.4 can be used for this purpose, as the following illustrates.

COROLLARY 11.4.5. *Let $\mathfrak{B}$ be an ω-categorical ordered Ramsey structure and $f \in \mathrm{Pol}(\mathfrak{B})^{(k)}$ be injective. Then $\mathfrak{B}$ also has an injective polymorphism g of arity k which is canonical with respect to $\mathfrak{B}$.*

This corollary follows in a straightforward way from Proposition 11.4.4 and a compactness argument, which we do not present here since we will present a proper generalisation of Corollary 11.4.5 in full detail, namely Theorem 11.4.7. In our generalisation, we would like to drop the injectivity condition on f, and thus also on g. But if this is all we do we arrive at a triviality, since local generation—or, for that matter, generation—produces projections, and projections are canonical. To formulate an appropriate generalisation, we apply the operation of local closure to a subset of the set of all operations generated by f.

DEFINITION 11.4.6. Let $\mathfrak{B}$ be an ω-categorical structure and $f, g \colon B^d \to B$. Then f *interpolates g modulo* $\mathrm{Aut}(\mathfrak{B})$ if for every finite $S \subseteq B$ there are $\alpha_0, \alpha_1, \ldots, \alpha_d \in \mathrm{Aut}(\mathfrak{B})$ such that $g(x_1, \ldots, x_d) = \alpha_0 f(\alpha_1 x_1, \ldots, \alpha_d x_d)$ for all $x_1, \ldots, x_d \in B$, i.e.,

$$g \in \overline{\{(x_1, \ldots, x_d) \mapsto \alpha_0 f(\alpha_1(x_1), \ldots, \alpha_d(x_d)) \mid \alpha_0, \alpha_1, \ldots, \alpha_d \in \mathrm{Aut}(\mathfrak{B})\}}.$$

THEOREM 11.4.7. *Let $\mathfrak{B}$ be an ω-categorical ordered Ramsey structure and let $f \colon B^d \to B$ be an operation. Then f interpolates an operation modulo* $\mathrm{Aut}(\mathfrak{B})$ *that is canonical with respect to $\mathfrak{B}$.*

Note that Theorem 11.4.7 is indeed a generalisation of Corollary 11.4.5, because operations that are interpolated by injective operations modulo automorphisms are again injective, and operations that are interpolated by polymorphisms modulo automorphisms are again polymorphisms. However, Theorem 11.4.7 is still not in its most general and most useful form. For this, we need a further generalisation of the notion of interpolation modulo automorphisms to the situation where f is a function between different structures $\mathfrak{A}$ and $\mathfrak{B}$.

DEFINITION 11.4.8. Let $\mathfrak{A}$ and $\mathfrak{B}$ be structures and let $f, g \colon A \to B$ be functions. Then f *interpolates* g *modulo* $(\mathrm{Aut}(\mathfrak{A}), \mathrm{Aut}(\mathfrak{B}))$ if for every finite $S \subseteq A$ there are $\alpha \in \mathrm{Aut}(\mathfrak{A})$ and $\beta \in \mathrm{Aut}(\mathfrak{B})$ such that $g(x) = \beta f(\alpha x)$ for every $x \in S$.

The following is the central statement about Ramsey structures and interpolation modulo automorphisms; it has many applications. We present two proofs of this result, the first making direct use of the Ramsey property of $\mathfrak{A}$ and the compactness principle of Lemma 4.1.10, the second based on the extreme amenability of $\mathrm{Aut}(\mathfrak{A})$.

LEMMA 11.4.9 (Canonisation Lemma). *Let $\mathfrak{A}$ be an ω-categorical ordered Ramsey structure and let $\mathfrak{B}$ be ω-categorical. Then every $f : A \to B$ interpolates an operation $g \colon A \to B$ modulo* $(\mathrm{Aut}(\mathfrak{A}), \mathrm{Aut}(\mathfrak{B}))$ *that is canonical with respect to* $(\mathfrak{A}, \mathfrak{B})$.

FIRST PROOF. By Lemma 11.4.2, for every finite $S \subseteq A$ there is a function from $A \to B$ that is canonical on S with respect to $(\mathfrak{A}, \mathfrak{B})$ and interpolated by f modulo $(\mathrm{Aut}(\mathfrak{A}), \mathrm{Aut}(\mathfrak{B}))$. Now the statement follows from Lemma 4.1.10, because the property to be canonical on S with respect to $(\mathfrak{A}, \mathfrak{B})$ is a universal first-order statement about f in the signature of $\mathfrak{A}$ and $\mathfrak{B}$. ⊣

SECOND PROOF. Our second proof, following [92], uses the extreme amenability of $\mathrm{Aut}(\mathfrak{A})$. Recall the definition of the space $B^A /\!\!/ \mathrm{Aut}(\mathfrak{B})$ (Definition 10.1.7) which is compact and Hausdorff (Proposition 10.1.8), and the notation $f \sim g$ meaning that f is in the closure of the orbit of g under $\mathrm{Aut}(\mathfrak{B})$. The space

$$\overline{\mathrm{Aut}(\mathfrak{B})\, f\ \mathrm{Aut}(\mathfrak{A})}/_{\sim}$$

is a closed subspace of $B^A /\!\!/ \mathrm{Aut}(\mathfrak{B})$, and hence compact (Proposition 9.1.18) and Hausdorff as well. We define a continuous action of $\mathrm{Aut}(\mathfrak{A})$ on

$$\overline{\mathrm{Aut}(\mathfrak{B})\, f\ \mathrm{Aut}(\mathfrak{A})}/_{\sim}$$

by

$$(\alpha, g/_{\sim}) \mapsto g\,\alpha^{-1}/_{\sim}.$$

This assignment is well defined, it is a group action, and it is continuous. Since $\mathrm{Aut}(\mathfrak{A})$ is extremely amenable, this action has a fixed point. Any element g of this fixed point is canonical: whenever $\alpha \in \mathrm{Aut}(\mathfrak{A})$, then $g\,\alpha/_{\sim} = g/_{\sim}$, which is the definition of canonicity. ⊣

PROOF OF THEOREM 11.4.7. We apply Lemma 11.4.9 to the structure

$$\mathfrak{A} := \mathfrak{B}^{(d)},$$

which is Ramsey if $\mathfrak{B}$ is Ramsey (Theorem 11.2.14). As in the proof of Proposition 11.4.4, the canonicity of a function g with respect to $(\mathfrak{B}^{(d)}, \mathfrak{B})$ translates into the canonicity of the d-ary function g with respect to $\mathfrak{B}$, and interpolation of operations modulo $(\mathrm{Aut}(\mathfrak{B}^{(d)}), \mathrm{Aut}(\mathfrak{B}))$ translates to interpolation modulo $\mathrm{Aut}(\mathfrak{B})$. ⊣

11.4.3. Canonisation with respect to constants. 'Canonisation' of operations as exhibited in Theorem 11.4.7 becomes particularly powerful when we combine it with expansions by constants.

THEOREM 11.4.10. *Let $\mathfrak{A}$ be an ω-categorical ordered Ramsey structure and let $\mathfrak{B}$ be ω-categorical. Let $f : A^k \to B$ be a function and let $c^1, \dots, c^m \in A^k$. Then*

$$\overline{\{(x_1, \dots, x_k) \mapsto \beta f(\alpha_1(x_1), \dots, \alpha_k(x_k)) \mid \beta \in \mathrm{Aut}(\mathfrak{B}),} \\ \overline{\alpha_i \in \mathrm{Aut}(\mathfrak{A}, c^1_i, \dots, c^m_i)\}}$$

contains an operation which is canonical with respect to $((\mathfrak{A}^{(k)}, c^1, \dots, c^m), \mathfrak{B})$, and which is identical with f on $c^1, \dots, c^m$.

PROOF. The ω-categorical structure $\mathfrak{A}^{(k)}$ is Ramsey (Theorem 11.2.14) and a linear order $<$ is first-order definable in it. The expansion $(\mathfrak{A}^{(k)}, <)$ is ordered and also ω-categorical and Ramsey. It follows that $\mathfrak{C} := (\mathfrak{A}^{(k)}, <, c^1, \dots, c^m)$ is Ramsey and ω-categorical by Corollary 11.2.16. Note that $\mathrm{Aut}(\mathfrak{C}) = \mathrm{Aut}(\mathfrak{A}, c^1_1, \dots, c^m_1) \times \cdots \times \mathrm{Aut}(\mathfrak{A}, c^1_k, \dots, c^m_k)$. Then Lemma 11.4.9 shows that f interpolates modulo $(\mathrm{Aut}(\mathfrak{C}), \mathrm{Aut}(\mathfrak{B}))$ an operation g that satisfies the requirements. ⊣

11.4.4. Behaviours of functions. To work with canonical functions the definition of a *behaviour* of a canonical function. is important. It is sometimes necessary to work with functions that share some properties with canonical functions. The following definition from [88] gives us some flexibility in specifying such properties.

DEFINITION 11.4.11. Let $\mathfrak{A}$ and $\mathfrak{B}$ be structures. An *(n-)type condition* between $\mathfrak{A}$ and $\mathfrak{B}$ is a pair (r, s) where r is a complete n-type in $\mathfrak{A}$ and s is a complete n-type in $\mathfrak{B}$. A function $f : A \to B$ *satisfies the n-type condition (r, s) on $X \subseteq A$*, if for every $t \in X^n$ of type r in $\mathfrak{A}$ the n-tuple $f(t)$ has type s in $\mathfrak{B}$. It satisfies a set Λ of type conditions on X if it satisfies every type condition in Λ on X; if $X = A$ we may also drop 'on X', and simply write that f satisfies Λ. A set Λ of type conditions is called *complete* if for every $n \geq 1$ and and every complete n-type r in $\mathfrak{A}$ there exists a complete n-type s in $\mathfrak{B}$ such that f satisfies the type condition (r, s).

Note that $f : \mathfrak{A} \to \mathfrak{B}$ is canonical if and only if it satisfies a complete set of type conditions. The *behaviour* of a canonical function $f : \mathfrak{A} \to \mathfrak{B}$ is the function that maps each complete n-type r in $\mathfrak{A}$ to the complete n-type s in $\mathfrak{B}$ such that f satisfies the type condition (r, s).

LEMMA 11.4.12. *Let $\mathfrak{A}$ and $\mathfrak{B}$ be ω-categorical structures.*

1. *If $f : A \to B$ interpolates g modulo $(\mathrm{Aut}(\mathfrak{A}), \mathrm{Aut}(\mathfrak{B}))$ and satisfies a type condition, then g satisfies the type condition as well.*
2. *Two functions $f, g : A \to B$ that are canonical with respect to $(\mathfrak{A}, \mathfrak{B})$ have the same behaviour if and only if f interpolates g modulo $(\mathrm{Aut}(\mathfrak{A}), \mathrm{Aut}(\mathfrak{B}))$ and g interpolates f modulo $(\mathrm{Aut}(\mathfrak{A}), \mathrm{Aut}(\mathfrak{B}))$.*
3. *If $\mathfrak{B}$ is homogeneous with a finite relational signature of maximal arity m, then two canonical functions with respect to $(\mathfrak{A}, \mathfrak{B})$ have the same behaviour if and only if they satisfy the same m-type conditions.*

PROOF. The first statement follows directly from the definitions. For the second statement, the backwards implication follows from the first statement. The forward implication can be shown by a compactness argument and is left to the reader. For the third statement, suppose that $f, g : A \to B$ are canonical with respect to $(\mathfrak{A}, \mathfrak{B})$ and satisfy the same m-type conditions, and let $s, t \in A^n$, for $n \in \mathbb{N}$, be tuples that have the same n-type in $\mathfrak{A}$. We have to show that $f(s)$ and $g(t)$ have the same n-type in $\mathfrak{B}$. Since every atomic formula over the signature of $\mathfrak{B}$ has at most m variables, $f(s)$ and $g(t)$ satisfy the same atomic formulas, so the statement follows from the homogeneity of $\mathfrak{B}$. ⊣

Definition 11.4.11 can also be applied to operations of arbitrary arity over a structure $\mathfrak{B}$. A *type condition* for k-ary operations from $B^k \to B$ is a type condition between the algebraic power $\mathfrak{B}^{(k)}$ and $\mathfrak{B}$. Recall from Section 4.2.5.2 that a complete n-type in $\mathfrak{B}^{(k)}$ can be specified by specifying k complete n-types in $\mathfrak{B}$. So an n-type condition between $\mathfrak{B}^{(k)}$ and $\mathfrak{B}$ can be specified by a tuple $(r_1, \ldots, r_k, s)$ of complete n-types in $\mathfrak{B}$, and $f : B^k \to B$ satisfies this type condition if for all $t_1, \ldots, t_k \in X^n$, if t_i has type r_i for all $i \in \{1, \ldots, k\}$, then $f(t_1, \ldots, t_k)$ has type s.

Not for every set Λ of type conditions between $\mathfrak{A}$ and $\mathfrak{B}$ there exists a function that satisfies Λ, as we will see in the following example.

Example 11.4.13. We determine the possible behaviours of binary injective polymorphisms f of $(\mathbb{Q}; <)$. We write $<$ and $>$ stand for the complete 2-types in $(\mathbb{Q}; <)$ given by $x_1 < x_2$ and by $x_2 < x_1$. Note that every Λ must contain the 2-type condition $(<, <, <)$ because f preserves $<$. Moreover, there are complete 2-types $s_1, s_2, s_3 \in \{<, >\}$ such that Λ contains $(<, =, s_1)$, $(=, <, s_1)$, and $(<, >, s_3)$. The eight resulting sets of type conditions are illustrated in Figure 1. However, only four of these sets, namely the ones depicted in the first row, are satisfied by binary injective polymorphisms of $(\mathbb{Q}; <)$. The other diagrams contain directed cycles of length three; so the existence of operations satisfying these type conditions would imply the existence of three points x, y, z in $\mathbb{Q}$ such that $x < y < z < x$, a contradiction. The type conditions that are

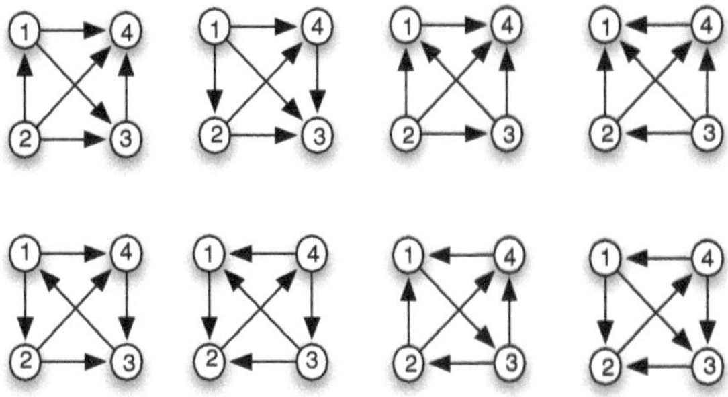

FIGURE 1. Complete sets of type conditions between $(\mathbb{Q}; <)^{(2)}$ and $(\mathbb{Q}; <)$ illustrated on a $[2]^2$ grid. Only the top row can be satisfied by operations $f : \mathbb{Q}^2 \to \mathbb{Q}$.

illustrated in the top row, on the other hand, are satisfied by the operations $\text{lex}(x, y)$, $\text{lex}(y, x)$, $\text{lex}(x, -y)$, and $\text{lex}(y, -x)$, respectively.

The analysis in Example 11.4.13 in combination with Corollary 11.4.5 shows that every binary injective polymorphism of $(\mathbb{Q}; <)$ interpolates modulo $\text{Aut}(\mathbb{Q}; <)$ at least one of the operations $\text{lex}(x, y)$, $\text{lex}(y, x)$, $\text{lex}(x, -y)$, or $\text{lex}(y, -x)$. The not necessarily injective case can be analysed similarly, and for later reference we state the following.

LEMMA 11.4.14. *Let f be a canonical binary polymorphism of $(\mathbb{Q}; <)$. Then f has the same behaviour as one of the following seven operations.*

- $\text{lex}(x, y)$ *or* $\text{lex}(y, x)$ (*Example* 10.5.4);
- $\text{lex}(x, -y)$ *or* $\text{lex}(y, -x)$.
- $(x, y) \mapsto x$ *or* $(x, y) \mapsto y$.
- *a constant operation.*

11.4.5. Canonical non-preservation. Let $\mathfrak{C}$ be an ω-categorical ordered Ramsey structure and let $\mathfrak{B}$ be a first-order reduct of $\mathfrak{C}$. Suppose that a relation R does not have a primitive positive definition in $\mathfrak{B}$, so that there is an f in $\text{Pol}(\mathfrak{C})$ that does not preserve R (Theorem 6.1.12). It is natural to ask whether f may be taken to be canonical. The following example shows that in general, there may not be a canonical polymorphism that does not preserve R.

Example 11.4.15. Let $\mathfrak{C} := (\mathbb{Q}; <)$, and let $\mathfrak{B}$ the first-order reduct of $\mathfrak{C}$ that contains all relations that are first-order definable in $\mathfrak{C}$ and are preserved by the operation $(x, y) \mapsto \min(x, y)$. Clearly, min does not preserve the relation

$$\text{Betw} := \{(a, b, c) \in \mathbb{Q}^3 \mid a < b < c \vee c < b < a\}$$

and hence Betw is not primitively positively definable in $\mathfrak{C}$. On the other hand, it can be shown that every operation that is locally generated by min and the automorphisms of $\mathfrak{C}$ and that is canonical with respect to $\mathfrak{C}$ is a projection, and hence preserves Betw.

On the other hand, the results from Section 11.2.4 suggest a sense in which we can still find canonical polymorphisms that do not preserve R. To illustrate the basic idea, we first discuss the unary case, with existential positive definability instead of primitive positive definability.

THEOREM 11.4.16. *Let $\mathfrak{C}$ be an ω-categorical ordered Ramsey structure and let $\mathfrak{B}$ be a first-order reduct of $\mathfrak{C}$. Let R be an m-ary relation that does not have an existential positive definition in $\mathfrak{B}$. Then there exists an endomorphism $e \in \text{End}(\mathfrak{B})$ and an m-tuple $c \in R$ such that*

- *$e(c) \notin R$;*
- *e is canonical with respect to $((\mathfrak{C}, c), \mathfrak{C})$.*

PROOF. The structure $\mathfrak{B}$ is ω-categorical. If R does not have an existential definition in $\mathfrak{B}$, then by Theorem 4.4.1 there is an endomorphism e' of $\mathfrak{B}$ which does not preserve R, that is, there is a m-tuple $c \in R$ such that $e'(c) \notin R$. By Theorem 11.4.10 applied with $k = 1$, the operation e' interpolates an operation e modulo $(\text{Aut}(\mathfrak{C}, c), \text{Aut}(\mathfrak{C}))$ that is canonical with respect to $((\mathfrak{C}, c), \mathfrak{C})$ and that has the same restriction to $\{c_1, \ldots, c_m\}$ as e'. Since e is an endomorphism of $\mathfrak{B}$ it therefore satisfies the stated properties. ⊣

We now give the analog of Theorem 11.4.16 for operations of higher arity and whose proof is analogous to the proof of the previous theorem.

THEOREM 11.4.17. *Let $\mathfrak{C}$ be an ω-categorical ordered Ramsey structure, let $\mathfrak{B}$ be a first-order reduct of $\mathfrak{C}$, and let R be an m-ary relation that does not have a primitive positive definition in $\mathfrak{B}$. Then there exist $k \in \mathbb{N}$, a k-ary polymorphism f of $\mathfrak{B}$, and m-tuples $c_1 = (c_1^1, \ldots, c_1^m), \ldots, c_k = (c_k^1, \ldots, c_k^m) \in R$ such that*

- *$f(c_1, \ldots, c_k) \notin R$*
- *f is canonical with respect to $((\mathfrak{C}^{(k)}, (c_1^1, \ldots, c_k^1), \ldots, (c_1^m, \ldots, c_k^m)), \mathfrak{C})$.*

PROOF. If R does not have a primitive positive definition in $\mathfrak{B}$, then there is $f' \in \text{Pol}(\mathfrak{B})$ which does not preserve R (by Theorem 6.1.12 since $\mathfrak{B}$ is ω-categorical). Let k be the arity of f'. Then there are m-tuples $c_1, \ldots, c_k \in R$ such that $f'(c_1, \ldots, c_k) \notin R$. By Theorem 11.4.10 there exists some operation f in the local closure of the set

$$\{(x_1, \ldots, x_k) \mapsto \beta f'(\alpha_1(x_1), \ldots, \alpha_k(x_k)) \mid \beta \in \text{Aut}(\mathfrak{C}),$$
$$\alpha_i \in \text{Aut}(\mathfrak{C}, c_i) \text{ for } i \in \{1, \ldots, k\}\}$$

which is canonical with respect to $((\mathfrak{C}^{(k)}, (c_1^1, \ldots, c_k^1), \ldots, (c_1^m, \ldots, c_k^m)), \mathfrak{B})$. Since f is locally generated by polymorphisms of $\mathfrak{B}$, it is itself a polymorphism of $\mathfrak{B}$, and satisfies our requirements. ⊣

We are now ready to prove the following result, which was already stated in Section 6.1.8.

THEOREM 11.4.18 (Proposition 24 in [96]). *Let $\mathfrak{B}$ be a reduct of a homogeneous ordered Ramsey structure $\mathfrak{C}$ with a finite relational signature. Then there are finitely many minimal closed clones above* Pol($\mathfrak{B}$).

PROOF. Every minimal closed clone above Pol($\mathfrak{B}$) is locally generated by a minimal operation f (Proposition 6.1.35), and if k is the arity of f then by Theorem 6.1.13 there must be a relation R in $\mathfrak{B}$ that is not preserved by f, that is, there are $c_1 = (c_1^1, \dots, c_1^m), \dots, c_k = (c_k^1, \dots, c_k^m) \in R$ such that $f(c_1, \dots, c_k) \notin R$. Since f is a minimal operation, Theorem 11.4.17 implies that f must be canonical with respect to $((\mathfrak{C}^{(k)}, (c_1^1, \dots, c_k^1), \dots, (c_1^m, \dots, c_k^m)), \mathfrak{B})$. There are only finitely many behaviours of such operations, because $\mathfrak{C}$ is homogeneous is a finite relational signature. As two minimal operations with the same behaviour locally generate the same closed clone above Pol($\mathfrak{B}$), we are done. ⊣

11.5. Application: preservation of pseudo-minor conditions

Minor-preserving maps preserve minor conditions (Section 6.7.3), but in general there is no reason why they should also preserve pseudo-minor conditions (Section 10.1). However, if we have a minor-preserving map from Pol($\mathfrak{B}$) to Pol($\mathfrak{A}$) where $\mathfrak{A}$ is a finite structure, then more can be said under some Ramsey-theoretic assumption on $\mathfrak{B}$. The following statement is only implicit in the literature [20]; it follows from the second proof of Theorem 1.9 in [20] (the author is grateful to Jakub Rydval for this observation).

THEOREM 11.5.1. *Let $\mathfrak{B}$ be an ω-categorical ordered Ramsey structure which is a model-complete core and suppose that* Pol($\mathfrak{B}$) *satisfies a pseudo-minor condition ϕ. Let ξ be a uniformly continuous minor-preserving map from* Pol($\mathfrak{B}$) *to* Pol($\mathfrak{A}$) *for some finite structure $\mathfrak{A}$. Then* Pol($\mathfrak{A}$) *also satisfies ϕ.*

PROOF. Choose $m \in \mathbb{N}$ larger than all the arities of all operations that occur in ϕ. Let S_m be the set of all functions from Pol$(\mathfrak{B})^{(m)}$ to Pol$(\mathfrak{A})^{(m)}$. Note that S_m can be equipped with the product topology, and since A is finite and discrete S_m is compact (Theorem 9.1.17, for the first time in this text applied to a product with an uncountable exponent). Also note that the space S_m is Hausdorff (Proposition 9.1.3). We define an action of Aut($\mathfrak{B}$) on S_m as follows. For $\alpha \in$ Aut($\mathfrak{B}$) and $\eta \in S_m$, define $\alpha \cdot \eta$ as the map in S_m given by

$$f \mapsto \eta(\alpha^{-1} \circ f).$$

Let $\xi_m \in S_m$ be the restriction of ξ to Pol$^{(m)}(\mathfrak{B})$. Then

$$C := \overline{\{\alpha \cdot \xi_m \mid \alpha \in \operatorname{Aut}(\mathfrak{B})\}}$$

is a compact subspace of S_m and all maps in C are minor-preserving.

Claim. The restriction of the action to C is continuous. By the uniform continuity of ξ, there exists a finite $F \subseteq B^m$ such that for all $f, g \in \mathrm{Pol}^{(m)}(\mathfrak{B})$ we have that $f|_F = g|_F$ implies $\xi(f) = \xi(f')$. Now let $\psi \in C$ and suppose that $f, g \in \mathrm{Pol}^{(m)}(\mathfrak{B})$ are such that $f|_F = g|_F$. Observe that then $\psi(f) = \psi(g)$: since $\psi \in C$ there exists $\alpha \in \mathrm{Aut}(\mathfrak{B})$ such that $\psi(f) = \alpha \cdot \xi_m(f)$ and $\psi(g) = \alpha \cdot \xi_m(g)$. Hence, $\psi(f) = \alpha \cdot \xi_m(f) = \alpha \cdot \xi_m(g) = \psi(g)$.

Now consider for some $\psi \in C$ and $f \in \mathrm{Pol}^{(m)}(\mathfrak{B})$ the basic open subset $O_f(\psi) \subseteq C$ given by

$$O_f(\psi) := \{\psi' \in C \mid \psi'(f) = \psi(f)\}.$$

Then the open set

$$\{(\alpha, \psi') \in \mathrm{Aut}(\mathfrak{B}) \times C \mid \alpha \in \mathrm{Aut}(\mathfrak{B})_{f(F)} \text{and } \psi'(f) = \psi(f)\}$$

contains (id, ψ) and is mapped into $O_f(\psi)$ by the action. To see this, let $\alpha \in \mathrm{Aut}(\mathfrak{B})_{f(F)}$ and $\psi' \in C$ such that $\psi'(f) = \psi(f)$. Then by the observation above

$$\begin{aligned}
(\alpha \cdot \psi')(f) &= \psi'(\alpha^{-1} \circ f) && \text{(by definition)}\\
&= \psi'(f) && (\text{since } \alpha^{-1} \circ f|_F = f|_F)\\
&= \psi(f) && \text{(by assumption)}
\end{aligned}$$

which shows that $\alpha \cdot \psi \in O_f(\psi)$. We have now proved the claim.

Since $\mathrm{Aut}(\mathfrak{B})$ is extremely amenable by Theorem 11.2.2, the action on C has a fixed point ξ'_m, i.e., $\alpha \cdot \xi'_m = \xi'_m$ for every $\alpha \in \mathrm{Aut}(\mathfrak{B})$. In other words, ξ'_m preserves left composition with $\mathrm{Aut}(\mathfrak{B})$, and by continuity even with $\overline{\mathrm{Aut}(\mathfrak{B})}$. Moreover, ξ'_m is minor-preserving and hence $\mathrm{Pol}(\mathfrak{A})$ must satisfy ϕ as well. ⊣

We obtain the following consequence [20]; this is the implication from (5) to (2) in Theorem 10.3.5, however, under very different assumptions: there we required that $\mathfrak{C}$ is a reduct of a finitely bounded homogeneous structure, and here we require that $\mathfrak{C}$ is an ω-categorical ordered Ramsey structure.

COROLLARY 11.5.2. *Let $\mathfrak{B}$ be an ω-categorical ordered Ramsey structure which is a model-complete core and has a pseudo-Siggers polymorphism. Then* $\mathrm{Pol}(\mathfrak{B})$ *does not have a uniformly continuous minor-preserving map to* **Proj**.

Recall that the atomless Boolean algebra both has a pseudo-Siggers polymorphism and a uniformly continuous minor-preserving map to the projections (Example 10.3.1). This is consistent with Corollary 11.5.2, because $\mathfrak{B}$ is not ordered. And its ordered Ramsey expansion $(\mathfrak{B}, <)$ (see Example 11.2.7), on the other hand, no longer has a pseudo-Siggers polymorphism.

11.6. Application: decidability results for meta-problems

Finitely bounded homogeneous structures $\mathfrak{B}$, though typically infinite, are determined by a finite amount of data: the bounds determining their age, which determines $\mathfrak{B}$ up to isomorphism. Since these bounds can be input to a computer, many questions about finitely bounded homogeneous structures can be viewed as algorithmic questions. The same applies to finite-signature first-order reducts $\mathfrak{A}$ of $\mathfrak{B}$, because it suffices to additionally specify (quantifier-free) first-order formulas that define the relations of $\mathfrak{A}$ in $\mathfrak{B}$. Many of the questions about finitely bounded homogeneous structures $\mathfrak{B}$ and their first-order reducts are not known to be decidable, and listed as open problems in Section 14.2.11. Some of these questions are also of interest when the structure $\mathfrak{B}$ is fixed. In this section we discuss applications of the techniques seen in this chapter to decidability results for finite-signature first-order reducts of homogeneous finitely bounded structures.

11.6.1. Deciding the existence of maps with specified behavior. Let $\mathfrak{A}$ and $\mathfrak{B}$ be homogeneous finitely bounded structures and let Λ be a finite set of type conditions between $\mathfrak{A}$ and $\mathfrak{B}$. We are interested in the question whether Λ can be realised by a function from $\mathfrak{A}$ to $\mathfrak{B}$. The following decidability result will be the basis for several other decidability results.

THEOREM 11.6.1. *Let $\mathfrak{A}$ and $\mathfrak{B}$ be homogeneous finitely bounded structures and suppose that $\mathfrak{B}$ is Ramsey. Let Λ be a finite set of type conditions between $\mathfrak{A}$ and $\mathfrak{B}$. Then the question whether there is a function from $\mathfrak{A}$ to $\mathfrak{B}$ that satisfies Λ is decidable.*

PROOF. Suppose that there exists a function $f: A \to B$ that satisfies Λ. We know from the canonisation lemma (Lemma 11.4.9) f interpolates a function g modulo $(\mathrm{Aut}(\mathfrak{A}), \mathrm{Aut}(\mathfrak{B}))$ which is canonical with respect to $(\mathfrak{A}, \mathfrak{B})$. Moreover, g satisfies Λ as well (Lemma 11.4.12); we call such a function g a *witness* for Λ. We decide the existence of a witness by an effective reduction to a finite-domain constraint satisfaction problem (similarly as in Section 10.5.3), which proves the decidability result.

Let $\mathcal{F}$ be a finite set of finite structures such that $\mathrm{Forb}^{\mathrm{emb}}(\mathcal{F}) = \mathrm{Age}(\mathfrak{B})$. It will be convenient to make the assumption that $\mathcal{F}$ is *minimal* in the sense that it does not contain structures $\mathfrak{F}_1, \mathfrak{F}_2$ such that $\mathfrak{F}_1$ is an induced substructure of $\mathfrak{F}_2$. This assumption is without loss of generality since $\mathcal{F}$ is finite; otherwise we replace $\mathcal{F}$ by a set $\mathcal{F}'$ of minimal cardinality with $\mathrm{Forb}^{\mathrm{emb}}(\mathcal{F}) = \mathrm{Forb}^{\mathrm{emb}}(\mathcal{F}')$, then $\mathcal{F}'$ we be minimal also in the desired sense.

Let $n = 2$ if all relations of $\mathfrak{A}$ are unary, and otherwise let n be the maximal arity of the relations in $\mathfrak{A}$ and in $\mathfrak{B}$. The domain of the CSP is the set of all complete n-types of $\mathfrak{B}$. The instance of the CSP has a variable for every complete n-type of $\mathfrak{A}$. Hence, every solution to the CSP describes a complete set Δ of type conditions between $\mathfrak{A}$ and $\mathfrak{B}$. The constraints are described below.

Compatibility. Let $r, t \in S_n^{\mathfrak{A}}$ and $I \subset [n]$ be such that the subtype of r induced on I and the subtype of t induced on I coincide. Let $r', t' \in S_n^{\mathfrak{B}}$ be such that $(r, r'), (t, t') \in \Delta$. Then we impose the binary constraint that I induces the same subtype in r' and in t'.

Satisfiability. To ensure that Δ is satisfiable by a function $g: A \to B$ (recall Example 11.4.13), we need to take care of two things.

- Δ should not force the existence of one of the forbidden substructures from $\mathcal{F}$ in the image of g. For each structure $\mathfrak{F} \in \mathcal{F}$ with $|F| =: s > n$ and each $t \in S_s^{\mathfrak{A}}$ with $(t, t') \in \Delta$ we have a constraint of arity $r := \binom{s}{n}$. Let $a_1, \ldots, a_s$ be the elements of $\mathfrak{F}$. Observe that for every subset $I \subseteq [s]$ with $|I| = n$ the structure induced on $\{a_i \mid i \in I\}$ in $\mathfrak{F}$ is an induced substructure of $\mathfrak{B}$, by the minimality assumption on $\mathcal{F}$. Let $\phi_I^{\mathfrak{F}}$ be the formula with variables $x_1, \ldots, x_n$ that contains for $i_1, \ldots, i_m \in I$ the conjunct $R(x_{i_1}, \ldots, x_{i_m})$ if and only if $(a_{i_1}, \ldots, a_{i_m}) \in R^{\mathfrak{F}}$. By the observation we just made and the homogeneity of $\mathfrak{B}$ in a finite relational signature, $\phi_I^{\mathfrak{F}}$ is contained in a unique complete n-type of $\mathfrak{B}$. We then impose the constraint of arity r that requires that for some $I \subseteq [s]$ with $|I| = n$ the subtype of t' induced on I does not contain $\phi_I^{\mathfrak{F}}$.
- Since equality is transitive[28], we also add for every $t \in S_3^{\mathfrak{A}}$ the ternary constraint which makes sure that if the subtype of t induced on $\{1, 2\}$ contains $x_1 = x_2$ and if the subtype of S induced on $\{2, 3\}$ contains $x_1 = x_2$, then the subtype of t induced on $\{1, 3\}$ contains $x_1 = x_2$.

Behaviour. If $(t, t') \in \Lambda$, then then we add a constraint that ensures that $(t, t') \in \Delta$.

We verify that there is a witness g if and only if the CSP instance described above has a satisfying assignment in the set of all n-types of $\mathfrak{B}$. For the easy direction, suppose that there exists a witness. Then the behaviour of the witness provides a solution to the instance that clearly satisfies the compatibility, satisfiability, and behaviour constraints.

For the opposite direction, suppose that α is a solution to the described CSP, i.e., a mapping $\Delta: S_n^{\mathfrak{A}} \to S_n^{\mathfrak{B}}$ that satisfies compatibility, satisfiability, and behaviour constraints. Clearly, if there is a map $g: \mathfrak{A} \to \mathfrak{B}$ with behaviour Δ then g also has behaviour Λ and we are done. To show the existence of such a map, by the ω-categoricity of $\mathfrak{B}$ and Lemma 4.1.10 it suffices to show that for every finite substructure $\mathfrak{S}$ of $\mathfrak{A}$ there exists a function $h: S \to B$ such that for every $a \in S^n$ we have that $\mathrm{tp}^{\mathfrak{B}}(h(a)) = \Delta(\mathrm{tp}^{\mathfrak{A}}(a))$. The existence of such a function can be shown similarly as in the proof of Theorem 10.5.10 because Δ respects the compatibility and satisfiability constraints. ⊣

[28] An alternative presentation would have been to add a binary relation symbol for equality and to add structures to $\mathcal{F}$ that code the Leibniz' laws for equality.

11.6.2. Decidability of definability. Let $\mathfrak{B}$ be a finitely bounded homogeneous structure. We will be interested in the following computational problem.

FO-Def
Instance: A finitely bounded homogeneous structure $\mathfrak{B}$ and two finite-signature first-order reducts $\mathfrak{A}_1$ and $\mathfrak{A}_2$ of $\mathfrak{B}$.
Question: Is $\mathfrak{A}_1$ a first-order reduct of $\mathfrak{A}_2$?

We are also interested in the variants of this problem where we replace first-order definability by other syntactically restricted versions of definability, in particular by primitive positive definability. The corresponding computational problem for primitive positive definability is denoted by PP-Def($\mathfrak{B}$), and the problem for existential and existential positive definability by EX-Def($\mathfrak{B}$) and EP-Def($\mathfrak{B}$), respectively.

Example 11.6.2. We can use an algorithm for PP-Def to decide whether all polymorphisms of a first-order reduct $\mathfrak{A}_1$ of a homogeneous finitely bounded structure $\mathfrak{B}$ are essentially unary. For that, we simply apply the algorithm to $\mathfrak{A}_1$ and $\mathfrak{A}_2$ where $\mathfrak{A}_2$ only contains the ternary relation defined by $x = y \vee y = z$ (Proposition 6.1.19).

For *finite* structures $\mathfrak{B}$ the problem PP-Def is co-NEXPTIME-complete [343]. To decide the problem FO-Def for finite structures $\mathfrak{B}$ it suffices to test for every relation of $\mathfrak{A}_1$ whether it is preserved by the automorphisms of $\mathfrak{A}_2$ (Theorem 4.2.9); since a finite structure has only finitely many automorphisms, this task can be performed algorithmically. Note that a relation has a first-order definition in a finite structure $\mathfrak{A}_2$ if and only if it has an existential definition in $\mathfrak{A}_2$ (since all finite structures are model complete) and hence FO-Def and EX-Def are the same problems for finite $\mathfrak{B}$. The problem EP-Def is for finite $\mathfrak{B}$ also easy to decide, replacing the automorphisms of $\mathfrak{B}$ with the endomorphisms of $\mathfrak{B}$ (Proposition 4.4.1).

Example 11.6.3. We can use an algorithm for EP-Def to test whether a first-order reduct $\mathfrak{A}$ of a finitely bounded homogeneous structure $\mathfrak{B}$ is a core. Recall from Section 2.6 and in particular from Proposition 2.6.10 that a τ-structure $\mathfrak{A}$ is a core if and only if for every atomic τ-formula ψ the formula $\neg\psi$ has an existential positive definition in $\mathfrak{A}$. We apply the algorithm for EP-Def to the structures $\mathfrak{A}_1 := \mathfrak{B}$ and the first-order reduct $\mathfrak{A}_2$ of $\mathfrak{B}$ that contains for every atomic formula ϕ of $\mathfrak{A}$ the relation defined by $\neg\phi$.

The main result of this section is the decidability of PP-Def if $\mathfrak{B}$ is a first-order reduct of a homogeneous finitely bounded ordered Ramsey structure. Even if we restrict $\mathfrak{B}$ to be the simplest of all countable structures $\mathfrak{B} = (\mathbb{N}; =)$, which is a first-order reduct of the homogeneous finitely bounded ordered Ramsey structure $(\mathbb{Q}; <)$, the decidability of PP-Def is not obvious (and has been posed as an open problem [54]).

THEOREM 11.6.4 (Theorem 1 in [96]). PP-Def *is decidable if* $\mathfrak{B}$ *is a first-order reduct of a homogeneous finitely bounded ordered Ramsey structure.*

PROOF. It clearly suffices to show the statement for the situation that $\mathfrak{B}$ itself is a homogeneous finitely bounded ordered Ramsey structure; let $\mathcal{F}$ be a finite set of finite structures such that $\text{Forb}^{\text{emb}}(\mathcal{F}) = \text{Age}(\mathfrak{B})$. Let $\mathfrak{A}_1$ and $\mathfrak{A}_2$ be two first-order reducts of $\mathfrak{B}$; we have to decide whether every relation of $\mathfrak{A}_2$ is primitively positively definable in $\mathfrak{A}_1$. Let R be a relation of $\mathfrak{A}_2$ and let ϕ be a first-order definition of R in $\mathfrak{B}$; since $\mathfrak{B}$ is homogeneous we may assume that ϕ is quantifier-free (Corollary 4.3.3). By Theorem 11.4.17, the relation R does not have a primitive positive definition in $\mathfrak{B}$ if and only if there exists a $d \in \mathbb{N}$, a d-ary $f \in \text{Pol}(\mathfrak{A}_1)$, and m-tuples $t^1, \dots, t^d \in R$ such that $f(t^1, \dots, t^d) \notin R$, and f is canonical with respect to $((\mathfrak{B}^{(d)}, (t_1^1, \dots, t_1^d), \dots, (t_m^1, \dots, t_m^d)), \mathfrak{B})$. It can be expressed as a behaviour between $(\mathfrak{B}^{(d)}, (t_1^1, \dots, t_1^d), \dots, (t_m^1, \dots, t_m^d))$ and $\mathfrak{B}$ whether a map $f : B^d \to B$ satisfies these conditions.

Recall that the algebraic power $\mathfrak{B}^{(d)}$ is homogeneous (Proposition 4.2.19) and Ramsey (Theorem 11.2.14) and a linear order $<$ is first-order definable in $\mathfrak{B}^{(d)}$. It follows that $(\mathfrak{B}^{(d)}, <, (t_1^1, \dots, t_1^d), \dots, (t_m^1, \dots, t_m^d))$ is homogeneous and Ramsey (Corollary 11.2.16). There exists a structure with finite relational signature and the same automorphism group as this structure (Proposition 4.3.12). Therefore, we can use the algorithm from Theorem 11.6.1 to test whether this behaviour can be realised. ⊣

While the assumptions for the structure $\mathfrak{B}$ in Theorem 11.6.4 might appear rather restrictive at first sight, they actually are quite general: we want to point out that we only require that $\mathfrak{B}$ is a *first-order reduct* of a homogeneous finitely bounded ordered Ramsey structure. If Conjecture 11.1 is true then every homogeneous finitely bounded structure satisfies the assumptions.

COROLLARY 11.6.5 ([96]). *The problems* EP-Def *and* EX-Def *are decidable if the input structure* $\mathfrak{B}$ *is a homogeneous finitely bounded ordered Ramsey structure.*

PROOF. Suppose that $\mathfrak{A}_1$ and $\mathfrak{A}_2$ of $\mathfrak{B}$. The result for EP-Def follows by applying the algorithm from Theorem 11.6.4 to the first-order reduct $(\mathfrak{A}_1, P^3_{A_1})$ and $\mathfrak{A}_2$ (see Proposition 6.1.19). The result for EX-Def can be reduced to EP-Def by expanding $\mathfrak{A}_1$ with relations for the negation of each atomic relation in $\mathfrak{A}_1$ (the same idea has been used in Example 11.6.3). ⊣

An important open problem is whether the method can be extended to show decidability of FO-Def for homogeneous finitely bounded Ramsey structures $\mathfrak{B}$. First-order definability is characterised by preservation under automorphisms (Theorem 4.1.6), i.e., surjective self-embeddings. But the requirement of surjectivity is difficult to deal with in our approach.

In order to formulate some negative results, we consider the restriction of the problems FO-Def, EX-Def, EP-Def, and FO-Def where we *fix* the homogeneous finitely bounded structure $\mathfrak{B}$ in the input to these problems, and study the decidability of the problems depending on $\mathfrak{B}$. The resulting computational problems will be denoted by FO-Def($\mathfrak{B}$), EX-Def($\mathfrak{B}$), EP-Def($\mathfrak{B}$), and FO-Def($\mathfrak{B}$), respectively.

We mention that Szymon Toruńczyk (personal communication) has observed that FO-Def($\mathfrak{B}$) is computationally equivalent to the problem of deciding for given first-order reducts $\mathfrak{A}_1$ and $\mathfrak{A}_2$ of $\mathfrak{B}$ whether $\mathfrak{A}_1$ and $\mathfrak{A}_2$ are isomorphic. Note that it follows from Theorem 11.6.1 that it is decidable whether there exists an embedding from $\mathfrak{A}_1$ to $\mathfrak{A}_2$ and an embedding from $\mathfrak{A}_2$ to $\mathfrak{A}_1$; but for infinite structures this of course does not imply that $\mathfrak{A}_1$ and $\mathfrak{A}_2$ are isomorphic.

The assumption that $\mathfrak{B}$ is finitely bounded is motivated by the following result.

PROPOSITION 11.6.6. *There exists a homogeneous ordered Ramsey structure $\mathfrak{B}$ with finite relational signature such that each of* PP-Def($\mathfrak{B}$), EP-Def($\mathfrak{B}$), EX-Def($\mathfrak{B}$), *and* FO-Def($\mathfrak{B}$) *are undecidable.*

PROOF. Recall the definition of *Henson digraphs* from Example 2.3.12. We first show that non-isomorphic Henson digraphs $\mathfrak{C}_1$ and $\mathfrak{C}_2$ have distinct PP-Def problems. In fact, we show the existence of a first-order formula ϕ_1 over digraphs such that the input with $\phi_0 := E(x, y)$ and ϕ_1 is a yes-instance of PP-Def($\mathfrak{C}_1$) and a no-instance of PP-Def($\mathfrak{C}_2$), or vice versa. Since there are uncountably many Henson digraphs, but only countably many algorithms, this clearly shows the existence of Henson digraphs $\mathfrak{C}$ such that PP-Def($\mathfrak{C}$) is undecidable.

Since $\mathfrak{C}_1$ and $\mathfrak{C}_2$ are non-isomorphic, there must be a structure $\mathfrak{A}$ that embeds to $\mathfrak{C}_1$ but not to $\mathfrak{C}_2$, or that embeds to $\mathfrak{C}_2$ but not to $\mathfrak{C}_1$. Assume the former is the case; in the latter, simply exchange $\mathfrak{C}_1$ and $\mathfrak{C}_2$. Let s be the number of elements of $\mathfrak{A}$, and denote the elements by $a_1, \dots, a_s$. Let ψ be the formula with variables $x_1, \dots, x_s$ that has for distinct $i, j \leq s$ a conjunct $E(x_i, x_j)$ if $E(a_i, a_j)$ holds in $\mathfrak{A}$, and a conjunct $\neg E(x_i, x_j) \wedge x_i \neq x_j$ otherwise. Let ϕ be the formula $\psi \Rightarrow E(x_{s+1}, x_{s+2})$. Consider the relation $R_1 \subseteq (C_1)^{s+2}$ defined by ϕ in $\mathfrak{C}_1$. Let R be a relation symbol of arity $s + 2$, and $\mathfrak{B}$ be the structures with signature $\{R\}$, domain C_1, and where R denotes the relation R_1. It is clear that $\exists x_1, \dots, x_s \colon R(x_1, \dots, x_s, x, y)$ is a primitive positive definition of $E(x, y)$ in $\mathfrak{B}$. Now consider the relation R_2 defined by ϕ in $\mathfrak{C}_2$. Since $\mathfrak{A}$ does not embed into $\mathfrak{C}_2$, the precondition of ϕ is never satisfied, and the relation R_2 is empty. Hence, the structure $(C_2; R_2)$ is preserved by all permutations. But the relation $E(x, y)$ is certainly not first-order definable over a structure that is preserved by all permutations.

Note that every Henson graph has a generic superposition $\mathfrak{C}$ with $(\mathbb{Q}; <)$ (Definition 2.3.22), and that the generic superposition is (ordered) Ramsey by the results described in Example 11.1.10. Clearly, PP-Def($\mathfrak{C}$) is undecidable as well. The same proof shows that EP-Def($\mathfrak{C}$), EX-Def($\mathfrak{C}$), and FO-Def($\mathfrak{C}$) are undecidable as well. ⊣

We mention that the application of the Henson digraphs in the proof of Proposition 11.6.6 is quite close to the original motivation of Henson for introducing his digraphs, namely proving the existence of an undecidable, ω-categorical theory in the signature of digraphs [200].

11.6.3. Testing the existence of pseudo-Siggers polymorphisms. Another application of the techniques from this chapter concern the existence of pseudo-Siggers polymorphisms of homogeneous finitely bounded ordered Ramsey structures. The same technique can be used to test the existence of polymorphisms satisfying a given pseudo-minor condition.

THEOREM 11.6.7. *There is an algorithm that decides whether a given homogeneous finitely bounded Ramsey structure $\mathfrak{B}$ has a pseudo-Siggers polymorphism.*

PROOF. Recall that Lemma 10.1.14 implies that if $\mathfrak{B}$ has a pseudo-Siggers polymorphism s then every operation in

$$S := \overline{\{(x_1,\dots,x_6) \mapsto \gamma s(\beta(x_1),\dots,\beta(x_6)) \mid \beta,\gamma \in \operatorname{End}(\mathfrak{B})\}}$$

is a pseudo-Siggers operation, too. Let $\mathfrak{D} := \mathfrak{B}^{[d]}$ be the finite-signature structure defined as in the proof of Lemma 3.5.4, which is Ramsey by Corollary 11.2.4. The canonisation lemma (Lemma 11.4.9) applied to $\mathfrak{D}$ implies that S contains a diagonally canonical operation. Hence, $\mathfrak{B}$ has a pseudo-Siggers polymorphism if and only if it has a pseudo-Siggers polymorphism which is diagonally canonical with respect to $\mathfrak{B}$. Whether a function $B^d \to B$ is diagonally canonical with respect to $\mathfrak{B}$ and whether it is a polymorphism of $\mathfrak{B}$ can be formulated by some behaviour Λ, and so we may use the algorithm from Theorem 11.6.1 (where $\mathfrak{B}$ is part of the input) to test whether such an operation exists. ⊣

See Section 14.2.11 for a series of related meta-problems whose decidability is open.

Chapter 12

TEMPORAL CONSTRAINT SATISFACTION PROBLEMS

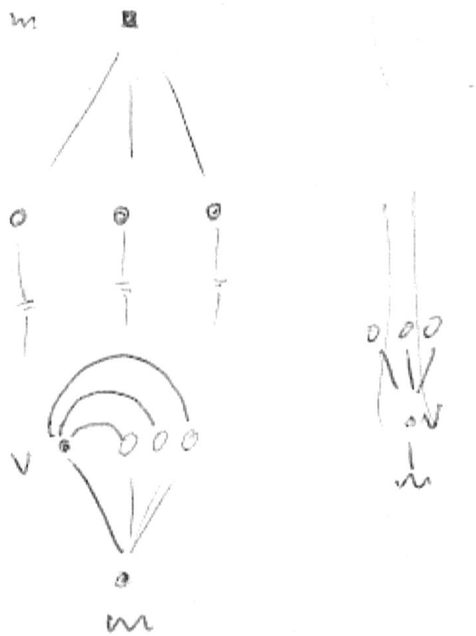

Martin Kutz, 2001

This chapter treats first-order reducts of $(\mathbb{Q}; <)$. When studying the constraint satisfaction problems for such reducts, we think of the set of rational numbers $\mathbb{Q}$ as time points, and therefore refer to such reducts as *temporal constraint languages*, to their relations as *temporal relations*, and to the corresponding CSPs as *temporal CSPs*. We have already discussed some temporal constraint languages earlier in this text: for instance and/or precedence constraints from scheduling (Section 1.6.8), and Ord-Horn constraints on time points (Section 1.6.9).

There are also several famous NP-complete temporal CSPs. For example the Betweenness Problem [176], which has been introduced in Example 1.1.3

as a CSP with domain $\mathbb{Z}$, can also be formulated as CSP($\mathbb{Q}$; Betw) where

$$\text{Betw} := \{(x, y, z) \in \mathbb{Q}^3 \mid (x < y < z) \vee (z < y < x)\}.$$

We have seen in Proposition 3.1.10 that this CSP is NP-hard. Similarly, the Cyclic Ordering Problem [176] can be formulated as the CSP for ($\mathbb{Q}$; Cycl) where

$$\text{Cycl} := \{(x, y, z) \mid (x < y < z) \vee (y < z < x) \vee (z < x < y)\}$$

and is also NP-complete [174] (an NP-hardness proof using primitive positive interpretations can be found in Section 12.2).

A subclass of temporal CSPs called *ordering CSPs* has been introduced in [190]. An ordering CSP is a temporal CSP whose constraint language only contains injective relations (i.e., the entries of all tuples of relations in the language are pairwise distinct; for example, CSP($\mathbb{Q}$; $\leq, \neq$) is *not* an ordering CSP). Satisfiability thresholds for *random instances* of ordering CSPs have been studied in [181], and *approximability* of ordering CSPs has been studied in [189].

The class of temporal constraint languages is of fundamental importance for infinite domain constraint satisfaction, since CSPs for such languages appear as important special cases in several other classes of CSPs that have been studied, e.g., constraint languages about branching time, partially ordered time, spatial reasoning, and set constraints [153, 220, 104]. Moreover, several polynomial-time solvable classes of constraint languages on *time intervals* [288, 248, 154] can be solved by translation into polynomial-time solvable temporal constraint languages (see Section 3.4).

The class of first-order reducts of ($\mathbb{Q}$; $<$) plays a fundamental role in the theory of infinite permutation groups: we mention the result of Cameron [121] that the highly set-transitive closed subgroups of the full symmetric group on a countably infinite set are precisely those permutation groups that are isomorphic (as permutation groups) to the automorphism groups of first-order reducts of ($\mathbb{Q}$; $<$). The structure ($\mathbb{Q}$; $<$) is also of fundamental importance in model theory. We mention for example the recent result of Pierre Simon [331] that an ω-categorical *NIP structure* $\mathfrak{B}$ is unstable if and only if ($\mathbb{Q}$; $<$) has a first-order interpretation in $\mathfrak{B}$.

In this chapter we prove a complete classification of the computational complexity of CSP($\mathfrak{B}$) for temporal constraint languages $\mathfrak{B}$. The classification confirms the general infinite-domain tractability conjecture for first-order reducts of finitely bounded homogeneous structures (Conjecture 3.1 and Conjecture 4.1).

THEOREM 12.0.8. *Let $\mathfrak{B}$ be a structure with a first-order definition in* ($\mathbb{Q}$; $<$). *Then exactly one of the following two cases applies.*

1. *$\mathfrak{B}$ has an at most ternary weak near-unanimity polymorphism modulo endomorphisms. In this case,* CSP($\mathfrak{B}'$) *is in P for every finite-signature reduct $\mathfrak{B}'$ of $\mathfrak{B}$.*
2. *All finite structures have a primitive positive interpretation with parameters in $\mathfrak{B}$. In this case, $\mathfrak{B}$ has a finite-signature reduct $\mathfrak{B}'$ such that* CSP($\mathfrak{B}'$) *is NP-hard by Corollary* 3.1.6.

This classification is a slight strengthening of the statement of the infinite-domain tractability conjecture, in two respects:

- In item (2), instead of primitive positive interpretability with parameters in the model-complete core of $\mathfrak{B}$ we directly get primitive positive interpretability with parameters in $\mathfrak{B}$.
- Instead of a pseudo-Siggers operation, we even get an at most ternary pseudo weak near-unanimity operation.

On the other hand, we mention that the pseudo weak near-unanimity operations that describe the polynomial-time tractable cases cannot be chosen to be canonical with respect to $(\mathbb{Q}; <)$. For the same reason, the polynomial-time algorithms in this section cannot be obtained by the reduction to finite-domain CSPs from Section 10.5. Our proof is based on the universal-algebraic approach and Ramsey theory as described in Chapter 6 and Chapter 11. This chapter contains and extends results from [69, 70, 88, 55].

Notation. All clones considered in this chapter contain all automorphisms of $(\mathbb{Q}; <)$; hence, for $f, g \in \mathcal{O}_{\mathbb{Q}}$ say that f *locally generates* g if g is in the local closure of $\langle \text{Aut}(\mathbb{Q}; <) \cup \{f\} \rangle$. Similarly, if $S \subseteq \mathcal{O}_{\mathbb{Q}}$ we say that S *locally generates* g if g is in the local closure of $\langle \text{Aut}(\mathbb{Q}; <) \cup S \rangle$.

Let $\mathbb{Q}^+$ denote the set of all positive rational numbers, and let $\mathbb{Q}_0^-$ denote $\mathbb{Q} \setminus \mathbb{Q}^+$. It will be convenient to use the notation $U < V$ if $U, V \subseteq \mathbb{Q}$ are such that $u < v$ for all $u \in U$ and $v \in V$. Likewise, if $u \in U$ we use the notation $u < V$ if $u < v$ for all $v \in V$. The notation $U < v$ is defined analogously.

The unary operation $-$ is defined as $-(x) := -x$ in the usual sense. Note that if $f : \mathbb{Q} \to \mathbb{Q}$ is canonical with respect to $(\mathbb{Q}; <)$ then f must have the same behaviour as $-$, as $\text{id}_{\mathbb{Q}}$, or as a constant map. This holds because $(\mathbb{Q}; <)$ is homogeneous over a binary signature; hence, the canonical functions are determined by their 2-type conditions, and only finitely many different behaviours need to be checked. For an introduction to canonical functions, see Section 11.4.4.

DEFINITION 12.0.9. Let f be a k-ary operation on $\mathbb{Q}$. Then the operation given by $-f(-x_1, \ldots, -x_k)$ is called the *dual* of f.

Note that if f preserves an m-ary relation R, then the dual of f preserves the relation $-R$ defined as

$$\{(-a_1, \ldots, -a_m) \mid (a_1, \ldots, a_m) \in R\}.$$

Clearly, CSP($\mathbb{Q}; R_1, \dots, R_k$) and CSP($\mathbb{Q}; -R_1, \dots, -R_k$) are exactly the same computational problem.

12.1. Endomorphisms and Cameron's theorem

Cameron [121] classified temporal constraint languages up to first-order interdefinability. In this section we present a refinement of this result, Theorem 12.1.7, which provides a classification of the model-complete cores of temporal constraint languages, up to interdefinability (and isomorphism). Our proof is based on canonical functions and the tools from Chapter 11. We follow a bottom-up strategy and use canonical functions with respect to expansions of $(\mathbb{Q}; <)$ by constants. The same strategy has been used successfully to classify the first-order reducts of

- the random graph [89], rederiving a result of Simon Thomas [339];
- the countable universal homogeneous poset [297] (Example 2.3.11);
- the expansion of the Henson graphs by a constant [307];
- the countable universal homogeneous ordered graph [93];
- the random permutation [265] (Example 2.3.23);
- the countable universal homogeneous binary branching C-relation [63] (Section 5.1.2);
- the countable model of the model companion theory of the theory of semilinear orders [49] (Section 5.2.2);
- the Henson digraphs [7] (Example 2.3.12).

For $x_1, \dots, x_n \in \mathbb{Q}$ write $\overrightarrow{x_1 \cdots x_n}$ if $x_1 < \cdots < x_n$.

THEOREM 12.1.1. (Relational version of Cameron's theorem; see e.g. [223]). *Let $\mathfrak{B}$ be a first-order reduct of $(\mathbb{Q}; <)$. Then $\mathfrak{B}$ is first-order interdefinable with exactly one out of the following five homogeneous structures.*

1. *The dense linear order $(\mathbb{Q}; <)$ itself,*
2. *The structure $(\mathbb{Q}; \mathrm{Betw})$, where* Betw *is the ternary relation*
$$\{(x, y, z) \in \mathbb{Q}^3 \mid \overrightarrow{xyz} \vee \overrightarrow{zyx}\},$$
3. *The structure $(\mathbb{Q}; \mathrm{Cycl})$, where* Cycl *is the ternary relation*
$$\{(x, y, z) \mid \overrightarrow{xyz} \vee \overrightarrow{yzx} \vee \overrightarrow{zxy}\},$$
4. *The structure $(\mathbb{Q}; \mathrm{Sep})$, where* Sep *is the 4-ary relation*
$$\{(x_1, y_1, x_2, y_2) \mid \overrightarrow{x_1x_2y_1y_2} \vee \overrightarrow{x_1y_2y_1x_2} \vee \overrightarrow{y_1x_2x_1y_2} \vee \overrightarrow{y_1y_2x_1x_2} \vee \overrightarrow{x_2x_1y_2y_1} \vee \overrightarrow{x_2y_1y_2x_1} \vee \overrightarrow{y_2x_1x_2y_1} \vee \overrightarrow{y_2y_1x_2x_1}\},$$
5. *The structure $(\mathbb{Q}; =)$.*

The relation Sep is the so-called *Separation Relation*; note that $\text{Sep}(x_1, y_1, x_2, y_2)$ holds for elements $x_1, y_1, x_2, y_2 \in \mathbb{Q}$ iff all four points x_1, y_1, x_2, y_2 are distinct and the smallest interval over $\mathbb{Q}$ containing x_1, y_1 properly overlaps with the smallest interval containing x_2, y_2 (where properly overlaps means that the two intervals have a non-empty intersection, but none of the intervals contains the other). Note that $-$ preserves Betw and Sep, but does not preserve $<$ and Cycl.

We now define an important operation on $\mathbb{Q}$. Let c be any irrational number, and let e be any order-preserving bijection between $(-\infty, c)$ and (c, ∞). Then the operation $\circlearrowright$ is defined by $e(x)$ for $x < c$ and by $e^{-1}(x)$ for $x > c$. Note that $\circlearrowright$ preserves Cycl and Sep, but does not preserve $<$ and Betw. With this operations and the observations above, Cameron's theorem follows from the following.

THEOREM 12.1.2. (Operational version of Cameron's theorem; see e.g. [223]). *Let $\mathfrak{B}$ be a temporal constraint language. Then exactly one of the following holds.*

1. $\text{Aut}(\mathfrak{B}) = \text{Aut}(\mathbb{Q}; <)$;
2. $\text{Aut}(\mathfrak{B}) = \overline{\langle \text{Aut}(\mathbb{Q}; <) \cup \{-\} \rangle}$;
3. $\text{Aut}(\mathfrak{B}) = \overline{\langle \text{Aut}(\mathbb{Q}; <) \cup \{\circlearrowright\} \rangle}$;
4. $\text{Aut}(\mathfrak{B}) = \overline{\langle \text{Aut}(\mathbb{Q}; <) \cup \{-, \circlearrowright\} \rangle}$;
5. $\text{Aut}(\mathfrak{B}) = \text{Sym}(\mathbb{Q})$.

Our bottom-up strategy starts with an analysis of first-order reducts $\mathfrak{B}$ of $(\mathbb{Q}; <)$ such that $<$ is *not* existentially positively definable in $\mathfrak{B}$.

PROPOSITION 12.1.3. *Let $\mathfrak{B}$ be a first-order reduct of $(\mathbb{Q}; <)$ such that the relation $<$ is not existentially positively definable in $\mathfrak{B}$. Then $\mathfrak{B}$ is preserved by $-$, by $\circlearrowright$, or by a constant operation.*

PROOF. Theorem 11.4.16 implies that there are $s, t \in \mathbb{Q}$, $e \in \text{End}(\mathfrak{B})$ such that

- $s < t$ and $e(t) \leq e(s)$;
- e is canonical with respect to $((\mathbb{Q}; <, s, t), (\mathbb{Q}; <))$.

If e is not injective then Lemma 4.4.6 implies that $\mathfrak{B}$ has a constant endomorphism and we are done; so we assume in the following that e is injective and in particular that $e(t) < e(s)$. Note that the substructure of $(\mathbb{Q}; <)$ induced on each of the three infinite orbits O_1, O_2, O_3 under $\text{Aut}(\mathbb{Q}; <, s, t)$ is an isomorphic copy of $(\mathbb{Q}; <)$. Also note that the canonicity of the operation e with respect to $((\mathbb{Q}; <, s, t), (\mathbb{Q}; <))$ implies that the restriction of e to O_i for some $i \in \{1, 2, 3\}$ is either order-preserving or order-reversing. If e is order-reversing on O_i, then e interpolates $-$ modulo $\text{Aut}(\mathbb{Q}; <)$ and hence $\mathfrak{B}$ is preserved by $-$. So we may suppose that e is order-preserving on each of O_1, O_2, O_3.

If $i, j \in \{1, 2, 3\}$ are such that $O_i < O_j$ and $e(O_j) < e(O_i)$, then e interpolates $\circlearrowright$ modulo $\mathrm{Aut}(\mathbb{Q}; <)$ and hence $\mathfrak{B}$ is preserved by $\circlearrowright$. So we may assume that $e(O_1) < e(O_2) < e(O_3)$, in which case the canonicity of e implies that e is order-preserving on $O_1 \cup O_2 \cup O_3$.

If the restriction of e to $S := \{x \in \mathbb{Q} \mid s < x\}$ were order-preserving and the restriction of e to $T := \{x \in \mathbb{Q} \mid x < t\}$ were order-preserving, then we would have $e(s) < e(t)$, contrary to our assumptions. So suppose without loss of generality that the restriction to S is not order-preserving. By canonicity $e(t) < e(S \setminus \{t\})$ or $e(S \setminus \{t\}) < e(t)$. We only consider the first case since the second case can be treated analogously.

Claim. $\circlearrowright \in \mathrm{Aut}(\mathfrak{B})$. We use the fact that $\mathrm{Aut}(\mathfrak{B})$ is closed in $\mathrm{Sym}(\mathbb{Q})$ and verify that for every finite $F \subseteq \mathbb{Q}$ the operation e and the automorphisms of $(\mathbb{Q}; <)$ generate an operation f such that $f|_F = \circlearrowright|_F$ (here, local interpolation is not enough). Let $F = \{u_1, \ldots, u_n\}$ be such that $u_1 < \cdots < u_n$. Let c be the constant from the definition of $\circlearrowright$. If $u_i > c$ for all $i \in \{1, \ldots, n\}$ then we can pick an automorphism of $(\mathbb{Q}; <)$ whose restriction to F equals $\circlearrowright|_F$. Otherwise, let $i \in \{1, \ldots, n\}$ be maximal such that $u_i < c$. Choose $\alpha_1 \in \mathrm{Aut}(\mathbb{Q}; <)$ that maps F to S and $\max(F)$ to t; such an α_1 exists by the homogeneity of $(\mathbb{Q}; <)$. Define $F_1 := e(\alpha_1(F))$. Now suppose inductively that F_j has already been defined for $j \geq 1$. Choose $\alpha_{j+1} \in \mathrm{Aut}(\mathbb{Q}; <)$ that maps F_j to S and $\max(F_j)$ to t. Let $f' := e \circ \alpha_{n-i} \cdots e \circ \alpha_1$. Then

$$f'(u_{i+1}) < f'(u_{i+2}) < \cdots < f'(u_n) < f'(u_1) < f'(u_2) < \cdots < f'(u_i)$$

and hence we find $\alpha \in \mathrm{Aut}(\mathbb{Q}; <)$ such that $\alpha f'(u_j) = \circlearrowright(u_j)$ for all $j \in \{1, \ldots, n\}$. ⊣

PROPOSITION 12.1.4. *Let $\mathfrak{B}$ be a first-order reduct of $(\mathbb{Q}; <)$ preserved by $-$. Then $\mathrm{End}(\mathfrak{B}) = \overline{\langle \mathrm{Aut}(\mathfrak{B}) \cup \{-\} \rangle}$ or $\mathfrak{B}$ is preserved by $\circlearrowright$ or by a constant operation.*

PROOF. If $\mathrm{End}(\mathfrak{B}) \neq \overline{\langle \mathrm{Aut}(\mathfrak{B}) \cup \{-\} \rangle}$ then there exists an $e \in \mathrm{End}(\mathfrak{B})$ and $t \in B^n$ for some $n \in \mathbb{N}$ such that $e(t)$ lies in a different orbit than t and than $-(t)$ under $\mathrm{Aut}(\mathbb{Q}; <)$. Again we may assume by Lemma 4.4.6 that e is injective. Theorem 11.4.16 implies that we may assume that e is canonical with respect to $((\mathbb{Q}; <, t_1, \ldots, t_n), (\mathbb{Q}; <))$.

If for some $i, j \in \{1, \ldots, n\}$ we have $O_i < O_j$ and $e(O_j) < e(O_i)$ and e is order-preserving on O_i or on O_j then one can show that $\mathfrak{B}$ is preserved by $\circlearrowright$ by a similar argument as in the final paragraph of the proof of Proposition 12.1.3; so in this case we are done. Similarly, we are done if for some $i \in \{1, \ldots, n\}$ there exists an infinite orbit O under $\mathrm{Aut}(\mathbb{Q}; <, t_1, \ldots, t_n)$ such O is preserved by e and ($O < t_i$ and $e(t_i) < e(O)$) or ($t_i < O$ and $e(O) < e(t_i)$). We are also done if $- \circ e \in \mathrm{End}(\mathfrak{B})$ satisfies the conditions above. Otherwise, one can conclude that e and $-$ have the same behaviour on all of $\mathbb{Q}$, in contradiction to the assumption that $e(t)$ lies in a different orbit than t and than $-(t)$. ⊣

PROPOSITION 12.1.5. *Let $\mathfrak{B}$ be a first-order reduct of $(\mathbb{Q};<)$ preserved by $\circlearrowright$. Then* $\mathrm{End}(\mathfrak{B}) = \overline{\langle \mathrm{Aut}(\mathfrak{B}) \cup \{\circlearrowright\} \rangle}$ *or $\mathfrak{B}$ is preserved by $-$ or by a constant operation.*

PROOF. If $\mathrm{End}(\mathfrak{B}) \neq \overline{\langle \mathrm{Aut}(\mathfrak{B}) \cup \{\circlearrowright\} \rangle}$ there exists an $e \in \mathrm{End}(\mathfrak{B})$ and $t \in B^n$ for some $n \in \mathbb{N}$ such that $e(t)$ lies in a different orbit than t and than $\circlearrowright(t)$ under $\mathrm{Aut}(\mathbb{Q};<)$. Again we may assume by Lemma 4.4.6 that e is injective. Theorem 11.4.16 implies that we also may assume that e is canonical with respect to $((\mathbb{Q};<,t_1,\dots,t_n),(\mathbb{Q};<))$. In the following, orbits always refer to the permutation group $\mathrm{Aut}(\mathbb{Q};<,t_1,\dots,t_n)$. If for some $i \in \{1,\dots,n\}$ there exists an infinite orbit O such that e is order-reversing on O then e interpolates $-$ modulo $\mathrm{Aut}(\mathbb{Q};<)$. Hence, $\mathfrak{B}$ is preserved by $-$ and we are done. So we may assume that e preserves the order on each infinite orbit.

If for every $i \in \{1,\dots,n\}$ and every infinite orbit O it holds that $t_i < O$ if and only if $e(t_i) < e(O)$ then $e(t)$ lies in the same orbit as t, contrary to our assumptions. Otherwise, by composing e with $\circlearrowright$ and automorphisms of $(\mathbb{Q};<)$ we may assume that there are $i \in \{1,\dots,n\}$, infinite orbits O_1, O_2 such that $O_1 < t_i < O_2$ and $e(t_i) < e(O_1) < e(O_2)$.

Claim. $- \in \mathrm{Aut}(\mathfrak{B})$. We use that $\mathrm{Aut}(\mathfrak{B})$ is closed in $\mathrm{Sym}(\mathbb{Q})$ and verify for every finite $F \subseteq \mathbb{Q}$ there exists $f \in \overline{\langle \mathrm{Aut}(\mathbb{Q};<) \cup \{e\} \rangle}$ such that $f|_F = -|_F$. Let $\{u_1,\dots,u_n\} = F$ be such that $u_1 < \dots < u_n$. Choose $\alpha_1 \in \mathrm{Aut}(\mathbb{Q};<)$ that maps F to $O_1 \cup \{u_i\} \cup O_2$ such that $\alpha_1(u_2) = u_i$. Define $e_1 := e \circ \alpha_1$ and $F_1 := e_1(F)$. Suppose inductively that e_j and F_j has already been defined for $j \geq 1$. Choose $\alpha_{j+1} \in \mathrm{Aut}(\mathbb{Q};<)$ that maps F_j to S and $e_j(u_{j+1})$ to u_i. Then $e_{n-1}(u_1) > e_{n-1}(u_2) > \dots > e_{n-1}(u_n)$ and hence we can find an $\alpha \in \mathrm{Aut}(\mathbb{Q};<)$ such that $\alpha(e_{n-1}(u_j)) = -(u_j)$ for all $j \in \{1,\dots,n\}$. $\dashv$

The following proposition is the final (and most complex) step in our bottom-up strategy where we 'reach $\mathrm{Sym}(\mathbb{Q})$'.

PROPOSITION 12.1.6. *Let $\mathfrak{B}$ be a first-order reduct of $(\mathbb{Q};<)$ preserved by $\{\circlearrowright,-\}$. Then* $\mathrm{End}(\mathfrak{B}) = \overline{\langle \mathrm{Aut}(\mathfrak{B}) \cup \{\circlearrowright,-\} \rangle}$ *or $\mathfrak{B}$ is preserved by* $\mathrm{Sym}(\mathbb{Q})$ *or by a constant.*

PROOF. If $\mathrm{End}(\mathfrak{B}) \neq \overline{\langle \mathrm{Aut}(\mathfrak{B}) \cup \{\circlearrowright,-\} \rangle}$ there exists an $e \in \mathrm{End}(\mathfrak{B})$ and $t \in B^n$ for some $n \in \mathbb{N}$ such that $e(t)$ lies in a different orbit than t, $\circlearrowright(t)$, and $-(t)$ with respect to $\mathrm{Aut}(\mathbb{Q};<)$. Again we may assume by Lemma 4.4.6 that e is injective. Theorem 11.4.16 implies that we may also assume that e is canonical with respect to $((\mathbb{Q};<,t_1,\dots,t_n),(\mathbb{Q};<))$. Orbits will in the following be always with respect to the permutation group $\mathrm{Aut}(\mathbb{Q};<,t_1,\dots,t_n)$.

To show that $\mathrm{Aut}(\mathfrak{B}) = \mathrm{Sym}(\mathbb{Q})$ it suffices to show that for every finite $F \subseteq \mathbb{Q}$ and every permutation of F there is an operation generated by $\mathrm{Aut}(\mathbb{Q}) \cup \{e,-,\circlearrowright\}$ whose restriction to F coincides with that permutation, because $\mathrm{Aut}(\mathfrak{B})$ is closed. Since every permutation of a finite set is generated by transpositions, it suffices to show this for a transposition of F, i.e., a permutation α such that there exist two distinct elements $u, v \in F$ with $\alpha(u) = v$, $\alpha(v) = u$, and $\alpha(w) = w$ for all other $w \in F$. In each of the following three claims we verify this condition. We always assume without loss of generality that $u < v$ and use the following notation:

$$F_1 := \{x \in F \mid x < u\}$$
$$F_2 := \{x \in F \mid u < x < v\}$$
$$F_3 := \{x \in F \mid v < x\}.$$

So we have $F_1 < u < F_2 < v < F_3$.

Claim 1. If there are infinite orbits O_1 and O_2 such that e reverses the order on O_1 and preserves the order on O_2 then $\text{Aut}(\mathfrak{B}) = \text{Sym}(\mathbb{Q})$. By composing e with $-$, automorphisms of $(\mathbb{Q}; <)$, and $\circlearrowright$ we may suppose that $O_1 < O_2$ and $e(O_1) < e(O_2)$. Let $\beta_1 \in \text{Aut}(\mathbb{Q}; <)$ be such that $\beta_1(F_1 \cup \{u\} \cup F_2) \subseteq O_1$ and $\beta_1(\{v\} \cup F_3) \subseteq O_2$. Setting $e_1 := e \circ \beta_1$ we have

$$e_1(F_2) < e_1(u) < e_1(F_1) < e_1(v) < e_1(F_3).$$

We may choose $\beta_2 \in \langle \text{Aut}(\mathbb{Q}; <) \cup \{\circlearrowright\} \rangle$ such that $\beta_2(F_1 \cup \{v\} \cup F_3) \subseteq O_1$ and $\beta_1(F_2 \cup \{u\}) \subseteq O_2$. Setting $e_2 := - \circ e \circ \beta_2$ we have

$$e_2(u) < e_2(F_2) < e_2(F_1) < e_2(v) < e_2(F_3).$$

We may then choose $\beta_3 \in \text{Aut}(\mathbb{Q}; <)$ such that $\beta_3(\{u\} \cup F_2 \cup F_1) \subseteq O_1$ and $\beta_3(\{v\} \cup F_3) \subseteq O_2$. Setting $e_3 := e \circ \beta_3$ we have

$$e_3(F_1) < e_3(v) < e_3(F_2) < e_3(u) < e_3(F_3).$$

Moreover,

- F_1 and F_2 are reversed by e_1, reversed by e_2, and preserved by e_3;
- F_3 is preserved by e_1, by e_2, and by e_3.

So there exists $\gamma \in \text{Aut}(\mathbb{Q}; <)$ such that the restriction of $\gamma \circ e_3$ to F coincides with the given transposition α.

The canonicity of e implies that e is either order-reversing on each infinite orbits, or order-preserving on each infinite orbit. By composing e with $-$ we therefore may now assume that e is order-preserving on each infinite orbit.

Claim 2. If there are infinite orbits O_1, O_2, O_3 such that $O_1 < O_2 < O_3$ and $e(O_2) < e(O_1) < e(O_3)$ then $\text{Aut}(\mathfrak{B}) = \text{Sym}(\mathbb{Q})$. Let $\beta_1 \in \text{Aut}(\mathbb{Q}; <)$ be such that $\beta_1(F_1) \subseteq O_1$, $\beta_1(u) \in O_2$, and $\beta_1(F_2 \cup \{v\} \cup F_3) \subseteq O_3$. Setting $e_1 := e \circ \beta_1$ we have

$$e_1(u) < e_1(F_1) < e_1(F_2) < e_1(v) < e_1(F_3).$$

Let $\beta_2 \in \text{Aut}(\mathbb{Q}; <)$ be such that $\beta_2(e_1(u)) \subseteq O_1$, $\beta_2(e_1(F_1 \cup F_2)) \subseteq O_2$, and $\beta_2(e_1(\{v\} \cup F_3)) \subseteq O_3$. Setting $e_2 := e \circ \beta_2 \circ e_1$ we have

$$e_2(F_1) < e_2(F_2) < e_2(u) < e_2(\{v\} \cup F_3).$$

Let $\beta_3 \in \mathrm{Aut}(\mathbb{Q};<)$ be such that $\beta_3(e_2(F_1)) \subseteq O_1$, $\beta_3(e_2(F_2 \cup \{u\})) \subseteq O_2$, and $\beta_3(e_2(\{v\} \cup F_3)) \subseteq O_3$. Setting $e_3 := e \circ \beta_3$ we have

$$e_3(F_2) < e_3(u) < e_3(F_1) < e_3(v) < e_3(F_3).$$

Let $\beta_4 \in \mathrm{Aut}(\mathbb{Q};<)$ be such that $\beta_4(e_3(F_2 \cup \{u\})) \subseteq O_1$, $\beta_4(e_3(F_1 \cup \{v\})) \subseteq O_2$, and $\beta_4(e_3(F_3))$. Setting $e_4 := e \circ \beta_4$ we have

$$e_4(F_1) < e_4(v) < e_4(F_2) < e_4(u) < e_4(F_3).$$

Moreover, e_4 is order-preserving on each of F_1, F_2, F_3, and hence we can find $\gamma \in \mathrm{Aut}(\mathbb{Q};<)$ such that the restriction of $\gamma \circ e_4$ to F equals the transposition α, which finishes the proof of the claim.

By composing e with $-$ if necessary, we may suppose that for any two distinct infinite orbits O_1 and O_2 we have $O_1 < O_2$ if and only if $e(O_1) < e(O_2)$.

Claim 3. If there are infinite orbits O_1 and O_2 and $i \in \{1, \ldots, n\}$ such that $O_1 < t_i < O_2$ and $e(t_i) < e(O_1) < e(O_2)$ or $e(O_1) < e(O_2) < e(t_i)$ then $\mathfrak{B}$ is preserved by $\mathrm{Sym}(\mathbb{Q})$. Note that $\langle \mathrm{Aut}(\mathbb{Q};<) \cup \{e, \circlearrowright\} \rangle$ contains an operation f such that $f(O_2) < f(t_i) < f(O_1)$. Let $\beta_1 \in \mathrm{Aut}(\mathbb{Q};<)$ be such that $\beta_1(F_1) \subseteq O_1$, $\beta_1(u) = t_i$, and $\beta_1(F_2 \cup \{v\} \cup F_3) \subseteq O_2$. Setting $f_1 := f \circ \beta_1$ we have

$$f_1(F_2) < f_1(v) < f_1(F_3) < f_1(u) < f_1(F_1).$$

Let $\beta_2 \in \mathrm{Aut}(\mathbb{Q};<)$ be such that $\beta_2(F_2) \subseteq O_1$, $\beta_2(v) = t_i$, and $\beta_2(F_3 \cup \{u\} \cup F_1) \subseteq O_2$. Setting $f_2 := f \circ \beta_2$ we have

$$f_2(F_3) < f_2(u) < f_2(F_1) < f_2(v) < f_2(F_2).$$

Let $\beta_3 \in \mathrm{Aut}(\mathbb{Q};<)$ be such that $\beta_3(F_3) \subseteq O_1$, $\beta_3(u) = t_i$, $\beta_3(F_1 \cup \{v\} \cup F_2) \subseteq O_2$. Setting $f_3 := f \circ \beta_3$ be have

$$f_3(F_1) < f_3(v) < f_3(F_2) < f_3(u) < f_3(F_3).$$

Since f_3 is order-preserving on each of F_1, F_2, F_3 we can find $\gamma \in \mathrm{Aut}(\mathbb{Q};<)$ such that the restriction of $\gamma \circ f_3$ to F equals the transposition α, which finishes the proof of the claim.

Finally we argue that the assumptions from Claim 3 must apply: otherwise, for all infinite orbits O_1 and O_2 and $i \in \{1, \ldots, n\}$ such that $O_1 < t_i < O_2$ we have $e(O_1) < e(t_i) < e(O_2)$, and hence $e(t)$ lies in the same orbit as t, contradicting our assumptions. ⊣

Note that the following theorem immediately implies Theorem 12.1.2 and hence also implies Cameron's theorem.

THEOREM 12.1.7. *Let $\mathfrak{B}$ be a first-order reduct of $(\mathbb{Q};<)$. Then exactly one of the following cases applies.*

1. *$\mathfrak{B}$ has a constant endomorphism;*
2. $\mathrm{End}(\mathfrak{B}) = \overline{\mathrm{End}(\mathbb{Q};<)}$;
3. $\mathrm{End}(\mathfrak{B}) = \overline{\langle \mathrm{Aut}(\mathbb{Q};<) \cup \{-\} \rangle}$;

4. $\mathrm{End}(\mathfrak{B}) = \overline{\langle \mathrm{Aut}(\mathbb{Q};<) \cup \{\circlearrowleft\}\rangle}$;
5. $\mathrm{End}(\mathfrak{B}) = \overline{\langle \mathrm{Aut}(\mathbb{Q};<) \cup \{-,\circlearrowleft\}\rangle}$;
6. $\mathrm{End}(\mathfrak{B})$ *equals the set of all injective unary operations.*

PROOF. The result is an immediate consequence of the Propositions 12.1.3, 12.1.4, 12.1.5, and 12.1.6.

We now present an alternative proof based on Cameron's theorem. First note that all the cases are indeed disjoint: a constant endomorphism does not preserve $<$, and cannot be generated by a set of injective unary operations; this shows that the first case is distinct from all others. Disjointness of the remaining cases follows from Theorem 12.1.2. If $\mathfrak{B}$ has a non-injective endomorphism, then Corollary 6.1.27 shows that there is also a constant endomorphism. Otherwise all endomorphisms of $\mathfrak{B}$ are injective. We show that then all endomorphisms e of $\mathfrak{B}$ are locally invertible: for any $a_1,\dots,a_l \in \mathbb{Q}$ there exists a self-embedding f of $\mathfrak{B}$ into $\mathfrak{B}$ such that $f(e(a_i)) = a_i$ for all $i \in \{1,\dots,l\}$. Because e is injective, there is an $\alpha \in \mathrm{Aut}(\mathbb{Q};<)$ such that $\alpha e(\{a_1,\dots,a_l\}) = \{a_1,\dots,a_l\}$. Then $(\alpha e)^{l!}$, i.e., the composition of $(\alpha e)\dots(\alpha e)$ with l-factorial many terms of the form (αe), maps a_i to itself for all $1 \leq i \leq l$. Then $(\alpha e)^{l!-1}\alpha$ is also an endomorphism of $\mathfrak{B}$, and we have $((\alpha e)^{l!-1}\alpha)(e(a_1),\dots,e(a_l)) = (\alpha e)^{l!}(a_1,\dots,a_l) = (a_1,\dots,a_l)$. This proves that e is locally invertible.

Theorem 4.5.1 (6) $\Rightarrow$ (5) shows that the automorphisms of $\mathfrak{B}$ lie dense in the endomorphisms of $\mathfrak{B}$. The claim of the statement then follows directly from Theorem 12.1.2. ⊣

Note that Theorem 12.1.7 can be understood as a classification of the model-complete cores of first-order reducts $\mathfrak{B}$ of $(\mathbb{Q};<)$, considered up to first-order interdefinability and isomorphism: in case (1), the model-complete core of $\mathfrak{B}$ has just one element, and in all other cases, $\mathfrak{B}$ is already a model-complete core.

The fact that this classification can be derived from a classification for first-order reducts of $\mathfrak{B}$ (Cameron's theorem) is quite particular for $(\mathbb{Q};<)$; we make essential use of high set-transitivity of $\mathrm{Aut}(\mathbb{Q};<)$ in our proof. Already for the Random graph, where a similar classification of the model-complete cores of the first-order reducts is known [89], it is no longer clear how to obtain this result from a classification of the first-order reducts of the random graph, proved much earlier by Simon Thomas [339].

12.2. Hard temporal CSPs

In this section we show for certain fundamental first-order reducts of $(\mathbb{Q};<)$ that they can interpret all finite structures with parameters; it follows that their CSP is NP-complete. We have already mentioned in the introduction that the Betweenness and the Cyclic Ordering Problem in [176] can be formulated as temporal CSPs, and that these problems are NP-complete. The corresponding relations Betw and Cycl re-appeared in Cameron's theorem (Theorem 12.1.1). Another important relation for our classification is the relation T_3 from Definition 3.1.8 (also see Example 9.6.3). In fact, if $\mathfrak{B}$ is $(\mathbb{Q}; R)$ for one of the relations R mentioned above, then we give primitive positive interpretations of $(\{0,1\}; \text{1IN3})$ with finitely many constants in $\mathfrak{B}$. Thus, hardness of temporal CSPs can always be shown with Proposition 3.1.7. We have already seen this for Betw and T_3 and complete this here by showing it for Cycl and Sep. We thank Trung Van Pham for pointing out a simpler proof for Cycl than our original proof, which was inspired by the NP-hardness proof of [174] for the 'Cyclic ordering problem' (see [176]).

THEOREM 12.2.1. *The structure* $(\mathbb{Q}; <, \text{Betw})$ *has a primitive positive interpretation in* $(\mathbb{Q}; \text{Cycl}, 0, 1)$. *The structure* $(\{0,1\}; \text{1IN3})$ *has a primitive positive interpretation with parameters in* $(\mathbb{Q}; \text{Cycl})$, *and* $\text{CSP}(\mathbb{Q}; \text{Cycl})$ *is NP-hard.*

PROOF. Our interpretation I of $(\mathbb{Q}; <, \text{Betw})$ in $(\mathbb{Q}; \text{Cycl}, 0, 1)$ is 1-dimensional. The domain formula $\top_I(x)$ is $\text{Cycl}(0, x, 1)$, and defines the open interval $(-1, 1) \subseteq \mathbb{Q}$. The coordinate map is any isomorphism between $(\mathbb{Q}; <)$ and the substructure induced by these numbers. The interpreting relation for $x < y$ is $\text{Cycl}(0, x, y)$. It is easy to verify that the relation Cycl is not preserved by any of the relations listed in Lemma 12.6.2. Hence, Betw has a primitive positive definition in $(\mathbb{Q}; \text{Cycl}, <)$, which is the interpreting formula for Betw in the interpretation.

Since $(\mathbb{Q}; \text{Betw})$ can in turn interpret primitively positively $(\{0,1\}; \text{1IN3})$ with parameters by Proposition 3.1.10, the desired interpretation can be obtained by composing interpretations (Remark 3.1.5). We can then apply Proposition 3.1.7, and the NP-hardness of $\text{CSP}(\mathbb{Q}; \text{Cycl})$ follows from the NP-hardness of $(\{0,1\}; \text{1IN3})$. ⊣

Another relation that appeared in Theorem 12.1.1 is the separation relation Sep. The corresponding CSP is again NP-complete.

PROPOSITION 12.2.2. *There is a primitive positive interpretation of* $(\mathbb{Q}; \text{Betw})$ *in* $(\mathbb{Q}; \text{Sep}, -1, 1, 2)$, *and therefore also a primitive positive interpretation of* $(\{0,1\}; \text{1IN3})$ *in* $(\mathbb{Q}; \text{Sep})$ *with parameters. The problem* $\text{CSP}(\mathbb{Q}; \text{Sep})$ *is NP-hard.*

PROOF. Our interpretation I of $(\mathbb{Q}; \text{Betw})$ in $(\mathbb{Q}; \text{Sep}, -1, 1, 2)$ is 1-dimensional. The domain formula $\top_I(x)$ is $\text{Sep}(-1, 1, x, 2)$, and defines the open interval $(-1, 1) \subseteq \mathbb{Q}$. The coordinate map is any isomorphism between $(\mathbb{Q}; <)$ and the substructure of $(\mathbb{Q}; <)$ induced on these numbers. Then the formula $\text{Sep}(x, z, y, 1)$ interprets Betw: x and y must satisfy $\top_I$, and so $-1 < x, y < 1$. Therefore,

$$\begin{aligned} \text{Betw}(I(x), I(y), I(z)) &\Leftrightarrow -1 < x < y < z < 1 \text{ or } -1 < z < y < x < 1 \\ &\Leftrightarrow \text{Sep}(x, z, y, 1) \end{aligned}$$

A primitive positive interpretation of $(\{0,1\}; \mathrm{1IN3})$ can be obtained as follows. The argument above shows that the structure $(\mathbb{Q}; \mathrm{Sep})$ can interpret primitively positively $(\mathbb{Q}; \mathrm{Betw}, 0)$ with parameters, which in turn can interpret primitively positively $(\{0,1\}; \mathrm{1IN3})$ by Proposition 3.1.10. Then the desired interpretation can be obtained by composing interpretations (see Section 3.4). We can then apply Proposition 3.1.7, and the NP-hardness of CSP$(\mathbb{Q}; \mathrm{Sep})$ follows from the NP-hardness of $(\{0,1\}; \mathrm{1IN3})$. ⊣

12.3. Definability of the order

As an application of our classification of the model-complete cores of first-order reducts of $(\mathbb{Q}; <)$ from Theorem 12.1.7 and the hardness results in Section 12.2 we can simplify the complexity classification task for temporal CSPs to the classification for first-order *expansions* of $(\mathbb{Q}; <)$.

THEOREM 12.3.1. *Let $\mathfrak{B}$ be a temporal constraint language. Then it satisfies at least one of the following*:

(*a*) *There is a primitive positive definition of* Cycl, Betw, *or* Sep *in* $\mathfrak{B}$.
(*b*) Pol$(\mathfrak{B})$ *contains a constant operation.*
(*c*) *Aut*$(\mathfrak{B})$ *contains all permutations of* $\mathbb{Q}$.
(*d*) *There is a primitive positive definition of* $<$ *in* $\mathfrak{B}$.

PROOF. If there is a primitive positive definition of $<$ in $\mathfrak{B}$, we are in case (d). Otherwise, Proposition 12.1.7 shows that $\mathfrak{B}$ is preserved by a constant, $-$, or $\circlearrowright$. For each of these three operations we show the claim of the statement separately in the following three paragraphs.

If $\mathfrak{B}$ is preserved by a constant we are in case (b), so we assume in the following that $\mathfrak{B}$ is not preserved by a constant.

If $\mathfrak{B}$ is preserved by $-$, the relation Betw consists of only one orbit of triples. If there is a primitive positive definition of Betw in $\mathfrak{B}$ we are in case (a). Otherwise, Lemma 6.1.24 shows that there is an endomorphism that does not preserve Betw. Proposition 12.1.7 then implies that $\mathfrak{B}$ is also preserved by $\circlearrowright$. Thus, the relation Sep consists of only one orbit of 4-tuples. Again, either Sep has a primitive positive definition, and we are in case (a), or there is an endomorphism that does not preserve Sep. Proposition 12.1.7 now shows that $\mathfrak{B}$ is preserved by all unary injections and we are in case (c).

If $\mathfrak{B}$ is preserved by $\circlearrowright$, then the relation Cycl consists of only one orbit of triples. If Cycl has a pp definition in $\mathfrak{B}$, we are in case (a). Otherwise, Lemma 6.1.24 shows that there is an endomorphism that does not preserve Cycl. Proposition 12.1.7 then shows that $\mathfrak{B}$ is also preserved by $-$. But the statement of the lemma has already been shown in the case that $\mathfrak{B}$ is preserved by both $-$ and $\circlearrowright$ in the previous paragraph, so we are done. ⊣

In case (a), there is a finite-signature first-order reduct $\mathfrak{B}'$ of $\mathfrak{B}$ such that CSP($\mathfrak{B}'$) is NP-hard, as we have seen in Section 12.2. In case (b), for all finite-signature first-order reducts $\mathfrak{B}'$ of $\mathfrak{B}$ the problem CSP($\mathfrak{B}$) is trivially in P (see Proposition 1.1.12). In case (c) $\mathfrak{B}$ is an equality constraint language and the complexity of the respective CSPs has been classified in Chapter 7. In the following, we therefore study only those temporal constraint languages where $<$ is primitively positively definable.

12.4. Lex-closed constraints

An important class of temporal constraint languages are the languages preserved by the operation lex, introduced in Section 11.4.4. Recall that lex is a binary injective operation on $\mathbb{Q}$ such that $\text{lex}(a, b) < \text{lex}(a', b')$ if either $a < a'$, or $a = a'$ and $b < b'$. Note that lex is canonical with respect to $\text{Aut}(\mathbb{Q}; <)$ and that the conditions above completely describe the behaviour of lex. It follows from Lemma 11.4.12 that all operations locally generate the same clone. We also write

- $\text{lex}_{y,x}$ for the operation $(x, y) \mapsto \text{lex}(y, x)$,
- $\text{lex}_{y,-x}$ for the operation $(x, y) \mapsto \text{lex}(y, -x)$,
- $\text{lex}_{x,-y}$ for the operation $(x, y) \mapsto \text{lex}(x, -y)$, and
- $\text{lex}_{x,y}$ for the operation $(x, y) \mapsto \text{lex}(x, y)$.

In diagrams for binary operations f as in Figure 1, we draw a directed edge from (a, b) to (a', b') if $f(a, b) < f(a', b')$. Unoriented lines in rows and columns of picture for an operation f relate pairs of values that get the same value under f.

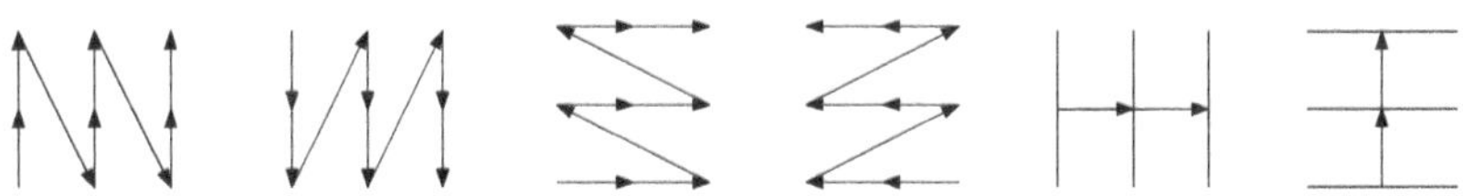

FIGURE 1. Illustrations of the six basic operations $\text{lex}_{x,y}$, $\text{lex}_{x,-y}$, $\text{lex}_{y,x}$, $\text{lex}_{y,-x}$, π_1, π_2.

A k-ary operation $f : \mathbb{Q}^k \to \mathbb{Q}$ is *dominated* by the i-th argument if for all $a, b \in \mathbb{Q}^k$ it holds that $f(a_1, \ldots, a_k) \leq f(b_1, \ldots, b_k)$ if and only if $a_i \leq b_i$. Examples of operations dominated by the first argument are π_1, $\text{lex}_{x,y}$, and $\text{lex}_{x,-y}$, and examples of operations dominated by the second argument are π_y, $\text{lex}_{y,x}$, $\text{lex}_{y,-x}$. It is easy to see that the relation Betw is preserved by lex, and more generally by all operations that are dominated by one argument. Therefore, we are interested in further restrictions of languages preserved by lex that imply polynomial-time tractability of the corresponding CSPs.

12.4.1. The operations lex **and** ll. A large class of polynomial-time tractable temporal CSPs has been introduced in [70]. The class is defined in terms of a binary polymorphism, denoted by ll. We will see in Proposition 12.4.2 that this class contains the class of Ord-Horn constraint languages (Section 1.6.9).

DEFINITION 12.4.1. Let ll: $\mathbb{Q}^2 \to \mathbb{Q}$ be such that $\text{ll}(a,b) < \text{ll}(a',b')$ if

- $a \leq 0$ and $a < a'$, or
- $a \leq 0$ and $a = a'$ and $b < b'$, or
- $a, a' > 0$ and $b < b'$, or
- $a > 0$ and $b = b'$ and $a < a'$.

All operations satisfying these conditions are by definition injective, and they all generate the same clone. The function ll is not canonical with respect to $(\mathbb{Q}; <)$, but it is canonical with respect to $(\mathbb{Q}; <, 0)$. Clearly, ll interpolates lex modulo $\text{Aut}(\mathbb{Q}; <)$. The function ll has a *dual* that locally generates a different clone. For an illustration of ll and its dual, see Figure 2.

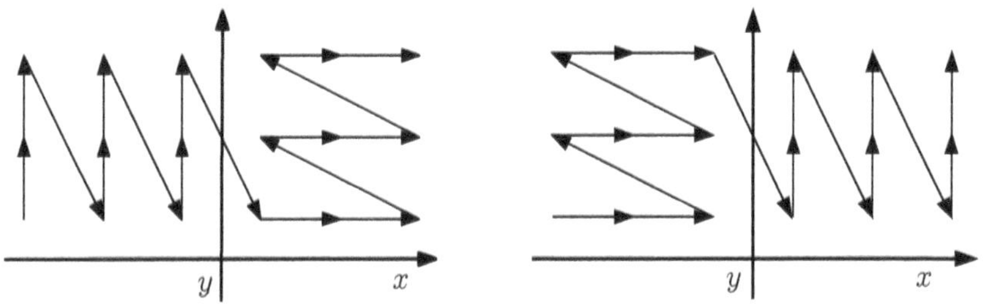

FIGURE 2. A visualization of ll (left) and dual-ll (right).

PROPOSITION 12.4.2. *All relations in Ord-Horn are preserved by ll and dual ll.*

PROOF. We give the argument for ll only; the argument for dual ll is analogous. It suffices to show that every relation that can be defined by a formula ϕ of the form $(x_1 = y_1 \wedge \cdots \wedge x_{k-1} = y_{k-1}) \to x_k \ O \ y_k$ is preserved by ll, where $O \in \{=, <, \leq, \neq\}$. Let t_1 and t_2 be two $2k$-tuples that satisfy ϕ. Consider a $2k$-tuple t_3 obtained by applying ll componentwise to t_1 and t_2. Suppose first that there is an $i \leq k-1$ such that one of the tuples does not satisfy $x_i = y_i$. Then $x_i = y_i$ is not satisfied in t_3 as well, by injectivity of ll, and therefore the tuple t_3 satisfies ϕ. Now consider the case that $x_i = y_i$ holds for all $i \leq k-1$ in both tuples t_1 and t_2. Since t_1 and t_2 satisfy ϕ, the literal $x_k O y_k$ holds in both t_1 and t_2. Because ll preserves all relations in $\{=, <, \leq, \neq\}$, the literal $x_k O y_k$ holds in t_3, and therefore t_3 satisfies ϕ as well. ⊣

Since the relation R^{min} defined by $(x > y) \vee (x > z)$ (see Section 1.6.8) is preserved by ll but not by dual ll, the class of ll-closed constraints is *strictly* larger than Ord-Horn.

12.4.2. Operations locally generating ll, dual-ll, **or** lex. In this section we present conditions that imply that an operation locally generates ll, dual-ll, or lex.

DEFINITION 12.4.3. Let f, g be from $\mathbb{Q}^2 \to \mathbb{Q}$. Then $[f|g]$ denotes an arbitrary operation from $\mathbb{Q}^2 \to \mathbb{Q}$ with the following properties. For all $x, x', y, y' \in \mathbb{Q}$,

- if $x \leq 0$ and $x' > 0$ then $[f|g](x, y) < [f|g](x', y')$;
- $[f|g]$ and f satisfy the same type conditions on $\mathbb{Q}_0^- \times \mathbb{Q}$;
- $[f|g]$ and g satisfy the same type conditions on $\mathbb{Q}^+ \times \mathbb{Q}$.

For example, if $f = lex_{x,y}$ and $g = \text{lex}_{y,x}$, then $[f|g]$ and ll satisfy the same type conditions.

LEMMA 12.4.4. *Let* $f, g \in \{\text{lex}_{x,y}, \text{lex}_{x,-y}, \text{lex}_{y,x}, \text{lex}_{y,-x}, \pi_1, \pi_2\}$, *and let* f' (g') *be* $\text{lex}_{x,y}$ *if* f (g) *is dominated by the first argument, and* $\text{lex}_{y,x}$ *otherwise. Then* $\{\text{lex}, [f|g]\}$ *locally generates* $[f'|g']$.

PROOF. Let $r, s \in \mathbb{Q}^k$. Let $\alpha \in \text{Aut}(\mathbb{Q}; <)$ be such that for each entry x of r and for each entry y of s, the value of $\alpha\text{lex}(x, y)$ is negative if $x \leq 0$, and positive otherwise. We claim that

$$u := [f|g](\alpha\text{lex}(r, s), \text{lex}(s, r))$$

lies in the same orbit as $t := [f'|g'](r, s)$. By the homogeneity of $\text{Aut}(\mathbb{Q}; <)$ it suffices to show for $i, j \in \{1, \ldots, k\}$ that

$$u_i \leq u_j \text{ if and only if } t_i \leq t_j. \tag{1}$$

We can assume that $r_i \leq r_j$ by exchanging the name of i and j if necessary, and distinguish three cases:

- $r_i \leq 0 < r_j$. Then $t_i < t_j$ by the definition of $[f'|g']$. Since for $l \in \{1, \ldots, k\}$ the value of $\alpha\text{lex}(r_l, s_l)$ is positive if and only if $r_l > 0$, we have $u_i < u_j$ by the definition of $[f|g]$. Thus, we have verified (1) in this case.
- $r_j \leq 0$. Note that $f(\text{lex}(x, y), \text{lex}(y, x))$ and $f'(x, y)$ have the same behavior. Moreover, $f(x, y)$ and $f(\alpha(x), y)$ have the same behaviour since f is canonical with respect to $(\mathbb{Q}; <)$. Also, $\alpha\text{lex}(r_i, s_i), \alpha\text{lex}(r_j, s_j) < 0$ and thus we have the following equivalences.

$$\begin{aligned} t_i \leq t_j &\text{ iff } f'(r_i, s_i) \leq f'(r_j, s_j) \\ &\text{ iff } f(\text{lex}(r_i, s_i), \text{lex}(s_i, r_i)) \leq f(\text{lex}(r_j, s_j), \text{lex}(s_j, r_j)) \\ &\text{ iff } f(\alpha\text{lex}(r_i, s_i), \text{lex}(s_i, r_i)) \leq f(\alpha\text{lex}(r_j, s_j), \text{lex}(s_j, r_j)) \\ &\text{ iff } u_i \leq u_j. \end{aligned}$$

- $0 < r_j$. This case is analogous to the previous one and left to the reader.

Since $r, s \in \mathbb{Q}^k$ were chosen arbitrarily, the claim implies that $\{\text{lex}, [f|g]\}$ locally generates $[f'|g']$. ⊣

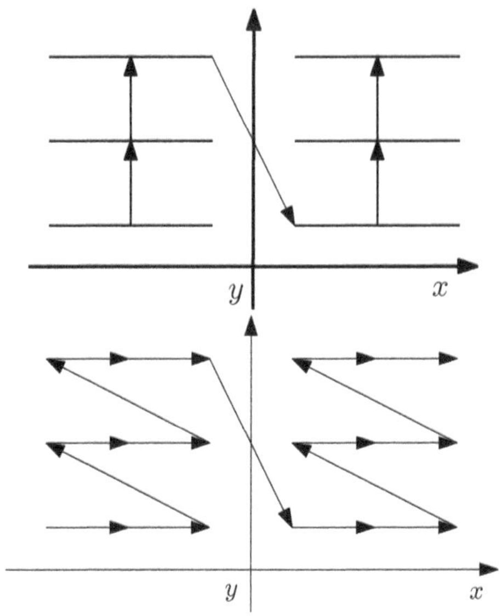

FIGURE 3. An illustration of the operation $[\pi_2|\pi_2]$ (on the left) and the operation $[\text{lex}_{y,x}|\text{lex}_{y,x}]$ (on the right).

LEMMA 12.4.5. *For* $f, g \in \{\pi_2, \text{lex}_{y,x}\}$ *the operation* $[f|g]$ *locally generates* $[\text{lex}_{x,y}|g]$.

In particular, for $f = g = lex_{y,x}$ the lemma shows that $[f|g]$ generates ll. For $f = g = \pi_2$, the lemma shows that $[f|g]$ generates $[\text{lex}_{x,y}|\pi_2]$ and in particular $\text{lex}_{x,y}$. See Figure 3 for illustrations of those two cases.

PROOF OF LEMMA 12.4.5. Let $r, s \in \mathbb{Q}^k$. Let l denote the number of non-positive values in r. We take $\alpha_1, \dots, \alpha_l$ from $\text{Aut}(\mathbb{Q}; <)$ such that α_i maps exactly the i smallest values in r to non-positive values. We define a sequence of tuples $u_1, \dots, u_l$ as follows: $u_1 = s$, and for $m \geq 2$

$$u_m := [f|g](\alpha_m r, u_{m-1}).$$

We claim that u_l lies in the same orbit as $t := [\text{lex}_{x,y}|g](r, s)$. By the homogeneity of $(\mathbb{Q}; <)$ it is enough to show for all $i, j \in \{1, \dots, k\}$ with $r_i \leq r_j$ that

$$(u_l)_i \leq (u_l)_j \text{ if and only if } t_i \leq t_j. \tag{2}$$

We distinguish three cases:

- $r_i = r_j \leq 0$. Since $\alpha_m r_i = \alpha_m r_j$ for all $m \leq l$, we have $(u_l)_i \leq (u_l)_j$ if and only if $(u_1)_I \leq (u_1)_j$. Since $u_1 = s$ and $r_i \leq 0$, and as f is dominated by the second argument, $(u_1)_i \leq (u_1)_j$ if and only if $t_i \leq t_j$, which proves (2).

- $r_i < r_j, r_i \leq 0$. Let $m \in l$ be such that $\alpha_m r_i \leq 0$ and $\alpha_m r_j > 0$. By definition of $[f|g]$ we see that $(u_m)_i < (u_m)_j$. Because $\alpha_m r_i < \alpha_m r_j$, and because $[f|g]$ preserves $<$, by induction on $n \geq m$ we have that $(u_n)_i < (u_n)_j$. In particular, $(u_l)_i < (u_l)_j$. On the other hand, $t_i < t_j$ by definition of $\text{lex}_{x,y}$ and $[\text{lex}_{x,y}|g]$, and so (2) also holds in this case.
- $r_i > 0$. Observe that by the choice of l we have $\alpha_m r_i > 0$ for all $m \leq l$. Thus (2) holds, because both $[f|g]$, $[\text{lex}_{x,y}|g]$, and g satisfy the same type conditions on $\mathbb{Q}^+ \times \mathbb{Q}$.

Since $r, s \in \mathbb{Q}^k$ were chosen arbitrarily the claim implies that $[f|g]$ locally generates $[\text{lex}_{x,y}|g]$. ⊣

12.5. Shuffle-closed constraints

Another important subclass of the class of all temporal constraint languages is the class of languages that only contain *shuffle-closed* temporal relations. As we will see, there are shuffle-closed temporal relations giving rise to NP-complete CSPs. However, in this section we present three additional restrictions that imply that the corresponding CSPs can be solved in polynomial time.

12.5.1. Shuffle closure. We define shuffle closure, and show how shuffle closure can also be described by a certain binary operation on $\mathbb{Q}$.

DEFINITION 12.5.1. $R \subseteq \mathbb{Q}^k$ is called *shuffle-closed* iff for all tuples $r, s \in R$ and every $l \in \{1, \dots, k\}$ there is a tuple $t \in R$ such that for all $i, j \in \{1, \dots, k\}$ we have $t_i \leq t_j$ if and only if

- $r_i \leq r_l$ and $r_i \leq r_j$, or
- $r_l < r_i$, $r_l < r_j$, and $s_i \leq s_j$.

Let pp be an arbitrary binary operation on $\mathbb{Q}$ such that $\text{pp}(a, b) \leq \text{pp}(a', b')$ iff one of the following cases applies:

- $a \leq 0$ and $a \leq a'$, or
- $0 < a, 0 < a'$, and $b \leq b'$.

Clearly, such an operation exists. For an illustration, see the left diagram in Figure 4. The right diagram of Figure 4 is an illustration of the dual-pp operation. The name of the operation pp is derived from the word *'projection-projection'*, since the operation satisfies the same type conditions as the projection to the second argument if the first argument is positive, and satisfies the same type conditions as the projection to the first argument otherwise.

PROPOSITION 12.5.2. *Let $R \subseteq \mathbb{Q}^k$ be a temporal relation. Then R is shuffle-closed if and only if it is preserved by* pp.

PROOF. Let $r, s \in R$. We first suppose that R is shuffle-closed and prove that $t := \text{pp}(r, s) \in R$. If r only contains positive values, then there clearly exists $\alpha \in \text{Aut}(\mathbb{Q}; <)$ such that $t = \alpha s$, and since R is preserved by $\text{Aut}(\mathbb{Q}; <)$ we are

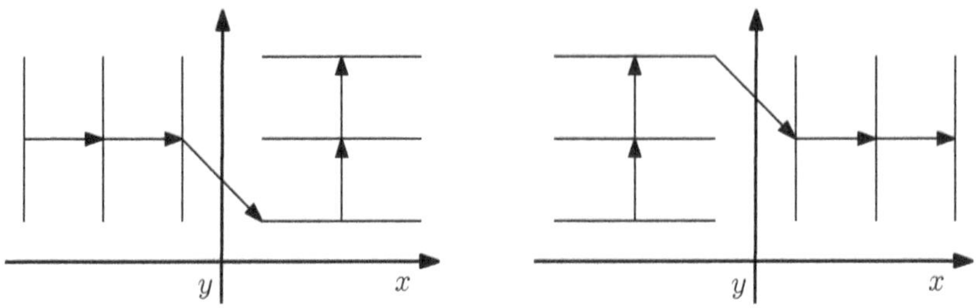

FIGURE 4. A visualisation of pp (left) and dual-pp (right).

done. Otherwise, let $l \in \{1, \ldots, k\}$ be such that r_l is the largest non-positive entry in r. Since R is shuffle-closed, there exists $t' \in R$ such that $t'_i \leq t'_j$ iff ($r_i \leq r_l$ and $r_i \leq r_j$) or ($r_l < r_i, r_l < r_j$, and $s_i \leq s_j$) for all $i, j \in \{1, \ldots, k\}$. By the definition of pp, and the choice of l, the tuple t satisfies the same property, $t \in R$ follows from the homogeneity of $\text{Aut}(\mathbb{Q}; <)$.

For the opposite direction, we assume that R is preserved by pp, and have to show shuffle closure of R. Let $l \in \{1, \ldots, k\}$. Choose $\gamma \in \text{Aut}(\mathbb{Q}; <)$ such that γ maps r_l to 0. Then $t = \text{pp}(\gamma r, s)$ is a tuple that satisfies the conditions specified in the definition of shuffle-closure. ⊣

Due to Proposition 12.5.2, we use the phrase '$\mathfrak{B}$ is shuffle-closed' interchangeably with '$\mathfrak{B}$ is preserved by pp'. The following lemma states an important property of shuffle-closed languages that will be used several times in the next sections.

LEMMA 12.5.3. *Let $t_1, \ldots, t_l$ be tuples from a k-ary shuffle-closed relation R, and let $M_1, \ldots, M_l \subset [k]$ be disjoint sets of indices such that $\bigcup_{i=1}^{l} M_i = \{1, \ldots, k\}$ and such that for all $i, j \in [l]$ with $i < j$ and for all $i' \in M_i$, $j' \in M_j$ it holds that $t_i[i'] < t_i[j']$. Then there is a tuple $t \in R$ such that*

- *$t[i'] < t[j']$ for all $i, j \in \{1, \ldots, l\}$ with $i < j$ and for all $i' \in M_i$, $j' \in M_j$;*
- *$t[i'] \leq t[i'']$ iff $t_i[i'] \leq t_i[i'']$ for all $i \in \{1, \ldots, l\}$ and all $i', i'' \in M_i$.*

PROOF. Let $\beta_1, \ldots, \beta_{l-1} \in \text{Aut}(\mathbb{Q}; <)$ be such that β_i maps $\max\{t_i[i'] \mid i' \in M_i\}$ to 0. We set

$$t := \text{pp}(\beta_1 t_1, \text{pp}(\beta_2 t_2, \ldots, \text{pp}(\beta_{l-1} t_{l-1}, t_l) \ldots)).$$

The tuple t clearly belongs to R.

We prove by induction on l that t satisfies the other conditions of the lemma. Observe that β_1 maps all the entries of t_1 at M_1 to non-positive values. Thus for $l = 2$, it is easy to check from the properties of pp that for each $i \in M_1$ and $i' \in M_2$ we have $t[i] < t[i']$ as required by the statement of the lemma. Also the second condition is immediate. For $l > 2$ let t' be defined by

$$t' := \text{pp}(\beta_2 t_2, \text{pp}(\beta_3 t_3, \ldots, \text{pp}(\beta_{l-1} t_{l-1}, t_l) \ldots)).$$

Then we have $t = \text{pp}(\beta_1 t_1, t')$. Now we apply the same argument as for $l = 2$. Because the order on $\{1, \ldots, k\} \setminus M_1$ is preserved by the application of pp, we know that the conditions are satisfied for the sets $M_2, \ldots, M_l$. The argument also shows that the entries at M_1 are smaller than the entries at $\{1, \ldots, k\} \setminus M_1$ and that their order is the same as in t_1. ⊣

We give a simple criterion for showing that certain operations generate pp; the proof is immediate from the definitions.

LEMMA 12.5.4. *Let $f \in \text{Pol}^{(2)}(\mathbb{Q}; <)$ be such that for every finite $F \subseteq \mathbb{Q}$ and $p \in F$ there exist $\alpha, \beta \in \text{Aut}(\mathbb{Q}; <)$ such that $f(x, y) = \alpha x$ for all $x, y \in F$ with $x \leq p$, and $f(x, y) = \beta y$ for all $x, y \in F$ with $x > p$. Then f interpolates* pp *modulo* $\text{Aut}(\mathbb{Q}; <)$. *In particular, f locally generates* pp.

It is easy to verify that the relation T_3 (Definition 3.1.8) is shuffle-closed. Proposition 3.1.9 shows that $\text{CSP}(\mathbb{Q}; T_3)$ is NP-complete, and thus the property of shuffle-closure is not strong enough to guarantee tractability.

12.5.2. Min-union closure. This section introduces and studies a stronger property than shuffle-closure, namely preservation under the binary operation min that maps two values x and y to the smaller of the two values; see Figure 5 for an illustration of the operation min.

For constraint languages over a finite domain, min- and max-closed relations were studied in [215]. An equivalent clausal description of such constraints is known; however, the equivalence only holds for *finite* domains. The tractability of the CSP where the constraint language has such a clausal description has also been shown for infinite domains [134]. But the algorithm presented in [134] cannot be applied to all min-closed constraint languages over an infinite domain; it is already not clear how to adapt their approach to deal with the relation $R_<^{\min} = \{(x, y, z) \mid y < x \vee z < x\}$ which is preserved by min. We give another example of a relation that is preserved by min.

Example 12.5.5. The relation U defined below is preserved by min.

$$\begin{aligned} U := \big\{(x, y, z) \in \mathbb{Q}^3 \mid\ & (x = y \wedge y < z) \\ & \vee (x = z \wedge z < y) \\ & \vee (x = y \wedge y = z)\big\}. \end{aligned}$$

In Section 12.8.3 we describe an algorithm that efficiently solves the CSP for temporal constraint languages that are preserved by min; a fundamentally different algorithm can be found in [74].

DEFINITION 12.5.6. Let $t \in \mathbb{Q}^k$. The i-th entry in $t \in \mathbb{Q}^k$ is called *minimal* if $t_i \leq t_j$ for every $j \in [k]$. The *min-set of t* is the set $M(t)$ of all indices with minimal entries, i.e.,

$$M(t) := \{i \in [k] \mid t_i \leq t_j \text{ for every } j \in [k]\}.$$

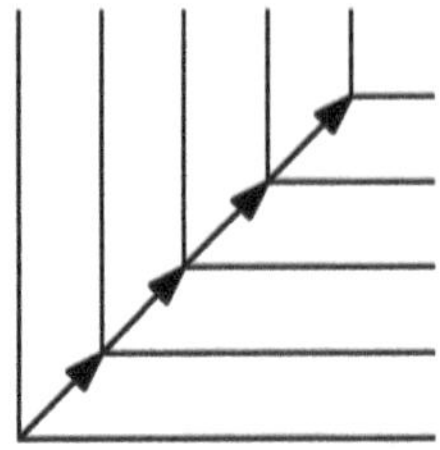

FIGURE 5. Illustration of the operation min.

DEFINITION 12.5.7. A relation $R \subseteq \mathbb{Q}^k$ is called *min-union closed* if for all $r, s \in R$ there exists $t \in R$ such that $M(t) = M(r) \cup M(s)$.

Min-union closure of a relation is linked to the existence of certain polymorphisms.

DEFINITION 12.5.8. We say that $f \in \mathrm{Pol}^{(2)}(\mathbb{Q}; <)$ *provides min-union closure* if $f(0,0) = f(0,x) = f(x,0)$ for all positive $x \in \mathbb{Q}$.

The operation min is an example of an operation providing min-union closure. The following lemma connects Definition 12.5.7 and Definition 12.5.8.

LEMMA 12.5.9. *Let R be a temporal relation preserved by an operation f providing min-union closure. Then R is min-union closed.*

PROOF. Let $t_1, t_2 \in R$ and let a_1 and a_2 be the minimal values among the entries of t_1 and t_2, respectively. Then there are $\alpha_1, \alpha_2 \in \mathrm{Aut}(\mathbb{Q}; <)$ such that $\alpha_1 a_1 = \alpha_2 a_2 = 0$. Observe that in the tuple $t_3 = f(\alpha_1 t_1, \alpha_2 t_2)$ all entries at indices from $M(t_1) \cup M(t_2)$ have the same value. Since f preserves $<$ this value is strictly smaller than the values at all other entries in t_3. Hence, $M(t_3) = M(t_1) \cup M(t_2)$. ⊣

The following proposition implies that $\{f, \mathrm{pp}\}$ generates min for every operation f that provides min-union closure.

PROPOSITION 12.5.10. *A temporal relation R is preserved by* pp *and an operation providing min-union closure if and only if R is preserved by* min.

PROOF. Clearly, min provides min-union closure. Also observe that min satisfies the conditions of Lemma 12.5.4 and thus locally generates pp.

For the opposite direction, suppose that R is k-ary and preserved by pp and an operation f providing min-union closure. We show that for any two tuples $r, s \in R$ the tuple $t = \min(r, s)$ is in R as well. Let l be the number of distinct values in t and $v_1 < v_2 < \cdots < v_l$ be these values. We define M_i for $i \in \{1, \ldots, l\}$ to be the set of indices of t with the i-th lowest value, i.e., $M_i = \{j \in \{1, \ldots, k\} \mid t_j = v_i\}$.

Now let $\alpha_1, \ldots, \alpha_l \in \mathrm{Aut}(\mathbb{Q}; <)$ be such that $\alpha_i v_i = 0$. Using these automorphisms we define the tuples $s_1, \ldots, s_l$ by $u^i := f(\alpha_i r, \alpha_i s)$. Clearly, these tuples belong to R. It also holds that u^i is constant at M_i because for each $j \in M_i$ at least one of the entries r_j, s_j is equal to v_i (the other one can be only greater) which is subsequently mapped to 0 by α_i and f maps all such pairs to the same value. Furthermore, for each $j' \in M_{i'}$ for $i < i' \leq l$ we have that $u^i_{j'}$ is greater than the value of u^i at M_i, because $\min(r_{j'}, s_{j'}) = v_{i'}$ is greater than v_i and f preserves $<$.

Now we can apply Lemma 12.5.3 to $u^1, \dots, u^l$ and $M_1, \dots, M_l$. The lemma gives us some tuple $u \in R$ which is constant at each set M_i for $i \in \{1, \dots, l\}$ and such that for each $i < j \leq l$ the value of u at M_i is lower than the value of u at M_j. Thus u has the same order of entries as t which shows that $t \in R$ as well. ⊣

12.5.3. Min-intersection closure. In this section, we study a different restriction of shuffle-closed constraint languages.

DEFINITION 12.5.11. A relation $R \subseteq \mathbb{Q}^k$ is called *min-intersection closed* if for all $r, s \in R$, if $M(r) \cap M(r) \neq \emptyset$, then there exists $t \in R$ such that $M(t) = M(r) \cap M(s)$.

DEFINITION 12.5.12. We say that $f \in \mathrm{Pol}^{(2)}(\mathbb{Q}; <)$ *provides min-intersection closure* if $f(0,0) < f(0,x)$ and $f(0,0) < f(x,0)$ for all positive $x \in \mathbb{Q}$.

LEMMA 12.5.13. *Let R be a temporal relation that is preserved by an operation f that provides min-intersection closure. Then R is min-intersection closed.*

PROOF. Let $r, s \in R$ be such that $M(r) \cap M(s) \neq \emptyset$; choose $i \in M(r) \cap M(s)$. Let $\alpha, \beta \in \mathrm{Aut}(\mathbb{Q}; <)$ be such that $\alpha r_i = \beta s_i = 0$. Consider the tuple $t = f(\alpha r, \beta s)$. At the entries from $M(r)$ the tuple αr equals 0 and likewise at the entries from $M(s)$ the tuple βs equals 0. Since $f(0,0) < f(0,x)$ and $f(0,0) < f(x,0)$ for all positive x it follows that in t all entries at $M(r) \cap M(s)$ have a strictly smaller value than all values at the symmetric difference $M(r) \Delta M(s)$. As f preserves $<$, all entries at $M(r) \cap M(s)$ have a strictly smaller value than the entries not at $M(r) \cup M(s)$. We conclude that $M(t) = M(r) \cap M(s)$. ⊣

An important example of an operation that provides min-intersection closure is given in the following definition.

DEFINITION 12.5.14. The operation $\mathrm{mi}\colon \mathbb{Q}^2 \to \mathbb{Q}$ is defined by

$$\mathrm{mi}(x,y) := \begin{cases} a(x) & \text{if } x < y \\ b(x) & \text{if } x = y \\ c(y) & \text{if } x > y \end{cases}$$

where $a, b, c \in \mathrm{End}(\mathbb{Q}; <)$ are such that

$$b(x) < c(x) < a(x) < b(x + \varepsilon)$$

for all $x \in \mathbb{Q}$ and all $0 < \varepsilon \in \mathbb{Q}$ (see Figure 6).

Operations a, b, c with the properties described in Definition 12.5.14 can be constructed as follows. Let $q_1, q_2, \dots$ be an enumeration of $\mathbb{Q}$. Inductively assume that we have already defined a, b, c on $\{q_1, \dots, q_n\}$ such that $b(q_i) < c(q_i) < a(q_i) < b(q_j)$ whenever $q_i < q_j$, for $i, j \in [n]$. Clearly, this is possible for $n = 1$. If $q_{n+1} > q_i$ for all $i \in [n]$, let q_j be the maximum of $\{q_1, \dots, q_n\}$, and define $a(q_j) < b(q_{n+1}) < c(q_{n+1}) < a(q_{n+1})$. In the case that $q_{n+1} < q_i$ for all $i \in [n]$ we proceed analogously. Otherwise, let $i, j \in [n]$ such that q_i is the largest possible and q_j is smallest possible such that $q_i < q_{n+1} < q_j$. In this case, define $a(q_i) < b(q_{n+1}) < c(q_{n+1}) < a(q_{n+1}) < b(q_j)$. In this way we define unary operations a, b, c on all of $\mathbb{Q}$ with the desired properties.

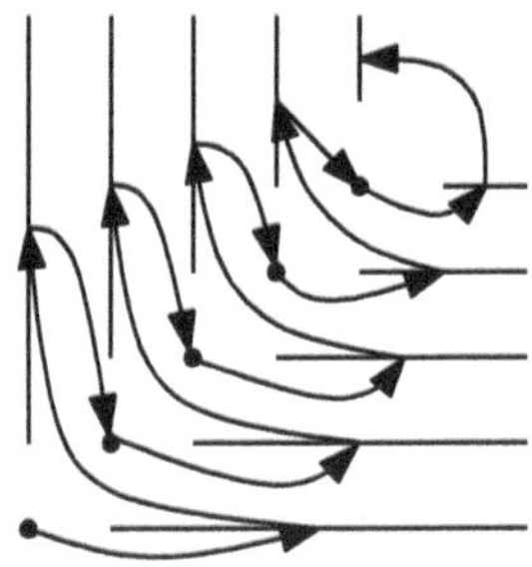

FIGURE 6. Illustration of the operation mi.

The operation mi will be of special importance, because the following proposition shows that pp together with any operation providing min-intersection closure generates the operation mi.

PROPOSITION 12.5.15. *A temporal relation R is preserved by* pp *and an operation f providing min-intersection closure if and only if R is preserved by* mi.

PROOF. It is clear that mi provides min-intersection closure. Lemma 12.5.4 shows that mi interpolates pp modulo $\mathrm{Aut}(\mathbb{Q}; <)$.

For the opposite direction, suppose R is k-ary and preserved by pp and an operation f providing min-intersection closure. We show that for any two tuples $r, s \in R$ the tuple $t = \mathrm{mi}(r, s)$ is in R as well. Let a, b, c be the mappings from the definition of the operation mi. Let $v_1 < \dots < v_l$ be the minimal-length sequence of rational numbers such that for each $i' \in [k]$ it holds that $t_{i'} \in \bigcup_{j \in [l]} \{a(v_j), b(v_j), c(v_j)\}$. Let M_i be

$$\{i' \in [k] \mid t_{i'} \in \{a(v_i), b(v_i), c(v_i)\}\}.$$

Observe that for each $i' \in M_i$ at least one of $r_{i'}$ and $s_{i'}$ is equal to v_i and the other value is greater or equal to v_i. Let M_i^a be the set of those $i' \in M_i$ where

$v_i = r_{i'} < s_{i'}$, M_i^b the set of those $i' \in M_i$ where $v_i = r_{i'} = s_{i'}$, and M_i^c the set of those $i' \in M_i$ where $v_i = s_{i'} < r_{i'}$.

Let $\alpha_1, \dots, \alpha_l \in \text{Aut}(\mathbb{Q}; <)$ be such that α_i maps v_i to 0. Let $\beta \in \text{Aut}(\mathbb{Q}; <)$ be such that $\beta f(0,0) = 0$. For each $i \in [l]$ we define

$$u^i := \text{pp}(\beta f(\alpha_i r, \alpha_i s), \text{pp}(\alpha_i s, r)). \tag{3}$$

We verify that for all $i \in [l]$ the tuple u^i is constant on each of the sets M_i^a, M_i^b, M_i^c, the value at M_i^b is lower than the value at M_i^c which is lower than the value at M_i^a. Furthermore, for each $j \in [l], j > i$, and each $i' \in M_i$ and $j' \in M_j$ it holds that $u^i_{i'} < u^i_{j'}$. Having this, we can apply Lemma 12.5.3 and obtain a tuple from R with the same ordering of entries as in t, which will prove the lemma.

Because α_i maps v_i to 0, the properties of pp imply that the tuple $w^i = pp(\alpha_i s, r)$ is constant at $M_i^b \cup M_i^c$ and at M_i^a, and the value at the first set is smaller than the value at the second set. Because the values of s at $M_i^a \cup \bigcup_{j=i+1}^l M_j$ are greater than v_i and the values of r at $\bigcup_{j=i+1}^l M_j$ are also greater than v_i (recall that for each $j \in [l]$, $j' \in M_j$ it holds that $\min(r_{j'}, s_{j'}) = v_j$) we conclude that the values of w^i at $\bigcup_{j=i+1}^l M_j$ are greater than those at M_i. The application of f in (3) yields a tuple which is constant on M_i^b and its value there (which is consequently mapped to 0 by β) is smaller than the values at $M_i^a \cup M_i^c \cup \bigcup_{j=i+1}^l M_j$. Thus it is easy to verify from the properties of pp that the outer application of pp in (3) yields a tuple with the desired properties. ⊣

A syntactic description of temporal relations preserved by mi can be found in Section 12.7.4. An algorithm that solves temporal constraint languages preserved by mi can be found in Section 12.8.4.

Example 12.5.16. We present a relation $I \subseteq \mathbb{Q}^4$ which is preserved by mi but not by min.

$$\begin{aligned} I := \{(a,b,c,d) \mid\ & (a = b \wedge b < c \wedge c = d) \\ & \vee (a = b \wedge b > c \wedge c = d) \\ & \vee (a = b \wedge b < c \wedge c < d) \\ & \vee (a > b \wedge b > c \wedge c = d)\}. \end{aligned}$$

It can be verified that I is preserved by mi; this is tedious, but it is easy to write a computer program that does the verification. Alternatively, note that

$$\begin{aligned} I(a,b,c,d) \Leftrightarrow ((a \geq b) \wedge (b \neq c) \wedge (c \leq d) \wedge (a = b \vee b > c) \wedge \\ (b < c \vee c = d)), \end{aligned}$$

and use the syntactic characterisation that we present in Section 12.7.4. The relation I is not preserved by min because

$$\min((0,0,1,2),(2,1,0,0)) = (0,0,0,0) \notin I$$

but $(0,0,1,2) \in I$ and $(2,1,0,0) \in I$.

Example 12.5.17. The following ternary temporal relation U is preserved by min (we omit the easy proof), but not preserved by mi.

$$\begin{aligned} U := \{(x,y,z) \mid & (x = y \wedge y < z) \\ & \vee (x = z \wedge z < y) \\ & \vee (x = y \wedge y = z)\} \end{aligned}$$

To see that U is not preserved by mi, note that the tuple $\text{mi}((0,0,1),(0,1,0))$ has three distinct values and hence is not in U, but $(0,0,1),(0,1,0) \in U$.

12.5.4. Min-xor closure. We introduce the third and final closure condition that is important when studying shuffle-closed temporal relations.

DEFINITION 12.5.18. A relation $R \subseteq \mathbb{Q}^k$ is called *min-xor closed* if for all $t_1, t_2 \in R$ such that $M(t_1)\Delta M(t_2)$ is non-empty there exists $t_3 \in R$ such that $M(t_3) = M(t_1)\Delta M(t_2)$.

DEFINITION 12.5.19. We say that $f \in \text{Pol}^{(2)}(\mathbb{Q};<)$ *provides min-xor closure* if $f(0,0) > f(0,x) = f(y,0)$ for all positive $x, y \in \mathbb{Q}$.

For an example of a binary operation that provides min-xor closure, consider the following binary operation, which we denote by mx.

$$\text{mx}(x,y) := \begin{cases} a(\min(x,y)) & \text{if } x \neq y \\ b(x) & \text{if } x = y \end{cases}$$

where b and b are unary operations that preserve $<$ such that $a(x) < b(x) < a(x+\varepsilon)$ for all $x \in \mathbb{Q}$ and all $0 < \varepsilon \in \mathbb{Q}$ (see Figure 7). Similarly as in the definition of mi, such operations a, b can be easily constructed. Note that the operation mx neither preserves the relation I nor the relation U introduced in Section 12.5.3.

LEMMA 12.5.20. *Let R be a temporal relation that is preserved by an operation f providing min-xor closure. Then R is min-xor closed.*

PROOF. Let $r, s \in R$ and suppose that the symmetric difference $M(t_1)\Delta M(t_2)$ is non-empty. Let u and v be the minimal values of the entries of r and of s, respectively. Then there are $\alpha, \beta \in \text{Aut}(\mathbb{Q};<)$ such that $\alpha u = 0$ and $\beta v = 0$. Consider the tuple $t = f(\alpha r, \beta s)$. Because αr is 0 for all entries at $M(r)$, βs is 0 for all entries at $M(s)$, and $f(0,0) > f(0,x) = f(y,0)$ for all $x, y > 0$, it follows that all entries of t at $M(r) \cap M(s)$ have a strictly larger value than all entries at $M(r)\Delta M(s)$, which all have the same value. As f preserves $<$, all

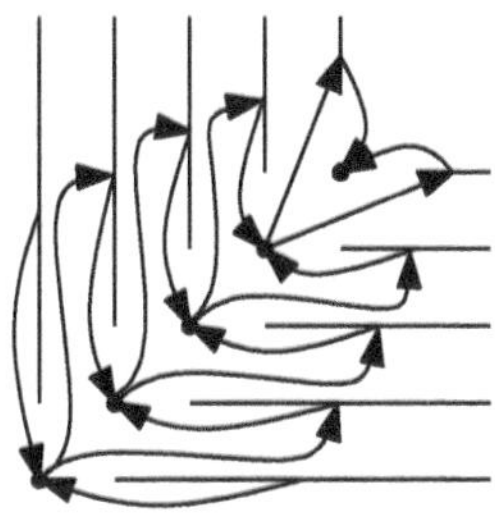

FIGURE 7. Illustration of the operation mx.

entries of t at $M(r) \cap M(s)$ are smaller than all entries not at $M(r) \cup M(s)$. We conclude that $M(t) = M(r)\Delta M(s)$. ⊣

The following lemma implies that $\{f, \text{pp}\}$ generates mx for any operation f that provides min-xor closure.

PROPOSITION 12.5.21. *A temporal relation R is preserved by* pp *and an operation f providing min-xor closure if and only if R is preserved by* mx.

PROOF. Clearly, mx provides min-xor closure. Lemma 12.5.4 shows that mx locally generates pp.

For the opposite direction, suppose that R is k-ary and preserved by pp and an operation f providing min-xor closure. We show that for any two tuples $r, s \in R$ the tuple $t = \text{mx}(r, s)$ is in R as well. Let a, b be the mappings as in the definition of the operation mx. Let $v_1 < \cdots < v_l$ be minimal set of rational numbers such that $t_i \in \bigcup_{j\in[l]}\{a(v_j), b(v_j)\}$ for all $i \in [k]$, and let M_i be the set of indices $\{i' \in [k] \mid t_{i'} \in \{a(v_i), b(v_i)\}\}$. Observe that for each $i' \in M_i$ at least one of $r_{i'}$ and $s_{i'}$ is equal to v_i and the other value is greater or equal to v_i. Let M_i^a be the set of those $i' \in M_i$ where $r_{i'} \neq s_{i'}$ and M_i^b the set of those $i' \in M_i$ where $v_i = r_{i'} = s_{i'}$.

Let $\alpha_1, \dots, \alpha_l \in \text{Aut}(\mathbb{Q}; <)$ be such that α_i maps v_i to 0. For each $i \in [l]$ we define $u^i := f(\alpha_i r, \alpha_i s)$. It is easy to see from the choice of α_i and properties of f that for each $i \in [l]$ the tuple u^i is constant at M_i^a, M_i^b, and that the value at M_i^a is lower than the value at M_i^b. Furthermore, because f preserves $<$, because the values of r at $\bigcup_{j=i+1}^{l} M_j$ are greater than v_i, and because the values of s at $\bigcup_{j=i+1}^{l} M_j$ are greater than v_i, we see that for each $j \in [l], j > i$ and each $i' \in M_i, j' \in M_j$ it holds that $u^i_{i'} < u^i_{j'}$. Having this, we can apply Lemma 12.5.3 and obtain a tuple from R with the same ordering of entries as in t, which proves the lemma. ⊣

Example 12.5.22. An interesting example of a temporal relation that is preserved by mx is the ternary relation X defined as follows.

$$X := \{(x, y, z) \mid (x = y \wedge y < z) \vee (x = z \wedge z < y) \vee (y = z \wedge y < x)\}.$$

The relation X is neither preserved by min nor by mi: the tuples $r := (0, 0, 1)$ and $s = (0, 1, 0)$ are in X, but $\min(r, s) = (0, 0, 0) \notin R$, and $\mathrm{mi}(r, s)$ has three distinct entries and hence is not in X as well.

A syntactic description of the relations preserved by mx can be found in Section 12.7.5, and an algorithm that solves constraint languages preserved by mx can be found in Section 12.8.5.

12.5.5. Operations generating min, mi, and mx. As we have seen in Proposition 3.1.9, if the relation T_3 has a primitive positive definition in $\mathfrak{B}$, then $\mathrm{CSP}(\mathfrak{B})$ is NP-hard. In this section we show that if a temporal constraint language is shuffle-closed and does not admit a primitive positive definition of T_3, then it is preserved by min, mi, or mx.

LEMMA 12.5.23. *Let $f \in \mathrm{Pol}^{(2)}(\mathbb{Q}; <)$ and $u_1, u_2, v_1, v_2 \in \mathbb{Q}$ be such that $u_1 < u_2$, $v_1 < v_2$, and $f(u_1, v_1) < f(u, v_1)$ and $f(u_1, v_1) < f(u_1, v)$ for all $u, v \in \mathbb{Q}$ with $u_2 < u$ and $v_2 < v$. Then f locally generates an operation providing min-intersection closure.*

PROOF. Let $a \in \mathrm{End}(\mathbb{Q}; <) = \overline{\mathrm{Aut}(\mathbb{Q}; <)}$ be such that $a(u_1) = 0$ and $u_2 < a(\mathbb{Q}^+)$. Let $b \in \mathrm{End}(\mathbb{Q}; <)$ be such that $b(v_1) = 0$ and $v_2 < b(\mathbb{Q}^+)$. Then the operation $(x, y) \mapsto f(a(x), b(y))$ provides min-union closure. ⊣

LEMMA 12.5.24. *Let $f \in \mathrm{Pol}^{(2)}(\mathbb{Q}; <)$ and $u_1, u_2, v_1 \in \mathbb{Q}$ be such that $u_1 < u_2$, $f(u_2, v_1) \leq f(u_1, v_1)$, and for every $u \in \mathbb{Q}$ with $u_2 < u$ we have $f(u_2, v_1) < f(u, v_1)$. Then f locally generates an operation providing min-intersection closure.*

PROOF. Note that for all $u, v \in \mathbb{Q}$ with $u_2 < u$ and $v_1 < v$ we have $f(u_2, v_1) < f(u, v_1)$ and $f(u_2, v_1) \leq f(u_1, v_1) < f(u_2, v)$ since f preserves $<$, so the statement follows from Lemma 12.5.23. ⊣

LEMMA 12.5.25. *Let $f \in \mathrm{Pol}^{(2)}(\mathbb{Q}; <)$ and $u_1, u_2, v_1, v_2 \in \mathbb{Q}$ be such that $u_1 < u_2$, $v_1 < v_2$, and $f(u_1, v_1) = f(u, v_1)$ and $f(u_1, v_1) = f(u_1, v)$ for all $u, v \in \mathbb{Q}$ with $u_1 \leq u < u_2$ and $v_1 \leq v < v_2$. Then f locally generates an operation providing min-union closure.*

PROOF. Let $a \in \mathrm{End}(\mathbb{Q}; <)$ be such that $a(0) = u_1$ and $a(\mathbb{Q}^+) < u_2$. Let $b \in \mathrm{End}(\mathbb{Q}; <)$ be such that $b(0) = v_1$ and $b(\mathbb{Q}^+) < v_2$. Then the operation $(x, y) \mapsto f(a(x), b(y))$ provides min-union closure. ⊣

LEMMA 12.5.26. *Let $f \in \mathrm{Pol}^{(2)}(\mathbb{Q}; <)$ and $u_1, u_2, v_1 \in \mathbb{Q}$ be such that for all $u, v \in \mathbb{Q}$ with $u_2 < u$ and $v_1 < v$ we have $f(u, v_1) < f(u_1, v_1)$ and $f(u, v_1) = f(u_1, v)$. Then f locally generates an operation providing min-xor closure.*

PROOF. Let $a \in \mathrm{End}(\mathbb{Q}; <)$ be such that $a(0) = u_1$ and $u_2 < a(\mathbb{Q}^+)$. Let $b \in \mathrm{Aut}(\mathbb{Q}; <)$ be such that $b(0) = v_1$. Then the operation $(x, y) \mapsto f(a(x), b(y))$ provides min-xor closure. ⊣

LEMMA 12.5.27. *Let $f \in \mathrm{Pol}^{(2)}(\mathbb{Q};<)$ and $u_1, u_2, v_1, v_2 \in \mathbb{Q}$ be such $u_1 < u_2$, $v_1 < v_2$, and for all $u, v \in \mathbb{Q}$ with $u_2 < u$ and $v_2 < v$ we have $f(u, v_1) < f(u_1, v_1)$ and $f(u, v_1) < f(u_1, v_1)$. Then f locally generates an operation providing min-intersection closure.*

PROOF. Let $a, b \in \mathrm{End}(\mathbb{Q};<)$ be such that $a(0) = u_1$, $u_2 < a(\mathbb{Q}^+)$, $b(0) = v_1$, and $v_2 < b(\mathbb{Q}^+)$. Note that $(x, y) \mapsto f(a(x), b(y))$ satisfies the assumptions on f from the statement so we may from now on assume that $u_1 = v_1 = 0$. Let $c \in \mathrm{End}(\mathbb{Q};<)$ be such that $c(\mathbb{Q}) \subseteq \mathbb{Q}^+$ and define

$$g(x, y) := f(c(f(x, y)), y).$$

It follows from the assumptions on f that for every $x \in \mathbb{Q}^+$ we have $f(0,0) > f(x, 0)$, and hence

$$g(0,0) = f(c(f(0,0)), 0) < f(c(f(x,0)), 0) = g(x, 0).$$

Moreover, for every $y \in \mathbb{Q}^+$ we have $f(0,0) < f(0, y)$, and as f preserves $<$ we get

$$g(0,0) = f(c(f(0,0)), 0) < f(c(f(0, y)), y) = g(0, y)$$

showing that g provides min-intersection closure. ⊣

LEMMA 12.5.28. *Suppose that $f \in \mathrm{Pol}^{(2)}(\mathbb{Q};<)$ does not preserve $\leq$. Then f locally generates an operation providing min-intersection or min-xor closure.*

PROOF. Without loss of generality we may assume that there are $u_1, u_2, v_1 \in \mathbb{Q}$ such that $u_1 < u_2$ and $f(u_2, v_1) < f(u_1, v_1)$; otherwise, exchange $f(x, y)$ by $f(y, x)$. By Theorem 11.4.17 we may suppose that the operation f is canonical with respect to $(((\mathbb{Q};<)^{[2]}, (u_1, v_1), (u_2, v_1)), (\mathbb{Q};<))$. First suppose that $f(u_2, v_1) < f(u, v_1)$ for all $u \in \mathbb{Q}$ such that $u_2 < u$. Then Lemma 12.5.24 implies that f locally generates an operation providing min-intersection closure. So by canonicity we may suppose that $f(u, v_1) \leq (u_2, v_1)$ for all $u \in \mathbb{Q}$ such that $u_2 < u$. Also by canonicity, one of the following cases must apply.

- $f(u_1, v) = f(u, v_1)$ for all $u, v \in \mathbb{Q}$ such that $v_1 < v$ and $u_2 < u$. Then Lemma 12.5.26 implies that f locally generates an operation providing min-xor closure.
- $f(u_1, v) < f(u, v_1)$ for all $u, v \in \mathbb{Q}$ such that $v_1 < v$ and $u_2 < u$. In this case we can apply Lemma 12.5.27 and obtain that f locally generates an operation providing min-intersection closure.
- $f(u_1, v) > f(u, v_1)$ for all $u, v \in \mathbb{Q}$ such that $v_1 < v$ and $u_2 < u$. In this case we can apply Lemma 12.5.27 to $(x, y) \mapsto f(y, x)$ and again obtain that f locally generates an operation providing min-intersection closure. ⊣

Recall that the relation T_3 was defined in Definition 3.1.8 to be

$$\{(x, y, z) \in \mathbb{Q}^3 \mid (x = y < z) \vee (x = z < y)\}.$$

LEMMA 12.5.29. *Suppose that* $f \in \mathrm{Pol}^{(2)}(\mathbb{Q}; <)$ *does not preserve* T_3. *Then* f *locally generates an operation providing min-union, min-intersection, or min-xor closure.*

PROOF. If f does not preserve $\leq$, then we are immediately done by Lemma 12.5.28. So we assume in the following that f preserves $\leq$. As f preserves $<$ and does not preserve T_3 we may assume without loss of generality (possibly after swapping arguments) that there are $x_1, x_2, y_1, y_2 \in \mathbb{Q}$ such that $x_1 < x_2$, $y_1 < y_2$ and

$$t := (f(x_1, y_1), f(x_2, y_1), f(x_1, y_2)) \notin T_3.$$

Because f preserves $\leq$ we have that $f(x_1, y_1) \leq f(x_2, y_1)$ and $f(x_1, y_1) \leq f(x_1, y_2)$. Since $t \notin T_3$, there are only two possibilities:

1. $t_1 = t_2 = t_3$. In this case Lemma 12.5.25 applied to $u_1 := x_1$, $u_2 := x_2$, $v_1 := y_1$, and $v_2 := y_2$ shows that f locally generates an operation providing min-union closure.
2. $t_1 < t_2$ and $t_1 < t_3$. In this case Lemma 12.5.23 shows that f locally generates an operation providing min-intersection closure. ⊣

COROLLARY 12.5.30. *Let* $\mathfrak{B}$ *be a first-order expansion of* $(\mathbb{Q}; <)$ *preserved by* pp *such that* T_3 *is not primitively positively definable in* $\mathfrak{B}$. *Then* $\mathfrak{B}$ *is preserved by* min, mi, *or* mx.

PROOF. If T_3 is not primitively positively definable, the by Lemma 6.1.24 there exists an $f \in \mathrm{Pol}^{(2)}(\mathfrak{B})$ that does not preserve T_3. Lemma 12.5.29 implies that f locally generates an operation providing min-union, min-intersection, or min-xor closure, and so the statement follows from Proposition 12.5.10, 12.5.15, or 12.5.21. ⊣

12.6. The fundamental case distinction

In this section we prove that every first-order expansion $\mathfrak{B}$ of $(\mathbb{Q}; <)$ where Betw is not primitively positively definable is preserved by pp, dual-pp, ll, or dual-ll.

LEMMA 12.6.1. *Let* $\mathfrak{B}$ *be a first-order expansion of* $(\mathbb{Q}; <)$ *such that* Betw *is not primitively positively definable in* $\mathfrak{B}$. *Then there exists a binary* $f \in \mathrm{Pol}(\mathfrak{B})$ *and* $r, s \in$ Betw *such that* $f(r, s)$ *has three distinct entries and* $f(r, s) \notin$ Betw.

PROOF. If Betw is not primitively positively definable in $\mathfrak{B}$ then Theorem 6.1.12 implies that there exists a polymorphism f of $\mathfrak{B}$ that does not preserve Betw. We may assume that f is binary because Betw consists of two orbits of triples under $\mathrm{Aut}(\mathbb{Q}; <)$ (Lemma 6.1.24). Choose $o, p \in$ Betw such that $q := f(o, p) \notin$ Betw. Since f preserves $<$, we can assume without loss of generality that $o_1 < o_2 < o_3$ and $p_1 > p_2 > p_3$. If q has three distinct entries, then we are done. Otherwise we distinguish two cases:

1. $q_1 = q_2 = q_3$: in that case, choose $r, s \in \mathbb{Q}^3$ such that $r_1 < o_1$, $r_2 = o_2$, and $r_3 = o_3$, and $p_2 < s_1 < p_1$, $s_2 = p_2$, and $s_3 = p_3$. It is straightforward to check that $r_1 < r_2 < r_3$ and $s_1 > s_2 > s_3$ and thus both triples belong to Betw. Now, consider $t := f(r, s)$. We have that $t_2 = q_2$, $t_3 = q_3$, and $t_1 < q_1 = t_2 = t_3$ because f preserves $<$. Therefore $t \notin$ Betw. Take r instead of o, s instead of p and proceed with case (2).
2. If exactly two entries in q have the same value, let i, j be their indices and let k be the index of the entry with the unique value. We assume that $q_k > q_i$ (the other case is symmetric). It is straightforward to verify that there is an entry in q such that making the value of this entry smaller would make q injective and it would still not be in Betw. We can assume without loss of generality that i is an index of such an entry. We choose r so that $r_i < o_i$, $r_j = o_j$, $r_k = o_k$, and $r_1 < r_2 < r_3$. We choose s such that $s_i < p_i$, $s_j = p_j$, $s_k = p_k$, and $s_1 > s_2 > s_3$.

Note that $r, s \in$ Betw. The tuple $t := f(r, s)$ satisfies $t_i < q_i$, $t_j = q_j$, and $t_k = q_k$. By the choice of i we conclude that t is injective, $t \notin$ Betw and we are done. ⊣

We use Ramsey theory via Theorem 11.4.10 to prove the following.

LEMMA 12.6.2. *Let $\mathfrak{B}$ be a first-order expansion of $(\mathbb{Q}; <)$ where* Betw *is not primitively positively definable. Then $\mathfrak{B}$ is preserved by* ll, dual-ll, pp, *or* dual-pp.

PROOF. If f does not preserve Betw and preserves $<$, then Lemma 12.6.1 asserts that there are $r, s \in$ Betw such that $t := f(r, s) \notin$ Betw and t is injective. As f preserves $<$, we can assume without loss of generality that $r_1 < r_2 < r_3$ and $s_1 > s_2 > s_3$ (otherwise, we apply the argument to $f(y, x)$). By Theorem 11.4.17 we may suppose that the operation f is canonical with respect to $(((\mathbb{Q}; <)^{[2]}, (r_1, s_1), (r_2, s_2), (r_3, s_3)), (\mathbb{Q}; <))$.

Note that either $t_1 > t_2 < t_3$ or $t_1 < t_2 > t_3$; see Figure 8 for an illustration. In the first case, let

$$R_1 := \{x \in \mathbb{Q} \mid r_1 < x < r_2\}, \qquad S_1 := \{y \in \mathbb{Q} \mid s_3 < y < s_2\},$$
$$R_2 := \{x \in \mathbb{Q} \mid r_3 < x\}, \qquad S_2 := \{y \in \mathbb{Q} \mid s_1 < y\}.$$

In the second case, we choose

$$R_1 := \{x \in \mathbb{Q} \mid r_2 < x < r_3\}, \qquad S_1 := \{y \in \mathbb{Q} \mid s_2 < y < s_1\},$$
$$R_2 := \{x \in \mathbb{Q} \mid x < r_1\}, \qquad S_2 := \{y \in \mathbb{Q} \mid y < s_3\}.$$

and the proof will be similar. Since f preserves $<$, we have

$$f(R_1, S_1) < f(r_2, s_2) < f(r_1, s_1) < f(R_1, S_2)$$
$$\text{and } f(R_1, S_1) < f(r_2, s_2) < f(r_3, s_3) < f(R_2, S_1).$$

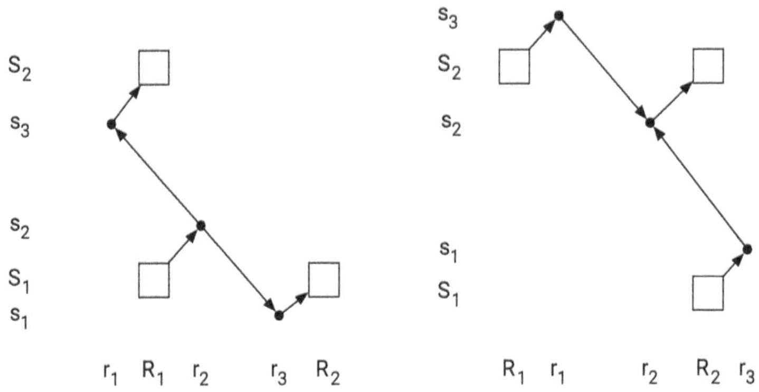

FIGURE 8. Illustration for the proof of Lemma 12.6.2.

First suppose that f is dominated by the same argument on $R_1 \times S_1$, $R_1 \times S_2$, and $R_2 \times S_1$. We can assume that f is dominated on these grids by the second argument; otherwise we swap the arguments of f. Let $g, h \in \{\text{lex}_{y,x}, \text{lex}_{y,-x}, p_y\}$ be such that f satisfies the same type conditions as g on $R_1 \times S_1$ and the same type conditions as h on $R_2 \times S_1$. Then by the above observations f locally generates $[g|h]$. Moreover, we show that f also generates lex.

- If g or h is $\text{lex}_{x,y}$ or $\text{lex}_{y,x}$, then f clearly generates lex.
- If g or h is $\text{lex}_{x,-y}$ or $\text{lex}_{y,-x}$, then f generates lex as well, because $\text{lex}(x, -\text{lex}(x, -y))$ and $\text{lex}(x, y)$ have the same behaviour.
- If g is p_y and h is p_y, then f generates lex by Lemma 12.4.5.

Note that the operation $[g|h]$ satisfies the conditions in Lemma 12.4.4, and hence $\{\text{lex}, [g|h]\}$ locally generates $[\text{lex}_{y,x}|\text{lex}_{y,x}]$. Lemma 12.4.5 implies that f locally generates ll.

Now we consider the case that f is dominated by different arguments on the grids $R_1 \times S_1$ and $R_1 \times S_2$, or by different arguments on the grids $R_1 \times S_1$ and $R_2 \times S_1$. We only consider the first case; the second case is symmetric under swapping the arguments of f. Let g, h be from $\{\text{lex}_{x,y}, \text{lex}_{x,-y}, \text{lex}_{y,x}, \text{lex}_{y,-x}, p_x, p_y\}$ such that f satisfies the same type conditions as h on $R_1 \times S_1$ and the same type conditions as g on $R_2 \times S_1$. Again, f locally generates $[h|g]$. If h is p_y and g is p_x, then $[h|g]$ has the same behaviour as pp; hence f locally generates pp and we are done. Dually, if h is p_x and g is p_y, then f locally generates dual-pp. In all other cases, either h or g is from $\text{lex}_{x,y}$, $\text{lex}_{y,x}$, $\text{lex}_{x,-y}$, or $\text{lex}_{y,-x}$, and thus f locally generates lex as we have already seen before. But then Lemma 12.4.4 shows that f locally generates ll or dual-ll. ⊣

12.7. Syntactic descriptions

A *temporal formula* is a quantifier-free formula that will be interpreted over $(\mathbb{Q};<)$. In this section we present syntactic characterisations of temporal relations that are preserved by ll and by pp. The second characterisation will be refined to characterisations of min-, mi-, and mx-closed relations. All of these characterisations will be *syntactical*, i.e., we present a restricted class of temporal formulas that precisely define the respective relations over $(\mathbb{Q};<)$. As a consequence, we also obtain a better understanding of the clones locally generated by ll, pp, min, mi, and mx, and their duals. This will be used in Section 12.9 to show that these clones satisfy certain pseudo-minor conditions. We also prove that each of these clones $\mathcal{C}$ is finitely related (Definition 6.1.2). The algorithms presented in the next section, however, can be presented without knowing about the syntactic characterisation.

12.7.1. A syntactic description of ll-closed constraints. In this section we present a family of syntactically restricted quantifier-free formulas over $(\mathbb{Q};<)$ such that the relations that are defined by these formulas are precisely the temporal relations preserved by ll.

DEFINITION 12.7.1. A temporal formula is called *ll-Horn* if it is a conjunction of ll-*Horn clauses*, which are formulas of the following form

$$(x_1 = y_1 \wedge \cdots \wedge x_k = y_k) \Rightarrow (z_1 < z_0 \vee \cdots \vee z_l < z_0) \text{ , or}$$
$$(x_1 = y_1 \wedge \cdots \wedge x_k = y_k) \Rightarrow (z_1 < z_0 \vee \cdots \vee z_l < z_0 \vee (z_0 = z_1 = \cdots = z_l))$$

where $0 \leq k, l$.

Note that k or l might be 0: if $k = 0$, we obtain a formula of the form $z_1 < z_0 \vee \cdots \vee z_l < z_0$ or $(z_1 < z_0 \vee \cdots \vee z_l < z_0 \vee (z_0 = z_1 = \cdots = z_l))$, and if $l = 0$ we obtain a disjunction of disequalities. Also note that the variables $x_1, \ldots, x_k, y_1, \ldots, y_k, z_0, \ldots, z_l$ need not be pairwise distinct. On the other hand, the clause $z_1 < z_2 \vee z_3 < z_4$ is an example of a formula that is *not* ll-Horn. Note that every Ord-Horn formula is also ll-Horn. The following result is from [70], but Antoine Mottet found a mistake in the proof presented there; the new proof presented below was found by him.

PROPOSITION 12.7.2. *A temporal relation is preserved by* ll *if and only if it can be defined by an ll-Horn formula.*

PROOF. The proof that every relation defined by an ll-Horn formula is ll-closed is similar to the proof of Proposition 12.4.2. We just need to additionally check that the relation defined by $z_1 < z_0 \vee \cdots \vee z_l < z_0$ and the relation defined by $z_1 < z_0 \vee \cdots \vee z_l < z_0 \vee (z_0 = \cdots = z_l)$ are preserved by ll. So let s and t be two assignments that satisfy $\phi := z_1 < z_0 \vee \cdots \vee z_l < z_0$, and let $r := \text{ll}(s,t)$. Let $i \in \{1, \ldots, l\}$ be such that $s(z_i) = \min(s(z_1), \ldots, s(z_l))$. Note that $s(z_i) < s(z_0)$. Let $j \in \{1, \ldots, l\}$ be such that $t(z_j) < t(z_0)$.

- If $s(z_i) \leq 0$ then $\text{ll}(s(z_i), t(z_i)) < \text{ll}(s(z_0), s(z_0))$ since $s(z_i) < s(z_0)$, and hence r satisfies ϕ.
- If $s(z_i) > 0$, then $s(z_0) > s(z_i) > 0$ and $s(z_j) > s(z_i) > 0$, and hence $\text{ll}(s(z_j), t(z_j)) < \text{ll}(s(z_0), s(z_0))$ since $t(z_j) < t(z_0)$, and hence r satisfies ϕ.

If s and t are satisfying assignments of $z_1 < z_0 \vee \cdots \vee z_l < z_0 \vee (z_0 = \cdots = z_l)$ where one of the assignments satisfies the last clause, then the statement follows from the fact that ll is injective and preserves $\leq$.

Let R be a temporal relation and let ϕ be a quantifier-free formula in CNF that defines R over $(\mathbb{Q}; <)$. In this formula, we write replace literals of the form $\neg(y < x)$ by $x < y \vee x = y$, and we use $x \leq y$ as shortcut for those two literals. For reasons that will become clear later, we additionally allow that clauses contain 'clustered equations' which are expressions of the form $x_1 = x_2 = \cdots = x_n$ and which stand for $x_1 = x_2 \wedge \cdots \wedge x_1 = x_n$; such an expression will be treated as one literal. We describe four rewriting rules that yield a formula ψ that also defines R over $(\mathbb{Q}; <)$ such that R is preserved by ll if and only if ψ is ll-Horn.

1. Suppose that ϕ contains a clause θ of the form $x < y \vee u < v \vee \theta'$, let ϕ' be the other clauses of ϕ, and suppose that
$$(\phi' \wedge \neg\theta' \wedge x < y) \text{implies } (u \leq v \vee x \leq v)$$
$$\text{and } (\phi' \wedge \neg\theta' \wedge u < v) \text{implies } (x \leq y \vee u \leq y).$$
Then replace θ by
$$(u \leq v \vee x \leq v \vee \theta') \wedge (u \neq v \vee x < y \vee \theta')$$
$$\wedge\, (x \leq y \vee u \leq y \vee \theta') \wedge (x \neq y \vee u < v \vee \theta').$$
2. Suppose that ϕ contains a clause θ of the form $x < y \vee u < v \vee \theta'$, let ϕ' are the other clauses of ϕ, and suppose that
$$(\phi' \wedge \neg\theta' \wedge x < y) \text{ implies } u \leq v.$$
Then replace θ by
$$(u \leq v \vee \theta') \wedge (x < y \vee u \neq v \vee \theta').$$
3. Suppose that θ is a clause of ϕ of the form
$$x_1 \neq y_1 \vee \cdots \vee x_k \neq y_k \vee z_1 < z_0 \vee \cdots \vee z_l < z_0 \vee u = v\,,$$
let ϕ' be the other clauses of ϕ, and suppose that
$$\phi' \wedge x_1 = y_1 \wedge \cdots \wedge x_k = y_k \wedge z_0 \leq z_1 \wedge \cdots \wedge z_0 \leq z_l \wedge u = v$$
implies that $z_0 = z_1 = \cdots = z_l$. Then replace θ by
$$(x_1 \neq y_1 \vee \cdots \vee x_k \neq y_k \vee z_0 \neq z_1 \vee \cdots \vee z_0 \neq z_l \vee u = v)$$
$$\wedge\, (x_1 \neq y_1 \vee \cdots \vee x_k \neq y_k \vee z_1 < z_0 \vee \cdots \vee z_l < z_0 \vee z_0 = z_1 = \cdots = z_l).$$

4. If ϕ contains a literal such that removing this literal from ϕ results in an equivalent formula, then remove the literal.

We claim that for each of the four rewriting rules, the resulting formula ψ is equivalent to ϕ.This is obvious for rule (4). To see that ϕ implies the new clauses in rule (1), let s be a satisfying assignment to ϕ.If s satisfies θ', then s also satisfies the new clauses, so let us assume that θ' is false. Then s satisfies $x < y$ or $u < v$. The two cases are symmetric, so we only treat the case that s satisfies $x < y$ in the following. By assumption, s must then satisfy $u \leq v \vee y \leq v$, and hence the first new clause is satisfied by s. Since $x < y$, the other new clauses are satisfied as well.

Now suppose conversely that s is a solution to ϕ' and the four new clauses and that θ does not hold. Because of the second and fourth new clause, we then must have $u \neq v$ and $x \neq y$. Then the first new clause implies that $x \leq v$ and the third new clause implies that $u \leq y$. But then $x \leq v \leq u \leq y \leq x$, a contradiction to $x \neq y$.

For rule (2), let s be a solution to ϕ. Then s obviously satisfies the first new clause if $u < v$ or θ' holds; otherwise, s must satisfy $x < y$ because of θ. But then $u \geq v$ by assumption and hence the first new clause also holds in this case. The second new clause is weaker then θ, so it is also satisfied by s. Now suppose conversely that s satisfies ϕ' and the two new clauses, and suppose for contradiction that θ does not hold. Then in particular $v \leq u$ holds and the first new clause implies that $u = v$, and hence $x < y$ because of the second new clause, contradiction to the assumption that θ does not hold.

Finally, for rule (3), the first new clause is a weakening of θ, and the second new clause is a consequence of ϕ by assumption. Conversely, suppose that s satisfies all clauses of ϕ except for θ which is not satisfied. Then the first new clause implies that $z_1 < z_1 \vee \cdots \vee z_l < z_0$, and thus the second new clause implies that $u = v$, and hence θ holds, contradiction. Hence, ψ is indeed equivalent to ϕ.

Note that rules (1) and (2) strictly reduce the number of pairs of literals $x < y$ and $u < v$ in the same clause where y and v are distinct variables. Rule (3) leaves this number invariant, but strictly reduces the number of literals of the form $u = v$ or of the form $u < v$ in the clause (here, we do not count complex equations). Rule (4) does not increase these numbers, and strictly reduces the total number of literals. Hence, when we repeatedly apply these rules, the procedure will eventually terminate.

Claim 1. The formula ψ cannot contain a clause θ of the form $x < y \vee u < v \vee \theta'$ where x and u are distinct variables. Since rule (1) is not applicable, there must exist a solution s to $\phi' \wedge \neg\theta' \wedge x < y \wedge v < u \wedge v < x$ or to $\phi' \wedge \neg\theta' \wedge u < v \wedge y < x \wedge y < u$. Suppose the former is the case, since the latter case can be treated similarly. Since rule (2) is not applicable, there exists a solution t to $\phi' \wedge \neg\theta' \wedge u < v \wedge y < x$. Let $\alpha \in \mathrm{Aut}(\mathbb{Q}; <)$ be such that $\alpha s(v) = 0$. We claim that $r = \mathrm{ll}(\alpha s, t)$ does not satisfy θ:

- we have $r(y) < r(x)$ since $0 < s(x), s(y)$ and $t(y) < t(x)$;
- we have $r(v) < r(u)$ since $s(v) = 0$ and $s(u) > 0$;
- finally, r does not satisfy θ' since neither s nor t satisfy θ'.

Hence, r does not satisfy ψ, in contradiction to the assumption that ll preserves R.

Claim 2. The formula ψ cannot contain a clause with two distinct literals $x = y$ and $u = v$. This is because rule (4) and since ϕ is preserved by the injection ll.

Claim 3. If ψ contains a clause with a literal $z_1 < z_0$ and a literal $u = v$, then $\{u, v\} = \{x, y\}$. This is because of Claim 1 and Claim 2, any such clause must be of the form $x_1 \neq y_1 \vee \cdots \vee x_k \neq y_k \vee z_1 < z_0 \vee \cdots \vee z_l < z_0 \vee u = v$. Since rule (3) does not apply, there exists a solution s to

$$\phi' \wedge x_1 = y_1 \wedge \cdots \wedge x_k = y_k \wedge z_0 \leq z_1 \wedge \cdots \wedge z_0 \leq z_l \wedge u = v$$
$$\wedge\, z_0 \neq z_1 \vee \cdots \vee z_0 \neq z_l.$$

Hence, there exists an $i \in \{1, \ldots, l\}$ such that $s(z_0) \neq s(z_i)$. Because the literal $z_i < z_0$ cannot be removed from ψ with rule (4), there exists a solution t to ϕ that such that $z_i < z_0$ is the only literal in θ satisfied by t. Let $\alpha \in \mathrm{Aut}(\mathbb{Q}; <)$ be such that $\alpha t(z_i) = 0$. Then $r := \mathrm{ll}(\alpha t, s)$ does not satisfy θ:

- r satisfies $\theta' \wedge x_1 = y_1 \wedge \cdots \wedge x_k = y_k$ since both t and s satisfy this formula.
- $r(z_i) < r(z_0)$ since $0 = \alpha t(z_i) < \alpha t(z_0)$.
- $r(z_j) \leq r(z_0)$ for all $i \in \{1, \ldots, k\} \setminus \{i\}$ since $t(z_0) \leq t(z_j)$ and $s(z_0) \leq s(z_j)$.
- $r(u) \neq r(v)$ since $t(u) \neq t(v)$ and ll is injective.

The three claims imply that each of the clauses of ψ must be logically equivalent to an implication as in Definition 12.7.1, and this concludes the proof. ⊣

There is yet another equivalent description of the temporal relations that are preserved by ll.

THEOREM 12.7.3. *Let $R \subseteq \mathbb{Q}^k$ be a relation with a first-order definition in $(\mathbb{Q}; <)$. The following are equivalent.*

1. *R is preserved by* ll.
2. *R has an* ll*-Horn definition.*
3. *R has a primitive positive definition in $(\mathbb{Q}; \neq, I_4, L)$ where*

$$L := \{(z_0, z_1, z_2) \mid z_1 < z_0 \vee z_2 < z_0 \vee z_0 = z_1 = z_2\}.$$

PROOF. (1) $\Leftrightarrow$ (2) has been shown in Proposition 12.7.2. For (3) $\Rightarrow$ (1) it suffices to show that each of the relations I_4, $\neq$, and L is preserved by ll, which is a straightforward exercise. For a proof sketch of (2) $\Rightarrow$ (3), let ϕ be a ll-Horn formula. It suffices to prove that each of the conjuncts of ϕ defines a relation which has a primitive positive definition in $(\mathbb{Q}; I_4, \neq, L)$. First observe that formulas of the form

$$z_1 < z_0 \vee \cdots \vee z_l < z_0 \vee (z_0 = z_1 = \cdots = z_l)$$

have a primitive positive definition in $(\mathbb{Q}; I_4, L)$, using a similar idea as presented in Section 1.6.8. Moreover, disjunctions of disqualities

$$z_1 \neq z_0 \vee \cdots \vee z_l \neq z_0$$

have a primitive positive definition in $(\mathbb{Q}; \neq, I_4)$, and hence formulas of the form

$$z_1 < z_0 \vee \cdots \vee z_l < z_0$$

have a primitive positive definition in $(\mathbb{Q}; \neq, I_4, L)$. Finally, suppose that $\phi(z_0, \ldots, z_l)$ has a primitive positive definition $\psi(z_0, \ldots, z_l)$ in $(\mathbb{Q}; \neq, I_4, L)$. Then the formula $x = y \Rightarrow \phi(z_0, z_1, \ldots, z_l)$ can be defined by

$$\psi(z'_0, \ldots, z'_l) \wedge \bigwedge_{0 \leq i \leq l} I_4(x, y, z'_i, z_i).$$

⊣

12.7.2. Syntax of shuffle-closed relations. Shuffle-closed temporal relations can be characterised in many different ways; the equivalence between (1) and (3) is from [55].

THEOREM 12.7.4. *Let $R \subseteq \mathbb{Q}^k$ be a relation with a first-order definition in $(\mathbb{Q}; <)$. The following are equivalent.*

1. *R is preserved by* pp.
2. *R is shuffle-closed.*
3. *R can be defined by a conjunction of clauses of the form*

$$y_1 \neq x \vee \cdots \vee y_k \neq x \vee z_1 \leq x \vee \cdots \vee z_l \leq x. \quad (4)$$

4. *R has a primitive positive definition in $(\mathbb{Q}; \neq, R^{\text{min}}_{\leq}, S^{\text{mi}})$ where*

$$R^{\text{min}}_{\leq} := \{(x, y, z) \in \mathbb{Q}^3 \mid y \leq x \vee z \leq x\};$$
$$\textit{and } S^{\text{mi}} := \{(x, y, z) \in \mathbb{Q}^3 \mid y \neq x \vee z \leq x\}.$$

We even may require that all variables in (4) are pairwise distinct. It is permitted that $l = 0$ or $k = 0$, in which cases the clause is a disjunction of disequalities or contains no disequalities, respectively.

PROOF. (1) ⇔ (2) has already been shown in Proposition 12.5.2.

(1) ⇒ (3) Let Φ_R be the set containing all temporal formulas $\phi(v_1, \ldots, v_n)$

- that are implied by $R(v_1, \ldots, v_n)$,
- that are of the form

$$\neg(x_1 \circ_1 x_2 \wedge x_2 \circ_2 x_3 \wedge \cdots \wedge x_{k-1} \circ_{k-1} x_k), \quad (5)$$

where $x_1, \ldots, x_k \in \{v_1, \ldots, v_n\}$ are pairwise different variables and every $\circ_i$ is in $\{<, =\}$, and

- that satisfy the following *unambiguity condition*: if $(v_i = v_j)$ occurs in ϕ, then $i < j$.

For each orbit of n-tuples under $\mathrm{Aut}(\mathbb{Q}; <)$ there is a formula of this form which is false precisely on tuples from this orbit. Since every temporal relation is a union of orbits of n-tuples, it follows that Φ_R defines R. We define an order on Φ_R. For $\phi_1, \phi_2 \in \Phi_R$ of the form $\neg(x_1 \circ_1 x_2 \wedge \cdots \wedge x_{k-1} \circ_{k-1} x_k)$ and $\neg(y_1 \circ_1 y_2 \wedge \cdots \wedge y_{l-1} \circ_{l-1} y_l)$, respectively, we will say that ϕ_1 is less than ϕ_2, in symbols $\phi_1 \preceq \phi_2$, if $x_1, \ldots, x_k$ is a subsequence of $y_1, \ldots, y_l$ and ϕ_1 entails ϕ_2. It is easy to see that this relation is reflexive and transitive. By the definition of Φ_R, we have that two formulas $\phi_1, \phi_2 \in \Phi_R$ with the same set of variables entail each other if and only if they are the same formula. Hence, $(\Phi_R; \preceq)$ is also antisymmetric, and we have defined a partial order on Φ_R. Let Φ_R^m be the set of minimal elements of $(\Phi_R; \preceq)$. Again it is clear that $\bigwedge \Phi_R^m$ defines R.

We obtain the desired definition of R by replacing every member ϕ of Φ_R^m, which is of the form (5), with an equivalent clause of the form (4). If ϕ is of the form $\neg(y_1 = y_2 \wedge \cdots \wedge y_{k-1} = y_k)$, then we replace it by the equivalent formula $(y_1 \neq y_2 \vee \cdots \vee y_{k-1} \neq y_k)$. Otherwise, ϕ contains at least one occurrence of $<$ and hence we can assume that ϕ is of the form:

$$\neg(y_1 = y_2 \wedge \cdots \wedge y_{k-1} = y_k \wedge y_k < z_1 \wedge z_1 \circ_1 z_2 \wedge \cdots \wedge z_{l-1} \circ_{l-1} z_l),$$

where every $\circ_i$ is in $\{=, <\}$. We consider two cases. The first is that R does not contain a tuple t that satisfies[29] the formula $y_1 = \cdots = y_k < z_i$ for every $i \in [l]$. Observe that in this case R entails

$$\psi := (y_2 \neq y_1 \vee \cdots \vee y_k \neq y_1 \vee z_1 \leq y_1 \vee \cdots \vee z_l \leq y_1)$$

and that ψ implies ϕ. Hence, we can safely replace ϕ by ψ in Φ_R^m.

If R does contain a tuple t that satisfies $x = y_1 = \cdots = y_k < z_i$ for every $i \in [l]$, then, as we show, R is not preserved by pp, in contradiction with the assumptions. Let θ be the formula $\neg(z_1 \circ_1 z_2 \wedge \cdots \wedge z_{l-1} \circ_{l-1} z_l)$. We have $\theta \preceq \phi$. Since ϕ is in Φ_R^m, it follows that θ is not in Φ_R, and hence not entailed by R. This implies that R contains a tuple r satisfying $(z_1 \circ_1 z_2 \wedge \cdots \wedge z_{l-1} \circ_{l-1} z_l)$. Let $\alpha \in \mathrm{Aut}(\mathbb{Q}; <)$ be such that $\alpha(r(x)) < 0 < \alpha(r(z_i))$ for all $i \in [l]$. Then $s := \mathrm{pp}(\alpha(r), t)$ satisfies $\neg\phi$. Since ϕ is entailed by R, it follows $s \notin R$, and thus that R is not preserved by pp.

(3) $\Rightarrow$ (4). If suffices to show that every relation defined by a formula ϕ as in (4) has a primitive positive definition in $(\mathbb{Q}; \neq, R_{\leq}^{\mathrm{min}}, S^{\mathrm{mi}})$ We prove this by induction on the number of disjuncts in ϕ. The statement is clearly true if $l = 0$, i.e., if ϕ is a disjunction of disequalities, since I_4 has a primitive positive definition in $(\mathbb{Q}; S^{\mathrm{mi}})$ and hence ϕ is primitively positively definable over $(\mathbb{Q}; \neq, S^{\mathrm{mi}})$, too (see the proof of Theorem 7.5.2). Now suppose inductively

[29] In this proof we view l-tuples of elements of $\mathbb{Q}$ as assignments to the variables $v_1, \ldots, v_l$.

that for $l > 0$ the formula

$$\phi := (y_1 \neq x \vee \cdots \vee y_k \neq x \vee z_1 \leq x \vee \cdots \vee z_{l-1} \leq x)$$

defines a relation with a primitive positive definition $\theta(x, y_1, \ldots, y_k, z_1, \ldots, z_{l-1})$ over $(\mathbb{Q}; \neq, R^{\min}_{\leq}, S^{\mathrm{mi}})$. Then the formula

$$\exists h \big(\theta(x, y_1, \ldots, y_k, h, z_2, \ldots, z_{l-1}) \wedge S^{\mathrm{mi}}(h, z_1, z_l)\big) \tag{6}$$

defines over $(\mathbb{Q}; \neq, R^{\min}_{\leq}, S^{\mathrm{mi}})$ the same relation as $\phi \vee z_l \leq x$. Suppose first that

$$(y_1 \neq x \vee \cdots \vee y_k \neq x \vee h \leq x \vee z_2 \leq x \vee \cdots \vee z_{l-1} \leq x) \wedge \\ (z_1 \leq h \vee z_l \leq h);$$

then $y_1 \neq x \vee \cdots \vee y_k \neq x \vee z_1 \leq x \vee \cdots \vee z_{l-1} \leq x \vee z_l \leq x$. Conversely, suppose that the latter formula holds. If $y_1 = \cdots = y_k = x$ and $x > z_i$ for every $i \in [l-1]$ then we must have $x \leq z_l$ and choose $h := z_l$; this clearly satisfies the quantifier-free part of the formula given in 6. If $y_i \neq x$ for some $i \in [k]$ or $x \leq z_i$ for some $i \in [l-1]$ then the first conjunct of the formula given in 6 is satisfied and we choose $h := \max(z_0, z_{l-1})$, which clearly satisfies the second conjunct as well.

(4) ⇒ (1). It suffices to verify that the relations $R^{\min}_{\leq}$ and S^{mi} are preserved by pp, which is an easy exercise. ⊣

12.7.3. Syntax of min**-closed temporal relations.** We present a syntactic characterisation of temporal relations preserved by min from [55], and a finite set of relations that can pp-define all of those relations. Both results have later been generalised to the larger class of semilinear relations over $\mathbb{Q}$ in [77] (presented there for max-closed rather than min-closed relations, but the results of course dualise).

THEOREM 12.7.5. *Let $R \subseteq \mathbb{Q}^n$ be first-order definable in $(\mathbb{Q}; <)$. Then the following are equivalent.*

1. *R is preserved by* min.
2. *R can be defined as a conjunction of so-called* min-clauses *of the form*

$$z_1 \circ_1 x \vee \cdots \vee z_l \circ_l x, \tag{7}$$

where for all $i \in [l]$, we have that $\circ_i$ is in $\{\leq, <\}$.

3. *R is primitively positively definable over $(\mathbb{Q}; <, R^{\min}_{\leq})$ where*

$$R^{\min}_{\leq} := \{(x, y, z) \in \mathbb{Q}^3 \mid y \leq x \vee z \leq x\}.$$

PROOF. (1) ⇒ (2). Suppose that R is preserved by min. Since min locally generates pp, Theorem 12.7.4 implies that R can be defined by a conjunction of clauses of the form

$$y_1 \neq x \vee \cdots \vee y_k \neq x \vee z_1 \circ_1 x \vee \cdots \vee z_l \circ_l x, \tag{8}$$

where $\circ_i \in \{<, \leq\}$ for every $i \in [l]$. Consider the set of such definitions of R with a minimal number of disequalities and from this set choose one with a minimal number of literals. Denote that formula by ϕ_R.

If ϕ_R does not have any disequalities, then we are done. Otherwise it contains a clause C of the form (8) with at least one disequality $(y_1 \neq x)$. Since ϕ_R does not contain any unnecessary literals, there is a tuple in R that falsifies all literals in C except for $(y_1 \neq x)$. Suppose that R contains both tuples t_1, t_2 that falsify all literals in C except for $(y_1 \neq x)$ and satisfy $(y_1 > x)$ and $(y_1 < x)$, respectively. Let α_1 be an automorphism of $(\mathbb{Q}; <)$ that sends $(t_1(x), t_1(y_1))$ to $(0, 1)$, and α_2 an automorphism of $(\mathbb{Q}; <)$ that sends $(t_2(x), t_2(y_1))$ to $(1, 0)$. Observe that $t_3 = \min(\alpha_1(t_1), \alpha_2(t_2))$ satisfies $(y_1 = x)$. Since min preserves $=$, $\leq$, and $<$, we have that t_3 falsifies all literals in C, hence we have a contradiction with the fact that R is preserved by min. It follows that R does not have either t_1 or t_2. If R does not contain t_1, then in ϕ_R we can replace C by $(y_1 < x \vee y_2 \neq x \vee \cdots \vee y_k \neq x \vee z_1 \circ_1 x \vee \cdots \vee z_l \circ_l x)$ obtaining a definition of R with a smaller number of disequalities. From now on we can therefore assume that R contains t_1.

Consider now the formula ϕ'_R obtained from ϕ_R by replacing the clause C by the clause $C' := (y_1 \neq x \vee y_1 \neq y_2 \vee \cdots \vee y_1 \neq y_k \vee y_1 \circ_1 z_1 \vee \cdots \vee y_1 \circ_l z_l)$. Observe that C and C' entail each other. Hence ϕ'_R also defines R. Consider C' as a part of ϕ'_R and observe that no literal from C' can be removed. Indeed, the new definition would have less disequalities or the same number of disequalities and less literals than ϕ_R. Thus R contains a tuple t' that satisfies all disjuncts of C' except for $(y_1 \neq x)$. As in the previous paragraph, we argue that R cannot have both tuples t'_1, and t'_2 that falsify all literals in C' except for $(x \neq y_1)$ and satisfy $(y_1 < x)$ and $(y_1 > x)$, respectively. If R does not contain t'_1, then in ϕ'_R we can replace C' by $(y_1 > x \vee y_1 \neq y_2 \vee \cdots \vee y_1 \neq y_k \vee y_1 \circ_1 z_1 \vee \cdots \vee y_1 \circ_l z_l)$ obtaining a definition of R with a smaller number of disequalities. Thus, we may assume that R contains t'_1.

Now, suppose towards the contradiction that R has both: t_1 that falsifies all disjuncts of C except for $(x \neq y_1)$ and satisfies $(x < y_1)$; and t'_1 that falsifies all disjuncts of C' except for $(y_1 \neq x)$ and satisfies $(y_1 < x)$. Let α, α' be automorphisms of $\text{Aut}(\mathbb{Q}; <)$ such that α sends $(t_1(x), t_1(y_1))$ to $(0, 1)$ and α' sends $(t'_1(x), t'_1(y_1))$ to $(1, 0)$. To complete the proof we will show that $t_4 = \min(\alpha(t_1), \alpha'(t'_1))$ falsifies all disjuncts of C. Since both t_1 and t'_1 are in R, we obtain a contradiction to the assumption that R is preserved by min. Since $\alpha(t_1(x)) = \alpha(t_1(y_2)) = \cdots = \alpha(t_1(y_k)) = 0$ and $\alpha'(t'_1(y_1)) = \alpha'(t'_1(y_2)) = \cdots = \alpha'(t'_1(y_k)) = 0$, it follows that $t_4(x) = t_4(y_1) = \cdots = t_4(y_k) = 0$ and hence t_4 falsifies all disequalities in C. Now, the clause C contains a disjunct $(x \circ_i z_i)$ with $\circ_i \in \{>, \geq\}$ and $i \in [k]$ if and only if C' contains $(y_1 \circ_i z_i)$. Assume that $\circ_i$ is $>$. The same argument will work for $\geq$. Observe that $\alpha(t_1)$ satisfies $(x \leq z_i)$ and sends x to 0; and that $\alpha'(t'_1)$ satisfies $(y_1 \leq z_i)$ and sends y_1 to 0. It follows that $t_4(x) \leq 0 \leq t_4(z_i)$. Thus t_4 falsifies $(x < z_i)$ and we are done.

(2) $\Rightarrow$ (3). Clauses of the form $z_1 \leq x \vee \cdots \vee z_l \leq x$ have a primitive positive definition over $(\mathbb{Q}; R^{\min}_{\leq})$; This can be shown as in the proof of the implication (3) $\Rightarrow$ (4) in Theorem 12.7.4. It is straightforward to modify the primitive positive definition, additionally using $<$, to define general min-clauses as in (7) primitively positively over $(\mathbb{Q}; <, R^{\min}_{\leq})$.

(3) $\Rightarrow$ (1). It suffices to verify that $<$ and $R^{\min}_{\leq}$ are preserved by min. This is straightforward for $<$ because if the arguments of min are strictly increased, then so is the output of min. To verify that $R^{\min}_{\leq}$ is preserved, let $r, s \in R^{\min}_{\leq}$. We want to show that $t := \min(r, s) \in R^{\min}_{\leq}$. There are $i, j \in \{2, 3\}$ such that $r_i \leq r_1$ and $s_j \leq s_1$. We assume that $r_i \leq s_j$; this is without loss of generality since otherwise we may change the roles of r and s since min is symmetric. Then $r_i \leq s_j \leq s_1$ and $r_i \leq r_1$, and hence $t_i = \min(r_i, s_i) \leq r_i \leq \min(r_1, s_1) = t_1$ showing that $t \in R^{\min}_{\leq}$. ⊣

Example 12.7.6. According to Theorem 12.7.5, the relation U from Example 12.5.5 can be defined by a conjunction of clauses of the form (7). Indeed, observe that $U(x, y, z)$ is equivalent to $(y \leq x \vee z \leq x) \wedge x \leq y \wedge x \leq z$.

12.7.4. Syntax of mi-closed temporal relations. We present various equivalent characterisations of temporal relations preserved by mi.

DEFINITION 12.7.7. A temporal formula is called mi-*Horn* if it is a conjunction of mi-*Horn clauses*, i.e., formulas of the form

$$\begin{aligned} &(y_1 \neq x) \vee \cdots \vee (y_k \neq x) \\ &\vee (z_0 \leq x) \\ &\vee (z_1 < x) \vee \cdots \vee (z_l < x) \end{aligned} \tag{9}$$

where each of the disjuncts may also be omitted.

The equivalence (1) $\Leftrightarrow$ (2) is an unpublished result of Michał Wrona, and the addition of item (3) is from Johannes Greiner and Jakub Rydval.

THEOREM 12.7.8. *Let* $R \subseteq \mathbb{Q}^n$ *be first-order definable over* $(\mathbb{Q}; <)$. *Then the following are equivalent.*

1. *R is preserved by the binary operation* mi.
2. *R can be defined by a* mi-*Horn formula.*
3. *R has a primitive positive definition in* $(\mathbb{Q}; R^{\mathrm{mi}}, S^{\mathrm{mi}})$ *where*

$$R^{\mathrm{mi}} := \{(x, z_0, z_1) \in \mathbb{Q}^3 \mid z_0 \leq x \vee z_1 < x\};$$
$$S^{\mathrm{mi}} := \{(x, y, z) \in \mathbb{Q}^3 \mid y \neq x \vee z \leq x\}.$$

PROOF. For (1) $\Rightarrow$ (2), recall that mi generates pp, and hence R can be defined by a conjunction ϕ of clauses of the form

$$y_1 \neq x \vee \cdots \vee y_k \neq x \vee z_1 \leq x \vee \cdots \vee z_l \leq x$$

for pairwise distinct variables $x_1, \ldots, x_k, y_1, \ldots, y_l, z$ (Theorem 12.7.4). Replace in ϕ a symbol $\leq$ by the symbol $<$ if the resulting formula is logically equivalent. Continue with such replacement steps until this is no longer possible, and let ψ be the resulting formula. Suppose for contradiction that ψ contains a clause C with two literals $z_1 \leq x$ and $z_2 \leq x$ for two distinct variables y_1, y_2. Write V for the set of variables that appear in ϕ.Then there exists an assignment $s_1 : V \to \mathbb{Q}$ that satisfies ψ, satisfies $z_1 = x$, and falsifies all other literals of C. Likewise, there exists an assignment $s_1 : V \to \mathbb{Q}$ that satisfies ψ, satisfies $z_2 = x$ and falsifies all other literals of C. Let $\alpha \in (\mathbb{Q}; <)$ be such that $\alpha(s_1(x)) = s_2(x)$. We claim that the map $s : V \to \mathbb{Q}$ given by $v \mapsto \text{mi}(s_1(v), s_2(v))$ does not satisfy C:

- s preserves $=$, $\leq$, and $<$ and hence does not satisfy literals of C that are falsified by both s_1 and s_2;
- $s(x) < s(z_2) < s(z_1)$ and hence s satisfies neither $z_1 \leq x$ nor $z_2 \leq x$.

This contradicts the assumption that R is preserves by mi.

The implication (2) $\Rightarrow$ (3) can be shown similarly to the implication (2) $\Rightarrow$ (3) in Theorem 12.7.4, and is left to the reader.

For the implication (3) $\Rightarrow$ (1) is suffices to verify that R^{mi} and S^{mi} are preserved by mi. Let $r, s \in R^{\text{mi}}$ and $t := \text{mi}(r, s)$. The following four cases are exhaustive.

1. $r_2 \leq r_1$ and $s_2 \leq s_1$. Then $t_2 \leq t_1$.
2. $r_3 < r_1$ and $s_3 < s_1$. Then $t_3 < t_1$.
3. $r_2 \leq r_1$ and $s_3 < s_1$. If $s_3 < r_2$ then we have $s_3 < r_2 \leq r_1$ and $s_3 < s_1$, so $s_3 < \min(r_1, s_1)$. It follows that $t_3 = \text{mi}(r_3, s_3) < \text{mi}(r_1, s_1) = t_1$. Otherwise, $r_2 \leq s_3$. Then $r_2 \leq s_3 < s_1$ and $r_2 \leq r_1$. Since $\text{mi}(r_2, x) \leq \text{mi}(r_1, s_1)$ for all $x \in \mathbb{Q}$ we have in particular that $t_2 = \text{mi}(r_1, r_2) \leq \text{mi}(r_1, s_1) = t_1$.
4. $r_3 < r_1$ and $s_2 \leq s_1$. If $r_3 < s_2$ then we have $r_3 < s_2 \leq s_1$ and $r_3 < r_1$, so $r_3 < \min(r_1, s_1)$. It follows that $t_3 = \text{mi}(r_3, s_3) < \text{mi}(r_1, s_1) = t_1$. Otherwise, $s_2 \leq r_3$. Then $s_2 \leq r_3 < r_1$ and $s_2 \leq s_1$. Since $\text{mi}(x, s_2) \leq \text{mi}(r_1, s_1)$ for every $x \in \mathbb{Q}$ we have in particular that $t_2 = \text{mi}(r_2, s_2) \leq \text{mi}(r_1, s_1) = t_1$.

In each of the cases we have that $t \in R^{\text{mi}}$. The verification that mi preserves S^{mi} is similarly straightforward and left to the reader. ⊣

12.7.5. Characterisation of mx**-closed relations.** The syntactic description of mx-closed relations is more involved than the syntactic descriptions in the previous sections. For $t \in \{0,1\}^k$ we write $\phi_t(x_1, \ldots, x_k)$ for the temporal formula

$$\Big(\bigwedge_{i,j:t_i=t_j=1} x_i = x_j\Big) \wedge \Big(\bigwedge_{i,j:t_i=1,t_j=0} x_i < x_j\Big).$$

A temporal formula $\phi(x_1, \dots, x_k)$ is called an *min-affine clause* if there exists $I \subseteq [k]$ such that ϕ is of the form

$$\phi_I^k := \bigvee \Big\{ \phi_t \mid t \in \{0,1\}^k, \sum_{i \in I} t_i = 0 \Big\}.$$

We additionally permit $\perp$ (the empty disjunction) as a min-affine clause.

Example 12.7.9. The formula $\phi_{[3]}^3$ defines X from Example 12.5.22.

DEFINITION 12.7.10. Let $t \in \mathbb{Q}^k$. Then the *min-tuple of* t is the Boolean vector $\chi(t) \in \{0,1\}^k$ such that $\chi(t)_i = 1$ if and only if t_i is minimal in t, i.e., $t_i \leq t_j$ for all $j \in [k]$.

Note that the min-tuple of $t \in \mathbb{Q}^n$ always contains at least one entry equal to 1. Observe also that a tuple $t \in \mathbb{Q}^n$ is in the relation defined by a min-affine clause if and only if $s := \chi(t)$ satisfies $\sum_{i \in I} s_i = 0$.

The equivalence of the first two items in Theorem 12.7.11 below is from [55]. The addition of the third item is due to Jakub Rydval, and substantially more difficult to prove than the corresponding implications in Theorem 12.7.4, Theorem 12.7.5, and Theorem 12.7.8.

THEOREM 12.7.11. *Let $R \subseteq \mathbb{Q}^k$ be first-order definable over $(\mathbb{Q}; <)$. Then the following are equivalent.*

1. *R is preserved by* mx.
2. *R is definable by a conjunction (possibly empty, which corresponds to $\top$) of min-affine clauses.*
3. *R is primitively positively definable in $(\mathbb{Q}; X)$.*

PROOF. (1) $\Rightarrow$ (2). We proceed by induction on the arity k of the relation $R \subseteq \mathbb{Q}^k$. The case $k = 0$ is verified as follows: if R is empty, then take the empty conjunction; if R is non-empty, then take the conjunction consisting of the single min-affine clause on no variables with $T = \emptyset$. In what follows, we assume that $k > 0$. Let ϕ be the conjunction of all min-affine clauses that are entailed by $R(v_1, \dots, v_k)$. Suppose that $b = (b_1, \dots b_k) \in \mathbb{Q}^k$ satisfies ϕ. Our goal is to show $b \in R$. Let $t \in \{0,1\}^k$ be the min-tuple of b. We claim that there is a tuple $c \in R$ with min-tuple t. Let $T \in \{0,1\}^k$ be the set of min-tuples of tuples in R. Since mx provides min-xor closure, it follows from the observations in Section 12.5.4 that $T \cup \{(0, \dots, 0)\}$ is preserved by the operation $\text{xor}(s,t) := \text{minority}(s, s', (0, \dots, 0))$. Since for $r, s, t \in \{0,1\}^k$ we have that $\text{minority}(r,s,t) = \text{minority}(\text{minority}(r,s,(0,\dots,0)), t, (0,\dots,0))$, the relation $T \cup \{(0, \dots, 0)\}$ is also preserved by minority and hence can be defined by a system of Boolean linear equations (Proposition 6.2.3). As $(0, \dots, 0)$ must satisfy this system, the system is homogeneous, i.e., every linear equation in the system is of the form $\sum_{i \in I} t_i = 0$ for some $I \subseteq [k]$. The formula ϕ consists of precisely the conjuncts of the form $\phi_I^{[k]}$ for such $I \subseteq [k]$, and the min-tuple of b must satisfy all of them. This shows that there exists $c \in R$ with min-tuple t.

If $t = (1, \dots, 1)$, then $c \in R$ and b are both constant tuples. Since b is equal to c under an automorphism of $(\mathbb{Q}; <)$, we have $b \in R$. So, we suppose that t contains 0 as an entry; for the sake of notation, we assume that t has the form $(0, \dots, 0, 1, \dots, 1)$, where the first m entries are 0; here, $0 < m < k$. We have that $(b_1, \dots, b_m)$ satisfies all min-affine clauses on variables from $v_1, \dots, v_m$ which are entailed by $R(v_1, \dots, v_k)$. Hence, by induction, it holds that $(b_1, \dots, b_m) \in \pi^k_{\{1,\dots,m\}} R$ (we use the notation from Definition 6.8.3), and that there exists a tuple of the form $(b_1, \dots, b_m, d)$ in R, where d is a tuple of length $(k - m)$. We can apply an automorphism to the tuple $(b_1, \dots, b_m, d)$ to obtain a tuple $(b'_1, \dots, b'_m, d') \in R$ where all entries are positive. Also, by applying an automorphism to $c \in R$, we can obtain a tuple of the form $(c'_1, \dots, c'_m, 0, \dots, 0)$ where, for all coordinates i from 1 to m, it holds that $c'_i > b'_i$. Applying mx to the tuples $(b'_1, \dots, b'_m, d')$ and $(c'_1, \dots, c'_m, 0, \dots, 0)$, one obtains the tuple $\big(\mathrm{mx}(b'_1, c'_1), \dots, \mathrm{mx}(b'_m, c'_m), \mathrm{mx}(d', (0, \dots, 0))\big)$, which is equivalent to b under an automorphism.

(2) $\Rightarrow$ (3). We refer to [87].

(3) $\Rightarrow$ (1) is straightforward to verify and has already been mentioned in Example 12.5.22. ⊣

12.8. Polynomial-time algorithms

We present polynomial-time algorithms for temporal CSPs that are preserved by at least one of ll, min, mi, mx (or their duals). All the algorithms also work for *infinite* signatures if the relations are appropriately represented. For example, we may represent an n-ary temporal relation by the set of orbits of n-tuples in the relation where each orbit is represented by the respective weak linear order induced on the entries of the tuples in that orbit. Another possibility is to represent a temporal relation by the first-order definition in the syntactically restricted forms that we presented in the previous section. Note that each of the four syntactically restricted forms that we presented may be viewed as a temporal formula in *disjunctive normal form* (*DNF*); we therefore often work with temporal formulas in DNF in general. As always, the representation does not matter if the signature of the template is finite.

We need an operation on conjunctions of temporal relations in DNF which is important operation in all of our algorithms. If ϕ is a temporal formula, we write $V(\phi)$ for the set of free variables of ϕ.

DEFINITION 12.8.1. If $\phi = (\phi_1 \wedge \dots \wedge \phi_m)$ is a conjunction of temporal formulas with $V(\phi) = \{x_1, \dots, x_n, y_1, \dots, y_k\}$ then we write $\phi[\{x_1, \dots, x_n\}]$

for the formula

$$(\exists y_1, \dots, y_k \colon \phi_1) \wedge \cdots \wedge (\exists y_1, \dots, y_k \colon \phi_m)$$

(which is in general of course not equivalent to $\exists y_1, \dots, y_k \colon \phi$).

Note that if ϕ_i for $i \in \{1, \dots, m\}$ is represented by a quantifier-free formula in DNF, then a quantifier-free DNF representation of $\exists y_1, \dots, y_k \colon \phi_i$ can be computed in linear time in the representation size of ϕ_i. Another important concept for all of the algorithms in this section are min-sets.

DEFINITION 12.8.2. Let $\psi(x_1, \dots, x_k)$ be a temporal formula. A *min-set of* $\psi(x_1, \dots, x_k)$ is a set $L \subseteq \{x_1, \dots, x_k\}$ such that there exists a tuple $t \in \mathbb{Q}^k$ with $M(t) = \{i \in [k] \mid x_i \in L\}$. For $S \subseteq \{x_1, \dots, x_k\}$, we write $\uparrow^{\psi}(S)$ for the set of all min-sets of ψ which contain S, and $\downarrow^{\psi}(S)$ for the set of all min-sets of ψ which are contained in S.

Note that if ψ is preserved by an operation providing min-intersection closure then $\bigcap \uparrow^{\psi}(S)$ is itself a min-set that contains S, and if ψ is preserved by an operation providing min-union closure then $\bigcup \downarrow^{\psi}(S)$ is itself a min-set that is contained in S.

12.8.1. An algorithm for ll-closed constraints. In this section we present an algorithm to decide the satisfiability of a conjunction of temporal formulas that are preserved by ll and given in DNF. One of the underlying ideas of the algorithm is to use a subroutine that tries to find a solution where every variable has a different value. If this is impossible, the subroutine must return a set of at least two variables that denote the same value in all solutions – since the constraints are preserved by a binary injective operation, such a set must exist (Proposition 7.1.6).

Let $\psi(x_1, \dots, x_k)$ be a temporal formula preserved by ll, and hence in particular by lex. Note that lex is an operation providing min-intersection closure, and thus, as we have mentioned in the previous section, for any $i \in [k]$ the set $\uparrow^{\psi}(\{x_i\})$ is a min-set of ψ.

LEMMA 12.8.3. *Let* $\psi(x_1, \dots, x_k)$ *be a temporal relation preserved by* lex. *Let* $i \in [k]$ *and* $S := \bigcap \uparrow^{\psi}(\{x_i\})$, *i.e.,* S *is the minimal min-set that contains* x_i. *Then*

$$\psi \wedge \bigwedge_{x \in S} x_i \leq x \text{ implies } \bigwedge_{x \in S} x_i = x.$$

PROOF. We prove the contraposition. Suppose that $r \in \mathbb{Q}^k$ satisfies $\psi \wedge x_i \neq x_\ell$ for some $x_\ell \in S$. Since S is a min-set of ψ there exists $s \in \mathbb{Q}^k$ satisfying ψ such that $S = \{x_j \mid j \in M(s)\} =: P$. Consider the tuple $t := \text{lex}(s, r)$ which satisfies ψ, too. Then $t_i < t_j$ for every $j \in [k] \setminus P$, because lex preserves $<$. If $r_i < r_\ell$ then $t_i < t_l$ because $s_i = s_\ell$, which is in contradiction to the minimality of S. Hence, $r_i > r_\ell$ and thus $t_i > t_\ell$. ⊣

To develop our algorithm, we use a specific notion of *constraint graph* of a conjunction of temporal formulas, defined as follows.

DEFINITION 12.8.4. The *constraint graph* G_ϕ of a conjunction ϕ of temporal formulas is a directed graph $(V; E)$ defined on the variables V of ϕ. For each conjunct $\psi(x_1, \ldots, x_k)$ of ϕ we add a directed edge (x_i, x_j) to E if in every satisfying assignment for ψ where x_i is minimal, x_j is minimal as well.

DEFINITION 12.8.5. If a conjunction of temporal formulas contains a conjunct ϕ such that ϕ does not admit a solution where the variable y denotes the minimal value, the we say that $y \in V(\phi)$ is *blocked* (*by* ϕ).

Note that if a temporal formula ϕ is represented as a list of orbits or even by a formula in disjunctive normal form then we can efficiently determine which variables are blocked by ϕ. Thus, by inspecting all the constraints in an instance it is possible to compute the blocked variables in linear time in the input size. We want to use the constraint graph to identify variables that have to denote the same value in all solutions, and therefore introduce the following concepts.

DEFINITION 12.8.6. A strongly connected component K of the constraint graph G_ϕ for a conjunction ϕ of temporal formulas is called a *sink component* if no edge in G_ϕ leaves K. We call K *unblocked* if all elements of K are unblocked. A vertex of G that belongs to a sink component of size one is called a *sink*.

LEMMA 12.8.7. *Let $\phi(x_1, \ldots, x_n)$ be a conjunction of temporal formulas that are preserved by* lex *and let $K \subseteq \{x_1, \ldots, x_n\}$ be an unblocked sink component of G_ϕ. Then in every satisfying assignment to ϕ all variables from K must have equal values.*

PROOF. Let $s \colon V \to \mathbb{Q}$ be a satisfying assignment to ϕ.Suppose that K has at least two vertices; otherwise the statement is trivial. Define

$$M := \{x \in K \mid s(x) \leq s(y) \text{ for all } y \in K\}.$$

We want to show that $M = K$. Otherwise, since K is a strongly connected component, there is an edge in G_ϕ from some variable $u \in M$ to some variable $v \in K \setminus M$. By the definition of G_ϕ there is a conjunct ψ of ϕ such that if u denotes the minimal value in a satisfying assignment for ψ, then v has to denote the minimal value as well. Note that

$$S := \bigcap \uparrow^{\psi}(\{u\})$$

is non-empty, because $u \in K$ is by assumption unblocked. Also note that G_ϕ contains an edge from u to w for all $w \in S$. Since K is a sink component, all these variables w are in K. There is no variable $w \in S$ such that $s(w) < s(u)$, because $s(u) \leq s(y)$ for all $y \in K$. This contradicts Lemma 12.8.3, because $s(u) \neq s(v)$. ⊣

```
Spec(ϕ)
// Input: A conjunction ϕ of ll-closed temporal formulas in DNF.
// Output: If Spec returns false then ϕ has no solution.
// If ϕ has an injective solution, then Spec returns true.
// Otherwise return S ⊆ V(ϕ), |S| ≥ 2, such that
// for all x, y ∈ S we have x = y in all solutions to ϕ.
Set X := ∅;
While G_ϕ contains an unblocked sink s
    X := X ∪ {s};
    If X = V(ϕ) then return true
    else ϕ := ϕ[V(ϕ) \ X].
If G_ϕ has an unblocked sink component S then return S
else return false.
```

FIGURE 9. A polynomial-time algorithm to decide satisfiability of conjunctions of temporal formulas each preserved by ll: the sub-procedure Spec.

Lemma 12.8.7 immediately implies that for checking satisfiability we may replace ϕ by the conjunction of temporal formulas obtained from ϕ where all the variables in K are *contracted*, i.e., where all variables from K are replaced by the same variable.

LEMMA 12.8.8. *Let ϕ be a conjunction of temporal formulas that are preserved by* ll *and let $x \in V := V(\phi)$ be an unblocked sink in G_ϕ. If $\phi[V \setminus \{x\}]$ has an injective solution, then ϕ has an injective solution as well.*

PROOF. Let $s \colon V \to \mathbb{Q}$ be an injective solution to $\phi[V \setminus \{x\}]$. We claim that any extension s' of s to x such that $s'(x) < s(y)$ for all $y \in V \setminus \{x\}$ is injective and satisfies ϕ. If x does not appear in ϕ, then the statement is trivial. Consider a constraint $\psi(x_1, \ldots, x_k)$ from ϕ that is imposed on x; without loss of generality, $x = x_1$. By the definition of $\phi[V \setminus \{x\}]$ there exists $t \in \mathbb{Q}^k$ satisfying ψ such that $t_i = s(x_i)$ for all $i \in \{2, \ldots, k\}$. Because x is unblocked, $\bigcap \uparrow^{\psi}(\{x\})$ is non-empty, and hence there exists a t' satisfying ψ such that $M(t') = \{i \in [k] \mid x_i \in S\}$. Let $\alpha \in \mathrm{Aut}(\mathbb{Q}; <)$ be such that $\alpha t'_1 = 0$. Then $r := \mathrm{ll}(\alpha t', t)$ satisfies ψ. Note that for all $i, j \in [k]$ we have that $r_i \leq r_j$ if and only if $s'(x_i) \leq s'(x_j)$. Hence, s' satisfies all constraints from ϕ, which is what we had to show. ⊣

Our algorithm for ll-closed constraints can be found in Figure 10; we are now ready to prove its correctness.

THEOREM 12.8.9. *The procedure Solve in Algorithm* 10 *decides satisfiability of a conjunction of ll-closed temporal formulas ϕ given in disjunctive normal form.*

```
Solve(φ)
// Input: A conjunction φ of ll-closed temporal formulas in DNF.
// Output: accept if φ is satisfiable in (ℚ; <), reject otherwise.
S := Spec(φ)
If S = false then reject
else if S = true then accept
else
   Let φ' be contraction of S in φ.
   Return Solve(φ').
end if
```

FIGURE 10. The main procedure to decide satisfiability of conjunctions of temporal formulas preserved by ll in polynomial time.

There is an implementation of the algorithm that runs in time $O(nm)$ where n is the number of variables of ϕ and m is the size of the input.

PROOF. The correctness of the procedure Spec immediately implies the correctness of the procedure Solve. In the procedure Spec, after iterated deletion of sinks in G_ϕ, we have to distinguish three cases.

First, consider the case $V(\phi) = X$. In this case it follows from Lemma 12.8.8 by a straightforward induction that ϕ has an injective solution. Otherwise, consider the case that G_ϕ contains an unblocked sink component S with $|S| \geq 2$. Lemma 12.8.7 applied to $\phi[V(\phi) \setminus X]$ implies that all variables in S must have the same value in every solution, and hence the output is correct in this case as well.

In the third case we have $X \neq V(\phi)$ but G_ϕ does not contain an unblocked sink component. Note that in every solution to ϕ some variable must take the minimal value. However, since each strongly connected component without outgoing edges contains a blocked vertex, there is no variable that can denote the minimal element, and hence ϕ has no solution. Because ϕ is at all times of the execution of the algorithm implied by the original input constraints, the algorithm correctly rejects.

In each recursive call of Solve the instance in the argument has at least one variable less, and therefore Solve is executed at most n times. It is not difficult to implement the algorithm such that the total running time is cubic in the input size. However, it is possible to implicitly represent the constraint graph and to implement all sub-procedures such that the total running time is in $O(nm)$; for the details, we refer to [70]. ⊣

COROLLARY 12.8.10. *If $\mathfrak{B}$ is a finite-signature first-order reduct of $(\mathbb{Q}; <)$ which is preserved by min then there is an algorithm that solves* CSP($\mathfrak{B}$) *in time $O(nm)$ where n is the number of variables and m is the number of constraints.*

12.8.2. Algorithms for shuffle-closed languages. In the following we present three algorithms for shuffle-closed temporal relations, namely for constraints preserved by mi, by min, and by mx, respectively. All three algorithms follow a common strategy. They are searching for a subset of the variables that can have the minimal value in a solution. If they have found such a subset S, the algorithms add equalities and inequalities that are implied by all constraints under the assumption that the variables in S denote the minimal value in *all* solutions. Next, the algorithms recursively solve the instance consisting of the projections of all constraints to the variables that do not denote the minimal value in all solutions. We later show that if the instance has a solution and all constraints are shuffle-closed, then the instance also has a solution that satisfies all the additional constraints.

DEFINITION 12.8.11. Let ϕ be a conjunction of temporal formulas. Then $F \subseteq V(\phi)$ is called *free* if it is non-empty and for every conjunct $\psi(x_1, \ldots, x_k)$ of ϕ the set $F \cap \{x_1, \ldots, x_k\}$ is either empty or a min-set of the temporal relation defined by ψ (Definition 12.8.2).

Free sets can be used to decide the satisfiability of conjunctions of shuffle-closed temporal formulas. The following is a strengthening of a lemma from [69] due to Jakub Rydval [87], which allows a simplification in the later presentation.

LEMMA 12.8.12. *Let ϕ be a conjunction of shuffle-closed temporal formulas and let $F \subseteq V := V(\phi)$ be a union of free sets. Then ϕ has a solution if and only if $\phi[V \setminus F]$ has a solution.*

PROOF. Let $F_1, \ldots, F_k$ be free sets of ϕ and let $F := F_1 \cup \cdots \cup F_k$. Clearly, if ϕ has a solution then so has $\phi[V \setminus F]$. For the converse, suppose that $\phi[V \setminus F]$ has a solution s. Let $S_i := F_i \setminus (F_1 \cup \cdots \cup F_{i-1})$ for every $i \in \{1, \ldots, k\}$ (so $S_1 = F_1$). We claim that any extension s' of s is a solution to ϕ if $s'(S_1) < s'(S_2) < \cdots < s'(S_k) < s'(V \setminus F)$ and $s'(x) = s'(y)$ whenever there exists $i \in \{1, \ldots, k\}$ such that $x, y \in S_i$.

To verify this, let $\psi(x_1, \ldots, x_m)$ be a conjunct of ϕ such that, without loss of generality,

$$\{x_1, \ldots, x_m\} \cap F = \{x_1, \ldots, x_\ell\} \neq \emptyset.$$

By the definition of $\phi[V \setminus F]$ there is a tuple $t \in \mathbb{Q}^m$ that satisfies ψ and such that $t_j = s(x_j)$ for every $j \in \{\ell + 1, \ldots, m\}$. Since $F_1, \ldots, F_k$ are free, there are tuples $t^1, \ldots, t^k \in \mathbb{Q}^m$ such that for every $i \in \{1, \ldots, k\}$ the tuple t^i satisfies $\psi(x_1, \ldots, x_m)$ and for every $j \in \{1, \ldots, m\}$

$$j \in M(t^i) \text{ if and only if } x_j \in F_i.$$

For every $i \in \{1, \ldots, k\}$ let $\alpha^i \in \mathrm{Aut}(\mathbb{Q}; <)$ be such that α^i maps the minimal entry of t^i to 0. The tuple $r^i := \mathrm{pp}(\alpha^i t^i, t)$ satisfies ψ, because ψ is preserved by pp. It follows from the definition of pp that $j \in M(r^i)$ if and only if $x_j \in F_i$

for all $j \in \{1, \dots, m\}$. Moreover, $(r^i_{\ell+1}, \dots, r^i_m)$ and $(t_{\ell+1}, \dots, t_m)$ lie in the same orbit under Aut($\mathbb{Q}$; <). Define $p^k, p^{k-1}, \dots, p^1 \in \mathbb{Q}^m$ in this order as follows. Define $p^k := r^k$ and for $i \in \{1, \dots, k-1\}$

$$p^i := \text{pp}(\beta^i r^i, p^{i+1})$$

where $\beta^i \in \text{Aut}(\mathbb{Q}; <)$ is chosen such that $\beta^i(r^i_j) = 0$ for all $j \in M(r^i)$. We verify by induction that for all $i \in \{1, \dots, k\}$

1. p^i satisfies ψ;
2. $(p^i_{\ell+1}, \dots, p^i_m)$ and $(t_{\ell+1}, \dots, t_m)$ lie in the same orbit under Aut($\mathbb{Q}$; <);
3. $j \in M(p^i)$ if and only if $x_j \in F_i$ for all $j \in \{1, \dots, m\}$;
4. $p^i_u = p^i_v$ for all $u, v \in \{1, \dots, m\}$ with $x_u, x_v \in S_a$ for some $a \in \{i+1, \dots, k\}$;
5. $p^i_u < p^i_v$ for all $a, b \in \{i, i+1, \dots, k\}$ with $a < b$ and $x_u \in S_a$, $x_v \in S_b$.

For $i = k$ the items (1), (2), and (3) follow from the respective property of r^k and items (4) and (5) are trivial. For the induction step and $i \in \{1, \dots, k-1\}$ we have that $p^i = \text{pp}(\beta^i r^i, p^{i+1})$ satisfies item (1) and (2) because p^{i+1} satisfies item (1) and (2) by inductive assumption. For item (3), note that $M(p^i) = M(r^i)$. Finally, if $x_u, x_v \in S_{i+1} \cup \dots \cup S_k$ then $p^i_u \leq p^i_v$ if and only if $p^{i+1}_u \leq p^{i+1}_v$. This implies items (4) and (5) by induction. Finally, note that $(s'(x_1), \dots, s'(x_m))$ lies in the same orbit as p^1 and hence satisfies ψ. ⊣

```
Solve(φ)
// Input: A conjunction φ of shuffle-closed temporal formulas given in DNF.
// Output: A solution s to φ, or false if there is no solution.
i := 0;
While V(φ) ≠ ∅ do
    F := FindFreeSets(φ);
    If F = ∅ then return false;
    For each x ∈ F do s(x) := i
    i := i + 1;
    φ := φ[V(φ) \ F].
Return s.
```

FIGURE 11. A polynomial-time algorithm that decides satisfiability of conjunctions of shuffle-closed temporal formulas if free sets can be computed efficiently.

The above lemma implies that if we are able to identify a union of free sets for conjunctions ϕ of shuffle-closed temporal formulas in polynomial time, then we also have a polynomial-time algorithm that decides the satisfiability of

```
FindFreeSetMin(φ)
// Input: A conjunction φ of min-closed temporal formulas given in
DNF.
// Output: ∅ or a free subset S of the variables of φ.
// If the algorithm returns ∅, then φ is unsatisfiable.
S := V(φ);
Do
    For every conjunct ψ of φ do
        If S ∩ V(ψ) ≠ ∅ then
            S := (S \ V(ψ)) ∪ ⋃↓^ψ(S ∩ V(ψ));
Loop while S does not change.
Return S.
```

FIGURE 12. A polynomial-time algorithm that computes a free set for a satisfiable conjunction of min-closed temporal formulas in DNF.

ϕ in polynomial time, namely the algorithm shown in Figure 11. The running time of the algorithm is $O(n \cdot (m + t(n, m)))$, where $n = |V|$ is the number of variables, m is the size of the input formula ϕ, and $t(n, m)$ is the running time of the procedure that computes a free set.

12.8.3. An algorithm for languages preserved by min. In order to show that temporal CSPs for min-closed constrains are in P, it suffices to find a polynomial-time algorithm that computes a free set for a given conjunction of temporal formulas that are preserved by min and given in DNF, by the results of the previous section.

Let $\psi(x_1, \ldots, x_k)$ be a conjunct of ϕ and let L be a subset of $\{x_1, \ldots, x_k\}$. Note that if $\downarrow^\psi(L)$ is non-empty, i.e., if ψ has a min-set contained in L, then $\bigcup\downarrow^\psi(L)$ is a min-set contained in L, too, because ψ is preserved by min, which provides min-union closure by Lemma 12.5.9. We call this min-set the *maximal min-set of ψ contained in L*.

Figure 12 shows our procedure for finding a free set for conjunctions ϕ of min-closed temporal formulas. The procedure FindFreeSetsMin can be implemented so that its running time is in $O(m)$ where m is the length of the formula ϕ.

LEMMA 12.8.13. *Let ϕ be a conjunction of temporal formulas preserved by* mx *and given in DNF or as affine clauses. Then the procedure FindFreeSetMin in Figure* 12 *returns a free set of ϕ or rejects. If it rejects, then ϕ is unsatisfiable.*

PROOF. Suppose that the algorithm returns a non-empty set S. Therefore, for all conjuncts $\psi(x_1, \ldots, x_k)$ of ϕ such that $S \cap \{x_1, \ldots, x_k\} \neq \emptyset$ the maximal min-set of ψ contained in S equals $S \cap \{x_1, \ldots, x_k\}$. We conclude that S is free.

We now have to argue that if ϕ returns $\emptyset$, then ϕ is unsatisfiable. We show the contraposition, and suppose that ϕ has a solution. Then there is $S' \subseteq V(\phi)$ that have the minimal value in this solution. At the beginning of the procedure, $S = V(\phi)$ and therefore $S' \subseteq S$. We show that $S' \subseteq S$ during the entire execution of the procedure. Let $\psi(x_1, \ldots, x_k)$ be a conjunct of ϕ.Because $S' \cap \{x_1, \ldots, x_k\}$ is a min-set of ψ that is contained in S, the maximal min-set of ψ added to $S \setminus \{x_1, \ldots, x_k\}$ certainly contains $S' \cap \{x_1, \ldots, x_k\}$. Therefore, after the modification to S it still holds that $S \supseteq S' \neq \emptyset$ and hence the algorithm does not return $\emptyset$. ⊣

THEOREM 12.8.14. *There is an algorithm that decides the satisfiability of conjunctions ϕ of temporal formulas, each preserved by* min *and given in DNF, in time $O(nm)$, where n is the number of variables of ϕ and m the size of ϕ.*

PROOF. We use the procedure FindFreeSetMin in Figure 12 for the subroutine FindFreeSets in Figure 11. Then Lemma 12.8.12 and Lemma 12.8.13 imply the correctness of the resulting algorithm. ⊣

COROLLARY 12.8.15. *If $\mathfrak{B}$ is a finite-signature first-order reduct of* $(\mathbb{Q}; <)$ *which is preserved by* min *then there is an algorithm that solves* CSP$(\mathfrak{B})$ *in time $O(nm)$ where n is the number of variables and m the number of constraints.*

12.8.4. An algorithm for languages preserved by mi. In order to show that temporal CSPs for mi-closed constrains are in P, it suffices to find a polynomial-time algorithm that computes a free set for a given conjunction of temporal formulas that are preserved by mi and given in DNF, by the results of Section 12.8.2.

Let $\psi(x_1, \ldots, x_k)$ be a conjunct of ϕ and let $L \subseteq \{x_1, \ldots, x_k\}$. Note that if $\uparrow^{\psi}(L)$ is non-empty then $\bigcap \uparrow^{\psi}(L)$ is a min-set that contains, L, too, because ψ is preserved by mi, which provides min-intersection closure by Lemma 12.5.13. We call this min-set the *minimal min-set of R containing L*. A procedure for finding a union of free sets is given in Figure 13. It is straightforward to verify that the algorithm runs in time $O(n^2m)$ where n is the number of variables of ϕ and m is the size of ϕ.

LEMMA 12.8.16. *Let ϕ be a conjunction of temporal formulas preserved by* mi *and given in DNF. Then the procedure FindFreeSetsMi in Figure* 13 *returns a union of free set S of ϕ.If it returns $\emptyset$, then ϕ is unsatisfiable.*

PROOF. Suppose that y is an element of the set S returned by the algorithm. Then for some $x \in V(\phi)$, the set T computed by the inner loop must contain y. So for all conjuncts ψ of ϕ such that $V(\psi) \cap T \neq \emptyset$ the set T did not change. This implies that for all these constraints the minimal min-set of ψ containing $T \cap V(\psi)$ is equal to $T \cap V(\psi)$. Hence, T is a free set of ϕ; it follows that S is a union of free sets.

To show that if the set S returned by the algorithm is empty, then ϕ is unsatisfiable, we prove the contraposition. Suppose that ϕ has a solution. Then there

```
FindFreeSetsMi(φ)
// Input: a conjunction φ of mi-closed temporal formulas given in DNF.
// Output: a union of free sets of φ.
// If the algorithm returns ∅, then φ is unsatisfiable.
S := ∅;
For every x ∈ V(φ) do
     T := {x};
     Do
          For all conjuncts ψ(x_1, ..., x_k) of φ such that {x_1, ..., x_k}∩T ≠ ∅
do
               If ↑^ψ(S ∩ {x_1, ..., x_k}) = ∅ then T := ∅;
               else T := T ∪ ⋂↑^ψ(S ∩ {x_1, ..., x_k}).
     Loop until T does not change.
     S := S ∪ T.
Return S.
```

FIGURE 13. A polynomial-time algorithm that computes a union of free sets for a satisfiable conjunction of temporal formulas preserved by mi.

is some set S' of variables that have the minimal value in this solution. Consider a run of the while loop in the procedure FindFreeSetsMi for some variable $x \in S'$. In the beginning, it holds that $T = \{x\} \subseteq S'$. For each constraint ψ from Φ we have that $S' \cap V(\psi)$ is a min-set of ψ if $S' \cap V(\psi)$ is non-empty. Because we always only add to T variables of the *minimal* min-set of ψ containing $T \cap V(\psi)$, all the variables in T are contained in S'. Therefore, T remains a subset of S' all the time, and the output S of the algorithm is non-empty. ⊣

THEOREM 12.8.17. *There is an algorithm that decides the satisfiability of a given conjunction ϕ of temporal formulas, each preserved by* mi *and given in DNF, in time $O(n^3m)$, where $n = |V(\phi)|$ and m is the size of ϕ.*

PROOF. We use the procedure FindFreeSetsMi in Figure 13 for the subroutine FindFreeSet in Figure 11. Lemma 12.8.12 and Lemma 12.8.16 imply the correctness of these algorithms. ⊣

COROLLARY 12.8.18. *If $\mathfrak{B}$ is a finite-signature first-order reduct of $(\mathbb{Q};<)$ which is preserved by* mi, *then there is an algorithm solving* CSP($\mathfrak{B}$) *in time $O(n^3m)$ where n is the number of variables and m the number of constraints in an instance.*

12.8.5. An algorithm for languages preserved by mx. Finally, we consider the satisfiability problem for conjunctions ϕ of temporal formulas that are preserved by mx and given in DNF; alternatively (and even more succinctly), we also allow that the input is a finite disjunction of min-affine clauses (Section 12.7.5;

```
FindFreeSetsMx(φ)
// Input: a conjunction φ of mx-closed temporal formulas,
// given in DNF or as min-affine clauses.
// Output: ∅ or a free set of φ.
// If the algorithm returns ∅, then φ is unsatisfiable.
Let E = ∅, S = ∅.
For every ψ = φ_I^k(x_1, ..., x_k) ∈ φ do
    E := E ∪ {∑_I x_i = 0}.
If there exists a nontrivial s: V(φ) → {0, 1} satisfying E
    return {x ∈ V(φ) | s(x) = 1}
else return ∅.
```

FIGURE 14. A polynomial-time algorithm that computes a free set for a satisfiable conjunction of mx-closed temporal formulas.

note that the size of a temporal formula in DNF which is equivalent to ϕ_I^k has exponential size in k).

Let $\psi(x_1, \ldots, x_k)$ be a conjunct of ϕ. We define

$$\chi(\psi) := \{\chi(t) \mid t \in \mathbb{Q}^k \text{ satisfies } \psi\}.$$

Since ψ is preserved by mx and mx provides min-xor-closure by Lemma 12.5.20, the set $\chi(\psi) \cup \{(0, \ldots, 0)\}$ is closed under componentwise addition of vectors modulo 2, and hence in particular closed under the Boolean minority operation $\text{minority}(x, y, z) = x \oplus y \oplus z$. By Theorem 6.2.7, the Boolean relation $\chi(\psi) \cup \{0^k\}$ is exactly the set of solutions of a system of linear equations over $\{0, 1\}$.

LEMMA 12.8.19. *Let ϕ be a conjunction of temporal formulas preserved by* mx *and given in DNF or as affine clauses. Then the procedure FindFreeSetsMx in Figure* 14 *returns $\emptyset$ or a free set of ϕ.If it returns $\emptyset$, then ϕ is unsatisfiable.*

PROOF. It is well known that a solution of S that is not constant to 0 can be computed in cubic time (by Gaussian elimination). If there is such a solution, then the set of variables mapped to 1 is a free set of ϕ. If the system has no such solution, then there is no free set of variables, and there is no solution for ϕ. ⊣

THEOREM 12.8.20. *There is an algorithm that decides the satisfiability of a given conjunction $\phi(x_1, \ldots, x_n)$ of temporal formulas, each mx-closed and given in DNF or as min-affine clauses, in time $O(n^4)$.*

PROOF. The statement follows from Lemma 12.8.12 and Lemma 12.8.19. ⊣

COROLLARY 12.8.21. *If $\mathfrak{B}$ is a finite-signature first-order reduct of $\mathfrak{B}$ preserved by* mx *then there is an algorithm solving* CSP($\mathfrak{B}$) *in time $O(n^4)$.*

12.9. Equational descriptions

The complexity classification for temporal CSPs from Theorem 12.0.8 (which we have not yet proved) states that the existence of an at most ternary pseudo weak near-unanimity (PWNU) polymorphism implies polynomial-time tractability. The operations min and mx are binary symmetric polymorphisms and hence weak near-unanimity operations. On the other hand, both ll and mi do not locally generate symmetric operations, not even modulo endomorphisms (Section 12.9.1). However, we show that they locally generate ternary PWNU operations (Sections 12.9.2 and 12.9.3).

12.9.1. Pseudo-cyclic polymorphisms. The clone locally generated by ll does not contain cyclic polymorphisms modulo endomorphisms. In fact, this holds for all clones $\mathcal{C}$ over the domain $\mathbb{Q}$ that preserve $<$ and the relation

$$I_4 := \{(x, y, u, v) \in \mathbb{Q}^4 \mid x = y \Rightarrow u = v\}.$$

Recall from Lemma 7.5.1 that every operation $f \in \mathcal{C}$ that depends on all arguments is injective. Let f be such an operation of arity $n \geq 2$, and consider

$$d_0 := f(1, 0, \dots, 0), d_1 := f(0, 1, 0, \dots, 0), \dots, d_{n-1} := f(0, \dots, 0, 1).$$

We claim that there is no $\in \mathrm{End}(\mathbb{Q}; <)$ such that $d_i = \alpha(d_{i+1})$ for all indices $i \in \mathbb{Z}/n\mathbb{Z}$. If $d_0 < d_1$, we would obtain that $d_0 < d_1 < d_2 < \cdots < d_{n-1}$, a contradiction. Similarly, we obtain a contradiction if $d_0 > d_1$.

12.9.2. An injective ternary pseudo weak near-unanimity operation. We introduce the shortcut

$$\mathrm{lex}_n(x_1, \dots, x_n) := \mathrm{lex}(x_1, \mathrm{lex}(x_2, \dots \mathrm{lex}(x_{n-1}, x_n) \dots)).$$

Also recall from Example 8.5.5 the definition of the median operation

$$\mathrm{median}(x, y, z) := \max(\min(x, y), \min(x, z), \min(y, z)).$$

Finally, define $\mathrm{ll}_3 : \mathbb{Q}^3 \to \mathbb{Q}$ by

$$\mathrm{ll}_3(x, y, z) := \mathrm{lex}_5(\min(x, y, z), \mathrm{median}(x, y, z), x, y, z).$$

Note that ll_3 is injective and preserves the relation $\leq$.

PROPOSITION 12.9.1. *There are $a, b, c \in \mathrm{End}(\mathbb{Q}; <)$ such that for all $x, y \in \mathbb{Q}$*

$$a(\mathrm{ll}_3(x, x, y)) = b(\mathrm{ll}_3(x, y, x)) = c(\mathrm{ll}_3(y, x, x)).$$

That is, ll_3 is a weak near-unanimity modulo endomorphisms of $(\mathbb{Q}; <)$.

PROOF. By Lemma 10.1.5, it suffices to show that for every finite $S \subset \mathbb{Q}$ there are $\alpha, \beta \in \mathrm{Aut}(\mathbb{Q}; <)$ such that for all $x, y \in S$

$$\mathrm{ll}_3(y, x, x) = \alpha_1 \mathrm{ll}_3(x, y, x) = \alpha_2 \mathrm{ll}_3(x, x, y).$$

By the properties of f we have that $\mathrm{ll} - 3(y, x, x) \leq \mathrm{ll}_3(y', x', x')$ if and only if one of the following holds:

- $\min(x, y) < \min(x', y')$;
- $\min(x, y) = \min(x', y')$ and $x < x'$;
- $\min(x, y) = \min(x', y')$, $x = x'$, and $y < y'$;
- $x = x'$ and $y = y'$.

Note that this is the case if and only if $\text{ll}_3(x, y, x) < \text{ll}_3(x', y', x')$, and if and only if $\text{ll}_3(x, x, y) < \text{ll}_3(x', x', y')$. Hence, the existence of α_1 and α_2 follows from the homogeneity of $(\mathbb{Q}; <)$. ⊣

THEOREM 12.9.2. *Let $R \subseteq \mathbb{Q}^n$ be first-order definable over $(\mathbb{Q}; <)$. Then the following are equivalent.*

1. *R is preserved by the ternary pseudo weak near-unanimity operation* ll_3.
2. *R is preserved by* ll.
3. *R has an ll-Horn definition.*

PROOF. The implication from (1) to (2) follows from the observation that $\text{ll}_3(x, x, y)$ interpolates ll modulo $\text{Aut}(\mathbb{Q}; <)$. The implication from (2) to (3) has already been shown in Lemma 12.7.2. For the implication from (3) to (1), it suffices to verify that ll_3 preserves all ll-Horn formulas. Since ll_3 is injective, it suffices to show that ll_3 preserves formulas ϕ of the form

$$(z_1 < z_0) \vee \cdots \vee (z_l < z_0)$$

and of the form

$$(z_1 < z_0) \vee \cdots \vee (z_l < z_0) \vee (z_0 = z_1 = \cdots = z_l).$$

Suppose that s_1, s_2, s_3 are assignments that satisfy ϕ; we have to show that the assignment s defined by $s(x) := \text{ll}_3(s_1(x), s_2(x), s_3(x))$ satisfies ϕ. Let $j \in \{1, 2, 3\}$ be such that $s_j(z_0) = \min(s_1(z_0), s_2(z_0), s_3(z_0))$.

Suppose first that s_j satisfies $z_1 < z_0 \vee \cdots \vee z_l < z_0$. Let i be such that $s_j(z_i) = \min(s_j(z_1), \ldots, s_j(z_l))$. Then $s_j(z_i) < s_j(z_0)$ by assumption, and hence

$$\min(s_1(z_i), s_2(z_i), s_3(z_i)) < \min(s_1(z_0), s_2(z_0), s_3(z_0)).$$

Therefore, $\text{ll}_3(s_1(z_i), s_2(z_i), s_3(z_i)) < \text{ll}_3(s_1(z_0), s_2(z_0), s_3(z_0))$ by the properties of ll_3, and s satisfies $z_1 < z_0 \vee \cdots \vee z_l < z_0$.

Otherwise, s_j must satisfy $z_0 = z_1 = \cdots = z_l$. We next consider the case that there exists $c \in \{1, 2, 3\}$ and $p \in \{1, \ldots, l\}$ such that $s_c(z_p) < s_j(z_0)$. Then $\min(s_1(z_p), s_2(z_p), s_3(z_p)) \leq s_c(z_p) < s_j(z_0) = \min(s_1(z_0), s_2(z_0), s_3(z_0))$ and hence, by the definition of ll_3, we have $s(z_p) < s(z_0)$ and s satisfies ϕ. Otherwise, $s_c(z_p) \geq s_j(z_0)$ for all $c \in \{1, 2, 3\}$ and $p \in \{1, \ldots, l\}$. Let $a, b \in \{1, 2, 3\}$ be such that $a < b$ and $\{a, b\} = \{1, 2, 3\} \setminus \{j\}$. The definition of ll_3 then implies for all $i \in \{1, \ldots, l\}$ that $\text{ll}_3(s_1(z_i), s_2(z_i), s_3(z_i)) < \text{ll}_3(s_1(z_0), s_2(z_0), s_3(z_0))$ if and only if $s_a(z_i) < s_a(z_0)$. If there exists an $i \in \{1, \ldots, l\}$ such that $s_a(z_i) < s_a(z_0)$, we therefore have $s(z_i) < s(z_0)$ and s satisfies $z_1 < z_0 \vee \cdots \vee z_l < z_0$. Otherwise, we must have that

$$s_a(z_0) = s_a(z_1) = \cdots = s_a(z_l).$$

If also $s_b(z_0) = s_b(z_1) = \cdots = s_b(z_l)$ then s satisfies $z_0 = z_1 = \cdots = z_l$, too. So suppose that there exists $p \in [l]$ such that $s_b(z_p) < s_b(z_0)$. Since $s_j(z_p) = s_j(z_0)$ and $s_a(z_p) = s_a(z_0)$ we then have $s(z_p) < s(z_0)$ as s is injective and preserves $\leq$. Hence, s satisfies ϕ also in this case. ⊣

12.9.3. A non-injective ternary pseudo weak near-unanimity operation. In this section we present an equivalent description of the temporal relations preserved by mi in terms of a certain ternary pseudo weak near-unanimity mi_3; the way we introduce mi_3 is taken from [87]. Let $\text{mi}_3 \colon \mathbb{Q}^3 \to \mathbb{Q}$ be the operation defined by

$$\text{mi}_3(x, y, z) := \text{lex}_5(\min(x, y, z), \text{median}(-\chi(x, y, z)), -\chi(x, y, z))$$

(where $\chi(t)$ denotes the min-tuple of t introduced in Definition 12.7.10). We first present an equivalent description of mi_3. Recall that if $A, B \subseteq \mathbb{Q}^3$ and $f \colon \mathbb{Q}^3 \to \mathbb{Q}$, then we write $f(A) < f(B)$ if for all $(x, y, z) \in A$ and $(x', y', z') \in B$ we have that $f(x, y, z) < f(x', y', z')$.

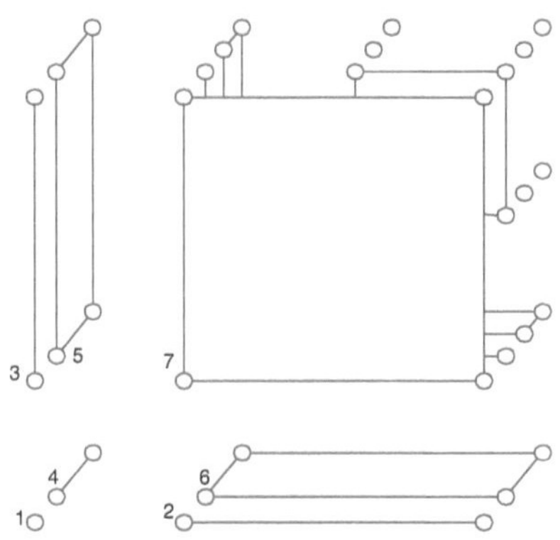

FIGURE 15. Illustration of the function mi_3; see Proposition 12.9.3.

PROPOSITION 12.9.3. *There kernel of* mi_3 *has the following classes: for each* $u \in \mathbb{Q}$

1. $x(u) := \{(a, b, c) \mid u = b = c, a > u\}$;
2. $y(u) := \{(a, b, c) \mid u = a = c, b > u\}$;
3. $z(u) := \{(a, b, c) \mid u = a = b, c > u\}$;
4. $X(u) := \{(a, b, c) \mid u = a, b > u, c > u\}$;
5. $Y(u) := \{(a, b, c) \mid u = b, a > u, c > u\}$;
6. $Z(u) := \{(a, b, c) \mid u = c, a > u, b > u\}$;

7. $D(u) := \{(u, u, u)\}$.

Moreover, for $u < v$, we have

$$\begin{aligned}\text{mi}_3(D(u)) &< \text{mi}_3(x(u)) < \text{mi}_3(y(u)) < \text{mi}_3(z(u)) \\ &< \text{mi}_3(Z(u)) < \text{mi}_3(Y(u)) < \text{mi}_3(X(u)) < \text{mi}_3(D(v)).\end{aligned}$$

PROOF. The specified countable family of subsets of $\mathbb{Q}^3$ indeed forms a partition of $\mathbb{Q}^3$. To see this, note that we distinguish which entries of the tuple are equal to the minimum u of the entries of the tuple. This splits $\mathbb{Q}^3$ into seven different classes for a given u, all of them pairwise disjoint. It is then straightforward to verify that the images of mi_3 are ordered as specified; see Figure 15. ⊣

PROPOSITION 12.9.4. *There are $a, b, c \in \text{End}(\mathbb{Q}; <)$ such that for all $x, y \in \mathbb{Q}$*

$$a(\text{mi}_3(y, x, x)) = b(\text{mi}_3(x, y, x)) = c(\text{mi}_3(x, x, y)).$$

That is, mi_3 is a weak near-unanimity modulo endomorphisms of $(\mathbb{Q}; <)$.

PROOF. By Lemma 10.1.5, it suffices to show that for all finite $S \subset \mathbb{Q}$ there are $\alpha, \beta \in \text{Aut}(\mathbb{Q}; <)$ such that for all $x, y \in S$

$$\text{mi}_3(y, x, x) = \alpha(\text{mi}(y, x)) \quad (10)$$

$$\text{mi}_3(x, y, x) = \beta(\text{mi}(y, x)) \quad (11)$$

$$\text{mi}_3(x, x, y) = \gamma(\text{mi}(y, x)) \quad (12)$$

Observe that for all $u, v, u', v' \in \mathbb{Q}$ we have $\text{mi}_3(v, u, u) \leq \text{mi}_3(v', u', u')$ iff one of the following cases applies:

- $\min(u, v) < \min(u', v')$;
- $u = v = \min(u', v')$;
- $u < v$ and $u = \min(u', v') < \max(u', v')$;
- $v < u$ and $v = v' < u'$.

This in turn is the case if and only if $\text{mi}(v, u) \leq \text{mi}(v', u')$. Then the statement for (10) follows from homogeneity of $(\mathbb{Q}; <)$. The proof for (11) and for (12) is analogous. ⊣

THEOREM 12.9.5. *Let $R \subseteq \mathbb{Q}^n$ be first-order definable over $(\mathbb{Q}; <)$. Then the following are equivalent.*

1. *R is preserved by the ternary pseudo weak near-unanimity operation mi_3.*
2. *R is preserved by mi.*
3. *R is primitively positively definable over $(\mathbb{Q}; R^{\text{mi}}, S^{\text{mi}})$.*

PROOF. The implication from (1) to (2) follows from the observation that $(x, y) \mapsto \text{mi}_3(x, x, y)$ interpolates mi modulo $\text{Aut}(\mathbb{Q}; <)$. The implication from (2) to (3) follows from Theorem 12.7.8. For the implication from (3) to (1) it suffices to verify that mi_3 preserves R^{mi} and S^{mi}. Let $t^1, t^2, t^3 \in$

$R^{\mathrm{mi}} = \{(x, z_0, z_1) \mid z_0 \leq x \vee z_1 < x\}$ and suppose for contradiction that $s := \mathrm{mi}_3(t^1, t^2, t^3) \notin R^{\mathrm{mi}}$.

Let $i \in \{1, 2, 3\}$ be such that $\min(t_1^1, t_1^2, t_1^3) = t_1^i$. Suppose that there exists $j \in \{1, 2, 3\}$ such that $t_3^j < t_1^i$. Then

$$\min(t_3^j, t_3^j, t_3^j) \leq t_3^j < t_1^i = \min(t_1^1, t_1^2, t_1^3)$$

and hence $s_3 < s_1$. Thus, $s \in R^{\mathrm{mi}}$ contrary to our assumptions. So suppose that $t_1^i \leq t_3^j$ for every $j \in \{1, 2, 3\}$.

Since $s \notin R^{\mathrm{mi}}$ we have $s_1 > s_2$. Then by the definition of mi_3 there are three cases to consider.

- $t_1^i < \min(t_2)$. Then $t_1^i \leq t_3^i$ and $t_1^i < t_3^i$ and hence $t^i \notin R^{\mathrm{mi}}$ contrary to our assumptions.
- $t_1^i = \min(t_2)$ and $\mathrm{median}(-\chi(t_1)) < \mathrm{median}(-\chi(t_2))$. Then there must exist $\ell \in [3]$ such that t_1^k is minimal in t_1 and t_3^ℓ is not minimal in t_3. But then $t_1^\ell = t_1^i \leq t_3^\ell$ and $t_1^\ell = t_1^i = \min(t_2) < t_2^\ell$, in contradiction to $t^\ell \in R^{\mathrm{mi}}$.
- $t_1^i = \min(t_2^1, t_2^2, t_2^3)$, $\mathrm{median}(-\chi(t_1)) = \mathrm{median}(-\chi(t_2))$, and $\mathrm{lex}_3(-\chi(t_1)) < \mathrm{lex}_3(-\chi(t_2))$. In this case there must again exist $\ell \in [3]$ such that t_1^ℓ is minimal in t_1 and t_3^ℓ is not minimal in t_3, and the argument is as in the previous case.

This concludes the proof that mi_3 preserves R^{mi}.

Suppose for contradiction that $a, b, c \in S^{\mathrm{mi}} = \{(x, y, z) \in \mathbb{Q}^3 \mid y \neq x \vee z \leq x\}$ and $d := f(a, b, c) \notin S^{\mathrm{mi}}$. In particular, $d_1 = d_2$. By the definition of f, there exists a $u \in \mathbb{Q}$ such that

$$I := \{(a_1, b_1, c_1), (a_2, b_2, c_2)\}$$

is contained in one of the sets $x(u)$, $y(u)$, $z(u)$, $X(u)$, $Y(u)$, $Z(u)$, or $D(u)$ from the definition of f. It follows that $a_1 = a_2 = u$ or $b_1 = b_2 = u$ or $c_1 = c_2 = u$. Suppose without loss of generality that the first case applies (the argument in the other cases is symmetric), and hence I is contained in either $y(u)$, $z(u)$, $X(u)$, or $D(u)$.

- If $I \subseteq D(u)$ then we must have $a_3 \leq a_1 = u$, $b_3 \leq b_1 = u$, and $c_3 \leq c_1 = u$ because $a, b, c \in S^{\mathrm{mi}}$. Since f preserves $\leq$ we have that $d_3 \leq d_1$.
- If $I \subseteq y(u)$ then we must have $a_z \leq a_1 = u$ and $c_3 \leq c_1 = u$. Then

$$\begin{aligned} f(a_3, b_3, c_3) &\leq f(u, b_3, u) && \text{(since } f \text{ preserves } \leq) \\ &\leq f(u, b_1, u) && \text{(by the properties of } f) \\ &= f(a_1, b_1, c_1) \end{aligned}$$

  and hence $d_3 \leq d_1$.
- The case that $I \subseteq z(u)$ is similar to the previous one.

- If $I \subseteq X(u)$ then $a_3 \leq a_1 = u$, and
$$\begin{aligned} f(a_3, b_3, c_3) &\leq f(u, b_3, c_3) && \text{(since } f \text{ preserves } \leq) \\ &\leq f(u, b_1, c_1) && \text{(by the properties of } f) \\ &= f(a_1, b_1, c_1) \end{aligned}$$
and hence $d_3 \leq d_1$.

In each of the cases we have that $d \in S^{\text{mi}}$, a contradiction. ⊣

12.10. The classification

This section combines the previous results to show that every finite temporal constraint language has a constraint satisfaction problem which can be solved in polynomial time or is NP-complete.

THEOREM 12.10.1. *Let $\mathfrak{B}$ be a first-order reduct of $(\mathbb{Q}; <)$. Then one of the following applies.*

- *$\mathfrak{B}$ is preserved by at least one of the following nine operations*: ll, min, mi, mx, *their duals, or a constant operation.*
- Betw, Cycl, Sep, T_3, $-T_3$, *or* P_B^4 *is primitively positively definable in* $\mathfrak{B}$.

PROOF. Theorem 12.3.1 asserts that one of the following cases holds:

1. There is a primitive positive definition of Cycl, Betw, or Sep in $\mathfrak{B}$.
2. $\text{Pol}(\mathfrak{B})$ contains a constant operation.
3. $\text{Pol}(\mathfrak{B})$ contains all permutations of $\mathbb{Q}$. In this case, Theorem 7.4.2 shows that $\mathfrak{B}$ either has a binary injective polymorphism g, or the relation P_B^4 has a primitive positive definition in $\mathfrak{B}$. In the first case, by composing g with a permutation, we see that *all* binary injective operations preserve $\mathfrak{B}$, and hence in particular the operation ll is a polymorphism of $\mathfrak{B}$.
4. all $f \in \text{Pol}(\mathfrak{B})$ preserve $<$.

We are done in all cases except the fourth. Also, we can assume that $\mathfrak{B}$ has a polymorphism f that does not preserve Betw. By Lemma 6.1.29, we can assume that f is binary. Then Lemma 12.6.2 implies that the operation f generates pp, dual-pp, ll, or dual-ll. If f generates ll or dual-ll we are done. If f generates pp then Corollary 12.5.30 shows that either T_3 has a primitive positive definition in $\mathfrak{B}$, or $\mathfrak{B}$ is preserved by min, mi, or mx. Dually, if f generates dual-pp then either $-T_3$ has a primitive positive definition in $\mathfrak{B}$, or $\mathfrak{B}$ is preserved by one of the duals of min, mi, or mx, which completes the proof. ⊣

With the previous theorem it is easy to obtain the full complexity classification for temporal constraint satisfcation problems, and finally show Theorem 12.0.8.

PROOF OF THEOREM 12.0.8. If $\mathfrak{B}$ is preserved by ll, min, mi, mx, one of their duals, or the constant operation, then $\mathfrak{B}$ has an at most ternary weak near-unanimity polymorphism modulo endomorphisms; this is immediate for the commutative binary functions mx, min, their duals, and for the constant function. For ll, this has been shown in Theorem 12.9.2, and for mi in Theorem 12.9.5. For dual mi and dual ll the dual argument works.

Now let $\mathfrak{B}'$ be a finite signature reduct of $\mathfrak{B}$. If $\mathfrak{B}'$ is preserved by a constant operation, then tractability of CSP($\mathfrak{B}'$) follows from Proposition 1.1.12. For the case that $\mathfrak{B}'$ is preserved by ll or dual-ll we have presented a polynomial-time algorithm for CSP($\mathfrak{B}'$) in Theorem 12.8.9. If $\mathfrak{B}'$ is preserved by min, mi, mx, or one of their duals, tractability of CSP($\mathfrak{B}'$) is shown in Section 12.8.2.

Now suppose that $\mathfrak{B}$ is not preserved by one of the listed operations. Then by Theorem 12.10.1 we know that one of the relations Betw, Cycl, Sep, T_3, $-T_3$, or P_B^4 has a primitive positive definition in $\mathfrak{B}$. Each of those relations together with finitely many constants primitively positively interprets all finite structures, each time using Theorem 3.2.2:

- For ($\mathbb{Q}$; Betw, 0) a primitive positive interpretation of ($\{0, 1\}$; NAE) has been shown in Proposition 3.1.10 via Theorem 3.2.2.
- a primitive positive interpretation of ($\{0, 1\}$; 1IN3) in ($\mathbb{Q}$; Cycl) with parameters has been given in Theorem 12.2.1.
- The structure ($\mathbb{Q}$; Sep, 0, 1) primitively positively interprets ($\{0, 1\}$; 1IN3) by Proposition 12.2.2.
- The structure ($\mathbb{Q}$; T_3, 0) primitively positively interprets ($\{0, 1\}$; 1IN3) by Proposition 3.1.9; the proof for $-T$ is dual.
- The structure ($\mathbb{Q}$; P_B^4, $\neq$) primitively positively interprets ($\{0, 1\}$; 1IN3) by Theorem 7.4.1.

Finally, recall from Theorem 12.1.7 that if $\mathfrak{B}$ does not have a constant endomorphism, then it is a model-complete core, and hence Theorem 10.3.5 shows that the two cases in the statement of Theorem 12.0.8 are distinct. ⊣

See Figure 16 for an overview over the nine largest tractable temporal constraint languages; the first four rows in the table list the classes that have a dual class which is not listed in the table. The second column of the table shows relations that can be used to primitively positively define all other relations in the respective language; these relations and their polymorphisms can also be used to show that all the listed classes are distinct.

12.10.1. Decidability of tractability. We want to remark that the so-called *meta-problem* for tractability is decidable; this is formally stated in the following corollary.

COROLLARY 12.10.2. *There is an algorithm that, given quantifier-free first-order formulas* $\phi_1, \ldots, \phi_n$ *that define over* ($\mathbb{Q}$; $<$) *the relations* $R_1, \ldots, R_n$, *decides whether* CSP($\mathbb{Q}$; $R_1, \ldots, R_n$) *is tractable or NP-complete.*

Polymorphism	*Relational Generators*	*Complexity*	*Reference*
min	$<, R^{\min}_{\leq}$ (12.7.5)	$O(nm)$	Theorem 12.8.14
mi	$R^{\mathrm{mi}}, S^{\mathrm{mi}}$ (12.7.8)	$O(n^3m)$	Theorem 12.8.17
mx	X (12.7.11)	$O(n^4)$	Theorem 12.8.20
ll	$\neq, I_4, L$ (12.7.3)	$O(nm)$	Theorem 12.8.9
constant	$\leq, \mathrm{Betw}_=$	$O(m)$	Proposition 1.1.12

FIGURE 16. Overview of the polynomial-time tractable temporal CSPs.

PROOF. We can use Theorem 11.6.4 to test the primitive positive definability of one of the relations given in Theorem 12.10.1. Alternatively, we can algorithmically test the existence of a pseudo-Siggers polymorphism (Theorem 11.6.7), which characterises polynomial-time tractability by Theorem 12.0.8. ⊣

12.10.2. Discrete temporal CSPs. We close this section with some remarks on a related classification for *discrete temporal constraint satisfaction problems*, i.e., CSPs for first-order reducts of $(\mathbb{Z}; <)$. The computational complexity of CSP($\mathfrak{B}$) for first-order reducts of $(\mathbb{Z}; <)$ has been classified [79]; every such problem is either in P or NP-complete. Note that in particular the structure $(\mathbb{Z}; \mathrm{Succ})$ with $\mathrm{Succ} = \{(x, y) \mid x = y + 1\}$ is first-order definable in $(\mathbb{Z}; <)$. Clearly, first-order reducts of $(\mathbb{Z}; <)$ are in general not ω-categorical. However, polymorphisms (either of $\mathfrak{B}$, or of elementarily equivalent structures) still play a central role in the classification proof.

The complexity classification for discrete temporal CSPs depends on the dichotomy for finite-domain CSPs: Note that a first-order reduct $\mathfrak{B}$ of $(\mathbb{Z}; <)$ may be homomorphically equivalent to a finite structure. For instance, the undirected graph $(\mathbb{Z}; \{(x, y) \mid |x - y| \in \{1, 2\}\})$ is is a first-order reduct of $(\mathbb{Z}; <)$ and homomorphically equivalent to K_3. Finite cores that arise in this way have a transitive automorphism group; this follows easily from the fact that $\mathrm{Aut}(\mathbb{Z}; \mathrm{Succ})$ is transitive. The only proofs of the dichotomy conjecture for CSPs whose template is finite and has a transitive automorphism group are the proofs for the full dichotomy conjecture for finite templates from [111, 348].

Chapter 13

NON-DICHOTOMIES

There are basically two methods for proving that a subclass of NP does not have a complexity dichotomy. The first is to show that for every problem in NP there is a polynomial-time equivalent problem in the subclass. By polynomial-time equivalent we mean that there are polynomial-time Turing reductions between the two problems. The non-dichotomy result then follows from Ladner's theorem [252], which asserts that there are problems in NP that are neither in P nor NP-complete, unless P = NP (and if P = NP, there is of course also no dichotomy). This method has been applied to show that, for example, the class of monotone SNP does not exhibit a complexity dichotomy [170]. We will apply this technique in Section 13.1 and in Section 13.2 to give two different proofs of the fact that the class of all constraint satisfaction problems with infinite domains does not have a complexity dichotomy.

The second technique to show a non-dichotomy is to directly use Ladner's proof technique, which is sometimes called *delayed diagonalisation*. We will use this method in Section 13.3 to show that there are ω-categorical structures $\mathfrak{B}$ such that CSP($\mathfrak{B}$) is in coNP, but neither in P nor coNP-complete (unless P=coNP). The question whether there are ω-categorical structures $\mathfrak{B}$ such that CSP($\mathfrak{B}$) is in NP \ P but not NP-complete is still open (Question 48).

13.1. Arithmetical templates

In this section we show that for every decision problem there exists a polynomial-time equivalent constraint satisfaction problem with an infinite template $\mathfrak{B}$. This result was first shown in [59]. Here we present a new proof that uses Matiyasevich's theorem. In fact, we prove a stronger result, namely the existence of a single structure $\mathfrak{C}$ such that for every recursively enumerable problem $\mathcal{P}$ there is a structure $\mathfrak{B}$ with a first-order definition in $\mathfrak{C}$ such that CSP($\mathfrak{B}$) is polynomial-time equivalent to $\mathcal{P}$. A second proof, based on the results in Section 1.4.2, can be found in the next section.

Previously, Bauslaugh [30] showed that for every recursive function f there exists an infinite structure $\mathfrak{B}$ such that CSP($\mathfrak{B}$) is decidable, but has time complexity at least f. Later, Schwandtner gave upper and lower bounds in the exponential time hierarchy for some infinite-domain CSPs [327]; but these bounds leave an exponential gap.

We make essential use of the following theorem, which is due to Davis, Matiyasevich, Putnam, and Robinson.

THEOREM 13.1.1 (see e.g. [282]). *A subset of $\mathbb{Z}$ is recursively enumerable if and only if it has a primitive positive definition in* $(\mathbb{Z}; *, +, 1)$.

THEOREM 13.1.2 (Theorem 57 in [76]). *For every recursively enumerable language $\mathcal{L}$ there exists a relational structure $\mathfrak{B}$ with a first-order (in fact, a primitive positive) definition in* $(\mathbb{Z}; *, +, 1)$ *such that* CSP($\mathfrak{B}$) *is polynomial-time Turing equivalent to $\mathcal{L}$.*

PROOF. We code $\mathcal{L}$ as a set L of natural numbers, viewing the binary encodings of natural numbers as bit strings. More precisely, $s \in \mathcal{L}$ if and only if the number represented in binary by the string $1s$ is in L. That is, we prepend the symbol 1 at the front so that for instance $00 \in \mathcal{L}$ and $01 \in \mathcal{L}$ correspond to different numbers in L. Now consider the structure $\mathfrak{B} := (\mathbb{Z}; S, D, L', N)$

where

$$\begin{aligned} S &:= \{(x, y) \in \mathbb{Z}^2 \mid (y = x + 1 \wedge x \geq 0) \vee (x = y = -1)\}, \\ D &:= \{(x, y) \in \mathbb{Z}^2 \mid (y = 2x \wedge x \geq 0) \vee (x = y = -1)\}, \\ L' &:= L \cup \{-1\}, \text{ and} \\ N &:= \{0\}. \end{aligned}$$

Clearly, if $\mathcal{L}$ is recursively enumerable, then L and L' are recursively enumerable as well.

We have to verify that CSP($\mathfrak{B}$) is polynomial-time equivalent to $\mathcal{L}$. We first show that there is a polynomial-time reduction from $\mathcal{L}$ to CSP($\mathfrak{B}$). View an instance of $\mathcal{L}$ as a number $n \geq 0$ as above, and let $\eta(x)$ be a primitive positive definition for $x = n$ in $\mathfrak{B}$. It is possible to find such a definition in polynomial time by repeatedly doubling ($y = x + x$) and incrementing ($y = x + 1$) the value 0 (this also follows from the more general Lemma 1.6.1). It is clear that n codes a yes-instance of $\mathcal{L}$ if and only if $\exists x(\eta(x) \wedge L'(x))$ is true in $\mathfrak{B}$.

To reduce CSP($\mathfrak{B}$) to $\mathcal{L}$, we present a polynomial-time algorithm for CSP($\mathfrak{B}$) that uses an oracle for $\mathcal{L}$ (so our reduction will be a polynomial-time Turing reduction). Let ϕ be an instance of CSP($\mathfrak{B}$), and let H be the undirected graph whose vertices are the variables W of ϕ, and which has an edge between x and y if ϕ contains the constraint $S(x, y)$ or the constraint $D(x, y)$. Compute the connected components of H. If a connected component does not contain x with a constraint $N(x)$ in ϕ, then we can set all variables of that component to -1 and satisfy all constraints involving those variables.

Otherwise, suppose that we have a component C that does contain x_0 with a constraint $N(x_0)$. Observe that by connectivity, if there exists a solution, then all variables in C must take non-negative values. Consider the following linear system: for each constraint of the form $S(x, y)$ for $x, y \in C$ we add $y = x + 1$ and $x \geq 0$ to the system, and for each constraint of the form $D(x, y)$ for $x, y \in D$ we add $z = 2x$ and $x \geq 0$. Subject to $x_0 = 0$ this system has either one or no solution. We can check in polynomial time whether a linear system with 2 variables per constraint has no integer solution [86], and if there is no solution, the algorithm rejects. Otherwise, the algorithm assigns to each variable $x \in C$ its unique integer value, and if ϕ contains a constraint $L'(x)$, we call the oracle for $\mathcal{L}$ with the binary encoding of this value. If any of those oracle calls has a negative result, reject. Otherwise, we have found an assignment that satisfies all constraints, and accept. ⊣

The universal-algebraic approach fails badly when it comes to analyzing the computational complexity of CSP($\mathfrak{B}$): the semilattice operation $(x, y) \mapsto \max_<(x, y)$ preserves $\mathfrak{B}$ for all structures $\mathfrak{B}$ considered in the previous proof, and we cannot draw any consequences for the computational complexity of CSP($\mathfrak{B}$).

13.2. CSPs in SNP

Another proof that shows that every problem in NP is polynomial-time Turing equivalent to an infinite-domain CSP is based on a result by Feder and Vardi, and the results from Section 1.4.3.

THEOREM 13.2.1 (Theorem 3 in [170]). *Every problem in NP is equivalent to a problem in monotone SNP under polynomial-time reductions.*

We show the following.

PROPOSITION 13.2.2. *Every problem in monotone SNP is equivalent to a problem in monotone connected SNP under polynomial-time Turing reductions.*

PROOF. Let Φ be a monotone SNP sentence of the form

$$\exists R_1, \dots, R_k \, \forall x_1, \dots, x_l : \phi$$

for ϕ quantifier-free and in conjunctive normal form. The sentence Ψ that we are going to construct from Φ has an additional free relation symbol E, and an existentially quantified relation symbol T, and is defined by

$$\exists R_1, \dots, R_k, T \, \forall x_1, \dots, x_l : \psi$$

where ψ is the quantifier-free first-order formula with the following clauses.

1. $\neg E(x_1, x_2) \vee T(x_1, x_2)$;
2. $\neg T(x_1, x_2) \vee \neg T(x_2, x_3) \vee T(x_1, x_3)$;
3. $\neg T(x_1, x_2) \vee T(x_2, x_1)$;
4. for each clause ϕ' of ϕ with variables $x_1, \dots, x_q$, the clause

$$\phi' \vee \bigvee_{i<j<q} \neg T(x_i, x_j).$$

The sentence Ψ is clearly connected and monotone. We are therefore left with the task to verify that Φ and Ψ are equivalent under polynomial-time Turing reductions.

We start with the reduction from Φ to Ψ. When $\mathfrak{A}$ is a finite τ-structure, we expand $\mathfrak{A}$ to a $(\tau \cup \{E\})$-structure $\mathfrak{A}'$ by choosing for E the full binary relation. Then also T must denote the full binary relation (so that the clauses from item (1), (2), and (3) above are satisfied), and the clauses introduced in (4) are equivalent to ϕ'. Hence, Φ holds on $\mathfrak{A}$ if and only if Ψ holds on $\mathfrak{A}'$.

For the reduction from Ψ to Φ, let $\mathfrak{A}$ be an instance of Ψ. We can compute the connected components $C_1, \dots, C_k$ of the $\{E\}$-reduct of $\mathfrak{A}$ in polynomial time in the size $\mathfrak{A}$. For each of those connected components C, we evaluate Φ on the τ-reduct $\mathfrak{A}_C$ of $\mathfrak{A}[C]$. If for one component this evaluation is negative, then $\mathfrak{A}[C]$ and consequently $\mathfrak{A}$ do not satisfy Ψ. Otherwise, for each C there exists a $(\tau \cup \{R_1, \dots, R_k\})$-expansion of $\mathfrak{A}_C$ that satisfies ϕ. Let $\mathfrak{A}'$ be the expansion of the disjoint union of all those $(\tau \cup \{R_1, \dots, R_k\})$-structures by the relation T that denotes the equivalence relation with equivalence classes $C_1, \dots, C_k$.

Clearly, all clauses from items (1), (2), and (3) in the definition of Ψ are satisfied by $\mathfrak{A}'$. Each q-tuple $(a_1, \dots, a_q)$ from elements of $\mathfrak{A}'$ either contains entries from different components, and hence satisfies the disjunctions from item (4), or contains only entries from the same component C, but in this case the tuple also satisfies the disjunctions from item (4) since $\mathfrak{A}_C$ satisfies Φ. ⊣

COROLLARY 13.2.3. *For every problem in NP there is a structure* $\mathfrak{B}$ *such that the problem is polynomial-time Turing equivalent to* CSP($\mathfrak{B}$).

PROOF. By Theorem 13.2.1, every problem in NP is equivalent to a monotone SNP sentence Φ under polynomial-time reductions. We have shown in Proposition 13.2.2 that Φ is equivalent to a monotone connected SNP sentence Ψ, and by Theorem 1.4.12 there exists an infinite structure $\mathfrak{B}$ such that Ψ describes CSP($\mathfrak{B}$). ⊣

In Figure 1 the diagram about the fragments of SNP from Section 1.4 has been decorated with information about the complexity classification status.

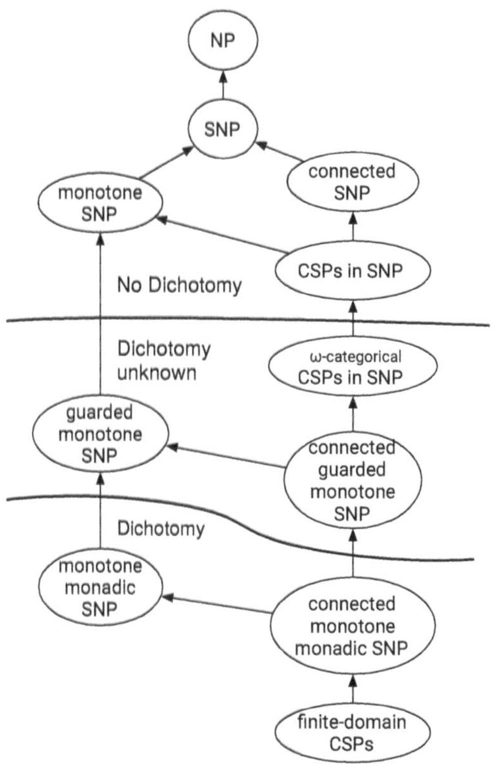

FIGURE 1. Dichotomies and non-dichotomies for fragments of SNP.

13.3. coNP-intermediate countably categorical templates

In this section we show that there exists an ω-categorical directed graph $\mathfrak{B}$ such that CSP($\mathfrak{B}$) is in coNP, but neither coNP-complete nor in P (unless coNP=P). All structures in this section will be Fraïssé limits of classes of directed graphs.

Let $\mathcal{N}$ be a class of finite tournaments, and recall that $\text{Forb}^{\text{emb}}(\mathcal{N} \cup \{\mathfrak{L}\})$, the class of all finite loopless digraphs that does not embed a tournament from $\mathcal{N}$, is an amalgamation class (Example 2.3.12). We write $\mathfrak{B}_\mathcal{N}$ for the Fraïssé-limit of $\text{Forb}^{\text{emb}}(\mathcal{N} \cup \{\mathfrak{L}\})$. Observe that for finite $\mathcal{N}$ the problem CSP($\mathfrak{B}_\mathcal{N}$) can be solved in deterministic polynomial time, because for a given instance $\mathfrak{A}$ of this problem an algorithm simply has to check whether there is a homomorphism from one of the structures in $\mathcal{N} \cup \{\mathfrak{L}\}$ to $\mathfrak{A}$, which is the case if and only if there is a homomorphism from $\mathfrak{A}$ to $\mathfrak{B}_\mathcal{N}$.

When proving that there are uncountably many homogeneous digraphs, Henson specified an infinite set $\mathcal{T}$ of tournaments $\mathfrak{T}_3, \mathfrak{T}_4, \dots$ with the property that $\mathfrak{T}_i$ does not embed into $\mathfrak{T}_j$ if $i \neq j$. The tournament $\mathfrak{T}_n$, for $n \geq 3$, in Henson's set $\mathcal{T}$ has vertices $0, \dots, n+1$, and the following edges:

- $(i, i+1)$ for $0 \leq i \leq n$;
- $(0, n+1)$;
- (j, i) for $j > i+1$ and $(i, j) \neq (0, n+1)$.

PROPOSITION 13.3.1. *The problem* CSP($\mathfrak{B}_\mathcal{T}$) *is coNP-complete.*

PROOF. The problem is contained in coNP, because we can efficiently test whether a sequence $v_1, \dots, v_k$ of distinct vertices of a given directed graph $\mathfrak{A}$ induces $\mathfrak{T}_k$ in $\mathfrak{A}$, i.e., whether (v_i, v_j) is an arc in $\mathfrak{A}$ if and only if (i, j) is an arc in $\mathfrak{T}_k$, for all $i, j \in \{1, \dots, k\}$. If for all such sequences of vertices this test is negative, we can be sure that $\mathfrak{A}$ is from $\text{Forb}^{\text{emb}}(\mathcal{T} \cup \{\mathfrak{L}\})$, and hence maps homomorphically to $\mathfrak{B}_\mathcal{T}$. Otherwise, $\mathfrak{A}$ embeds a structure from $\mathcal{T}$, and hence does not map homomorphically to $\mathfrak{B}_\mathcal{T}$.

The proof of coNP-hardness goes by reduction from the complement of the NP-complete 3SAT problem (see Example 1.2.2), and is inspired by a classical reduction from 3SAT to the clique problem. For a given 3SAT instance, we create an instance $\mathfrak{A}$ of CSP($\mathfrak{B}_\mathcal{T}$) as follows: If

$$(\ell_0^1 \vee \ell_0^2 \vee \ell_0^3), \dots, (\ell_{k+1}^1 \vee \ell_{k+1}^2 \vee \ell_{k+1}^3)$$

are all the clauses of the 3SAT formula for some positive or negative literals ℓ_j^i (we assume without loss of generality that the 3SAT instance has at least three clauses and that each clause has exactly three literals), then the vertex set of $\mathfrak{A}$ is

$$\{(0,1), (0,2), (0,3), \dots, (k+1,1), (k+1,2), (k+1,3)\},$$

and the arc set of $\mathfrak{A}$ consists of all pairs $((i, j), (p, q))$ of vertices such that $\ell_i^j \neq \neg\ell_p^q$ (where we identify literals of the form x and of the form $\neg\neg x$) and such that (i, p) is an arc in $\mathfrak{T}_k$.

We claim that a 3SAT instance is unsatisfiable if and only if the created instance $\mathfrak{A}$ maps homomorphically to $\mathfrak{B}_{\mathcal{T}}$. The 3SAT instance is satisfiable iff there is a mapping from the variables to true and false such that in each clause at least one literal, say $\ell_0^{j_0}, \ldots, \ell_{k+1}^{j_{k+1}}$, is true. This is the case if and only if the vertices $(0, j_1), \ldots, (k+1, j_{k+1})$ induce T_k in $\mathfrak{A}$, i.e., $((i, j_i), (p, j_p))$ is an edge if and only if (i, p) is an edge in $\mathfrak{T}_k$. This is the case if and only if T_k embeds into $\mathfrak{A}$. To conclude, it suffices to prove that $\mathfrak{T}_k$ embeds into $\mathfrak{A}$ if and only if $\mathfrak{A}$ does not map homomorphically to $\mathfrak{B}_{\mathcal{T}}$. It is clear that if T_k embeds into $\mathfrak{A}$, then $\mathfrak{A}$ does not map homomorphically to $\mathfrak{B}_{\mathcal{T}}$. Conversely, if $\mathfrak{A}$ does not homomorphically embed to $\mathfrak{B}_{\mathcal{T}}$, then there exists a j such that there is an embedding e of $\mathfrak{T}_j$ into $\mathfrak{A}$. Then for any $(i, j), (p, q)$ in the image of e we have that (i, p) is an edge of $\mathfrak{T}_k$. Therefore, the mapping that sends an element u of $\mathfrak{T}_j$ to the first component of $e(u)$ is an embedding of $\mathfrak{T}_j$ into $\mathfrak{T}_k$. Since $\mathfrak{T}_j$ and $\mathfrak{T}_k$ are homomorphically inequivalent for all distinct $j, k \geq 3$ we obtain that $j = k$ and that $\mathfrak{T}_k$ embeds into $\mathfrak{A}$, which finishes the proof. ⊣

We now modify the proof of Ladner's theorem given in [299] (which is basically Ladner's original proof) to create a subset $\mathcal{T}_0$ of $\mathcal{T}$ such that CSP($\mathfrak{B}_{\mathcal{T}_0}$) is in coNP, but neither in P nor coNP-complete (unless coNP=P). One of the ideas in Ladner's proof is to *'blow holes into SAT'*, so that the positive instances of the resulting problem are too sparse to be NP-complete and too dense to be in P. Our modification is that we do not blow holes into a computational problem itself, but that we *'blow holes into the obstruction set $\mathcal{T}$ of* CSP($\mathfrak{B}_{\mathcal{T}}$)'.

In the following, we fix one of the standard encodings of graphs as strings over the alphabet $\{0, 1\}$. Let $M_1, M_2, \ldots$ be an enumeration of all polynomial-time bounded Turing machines, and let $R_1, R_2, \ldots$ be an enumeration of all polynomial-time bounded reductions. We assume that these enumerations are effective; it is well known that such enumerations exist.

The definition of $\mathcal{T}_0$ uses a Turing machine F that computes a function $f: \mathbb{N} \to \mathbb{N}$, which is defined below. The set $\mathcal{T}_0$ is then defined as follows.

$$\mathcal{T}_0 = \{\mathfrak{T}_n \mid f(n) \text{ is even}\}.$$

The input number n is given to the machine F in unary representation. The computation of F proceeds in two phases. In the first phase, F simulates itself[30] on input 1, then on input 2, 3, and so on, until the number of computation steps of F in this phase exceeds n (we can always maintain a counter during the simulation to recognise when to stop). Let k be the value $f(i)$ for the last input i for which the simulation was completely performed by F.

[30]Note that by the fixpoint theorem of recursion theory we can assume that F has access to its own description.

In the second phase, the machine stops if phase two takes more than n computation steps, and F returns k. We distinguish whether k is even or odd. If k is even, all isomorphism types of directed graphs $\mathfrak{A}$ on $s = 1, 2, 3, \dots$ vertices are enumerated. For each directed graph $\mathfrak{A}$ in the enumeration the machine F simulates $M_{k/2}$ on the encoding of $\mathfrak{A}$. Moreover, F computes whether $\mathfrak{A}$ maps homomorphically to $\mathfrak{B}_{\mathcal{T}_0}$. This is the case if for all structures $\mathfrak{T}_l \in \mathcal{T}$ that embed into $\mathfrak{A}$ the value of $f(l)$ is even. So F tests for $l = 1, 2, \dots, s$ whether $\mathfrak{T}_l$ embeds to $\mathfrak{A}$ (F uses any straightforward exponential time algorithm for this purpose), and if it does, simulates itself on input l to find out whether $f(l)$ is even. If

(1) $M_{k/2}$ rejects and $\mathfrak{A}$ maps homomorphically to $\mathfrak{B}_{\mathcal{T}_0}$, or
(2) $M_{k/2}$ accepts and $\mathfrak{A}$ does not map homomorphically to $\mathfrak{B}_{\mathcal{T}_0}$,

then F returns $k + 1$ (and $f(n) = k + 1$).

The other case of the second phase is that k is odd. Again, F enumerates all isomorphism types of directed graphs $\mathfrak{A}$ with $s = 1, 2, 3, \dots$ vertices, and simulates the computation of $R_{\lfloor k/2 \rfloor}$ on the encoding of $\mathfrak{A}$. Then F computes whether the output of $R_{\lfloor k/2 \rfloor}$ encodes a directed graph $\mathfrak{A}'$ that maps homomorphically to $\mathfrak{B}_{\mathcal{T}_0}$. The digraph $\mathfrak{A}'$ maps homomorphically to $\mathfrak{B}_{\mathcal{T}_0}$ if and only if for all tournaments $\mathfrak{T}_l$ that embed into $\mathfrak{A}'$ the value $f(l)$ is even. Whether $\mathfrak{T}_l$ embeds into $\mathfrak{A}'$ is tested with a straightforward exponential-time algorithm. To test whether $f(l)$ is even, F simulates itself on input l. Finally, F tests with a straightforward exponential-time algorithm whether $\mathfrak{A}$ maps homomorphically to $\mathfrak{B}_{\mathcal{T}}$. If

(3) $\mathfrak{A}$ maps homomorphically to $\mathfrak{B}_{\mathcal{T}}$ and $\mathfrak{A}'$ does not map homomorphically to $\mathfrak{B}_{\mathcal{T}_0}$, or
(4) $\mathfrak{A}$ does not map homomorphically to $\mathfrak{B}_{\mathcal{T}}$ and $\mathfrak{A}'$ maps homomorphically to $\mathfrak{B}_{\mathcal{T}_0}$,

then F returns $k + 1$.

LEMMA 13.3.2. *The function f is a non-decreasing function, that is, for all n we have $f(n) \leq f(n + 1)$.*

PROOF. We inductively assume that $f(s - 1) \leq f(s)$ for all $s \leq n$, and have to show that $f(n) \leq f(n+1)$. Since F has more time to simulate itself when we run it on $n + 1$ instead of n, the value i computed in the first phase of F cannot become smaller. By inductive assumption, $k = f(i)$ cannot become smaller as well. In the second phase, we either return k or $k + 1$. Hence, if k becomes larger in the first phase, the output of F cannot become smaller. If k does not become larger, then the only difference between the second phase of F for input $n + 1$ compared to input n is that there is more time for the computations. Hence, if the machine F on input n verifies condition (1),(2),(3),(4) for some digraph $\mathfrak{A}$ (and hence returns $k + 1$), then F also verifies this condition for $\mathfrak{A}$

on input $n + 1$, and returns $k + 1$ as well. Otherwise, $f(n) = k$, and also here $f(n + 1) \geq f(n)$ holds. ⊣

LEMMA 13.3.3. *For every $n_0 \in \mathbb{N}$ there exists an $n > n_0$ such that $f(n) > f(n_0)$ (unless coNP $\neq$ P).*

PROOF. Assume for contradiction that there exists an n_0 such that $f(n)$ equals a constant k_0 for all $n \geq n_0$. Then there also exists an n_1 such that for all $n \geq n_1$ the value of k computed by the first phase of F on input n is k_0.

If k_0 is even, then on all inputs $n \geq n_1$ the second phase of F simulates $M_{k_0/2}$ on encodings of an enumeration of digraphs. Since the output of F must be k_0, for all digraphs neither (1) nor (2) can apply. Since this holds for all $n \geq n_1$, the polynomial-time bounded machine $M_{k_0/2}$ correctly decides CSP($\mathfrak{B}_{\mathcal{T}_0}$), and hence CSP($\mathfrak{B}_{\mathcal{T}_0}$) is in P. But then there is the following polynomial-time algorithm that solves CSP($\mathfrak{B}_{\mathcal{T}}$), a contradiction to coNP-completeness of CSP($\mathfrak{B}_{\mathcal{T}}$) (Proposition 13.3.1) and our assumption that coNP $\neq$ P.

```
// Input: A finite digraph 𝔄.
If 𝔄 maps homomorphically to 𝔅_{T_0} then accept.
Test whether one of the finitely many digraphs in T \ T_0 embeds into 𝔄.
Accept if none of them embeds into 𝔄.
Reject otherwise.
```

If k_0 is odd, then on all inputs $n \geq n_1$ the second phase of F does not find a digraph $\mathfrak{A}$ for which (3) or (4) applies, because the output of F must be k_0. Hence, $R_{\lfloor k_0/2 \rfloor}$ is a polynomial-time reduction from CSP($\mathfrak{B}_{\mathcal{T}}$) to CSP($\mathfrak{B}_{\mathcal{T}_0}$), and by Proposition 13.3.1 the problem CSP($\mathfrak{B}_{\mathcal{T}_0}$) is coNP-hard. But note that because $f(n)$ equals the odd number k_0 for all but finitely many n, the set $\mathcal{T}_0$ is finite. Therefore, CSP($\mathfrak{B}_{\mathcal{T}_0}$) can be solved in polynomial time, contradicting our assumption that coNP $\neq$ P. ⊣

THEOREM 13.3.4. CSP($\mathfrak{B}_{\mathcal{T}_0}$) *is in coNP, but neither in P nor coNP-complete (unless coNP=P).*

PROOF. It is easy to see that CSP($\mathfrak{B}_{\mathcal{T}_0}$) is in coNP. On input $\mathfrak{A}$ the algorithm non-deterministically chooses a sequence of l vertices, and checks in polynomial time whether this sequence induces a copy of $\mathfrak{T}_l$. If yes, the algorithm computes $f(l)$, which can be done in linear time by executing F on the unary representation of l. If $f(l)$ is even, the algorithm accepts. Recall that $\mathfrak{A}$ does not map homomorphically to $\mathfrak{B}_{\mathcal{T}_0}$ iff a tournament $\mathfrak{T}_l \in \mathcal{T}_0$ embeds into $\mathfrak{A}$, which is the case if and only if there is an accepting computation path for the above non-deterministic algorithm.

Suppose that CSP($\mathfrak{B}_{\mathcal{T}_0}$) is in P. Then for some i the machine M_i decides CSP($\mathfrak{B}_{\mathcal{T}_0}$). By Lemma 13.3.2 and Lemma 13.3.3 there exists an n_0 such that $f(n_0) = 2i$. Then there must also be an $n_1 > n_2$ such that the value k computed during the first phase of F on input n_1 equals $2i$. Since M_i correctly decides

CSP($\mathfrak{B}_{\mathcal{T}_0}$), the machine F returns $2i$ on input n_1. By Lemma 13.3.2, the machine F also returns $2i$ for all inputs from n_1 to n_2, and by induction it follows that it F returns $2i$ for *all* inputs larger than $n \geq n_0$, in contradiction to Lemma 13.3.3.

Finally, suppose that CSP($\mathfrak{B}_{\mathcal{T}_0}$) is coNP-complete. Then for some i the machine R_i is a valid reduction from CSP($\mathfrak{B}_{\mathcal{T}}$) to CSP($\mathfrak{B}_{\mathcal{T}_0}$). Again, by Lemma 13.3.2 and Lemma 13.3.3 there exists an n_1 such that the value k computed during the first phase of F on input n_1 equals $2i$. Since the reduction R_i is correct, the machine F returns $2i$ on input n_1, and in fact returns $2i$ on all inputs greater than n_1. This contradicts Lemma 13.3.3. ⊣

The same technique has been applied to prove several other non-dichotomy results, e.g. for infinite constraint languages [59], in the parameterised complexity setting [221], and for the complexity of planning in AI [127].

Chapter 14

CONCLUSION AND OUTLOOK

Constraint satisfaction problems provide a common framework for a variety of computational problems that appeared in temporal reasoning (Sections 1.6.1 and 1.6.3), phylogenetic analysis (Section 1.6.2), computational linguistics (Section 1.6.4), spatial reasoning (Section 1.6.6), scheduling (Section 1.6.8), and verification (Section 1.6.5). They generalise graph and digraph homomorphism problems that have been studied in combinatorics (see Section 1.1) and have strong links with complexity classification problems in finite model theory, e.g., for the complexity class MMSNP (see Section 1.4). CSPs also generalise network satisfaction problems that have been studied for finite relation algebras (see Section 1.5).

In spite of the diversity of the computational problems from these areas, they can all be formulated as constraint satisfaction problems for an appropriately chosen infinite-domain template structure. This common framework has the advantage that the same tools can be applied to study their computational complexity, for instance the concepts of *primitive positive interpretability* and *homomorphic equivalence*, introduced in Chapter 3. Very often, we can even find templates that satisfy strong model-theoretic assumptions; the reader has learned the relevant model theory in Chapters 2 and 4.

If the template $\mathfrak{B}$ of a CSP can be chosen to be *ω-categorical*, then we have additional methods to study definability and interpretability in $\mathfrak{B}$ via the automorphism group of $\mathfrak{B}$. In Chapter 4 we have presented a beautiful dictionary to translate back-and-forth between concepts and results from model theory and from permutation groups. This dictionary can be extended to *primitive positive* definability and interpretability if we replace the automorphism group of $\mathfrak{B}$ by the polymorphism clone of $\mathfrak{B}$, which takes us into universal algebra in Chapter 6. The universal-algebraic perspective is also very fruitful for example to better understand one of the main algorithmic approach to infinite-domain CSPs, which is *local consistency*. Local consistency algorithms can often be formulated using Datalog programs from finite model theory; pebble games can be used to prove Datalog inexpressibility results; all this can be found in Chapter 8.

Operation clones carry a natural topology, the topology of pointwise convergence, which makes them *topological clones*, similarly as permutation groups may be viewed as topological groups. Chapter 9 shows that the computational complexity of the CSP of an ω-categorical structure only depends on the polymorphism clone, viewed as a topological clone. The most recent developments concerning polymorphism clones of ω-categorical structures have been presented in Chapter 10. Finally, one of the strongest combinatorial tools for obtaining complexity classification results with the universal-algebraic approach is *Ramsey theory*, via *canonical functions*, presented in Chapter 11; all this is used for the complexity classification of temporal constraint languages in Chapter 12.

The relations among the research areas mentioned are illustrated in Figure 1.

14.1. Future research directions

Infinite-domain constraint satisfaction, even when restricted to ω-categorical templates, is still in its infancy. The infinite-domain tractability conjecture for reducts of finitely bounded homogeneous structures remains wide open. There are several important directions for future research.

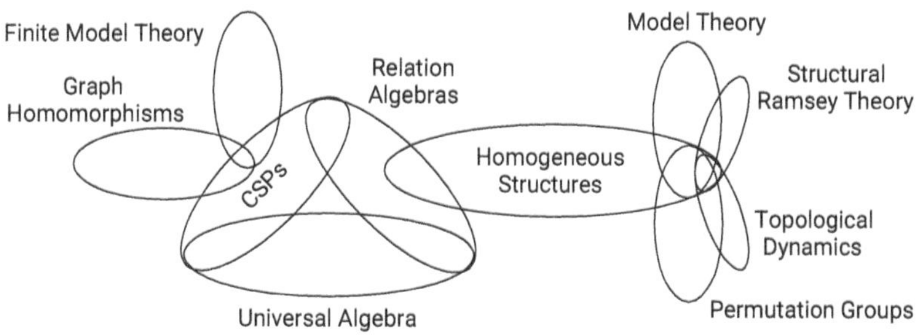

FIGURE 1. Landscape of CSP-relevant research areas around the year 2003: universal algebra and model theory had long separated into different research communities [206]; the Kechris-Pestov-Todorčević connection [229] had just been announced, creating the intersection of topological dynamics and structural Ramsey theory for homogeneous structures. The intersection of model theory and infinite permutation groups for ω-categorical structures was recognised as a fruitful field of research already since the 80s [226]. Homogeneous structures had been used to construct representations of relation algebras [202, 204], and relation algebras had been used to model CSPs [251, 203]. The finite-domain CSP research community with its big intersection of universal algebra, finite model theory, and graph homomorphisms [142] had not yet been fully formed; however, the connection between CSPs and universal algebra was already discovered in the late 90s [217], and the field of CSPs as a common topic in graph homomorphisms and finite model theory was basically created by Feder and Vardi [170].

14.1.1. Universal algebra. Some of the important results that hold for operation clones on finite domains have not yet been generalised to oligomorphic clones; outstanding examples are the equivalence of the existence of a Siggers operation, the existence of a weak near-unanimity operation, and the existence of a 4-ary Siggers operation mentioned in Section 6.9 (cf. Questions 21 and 22). A main obstruction seems to be that absorption theory [24] has only been developed for finite-domain algebras. See Section 14.2.6 for a list of concrete open research questions in this direction.

14.1.2. Polynomial-time tractability. We already know that ω-categorical model-complete cores $\mathfrak{B}$ without a pseudo-Siggers polymorphism have an NP-hard CSP. Therefore, if the infinite-domain tractability conjecture (Conjecture 3.1) is true, then what is left to classify the computational complexity of all finite-signature first-order reducts of finitely bounded homogeneous structures is to prove polynomial-time tractability of CSP($\mathfrak{B}$) if $\mathfrak{B}$ has a pseudo-Siggers polymorphism. The main challenges here are

- finding suitable polynomial-time algorithms, and
- proving that if $\mathfrak{B}$ has a pseudo-Siggers polymorphism then one of these algorithms applies.

This suggests a bottom-up approach: prove that if the polymorphism clone of $\mathfrak{B}$ satisfies identities that are stronger than the pseudo-Siggers identity, CSP($\mathfrak{B}$) can be solved in polynomial time. Examples of such identities for which this has been successful are

- quasi near-unanimity identities from Section 8.5.2, in which case we can use Datalog to solve CSP($\mathfrak{B}$) in polynomial time;
- canonical pseudo-Siggers polymorphisms from Section 10.5.5, in which case we can reduce CSP($\mathfrak{B}$) to a polynomial-time tractable finite-domain CSP.

A next step might be to find stronger conditions that imply that CSP($\mathfrak{B}$) can be expressed in fixed-point logic (Section 8.7). Some strong polymorphism conditions that should imply polynomial-time tractability can be found in the open problem list, Section 14.2.6.

14.1.3. Classification results. With the techniques developed in this text, several complexity classification projects now appear to be feasible. We expect interesting polynomial-time tractable CSPs for the class of first-order reducts of ω-categorical semilinear orders (these were classified by Droste [155]), so classifying the CSP for finite-signature first-order reducts of the structure $(\mathbb{S}; \leq)$ from Section 5.2 appears to be an attractive goal. Another challenge would be a full classification for guarded monotone SNP (Section 5.6.3). Several other promising concrete classification tasks are listed in Section 14.2.9.

14.1.4. CSPs over numerical domains. Many constraint satisfaction problems cannot be formulated with ω-categorical templates; this is typically the case if the CSP involves some sort of numeric reasoning, e.g., if it can express addition over some infinite set. These problems often occur in practice. Some of the techniques developed here for ω-categorical structures can be applied by working with sufficiently saturated templates [61]. In this way, a complete complexity classification for CSPs of first-order reducts of $(\mathbb{Z}; <)$ has been obtained [79]. However, even the CSPs for first-order reducts of $(\mathbb{Q}; <, \{(x, y) \mid x = y + 1\})$ or of $(\mathbb{Z}; \{(x, y) \mid x = 2y\}, \{(x, y) \mid x = y + 1\})$ have not yet been fully classified. Unlike ω-categorical templates, there are even concrete structures over numerical domains whose CSP is neither known to be in P nor known to be NP-hard, such as

- CSP$(\mathbb{Z}; \{(x, y) \mid x = 2y\}, \{(x, y) \mid x = y + 1\}, <)$,
- CSP$(\mathbb{Q}; \{(x, y) \mid x = 2y\}, \{(x, y) \mid x = y + 1\}, R^{\min})$,

- CSP($\mathbb{R}$; $\{(x, y, z) \mid x = y + z\}, \{1\}, \{(x, y) \mid y \geq x^2\}$).

A survey article about the complexity of numeric CSPs is available [76].

14.2. Open problem list

We list concrete open research questions whose answer would help to improve our understanding of the infinite-domain tractability conjecture. Luckily, solving all of these problems might not be necessary to prove the conjecture.

14.2.1. Classical model theory.

1. (Cherlin [130], Problem D) Let $\mathfrak{B}$ be countably infinite and homogeneous in a finite binary relational signature. If $\mathfrak{B}$ has a primitive automorphism group, is $\text{acl}(A) = A$ for all finite A? Addition: Can we show this under the additional assumption that $\mathfrak{B}$ is finitely bounded?

2. Does every primitive oligomorphic permutation group have the orbital extension property (Definition 6.1.28)?

3. (Macpherson [270], Question 2.2.7 (4)) Is the age of a homogeneous structure well-quasi-ordered by embeddings if and only if the growth of the number of orbits of n-element subsets is bounded from above by an exponential function in n?

14.2.2. Existential positive model theory.

4. Thomas' conjecture [339] states that every countable homogeneous structure with finite relational signature $\mathfrak{B}$ has only finitely many first-order reducts, up to first-order interdefinability. We ask the following strengthening: under the same assumptions on $\mathfrak{B}$, are there only finitely many endomorphism monoids of model-complete cores of first-order reducts of $\mathfrak{B}$?

5. If $\mathfrak{B}$ is a reduct of a finitely bounded homogeneous structure, is the model-complete core of $\mathfrak{B}$ also the reduct of a finitely bounded homogeneous structure?

14.2.3. Automorphism groups.

6. (MacPherson and Praeger [272]) Is it true that whenever $\mathfrak{B}$ is a countably infinite structure which is homogeneous over a finite relational signature, then $\text{Aut}(\mathfrak{B})$ has just finitely many closed normal subgroups?

7. Is the property to be a first-order reduct of a finitely bounded homogeneous structure a property of the topological group?

8. We have seen that the property of an oligomorphic clone to be finitely related is a property of the topological clone (Proposition 9.5.19). There is an analogous statement for groups instead of clones. Can this property be expressed naturally in terms of the topological group? We mention that the

decision question to decide whether a given finite algebra is finitely related (Definition 6.1.2) is undecidable [286].

14.2.4. Polymorphism clones.

9. (Question 9.1) Does there exist a reduct of a finitely bounded homogeneous structure $\mathfrak{B}$ such that $\mathrm{Pol}(\mathfrak{B})$ has a homomorphism to the clone projections **Proj**, but no continuous one?

10. Let ξ be an isomorphism between the polymorphism clones of two (reducts of) finitely bounded homogeneous structures. Is ξ always a homeomorphism?

11. Suppose that there exists an isomorphism between the polymorphism clones of two (reducts of) finitely bounded homogeneous structures. Does there also exist an isomorphism which is additionally a homeomorphism?

12. Is it consistent with ZF that every homomorphism from an oligomorphic clone to **Proj** is continuous?

13. Is it consistent with ZF that every isomorphism between polymorphism clones of countably infinite structures is a homeomorphism?

14.2.5. Ramsey theory.

14. Can every finitely bounded homogeneous structure be expanded to a homogeneous finitely bounded structure with the Ramsey property? (Conjecture 11.1).

15. Can every homogeneous structure with finite relational signature be expanded to a homogeneous structure with finite relational signature and additionally the Ramsey property?

16. Can the model-complete core of a reduct of a finitely bounded homogeneous Ramsey structure be expanded to a finitely bounded homogeneous Ramsey structure? A solution to this question has been announced recently by Mottet and Pinsker [287].

14.2.6. Universal algebra.

17. Is every algebra with few subpowers (see Section 6.9.2) finitely related (Definition 6.1.2)?

18. Does every ω-categorical structure with finite relational signature and a quasi Jónsson polymorphism also have a quasi near-unanimity polymorphism?

19. Does every ω-categorical structure with a quasi Jónsson polymorphism also have a quasi directed Jónsson operation?

20. Does every ω-categorical structure with finite relational signature and an (idempotent) Jónsson polymorphism also have a near-unanimity polymorphism?

21. Does every ω-categorical model-complete core with a 6-ary pseudo-Siggers polymorphism also have some pseudo weak near-unanimity polymorphism?

22. Does every ω-categorical model-complete core with a 6-ary pseudo-Siggers polymorphism also have a 4-ary pseudo-Siggers polymorphism? This would follow from a positive answer to the following question of Barto and Pinsker.

23. If $\mathfrak{B}$ is an ω-categorical digraph without sources and sinks of algebraic length 1, it is true that either $K_3 \in \mathrm{HI}(\mathfrak{B})$ or $\mathfrak{B}$ contains a pseudo-loop? (Conjecture 3.7 in [28]; it generalises Theorem 6.9.4 in the same way as Lemma 10.2.3 generalises Lemma 10.2.2.)

24. Does every countably infinite ω-categorical core with an essential polymorphism also have a binary essential polymorphism?

25. Consider the following equivalence relation on finite structures [98]: put $\mathfrak{A} \sim \mathfrak{B}$ if $\mathfrak{A} \in \mathrm{HI}(\mathfrak{B})$ and $\mathfrak{B} \in \mathrm{HI}(\mathfrak{A})$. Does this equivalence relation have countably many equivalence classes? Note that there are uncountably many structures with the domain $\{0, 1, 2\}$ up to primitive positive interdefinability [346, 347]; however, the equivalence relation defined above is coarser. It is already open whether the restriction of the equivalence relation described above to three-element structures has countably many equivalence classes.

26. Does the poset from the previous question have uncountably many equivalence classes for countable structures that are preserved by all permutations? (This would be a strengthening of the result from [54] which states that there are uncountably many closed clones over $\mathbb{N}$ that are preserved by all permutations.)

27. Does every ω-categorical structure without algebraicity that can be solved by Datalog also have a binary injective polymorphism?

28. Is every oligomorphic clone locally generated by its diagonally canonical operations? Equivalently, is it true that a relation R is primitively positively definable in an ω-categorical structure $\mathfrak{B}$ if and only if $\mathfrak{B}$ is preserved by all *diagonally canonical* polymorphisms of $\mathfrak{B}$?

29. Can expressibility in fixed point logic of the CSP for a model-complete core first-order reduct $\mathfrak{B}$ of a finitely bounded homogeneous structure be characterised by a pseudo height-one condition that must hold in the polymorphism clone of $\mathfrak{B}$? A candidate for such a condition is proposed in [87].

14.2.7. Finite model theory.

30. Is there a Henson digraph (Example 2.3.12) whose CSP is in monotone SNP (Section 1.4.3) but not in FO?

31. Is every CSP in existential MSO also in SNP?

32. Is every CSP which is both in SNP (and hence in monotone SNP, see Section 5.6.2) and in USO (universal second-order logic) in P?

33. Is every CSP in MSO the CSP for an ω-categorical structure?

14.2.8. Polynomial-time tractability.

34. Suppose that $\mathfrak{B}$ is a reduct of a finitely bounded homogeneous structure with a quasi edge polymorphism. Is CSP($\mathfrak{B}$) in P? The same question is open even if $\mathfrak{B}$ has an (idempotent) edge polymorphism.

35. Suppose that $\mathfrak{B}$ is a reduct of a finitely bounded homogeneous structure with a chain of quasi Pixley polymorphisms (cf. Section 8.5.5). Can CSP($\mathfrak{B}$) be solved in Datalog? The question is already open for chains of (idempotent) Pixley polymorphisms. Note that a positive answer to Question 18 also implies a positive answer to this question, because chains of quasi Pixley operations provide chains of quasi Jonsson operations (Proposition 6.9.12).

36. Suppose that $\mathfrak{B}$ is a reduct of a finitely bounded homogeneous structure and $\mathfrak{B}$ is preserved by totally symmetric operations of all arities. Can CSP($\mathfrak{B}$) be expressed in fixed point logic?

14.2.9. Open classifications. In each of the following questions we describe classes of computational problems that can be formulated as CSPs for reducts of finitely bounded homogeneous structures where the computational complexity has not yet been classified. Does Conjecture 3.1 hold for all

37. CSPs for first-order reducts of RCC5 (Section 5.4; [62])?

38. CSPs for first-order reducts of Allen's Interval Algebra[62]?

39. CSPs for first-order reducts of finitely bounded homogeneous structures whose age has *free* amalgamation (this has also been raised in [80])?

40. CSPs for structures with a first-order interpretation over $(\mathbb{N}; \neq)$? This question was also posed in [81].

41. CSPs for structures where the number of orbits of n-element subsets grows at most polynomially? For some partial results in this direction, see [74, 169, 48].

42. CSPs for structures with a first-order interpretation over $(\mathbb{Q}; <)$ (harder than the previous two, mentioned already in [43])?

43. CSPs for finite-signature structures that are first-order reducts of finitely bounded homogeneous structures that are additionally ω-stable [337][31].

44. general network satisfaction problems for finite relational algebras with a normal representation (see [47])?

45. CSPs that can be expressed in guarded monotone SNP (Section 5.6.3)?

Moreover, in each of the classes above, we might ask for classifications whether the respective problems can be solved by Datalog or expressed in fixed point logic. We pick out two of the resulting classification problems that appear to be particularly interesting.

46. For which structures $\mathfrak{B}$ with a first-order interpretation over $(\mathbb{Q}; <)$ can CSP($\mathfrak{B}$) be expressed in fixed point logic?

47. Is it true that a problem in connected MMSNP can be expressed in Datalog if and only if the ω-categorical model-complete core template of the corresponding CSP (see Section 5.6.2) has polymorphisms satisfying the

[31] Note that ω-stability alone is not strong enough to hope for a complexity classification: Bodirsky and Grohe [59] constructed for every computational problem a constraint satisfaction problem which is equivalent under polynomial-time Turing reductions and whose template is ω-stable.

pseudo-variants (cf. Section 10.1) of the identities presented in Theorem 8.8.2 (which capture Datalog expressibility over finite domains; this question has already been asked in [75], aiming at a solution to an open problem from [40])?

14.2.10. Complexity theory.

48. Are there ω-categorical structures whose CSP is in NP, but neither NP-hard nor in P?

49. Is there a complexity non-dichotomy for ω-categorical structures $\mathfrak{B}$ if $\mathrm{CSP}(\mathfrak{B})$ is in SNP?

50. Is every ω-categorical CSP equivalent to the CSP of an ω-categorical (directed or even undirected) graph?

14.2.11. Decidability of meta questions.

51. Given a first-order reduct $\mathfrak{B}$ of a finitely bounded homogeneous structure (represented as discussed in Section 11.6), can we decide whether $\mathrm{Pol}(\mathfrak{B})$ has a uniformly continuous homomorphism to the projections? Equivalently, can we decide whether the model-complete core of $\mathfrak{B}$ has a pseudo-Siggers polymorphism?

52. Given a first-order reduct $\mathfrak{B}$ of a finitely bounded homogeneous structure, is the following decidable: Is $\mathrm{CSP}(\mathfrak{B})$ in Datalog? Is it in (ℓ, k)-Datalog for some fixed $\ell, k \in \mathbb{N}$?

53. Can we effectively decide for a given finite set of finite τ-structures whether $\mathrm{Forb}^{\mathrm{emb}}(\mathcal{F})$ is an amalgamation class (a positive answer is known if all symbols in τ are binary; see Proposition 2.3.19)?

54. Can we effectively decide first-order definability in first-order reducts of homogeneous finitely bounded Ramsey structures?

14.2.12. Theory of relation algebras.

55. Is there a finite relation algebra with a fully universal square representation, but without an ω-categorical fully universal square representation?

56. Is there a finite relation algebra with an ω-categorical fully universal square representation but without a fully universal square representation which is not a first-order reduct of a finitely bounded homogeneous structure?

57. Find a finite relation algebra $\mathbf{A}$ such that there is no ω-categorical structure $\mathfrak{B}$ such that the general network satisfaction problem for $\mathbf{A}$ equals the constraint satisfaction problem for $\mathfrak{B}$ (note that we do not insist on $\mathfrak{B}$ being a representation of $\mathbf{A}$).

58. Find a finite relation algebra with an ω-categorical fully universal square representation which is not the orbital relation algebra of an ω-categorical structure.

14.2.13. Other.

59. Let τ be a finite relational signature. Is there for all countable ω-categorical τ-structures $\mathfrak{A}, \mathfrak{B}$ a countable ω-categorical structure $\mathfrak{C}$ such that

for all finite (equivalently, for all countable) $\mathfrak{H}$

$$\mathfrak{A} \times \mathfrak{H} \to \mathfrak{B} \;\Leftrightarrow\; \mathfrak{H} \to \mathfrak{C}?$$

This would turn the lattice arising from ordering ω-categorical τ-structures by the existence of homomorphisms (Remark 4.2.21) into a Heyting algebra. The corresponding statement is true for the class of all finite τ-structures; the so-called $\mathfrak{A}$-*th power of* $\mathfrak{B}$ satisfies the given condition for $\mathfrak{C}$ and is finite if $\mathfrak{A}$ and $\mathfrak{B}$ are finite; see [197, 254].

REFERENCES

[1] Fred G. Abramson and Leo Harrington, *Models without indiscernibles*, ***The Journal of Symbolic Logic***, vol. 43 (1978), no. 3, pp. 572–600.

[2] Dimitris Achlioptas, Amin Coja-Oghlan, and Federico Ricci-Tersenghi, *On the solution-space geometry of random constraint satisfaction problems*, ***Random Structures and Algorithms***, vol. 38 (2011), no. 3, pp. 251–268.

[3] Samson Adepoju Adeleke and Peter M. Neumann, *Primitive permutation groups with primitive Jordan sets*, ***Journal of the London Mathematical Society***, vol. 53 (1994), no. 2, pp. 209–229.

[4] ———, ***Relations Related to Betweenness: Their Structure and Automorphisms***, Memoirs of the AMS, vol. 623, American Mathematical Society, 1998.

[5] Hans Adler, *A geometric introduction to forking and thorn-forking*, ***Journal of Mathematical Logic***, vol. 9 (2009), no. 1, pp. 1–20.

[6] Lovkush Agarwal, *The reducts of the generic digraph*, ***Annals of Pure and Applied Logic***, vol. 167 (2016), pp. 370–391.

[7] Lovkush Agarwal and Michael Kompatscher, $2^{\aleph_0}$ *pairwise nonisomorphic maximal-closed subgroups of Sym*($\mathbb{N}$) *via the classification of the reducts of the Henson digraphs*, ***The Journal of Symbolic Logic***, vol. 83 (2018), no. 2, pp. 395–415.

[8] Gisela Ahlbrandt and Martin Ziegler, *Quasi-finitely axiomatizable totally categorical theories*, ***Annals of Pure and Applied Logic***, vol. 30 (1986), no. 1, pp. 63–82.

[9] Alfred V. Aho, Yehoshua Sagiv, Thomas G. Szymanski, and Jeffrey D. Ullman, *Inferring a tree from lowest common ancestors with an application to the optimization of relational expressions*, ***SIAM Journal on Computing***, vol. 10 (1981), no. 3, pp. 405–421.

[10] James F. Allen, *Maintaining knowledge about temporal intervals*, ***Communications of the ACM***, vol. 26 (1983), no. 11, pp. 832–843.

[11] Eric Allender, Michael Bauland, Neil Immerman, Henning Schnoor, and Heribert Vollmer, *The complexity of satisfiability problems:*

Refining Schaefer's theorem, ***Journal of Computer and System Sciences***, vol. 75 (2009), no. 4, pp. 245–254.

[12] BENGT ASPVALL, MICHAEL F. PLASS, and ROBERT ENDRE TARJAN, *A linear-time algorithm for testing the truth of certain quantified boolean formulas*, ***Information Processing Letters***, vol. 8 (1979), no. 3, pp. 121–123.

[13] ALBERT ATSERIAS, *On digraph coloring problems and treewidth duality*, ***Proceedings of the Symposium on Logic in Computer Science (LICS)***, 2005, pp. 106–115.

[14] ALBERT ATSERIAS, ANDREI A. BULATOV, and ANUJ DAWAR, *Affine systems of equations and counting infinitary logic*, ***Theoretical Computer Science***, vol. 410 (2009), no. 18, pp. 1666–1683.

[15] FRANZ BAADER and TOBIAS NIPKOW, ***Term Rewriting and All That***, Cambridge University Press, 1999.

[16] FRANZ BAADER and KLAUS U. SCHULZ, *Combining constraint solving*, ***Constraints in Computational Logics*** (H. Comon, C. Marché, and R. Treinen, editors), Lecture Notes in Computer Science, vol. 2002, Springer-Verlag, 2001.

[17] LIBOR BARTO, *The dichotomy for conservative constraint satisfaction problems revisited*, ***Proceedings of the Symposium on Logic in Computer Science (LICS)*** (Toronto, Canada), 2011.

[18] ———, *Finitely related algebras in congruence distributive varieties have near unanimity terms*, ***Canadian Journal of Mathematics***, vol. 65 (2013), no. 1, pp. 3–21.

[19] ———, *Finitely related algebras in congruence modular varieties have few subpowers*, ***Journal of the European Mathematical Society***, vol. 20 (2018), no. 6, pp. 1439–1471.

[20] LIBOR BARTO, MICHAEL KOMPATSCHER, MIROSLAV OLŠÁK, TRUNG VAN PHAM, and MICHAEL PINSKER, *Equations in oligomorphic clones and the constraint satisfaction problem for ω-categorical structures*, ***Journal of Mathematical Logic***, vol. 19 (2019), no. 2, p. #1950010, An extended abstract appeared at the Proceedings of the 32nd Annual ACM/IEEE Symposium on Logic in Computer Science – LICS'17.

[21] LIBOR BARTO and MARCIN KOZIK, *Constraint satisfaction problems of bounded width*, ***Proceedings of the Annual Symposium on Foundations of Computer Science (FOCS)***, 2009, pp. 595–603.

[22] ———, *New conditions for Taylor varieties and CSP*, ***Proceedings of the Symposium on Logic in Computer Science (LICS)***, 2010, pp. 100–109.

[23] ———, *Absorbing subalgebras, cyclic terms and the constraint satisfaction problem*, ***Logical Methods in Computer Science***, vol. 8/1 (2012), no. 07, pp. 1–26.

[24] ———, *Absorption in universal algebra and CSP*, ***The Constraint Satisfaction Problem: Complexity and Approximability***, Dagstuhl Follow-Ups, vol. 7, 2017, pp. 45–77.

[25] LIBOR BARTO, MARCIN KOZIK, and TODD NIVEN, *The CSP dichotomy holds for digraphs with no sources and no sinks (a positive answer to a conjecture of Bang-Jensen and Hell)*, ***SIAM Journal on Computing***, vol. 38 (2009), no. 5.

[26] LIBOR BARTO, ANDREI A. KROKHIN, and ROSS WILLARD, *Polymorphisms, and how to use them*, ***The Constraint Satisfaction Problem: Complexity and Approximability***, Schloss Dagstuhl – Leibniz-Zentrum für Informatik, 2017, pp. 1–44.

[27] LIBOR BARTO, JAKUB OPRŠAL, and MICHAEL PINSKER, *The wonderland of reflections*, ***Israel Journal of Mathematics***, vol. 223 (2018), no. 1, pp. 363–398.

[28] LIBOR BARTO and MICHAEL PINSKER, *The algebraic dichotomy conjecture for infinite domain constraint satisfaction problems*, ***Proceedings of the 31th Annual IEEE Symposium on Logic in Computer Science – LICS'16***, 2016, Preprint arXiv:1602.04353, pp. 615–622.

[29] ———, *Topology is irrelevant*, ***SIAM Journal on Computing***, vol. 49 (2020), no. 2, pp. 365–393.

[30] BRUCE L. BAUSLAUGH, *The complexity of infinite H-coloring*, ***Journal of Combinatorial Theory, Series B***, vol. 61 (1994), no. 2, pp. 141–154.

[31] ———, *Core-like properties of infinite graphs and structures*, ***Discrete Mathematics***, vol. 138 (1995), no. 1, pp. 101–111.

[32] ———, *Cores and compactness of infinite directed graphs*, ***Journal of Combinatorial Theory, Series B***, vol. 68 (1996), no. 2, pp. 255–276.

[33] HOWARD BECKER and ALEXANDER KECHRIS, ***The Descriptive Set Theory of Polish Group Actions***, LMS Lecture Note Series, no. 232, Cambridge University Press, 1996.

[34] MIKE BEHRISCH, JOHN K. TRUSS, and EDITH VARGAS-GARCIA, *Reconstructing the topology on monoids and polymorphism clones of the rationals*, ***Studia Logica***, vol. 105 (2017), pp. 65–91.

[35] ITAÏ BEN YAACOV and BRUNO POIZAT, *Fondements de la logique positive*, ***The Journal of Symbolic Logic***, vol. 72 (2007), no. 4, pp. 1141–1162.

[36] ITAY BEN YAACOV, *Positive model theory and compact abstract theories*, ***Journal of Mathematical Logic***, vol. 3 (2003), no. 1, pp. 85–118.

[37] BRANDON BENNETT, *Spatial reasoning with propositional logics*, ***Proceedings of the International Conference on Knowledge Representation and Reasoning***, Morgan Kaufmann, 1994, pp. 51–62.

[38] JOEL BERMAN, PAWEL IDZIAK, PETAR MARKOVIC, RALPH MCKENZIE, MATTHEW VALERIOTE, and ROSS WILLARD, *Varieties with few subalgebras of powers*, ***Transactions of the American Mathematical Society***, vol. 362 (2010), no. 3, pp. 1445–1473.

[39] MARC BEZEM, ROBERT NIEUWENHUIS, and ENRIC RODRÍGUEZ-CARBONELL, *The max-atom problem and its relevance*, ***Proceedings of the International Conference on Logic for Programming, Artificial Intelligence, and Reasoning (LPAR)***, 2008, pp. 47–61.

[40] MEGHYN BIENVENU, BALDER TEN CATE, CARSTEN LUTZ, and FRANK WOLTER, *Ontology-based data access: A study through disjunctive Datalog, CSP, and MMSNP*, ***ACM Transactions of Database Systems***, vol. 39 (2014), no. 4, p. 33.

[41] NORMAN BIGGS, *Constructions for cubic graphs with large girth*, ***Electronic Journal of Combinatorics***, vol. 5 (1998).

[42] GARRETT BIRKHOFF, *On the structure of abstract algebras*, ***Mathematical Proceedings of the Cambridge Philosophical Society***, vol. 31 (1935), no. 4, pp. 433–454.

[43] MANUEL BODIRSKY, *Constraint satisfaction with infinite domains*, Dissertation, Humboldt-Universität zu Berlin, 2004.

[44] ———, *Cores of countably categorical structures*, ***Logical Methods in Computer Science (LMCS)***, vol. 3 (2007), no. 1, pp. 1–16.

[45] ———, *New Ramsey classes from old*, ***Electronic Journal of Combinatorics***, vol. 21 (2014), no. 2, preprint arXiv:1204.3258.

[46] ———, *Ramsey classes: Examples and constructions*, ***Surveys in Combinatorics. London Mathematical Society Lecture Note Series 424***, Cambridge University Press, 2015, invited survey article for the British Combinatorial Conference; ArXiv:1502.05146.

[47] ———, *Finite relation algebras with normal representations*, ***Relational and Algebraic Methods in Computer Science – 17th International Conference, RAMiCS 2018, Groningen, The Netherlands, October 29-November 1, 2018, Proceedings***, 2018, pp. 3–17.

[48] MANUEL BODIRSKY and BERTALAN BODOR, *Structures with small orbit growth*, 2019, preprint available at ArXiv:1810.05657.

[49] MANUEL BODIRSKY, DAVID BRADLEY-WILLIAMS, MICHAEL PINSKER, and ANDRÁS PONGRÁCZ, *The universal homogeneous binary tree*, ***Journal of Logic and Computation***, vol. 28 (2018), no. 1, pp. 133–163.

[50] MANUEL BODIRSKY and HUBERT CHEN, *Oligomorphic clones*, ***Algebra Universalis***, vol. 57 (2007), no. 1, pp. 109–125.

[51] MANUEL BODIRSKY and HUBIE CHEN, *Qualitative temporal and spatial reasoning revisited*, ***Journal of Logic and Computation***, vol. 19 (2009), no. 6, pp. 1359–1383.

[52] ———, *Quantified equality constraints*, ***SIAM Journal on Computing***, vol. 39 (2010), no. 8, pp. 3682–3699, A preliminary version appeared in the proceedings of LICS'07.

[53] MANUEL BODIRSKY, HUBIE CHEN, JAN KÁRA, and TIMO VON OERTZEN, *Maximal infinite-valued constraint languages*, ***Theoretical Computer Science (TCS)***, vol. 410 (2009), pp. 1684–1693, a preliminary version appeared at ICALP'07.

[54] MANUEL BODIRSKY, HUBIE CHEN, and MICHAEL PINSKER, *The reducts of equality up to primitive positive interdefinability*, ***The Journal of Symbolic Logic***, vol. 75 (2010), no. 4, pp. 1249–1292.

[55] MANUEL BODIRSKY, HUBIE CHEN, and MICHAŁ WRONA, *Tractability of quantified temporal constraints to the max*, ***International Journal of Algebra and Computation***, vol. 24 (2014), no. 8, pp. 1141–1156.

[56] MANUEL BODIRSKY and VÍCTOR DALMAU, *Datalog and constraint satisfaction with infinite templates*, ***Journal on Computer and System Sciences***, vol. 79 (2013), pp. 79–100, a preliminary version appeared in the proceedings of the Symposium on Theoretical Aspects of Computer Science (STACS'05).

[57] MANUEL BODIRSKY, DAVID EVANS, MICHAEL KOMPATSCHER, and MICHAEL PINSKER, *A counterexample to the reconstruction of ω-categorical structures from their endomorphism monoids*, ***Israel Journal of Mathematics***, vol. 224 (2018), no. 1, pp. 57–82, preprint arXiv:1510.00356.

[58] MANUEL BODIRSKY and JOHANNES GREINER, *Complexity of combinations of qualitative constraint satisfaction problems*, ***9th International Joint Conference on Automated Reasoning (IJCAR)***, 2018, pp. 263–278.

[59] MANUEL BODIRSKY and MARTIN GROHE, *Non-dichotomies in constraint satisfaction complexity*, ***Automata, Languages and Programming. 35th International Colloquium, ICALP 2008, Reykjavik, Iceland, July 7-11, 2008, Proceedings, Part II*** (Luca Aceto, Ivan Damgård, Leslie Ann Goldberg, Magnús M. Halldórsson, Anna Ingólfsdóttir, and Igor Walukiewicz, editors), Lecture Notes in Computer Science, vol. 5126, Springer, 2008, pp. 184–196.

[60] MANUEL BODIRSKY and MARTIN HILS, *Tractable set constraints*, ***Journal of Artificial Intelligence Research***, vol. 45 (2012), pp. 731–759.

[61] MANUEL BODIRSKY, MARTIN HILS, and BARNABY MARTIN, *On the scope of the universal-algebraic approach to constraint satisfaction*, ***Logical Methods in Computer Science (LMCS)***, vol. 8 (2012), no. 3, an extended abstract that announced some of the results appeared in the proceedings of Logic in Computer Science (LICS'10).

[62] MANUEL BODIRSKY and PETER JONSSON, *A model-theoretic view on qualitative constraint reasoning*, ***Journal of Artificial Intelligence Research***, vol. 58 (2017), pp. 339–385.

[63] MANUEL BODIRSKY, PETER JONSSON, and TRUNG VAN PHAM, *The reducts of the homogeneous binary branching C-relation*, ***The Journal of Symbolic Logic***, vol. 81 (2016), no. 4, pp. 1255–1297, preprint arXiv:1408.2554.

[64] ———, *The Complexity of Phylogeny Constraint Satisfaction Problems*, ***ACM Transactions on Computational Logic (TOCL)***, vol. 18 (2017), no. 3, an extended abstract appeared in the conference STACS 2016.

[65] MANUEL BODIRSKY, PETER JONSSON, and TIMO VON OERTZEN, *Horn versus full first-order: complexity dichotomies for algebraic constraint satisfaction*, ***Journal of Logic and Computation***, vol. 22 (2011), no. 3, pp. 643–660.

[66] ———, *Essential convexity and complexity of semi-algebraic constraints*, ***Logical Methods in Computer Science***, vol. 8 (2012), no. 4, an extended abstract about a subset of the results has been published under the title *Semilinear Program Feasibility* at ICALP'10.

[67] MANUEL BODIRSKY and MARKUS JUNKER, *$\aleph_0$-categorical structures: interpretations and endomorphisms*, ***Algebra Universalis***, vol. 64 (2011), no. 3-4, pp. 403–417.

[68] MANUEL BODIRSKY and JAN KÁRA, *The complexity of equality constraint languages*, ***Theory of Computing Systems***, vol. 3 (2008), no. 2, pp. 136–158, a conference version appeared in the proceedings of Computer Science Russia (CSR'06).

[69] ———, *The complexity of temporal constraint satisfaction problems*, ***Journal of the ACM***, vol. 57 (2009), no. 2, pp. 1–41, an extended abstract appeared in the Proceedings of the Symposium on Theory of Computing (STOC).

[70] ———, *A fast algorithm and Datalog inexpressibility for temporal reasoning*, ***ACM Transactions on Computational Logic***, vol. 11 (2010), no. 3.

[71] MANUEL BODIRSKY, SIMON KNÄUER, and FLORIAN STARKE, *ASNP: a tame fragment of existential second-order logic*, ***Computing Research Repository (CoRR)***, vol. abs/2001.08190 (2020).

[72] MANUEL BODIRSKY and MARTIN KUTZ, *Pure dominance constraints*, ***STACS***, 2002, pp. 287–298.

[73] ———, *Determining the consistency of partial tree descriptions*, ***Artificial Intelligence***, vol. 171 (2007), pp. 185–196.

[74] MANUEL BODIRSKY, DUGALD MACPHERSON, and JOHAN THAPPER, *Constraint satisfaction tractability from semi-lattice operations on infinite sets*, ***Transaction of Computational Logic (ACM-TOCL)***, vol. 14 (2013), no. 4, pp. 1–30.

[75] MANUEL BODIRSKY, FLORENT MADELAINE, and ANTOINE MOTTET, *A universal-algebraic proof of the complexity dichotomy for Monotone Monadic SNP*, ***Proceedings of the Symposium on Logic in Computer Science (LICS)***, 2018, preprint arXiv:1802.03255, pp. 105–114.

[76] MANUEL BODIRSKY and MARCELLO MAMINO, *Constraint satisfaction problems over numeric domains*, ***The Constraint Satisfaction Problem: Complexity and Approximability***, Schloss Dagstuhl – Leibniz-Zentrum für Informatik, 2017, pp. 79–111.

[77] ———, *Tropically convex constraint satisfaction*, ***Theory of Computing Systems***, vol. 62 (2018), no. 3, pp. 481–509, an extended abstract of the paper appeared under the title "Max-Closed Semilinear Constraints" in the proceedings of CSR'16; preprint available under ArXiv:1506.04184.

[78] MANUEL BODIRSKY, MARCELLO MAMINO, and CATERINA VIOLA, *Submodular functions and valued constraint satisfaction problems over infinite domains*, ***Proceedings of Computer Science Logic (CSL)***, 2018, arXiv:1804.01710.

[79] MANUEL BODIRSKY, BARNABY MARTIN, and ANTOINE MOTTET, *Discrete temporal constraint satisfaction problems*, ***Journal of the ACM***, vol. 65 (2018), no. 2, pp. 9:1–9:41, preprint ArXiv:1503.08572.

[80] MANUEL BODIRSKY, BARNABY MARTIN, MICHAEL PINSKER, and ANDRÁS PONGRÁCZ, *Constraint satisfaction problems for reducts of homogeneous graphs*, ***SIAM Journal on Computing***, vol. 48 (2019), no. 4, pp. 1224–1264, A conference version appeared in the Proceedings of the 43rd International Colloquium on Automata, Languages, and Programming, ICALP 2016, pages 119:1–119:14.

[81] MANUEL BODIRSKY and ANTOINE MOTTET, *A dichotomy for first-order reducts of unary structures*, ***Logical Methods in Computer Science***, vol. 14 (2018), no. 2.

[82] MANUEL BODIRSKY, ANTOINE MOTTET, MIROSLAV OLŠÁK, JAKUB OPRŠAL, MICHAEL PINSKER, and ROSS WILLARD, *Topology is relevant (in the infinite-domain dichotomy conjecture for constraint satisfaction problems)*, ***Proceedings of the Symposium on Logic in Computer Science – LICS'19***, 2019, preprint arXiv:1901.04237.

[83] MANUEL BODIRSKY and JENS K. MUELLER, *Rooted phylogeny problems*, ***Logical Methods in Computer Science***, vol. 7 (2011), no. 4, An extended abstract appeared in the proceedings of ICDT'10.

[84] MANUEL BODIRSKY and JAROSLAV NEŠETŘIL, *Constraint satisfaction with countable homogeneous templates*, ***Proceedings of Computer Science Logic (CSL)*** (Vienna), 2003, pp. 44–57.

[85] ———, *Constraint satisfaction with countable homogeneous templates*, ***Journal of Logic and Computation***, vol. 16 (2006), no. 3, pp. 359–373.

[86] MANUEL BODIRSKY, GUSTAV NORDH, and TIMO VON OERTZEN, *Integer programming with 2-variable equations and 1-variable inequalities*, ***Information Processing Letters***, vol. 109 (2009), no. 11, pp. 572–575.

[87] MANUEL BODIRSKY, WIED PAKUSA, and JAKUB RYDVAL, *Temporal constraint satisfaction problems in fixed-point logic*, ***LICS '20: 35th Annual ACM/IEEE Symposium on Logic in Computer Science, Saarbrücken, Germany, July 8-11, 2020*** (Holger Hermanns, Lijun Zhang, Naoki Kobayashi, and Dale Miller, editors), ACM, 2020, pp. 237–251.

[88] MANUEL BODIRSKY and MICHAEL PINSKER, *Reducts of Ramsey structures*, ***AMS Contemporary Mathematics, vol. 558 (Model Theoretic Methods in Finite Combinatorics)***, (2011), pp. 489–519.

[89] ———, *Minimal functions on the random graph*, ***Israel Journal of Mathematics***, vol. 200 (2014), no. 1, pp. 251–296.

[90] ———, *Schaefer's theorem for graphs*, ***Journal of the ACM***, vol. 62 (2015), no. 3, p. 52 pages (article number 19), A conference version appeared in the Proceedings of STOC 2011, pages 655-664.

[91] ———, *Topological Birkhoff*, ***Transactions of the American Mathematical Society***, vol. 367 (2015), pp. 2527–2549.

[92] ———, *Canonical functions: a proof via topological dynamics*, Preprint arXiv:1610.09660, 2016.

[93] MANUEL BODIRSKY, MICHAEL PINSKER, and ANDRÁS PONGRÁCZ, *The 42 reducts of the random ordered graph*, ***Proceedings of the LMS***, vol. 111 (2015), no. 3, pp. 591–632, preprint available from arXiv:1309.2165.

[94] ———, *Reconstructing the topology of clones*, ***Transactions of the American Mathematical Society***, vol. 369 (2017), pp. 3707–3740.

[95] ———, *Projective clone homomorphisms*, ***The Journal of Symbolic Logic***, (2019), To appear. doi:10.1017/jsl.2019.23.

[96] MANUEL BODIRSKY, MICHAEL PINSKER, and TODOR TSANKOV, *Decidability of definability*, ***The Journal of Symbolic Logic***, vol. 78 (2013), no. 4, pp. 1036–1054, A conference version appeared in the Proceedings of LICS 2011.

[97] MANUEL BODIRSKY and FRIEDRICH MARTIN SCHNEIDER, *A topological characterisation of endomorphism monoids of countable structures*, ***Algebra Universalis***, vol. 77 (2017), no. 3, pp. 251–269, preprint available at arXiv:1508.07404.

[98] MANUEL BODIRSKY and ALBERT VUCAJ, *Two-element structures modulo primitive positive constructability*, ***Algebra Universalis***, vol. 81 (2020), no. 20, preprint available at ArXiv:1905.12333.

[99] MANUEL BODIRSKY and STEFAN WOELFL, *RCC8 is tractable on instances of bounded treewidth*, ***Proceedings of International Joint Conferences on Artificial Intelligence (IJCAI)*** (Toby Walsh, editor), AAAI, July 2011, pp. 756–761.

[100] MANUEL BODIRSKY and MICHAŁ WRONA, *Equivalence constraint satisfaction problems*, ***Proceedings of Computer Science Logic*** (Patrick Cégielski and Arnaud Durand, editors), LIPICS, vol. 16, Dagstuhl Publishing, September 2012, pp. 122–136.

[101] V. G. BODNARČUK, L. A. KALUŽNIN, V. N. KOTOV, and B. A. ROMOV, *Galois theory for Post algebras, part I and II*, ***Cybernetics***, vol. 5 (1969), pp. 243–539.

[102] FERDINAND BÖRNER, ANDREI A. BULATOV, HUBIE CHEN, PETER JEAVONS, and ANDREI A. KROKHIN, *The complexity of constraint satisfaction games and QCSP*, ***Information and Computation***, vol. 207 (2009), no. 9, pp. 923–944.

[103] SAMUEL BRAUNFELD, *Towards the undecidability of atomicity for permutation classes via the undecidability of joint embedding for hereditary graph classes*, 2019, preprint available at arXiv:1903.11932.

[104] MATHIAS BROXVALL and PETER JONSSON, *Point algebras for temporal reasoning: Algorithms and complexity*, ***Artificial Intelligence***, vol. 149 (2003), no. 2, pp. 179–220.

[105] MATHIAS BROXVALL, PETER JONSSON, and JOCHEN RENZ, *Disjunctions, independence, refinements*, ***Artificial Intelligence***, vol. 140 (2002), no. 1/2, pp. 153–173.

[106] ANDREI A. BULATOV, *Tractable conservative constraint satisfaction problems*, ***Proceedings of the Symposium on Logic in Computer Science (LICS)*** (Ottawa, Canada), 2003, pp. 321–330.

[107] ———, *H-coloring dichotomy revisited*, ***Theoretical Computer Science***, vol. 349 (2005), no. 1, pp. 31–39.

[108] ———, *A dichotomy theorem for constraint satisfaction problems on a 3-element set*, ***Journal of the ACM***, vol. 53 (2006), no. 1, pp. 66–120.

[109] ———, *The complexity of the counting constraint satisfaction problem*, ***Journal of the ACM***, vol. 60 (2013), no. 5, pp. 34:1–34:41.

[110] ———, *Conservative constraint satisfaction re-revisited*, ***Journal Computer and System Sciences***, vol. 82 (2016), no. 2, pp. 347–356, arXiv:1408.3690.

[111] ———, *A dichotomy theorem for nonuniform CSPs*, ***58th IEEE Annual Symposium on Foundations of Computer Science, FOCS 2017, Berkeley, CA, USA, October 15-17***, 2017, pp. 319–330.

[112] ANDREI A. BULATOV and VÍCTOR DALMAU, *Towards a dichotomy theorem for the counting constraint satisfaction problem*, ***Proceedings of the Annual Symposium on Foundations of Computer Science (FOCS)***, 2003, pp. 562–574.

[113] ———, *A simple algorithm for Mal'tsev constraints*, ***SIAM Journal on Computing***, vol. 36 (2006), no. 1, pp. 16–27.

[114] ANDREI A. BULATOV, VÍCTOR DALMAU, MARTIN GROHE, and DÁNIEL MARX, *Enumerating homomorphisms*, ***Journal of Computer and System Sciences***, vol. 78 (2012), no. 2, pp. 638–650.

[115] ANDREI A. BULATOV and PETER JEAVONS, *Algebraic structures in combinatorial problems*, Technical report MATH-AL-4-2001, Technische Universität Dresden, 2001.

[116] ANDREI A. BULATOV, ANDREI A. KROKHIN, and PETER G. JEAVONS, *Classifying the complexity of constraints using finite algebras*, ***SIAM Journal on Computing***, vol. 34 (2005), pp. 720–742.

[117] JAKUB BULÍN, ANDREI A. KROKHIN, and JAKUB OPRŠAL, *Algebraic approach to promise constraint satisfaction*, ***Proceedings of the 51st Annual ACM SIGACT Symposium on Theory of Computing, STOC 2019, Phoenix, AZ, USA, June 23-26, 2019***, 2019, pp. 602–613.

[118] HANS KLEINE BÜNING and THEODOR LETTMANN, ***Propositional Logic: Deduction and Algorithms***, Cambridge University Press, 1999.

[119] STANLEY N. BURRIS and HANAMANTAGOUDA P. SANKAPPANAVAR, ***A Course in Universal Algebra***, Springer Verlag, Berlin, 1981.

[120] CRISTIAN S. CALUDE, SANJAY JAIN, BAKHADYR KHOUSSAINOV, WEI LI, and FRANK STEPHAN, *Deciding parity games in quasipolynomial time*, ***Proceedings of the 49th Annual ACM SIGACT Symposium on Theory of Computing, STOC 2017, Montreal, QC, Canada, June 19-23, 2017***, 2017, pp. 252–263.

[121] PETER J. CAMERON, *Transitivity of permutation groups on unordered sets*, ***Mathematische Zeitschrift***, vol. 148 (1976), pp. 127–139.

[122] ———, *Some treelike objects*, ***Quarterly Journal of Mathematics, Oxford Series (2)***, vol. 38 (1987), no. 150, pp. 155–183.

[123] ———, ***Oligomorphic Permutation Groups***, Cambridge University Press, Cambridge, 1990.

[124] ———, *Homogeneous permutations*, ***Electronic Journal of Combinatorics***, vol. 9 (2002), no. 2.

[125] GEORG CANTOR, *Über unendliche, lineare Punktmannigfaltigkeiten*, ***Mathematische Annalen***, vol. 23 (1884), pp. 453–488.

[126] HUBIE CHEN and VÍCTOR DALMAU, *(Smart) look-ahead arc consistency and the pursuit of CSP tractability*, ***Proceedings of the International Conference on Principles and Practice of Constraint Programming (CP)***, 2004, pp. 182–196.

[127] HUBIE CHEN and OMER GIMÉNEZ, *Causal graphs and structurally restricted planning*, ***Journal of Computer and System Sciences***, vol. 76 (2010), no. 7, pp. 579–592.

[128] HUBIE CHEN and MARTIN GROHE, *Constraint satisfaction with succinctly specified relations*, ***Journal of Computer and System Sciences***, vol. 76 (2010), no. 8, pp. 847–860.

[129] GREGORY CHERLIN, SAHARON SHELAH, and NIANGDONG SHI, *Universal graphs with forbidden subgraphs and algebraic closure*, ***Advances in Applied Mathematics***, vol. 22 (1999), pp. 454–491.

[130] GREGORY L. CHERLIN, *Combinatorial problems connected with finite homogeneity*, ***Contemporary Mathematics***, vol. 131 (1993), pp. 3–30.

[131] ———, *The classification of countable homogeneous directed graphs and countable homogeneous n-tournaments*, ***AMS Memoir***, vol. 131 (1998), no. 621.

[132] DAVID COHEN, PETER JEAVONS, PETER JONSSON, and MANOLIS KOUBARAKIS, *Building tractable disjunctive constraints*, ***Journal of the ACM***, vol. 47 (2000), no. 5, pp. 826–853.

[133] DAVID A. COHEN, MARTIN C. COOPER, PETER G. JEAVONS, and STANISLAV ŽIVNÝ, *Binarisation via dualisation for valued constraints*, ***Proceedings of the Twenty-Ninth AAAI Conference on Artificial Intelligence, January 25-30, 2015, Austin, Texas, USA***, 2015, pp. 3731–3737.

[134] DAVID A. COHEN, PETER JEAVONS, PETER JONSSON, and MANOLIS KOUBARAKIS, *Building tractable disjunctive constraints*, ***Journal of the ACM***, vol. 47 (2000), no. 5, pp. 826–853.

[135] PAUL M. COHN, ***Universal Algebra***, Harper and Row, New York, NY, 1965.

[136] AMIN COJA-OGHLAN, *Random constraint satisfaction problems*, ***Proceedings Fifth Workshop on Developments in Computational Models – Computational Models From Nature, DCM 2009, Rhodes, Greece, 11th July 2009***, 2009, pp. 32–37.

[137] THOMAS H. CORMEN, CHARLES E. LEISERSON, RONALD L. RIVEST, and CLIFFORD STEIN, ***Introduction to Algorithms***, MIT Press, 2009, third edition.

[138] THOMAS CORNELL, *On determining the consistency of partial descriptions of trees*, ***Proceedings of the ACL***, 1994, pp. 163–170.

[139] MIGUEL COUCEIRO, LUCIEN HADDAD, and VICTOR LAGERKVIST, *Fine-grained complexity of constraint satisfaction problems through partial polymorphisms: A survey*, ***2019 IEEE 49th International Symposium on Multiple-Valued Logic (ISMVL), Fredericton, NB, Canada, May 21-23, 2019***, 2019, pp. 170–175.

[140] MIGUEL COUCEIRO and ERKKO LEHTONEN, *Generalizations of Świerczkowski's lemma and the arity gap of finite functions*, ***Discrete Mathematics***, vol. 309 (2009), no. 20, pp. 5905–5912.

[141] JACINTA COVINGTON, *Homogenizable relational structures*, ***Illinois Journal of Mathematics***, vol. 34 (1990), no. 4, pp. 731–743.

[142] Nadia Creignou, Phokion G. Kolaitis, and Heribert Vollmer (editors), ***Complexity of Constraints – An Overview of Current Research Themes [Result of a Dagstuhl Seminar]***, Lecture Notes in Computer Science, vol. 5250, Springer, 2008.

[143] MATTEO CRISTIANI and ROBIN HIRSCH, *The complexity of the constraint satisfaction problem for small relation algebras*, ***Artificial Intelligence Journal***, vol. 156 (2004), pp. 177–196.

[144] BÉLA CSÁKÁNY, *Minimal clones*, ***Algebra Universalis***, vol. 54 (2005), no. 1, pp. 73–89.

[145] VÍCTOR DALMAU, *Linear datalog and bounded path duality of relational structures*, ***Logical Methods in Computer Science***, vol. 1 (2005), no. 1.

[146] VICTOR DALMAU, PHOKION G. KOLAITIS, and MOSHE Y. VARDI, *Constraint satisfaction, bounded treewidth, and finite-variable logics*, ***Proceedings of the International Conference on Principles and Practice of Constraint Programming (CP)***, 2002, pp. 310–326.

[147] VÍCTOR DALMAU and JUSTIN PEARSON, *Closure functions and width 1 problems*, ***Proceedings of the International Conference on Principles and Practice of Constraint Programming (CP)***, 1999, pp. 159–173.

[148] RINA DECHTER, ***Constraint Processing***, Morgan Kaufmann, 2003.

[149] WALTER DEUBER, *A generalization of Ramsey's theorem for regular trees*, ***Journal of Combinatorial Theory, Series B***, vol. 18 (1975), pp. 18–23.

[150] REINHARD DIESTEL, ***Graph Theory***, Springer–Verlag, New York, 2005, third edition.

[151] JOHN DIXON, PETER M. NEUMANN, and SIMON THOMAS, *Subgroups of small index in infinite symmetric groups*, ***Bulletin of the London Mathematical Society***, vol. 18 (1986), no. 6, pp. 580–586.

[152] WILLIAM F. DOWLING and JEAN H. GALLIER, *Linear-time algorithms for testing the satisfiability of propositional Horn formulae*, ***The Journal of Logic Programming***, vol. 1 (1984), no. 3, pp. 267–284.

[153] THOMAS DRAKENGREN and PETER JONSSON, *Reasoning about set constraints applied to tractable inference in intuitionistic logic*, ***Journal of Logic and Computation***, vol. 8 (1998), no. 6, pp. 855–875.

[154] ———, *Computational complexity of temporal constraint problems*, ***Handbook of Temporal Reasoning in Artificial Intelligence*** (M. Fisher, D. Gabbay, and L. Vila, editors), Elsevier, 2005, pp. 197–218.

[155] MANFRED DROSTE, *Structure of partially ordered sets with transitive automorphism groups*, ***AMS Memoir***, vol. 57 (1985), no. 334.

[156] ———, *Partially ordered sets with transitive automorphism groups*, ***Proceedings of the London Mathematical Society***, vol. 54 (1987), pp. 517–543.

[157] MANFRED DROSTE, CHARLES W. HOLLAND, and DUGALD MACPHERSON, *Automorphism groups of infinite semilinear orders (II)*, ***Proceedings of the London Mathematical Society***, vol. 58 (1989), pp. 479–494.

[158] ———, *Automorphism groups of homogeneous semilinear orders: normal subgroups and commutators*, ***Canadian Journal of Mathematics***, vol. 43 (1991), pp. 721–737.

[159] IVO DÜNTSCH, *Relation algebras and their application in temporal and spatial reasoning*, ***Artificial Intelligence Review***, vol. 23 (2005), pp. 315–357.

[160] MARTIN E. DYER and CATHERINE S. GREENHILL, *The complexity of counting graph homomorphisms*, ***Random Structures and Algorithms***, vol. 17 (2000), pp. 260–289.

[161] MARTIN E. DYER and DAVID RICHERBY, *An effective dichotomy for the counting constraint satisfaction problem*, ***SIAM Journal on Computing***, vol. 42 (2013), no. 3, pp. 1245–1274.

[162] H.-D. EBBINGHAUS and J. FLUM, ***Finite Model Theory***, Springer, Berlin, Heidelberg, New York, 1999, second edition.

[163] H.-D. EBBINGHAUS, J. FLUM, and WOLFGANG THOMAS, ***Mathematical Logic***, Springer, Berlin, Heidelberg, New York, 1984.

[164] LÁSZLÓ EGRI, BENOIT LAROSE, and PASCAL TESSON, *Symmetric datalog and constraint satisfaction problems in logspace*, ***Proceedings of the Symposium on Logic in Computer Science (LICS)***, 2007, pp. 193–202.

[165] E. ALLEN EMERSON and CHARANJIT S. JUTLA, *Tree automata, mu-calculus and determinacy*, ***Proceedings of the 32nd annual Symposium on Foundations of Computer Science (FOCS'91)*** (Washington, DC, USA), IEEE Computer Society, 1991, pp. 368–377.

[166] DAVID M. EVANS, *Subgroups of small index in general linear groups*, ***Bulletin of the London Mathematical Society***, vol. 18 (1986), pp. 587–590.

[167] DAVID M. EVANS and PAUL R. HEWITT, *Counterexamples to a conjecture on relative categoricity*, ***Annals of Pure and Applied Logic***, vol. 46 (1990), no. 2, pp. 201–209.

[168] DAVID M. EVANS, JAN HUBIČKA, and JAROSLAV NEŠETŘIL, *Automorphism groups and Ramsey properties of sparse graphs*, ***Proceedings of the London Mathematical Society***, vol. 119 (2019), pp. 515–546.

[169] JUSTINE FALQUE and NICOLAS M. THIÉRY, *The orbit algebra of a permutation group with polynomial profile is cohen-macaulay*, preprint ArXiv:1804.03489, 2018.

[170] TOMÁS FEDER and MOSHE Y. VARDI, *The computational structure of monotone monadic SNP and constraint satisfaction: a study through Datalog and group theory*, ***SIAM Journal on Computing***, vol. 28 (1999), pp. 57–104.

[171] ———, *Homomorphism closed vs. existential positive*, ***Proceedings of the Symposium on Logic in Computer Science (LICS)***, 2003, pp. 311–320.

[172] ROLAND FRAÏSSÉ, *Sur l'extension aux relations de quelques propriétés des ordres*, ***Annales Scientifiques de l'École Normale Supérieure***, vol. 71 (1954), pp. 363–388.

[173] ———, ***Theory of Relations***, Elsevier Science Ltd, North-Holland, 1986.

[174] ZVI GALIL and NIMROD MEGIDDO, *Cyclic ordering is NP-complete*, ***Theoretical Computer Science***, vol. 5 (1977), no. 2, pp. 179–182.

[175] SU GAO, ***Invariant Descriptive Set Theory***, Pure and applied mathematics, Taylor and Francis, 2008.

[176] MICHAEL GAREY and DAVID JOHNSON, ***A Guide to NP-Completeness***, CSLI Press, Stanford, 1978.

[177] MAI GEHRKE and MICHAEL PINSKER, *Uniform Birkhoff*, ***Journal of Pure and Applied Algebra***, vol. 222 (2018), no. 5, pp. 1242–1250.

[178] DAVID GEIGER, *Closed systems of functions and predicates*, ***Pacific Journal of Mathematics***, vol. 27 (1968), pp. 95–100.

[179] PIERRE GILLIBERT, JULIUS JONUŠAS, MICHAEL KOMPATSCHER, ANTOINE MOTTET, and MICHAEL PINSKER, *Hrushovski's encoding and ω-categorical CSP monsters*, ***47th International Colloquium on Automata, Languages, and Programming, ICALP 2020, July 8-11, 2020, Saarbrücken, Germany (Virtual Conference)***, LIPIcs, vol. 168, Schloss Dagstuhl – Leibniz-Zentrum für Informatik, 2020, pp. 131:1–131:17.

[180] ANDREAS GOERDT, *On random betweenness constraints*, ***FCT***, 2009, pp. 157–168.

[181] ———, *On random ordering constraints*, ***Proceedings of Computer Science Russia (CSR)***, 2009, pp. 105–116.

[182] ERICH GRÄDEL, *Capturing complexity classes by fragments of second-order logic*, ***Theoretical Computer Science***, vol. 101 (1992), no. 1, pp. 35–57.

[183] ERICH GRÄDEL, WOLFGANG THOMAS, and THOMAS WILKE, ***Automata, Logics, and Infinite Games. A Guide to Current Research***, LNCS 2500, Springer-Verlag, 2002.

[184] RON L. GRAHAM and BRUCE L. ROTHSCHILD, *Ramsey's theorem for n-parameter sets*, ***Transactions of the American Mathematical Society***, vol. 159 (1971), pp. 257–292.

[185] RON L. GRAHAM, BRUCE L. ROTHSCHILD, and JOEL H. SPENCER, ***Ramsey Theory***, Wiley-Interscience Series in Discrete Mathematics and Optimization, John Wiley & Sons, Inc., New York, 1990, second edition.

[186] EDMOND E. GRANIRER, *Extremely amenable semigroups 2*, ***Mathematica Scandinavica***, vol. 20 (1967), pp. 93–113.

[187] MARTIN GROHE, *The structure of fixed-point logics*, PhD-thesis at the Albert-Ludwigs Universität, Freiburg i. Br., 1994.

[188] ———, *The complexity of homomorphism and constraint satisfaction problems seen from the other side*, ***Journal of the ACM***, vol. 54 (2007), no. 1.

[189] VENKATESAN GURUSWAMI, JOHAN HÅSTAD, RAJSEKAR MANOKARAN, PRASAD RAGHAVENDRA, and MOSES CHARIKAR, *Beating the random ordering is hard: Every ordering CSP is approximation resistant*, ***SIAM Journal on Computing***, vol. 40 (2011), no. 3, pp. 878–914.

[190] WALTER GUTTMANN and MARKUS MAUCHER, *Variations on an ordering theme with constraints*, ***Proceedings of TCS***, 2006, pp. 77–90.

[191] LUCIEN HADDAD and IVO G. ROSENBERG, *Finite clones containing all permutations*, ***Canadian Journal of Mathematics***, vol. 46 (1994), no. 5, pp. 951–970.

[192] GODFREY H. HARDY and EDWARD M. WRIGHT, ***An Introduction to the Theory of Numbers***, Oxford University Press, 2008, sixth edition.

[193] DEIRDRE HASKELL and DUGALD MACPHERSON, *Cell decompositions of C-minimal structures*, ***Annals of Pure and Applied Logic***, vol. 66 (1994), pp. 113–162.

[194] LUTZ HEINDORF, *The maximal clones on countable sets that include all permutations*, ***Algebra Universalis***, vol. 48 (2002), pp. 209–222.

[195] PAVOL HELL and JAROSLAV NEŠETŘIL, *On the complexity of H-coloring*, ***Journal of Combinatorial Theory, Series B***, vol. 48 (1990), pp. 92–110.

[196] ———, *The core of a graph*, ***Discrete Mathematics***, vol. 109 (1992), pp. 117–126.

[197] ———, ***Graphs and Homomorphisms***, Oxford University Press, Oxford, 2004.

[198] J. WILLIAM HELTON and JIAWANG NIE, *Sufficient and necessary conditions for semidefinite representability of convex hulls and sets*, ***SIAM Journal on Optimization***, vol. 20 (2009), no. 2, pp. 759–791.

[199] C. WARD HENSON, *A family of countable homogeneous graphs*, ***Pacific Journal of Mathematics***, vol. 38 (1971), pp. 69–83.

[200] ———, *Countable homogeneous relational systems and categorical theories*, ***The Journal of Symbolic Logic***, vol. 37 (1972), pp. 494–500.

[201] BERNHARD HERWIG, *Extending partial isomorphisms for the small index property of many ω-categorical structures*, ***Israel Journal of Mathematics***, vol. 107 (1998), pp. 93–123.

[202] ROBIN HIRSCH, *Relation algebras of intervals*, ***Artificial Intelligence Journal***, vol. 83 (1996), pp. 1–29.

[203] ———, *Expressive power and complexity in algebraic logic*, ***Journal of Logic and Computation***, vol. 7 (1997), no. 3, pp. 309–351.

[204] ROBIN HIRSCH and IAN HODKINSON, ***Relation Algebras by Games***, North Holland, 2002.

[205] DAVID HOBBY and RALPH MCKENZIE, ***The Structure of Finite Algebras***, Contemporary Mathematics, vol. 76, American Mathematical Society, 1988.

[206] WILFRID HODGES, ***Model Theory***, Cambridge University Press, 1993.

[207] ———, ***A Shorter Model Theory***, Cambridge University Press, Cambridge, 1997.

[208] WILFRID HODGES, IAN HODKINSON, DANIEL LASCAR, and SAHARON SHELAH, *The small index property for ω-stable ω-categorical structures and for the random graph*, ***Journal of the London Mathematical Society***, vol. S2-48 (1993), no. 2, pp. 204–218.

[209] JAN HUBIČKA and JAROSLAV NEŠETŘIL, *Universal structures with forbidden homomorphisms*, ***Logic Without Borders – Essays on Set Theory, Model Theory, Philosophical Logic and Philosophy of Mathematics***, De Gruyter, 2015, arXiv:0907.4079, pp. 241–264.

[210] ———, *Homomorphism and embedding universal structures for restricted classes*, ***Multiple-Valued Logic and Soft Computing***, vol. 27 (2016), no. 2-3, pp. 229–253, arXiv:0909.4939.

[211] PAWEL M. IDZIAK, PETAR MARKOVIC, RALPH MCKENZIE, MATTHEW VALERIOTE, and ROSS WILLARD, *Tractability and learnability arising from algebras with few subpowers*, ***Proceedings of the Symposium on Logic in Computer Science (LICS)***, 2007, pp. 213–224.

[212] ———, *Tractability and learnability arising from algebras with few subpowers*, ***SIAM Journal on Computing***, vol. 39 (2010), no. 7, pp. 3023–3037.

[213] NEIL IMMERMAN, ***Descriptive Complexity***, Graduate Texts in Computer Science, Springer, 1998.

[214] SVANTE JANSON, TOMASZ LUCZAK, and ANDRZEJ RUCINSKI, ***Random Graphs***, John Wiley and Sons, 2000.

[215] P. G. JEAVONS and M. C. COOPER, *Tractable constraints on ordered domains*, ***Artificial Intelligence***, vol. 79 (1995), no. 2, pp. 327–339.

[216] PETER JEAVONS, DAVID COHEN, and MARTIN COOPER, *Constraints, consistency and closure*, ***Artificial Intelligence***, vol. 101 (1998), no. 1-2, pp. 251–265.

[217] PETER JEAVONS, DAVID COHEN, and MARC GYSSENS, *Closure properties of constraints*, ***Journal of the ACM***, vol. 44 (1997), no. 4, pp. 527–548.

[218] BJARNI JÓNSSON, *Algebras whose congruence lattices are distributive*, ***Mathematica Scandinavica***, vol. 21 (1967), pp. 110–121.

[219] PETER JONSSON and CHRISTER BÄCKSTRÖM, *A unifying approach to temporal constraint reasoning*, ***Artificial Intelligence***, vol. 102 (1998), no. 1, pp. 143–155.

[220] Peter Jonsson and Thomas Drakengren, *A complete classification of tractability in RCC-5*, ***Journal of Artificial Intelligence Research***, vol. 6 (1997), pp. 211–221.

[221] Peter Jonsson, Victor Lagerkvist, and Gustav Nordh, *Constructing np-intermediate problems by blowing holes with parameters of various properties*, ***Theoretical Computer Science***, vol. 581 (2015), pp. 67–82.

[222] Jelena Jovanović, Petar Marković, Ralph McKenzie, and Matthew Moore, *Optimal strong Mal'cev conditions for congruence meet-semidistributivity in locally finite varieties*, ***Algebra Universalis***, vol. 76 (2016), pp. 305–325.

[223] Markus Junker and Martin Ziegler, *The* 116 *reducts of* $(\mathbb{Q}, <, a)$, ***The Journal of Symbolic Logic***, vol. 74 (2008), no. 3, pp. 861–884.

[224] Paris C. Kanellakis, Gabriel M. Kuper, and Peter Z. Revesz, *Constraint query languages*, ***Proceedings of Symposium on Principles of Database Systems (PODS)***, 1990, pp. 299–313.

[225] Marek Karpinski, Hans Kleine Büning, and Peter H. Schmitt, *On the computational complexity of quantified Horn clauses*, ***Proceedings of Computer Science Logic (CSL)*** (Egon Börger, Hans Kleine Büning, and Michael M. Richter, editors), Lecture Notes in Computer Science, 1987, 1st Workshop on Computer Science Logic, Karlsruhe, Germany, pp. 129–137.

[226] Richard Kaye and Dugald Macpherson, ***Automorphisms of First-Order Structures***, Oxford University Press, 1994.

[227] Alexandr Kazda, Marcin Kozik, Ralph McKenzie, and Matthew Moore, *Absorption and directed Jónsson terms*, ***Outstanding Contributions to Logic***, vol. 16 (2018), pp. 203–220.

[228] Keith A. Kearnes and Emil W. Kiss, ***The Shape of Congruence Lattices***, Memoirs of the American Mathematical Society, vol. 222 (1046), American Mathematical Society, 2013.

[229] Alexander Kechris, Vladimir Pestov, and Stevo Todorčević, *Fraïssé limits, Ramsey theory, and topological dynamics of automorphism groups*, ***Geometric and Functional Analysis***, vol. 15 (2005), no. 1, pp. 106–189.

[230] Alexander S. Kechris and Christian Rosendal, *Turbulence, amalgamation and generic automorphisms of homogeneous structures*, ***Proceedings of the London Mathematical Society***, vol. 94 (2007), no. 3, pp. 302–350.

[231] Jerome Keisler, *Reduced products and Horn classes*, ***Transactions of the American Mathematical Society***, vol. 117 (1965), pp. 307–328.

[232] Leonid Khachiyan, *A polynomial algorithm in linear programming*, ***Doklady Akademii Nauk SSSR***, vol. 244 (1979), pp. 1093–1097.

[233] Hirotaka Kikyo, *Model companions of theories with an automorphism*, ***The Journal of Symbolic Logic***, vol. 65 (2000), no. 3, pp. 1215–1222.

[234] Bartek Klin, Eryk Kopczynski, Joanna Ochremiak, and Szymon Toruńczyk, *Locally finite constraint satisfaction problems*, ***30th Annual***

ACM/IEEE Symposium on Logic in Computer Science, LICS 2015, Kyoto, Japan, 2015, pp. 475–486.

[235] BARTEK KLIN, SLAWOMIR LASOTA, JOANNA OCHREMIAK, and SZYMON TORUNCZYK, *Homomorphism Problems for First-Order Definable Structures*, ***36th IARCS Annual Conference on Foundations of Software Technology and Theoretical Computer Science (FSTTCS 2016)*** (Dagstuhl, Germany) (Akash Lal, S. Akshay, Saket Saurabh, and Sandeep Sen, editors), Leibniz International Proceedings in Informatics (LIPIcs), vol. 65, Schloss Dagstuhl – Leibniz-Zentrum für Informatik, 2016, pp. 14:1–14:15.

[236] PHOKION G. KOLAITIS and MOSHE Y. VARDI, *The decision problem for the probabilities of higher-order properties*, ***Proceedings of the Symposium on Theory of Computing (STOC)***, 1987, pp. 425–435.

[237] ———, *On the expressive power of Datalog: Tools and a case study*, ***Journal of Computer and System Sciences***, vol. 51 (1995), no. 1, pp. 110–134.

[238] ———, *Conjunctive-query containment and constraint satisfaction*, ***Proceedings of Symposium on Principles of Database Systems (PODS)***, 1998, pp. 205–213.

[239] VLADIMIR KOLMOGOROV, ANDREI A. KROKHIN, and MICHAL ROLÍNEK, *The complexity of general-valued CSPs*, ***IEEE 56th Annual Symposium on Foundations of Computer Science, FOCS 2015, Berkeley, CA, USA, 17-20 October, 2015***, 2015, pp. 1246–1258.

[240] VLADIMIR KOLMOGOROV, JOHAN THAPPER, and STANISLAV ŽIVNÝ, *The power of linear programming for general-valued CSPs*, ***SIAM Journal on Computing***, vol. 44 (2015), no. 1, pp. 1–36.

[241] MICHAEL KOMPATSCHER and TRUNG VAN PHAM, *A complexity dichotomy for poset constraint satisfaction*, ***IfCoLog Journal of Logics and their Applications (FLAP)***, vol. 5 (2018), no. 8, pp. 1663–1696.

[242] SABINE KOPPELBERG, *Projective boolean algebras*, ***Handbook of Boolean Algebras*** (J. D. Monk and R. Bonnet, editors), vol. 3, North Holland, Amsterdam – New York – Oxford – Tokyo, 1989, pp. 741–773.

[243] M. KOUBARAKIS, *Tractable disjunctions of linear constraints: Basic results and applications to temporal reasoning*, ***Theoretical Computer Science***, vol. 266 (2001), pp. 311–339.

[244] MARCIN KOZIK, ANDREI KROKHIN, MATT VALERIOTE, and ROSS WILLARD, *Characterizations of several Maltsev conditions*, ***Algebra Universalis***, vol. 73 (2015), no. 3, pp. 205–224.

[245] MARCIN KOZIK and JOANNA OCHREMIAK, *Algebraic properties of valued constraint satisfaction problem*, ***Automata, Languages, and Programming – 42nd International Colloquium, ICALP 2015, Kyoto, Japan, July 6-10, 2015, Proceedings, Part I***, 2015, pp. 846–858.

[246] MARC KRASNER, *Généralisations et analogues de la théorie de Galois*, ***Congrès de la Victoire, Association Française pour l'Avancement des Sciences***, 1945, pp. 54–58.

[247] STEPHAN KREUTZER, *Expressive equivalence of least and inflationary fixed-point logic*, ***17th IEEE Symposium on Logic in Computer Science (LICS 2002), 22-25 July 2002, Copenhagen, Denmark, Proceedings***, 2002, p. 403.

[248] ANDREI A. KROKHIN, PETER JEAVONS, and PETER JONSSON, *Reasoning about temporal relations: The tractable subalgebras of Allen's interval algebra*, ***Journal of the ACM***, vol. 50 (2003), no. 5, pp. 591–640.

[249] GÁBOR KUN, *Constraints, MMSNP, and expander relational structures*, ***Combinatorica***, vol. 33 (2013), no. 3, pp. 335–347.

[250] ALISTAIR H. LACHLAN and ROBERT E. WOODROW, *Countable ultrahomogeneous undirected graphs*, ***Transactions of the American Mathematical Society***, vol. 262 (1980), no. 1, pp. 51–94.

[251] PETER B. LADKIN and ROGER D. MADDUX, *On binary constraint problems*, ***Journal of the Association for Computing Machinery***, vol. 41 (1994), no. 3, pp. 435–469.

[252] RICHARD E. LADNER, *On the structure of polynomial time reducibility*, ***Journal of the ACM***, vol. 22 (1975), no. 1, pp. 155–171.

[253] JEFFREY C. LAGARIAS, *The computational complexity of simultaneous Diophantine approximation problems*, ***SIAM Journal on Computing***, vol. 14 (1985), no. 1, pp. 196–209.

[254] BENOIT LAROSE, CYNTHIA LOTEN, and CLAUDE TARDIF, *A characterisation of first-order constraint satisfaction problems*, ***Logical Methods in Computer Science***, vol. 3 (2007), no. 4:6.

[255] BENOIT LAROSE and PASCAL TESSON, *Universal algebra and hardness results for constraint satisfaction problems*, ***Theoretical Computer Science***, vol. 410 (2009), no. 18, pp. 1629–1647.

[256] BENOIT LAROSE, MATT VALERIOTE, and LÁSZLÓ ZÁDORI, *Omitting types, bounded width and the ability to count*, ***International Journal of Algebra and Computation***, vol. 19 (2009), no. 5.

[257] BENOIT LAROSE and LÁSZLÓ ZÁDORI, *Bounded width problems and algebras*, ***Algebra Universalis***, vol. 56 (2007), no. 3-4, pp. 439–466.

[258] DANIEL LASCAR, *Autour de la propriété du petit indice*, ***Proceedings of the London Mathematical Society***, vol. 62 (1991), no. 1, pp. 25–53.

[259] JEAN-LOUIS LASSEZ and KEN MCALOON, *Independence of negative constraints*, ***International Joint Conference on Theory and Practice of Software Development (TAPSOFT), Volume 1***, 1989, pp. 19–27.

[260] ———, *A constraint sequent calculus*, ***Proceedings of the Symposium on Logic in Computer Science (LICS)***, 1990, pp. 52–61.

[261] KLAUS LEEB, ***Vorlesungen über Pascaltheorie***, Arbeitsberichte des Instituts für Mathematische Maschinen und Datenverarbeitung, vol. 6, Friedrich-Alexander-Universität Erlangen-Nürnberg, 1973.

[262] JASON JINGSHI LI, TOMASZ KOWALSKI, JOCHEN RENZ, and SANJIANG LI, *Combining binary constraint networks in qualitative reasoning*, ***18th European Conference on Artificial Intelligence (ECAI)***, 2008, pp. 515–519.

[263] LEONID LIBKIN, ***Elements of Finite Model Theory***, Springer, 2004.

[264] GÉRARD LIGOZAT and JOCHEN RENZ, *What is a qualitative calculus? A general framework*, ***Proceedings of Pacific Rim International Conferences on Artificial Intelligence (PRICAI)***, 2004, pp. 53–64.

[265] JULIE LINMAN and MICHAEL PINSKER, *Permutations on the random permutation*, ***Electronic Journal of Combinatorics***, vol. 22 (2015), no. 2, pp. 1–22.

[266] JOHN W. LLOYD, ***Foundations of Logic Programming***, Springer, 1987, second edition.

[267] DEBORAH LOCKETT and JOHN K. TRUSS, *Some more notions of homomorphism-homogeneity*, ***Discrete Mathematics***, vol. 336 (2014), pp. 69–79.

[268] R. LYNDON, *The representation of relational algebras*, ***Annals of Mathematics***, vol. 51 (1950), no. 3, pp. 707–729.

[269] HAJIME MACHIDA and MICHAEL PINSKER, *The minimal clones above the permutations*, ***Semigroup Forum***, vol. 75 (2007), pp. 181–211.

[270] DUGALD MACPHERSON, *A survey of homogeneous structures*, ***Discrete Mathematics***, vol. 311 (2011), no. 15, pp. 1599–1634.

[271] DUGALD MACPHERSON and CHARLES STEINHORN, *On variants of o-minimality*, ***Annals of Pure and Applied Logic***, vol. 79 (1996), no. 2, pp. 165–209.

[272] H. DUGALD MACPHERSON and CHERYL E. PRAEGER, *Infinitary Versions of the O'Nan-Scott Theorem*, ***Proceedings of the London Mathematical Society***, vol. 68 (1994), no. 3, pp. 518–540.

[273] FLORENT MADELAINE, *Constraint satisfaction problems and related logic*, PhD-thesis, University of Leicester, 2003.

[274] ———, *Universal structures and the logic of forbidden patterns*, ***Logical Methods in Computer Science***, vol. 5 (2009), no. 2.

[275] FLORENT MADELAINE and IAIN A. STEWART, *Constraint satisfaction, logic and forbidden patterns*, ***SIAM Journal on Computing***, vol. 37 (2007), no. 1, pp. 132–163.

[276] FLORENT R. MADELAINE, *On the containment of forbidden patterns problems*, ***Proceedings of the International Conference on Principles and Practice of Constraint Programming (CP)***, 2010, pp. 345–359.

[277] FLORENT R. MADELAINE and IAIN A. STEWART, *Some problems not definable using structure homomorphisms*, ***Ars Combinatorica***, vol. 67 (2003), pp. 153–159.

[278] DAVID MARKER, ***Model Theory: An Introduction***, Springer, New York, 2002.

[279] MIKLÓS MARÓTI and RALPH MCKENZIE, *Existence theorems for weakly symmetric operations*, ***Algebra Universalis***, vol. 59 (2008), no. 3.

[280] KIM MARRIOTT and MARTIN ODERSKY, *Negative Boolean constraints*, ***Theoretical Computer Science***, vol. 160 (1996), no. 1&2, pp. 365–380.

[281] DÁNIEL MARX, *Tractable structures for constraint satisfaction with truth tables*, ***Theory of Computing Systems***, vol. 48 (2011), no. 3, pp. 444–464.

[282] YURI V. MATIYASEVICH, ***Hilbert's Tenth Problem***, MIT Press, Cambridge, Massachusetts, 1993.

[283] STEFAN MENGEL, *Some fixed-point logic gymnastics*, 2013, unpublished draft.

[284] KEITH R. MILLIKEN, *A Ramsey theorem for trees*, ***Journal of Combinatorial Theory, Series A***, vol. 26 (1979), no. 3, pp. 215–237.

[285] ROLF H. MÖHRING, MARTIN SKUTELLA, and FREDERIK STORK, *Scheduling with and/or precedence constraints*, ***SIAM Journal on Computing***, vol. 33 (2004), no. 2, pp. 393–415.

[286] MATTHEW MOORE, *Finite degree clones are undecidable*, Preprint: http://arxiv.org/abs/1811.05056, 2019.

[287] ANTOINE MOTTET and MICHAEL PINSKER, *Cores over Ramsey structures*, ***Computing Research Repository (CoRR)***, vol. abs/2004.05936 (2020).

[288] BERNHARD NEBEL and HANS-JÜRGEN BÜRCKERT, *Reasoning about temporal relations: A maximal tractable subclass of Allen's interval algebra*, ***Journal of the ACM***, vol. 42 (1995), no. 1, pp. 43–66.

[289] GREG NELSON and DEREK C. OPPEN, *Fast decision procedures based on congruence closure*, ***Journal of the ACM***, vol. 27 (1980), no. 2, pp. 356–364.

[290] JAROSLAV NEŠETŘIL, *Ramsey theory*, ***Handbook of Combinatorics***, (1995), pp. 1331–1403.

[291] ———, *Ramsey classes and homogeneous structures*, ***Combinatorics, Probability & Computing***, vol. 14 (2005), no. 1-2, pp. 171–189.

[292] JAROSLAV NEŠETŘIL and VOJTĚCH RÖDL, *Ramsey classes of set systems*, ***Journal of Combinatorial Theory, Series A***, vol. 34 (1983), no. 2, pp. 183–201.

[293] ———, *The partite construction and Ramsey set systems*, ***Discrete Mathematics***, vol. 75 (1989), no. 1-3, pp. 327–334.

[294] MIROSLAV OLŠÁK, *The weakest nontrivial idempotent equations*, ***Bulletin of the London Mathematical Society***, vol. 49 (2017), no. 6, pp. 1028–1047.

[295] ———, *Loop conditions*, ***Algebra Universalis***, vol. 81 (2020), no. 2, Preprint arXiv:1701.00260.

[296] JAROSLAV OPATRNY, *Total ordering problem*, ***SIAM Journal on Computing***, vol. 8 (1979), no. 1, pp. 111–114.

[297] PÉTER PÁL PACH, MICHAEL PINSKER, GABRIELLA PLUHÁR, ANDRÁS PONGRÁCZ, and CSABA SZABÓ, *Reducts of the random partial order*, ***Advances in Mathematics***, vol. 267 (2014), pp. 94–120.

[298] C. H. PAPADIMITRIOU and M. YANNAKAKIS, *Optimization, approximation, and complexity classes*, ***Journal of Computer and System Sciences***, vol. 43 (1991), pp. 425–440.

[299] CHRISTOS H. PAPADIMITRIOU, ***Computational Complexity***, Addison-Wesley, 1994.

[300] CHRISTIAN PECH and MAJA PECH, *Towards a Ryll-Nardzewski-type theorem for weakly oligomorphic structures*, ***Mathematical Logic Quarterly***, vol. 62 (2016), no. 1-2, pp. 25–34.

[301] ———, *Reconstructing the topology of the elementary self-embedding monoids of countable saturated structures*, ***Studia Logica***, vol. 106 (2018), no. 3, pp. 595–613.

[302] MICHAEL PINSKER, *Rosenberg's characterization of maximal clones*, diploma thesis, 2002.

[303] ———, *Maximal clones on uncountable sets that include all permutations*, ***Algebra Universalis***, vol. 54 (2005), no. 2, pp. 129–148.

[304] ———, *The number of unary clones containing the permutations on an infinite set*, ***Acta Scientiarum Mathematicarum (Szeged)***, vol. 71 (2005), pp. 461–467.

[305] MICHAEL PINSKER, PIERRE GILLIBERT, and J. JONUŠAS, *Pseudo-loop conditions*, ***Bulletin of the London Mathematical Society***, vol. 51 (2019), no. 5, pp. 917–936.

[306] BRUNO POIZAT, ***A Course in Model Theory: An Introduction to Contemporary Mathematical Logic***, Springer, 2000.

[307] ANDRÁS PONGRÁCZ, *Reducts of the Henson graphs with a constant*, ***Annals of Pure and Applied Logic***, vol. 168 (2017), no. 7, pp. 1472–1489.

[308] REINHARD PÖSCHEL, *A general Galois theory for operations and relations and concrete characterization of related algebraic structures*, ***Tech Report of Akademie der Wissenschaften der DDR***, (1980).

[309] REINHARD PÖSCHEL and LEV A. KALUŽNIN, ***Funktionen- und Relationenalgebren***, Deutscher Verlag der Wissenschaften, 1979.

[310] EMIL L. POST, *The two-valued iterative systems of mathematical logic*, ***Annals of Mathematics Studies***, vol. 5 (1941).

[311] EVGENIA B. RABINOVICH, *Embedding theorems and de Bruijn's problem for bounded symmetry groups*, ***Dokl. Akad. Nauk. Belor. S.S.R.***, vol. 21 (1977), no. 9, pp. 784–7, Russian.

[312] MOTAKURI V. RAMANA, *An exact duality theory for semidefinite programming and its complexity implications*, ***Mathematical Programming***, vol. 77 (1997), pp. 129–162.

[313] OMER REINGOLD, *Undirected connectivity in Log-Space*, ***Journal of the ACM***, vol. 55 (2008), no. 4.

[314] JOCHEN RENZ and BERNHARD NEBEL, *On the complexity of qualitative spatial reasoning: A maximal tractable fragment of the region connection calculus*, ***Artificial Intelligence***, vol. 108 (1999), no. 1-2, pp. 69–123.

[315] BORIS A. ROMOV, *Extendable local partial clones*, ***Discrete Mathematics***, vol. 308 (2008), no. 17, pp. 3744–3760.

[316] IVO G. ROSENBERG, *Minimal clones I: the five types*, ***Lectures in Universal Algebra*** (L. Szabó and Á. Szendrei, editors), Colloquia Mathematica Societatis Janos Bolyai, North-Holland, Amsterdam, 1986, pp. 405–427.

[317] BENJAMIN ROSSMAN, *Homomorphism preservation theorems*, ***Journal of the ACM***, vol. 55 (2008), no. 3.

[318] DAN SARACINO, *Model companions for $\aleph_0$-categorical theories*, ***Proceedings of the American Mathematical Society***, vol. 39 (1973), pp. 591–598.

[319] THOMAS J. SCHAEFER, *The complexity of satisfiability problems*, ***Proceedings of the Symposium on Theory of Computing (STOC)***, 1978, pp. 216–226.

[320] SERGEJ SCHECK, *Pseudo Jónsson operations in highly transitive clones*, Master thesis, TU Dresden, 2019.

[321] CLAUS SCHEIDERER, *Spectrahedral shadows*, ***SIAM Journal on Applied Algebra and Geometry***, vol. 2 (2018), pp. 26–44.

[322] JAMES H. SCHMERL, *Countable homogeneous partially ordered sets*, ***Algebra Universalis***, vol. 9 (1979), pp. 317–321.

[323] FRIEDRICH MARTIN SCHNEIDER, *A uniform Birkhoff theorem*, ***Algebra Universalis***, vol. 78 (2017), no. 3, pp. 337–354.

[324] UWE SCHÖNING, ***Logic for Computer Scientists***, Springer, 1989.

[325] JOSEPH SCHREIER and STANISŁAW MARCIN ULAM, *Über die Permutationsgruppe der natürlichen Zahlenfolge*, ***Studia Mathematica***, vol. 4 (1933), pp. 134–141.

[326] ALEXANDER SCHRIJVER, ***Theory of Linear and Integer Programming***, Wiley-Interscience Series in Discrete Mathematics and Optimization, John Wiley & Sons, 1998.

[327] GOETZ SCHWANDTNER, *Datalog on infinite structures*, Dissertation, Humboldt-Universität zu Berlin, 2008.

[328] STEPHEN W. SEMMES, *Endomorphisms of infinite symmetric groups*, ***Abstracts of the American Mathematical Society***, vol. 2 (1981), p. 426.

[329] MARK H. SIGGERS, *A strong Mal'cev condition for varieties omitting the unary type*, ***Algebra Universalis***, vol. 64 (2010), no. 1, pp. 15–20.

[330] HAROLD SIMMONS, *Large and small existentially closed structures*, ***The Journal of Symbolic Logic***, vol. 41 (1976), no. 2, pp. 379–390.

[331] PIERRE SIMON, *Linear orders in NIP structures*, arXiv:1310.8157v2, 2019.

[332] MICHAEL STEEL, *The complexity of reconstructing trees from qualitative characters and subtrees*, ***Journal of Classification***, vol. 9 (1992), pp. 91–116.

[333] LÁSZLÓ SZABÓ, *Concrete representation of relational structures of universal algebras*, ***Acta Scientiarum Mathematicarum (Szeged)***, vol. 40 (1978), pp. 175–184.

[334] ÁGNES SZENDREI, ***Clones in Universal Algebra***, Séminaire de Mathématiques Supérieures, Les Presses de l'Université de Montréal, 1986.

[335] WALTER TAYLOR, *Varieties obeying homotopy laws*, ***Canadian Journal of Mathematics***, vol. 29 (1977), pp. 498–527.

[336] ———, *Abstract clone theory*, ***Algebras and Orders*** (Ivo G. Rosenberg and G. Sabidussi, editors), NATO Science Series C, Mathematical and Physical Sciences, vol. 389, Kluwer Academic Publishers, 1993, pp. 507–530.

[337] KATRIN TENT and MARTIN ZIEGLER, ***A Course in Model Theory***, Lecture Notes in Logic, Cambridge University Press, 2012.

[338] JOHAN THAPPER and STANISLAV ŽIVNÝ, *The complexity of finite-valued CSPs*, ***Proceedings of the Symposium on Theory of Computing Conference (STOC), Palo Alto, CA, USA, June 1-4, 2013***, 2013, pp. 695–704.

[339] SIMON THOMAS, *Reducts of the random graph*, ***The Journal of Symbolic Logic***, vol. 56 (1991), no. 1, pp. 176–181.

[340] BORIS TRAKHTENBROT, *The impossibility of an algorithm for the decidability problem on finite classes*, ***Proceedings of the USSR Academy of Sciences***, vol. 70 (1950), no. 4, pp. 569–572, (in Russian).

[341] JOHN K. TRUSS, *Infinite permutation groups. II. Subgroups of small index*, ***Journal of Algebra***, vol. 120 (1989), no. 2, pp. 494–515.

[342] TODOR TSANKOV, *Unitary representations of oligomorphic groups*, ***Geometric and Functional Analysis***, vol. 22 (2012), no. 2, pp. 528–555.

[343] ROSS WILLARD, *Testing expressibility is hard*, ***Proceedings of the International Conference on Principles and Practice of Constraint Programming (CP)***, 2010, pp. 9–23.

[344] STEPHEN WILLARD, ***General Topology***, Dover Publications, 2004.

[345] HENRY WOLKOWICZ, ROMESH SAIGAL, and LIEVEN VANDENBERGHE, ***Handbook of Semidefinite Programming: Theory, Algorithms, and Applications***, Springer, 2000.

[346] YURI I. YANOV and ALBERT A. MUCHNIK, *On the existence of k-valued closed classes without a finite basis*, ***Doklady Akademii Nauk SSSR***, vol. 127 (1959), pp. 44–46, in Russian.

[347] DMITRIY N. ZHUK, *The lattice of all clones of self-dual functions in three-valued logic*, ***Multiple-Valued Logic and Soft Computing***, vol. 24 (2015), no. 1-4, pp. 251–316.

[348] ———, *A proof of CSP dichotomy conjecture*, ***58th IEEE Annual Symposium on Foundations of Computer Science, FOCS 2017, Berkeley, CA, USA, October 15-17***, 2017, https://arxiv.org/abs/1704.01914., pp. 331–342.

[349] ANDY ZUCKER, *Topological dynamics of automorphism groups, ultrafilter combinatorics, and the generic point problem*, ***Transactions of the American Mathematical Society***, vol. 368 (2016), no. 9, pp. 6715–6740.

INDEX